Lecture Notes in Computer Science 3624

Commenced Publication in 1973
Founding and Former Series Editors:
Gerhard Goos, Juris Hartmanis, and Jan van Leeuwen

Chandra Chekuri   Klaus Jansen
José D.P. Rolim   Luca Trevisan (Eds.)

Approximation, Randomization and Combinatorial Optimization

Algorithms and Techniques

8th International Workshop on Approximation Algorithms
for Combinatorial Optimization Problems, APPROX 2005
and 9th International Workshop on Randomization
and Computation, RANDOM 2005
Berkeley, CA, USA, August 22-24, 2005
Proceedings

 Springer

Volume Editors

Chandra Chekuri
Lucent Bell Labs
600 Mountain Avenue, Murray Hill, NJ 07974, USA
E-mail: chekuri@research.bell-labs.com

Klaus Jansen
University of Kiel, Institute for Computer Science
Olshausenstr. 40, 24098 Kiel, Germany
E-mail: kj@informatik.uni-kiel.de

José D.P. Rolim
Université de Genève, Centre Universitaire d'Informatique
24, Rue Général Dufour, 1211 Genève 4, Suisse
E-mail: jose.rolim@cui.unige.ch

Luca Trevisan
University of California, Computer Science Department
679 Soda Hall, Berkeley, CA 94720-1776, USA
E-mail: luca@cs.berkeley.edu

Library of Congress Control Number: 2005930720

CR Subject Classification (1998): F.2, G.2, G.1

ISSN   0302-9743
ISBN-10   3-540-28239-4 Springer Berlin Heidelberg New York
ISBN-13   978-3-540-28239-6 Springer Berlin Heidelberg New York

Springer is a part of Springer Science+Business Media

springeronline.com

© Springer-Verlag Berlin Heidelberg 2005
Printed in Germany

Typesetting: Camera-ready by author, data conversion by Olgun Computergrafik
Printed on acid-free paper   SPIN: 11538462   06/3142   5 4 3 2 1 0

Preface

This volume contains the papers presented at the 8th International Workshop on Approximation Algorithms for Combinatorial Optimization Problems (APPROX 2005) and the 9th International Workshop on Randomization and Computation (RANDOM 2005), which took place concurrently at the University of California in Berkeley, on August 22–24, 2005. APPROX focuses on algorithmic and complexity issues surrounding the development of efficient approximate solutions to computationally hard problems, and APPROX 2005 was the eighth in the series after Aalborg (1998), Berkeley (1999), Saarbrücken (2000), Berkeley (2001), Rome (2002), Princeton (2003), and Cambridge (2004). RANDOM is concerned with applications of randomness to computational and combinatorial problems, and RANDOM 2005 was the ninth workshop in the series following Bologna (1997), Barcelona (1998), Berkeley (1999), Geneva (2000), Berkeley (2001), Harvard (2002), Princeton (2003), and Cambridge (2004).

Topics of interest for APPROX and RANDOM are: design and analysis of approximation algorithms, hardness of approximation, small space and data streaming algorithms, sub-linear time algorithms, embeddings and metric space methods, mathematical programming methods, coloring and partitioning, cuts and connectivity, geometric problems, game theory and applications, network design and routing, packing and covering, scheduling, design and analysis of randomized algorithms, randomized complexity theory, pseudorandomness and derandomization, random combinatorial structures, random walks/Markov chains, expander graphs and randomness extractors, probabilistic proof systems, random projections and embeddings, error-correcting codes, average-case analysis, property testing, computational learning theory, and other applications of approximation and randomness.

The volume contains 20 contributed papers selected by the APPROX Program Committee out of 50 submissions, and 21 contributed papers selected by the RANDOM Program Committee out of 51 submissions.

We would like to thank all of the authors who submitted papers, the members of the program committees

APPROX 2005

Matthew Andrews, Lucent Bell Labs
Avrim Blum, CMU
Moses Charikar, Princeton University
Chandra Chekuri, Lucent Bell Labs (Chair)
Julia Chuzhoy, MIT
Uriel Feige, Microsoft Research and Weizmann Institute
Naveen Garg, IIT Delhi
Howard Karloff, AT&T Labs – Research
Stavros Kolliopoulos, University of Athens
Adam Meyerson, UCLA
Seffi Naor, Technion
Santosh Vempala, MIT

RANDOM 2005

Dorit Aharonov, Hebrew University
Boaz Barak, IAS and Princeton University
Funda Ergun, Simon Fraser University
Johan Håstad, KTH Stockholm
Chi-Jen Lu, Academia Sinica
Milena Mihail, Georgia Institute of Technology
Robert Krauthgamer, IBM Almaden
Dana Randall, Georgia Institute of Technology
Amin Shokrollahi, EPF Lausanne
Angelika Steger, ETH Zurich
Luca Trevisan, UC Berkeley (Chair)

and the external subreferees Scott Aaronson, Dimitris Achlioptas, Mansoor Alicherry, Andris Ambainis, Aaron Archer, Nikhil Bansal, Tugkan Batu, Gerard Ben Arous, Michael Ben-Or, Eli Ben-Sasson, Petra Berenbrink, Randeep Bhatia, Nayantara Bhatnagar, Niv Buchbinder, Shuchi Chawla, Joseph Cheriyan, Roee Engelberg, Lance Fortnow, Tom Friedetzky, Mikael Goldmann, Daniel Gottesman, Sam Greenberg, Anupam Gupta, Venkat Guruswami, Tom Hayes, Monika Henzinger, Danny Hermelin, Nicole Immorlica, Piotr Indyk, Adam Kalai, Julia Kempe, Claire Kenyon, Jordan Kerenidis, Sanjeev Khanna, Amit Kumar, Ravi Kumar, Nissan Lev-Tov, Liane Lewin, Laszlo Lovasz, Elitza Maneva, Michael Mitzenmacher, Cris Moore, Michele Mosca, Kamesh Munagala, Noam Nisan, Ryan O'Donnell, Martin Pal, Vinayaka Pandit, David Peleg, Yuval Rabani, Dror Rawitz, Danny Raz, Adi Rosen, Ronitt Rubinfeld, Cenk Sahinalp, Alex Samorodnitsky, Gabi Scalosub, Leonard Schulman, Roy Schwartz, Pranab Sen, Mehrdad Shahshahani, Amir Shpilka, Anastasios Sidiropoulos, Greg Sorkin, Adam Smith, Ronen Shaltiel, Maxim Sviridenko, Amnon-Ta-Shma, Emre Telatar, Alex Vardy, Eric Vigoda, Da-Wei Wang, Ronald de Wolf, David Woodruff, Hsin-Lung Wu, and Lisa Zhang.

We gratefully acknowledge the support from the Lucent Bell Labs of Murray Hill, the Computer Science Division of the University of California at Berkeley, the Institute of Computer Science of the Christian-Albrechts-Universität zu Kiel and the Department of Computer Science of the University of Geneva. We also thank Ute Iaquinto and Parvaneh Karimi Massouleh for their help.

August 2005   Chandra Chekuri and Luca Trevisan, Program Chairs
Klaus Jansen and José D.P. Rolim, Workshop Chairs

Table of Contents

Contributed Talks of APPROX

Contributed Talks of RANDOM

The Network as a Storage Device:
Dynamic Routing with Bounded Buffers

Stanislav Angelov*, Sanjeev Khanna**, and Keshav Kunal***

University of Pennsylvania, Philadelphia, PA 19104, USA
{angelov,sanjeev,kkunal}@cis.upenn.edu

Abstract. We study dynamic routing in store-and-forward packet networks where each network link has bounded buffer capacity for receiving
incoming packets and is capable of transmitting a fixed number of packets per unit of time. At any moment in time, packets are injected at
various network nodes with each packet specifying its destination node.
The goal is to maximize the *throughput*, defined as the number of packets
delivered to their destinations.

In this paper, we make some progress in understanding what is achievable
on various network topologies. For line networks, Nearest-to-Go (NTG),
a natural greedy algorithm, was shown to be $O(n^{2/3})$-competitive by
Aiello *et al* [1]. We show that NTG is $\tilde{O}(\sqrt{n})$-competitive, essentially
matching an identical lower bound known on the performance of *any*
greedy algorithm shown in [1]. We show that if we allow the online routing
algorithm to make centralized decisions, there is indeed a randomized
polylog(n)-competitive algorithm for line networks as well as rooted tree
networks, where each packet is destined for the root of the tree. For grid
graphs, we show that NTG has a performance ratio of $\tilde{\Theta}(n^{2/3})$ while no
greedy algorithm can achieve a ratio better than $\Omega(\sqrt{n})$. Finally, for an
arbitrary network with m edges, we show that NTG is $\tilde{\Theta}(m)$-competitive,
improving upon an earlier bound of $O(mn)$ [1].

1 Introduction

The problem of dynamically routing packets is central to traffic management
on large scale data networks. Packet data networks typically employ store-and-
forward routing where network links (or routers) store incoming packets and
schedule them to be forwarded in a suitable manner. Internet is a well-known
example of such a network. In this paper, we consider routing algorithms for
store-and-forward networks in a dynamic setting where packets continuously arrive in an arbitrary manner over a period of time and each packet specifies its

* Supported in part by NSF Career Award CCR-0093117, NSF Award ITR 0205456
and NIGMS Award 1-P20-GM-6912-1.
** Supported in part by an NSF Career Award CCR-0093117, NSF Award CCF-
0429836, and a US-Israel Binational Science Foundation Grant.
*** Supported in part by an NSF Career Award CCR-0093117 and NSF Award CCF-
0429836.

C. Chekuri et al. (Eds.): APPROX and RANDOM 2005, LNCS 3624, pp. 1–13, 2005.
© Springer-Verlag Berlin Heidelberg 2005

destination. The goal is to maximize the *throughput*, defined as the number of packets that are delivered to their destinations. This model, known as the *Competitive Network Throughput* (CNT) model, was introduced by Aiello *et al* [1]. The model allows for both *adaptive* and *non-adaptive* routing of packets, where in the latter case each packet also specifies a path to its destination node. Note that there is no distinction between the two settings on line and tree networks.

Aiello *et al* analyzed *greedy* algorithms, which are characterized as ones that accept and forward packets whenever possible. For the non-adaptive setting, they showed that Nearest-to-Go (NTG), a natural greedy algorithm that favors packets whose remaining distance is the shortest, is $O(mn)$-competitive[1] on any network with n nodes and m edges. On line networks, NTG was shown to be $O(n^{2/3})$-competitive provided each link has buffer capacity at least 2.

Our Results. Designing routing algorithms in the CNT model is challenging since buffer space is limited at any individual node and a good algorithm should use all nodes to buffer packets and not just nodes where packets are injected. It is like viewing the entire network as a storage device where new information is injected at locations chosen by the adversary. Regions in the network with high injection rate need to continuously move packets to intermediate network regions with low activity. However, if many nodes simultaneously send packets to the same region, the resulting congestion would lead to buffer overflows. Worse still, the adversary may suddenly raise the packet injection rate in the region of low activity. In general, an adversary can employ different attack strategies in different parts of the network. How should a routing algorithm coordinate the movement of packets in presence of buffer constraints? This turns out to be a difficult question even on simple networks such as lines and trees.

In this paper, we continue the thread of research started in [1]. Throughout the remainder of this paper, we will assume that all network links have a uniform buffer size B and bandwidth 1. We obtain the following results:

Line Networks: For line networks with $B > 1$, we show that NTG is $\tilde{O}(\sqrt{n})$-competitive, essentially matching an $\Omega(\sqrt{n})$ lower bound of [1]. A natural question is whether non-greedy algorithms can perform better. For $B = 1$, we show that any deterministic algorithm is $\Omega(n)$-competitive, strengthening a result of [1], and give a randomized $\tilde{O}(\sqrt{n})$-competitive algorithm. For $B > 1$, however, no super-constant lower bounds are known leaving a large gap between known lower bounds and what is achievable by greedy algorithms. We show that centralized decisions can improve performance exponentially by designing a randomized $O(\log^3 n)$-competitive algorithm, referred to as *Merge and Deliver*.

Tree Networks: For tree networks of height h, it was shown in [1] that any greedy algorithm has a competitive ratio of $\Omega(n/h)$. Building on the ideas used

---

[1] The exact bound shown in [1] is $O(mD)$ where D is the maximum length of any given path in the network. We state here the worst-case bound with $D = \Omega(n)$.

Table 1. Summary of results; the bounds obtained in this paper are in boldface.

Algorithm	Line	Tree	Grid	General
Greedy	$\Omega(\sqrt{n})$ [1]	$\Omega(n)$ [1]	$\pmb{\Omega(\sqrt{n})}$	$\Omega(m)$
NTG	$\pmb{\tilde{O}(\sqrt{n})}$	$\pmb{\tilde{O}(n)}$	$\pmb{\tilde{\Theta}(n^{2/3})}$	$\tilde{O}(m)$
Previous bounds	$O(n^{2/3})$ [1]	$O(nh)$ [1]	-	$O(mn)$ [1]
Merge and Deliver (M&D)	$\pmb{O(\log^3 n)}$	$\pmb{O(h\log^2 n)}$	-	-

in the line algorithm, we give a $O(\log^2 n)$-competitive algorithm when all packets are destined for the root. Such tree networks are motivated by the work on buffer overflows of merging streams of packets by Kesselman *et al* [2] and the work on information gathering by Kothapalli and Scheideler [3] and Azar and Zachut [4]. The result extends to a randomized $O(h\log^2 n)$-competitive algorithm when packet destinations are arbitrary.

Grid Networks: For adaptive setting in grid networks, we show that NTG with one-bend routing is $\tilde{\Theta}(n^{2/3})$-competitive. We establish a lower bound of $\Omega(\sqrt{n})$ on the competitive ratio for greedy algorithms with shortest path routing.

General Networks: Finally, for arbitrary network topologies, we show that any greedy algorithm is $\Omega(m)$-competitive and prove that NTG is, in fact, $\tilde{\Theta}(m)$-competitive. These results hold for both adaptive and non-adaptive settings, where NTG routes on a shortest path in the adaptive case.

Related Work. Dynamic store-and-forward routing networks have been studied extensively. An excellent survey of packet drop policies in communication networks can be found in [5]. Much of the earlier work has focused on the issue of stability with packets being injected in either probabilistic or adversarial manner. In stability analysis the goal is to understand how the buffer size at each link needs to grow as a function of the packet injection rate so that packets are never dropped. A *stable* protocol is one where the maximum buffer size does not grow with time. For work in the probabilistic setting, see [6–10]. Adversarial queuing theory introduced by Borodin *et al* [11] has also been used to study the stability of protocols and it has been shown in [11, 12] that certain greedy algorithms are unstable, that is, require unbounded buffer sizes. In particular, NTG is unstable.

The idea of using the entire network to store packets effectively has been used in [13] but in their model, packets can not be buffered while in transit and the performance measure is the time required to deliver all packets. Moreover, packets never get dropped because there is no limit on the number of packets which can be stored at their respective source nodes.

Throughput competitiveness was highlighted as a network performance measure by Awerbuch *et al* in [14]. They used adversarial traffic to analyze store-

and-forward routing algorithms in [15], but they compared their throughput to an adversary restricted to a certain class of strategies and with smaller buffers.

Kesselman *et al* [2] study the throughput competitiveness of work-conserving algorithms on line and tree networks when the adversary is a work-conserving algorithm too. Work-conserving algorithms always forward a packet if possible but unlike greedy, they need not accept packets when there is space in the buffer. However, they do not make assumptions about uniform link bandwidths or buffer sizes as in [1] and our model. They also consider the case when packets have different weights. Non-preemptive policies for packets with different values but with unit sized buffers have been analyzed in [16, 17].

In a recent parallel work, Azar and Zachut [4] also obtain centralized algorithms with polylog(n) competitive ratios on lines. They first obtain a $O(\log n)$-competitive deterministic algorithm for the special case when all packets have the same destination (which is termed information gathering) and then show that it can be extended to a randomized algorithm with $O(\log^2 n)$-competitive ratio for the general case. Their result can be extended to get $O(h \log n)$-competitive ratio on trees. Information gathering problem is similar to our notion of balanced instances (see Section 2.3) though their techniques are very different from ours – they construct an online reduction from the fractional buffers packet routing with bounded delay problem to fractional information gathering and the former is solved by an extension of the work of Awerbuch *et al* [14] for the discrete version. The fractional algorithm for information gathering is then transformed to a discrete one.

2   Preliminaries

2.1   Model and Problem Statement

We model the network as a directed graph on n nodes and m links. A node has at most I traffic ports where new packets can be injected, at most one at each port. Each link has an output port at its tail with a capacity $B > 0$ buffer, and an input port at its head that can store 1 packet. We assume uniform buffer size B at each link and bandwidth of each link to be 1.

Time is synchronous and each time step consists of *forwarding* and *switching* sub-steps. During the forwarding phase, each link selects at most one packet from its output buffer according to an *output scheduling policy* and forwards it to its input buffer. During the switching phase, a node clears all packets from its traffic ports and input ports at its incoming link(s). It delivers packets if the node is their destination or assigns them to the output port of the outgoing link on their respective paths. When more than B packets are assigned to a link's output buffer, packets are discarded based on a *contention-resolution policy*. We consider preemptive contention-resolution policies that can replace packets already stored at the buffer with new packets. A routing protocol specifies the output scheduling and contention-resolution policy. We are interested in *online* policies which make decisions with no knowledge of future packet arrivals.

Each injected packet comes with a specified destination. We assume that the destination is different from the source, otherwise the packet is routed optimally by any algorithm and does not interfere with other packets. The *goal* of the routing algorithm is to maximize throughput, that is, the total number of packets delivered by it. We distinguish between two types of algorithms, namely *centralized* and *distributed*. A centralized algorithm makes coordinated decisions at each node taking into account the state of the entire network while a distributed algorithm requires that each node make its decisions based on local information only. Distributed algorithms are of great practical interest for large networks. Centralized algorithms, on the other hand, give us insight into the inherent complexity of the problem due to the online nature of the input.

2.2   Useful Background Results and Definitions

The following lemma which gives an upper bound on the packets that can be absorbed over a time interval, will be a recurrent idea while analyzing algorithms.

Lemma 1. [1] In a network with m links, the number of packets that can be delivered and buffered in a time interval of length T units by *any* algorithm is $O(mT/d + mB)$, where $d > 0$ is a lower bound on the number of links in the shortest path to the destination for each injected packet.

The proof bounds the available bandwidth on all links and compares it against the minimum bandwidth required to deliver a packet injected during the time interval. The $O(mB)$ term accounts for packets buffered at the beginning and the end of the interval, which can be arbitrarily close to their destination.

An important class of distributed algorithms is *greedy algorithms* where each link always accepts an incoming packet if there is available buffer space and always forwards a packet if its buffer is non-empty. Based on how contention is resolved when receiving/forwarding packets, we obtain different algorithms. *Nearest-To-Go* (NTG) is a natural greedy algorithm which always selects a shortest path to route a packet and prefers a packet that has shortest remaining distance to travel, in both choosing packets to accept or forward.

Line and tree networks are of special interest to us, where there is a unique path between every source and destination pair. A *line network* on n nodes is a directed path with nodes labeled $1, 2, \ldots, n$ and a link from node i to $i + 1$, for $i \in [1, n)$. A *tree network* is a rooted tree with links directed towards the root. Note that there is a one-to-one correspondence between links and nodes (for all but one node) on lines and trees. For simplicity, whenever we refer to a node's buffer, we mean the output buffer of its unique outgoing link.

A simple useful property of any greedy algorithm is as follows:

Lemma 2. [1] *If at some time t, a greedy algorithm on a line network has buffered $k \leq nB$ packets, then it delivers at least k packets by time $t + (n - 1)B$.*

At times it will be useful to geometrically group packets into classes in the following manner. A packet belongs to **class** j if the length of the path on which it is routed by the optimal algorithm is in the range $[2^j, 2^{j+1})$. In the case when paths are unique or specified the class of a packet can be determined exactly.

2.3   Balanced Instances for Line Networks

At each time step, a node makes two decisions – which packets to accept (and which to drop if buffer overflows) and which packets to forward. When all packets have the same destination, the choice essentially comes down to whether a node should forward a packet or not, Motivated by this, we define the notion of a **balanced instance**, so that an algorithm can focus only on deciding when to forward packets. In a balanced instance, each released packet travels a distance of $\Theta(n)$. We establish a theorem that shows that with a logarithmic loss in competitive ratio, *any* algorithm on balanced instances can be converted to a one on arbitrary instances. Azar and Zachut [4] prove a similar theorem that it suffices to do so for information gathering.

Theorem 1. *An α-competitive algorithm for balanced instances gives an $O(\alpha \log n)$-competitive randomized algorithm for any instance on a line network.*

Proof. It suffices to randomly partition the line into disjoint networks of length ℓ each s.t. (i) online algorithm only accepts packets whose origin and destination are contained inside one of the smaller networks, (ii) each accepted packet travels a distance of $\Omega(\ell)$, and (iii) $\Omega(1/\log n)$-fraction of the packets routed by the optimal algorithm are expected to satisfy condition (i) and (ii).

We first pick a random integer $i \in [0, \log n)$. We next look at three distinct ways to partition the line network into intervals of length $3d$ each, starting from node with index 1, $d+1$, or $2d+1$, where $d = 2^i$. The online algorithm chooses one of the three partitionings at random and only routes class i packets whose end-points lie completely inside an interval. It is not hard to see that the process decomposes the problem into line routing instances satisfying (iii).

The above partitioning can be done by releasing $O(1)$ packets initially as a part of the network setup process. It can be used by distributed algorithms too.

3   Nearest-to-Go Protocol

In this section we prove asymptotically tight bounds (upto logarithmic factors) on the performance of Nearest-To-Go (NTG) on lines, show their extension to grids and then prove bounds on general graphs.

For a sequence σ of packets, let P_j denote packets of class j. Also, let OPT be the set of packets delivered by an optimal algorithm. Let class i be the class from which optimal algorithm delivers most packets, that is, $|P_i \cap \text{OPT}| = \Omega(|\text{OPT}|/\log n)$. We will compare the performance of NTG to an optimal algorithm on P_i, which we denote by OPT_i. It is easy to see that if NTG is α-competitive against OPT_i then it is $O(\alpha \log n)$-competitive against OPT.

3.1   Line Networks

We show that NTG is $O(\sqrt{n} \log n)$ competitive on line networks with $B > 1$. This supplements the lower bound $\Omega(\sqrt{n})$ on the competitive ratio of any greedy algorithm and improves the $O(n^{2/3})$-competitive ratio of NTG obtained in [1].

We prove that NTG maintains competitiveness w.r.t. to OPT_i on a suitable collection of instances, referred to as *virtual instances*.

Virtual Instances. Let class i be defined as above. Consider the three ways of partitioning defined in Section 2.3 with i as the chosen class. Instead of randomly choosing a partitioning, we pick one containing at least $|P_i \cap \mathrm{OPT}|/3$ packets.

For each interval in this optimal partitioning, we create a new *virtual instance* consisting of packets (of all classes) injected within the interval as well as any additional packets that travel some portion of the interval during NTG execution on the original sequence σ. Let us fix any such instance. W.l.o.g. we relabel the nodes by integers 1 through $3d$. We also add two specials nodes of index 0 (to model packets incoming from earlier instances) and $3d + 1$ (to model packets destined beyond the current instance). We divide the packets traveling through the instance as *good* and *bad* where a packet is good iff its destination is not the node labeled $3d + 1$. Due to the NTG property, bad packets never delay good packets or cause them to be dropped. Hence while analyzing our virtual instance, we will only consider good packets. We also let $n = 3d + 2$.

Analysis. For each virtual instance, we compare the number of good packets delivered by NTG to the number of class i packets contained within the interval that OPT_i delivered. The analysis (and not the algorithm) is in rounds where each round has 2 phases. Phase 1 terminates when OPT_i absorbs nB packets for the first time, and Phase 2 has a fixed duration of $(n - 1)B$ time units.

Packets are given a unique index and credits to account for dropped packets. We swap packets (meaning swap their index and credits) when either of the two events occurs: (i) a *new* packet causes another to be dropped, or (ii) a *moving* packet is delayed by another packet. The rules respectively ensure that a packet is (virtually) never dropped after being accepted and is (virtually) never delayed at a node once it starts moving. Recall that NTG drops packets at a node only when the node has B packets. For every node, mark the first time it overflows and give a charge of 2 to each of its packets. We do not charge that node for the next B time steps. Note that charge of 2 is enough since OPT_i could have moved away at most B packets from that node during that period and have B packets in its buffers. Repeat the process till the end of Phase 1.

Lemma 3. *NTG absorbs $\Omega(\sqrt{n}B)$ packets by the end of Phase 1.*

Proof. Let $2c$ be the maximum charge accumulated by a packet in Phase 1. NTG absorbed $\geq \frac{nB}{2c}$ packets since OPT_i absorbed at least nB packets. Now consider the rightmost packet with $2c$ charge. Split its charge into two parts, $2s$ – charge accumulated while static at its source node and $2m$ – charge accumulated while moving. Since the packet was charged $2s$, it stayed idle for at least $B(s - 1)$ time units and hence at least that many packets crossed over it. It is never in the same buffer with these packets again. On the other hand, while the packet was moving, it left behind $m(B - 1)$ packets which never catch up with it again. Therefore, NTG absorbs at least $\max\{\frac{nB}{2c}, (c - 1)(B - 1)\} = \Omega(\sqrt{n}B)$ packets.

By the end of the round, OPT_i buffered and delivered $O(nB)$ class i packets since it absorbed at most $2nB$ packets in Phase 1 and by Lemma 1, OPT_i absorbed $O(nB)$ packets during Phase 2. By Lemma 2 and 3, NTG delivers $\Omega(\sqrt{n}B)$ packets by the end of Phase 2. Thus we obtain an $O(\sqrt{n})$ competitive ratio. Note we can clear the charge of packets in NTG buffers and consider OPT_i buffers to be empty at the end of the round.

Theorem 2. *For $B > 1$, NTG is an $O(\sqrt{n}\log n)$-competitive algorithm on line networks.*

Remark 1. We note here that one can show that allowing greedy a buffer space that is α times larger can not improve the competitive ratio to $o(\sqrt{n}/\alpha)$.

3.2   Grid Networks

We consider 2-dimensional regular grids on n nodes, that is, with $\sqrt{n}$ rows and columns, where nodes are connected by directed links to the neighbors in the same row and column. Since it is no longer true that there is a unique shortest path between a source-destination pair on a grid, we restrict NTG to one-bend routing – every packet first moves along the row it is injected to its destination's column and then moves along that column. It is not hard to see that since distance between any two nodes is $O(\sqrt{n})$, any greedy algorithm that routes packets on shortest paths is $\Omega(\sqrt{n})$-competitive. An adversary releases $O(nB)$ packets at all links, all with the same destination. By time $O(\sqrt{n}B)$, the greedy algorithm will deliver $O(\sqrt{n}B)$ packets and drop the remaining while an optimal algorithm can deliver all packets. The theorem below gives tight bounds on the performance of NTG with one-bend routing on grid networks.

Theorem 3. *For $B > 1$, NTG with one-bend routing is $\tilde{\Theta}(n^{2/3})$-competitive algorithm on grid networks.*

3.3   General Networks

The following theorem improves the $O(mn)$ upper bound on the performance of NTG on general graphs obtained in [1].

Theorem 4. *NTG is $\tilde{\Theta}(m)$-competitive on any network for both adaptive and non-adaptive settings. Moreover, every greedy algorithm is $\Omega(m)$ competitive.*

4   The Merge and Deliver Algorithm

In this section, we present a randomized, centralized polylog(n)-competitive algorithm, referred to as *Merge and Deliver* (M&D), for line and tree networks.

4.1 Line Networks

By Theorem 1, it suffices to obtain a polylog(n)-competitive algorithm on balanced instances. A balanced instance allows us to treat all packets as equal and focus on the aspect of deciding if some buffered packet should be forwarded or not, which is where we feel the real complexity of the problem lies. Centralized algorithms make it possible to take a global view of the network and spread packets all around to overcome some of the challenges discussed in Section 1.

The M&D algorithm never drops accepted packets and forwards them in such a manner that it creates a small number of contiguous sequences of nodes with full buffers, which we call *segments*. A key observation is that once the adversary has filled buffers in a segment, it can absorb only one extra packet per time unit over the entire segment! This follows from the fact that there is a unit bandwidth out of the segment. The algorithm tries to decrease the number of such segments by merging segments of similar length. After a few rounds of merges, M&D accumulates enough packets without dropping too many. It then delivers them greedily knowing (from Lemma 1) that the adversary cannot get much leverage during that time.

We divide the execution of Merge and Deliver into rounds where each round has 2 phases and starts with empty buffers. Assume *w.l.o.g.* that OPT buffers are also empty, otherwise we view the stored packets as delivered.

1. Phase 1 is the *receiving phase*. The phase ends when the number of packets absorbed by M&D exceeds $nB/p(n)$ for a suitable $p(n) = \text{polylog}(n)$. We show that OPT absorbs $O(nB)$ packets in the phase.
2. Phase 2 is the *delivery phase* and consists of $(n-1)B$ time steps. In the phase, M&D greedily forwards packets without accepting any new packets. By Proposition 2 it delivers all packets that are buffered by the end of Phase 1. We show that OPT absorbs $O(nB)$ packets in this phase.

For clarity of exposition, we first describe and analyze the algorithm for buffers of size 2. We then briefly sketch extension to any $B > 2$.

Algorithm for Receiving Phase. At a given time step, a node can be *empty*, *single*, or *double*, where the terms denote the number of packets in its buffer. We define a *segment* to be a sequence (not necessarily maximal) of double nodes. Segments are assigned to classes, and a new double node (implying it received at least one new packet) is assigned to class 0. We ensure that a segment progresses to a higher class after some duration of time which depends on its current class.

It is instructive to think of the packets as forming two layers on the nodes. We call the bottom layer the *transport layer*. Now, consider a pair of segments s.t. there are only single nodes in between. It is possible to merge them, that is make the top layer of both segments adjacent to each other, using the transport layer, in a time proportional to the length of the left segment. Since segments might be separated by a large distance, it is not possible to physically move all those packets. Instead, we are only interested that the profile of packets changes as if the segments had actually become adjacent. M&D accomplishes this by forwarding at each time step a packet from the right end of the left segment and all the intermediate single nodes. We call this process *teleporting* and a teleport from node u to v is essentially forwarding a packet at each node in $[u, v-1]$.

We will only merge segments of the same class. The merged segment is then assigned to the next higher class. We ensure that segments of the same class have similar lengths or at least have enough credits as other segment of the same class. Thus we will be able to account for dropped packets in each class separately and will show that the class of a segment is bounded during the phase.

Note, that there could be merging segments of other classes in between two segments of a given class. To overcome this, we divide time into blocks of $\log n$ size. At the ith time unit of each block, only class i segments are *active*. Also, each class i starts its own clock at the beginning of Phase 1, with period 2^i. This clock ticks only during the time units of class i. Fix a class i, and consider its period and time steps, and the class i segments. At the beginning of the period, segments are numbered from left to right starting from 1. M&D merges each odd-number segment with the segment that follows it. The (at most one) unpaired segment is merged into the transport layer. At the end of the period, a merged pair becomes one segment of class $i+1$ and waits idle until the next class $i+1$ period begins. We now formally define the segment merging operation.

Segment Merging. Consider the pair of segments such that the first segment ends at node e and the second segment starts at $s+1$ at the time step. We teleport a packet from e to s. We distinguish the following special cases:

(i) If $e = s$ or one of the segments has length 0, the merge is completed.
(ii) If $s > e$ is already a double node we teleport the packet to s', where $s' \geq e$ is the minimum index s.t. nodes $[s', s]$ are double. The profile changes as if we moved all segments in $[s', s]$, which are of different class and therefore inactive, backward by one index and teleported the packet to s.
(iii) If the two segments are not linked by a transport layer, that is, there is an empty node to the right of the odd segment *or* the odd segment is the last segment of class i, we teleport the packet to the first empty node to its right.

Analysis. We first note that online can accept one packet (and drop at most one packet) at a single node. Hence the number of packets which online receives plus the number of packets dropped at double nodes is an upper bound on the number of packets dropped by online. We need to count only the packets dropped at double nodes to compare the packets in optimum to that in online. The movement strategy in Phase 1 ensures that there is never a net outflow of packets from the transport layer to the upper layer. Therefore we can treat packets in transport layer as good as delivered and devise a charging scheme which only draws credits from packets in top layer to account for dropped packets.

Since packets could join the transport layer during merging, we cannot ensure that segment lengths grow geometrically, but we show an upper bound on their lengths in this lemma which follows easily from induction.

Lemma 4. *At the start of class i period, the length of a class i segment is $\leq 2^i$.*

We note that each merge operation decreases the length of the left segment of a merging segment pair by 1. The next lemma follows.

Lemma 5. *The merge of two class i segments needs at most 2^i ticks and can be accomplished in a class i period.*

Corollary 1. *If a class i segment is present at the beginning of class i period, by the end of the period, its packets either merge with another class i segment to form a class $i + 1$ segment, or join the transport layer.*

We assign $O(\log^2 n)$ credits to each packet which are absorbed as follows:

- $\log^2 n + 3 \log n$ units of *merging* credits, which are distributed equally over all classes, that is a credit of $\log n + 3$ in each class.
- $\log^2 n + 2 \log n$ units of *idling* credits, which are distributed equally over all classes, that is a credit of $\log n + 2$ in each class.

When a new segment of class 0 is created it gets the credit of the new packet injected in the top layer. When two segments of class i are merged the unused credits of the segments are transferred to the new segment of class $i + 1$. If a packet gets delivered or joins the transport layer, it gives its credit to the segment it was trying to merge into. The following lemma establishes that a class i segment has enough credit irrespective of its length.

Lemma 6. *A class i segment has $2^i(\log n + 3)$ merging credits and $2^i(\log n + 2)$ idling credits. The class of a segment is bounded by $\log n$.*

Proof. The first part follows by induction. If a class i segment has $2^i(\log n + 3)$ merging credits, there must have been 2^i class 0 segments to begin with. Since we stop the phase right after $n/p(n)$ packets are absorbed, $i < \log n$.

We now analyze packets dropped by segments in each class. Consider a segment of length $l \leq 2^i$ in class i. It could have been promoted to class i in the middle of a class i period but since the period spans 2^i blocks of time, it stays idle for $\leq 2^i \log n$ time units. Number of packets dropped is bounded by $B = 2$ times the segment length plus the total bandwidth out of the segment which is $\leq 2 \cdot 2^i + 2^i \log n$. This can be paid by the idling credit of the segment. At the beginning of a new class i period, it starts merging with another segment of length, say $l' \leq 2^i$, to its right. The total number of packets dropped in both segments is bounded by $2(l + l + l') + 2 \cdot 2^i \log n \leq 6 \cdot 2^i + 2 \cdot 2^i \log n$, using a similar argument. This can be paid by the merging credit of the segments.

We let $p(n) = 2 \log^2 n + 5 \log n$ and note that the number of packets dropped by online during Phase 1 can be bounded by the credit assigned to all absorbed packets. Using Lemma 1 we show that OPT can only deliver and buffer $O(n)$ packets during Phase 2. Since M&D delivers at least $n/p(n)$ packets in the round, a competitive ratio of $O(p(n)) = O(\log^2 n)$ is established.

Generalization to $B > 2$. We can generalize M&D to $B > 2$ by redefining the concept of a double node and segment. Also, we increase the credits assigned to a packet by a factor of $B/(B-1) \leq 2$, scaling $p(n)$ accordingly.

Theorem 5. *For $B > 1$, Merge and Deliver is a randomized $O(\log^3 n)$-competitive algorithm on a line network.*

4.2   Tree Networks

The Merge and Deliver algorithm extends naturally to tree networks. We consider rooted tree networks where each edge is directed towards the root node. It is known that on trees with n nodes and height h, a greedy algorithm is $\Omega(n/h)$-competitive [1]. We show the following:

Theorem 6. *For $B > 1$, there is a $O(\log^2 n)$-competitive centralized deterministic algorithm for trees when packets have the same destination. When destinations are arbitrary, there is a randomized $O(h \log^2 n)$-competitive algorithm.*

5   Line Networks with Unit Buffers

For unit buffer sizes, Aiello *et al* [1] give a lower bound of $\Omega(n)$ on the competitive ratio of any greedy algorithm. We show that the lower bound holds for *any* deterministic algorithm. However, with randomization, we can beat this bound. We show that the following simple strategy is $\tilde{O}(\sqrt{n})$-competitive. The online algorithm chooses uniformly at random one of the two strategies, NTG and modified NTG, call it NTG', that forwards packets at every other time step. The intuition is that if in NTG there is a packet that collides with c other packets then $\Omega(c)$ packets are absorbed by NTG'.

Theorem 7. *For $B = 1$ any deterministic online algorithm is $\Omega(n)$-competitive on line networks. There is a randomized $\tilde{O}(\sqrt{n})$-competitive algorithm for the problem.*

6   Conclusion

Our results establish a strong separation between the performance of natural greedy algorithms and centralized routing algorithms even on the simple line topology. An interesting open question is to prove or disprove the existence of distributed polylog(n)-competitive algorithms for line networks.

References

1. Aiello, W., Ostrovsky, R., Kushilevitz, E., Rosén, A.: Dynamic routing on networks with fixed-size buffers. In: Proceedings of the 14th SODA. (2003) 771–780

2. Kesselman, A., Mansour, Y., Lotker, Z., Patt-Shamir, B.: Buffer overflows of merging streams. In: Proceedings of the 15th SPAA. (2003) 244–245
3. Kothapalli, K., Scheideler, C.: Information gathering in adversarial systems: lines and cycles. In: SPAA '03: Proceedings of the fifteenth annual ACM symposium on Parallel algorithms and architectures. (2003) 333–342
4. Azar, Y., Zachut, R.: Packet routing and information gathering in lines, rings and trees. In: Proceedings of 15th ESA. (2005)
5. Labrador, M., Banerjee, S.: Packet dropping policies for ATM and IP networks. IEEE Communications Surveys 2 (1999)
6. Broder, A.Z., Frieze, A.M., Upfal, E.: A general approach to dynamic packet routing with bounded buffers. In: Proceedings of the 37th FOCS. (1996) 390
7. Broder, A., Upfal, E.: Dynamic deflection routing on arrays. In: Proceedings of the 28th annual ACM Symposium on Theory of Computing. (1996) 348–355
8. Mihail, M.: Conductance and convergence of markov chains–a combinatorial treatment of expanders. In: Proceedings of the 30th FOCS. (1989) 526–531
9. Mitzenmacher, M.: Bounds on the greedy routing algorithm for array networks. Journal of Computer System Sciences 53 (1996) 317–327
10. Stamoulis, G., Tsitsiklis, J.: The efficiency of greedy routing in hypercubes and butterflies. IEEE Transactions on Communications 42 (1994) 3051–208
11. Borodin, A., Kleinberg, J., Raghavan, P., Sudan, M., Williamson, D.P.: Adversarial queueing theory. In: Proceedings of the 28th STOC. (1996) 376–385
12. Andrews, M., Mansour, Y., Fernéndez, A., Kleinberg, J., Leighton, T., Liu, Z.: Universal stability results for greedy contention-resolution protocols. In: Proceedings of the 37th FOCS. (1996) 380–389
13. Busch, C., Magdon-Ismail, M., Mavronicolas, M., Spirakis, P.G.: Direct routing: Algorithms and complexity. In: ESA. (2004) 134–145
14. Awerbuch, B., Azar, Y., Plotkin, S.: Throughput competitive on-line routing. In: Proceedings of 34th FOCS. (1993) 32–40
15. Scheideler, C., Vöcking, B.: Universal continuous routing strategies. In: Proceedings of the 8th SPAA. (1996) 142–151
16. Kesselman, A., Mansour, Y.: Harmonic buffer management policy for shared memory switches. Theor. Comput. Sci. 324 (2004) 161–182
17. Lotker, Z., Patt-Shamir, B.: Nearly optimal FIFO buffer management for DiffServ. In: Proceedings of the 21th PODC. (2002) 134–143

Rounding Two and Three Dimensional Solutions of the SDP Relaxation of MAX CUT

Adi Avidor[*] and Uri Zwick

School of Computer Science
Tel-Aviv University, Tel-Aviv 69978, Israel
{adi,zwick}@tau.ac.il

Abstract. Goemans and Williamson obtained an approximation algorithm for the MAX CUT problem with a performance ratio of $\alpha_{GW} \simeq 0.87856$. Their algorithm starts by solving a standard SDP relaxation of MAX CUT and then rounds the optimal solution obtained using a random hyperplane. In some cases, the optimal solution of the SDP relaxation happens to lie in a low dimensional space. Can an improved performance ratio be obtained for such instances? We show that the answer is yes in dimensions two and three and conjecture that this is also the case in any higher fixed dimension. In two dimensions an optimal $\frac{32}{25+5\sqrt{5}}$-approximation algorithm was already obtained by Goemans. (Note that $\frac{32}{25+5\sqrt{5}} \simeq 0.88456$.) We obtain an alternative derivation of this result using *Gegenbauer polynomials*. Our main result is an improved rounding procedure for SDP solutions that lie in $\mathbb{R}^3$ with a performance ratio of about 0.8818 . The rounding procedure uses an interesting yin-yan coloring of the three dimensional sphere. The improved performance ratio obtained resolves, in the negative, an open problem posed by Feige and Schechtman [STOC'01]. They asked whether there are MAX CUT instances with integrality ratios arbitrarily close to $\alpha_{GW} \simeq 0.87856$ that have optimal embedding, i.e., optimal solutions of their SDP relaxations, that lie in $\mathbb{R}^3$.

1 Introduction

An instance of the MAX CUT problem is an undirected graph $G = (V, E)$ with non-negative weights w_{ij} on the edges $(i, j) \in E$. The goal is to find a subset of vertices $S \subseteq V$ that maximizes the weight of the edges of the *cut* $(S, \bar{S})$. MAX CUT is one of Karp's original NP-complete problems, and has been extensively studied during the last few decades. Goemans and Williamson [GW95] used semidefinite programming and the *random hyperplane rounding* technique to obtain an approximation algorithm for the MAX CUT problem with a performance guarantee of $\alpha_{GW} \simeq 0.87856$. Håstad [Hås01] showed that MAX CUT does not have a $(16/17 + \varepsilon)$-approximation algorithm, for any $\varepsilon > 0$, unless $P = NP$. Recently, Khot *et al.* [KKMO04] showed that a certain plausible conjecture implies that MAX CUT does not have an $(\alpha_{GW} + \varepsilon)$-approximation, for any $\varepsilon > 0$.

The algorithm of Goemans and Williamson starts by solving a standard SDP relaxation of the MAX CUT instance. The value of the optimal solution obtained is an

---

[*] This research was supported by the ISRAEL SCIENCE FOUNDATION (grant no. 246/01).

C. Chekuri et al. (Eds.): APPROX and RANDOM 2005, LNCS 3624, pp. 14–25, 2005.

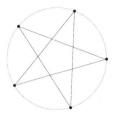

Fig. 1. An optimal solution of the SDP relaxation of C_5. Graph nodes lie on the unit circle

upper bound on the size of the maximum cut. The ratio between these two quantities is called the *integrality ratio* of the instance. The infimum of the integrality ratios of all the instances is the integrality ratio of the relaxation. The study of the integrality ratio of the standard SDP relaxation was started by Delorme and Poljak [DP93a, DP93b]. The worst instance they found was the 5-cycle C_5 which has an integrality ratio of $\frac{32}{25+5\sqrt{5}} \simeq 0.884458$. The optimal solution of the SDP relaxation of C_5, given in Figure 1, happens to lie in $\mathbb{R}^2$. Feige and Schechtman [FS01] showed that the integrality ratio of the standard SDP relaxation of MAX CUT is exactly α_{GW}. More specifically, they showed that for every $\varepsilon > 0$ there exists a graph with integrality ratio of at most $\alpha_{GW} + \varepsilon$. The optimal solution of the SDP relaxation of this graph lies in $\mathbb{R}^{\Theta(\sqrt{1/\varepsilon}\log 1/\varepsilon)}$. An interesting open question, raised by Feige and Schechtman, was whether it is possible to find instances of MAX CUT with integrality ratio arbitrarily close to $\alpha_{GW} \simeq 0.87856$ that have optimal embedding, i.e., optimal solutions of their SDP relaxations, that lie in $\mathbb{R}^3$. We supply a negative answer to this question.

Improved approximation algorithms for instances of MAX CUT that do not have large cuts were obtained by Zwick [Zwi99], Feige and Langberg [FL01] and Charikar and Wirth [CW04]. These improvements are obtained by modifying the *random hyperplane rounding* procedure of Goemans and Williamson [GW95]. Zwick [Zwi99] uses *outward rotations*. Feige and Langberg [FL01] introduced the RPR^2 rounding technique. Other rounding procedures, developed for the use in MAX 2-SAT and MAX DI-CUT algorithms, were suggested by Feige and Goemans [FG95], Matuura and Matsui [MM01a, MM01b] and Lewin *et al.* [LLZ02].

In this paper we study the approximation of MAX CUT instances that have a low dimensional optimal solution of their SDP relaxation. For instances with a solution that lies in the plane, Goemans (unpublished) achieved a performance ratio of $\frac{32}{25+5\sqrt{5}} = \frac{4}{5\sin^2(\frac{2\pi}{5})} \simeq 0.884458$. We describe an alternative way of getting such a performance ratio using a random procedure that uses *Gegenbauer polynomials*. This result is optimal in the sense that it matches the integrality ratio obtained by Delorme and Poljak [DP93a, DP93b].

Unfortunately, the use of the Gegenbauer polynomials does not yield improved approximation algorithms in dimensions higher than 2. To obtain an improved ratio in dimension three we use an interesting yin-yan like coloring of the unit sphere in $\mathbb{R}^3$. The question whether a similar coloring can be used in four or more dimensions remains an interesting open problem.

The rest of this extended abstract is organized as follows. In the next section we describe our notations and review the approximation algorithm of Goemans and

Williamson. In Section 3 we introduce the *Gegenbauer polynomials rounding* technique. In Section 4 we give our Gegenbauer polynomials based algorithm for the plane and compare it to the algorithm of Goemans. In Section 5 we introduce the *2-coloring rounding*. Finally, in Section 6 we present our 2-coloring based algorithm for the three dimensional case.

2   Preliminaries

Let $G = (V, E)$ be an undirected graph with non-negative weights attached to its edges. We let w_{ij} be the weight of an edge $(i, j) \in E$. We may assume, w.l.o.g., that $E = V \times V$. (If an edge is missing, we can add it and give it a weight of 0.) We also assume $V = \{1, \ldots, n\}$. We denote by S^{d-1} the unit sphere in $\mathbb{R}^d$. We define:

$$sgn(x) = \begin{cases} 1 & \text{if } x \geq 0 \\ -1 & \text{otherwise} \end{cases}$$

The Goemans and Williamson algorithm embeds the vertices of the input graph on the unit sphere S^{n-1} by solving a semidefinite program, and then rounds the vectors to integral values using random hyperplane rounding.

MAX CUT Semidefinite Program: MAX CUT may be formulated as the following integer quadratic program:

$$\max \sum_{i,j} w_{ij} \frac{1 - x_i \cdot x_j}{2}$$
$$s.t. \quad x_i \in \{-1, 1\}$$

Here, each vertex i is associated with a variable $x_i \in \{-1, 1\}$, where -1 and 1 may be viewed as the two sides of the cut. For each edge $(i, j) \in E$ the expression $\frac{1 - x_i x_j}{2}$ is 0 if the two endpoints of the edge are on the same side of the cut and 1 if the two endpoints are on different sides.

The latter program may be *relaxed* by replacing each variable x_i with a unit *vector* v_i in $\mathbb{R}^n$. The product between the two variables x_i and x_j is replaced by an inner product between the vectors v_i and v_j. The following semidefinite programming *relaxation* of MAX CUT is obtained:

$$\max \sum_{i,j} w_{ij} \frac{1 - v_i \cdot v_j}{2}$$
$$s.t. \quad v_i \in S^{n-1}$$

This semidefinite program can be solved in polynomial time up to any specified precision. The value of the program is an upper bound on the value of the maximal cut in the graph. Intuitively, the semidefinite programming embeds the input graph G on the unit sphere S^{n-1}, such that adjacent vertices in G are far apart in S^{n-1}.

Random Hyperplane Rounding: The *random hyperplane rounding* procedure chooses a uniformly distributed unit vector r on the sphere S^{n-1}. Then, x_i is set to $sgn(v_i \cdot r)$. We will use the following simple Lemma, taken from [GW95]:

Lemma 1. *If $u, v \in S^{n-1}$, then $Pr_{r \in S^{n-1}}[sgn(u \cdot r) \neq sgn(v \cdot r)] = \frac{\arccos(u \cdot v)}{\pi}$.*

3 Gegenbauer Polynomials Rounding

For any $m \geq 2$, and $k \geq 0$, the *Gegenbauer* (or ultraspherical) polynomials $G_k^{(m)}(x)$ are defined by the following recurrence relations:

$$G_0^{(m)}(x) = 1, \quad G_1^{(m)}(x) = x$$

$$G_k^{(m)}(x) = \frac{1}{k+m-3}\left((2k + m - 4)xG_{k-1}^{(m)}(x) - (k-1)G_{k-2}^{(m)}(x)\right) \quad \text{for } k \geq 2$$

$G_k^{(2)}$ are Chebyshev polynomials of the first kind, $G_k^{(3)}$ are Legendre polynomials of the first kind, and $G_k^{(4)}$ are Chebyshev polynomials of the second kind. The following interesting Theorem is due to Schoenberg [Sch42]:

Theorem 1. *A polynomial f is a sum of Gegenbauer polynomials $G_k^{(m)}$, for $k \geq 0$, with non-negative coefficients, i.e., $f(x) = \sum_{k \geq 0} \alpha_k G_k^{(m)}(x), \alpha_k \geq 0$, if and only if for any $v_1, \ldots, v_n \in S^{m-1}$ the matrix $(f(v_i \cdot v_j))_{ij}$ is positive semidefinite.*

Note that $G_k^{(m)}(1) = 1$ for all $m \geq 2$ and $k \geq 0$. Let $v_1, \ldots, v_n \in S^{m-1}$ and let f be a convex sum of $G_k^{(m)}$ with respect to k, i.e., $f(x) = \sum_{k \geq 0} \alpha_k G_k^{(m)}(x)$, where $\alpha_k \geq 0, \sum_k \alpha_k = 1$. Then, there exist **unit** vectors $v'_1, \ldots, v'_n \in S^{n-1}$ such that $v'_i \cdot v'_j = f(v_i \cdot v_j)$, for all $1 \leq i, j \leq n$.

In this setting, we define the *Gegenbauer polynomials rounding* technique with parameter f as:

1. Find vectors v'_i satisfying $v'_i \cdot v'_j = f(v_i \cdot v_j)$, where $1 \leq i, j \leq n$ (This can be done by solving a semidefinite program)
2. Let r be a vector uniformly distributed on the unit sphere S^{n-1}
3. Set $x_i = \text{sgn}(v'_i \cdot r)$, for $1 \leq i \leq n$

Note that the vectors $v'_1, \ldots, v'_n$ do not necessarily lie on the unit sphere S^{m-1}. Therefore, the vector r is chosen from S^{n-1}. Note also that the *random hyperplane rounding* of Goemans and Williamson is a special case of the *Gegenbauer polynomials rounding* technique as f can be taken to be $f = G_1^{(m)}$. The next corollary is an immediate result of Lemma 1:

Corollary 1. *If $v_1, \ldots, v_n$ are rounded using Gegenbauer polynomials rounding with polynomial f, then $Pr[x_i \neq x_j] = \frac{\arccos(f(v_i \cdot v_j))}{\pi}$, where $1 \leq i, j \leq n$.*

Alon and Naor [AN04] also studied a rounding technique which applies a polynomial on the elements of an inner products matrix. In their technique the polynomial may be arbitrary. However, the matrix must be of the form $(v_i \cdot u_j)_{ij}$, where v_i's and u_j's are distinct sets of vectors.

4 An Optimal Algorithm in the Plane

In this section we use *Gegenbauer polynomials* to round solutions of the semidefinite programming relaxation of MAX CUT that happen to lie in the plane. Note that as the solution lie in $\mathbb{R}^2$ the suitable polynomials to apply are only convex combinations of $G_k^{(2)}$'s.

Algorithm [MaxCut – Gege]:

1. Solve the MAX CUT semidefinite program
2. Run *random hyperplane rounding* with probability β and *Gegenbauer polynomials rounding* with the polynomial $G_4^{(2)}(x) = 8x^4 - 8x^2 + 1$ with probability $1 - \beta$

We choose $\beta = \frac{4}{5}\left(\frac{\pi}{5}\cot\frac{2\pi}{5} + 1\right) \simeq 0.963322$. The next Theorem together with the fact that the value of the MAX CUT semidefinite program is an upper bound on the value of the optimal cut, imply that our algorithm is a $\frac{32}{25+5\sqrt{5}}$ -approximation algorithm.

Theorem 2. *If the solution of the program lies in the plane, then the expected cut size produced by the algorithm is at least a fraction $\frac{32}{25+5\sqrt{5}}$ of the value of the MAX CUT semidefinite program.*

As the expected size of the cut produced by the algorithm equals $\sum_{i,j} w_{ij} Pr[x_i \neq x_j]$, we may derive Theorem 2 from the next lemma:

Lemma 2. $Pr[x_i \neq x_j] \geq \frac{32}{25+5\sqrt{5}} \frac{1 - v_i \cdot v_j}{2}$, *for all $1 \leq i, j \leq n$.*

Proof. Let $\theta_{ij} = \arccos(v_i \cdot v_j)$. By using Lemma 1, and Corollary 1 we get that

$$Pr[x_i \neq x_j] = \beta\frac{\theta_{ij}}{\pi} + (1 - \beta)\frac{\arccos(G_4^{(2)}(v_i \cdot v_j))}{\pi}.$$

As $G_k^{(2)}(x)$ are Chebyshev polynomials of the first kind, $G_k^{(2)}(\cos\theta) = \cos(k\theta)$. Therefore,

$$Pr[x_i \neq x_j] = \beta\frac{\theta_{ij}}{\pi} + (1 - \beta)\frac{\arccos(\cos(4\theta_{ij}))}{\pi}.$$

By using Lemma 3 in Appendix A, the latter value is greater or equals to $\frac{32}{25+5\sqrt{5}} \frac{1-\cos(\theta_{ij})}{2}$, and equality holds if and only if $\theta_{ij} = \frac{4\pi}{5}$ or $\theta_{ij} = 0$.

The last argument in the proof of Lemma 2 shows that the worst instance in our framework is when $\theta_{ij} = \frac{4\pi}{5}$ for all $(i, j) \in E$. The 5-cycle C_5, with an optimal solution $v_k = (\cos(\frac{4\pi}{5}k), \sin(\frac{4\pi}{5}k))$, for $1 \leq k \leq 5$, is such an instance. (See Figure 1.)

In our framework, *Gegenbauer polynomials rounding* performs better than the Goemans Williamson algorithm whenever the Gegenbauer polynomials transformation increases the expected angle between adjacent vertices. Denote the 'worst' angle of the Goemans-Williamson algorithm by $\theta_0 \simeq 2.331122$. In order to achieve an improved approximation, the Gegenbauer polynomial transformation should increase the angle between adjacent vertices with inner product $\cos\theta_0$. It can be shown that for any dimension $m \geq 3$, and any $k \geq 2$, $G_k^{(m)}(\cos\theta_0) > \cos\theta_0$. Hence, in higher dimensions other techniques are needed in order to attain an improved approximation algorithm.

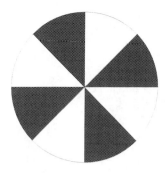

Fig. 2. $\mathcal{WINDMILL}_4$ 2-coloring

5 Rounding Using 2-Coloring of the Sphere

Definition 1. *A* 2-coloring $\mathcal{C}$ *of the unit sphere* S^{m-1} *is a function* $\mathcal{C} : S^{m-1} \rightarrow \{-1, +1\}$.

Let $\boldsymbol{v}_1, \ldots, \boldsymbol{v}_n \in S^{m-1}$ and let $\mathcal{C}$ be a 2-coloring of the unit sphere S^{m-1}. In the sphere *2-coloring rounding* method with respect to a 2-coloring $\mathcal{C}$, the 2-colored unit sphere is randomly rotated and then each variable x_i is set to $\mathcal{C}(\boldsymbol{v}_i)$. We give below a more formal definition of the *2-coloring rounding* with parameter $\mathcal{C}$. In the definition, it is more convenient to rotate $\boldsymbol{v}_1, \ldots, \boldsymbol{v}_n$ than to rotate the 2-colored sphere.

1. Choose a uniformly distributed *rotation* (orthonormal) matrix R
2. Set $x_i = \mathcal{C}(R\boldsymbol{v}_i)$, for $1 \le i \le n$

The matrix R in the first step may be chosen using the following procedure, which employs the Graham-Schmidt process:

1. Choose independently m vectors $\boldsymbol{r}_1, \ldots, \boldsymbol{r}_m \in \mathbb{R}^m$ according to the m-dimensional standard normal distribution
2. Iteratively, set $\boldsymbol{r}_i \leftarrow \boldsymbol{r}_i - \sum_{j=1}^{i-1} (\boldsymbol{r}_i \cdot \boldsymbol{r}_j)\boldsymbol{r}_j$, and then $\boldsymbol{r}_i \leftarrow \boldsymbol{r}_i / \|\boldsymbol{r}_i\|$
3. Let R be the matrix whose columns are the vectors $\boldsymbol{r}_i$

Note that when *2-coloring rounding* is used the probability $Pr[x_i \ne x_j]$ depends only on the angle between $\boldsymbol{v}_i$ and $\boldsymbol{v}_j$.

2-coloring rounding relates to *Gegenbauer polynomials rounding* by the following argument: Let $\mathcal{WINDMILL}_k$ be the 2-coloring of S^1 in which a point $(\cos\theta, \sin\theta) \in S^1$ is colored by black (-1) if $\theta \in [2\ell\frac{\pi}{k}, (2\ell+1)\frac{\pi}{k})$, and by white $(+1)$ otherwise (where $0 \le \ell < k$). The 2-coloring of $\mathcal{WINDMILL}_4$ for example is shown in Figure 2. Denote the angle between $\boldsymbol{v}_i$ and $\boldsymbol{v}_j$ by θ_{ij}. Also denote by $Pr_{WM}[x_i \ne x_j]$ the probability that $x_i \ne x_j$ if *2-coloring rounding* with $\mathcal{WINDMILL}_k$ is used, and by $Pr_{GEGE}[x_i \ne x_j]$ the probability that $x_i \ne x_j$ if *Gegenbauer polynomials rounding* with the polynomial $G_k^{(2)}$ is used. Then,

$$Pr_{WM}[x_i \ne x_j] = \begin{cases} \frac{\theta_{ij} - 2\ell\pi/k}{\pi/k} & \text{if} \quad \theta_{ij} \in [2\ell\frac{\pi}{k}, (2\ell+1)\frac{\pi}{k}) \quad \text{where } 0 \le \ell < k \\ \frac{(2\ell+1)\pi/k - \theta_{ij}}{\pi/k} & \text{if } \theta_{ij} \in [(2\ell+1)\frac{\pi}{k}, (2\ell+2)\frac{\pi}{k}) \quad \text{where } 0 \le \ell < k \end{cases}$$

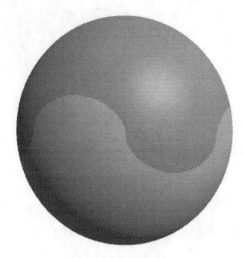

Fig. 3. $\mathcal{YINYAN}$ 2-coloring

In other words, $Pr_{WM}[x_i \neq x_j] = \frac{\arccos(\cos(k\theta_{ij}))}{\pi}$. On the other hand, by Corollary 1 $Pr_{GEGE}[x_i \neq x_j] = \frac{\arccos G_k^{(2)}(\cos(\theta_{ij}))}{\pi}$. As $G_k^{(2)}$ are Chebyshev polynomials of the first kind $G_k^{(2)}(\cos(\theta_{ij})) = \cos(k\theta_{ij})$. Hence, $Pr_{WM}[x_i \neq x_j] = Pr_{GEGE}[x_i \neq x_j]$. Goemans unpublished result was obtained in this way, with $k = 4$.

6   An Algorithm for Three Dimensions

Our improved approximation algorithm uses *2-coloring rounding*, combined with *random hyperplane rounding*. A 'good' 2-coloring would be one that helps the random hyperplane rounding procedure with the ratio of the so-called 'bad' angles.

In our approximation algorithm for three dimensions, we use a simple 2-coloring we call $\mathcal{YINYAN}$. We give a description of $\mathcal{YINYAN}$ in the spherical coordinates (θ, ϕ), where $0 \leq \theta \leq 2\pi$ and $-\frac{\pi}{2} \leq \phi \leq \frac{\pi}{2}$ (in our notations, the cartesian representation of (θ, ϕ) is $(\cos\theta\cos\phi, \sin\theta\cos\phi, \sin\phi)$.)

$$\mathcal{YINYAN}(\theta, \phi) = \begin{cases} +1 & \text{if } 4\phi > \sqrt[4]{\sin(4\theta)} \\ -1 & \text{otherwise} \end{cases}$$

Here $\sqrt[k]{x} = sgn(x)\sqrt{|x|}$. Figure 3 shows the $\mathcal{YINYAN}$ 2-coloring. Note that, the coloring of the equator in $\mathcal{YINYAN}$ is identical to the coloring $\mathcal{WINDMILL}_4$, which was used by Goemans in the plane.

Algorithm [MaxCut – $\mathcal{YINYAN}$]:

1. Solve the MAX CUT semidefinite program
2. Let CUT_1 be a cut obtained by running *random hyperplane rounding*
3. Let CUT_2 be a cut obtained by running *2-coloring rounding* with $\mathcal{YINYAN}$ 2-coloring
4. Return the largest of CUT_1 and CUT_2

We denote the weight of a cut CUT by $wt(CUT)$, the size of the optimal cut by OPT, and the expected weight of the cut produced by the algorithm by ALG. In addition, we denote by $Pr_{GW}[x_i \neq x_j]$ the probability that $x_i \neq x_j$ if the *random hyperplane rounding* of Goemans and Williamson is used, and by $Pr_{YY}[x_i \neq x_j]$ the probability that $x_i \neq x_j$ if *2-coloring rounding* with the 2-coloring $\mathcal{YINYAN}$ is used. If the angle between v_i and v_j is θ_{ij}, we write the latter also as $Pr_{YY}[\theta_{ij}]$. In these notations, for any $0 \leq \beta \leq 1$,

$$
\begin{aligned}
ALG &= E[\max\{wt(CUT_1), wt(CUT_2)\}] \\
&\geq \beta E[wt(CUT_1)] + (1 - \beta)E[wt(CUT_2)] \\
&= \beta \sum_{i,j} w_{ij} Pr_{GW}[x_i \neq x_j] + (1 - \beta) \sum_{i,j} w_{ij} Pr_{YY}[x_i \neq x_j]
\end{aligned}
$$

Thus, we have the following lower bound on the performance ratio of the algorithm, for any $0 \leq \beta \leq 1$:

$$
\begin{aligned}
\frac{ALG}{OPT} &\geq \frac{\beta \sum_{i,j} w_{ij} Pr_{GW}[x_i \neq x_j] + (1 - \beta) \sum_{i,j} w_{ij} Pr_{YY}[x_i \neq x_j]}{\sum_{i,j} w_{ij} \frac{1 - v_i \cdot v_j}{2}} \\
&\geq \inf_{0 < \theta \leq \pi} \left\{ \frac{\beta \frac{\theta}{\pi} + (1 - \beta) Pr_{YY}[\theta]}{\frac{1 - \cos \theta}{2}} \right\}
\end{aligned}
$$

The probability $Pr_{YY}(\theta)$ may be written as a three dimensional integral. However, this integral does not seem to have an analytical representation. We numerically calculated a lower bound on the latter integral. A lower bound of 0.8818 on the ratio may be obtained by choosing $\beta = 0.46$. It is possible to obtain a rigorous proof for the approximation ratio using a tool such as $\mathcal{REALSEARCH}$ (see [Zwi02]). However, this would require a tremendous amount of work. An easy way the reader can validate that using $\mathcal{YINYAN}$ yields a better than α_{GW} approximation is by computing $Pr_{YY}[\theta]$ for angles around the 'bad' angle $\theta_0 \simeq 2.331122$. This can be done by a simple easy to check MATLAB code given in Appendix B.

An improved algorithm is also obtained using, say, the coloring

$$
\mathcal{SINCOL}(\theta, \phi) = \begin{cases} +1 & \text{if } 2\phi > \sin(4\theta) \\ -1 & \text{otherwise} \end{cases}
$$

However, a better ratio is obtained by taking $\mathcal{YINYAN}$ instead of $\mathcal{SINCOL}$ or any version of $\mathcal{SINCOL}$ with adjusted constants. There is no reason to believe that the current function is optimal. We were mainly interested in showing that a ratio strictly larger than the ratio of Goemans and Williamson can be obtained.

7   Concluding Remarks

We presented two new techniques for rounding solutions of semidefinite programming relaxations. We used the first technique to give an approximation algorithm that matches the integrality gap of MAX CUT for instances embedded in the plane. We used the second method to obtain an approximation algorithm which is better than the Goemans-Williamson approximation algorithm for instances that have a solution in $\mathbb{R}^3$. The latter approximation algorithm rules out the existence of a graph with integrality ratio arbitrarily close to α_{GW} and an optimal solution in $\mathbb{R}^3$, thus resolving an open problem raised by Feige and Schechtman [FS01]. We conjecture that the *2-coloring rounding* method can be used to obtain an approximation algorithm with a performance ratio strictly larger than α_{GW} for any fixed dimension.

It can be shown that adding "triangle constraints" to the MAX CUT semidefinite programming relaxation rules out a non-integral solution in the plane. It would be interesting to derive similar approximation ratios for three (or higher) dimensional solutions that satisfy the "triangle constraints".

Acknowledgments

We would like to thank Olga Sorkine for her help in rendering Figure 3.

References

[AN04]   N. Alon and A. Naor. Approximating the cut-norm via Grothendieck's inequality. In *Proceedings of the 36th Annual ACM Symposium on Theory of Computing, Chicago, Illinois*, pages 72–80, 2004.

[CW04]   M. Charikar and A. Wirth. Maximizing Quadratic Programs: Extending Grothendieck's Inequality. In *Proceedings of the 45th Annual IEEE Symposium on Foundations of Computer Science, Rome, Italy*, pages 54–60, 2004.

[DP93a]   C. Delorme and S. Poljak. Combinatorial properties and the complexity of a max-cut approximation. *European Journal of Combinatorics*, 14(4):313–333, 1993.

[DP93b]   C. Delorme and S. Poljak. Laplacian eigenvalues and the maximum cut problem. *Mathematical Programming*, 62(3, Ser. A):557–574, 1993.

[FG95]   U. Feige and M. X. Goemans. Approximating the value of two prover proof systems, with applications to MAX-2SAT and MAX-DICUT. In *Proceedings of the 3rd Israel Symposium on Theory and Computing Systems, Tel Aviv, Israel*, pages 182–189, 1995.

[FL01]   U. Feige and M. Langberg. The RPR^2 rounding technique for semidefinite programs. In *Proceedings of the 28th Int. Coll. on Automata, Languages and Programming, Crete, Greece*, pages 213–224, 2001.

[FS01]   U. Feige and G. Schechtman. On the integrality ratio of semidefinite relaxations of MAX CUT. In *Proceedings of the 33th Annual ACM Symposium on Theory of Computing, Crete, Greece*, pages 433–442, 2001.

[GW95]   M. X. Goemans and D. P. Williamson. Improved Approximation Algorithms for Maximum Cut and Satisfiability Problems Using Semidefinite Programming. *Journal of the ACM*, 42:1115–1145, 1995.

[Hås01]   J. Håstad. Some optimal inapproximability results. *Journal of the ACM*, 48(4):798–859, 2001.

[KKMO04] S. Khot, G. Kindler, E. Mossel, and R. O'Donnell. Optimal inapproimability re-sutls for MAX-CUT and other 2-variable CSPs? In *Proceedings of the 45th Annual IEEE Symposium on Foundations of Computer Science, Rome, Italy*, pages 146–154, 2004.

[LLZ02]   M. Lewin, D. Livnat, and U. Zwick. Improved rounding techniques for the MAX 2-SAT and MAX DI-CUT problems. In *Proceedings of the 9th IPCO, Cambridge, Massachusetts*, pages 67–82, 2002.

[MM01a]   S. Matuura and T. Matsui. 0.863-approximation algorithm for MAX DICUT. In *Approximation, Randomization and Combinatorial Optimization: Algorithms and Techniques, Proceedongs of APPROX-RANDOM'01, Berkeley, California*, pages 138–146, 2001.

[MM01b]   S. Matuura and T. Matsui. 0.935-approximation randomized algorithm for MAX 2SAT and its derandomization. Technical Report METR 2001-03, Department of Mathematical Engineering and Information Physics, the University of Tokyo, Japan, September 2001.

[Sch42]   I. J. Schoenberg. Positive definite functions on spheres. *Duke Math J.*, 9:96–107, 1942.

[Zwi99]   U. Zwick. Outward rotations: a tool for rounding solutions of semidefinite program-ming relaxations, with applications to MAX CUT and other problems. In *Proceed-ings of the 31th Annual ACM Symposium on Theory of Computing, Atlanta, Georgia*, pages 679–687, 1999.

[Zwi02]   U. Zwick. Computer assisted proof of optimal approximability results. In *Proceed-ings of the 13th Annual ACM-SIAM Symposium on Discrete Algorithms, San Fran-cisco, California*, pages 496–505, 2002.

Appendix A

Lemma 3. *Let* $\beta = \frac{4}{5}\left(\frac{\pi}{5}\cot\frac{2\pi}{5} + 1\right)$ *and* $\alpha_{GEGE} = \frac{4}{5\sin^2(\frac{2\pi}{5})} = \frac{32}{25+5\sqrt{5}}$. *Then, for every* $\theta \in [0, \pi]$

$$\beta\frac{\theta}{\pi} + (1 - \beta)\frac{\arccos(\cos 4\theta)}{\pi} \geq \alpha_{GEGE}\frac{1 - \cos\theta}{2}$$

and equality holds if and only if $\theta = 0$ *or* $\theta = \frac{4\pi}{5}$.

Proof. Define $h(\theta) = \beta\frac{\theta}{\pi} + (1 - \beta)\frac{\arccos(\cos 4\theta)}{\pi} - \alpha_{GEGE}\frac{1-\cos\theta}{2}$.
   We show that the minimum of $h(\theta)$ is zero and is attained only at $\theta = 0$ and $\theta = \frac{4\pi}{5}$.
   Note that

$$\arccos(\cos 4\theta) = \begin{cases} 4\theta & \text{if } 0 \leq \theta < \frac{\pi}{4} \\ 2\pi - 4\theta & \text{if } \frac{\pi}{4} \leq \theta < \frac{\pi}{2} \\ 4\theta - 2\pi & \text{if } \frac{\pi}{2} \leq \theta < \frac{3\pi}{4} \\ 4\pi - 4\theta & \text{if } \frac{3\pi}{4} \leq \theta \leq \pi \end{cases}$$

Hence, for $\theta \in [0, \pi] \setminus \{0, \frac{\pi}{4}, \frac{\pi}{2}, \frac{3\pi}{4}, \pi\}$

$$h'(\theta) = \begin{cases} \frac{4-3\beta}{\pi} - \frac{1}{2}\alpha_{GEGE} \sin\theta & \text{if } 0 < \theta < \frac{\pi}{4} \text{ or } \frac{\pi}{2} < \theta < \frac{3\pi}{4} \\ \frac{5\beta-4}{\pi} - \frac{1}{2}\alpha_{GEGE} \sin\theta & \text{if } \frac{\pi}{4} < \theta < \frac{\pi}{2} \text{ or } \frac{3\pi}{4} < \theta < \pi \end{cases}$$

and $h''(\theta) = -\frac{1}{2}\alpha_{GEGE}\cos\theta$.

There are three cases:

Case I: $0 \le \theta \le \pi/2$

In this case $h''(\theta) < 0$ for $\theta \notin \{0, \pi/4, \pi/2\}$. Therefore, the minimum of $h(\theta)$ in the interval $0 \le \theta \le \pi/2$ may be attained at $\theta = 0$, $\theta = \pi/4$ or $\theta = \pi/2$. Indeed, $h(0) = 0$ and the minimum is attained. In addition, $h(\pi/4) = 1-3/4 \cdot \beta - \alpha_{GEGE}(1-\sqrt{2}/2)/2 = 0.14798\ldots > 0$, and $h(\pi/2) = (\beta - \alpha_{GEGE})/2 = 0.03942\ldots > 0$.

Case II: $\pi/2 \le \theta \le 3\pi/4$

As for all $\theta \in (\frac{\pi}{2}, \frac{3\pi}{4})$ the function $h''(\theta)$ is defined and positive, the minimum may be attained when $h'(\theta) = 0$, $\theta = \pi/2$ or $\theta = 3\pi/4$. The unique solution $\theta_0 \in (\frac{\pi}{2}, \frac{3\pi}{4})$ of the equation $0 = h'(\theta_0) = \frac{4-3\beta}{\pi} - \frac{1}{2}\alpha_{GEGE} \sin(\theta_0)$ is $\theta_0 = \arcsin\left(\frac{4-3\beta}{\pi}\frac{2}{\alpha_{GEGE}}\right)$. However, for this local minimum, $h(\theta_0) = 0.00146\ldots > 0$. In addition, as mentioned before $h(\frac{\pi}{2}) > 0$. Moreover, $h(\frac{3\pi}{4}) = \beta\frac{3}{4} + (1-\beta) - \frac{1}{2}\alpha_{GEGE}(1+\frac{\sqrt{2}}{2}) = 0.00423\ldots > 0$.

Case III: $3\pi/4 \le \theta \le \pi$

Again, for all $\theta \in (\frac{3\pi}{4}, \pi)$ the function $h''(\theta)$ is defined and positive. Hence, the minimum may be attained when $h'(\theta) = 0$, $\theta = 3\pi/4$ or $\theta = \pi$. The unique solution $\theta_0 \in (\frac{3\pi}{4}, \pi)$ of the equation $0 = h'(\theta_0) = \frac{5\beta-4}{\pi} - \frac{1}{2}\alpha_{GEGE} \sin(\theta_0)$ is $\theta_0 = \arcsin\left(\frac{5\beta-4}{\pi}\frac{2}{\alpha_{GEGE}}\right) = \arcsin\left(\frac{4}{5}\cot(\frac{2\pi}{5}) \cdot 2\frac{5\sin^2(\frac{2\pi}{5})}{4}\right) = \frac{4\pi}{5}$. For this local minimum, $h(\theta_0) = 0$. In addition, as mentioned before $h(\frac{3\pi}{4}) > 0$. Moreover, $h(\pi) = \beta - \alpha_{GEGE} > 0$.

Appendix B

A short MATLAB routine to compute the probability that two vectors on S^2 with a given *angle* between them are separated by rounding with $\mathcal{YINYAN}$ 2-coloring. The routine uses Monte-Carlo with *ntry* trials.

For example, for the angle 2.331122, by running $yinyan(2.331122, 10000000)$ we get that the probability is 0.7472. This is strictly larger than the random hyperplane rounding separating probability of $\frac{2.331122}{\pi} \simeq 0.7420$.

```
function p = yinyan(angle,ntry)

c = cos(angle); s = sin(angle);

cnt = 0;
for i=1:ntry

      % Choose two random vectors with the given angle.

      u = randn(1,3); u = u/sqrt(u*u');
      v = randn(1,3); v = v - (u*v')*u; v = v/sqrt(v*v');
      v = c*u + s*v;

      % Convert to polar coordinates

      [tet1,phi1] = cart2sph(u(1),u(2),u(3));
      [tet2,phi2] = cart2sph(v(1),v(2),v(3));

      % Check whether they are separated

      cnt = cnt + ((4*phi1>f(4*tet1))~=(4*phi2>f(4*tet2)));

end

p = cnt/ntry;


function y = f(x)
s = sin(x);
y = sign(s)*sqrt(abs(s));
```

What Would Edmonds Do? Augmenting Paths and Witnesses for Degree-Bounded MSTs

Kamalika Chaudhuri[1,*], Satish Rao[1,**],
Samantha Riesenfeld[1,***], and Kunal Talwar[2,†]

[1] UC Berkeley
{kamalika,satishr,samr}@cs.berkeley.edu
[2] Microsoft Research, Redmond, WA
kunal@microsoft.com

Abstract. Given a graph and degree upper bounds on vertices, the BDMST problem requires us to find the minimum cost spanning tree respecting the given degree bounds.Könemann and Ravi [10, 11] give bicriteria approximation algorithms for the problem using local search techniques of Fischer [5]. For a graph with a cost C, degree B spanning tree, and parameters $b, w > 1$, their algorithm produces a tree whose cost is at most wC and whose degree is at most $\frac{w}{w-1}bB + \log_b n$. We give a polynomial-time algorithm that finds a tree of *optimal* cost and with maximum degree at most $bB + 2(b+1)\log_b n$. We also give a quasi-polynomial algorithm which produces a tree of *optimal* cost C and maximum degree bounded by $B + O(\log n/\log\log n)$. Our algorithms work when there are upper as well as lower bounds on the degrees of the vertices.

1 Introduction

We study the problem of finding a minimum cost spanning tree in a graph $G = (V, E)$ such that the tree meets degree bounds on the vertices in V. A solution to this problem, called a degree-bounded minimum spanning tree (BDMST), can be used to find minimum cost travelling salesman path (TSPP) between two endpoints by finding the minimum cost spanning tree where the endpoints have degree bounds of one and the internal nodes have degree bounds of two. Since the Hamiltonian path problem can be solved given even a polynomial factor approximation algorithm for TSPP, and a Hamiltonian path is NP-hard to approximate to within any computable factor, we expect that it is necessary to relax the degree bounds to get any reasonable approximation algorithm. (Note the implicit assumption that the edge costs of the input graph do not define a metric.)

* Research partially supported by the Berkeley Fellowship and NSF Grant CCR-0121555
** Research partially supported by NSF Grant 0013128000
*** Research partially supported by NSF Graduate Fellowship
† Part of this research was performed when the author was a student at UC Berkeley and was supported by the NSF via grants CCR-0121555 and CCR-0105533

The BDMST problem, [14] also has several applications in efficient network routing protocols. For example, in multicast networks, one wishes to build a spanning tree that minimizes the total cost of the network as well as the maximum work done by (that is, the degree of) any vertex. The BDMST problem is a natural formulation of the duelling constraints.

For the sake of simplicity, we focus on a version of the problem in which each node has the same degree bound B, though our results generalize to the case of nonuniform degree bounds. We refer to this case as the uniform version of the problem. Bicriteria approximation algorithms for this problem were first given by Ravi et al. [14, 15]. The best previous results are by Könemann and Ravi [10, 11]. They give an algorithm that, for a graph with a BDMST of maximum degree B and cost C, and for any $w > 1$, finds a spanning tree with maximum degree $\frac{w}{w-1}bB + \log_b n$, for any constant b, and cost wC. While this expression yields a variety of tradeoffs, one can observe that the Könemann -Ravi algorithm always either violates the degree constraints or exceeds the optimal cost by a factor of two.

We give a polynomial-time algorithm that removes the error in the cost completely and also a quasi-polynomial-time algorithm that both obtains a solution of optimal cost and significantly improves the error in the degree bounds. Specifically, we give a polynomial-time algorithm that, given a graph with a BDMST of optimal cost C and maximum degree B, finds a spanning tree of cost exactly C and maximum degree $bB + 2(b+1)\log_b n$ for any constant b. The main technical contribution here is the use of a subroutine that finds minimum cost spanning trees while enforcing both lower and upper bounds on degrees. This allows for more natural and better performing cost-bounding techniques, which we discuss below. We also give a quasi-polynomial-time algorithm that finds a spanning tree of maximum degree $B + O(\frac{\log n}{\log \log n})$ and optimal cost C. If the degree is $\omega(\log n/ \log \log n)$, our algorithm gives a $1 + o(1)$ approximation in the degree while achieving optimal cost. The main technical contribution here is the use of augmenting paths to find low-degree minimum spanning trees, where Könemann and Ravi[10] used a local optimization procedure to solve this subproblem. We discuss this technique further below as well.

In a recent paper, Könemann and Ravi[11] improve their algorithms both in terms of implementation and applicability by providing an algorithm that can deal with nonuniform degree bounds and does not rely on explicitly solving a linear program. Our polynomial-time algorithm works as does theirs for nonuniform bounds, except we get cost optimality. Our quasi-polynomial-time algorithm extends easily to nonuniform degree bounds since our error is additive, rather than multiplicative, for this version of the algorithm. The algorithms that we give for enforcing lower-bound degree constraints require new ideas and may also be of independent interest. Our results extend to a more general version of the BDMST problem, in which upper bounds *and lower bounds* on the degree of vertices are given.

Previous Techniques

Fürer and Raghavachari[6] give a beautiful algorithm for un-weighted graphs that finds a spanning tree of degree $\Delta + 1$ in any graph that has a spanning tree of optimal degree Δ. This is optimal, in the sense that an exact algorithm would give an algorithm for Hamiltonian path. Their algorithm takes any spanning tree and relieves the degree of a high-degree node by an alternating sequence of edge additions and deletions. While there is a node in the current tree of degree at least $\Delta + 1$, their algorithm either finds such a way to lower its degree or certifies that the maximum degree of *any* spanning tree must be at least Δ. The certificate consists of a set of nodes S whose removal leaves at least $\Delta|S|$ connected components in the graph. This implies that the average degree of the set S must be at least Δ. This algorithm is reminiscent of Edmonds' algorithm for unweighted matching [3].

For the weighted version of the BDMST problem, we consult a different algorithm of Edmonds[2] that computes *weighted* matchings. This algorithm can be viewed as first finding a solution to the dual of the matching problem: it finds an assignment of penalties to nodes that lengthens adjacent edges and then finds a maximal packing of balls around subsets of nodes of the graph. Once the dual is known, one can ignore the values of the weights on the edges and rely solely on an un-weighted matching algorithm.

Könemann and Ravi begin along Edmond's path by examining a linear programming relaxation for the BDMST problem and its dual[1]. Unfortunately, in this case, the combinatorial subroutine must produce a low-degree *minimum cost* spanning tree in the graph where the edge costs have been modified by the dual potentials. To find an MST in the modified graph that has (approximately) minimum degree over all MSTs, Ravi and Könemann use an algorithm of Fischer. For a graph with an MST of degree Δ, Fischer's algorithm provides an MST of degree at most $b\Delta + \log_b n$ for any $b > 1$. They show how to get a good solution by solving a linear program with a relaxed (by a factor of $(w/w - 1)$) degree constraint and then applying Fischer's algorithm to the dual.

Our Results and Techniques

As mentioned above, Könemann and Ravi's methods introduce a tradeoff between degree and cost that always creates a constant-factor error somewhere. In this paper, we remedy this problem by following along the lines of Edmonds' approach to weighted matchings. We show that any MST in the modified graph in which the nodes with dual penalties have (relatively) high degree has low cost in the original graph. Thus, instead of just finding an MST of low maximum degree, we also wish to enforce that the nodes with penalties have high degree. This entails developing algorithms and certificates for enforcing degree lower bounds, in addition to our algorithms for enforcing degree upper bounds. Moreover, we have to combine the algorithms to simultaneously enforce the high- and low-degree constraints.

In addition to improving the cost of the solution, we improve the degree-bounds at the cost of a quasi-polynomial running time. Our approach relies on

---

[1] As in the case of Edmonds' non-bipartite matching algorithm, the linear program and its dual are of exponential size, though their solutions can be found in polynomial time.

our analogy with bipartite matching. As a guide to our approach, consider the problem on bipartite graphs of assigning each node on the "left" to a node on the "right" while minimizing the maximum degree of any node on the right. This problem can naturally be solved using matching algorithms. Consider instead finding a locally optimal solution where for each node on the left, all of its neighbors have degree within 1 of each other. One can then prove that no node on the right has degree higher than $b\Delta + \log_b n$ where Δ is the optimum value. The proof uses Hall's lower bound to show that any breadth first search in the graph of matched edges is shallow, and the depth of such a tree bounds the maximum degree of this graph. This is the framework of Fischer's algorithm for finding a minimum cost spanning tree of degree $b\Delta + \log_b n$.

In the case of the bipartite graph problem above, an augmenting or alternating path algorithm for matching gives a much better solution than the locally greedy algorithm. Based on this insight, we use augmenting paths to address Fischer's problem of finding a MST of minimum maximum degree. By doing so, we are able to find a minimum cost tree of degree at most $\Delta + O(\frac{\log n}{\log \log n})$. In contrast to Fisher's local (or single-swap) approach, our algorithm finds a sequence of moves to decrease the degree of one high degree node. This introduces significant complications since changing the structure of the tree also changes which swaps are legal.

We also remark that our augmenting path algorithms, though based on the most powerful methods in matching, fall short of the $\Delta + 1$ that Fürer and Raghavachari obtain. While we make some progress by removing all the leading constants, and by bounding the cost using degree lower bounds, $\Delta + 1$ is a tantalizing target.

2   Bounded Degree Minimum Spanning Trees

BDMST Problem: Given a weighted graph $G = (V, E, c)$, $c : E \to R^+$, and a positive integer $B \geq 2$, find the minimum cost spanning tree in the set $\{T : \forall v \in V, \deg_T(v) \leq B\}$.

Könemann and Ravi [11] show that an MST of minimum degree for a certain cost function is also a BDMST with analogous guarantees on degree and with cost within a constant factor of the optimal. We show in this section that an algorithm which obtains an MST with guarantees on both maximum and minimum degrees can be used to produce a BDMST with optimal cost and analogous degree bounds. We present an algorithm in Section 6 that obtains such guarantees.

Linear Programs for BDMST: An integer linear program for the BDMST problem is given by:

$$
\begin{aligned}
\text{OPT}_B = \quad & \min \ \sum_{e \in E} c_e x_e \\
& \text{s.t.} \ \sum_{e \in \delta(v)} x_e \leq B \ \forall v \in V; x \in \text{SP}_G; x_e \in \{0,1\}
\end{aligned}
\tag{1}
$$

where $\delta(v)$ is the set of edges of E that are incident to v, and SP_G is the convex hull of edge-incidence vectors of spanning trees of G. A tree defined by a vector $x \in \text{SP}_G$, the entries of which are not necessarily all integer, is called a *fractional spanning tree*. It can be written as a convex combination of spanning trees of G. For a fractional tree T with edge incidence vector $x \in \text{SP}_G$, let $\deg_T(v) = \sum_{e \in \delta(v)} x_e$. We can write the linear program relaxation of (1) as $\min\{c(T) : T \in \text{SP}_G, \forall v \in V \ \deg_T(v) \le B\}$. The approach used by Könemann and Ravi [11] is to take the Lagrangean dual of this LP, given by: $\max_{\lambda \ge 0} \min_{T \in \text{SP}_G}\{c(T) + \sum_{v \in V} \lambda_v(\deg_T(v) - B)\}$. We can think of the optimal solution to the dual as a vector λ^B of Lagrangean multipliers on the nodes and a set $\mathcal{O}^B$ of optimal trees, such that every tree $T^B \in \mathcal{O}^B$ minimizes $c(T^B) + \sum_{v \in V} \lambda_v^B(\deg_{T^B}(v) - B)$. The set of multipliers λ^B and $\mathcal{O}^B$ can be computed in polynomial time. The optimal value $\text{OPT}_{LD(B)}$ of this dual program is a lower bound on OPT_B and a tight lower bound on the optimal value of the LP relaxation.

Following the analysis of [11], let us define a new cost function c^{λ^B}, where $c^{\lambda^B}(e) = c_e + \lambda_u^B + \lambda_v^B$ for an edge $e = (u, v)$. Since $B\sum_{v \in V} \lambda_v^B$ is constant for a fixed choice of λ^B, every tree in $\mathcal{O}^B$ is an MST of G under the cost function c^{λ^B}. The optimal solution to the linear program is a *fractional MST* $T_B^f = \sum_{T \in \mathcal{O}^B} \alpha_T T$. Note that the degree of v in T_B^f is $\deg_{T_B^f}(v) = \sum_T \alpha_T \deg_T(v)$.

Let $L_B = \{v : \lambda_v^B > 0\}$ be the set of nodes with positive dual variables in the optimal solution. Complementary slackness conditions applied to the optimal solutions of the linear program and its dual imply the following claim.

Lemma 1 *For all $v \in V$, $\deg_{T_B^f}(v) \le B$, and for all $v \in L_B$, $\deg_{T_B^f}(v) = B$.*

So we are given the existence of the fractional MST T^f (under the cost function c^λ) that meets both upper and lower degree bounds (i.e. no node has degree more than B and every node in L_B has degree exactly B). Our approach to solving the BMST problem is to find an *integral* MST, under a slightly different cost function, that meets both the upper and lower degree bounds approximately. Then we use the dual program to show that this tree does not cost too much more than T^f.

Let $B^* = B + \omega$, for some $\omega > 0$ to be specified later. Let $T^{B^*} \in \mathcal{O}^{B^*}$, so T^{B^*} is an MST under the cost function $c^{\lambda^{B^*}}$. Since λ^{B^*} is a feasible solution for the dual LP, it is clear that $c(T^{B^*}) + \sum_{v \in V} \lambda_v^{B^*}(\deg_{T^{B^*}}(v) - B) \le \text{OPT}_{LD(B)}$. Further, if T^{B^*} has the property that for every node $v \in L_{B^*}$, $\deg_{T^{B^*}}(v)$ is at least $B = B^* - \omega$, the second term in the above expression is non-negative and hence $c(T^{B^*})$ is at most $\text{OPT}_{LD(B)}$.

Lemma 2 *Let T be an MST of G under the cost function $c^{\lambda^{B^*}}$ such that for every $v \in L_{B^*}$ $\deg_T(v) \ge B$. Then $c(T) \le \text{OPT}_{LD(B)}$.*

MSTs with Maximum and Minimum Degree Bounds

In Section 2, we observed that the problem of finding a BDMST can be reduced to the problem of finding an actual MST, under the cost function c^λ, which satisfies the upper and lower bounds ont the degrees. Our approach is to start with an arbitrary MST $T \in \mathcal{O}^{B^*}$ and try to decrease its maximum degree. So that we can bound the cost of T with respect to the optimal BDMST, we simultaneously increase to a relatively high degree of at least B the degree of each node in $L_{B^*} = \{v \in V : \lambda_v^{B^*} > 0\}$. We first consider meeting these dueling constraints one at a time. In Section 4 we give an algorithm for finding an MST in which the maximum degree bounds are approximately met. Our algorithm produces a witness $\mathcal{W}$ that certifies that the tree we find has near-optimal maximum degree. In Section 5, we show how to find an MST in which a specified subset L of the nodes meet minimum degree bounds approximately. Again, the algorithm produces a witness $\mathcal{W}_L$ that certifies near-optimality. Finally, in Section 6 we given an algorithm for achieving these goals simultaneously and an analogous witness.

3   Minimum Spanning Trees with Degree Bounds

MSTDB Problem: Given a graph $G = (V, E)$ with costs on edges, a degree upper bound B_H, a set of vertices L and a degree lower bound B_L, find, if one exists, a minimum spanning tree of G such that no vertex has degree more than B_H and all vertices in L have degree at least B_L.

Note that an MST with these degree bounds may not exist in the graph. If this happens, the algorithm is expected to provide a combinatorial proof of non-existence.

We present two algorithms for this problem. The first one, which is based on Fischer's algorithm for finding a minimum degree MST of a graph, runs in polynomial time and finds an MST such that (a) every vertex has degree at most $bB_H + \log_b n$ and (b) all vertices in L have degree at least $B_L/b - \log_b n$. If it fails, it finds a combinatorial witness to show that there exists no MST of the graph in which all vertices have degree at most B_H and the vertices in L have degree at least B_L. The second algorithm we present finds a MST with two properties: (a) every vertex has degree $B_H + O(\frac{\log n}{\log \log n})$ or less and (b) all vertices in L have degree $B_L - O(\frac{\log n}{\log \log n})$ or more. If it fails, it provides a combinatorial witness to shows that there exists no MST of G in which all vertices have degree at most B_H and the vertices in L have degree at least B_L. Our second algorithm runs in quasi-polynomial time.

Recall from Section 2 that our BDMST algorithm runs MSTDB with the following inputs: the cost function is $c'_{uv} = c_{uv} + \lambda_u + \lambda_v$, the degree upper bound B_H is B, the degree lower bound B_L is B, and L, the set of vertices on which the degree lower bounds should be enforced, is the set of vertices for which $\lambda_v > 0$. Lemma 1 guarantees that there always exists a fractional MST for this cost function in which the maximum degree of any vertex is B and in which all vertices in L have degree exactly B. The combinatorial witnesses produced by the algorithm certify non-existence of fractional trees, as well as integral trees, with the same degree bounds. It follows that the output of the first algorithm is an MST of maximum degree $bB + \log_b n$, where $b > 1$ is a constant, in which the minimum degree of L is also $B/b - \log_b n$, and the output of the second algorithm is an MST with maximum degree $B + O(\frac{\log n}{\log \log n})$ in which the minimum degree of L is $B - O(\frac{\log n}{\log \log n})$. Otherwise the witnesses produced by the algorithms provide a contradiction to Lemma 1.

Starting with an arbitrary MST, both of our MSTDB algorithms proceed in phases of improvement. Each phase is essentially a combination of two algorithms: MAXDMST and MINDMST. MAXDMST is essentially a subroutine that either decreases the number of "high degree" vertices or produces a witness to show that the maximum degree of the current tree is close to optimal. Similarly, MINDMST decreases the number of nodes in L which are of "low degree". If it fails, it finds a witness to guarantee that the minimum degree of L in the current tree is close to optimal.

In our first MSTDB algorithm, MAXDMST is essentially Fischer's algorithm, and MINDMST is a symmetric version of Fischer's algorithm that finds an MST in which a certain set of nodes have maximum minimum degree. Since these algorithms are largely based on Fischer's algorithm, we do not provide any more details on them in this paper. Instead we focus for the next two sections on our second MSTDB algorithm which uses an approach based on augmenting paths.

4   Minimum Spanning Trees with Degree Upper Bounds

Let $d_{\max}(T)$ denote the maximum degree over all vertices in the current MST T, and let $d_{\min}^L(T)$ denote the minimum degree over all vertices in L in T. In this section, we describe the MAXDMST algorithm, which solves the following problem.

Min Max Degree MST: Given a graph G and a degree bound $d > 0$, find an MST T of G such that $d_{\max}(T) \leq d$.

The main idea behind MAXDMST is to find a series of *edge swaps* that decreases in a controlled manner the degree of some high-degree vertex in the tree. We say that an edge e in tree T can be *swapped* for edge e' if the following hold: (a) $e' \notin T$, and (b) the unique cycle in $T \cup e'$ contains e. Performing the swap (e, e') involves removing e from T and adding e' to produce another tree T'. A swap (e, e') is called *feasible* if the edges e and e' have the same weight. Note that if T is an MST, then the tree T' produced after a feasible swap on T is an MST as well. MAXDMST attempts to find an augmenting path of swaps that decreases the degree of a high-degree vertex by one without creating any more vertices of the same or higher degree.

An important part of our algorithm is that when it terminates, it produces a proof that the resulting tree T has close-to-optimal degree. We call such a combinatorial proof a *witness*. The basic underlying structure of the witness produced by MAXDMST is a partition of the nodes in G into a center set W and k other sets $C_1, C_2, \ldots, C_k$ called *clusters*. The partition has the property that there are no tree edges between clusters. The use of this partition as a witness is based on the following idea: if there is no feasible swap involving an inter-cluster edge and an edge in T adjacent to a node in W, then *any* MST must connect the clusters $C_1, C_2, \ldots, C_k$ using edges adjacent to W. If it is also true that k is large, then the average degree of W in any MST must be high. Our version of the proof of the following lemma appears in the full version of the paper.

Lemma 3 ([6]) *Let V be partitioned into sets of nodes $W, C_1, C_2, \ldots, C_k$. If for any edge e between two clusters C_i and C_j, there is no MST on G containing e, then for any MST T of G, $d_{\max}(T) \geq \lceil \frac{|W|+k-1}{|W|} \rceil$.*

Definitions and Notation

Before describing our algorithm formally, we introduce some definitions and notation. Let T be the MST of G at the beginning of the current phase of the algorithm. We use W to refer generally to the center set, and W_i to denote the center set at some specific iteration i of this phase. For a node $u \in W$, the clusters connected to u by tree edges are called the *children clusters* of u. For any d, we denote by $S_{\geq d}$ the set of vertices that have degree d or more in T.

Let $\tau \subseteq T$ be a Steiner tree on the nodes of W_0 (the Steiner nodes are the nodes in $V - W_0$). We say that we *freeze* the edges of τ incident on W, meaning that the edges of τ that are incident on W are not allowed to be removed by any swap. Freezing these edges ensures that the edge swaps made at the end of the algorithm result in a tree (and not a graph containing a cycle).

Given T and a partition of the nodes into W and a set $\mathcal{C}$ of clusters, we call a non-tree inter-cluster edge e' *feasible with respect to W* if there exists a feasible swap (e, e') such that e is incident on W and e is not frozen. A feasible swap $(e = (u, v), e' = (u', v'))$ is called *good* if (a) $u \in W$ (b) $v \notin W$, (c) e is not frozen, and (d) e' is an inter-cluster edge. The algorithm uses good swaps to decrease the degrees of high-degree vertices in W.

The MAXDMST Algorithm

Given G and the current MST T on G, we begin each phase of the algorithm by choosing a degree d such that $|S_{\geq d-1}| \leq \frac{\log n}{\log \log n}|S_{\geq d}|$. Note that this inequality must be true for some d lying between $d_{\max}(T)$ and $d_{\max}(T) - \frac{\log n}{\log \log n}$. For the duration of this phase, we maintain a center set W of vertices, and a set $\mathcal{C}$ of clusters. The initial center set W_0 is $S_{\geq d-1}$, and the initial clusters are the connected components created by deleting W_0 from T. In general, though, a cluster is not necessarily internally connected. Consider the Steiner tree $\tau \subseteq T$ on W_0, and freeze the edges of τ incident on W_0. In each iteration i, we find a good swap $(e = (u, v), e' = (u', v'))$ where $u \in W_i$ and $v \notin W_i$. We then take the following steps:

1. Remove u from W_i.

2. Form a new cluster C_u by merging u with the cluster containing u', the cluster containing v', and all the children clusters of u.

3. Consider each feasible edge (u, w) such that w is a node that has been removed from the witness at some prior stage, and merge C_u with the cluster currently containing w.

Step 3 ensures that we never try to use a feasible edge between two nodes which initially have degree $d-1$ in T. This process is repeated until we either remove a vertex u^* with degree d or higher from W, or we run out of good swaps. Either event marks the end of the current phase.

Lemmas 4 and 5 show that in the first case, we can find a sequence of edge swaps to get a tree T' from T such that the degree of u^* decreases by one and no vertex has its degree raised to more than $d-1$. In the case that we run out of good swaps, Lemma 6 and Theorem 9 shows that we can interpret W_i as a witness to the fact that the maximum degree of T is close to optimal.

Lemma 4 *Suppose we remove a vertex v from the center set W_i in some iteration i. Then there is a sequence of feasible edge swaps which (a) decreases the degree of v by one and (b) does not increase the degree of any other node in T to d.*

Note that since each of $((u_i, v_i), (u'_i, v'_i))$ is a feasible swap when discovered, performing any one of them does not introduce a cycle in the tree. Yet it is not obvious that we can perform all of them together without disrupting the tree structure. Lemma 5 shows that the graph produced after performing all these edge swaps is still a tree. The frozen edges of the Steiner tree play a crucial role in its proof. Proofs form Lemmas 4 and 5 are omitted from this extended abstract.

Lemma 5 *The graph produced after performing all the edge swaps in Lemma 4 is a tree.*

The MAXDMST Witness

The actual witness $\mathcal{W}$ produced by the algorithm is slightly more complex than the one described at the beginning of Section 4. A witness $\mathcal{W}$ actually consists of a partition of the nodes into W and a set of clusters $C_1, C_2, \ldots, C_k$, and a set of edges R such that at least one endpoint of each edge in R is in W. The witness $\mathcal{W}$ must have the property that any MST that contains all of the edges of R cannot contain an inter-cluster edge. Now if the size of R is small and k is large, then any MST must use a large number of edges incident to W to connect the clusters. The proof of Lemma 6 is straightforward and is omitted from this extended abstract.

Lemma 6 (Witness to high optimal degree) *A high-degree witness $\mathcal{W} = (W, R, \{C_1, C_2, \ldots, C_k\})$ certifies that any MST of G has maximum degree at least $\lceil \frac{|W|+k-2|R|-1}{|W|} \rceil$.*

Now we interpret the structure produced by running the MAXDMST algorithm as a witness. Let R^* be the subset of frozen edges that have at least one endpoint incident on W^*, where W^* is the center set when the algorithm terminates. Let $C_1^*, C_2^*, \ldots, C_k^*$ be the clusters and T^* the tree when the algorithm terminates.

Lemma 7 $\mathcal{W}^* = (W^*, R^*, \{C_1^*, C_2^*, \ldots, C_k^*\})$ *is a high-degree witness.*

Lemma 8 *At most $2|W_0|$ edges are frozen by the algorithm, where W_0 is our initial witness.*

Theorem 9 *If the MAXDMST algorithm halts with an MST of maximum degree d, then $\Delta_{\text{OPT}} \geq d - 2 - O(\frac{\log n}{\log \log n})$.*

The proofs of Lemma 8, Lemma 7 and Theorem 9 are omitted from this extended abstract.

5 Minimum Spanning Trees with Degree Lower Bounds

Max Min Degree MST: Given a subset L of the nodes of G and a minimum degree bound $d > 0$, find an MST T of G such that $d_{\min}^L(T) \geq d$.

In this section, we consider the problem defined above. Note that this is an NP-hard decision problem, and our algorithm solves a gap version of it. Much like the algorithm of the previous section, our algorithm either finds a tree which respects the degree bounds up to an additive error of $\frac{\log n}{\log \log n}$ or gives a combinatorial proof showing that the decision problem is a no instance, i.e. there is no MST in G satisfying the degree constraints. While the algorithm and the witness here are somewhat analogous to those in the previous section, they are not symmetric, and new ideas are required for this case.

We first introduce the kind of witness we use for proving upper bounds on the minimum degree in L. A witness $\mathcal{W}_L$ consists of a set W of nodes in L, a set of clusters $C_1, C_2, \ldots, C_k, D$, that form a partition of $V \setminus W$, and a set R of inter-cluster edges. In any MST T of G, for all i, $1 \leq i \leq k$ the cluster C_i must be internally connected. For any MST T of G containing the edges of R, it must also be true that for all $v \in D$, there is some i, $1 \leq i \leq k$ such that there is a path from v to C_i in the forest created by deleting W from T.

Lemma 10 (Witness to low optimal degree) *A witness $\mathcal{W}_L = (W, R, \{C_1, \ldots, C_k\}, D)$ certifies that any MST of G has a node in L with degree at most $\lfloor \frac{2|W|+|R|+k-2}{|W|} \rfloor$.*

The proof of Lemma 10 is omitted from this extended abstract. Since it bounds the average degree of L, Corollary 11 follows.

Corollary 11 *Given a witness $\mathcal{W}_L$ as in the previous lemma, it holds that for any fractional MST T^f with edge incidence vector x, the minimum fractional degree in L, i.e. $\min_{v \in L} \sum_{e \in \delta(v)} x_e$, is at most $\frac{2|W|+|R|+k-2}{|W|}$.*

Recall that $d_{\min}^L(T) = \min_{v \in L} \deg_T(v)$ is the minimum degree over all nodes in L in tree T, and let Δ_{OPT}^L be the maximum possible value of $d_{\min}^L(T)$ attainable by any MST; the algorithm we describe in this section finds an MST in which $d_{\min}^L(T)$ is at least $\Delta_{\text{OPT}}^L - O(\frac{\log n}{\log \log n})$.

We now introduce the following notation: For a tree T, we denote by $S_{\leq d}^L$ the subset of nodes in L which have degree d or less in T.

Definition 1 *We call an edge e dotted with respect to an MST T and a set W, if e is a non-tree edge incident to a node in W and it is feasible with respect to some edge e' in T such that e' is not incident to a node in W.*

If the set W has low degree, we are interested in dotted edges with respect to W, since swapping in these edges and removing the edges they are feasible to, increases the degree of W.

The MINDMST Algorithm

Like the algorithm in Section 4, at the beginning of each phase, we pick a degree d such that in the current tree, $|S_{\leq d+1}| \leq \frac{\log n}{\log \log n} |S_{\leq d}|$.

Note that this inequality must be true for some d lying between $d^L_{\min}(T)$ and $d^L_{\min}(T) + \frac{\log n}{\log \log n}$. The aim is to improve the degree of some vertex of degree d or less without introducing more vertices of the same or lower degree. We begin with the center set $W_0 = S^L_{\leq d+1}$. Our initial clusters are the components obtained when we remove W_0 from T. Note that every node in L outside W_0 belongs to a cluster and has degree at least $d + 2$.

We maintain a set of special nodes N which is also initialized to $S^L_{\leq d+1}$. Consider the Steiner tree induced on N by the given tree T. We *freeze* the Steiner tree edges adjacent to N— that is, we disallow the algorithm from swapping out these edges— and put them in F. (Note that freezing edges reduces the set of dotted edges that can be used.)

In each iteration i, we look for a dotted edge $e = (u, v)$ adjacent to a node u in the witness W_i, and an intra-cluster tree edge $e' = (u', v')$ with which it can be swapped. If we find such a pair, we take the following steps:

1. Remove u from W_i.
2. Form a new cluster C_u by merging u along with all the children clusters of u, except for those which would involve a merge along an edge between u and a $(d + 1)$-degree node.
3. Split the cluster containing (u', v') along e' into two clusters $C_{u'}$ and $C_{v'}$.
4. Let u' be the endpoint of e' that has a tree path to u. Let w be the first node in N on the u'-u path. Add u' to N. Freeze the edges on this path that are incident on u' and on w, and add the new frozen edges to F. The u'-v path in T is called the *tail* of this swap.

Step 1 allows us to delete an edge incident on u at a later step. Steps 2 and 3 are needed to ensure that we never swap out more than one edge incident on any node. Finally, step 4 ensures that the sequence of swaps found by the end of the phase can be executed concurrently without resulting in non-tree structures.

We repeat the above process until we either remove a node of degree d or less from W, or there are no useful intra-cluster edges left. Either event ends the current phase. In the former case, Lemma 12 shows that we can find an improvement Once again the frozen edges play a crucial role in the proof, which appears in the full version of the paper.

In the latter case, Theorem 13 shows that we can turn the resulting structure into a witness W^*_L certifying that the tree has close to the optimal degree.

Lemma 12 *If we remove a vertex u from W, we can find a sequence of edge swaps which increases the degree of u by 1. Moreover, performing all these swaps results in a tree without introducing any new nodes of degree d or less.*

Theorem 13 *If the MINDMST algorithm halts with a tree with minimum degree d for a node in L, then $\Delta^L_{\text{OPT}} \leq d + 4 + O(\frac{\log n}{\log \log n})$.*

Proof. We begin by showing how to convert the structure produced by the MINDMST algorithm into a witness of the type in Lemma 10. Suppose the algorithm begins a phase with an MST T, chooses a degree d, sets $W_0 = S^L_{\leq d+1}$, but then cannot improve the degree of any vertex in L of degree d or lower. It terminates after $t > 0$ iterations with $W_t \supseteq S^L_{\leq d}$ as the center set and $C_1, C_2, \ldots, C_{k'}$, as the clusters.

Let $W^* = W_t$, and let I be the set of inter-cluster edges with respect to the clusters $C_1, \ldots, C_{k'}$. Recall that F is the set of edges frozen by the algorithm. Let $R^* = F \cup I$. We now describe how to modify the above clusters to get our witness. Let $\mathcal{C}^*$ be a collection of clusters, and let D^* be a collection of nodes. We begin with $\mathcal{C}^*$ and D^* empty. For $1 \leq i \leq k'$, let $C = C_i$ and do the following:

If cluster C is internally connected in every MST, then add C to $\mathcal{C}^*$. If not, let T' be an MST of G in which C is not internally connected, and let (u, v) be an edge of $T|_C \setminus T'|_C$.

Consider the clusters C_u and C_v that would result from splitting C about the edge (u, v). If both C_u and C_v have edges in T to W_t, then split C and recurse on each cluster C_u and C_v. If only one of the two, say C_u has a tree edge to W_t, then split C, recurse on C_u, and add the vertices in C_v to D^*.

The process terminates when every node outside W^* either belongs to D^* or to a cluster C that has been put in $\mathcal{C}^*$. Let us label the clusters in $\mathcal{C}^*$ by $C_1^*, C_2^*, \ldots C_k^*$, $k \geq k'$. Note that by construction, every vertex in D^* has a path in $T - W^*$ to a cluster in $\mathcal{C}^*$.

Lemma 14 $\mathcal{W}^*_L = (W^*, R^*, \mathcal{C}^* = \{C_1^*, \ldots C_k^*\}, D^*)$ *is a low-degree witness.*

The proof of Lemma 14 is omitted due to space constraints. To finish the proof of Theorem 13, we compute the bound that we get by using Lemma 10. First we count the number of clusters created by the algorithm. The number of clusters is the number of components formed by W_t from T, of which there are at most $(d+1)|W_t|$, plus the number of splits. There are at most $2|W_0 - W_t|$ additional clusters created by the splits during the algorithm. To see this, note that there are a total of $|W_0 - W_t|$ splits caused by finding a useable dotted edge, and each split creates one new cluster by splitting along the intra-cluster edge (u', v') involved in the swap. There are also some clusters split by edges adjacent to two $(d+1)$-degree nodes, but there can be at most $|W_0 - W_t|$ of these.

The number of clusters after the algorithm stops is thus at most $(d+1)|W_t| + 2|W_0 - W_t|$. In the process of breaking apart clusters with more than one edge to W_t, at most $|W_t|$ more splits may happen. Thus the total number k of clusters in $\mathcal{C}^*$ is at most $(d+2)|W_t| + 2|W_0 - W_t|$.

There at most $2|W_0| + 2|W_0 - W_t|$ edges in F, and at most $2|W_0 - W_t|$ inter-cluster edges with respect to the clusters $C_1, \ldots C_{k'}$ when the algorithm terminates. Therefore $|R^*| \leq 2\,|W_0| + 4\,|W_0 - W_t|$. According to Lemma 10, the witness $\mathcal{W}_L^*$ gives an upper bound of

$$\frac{2\,|W^*| + |R^*| + k - 2}{|W^*|} \leq \frac{(d+4)\,|W_t| + 6\,|W_0 - W_t| + 2\,|W_0| - 2}{|W_t|}$$

on the minimum degree of at least one node in L. Using the fact that $|S_{\leq d+1}| \leq \frac{\log n}{\log \log n}\,|S_{\leq d}|$, we conclude that $\Delta_{\mathrm{OPT}}^L \leq d + 4 + O(\frac{\log n}{\log \log n})$.

6   Combining Upper and Lower Bounds

Now we describe our MSTDB algorithm more formally. Recall from Section 3 that our MSTDB algorithm works in phases. Each phase employs algorithms MAXDMST and MINDMST to improve either a high degree vertex or a low degree vertex in L; when both improvements fail, their failure is justified by two combinatorial witnesses.

Let us now look into a phase of the algorithm in more detail. Each phase of MSTDB begins by picking a d such that

$$|S_{\leq B_L - d + 1}| \leq \frac{\log n}{\log \log n}|S_{\leq B_L - d}| \quad \text{and} \quad |S_{\geq B_H + d - 1}| \leq \frac{\log n}{\log \log n}|S_{\geq B_H + d}|.$$

It is easy to show that one can always find such a d between 0 and $2\frac{\log n}{\log \log n}$.

For the rest of the phase, vertices with degree $B_H + d$ or more are considered "high degree" vertices and those in L with degree $B_L - d$ are considered "low degree". We employ MAXDMST and MINDMST to reduce the degree of a high degree vertex and increase the degree of a low degree vertex respectively.

Before calling MAXDMST and MINDMST we need to ensure that the improvements they perform do not interfere. For this purpose, we look at all edges e such that one end point of e lies in $S_{\leq B_L - d + 1}$ and the other in $S_{\geq B_H + d - 1}$. Note that e might be a tree edge or a non-tree edge. We freeze all such edges, which means that we neither add any of the non-tree edges nor delete any of the tree edges while executing algorithms MAXDMST or MINDMST. Lemma 15 guarantees that this is enough to make the algorithm terminate. The proof of Lemma 15 is omitted from the extended abstract. Theorem 16 shows that when both MAXDMST and MINDMST fail with two combinatorial witnesses, at least one of the witnesses is good. The proof will appear in the full version of the paper.

Lemma 15 *Algorithm MSTDB terminates.*

Theorem 16 *If MSTDB fails, then in every MST T of the graph, either:*
  (1) The maximum degree of T is more than B_H, or
  (2) Some node in L has degree less than B_L.

References

1. T. M. Chan. Euclidean bounded-degree spanning tree ratios. In *Proceedings of the nineteenth annual symposium on Computational geometry*, pages 11–19. ACM Press, 2003.

2. J. Edmonds. Maximum matching and a polyhedron with 0–1 vertices. *Journal of Research National Bureau of Standards*, 69B:125–130, 1965.
3. J. Edmonds. Paths, trees, and flowers. *Canadian Journal of Mathematics*, 17:449–467, 1965.
4. S. Even and R. E. Tarjan. Network flow and testing graph connectivity. *SIAM Journal on Computing*, 4(4):507–518, Dec. 1975.
5. T. Fischer. Optimizing the degree of minimum weight spanning trees. Technical Report 14853, Dept of Computer Science, Cornell University, Ithaca, NY, 1993.
6. M. Fürer and B. Raghavachari. Approximating the minimum-degree Steiner tree to within one of optimal. *Journal of Algorithms*, 17(3):409–423, Nov. 1994.
7. J. Hopcroft and R. Karp. An n 5/2 algorithm for maximum matching in bipartite graphs. *SIAM Journal on Computing*, 2:225–231, 1973.
8. R. Jothi and B. Raghavachari. Degree-bounded minimum spanning trees. In *Proc. 16th Canadian Conf. on Computational Geometry (CCCG)*, 2004.
9. S. Khuller, B. Raghavachari, and N. Young. Low-degree spanning trees of small weight. *SIAM J. Comput.*, 25(2):355–368, 1996.
10. J. Könemann and R. Ravi. A matter of degree: improved approximation algorithms for degree-bounded minimum spanning trees. In *Proceedings of ACM STOC, 2000*.
11. J. Könemann and R. Ravi. Primal-dual meets local search: approximating MST's with nonuniform degree bounds. In ACM, editor, *Proceedings ACM STOC, 2003*.
12. R. Krishnan and B. Raghavachari. The directed minimum degree spanning tree problem. In *FSTTCS*, pages 232–243, 2001.
13. C. H. Papadimitriou and U. Vazirani. On two geometric problems related to the traveling salesman problem. *J. Algorithms*, 5:231–246, 1984.
14. R. Ravi, M. V. Marathe, S. S. Ravi, D. J. Rosenkrantz, and H. B. Hunt, III. Many birds with one stone: multi-objective approximation algorithms. In *Proceedings of ACM STOC,1993*.
15. R. Ravi, M. V. Marathe, S. S. Ravi, D. J. Rosenkrantz, and H. B. Hunt, III. Approximation algorithms for degree-constrained minimum-cost network-design problems. *Algorithmica*, 31, 2001.

A Rounding Algorithm for Approximating Minimum Manhattan Networks

(Extended Abstract)

Victor Chepoi, Karim Nouioua, and Yann Vaxès

Laboratoire d'Informatique Fondamentale de Marseille
Faculté des Sciences de Luminy, Universitée de la Méditerranée
F-13288 Marseille Cedex 9, France
{chepoi,nouioua,vaxes}@lif.univ-mrs.fr

Abstract. For a set T of n points (terminals) in the plane, a *Manhattan network* on T is a network $N(T) = (V, E)$ with the property that its edges are horizontal or vertical segments connecting points in $V \supseteq T$ and for every pair of terminals, the network $N(T)$ contains a shortest l_1-path between them. A *minimum Manhattan network* on T is a Manhattan network of minimum possible length. The problem of finding minimum Manhattan networks has been introduced by Gudmundsson, Levcopoulos, and Narasimhan (APPROX'99) and it is not known whether this problem is in P or not. Several approximation algorithms (with factors 8,4, and 3) have been proposed; recently Kato, Imai, and Asano (ISAAC'02) have given a factor 2 approximation algorithm, however their correctness proof is incomplete. In this note, we propose a rounding 2-approximation algorithm based on a LP-formulation of the minimum Manhattan network problem.

1 Introduction

A *rectilinear path* P between two points p, q of the plane $\mathbb{R}^2$ is a path connecting p and q and consisting of only horizontal and vertical line segments. More generally, a *rectilinear network* $N = (V, E)$ consists of a finite set V of points of $\mathbb{R}^2$ (the vertices of N) and of a finite set of horizontal and vertical segments connecting pairs of points of V (the edges of N). The *length* $l(P)$ (or $l(N)$) of a rectilinear path P (or of a rectilinear network N) is the sum of lengths of its edges. The l_1-distance between two points $p = (p^x, p^y)$ and $q = (q^x, q^y)$ in the plane $\mathbb{R}^2$ is $d(p, q) := ||p - q||_1 = |p^x - q^x| + |p^y - q^y|$. An l_1-*path* between two points $p, q \in \mathbb{R}^2$ is a rectilinear path connecting p, q and having length $d(p, q)$.

Given a set $T = \{t_1, \ldots, t_n\}$ of n points (*terminals*) in the plane, a *Manhattan network* [4] on T is a rectilinear network $N(T) = (V, E)$ such that $T \subseteq V$ and for every pair of points in T, the network $N(T)$ contains an l_1-path between them. A *minimum Manhattan network* on T is a Manhattan network of minimum possible length and the Minimum Manhattan Network problem (*MMN problem*) is to find such a network.

C. Chekuri et al. (Eds.): APPROX and RANDOM 2005, LNCS 3624, pp. 40–51, 2005.

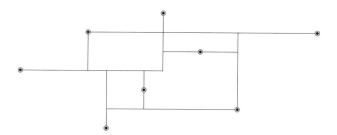

Fig. 1. A minimum Manhattan network

The minimum Manhattan network problem has been introduced by Gudmundsson, Levcopoulos, and Narasimhan [4]. It is not known whether this problem is in P or not. Gudmundsson et al. [4] proposed a factor 4 and a factor 8 approximation algorithms with different time complexity. They also conjectured that there exists a 2-approximation algorithm for this problem. Kato, Imai, and Asano [5] presented a factor 2 approximation algorithm, however, their correctness proof is incomplete (this fact was independently noticed by Benkert, Shirabe, and Wolf [1]). The algorithm in [5] proceeds in two steps. In the first step, a subnetwork of length at most the optimum is constructed. It connects by l_1-paths only certain pairs of terminals. In the second step, this partial network is augmented so that the remaining unconnected pairs are satisfied. It is claimed on page 355 of [5] that the length of this augmentation is again bounded from above by the length of a minimum Manhattan network. However, the details of how to perform this augmentation and the proof of its correctness are not provided. Following [5], Benkert et al. [1] outlined a factor 3 approximation algorithm and presented a mixed-integer programming formulation of the MMN problem. Notice that all four mentioned algorithms are geometric and some of them employ results from computational geometry. Nouioua [7] presented another factor 3 approximation algorithm based on the primal-dual method from linear programming. In this paper we present a rounding method applied to the optimal solution of the flow based linear program described in [1, 7] which leads to a factor 2 approximation algorithm for the minimum Manhattan network problem. For this, we define two subsets of pairs of terminals, called *strips* and *staircases*, and for each of them, we describe a specific rounding procedure. Each rounded up edge is paid by a group of parallel edges which together support at least one-half unit of fractional flow. Finally, we prove that a rectilinear network containing l_1-paths between all the pairs belonging to strips and staircases is a Manhattan network and thus, we end-up with an integer feasible solution whose cost is at most twice the fractional optimum.

Gudmundsson et al. [4] introduced the minimum Manhattan networks in connection with the construction of sparse geometric spanners. Given a set T of n points in the plane endowed with a norm $\|\cdot\|$, and a real number $t \geq 1$, a geometric network N is a *t-spanner* for T if for each pair of points $p, q \in T$, there exists a pq-path in N of length at t times the distance $\|p - q\|$ between p and q.

In the Euclidian plane (and more generally, for l_p-planes with $p \geq 2$), the linear segment is the unique shortest path between two endpoints, and therefore the unique 1-spanner of T is the trivial complete graph on T. On the other hand, if the unit ball of the norm is a polygon (in particular, for l_1 and l_∞), the points are connected by several shortest paths, therefore the problem of finding the sparsest 1-spanner becomes non trivial. In this connection, minimum Manhattan networks are precisely the optimal 1-spanners for the l_1 (or l_∞) plane. Sparse geometric spanners have applications in VLSI circuit design, network design, distributed algorithms and other areas, see for example the survey of [3]. Finally, Lam, Alexandersson, and Pachter [6] suggested to apply minimum Manhattan networks to design efficient search spaces for pair hidden Markov model (PHMM) alignment algorithms.

2   Properties and LP-Formulation

In this section, we present several properties of minimum Manhattan networks. First, we define some notations. Denote by $[p, q]$ the linear segment having p and q as end-points. The set of all points of $\mathbb{R}^2$ lying on l_1-paths between p and q constitute the smallest axis-parallel rectangle $R(p, q)$ containing the points p, q. For two terminals $t_i, t_j \in T$, set $R_{ij} := R(t_i, t_j)$. (This rectangle is degenerated if t_i and t_j have the same x- or y-coordinate.) We say that R_{ij} is an *empty rectangle* if $R_{ij} \cap T = \{t_i, t_j\}$. The *complete grid* is obtained by drawing in the smallest axis-parallel rectangle containing the set T a horizontal segment and a vertical segment through every terminal. Using standard methods for establishing Hanan grid-type results [10], it can be shown that the complete grid contains at least one minimum Manhattan network [4].

A point $p \in \mathbb{R}^2$ is said to be an *efficient* point of T [2, 9] if there does not exist any other point $q \in \mathbb{R}^2$ such that $d(q, t_i) \leq d(p, t_i)$ for all $t_i \in T$ and $d(q, t_j) < d(p, t_j)$ for at least one $t_j \in T$. Denote the set of all efficient points by $\mathcal{P}$, called the *Pareto envelope* of T. An optimal $O(n \log n)$ time algorithm to compute the Pareto envelope of n points in the l_1-plane is presented in [2] (for properties of $\mathcal{P}$ and an $O(n^2)$ time algorithm see also [9]). In particular, it is known that $\mathcal{P}$ is ortho-convex, i.e. the intersection of $\mathcal{P}$ with any vertical or horizontal line is convex, and that every two points of $\mathcal{P}$ can be joined in $\mathcal{P}$ by an l_1-path. $\mathcal{P}$, being ortho-convex, is a union of ortho-convex (possibly degenerated) rectilinear polygons (called *blocks*) glued together along vertices (they become cut points of $\mathcal{P}$); Fig. 2 presents two generic forms of the Pareto envelope of four points.

Lemma 1. *The Pareto envelope $\mathcal{P}$ contains at least one minimum Manhattan network on T.*

By this result, whose proof is omitted in this extended abstract, in order to solve the MMN problem on T it suffices to complete the set of terminals by adding to T the cut points of $\mathcal{P}$ and to solve a MMN problem on each block of $\mathcal{P}$ with respect to the new and old terminals located inside or on its boundary. Due

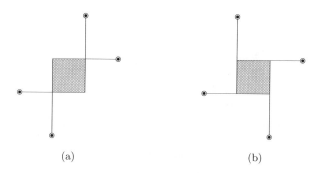

Fig. 2. Pareto envelope of four points

to this decomposition of the MMN problem into smaller subproblems, further we can assume without loss of generality that $\mathcal{P}$ consists of a single block with at least 3 terminals; denote by $\partial\mathcal{P}$ the boundary of this ortho-convex rectilinear polygon. Then every convex vertex of $\mathcal{P}$ is a terminal. Since the sub-path of $\partial\mathcal{P}$ between two consecutive convex vertices of $\partial\mathcal{P}$ is the unique l_1-path connecting these vertices inside $\mathcal{P}$ and $\partial\mathcal{P}$ is covered by such l_1-paths, from Lemma 2.1 we conclude that the edges of $\partial\mathcal{P}$ belong to any minimum Manhattan network inside $\mathcal{P}$.

From Lemma 2.1 and the result of [4] mentioned above we conclude that the part $\Gamma = (V, E)$ of the complete grid contained in $\mathcal{P}$ hosts at least one minimum Manhattan network. Two edges of Γ are called *twins* if they are opposite edges of a rectangular face of the grid Γ. Two edges e, f of Γ are called *parallel* if there exists a sequence $e = e_1, e_2, \ldots, e_{m+1} = f$ of edges such that for $i = 1, \ldots, m$ the edges e_i, e_{i+1} are twins. By definition, any edge e is parallel to itself and all edges parallel to e have the same length. Notice also that exactly two edges parallel to a given edge e belong to $\partial\mathcal{P}$.

We continue with the notion of generating set introduced in [5] and used in approximation algorithms from [1, 7]. A *generating set* is a subset F of pairs of terminals (or, more compactly, of their indices) with the property that a rectilinear network containing l_1-paths for all pairs in F is a Manhattan network on T. For example, $F_\varnothing$ consisting of all pairs ij with R_{ij} empty is a generating set. In the next section, we will describe a sparse generating set contained in $F_\varnothing$.

To give an LP-formulation of the minimum Manhattan network problem, let F be an arbitrary generating set; for each pair $ij \in F$, let $\Gamma_{ij} := \Gamma \cap R(t_i, t_j)$ and set $\Gamma_{ij} = (V_{ij}, E_{ij})$. We formulate the MMN problem as a cut covering problem using an exponential number of constraints, which we further convert into an equivalent formulation that employs only a polynomial number of variables and constraints. In both formulations, l_e will denote the length of an edge e of the network $\Gamma = (V, E)$ and x_e will be a 0-1 decision variable associated with e. A subset of edges C of E_{ij} is called a (t_i, t_j)-*cut* if every l_1-path between t_i and t_j in Γ_{ij} meets C. Let $\mathcal{C}_{ij}$ denote the collection of all (t_i, t_j)-cuts and set $\mathcal{C} := \cup_{ij \in F} \mathcal{C}_{ij}$. Then the minimum Manhattan networks can be viewed as the optimal solutions

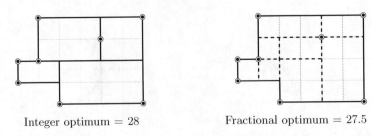

Integer optimum = 28   Fractional optimum = 27.5

Fig. 3. Integrality gap

of the following integer linear program (the dual of the relaxation of this program is a packing problem of the cuts from $\mathcal{C}$):

$$\text{minimize} \quad \sum_{e \in E} l_e x_e \tag{1}$$

$$\text{subject to} \quad \sum_{e \in C} x_e \geq 1, \qquad C \in \mathcal{C}$$

$$x_e \in \{0, 1\}, \qquad e \in E$$

Indeed, every Manhattan network is a feasible solution of (1). Conversely, let $x_e, e \in E$, be a feasible solution for (1). Considering x_e's as capacities of the edges e of Γ, and applying the covering constraints and the Ford-Fulkerson's theorem to each network $\Gamma_{ij}, ij \in F$, oriented as described below, we conclude the existence in Γ_{ij} of an integer (t_i, t_j)-flow of value 1, i.e., of an l_1-path between t_i and t_j. As a consequence, we obtain a Manhattan network of the same cost. This observation leads to the second integer programming formulation for the MMN problem (but this time, having a polynomial size). For each pair $ij \in F$ and each edge $e \in E_{ij}$ introduce a (flow) variable f_e^{ij}. Orient the edges of Γ_{ij} so that the oriented paths connecting t_i and t_j are exactly the l_1-paths between those terminals. For a vertex $v \in V_{ij} \setminus \{t_i, t_j\}$ denote by $\Gamma_{ij}^+(v)$ the oriented edges of Γ_{ij} entering v and by $\Gamma_{ij}^-(v)$ the oriented edges of Γ_{ij} out of v. We are lead to the following integer program:

$$\text{minimize} \quad \sum_{e \in E} l_e x_e \tag{2}$$

$$\text{subject to} \quad \sum_{e \in \Gamma_{ij}^+(v)} f_e^{ij} = \sum_{e \in \Gamma_{ij}^-(v)} f_e^{ij}, \qquad ij \in F, v \in V_{ij} \setminus \{t_i, t_j\}$$

$$\sum_{e \in \Gamma_{ij}^-(t_i)} f_e^{ij} = 1, \qquad ij \in F$$

$$0 \leq f_e^{ij} \leq x_e, \qquad ij \in F, \forall e \in E_{ij}$$

$$x_e \in \{0, 1\}, \qquad e \in E$$

Denote by (1′) and (2′) the LP-relaxation of (1) and (2) obtained by replacing the boolean constrains $x_e \in \{0,1\}$ by the linear constraints $x_e \geq 0$. Since (2′) contains a polynomial number of variables and inequalities, it can be solved in strongly polynomial time using the algorithm of Tardos [8]. The x-part of any optimal solution of (2′) is an optimal solution of (1′). Notice also that there exist instances of the MMN problem for which the cost of an optimal (fractional) solution of (1′) or (2′) is smaller than the cost of an optimal (integer) solution of (1) or (2). Fig. 3 shows such an example ($x_e = 1$ for bolded edges and $x_e = \frac{1}{2}$ for dashed edges). Finally observe that in any feasible solution of (1′) and (2′) for any edge $e \in \partial \mathcal{P}$ holds $x_e = 1$.

3 Strips and Staircases

A degenerated empty rectangle R_{ij} is called a *degenerated vertical* or *horizontal strip*. A non-degenerated empty rectangle R_{ij} is called a *vertical strip* if the x-coordinates of t_i and t_j take consecutive values in the sorted list of x-coordinates of the terminals and the intersection of R_{ij} with degenerated vertical strips is either empty or one of the points t_i or t_j. Analogously, a non-degenerated empty rectangle R_{ij} is called a *horizontal strip* if the y-coordinates of t_i and t_j take consecutive values in the sorted list of y-coordinates of the terminals and the intersection of horizontal sides of R_{ij} with degenerated horizontal strips is either empty or one of the points t_i or t_j; see Fig. 4. The *sides* of a vertical (resp., horizontal) strip R_{ij} are the vertical (resp., horizontal) sides of R_{ij}. We say that the strips $R_{ii'}$ and $R_{jj'}$ (degenerated or not) form a *crossing configuration* if they intersect and the Pareto envelope of the points $t_i, t_{i'}, t_j, t_{j'}$ is of type (a); see Fig. 2. The importance of such configurations resides in the following property whose proof is straightforward:

Lemma 2. *If the strips $R_{ii'}$ and $R_{jj'}$ form a crossing configuration as in Fig. 5, then from the l_1-paths between t_i and $t_{i'}$ and between t_j and $t_{j'}$ one can derive an l_1-path connecting t_i and $t_{j'}$ and an l_1-path connecting $t_{i'}$ and t_j.*

For a crossing configuration $R_{ii'}, R_{jj'}$, denote by o and o' the cut points of the rectangular block of the Pareto envelope of $t_i, t_{i'}, t_j, t_{j'}$, and assume that the

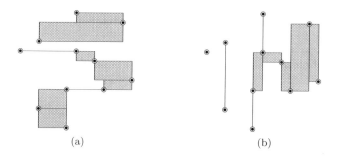

(a)   (b)

Fig. 4. Horizontal and vertical strips

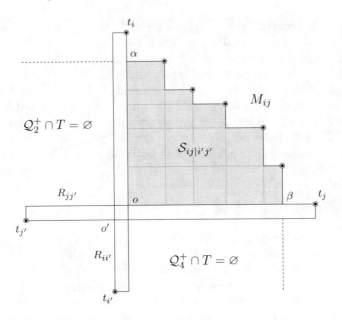

Fig. 5. Staircase $\mathcal{S}_{ij|i'j'}$

four tips of this envelope connect o with t_i, t_j and o' with $t_{i'}, t_{j'}$. Additionally, suppose without loss of generality, that t_i and t_j belong to the first quadrant $\mathcal{Q}_1$ with respect to the origin o (the remaining quadrants are labelled $\mathcal{Q}_2, \mathcal{Q}_3$, and $\mathcal{Q}_4$). Then $t_{i'}$ and $t_{j'}$ belong to the third quadrant with respect to the origin o'. Denote by T_{ij} the set of all terminals $t_k \in (T \setminus \{t_i, t_j\}) \cap \mathcal{Q}_1$ such that (i) $R(t_k, o) \cap T = \{t_k\}$ and (ii) the region $\{q \in \mathcal{Q}_2 : q^y \leq t_k^y\} \cup \{q \in \mathcal{Q}_4 : q^x \leq t_k^x\}$ does not contain any terminal of T. If T_{ij} is nonempty, then all its terminals belong to the rectangle R_{ij}, more precisely, they are all located on a common shortest rectilinear path between t_i and t_j. Denote by $\mathcal{S}_{ij|i'j'}$ the non-degenerated block of the Pareto envelope of the set $T_{ij} \cup \{o, t_i, t_j\}$ and call this rectilinear polygon a *staircase*; see Fig. 5 for an illustration. The point o is called the *origin* of this staircase. Analogously one can define the set $T_{i'j'}$ and the staircase $\mathcal{S}_{i'j'|ij}$ with origin o'. Two other types of staircases will be defined if t_i, t_j belong to the second quadrant and $t_{i'}, t_{j'}$ belong to the fourth quadrant. In order to simplify the presentation, further we will assume that after a suitable geometric transformation every staircase is located in the first quadrant. (Notice that our staircases are different from the staircase polygons occurring in the algorithms from [4].)

Let α be the leftmost highest point of the staircase $\mathcal{S}_{ij|i'j'}$ and let β be the rightmost lowest point of this staircase. Denote by M_{ij} the monotone boundary path of $\mathcal{S}_{ij|i'j'}$ between α and β and passing via the terminals of T_{ij}. By definition, $\mathcal{S}_{ij|i'j'} \cap T = T_{ij}$. By the choice of T_{ij}, there are no terminals of T located in the regions $\mathcal{Q}_2^+ := \{q \in \mathcal{Q}_2 : q^y \leq \alpha^y\}$ and $\mathcal{Q}_4^+ := \{q \in \mathcal{Q}_4 : q^x \leq \beta^x\}$. In

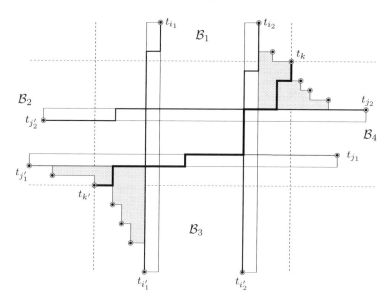

Fig. 6. To the proof of Lemma 3.2

particular, no strip traverses a staircase. From the definition of staircase immediately follows that two staircases either are disjoint or their intersection is a subset of terminals; in particular, every edge of the grid Γ belongs to at most one staircase.

Let F' be the set of all pairs ij such that R_{ij} is a strip. Let F'' be the set of all pairs $i'k$ such that there exists a staircase $\mathcal{S}_{ij|i'j'}$ such that t_k belongs to the set T_{ij}.

Lemma 3. $F := F' \cup F'' \subseteq F_\varnothing$ is a generating set.

Proof. Let N be a rectilinear network containing l_1-paths for all pairs in F. To prove that N is a Manhattan network on T, it suffices to establish that for an arbitrary pair $kk' \in F_\varnothing \setminus F$, the terminals t_k and $t_{k'}$ can be joined in N by an l_1-path. Assume without loss of generality that $t_{k'}^x \leq t_k^x$ and $t_{k'}^y \leq t_k^y$. The vertical and horizontal lines through the points t_k and $t_{k'}$ partition the plane into the rectangle $R_{kk'}$, four open quadrants and four closed unbounded half-bands labelled counterclockwise $\mathcal{B}_1, \mathcal{B}_2, \mathcal{B}_3$, and $\mathcal{B}_4$. Since $t_k \in \mathcal{B}_1 \cap T$ and $t_{k'} \in \mathcal{B}_3 \cap T$, there should exist at least one vertical strip between a terminal from $\mathcal{B}_1$ and a terminal from $\mathcal{B}_3$. Denote by $R_{i_1 i'_1}$ the leftmost strip and by $R_{i_2 i'_2}$ the rightmost strip traversing the rectangle $R_{kk'}$. These two strips may coincide (and one or both of them may be degenerated), however they are both different from $R_{kk'}$ because $kk' \notin F$. Suppose without loss of generality that $t_{i_1}, t_{i_2} \in \mathcal{B}_1$ and $t_{i'_1}, t_{i'_2} \in \mathcal{B}_3$. Analogously define the lowest horizontal strip $R_{j_1 j'_1}$ and the highest horizontal strip $R_{j_2 j'_2}$ traversing $R_{kk'}$. Again these strips may coincide and/or may be degenerated but they must be different from $R_{kk'}$. Let $t_{j_1}, t_{j_2} \in \mathcal{B}_4$ and $t_{j'_1}, t_{j'_2} \in \mathcal{B}_2$; see Fig. 6. From the choice of the strips in question, we conclude that each of four combinations of horizontal and vertical strips constitute crossing configurations. Moreover, $\mathcal{S}_{i_2 j_2 | i_2' j_2'}$ and $\mathcal{S}_{i_1' j_1' | i_1 j_1}$ must be staircases. The vertex t_k either belongs to $T_{i_2 j_2}$ or coincide with one of the vertices t_{i_2}, t_{j_2}. Analogously, $t_{k'} \in T_{i_1' j_1'} \cup \{t_{i_1'}, t_{j_1'}\}$. By Lemma 3.1 there is an l_1-path connecting each of the terminals $t_{i_1}, t_{i_2}, t_{j_1}, t_{j_2}$ to each of the terminals $t_{i_1'}, t_{i_2'}, t_{j_1'}, t_{j_2'}$. Also there exist l_1-paths between t_k and $t_{i_2'}, t_{j_2'}$ and between $t_{k'}$ and t_{i_1}, t_{j_2}. Combining certain pieces of these l_1-paths we will produce an l_1-path connecting t_k and $t_{k'}$. $\square$

4   The Rounding Algorithm

Let $(\mathbf{x}, \mathbf{f}) = ((x_e)_{e \in E}, (f_e^{ij})_{e \in E, ij \in F})$ be an optimal solution of the linear program $(2')$ (in general, this solution is not half-integral). The algorithm rounds up the solution $(\mathbf{x}, \mathbf{f})$ in three phases. In **Phase 0**, we insert all edges of $\partial \mathcal{P}$ in the integer solution. In **Phase 1**, the rounding is performed inside every strip $R_{ii'}$, in order to ensure the existence of an l_1-path $P_{ii'}$ between the terminals t_i and $t_{i'}$. In **Phase 2**, an iterative rounding procedure is applied to each staircase.

Let $R_{ii'}$ be a strip. If $R_{ii'}$ is degenerated, then $[t_i, t_{i'}]$ is the unique l_1-path between t_i and $t_{i'}$, yielding $x_e = f_e^{ii'} = 1$ for any edge $e \in [t_i, t_{i'}]$. If $R_{ii'}$ is not degenerated, then any l_1-path in Γ between t_i and $t_{i'}$ has a simple form: it goes along the side of $R_{ii'}$ containing t_i, then it makes a turn by following an edge of Γ traversing $R_{ii'}$ (called further a *switch* of $R_{ii'}$), and continues its way on the side containing $t_{i'}$ until it reaches $t_{i'}$. Although, it may happen that several such l_1-paths have been used by the fractional flow $f^{ii'}$ between t_i and $t_{i'}$, the cut condition ensures that $x_e + x_{e'} \geq f_e^{ii'} + f_{e'}^{ii'} \geq 1$ for any pair e, e' of twins on opposite sides of the strip $R_{ii'}$, yielding $\max\{x_e, x_{e'}\} \geq \frac{1}{2}$.

Let p be the furthest from t_i vertex on the side of $R_{ii'}$ containing t_i such that $x_e \geq \frac{1}{2}$ for every edge e of the segment $[t_i, p]$. Let pp' be the edge of Γ incident to p that traverses the strip $R_{ii'}$. By the choice of p we have $x_e \geq \frac{1}{2}$ for all edges e of the segment $[p', t_{i'}]$.

Phase 1 (procedure RoundStrip). For each strip $R_{ii'}$, if $R_{ii'}$ is degenerated, then take in the integer solution all edges of $[t_i, t_{i'}]$, otherwise round up the edges of $[t_i, p]$ and $[p', t_{i'}]$ and take the edge pp' as a switch of $R_{ii'}$; in both cases, denote by $P_{ii'}$ the resulting l_1-path between t_i and $t_{i'}$.

Let $\mathcal{S}_{ii' | jj'}$ be a staircase. Denote by ϕ the closest to t_i common point of the l_1-paths $P_{ii'}$ and $P_{jj'}$ (this point is a corner of the rectangular face of Γ containing the vertices o and o'). Let $P_{ii'}^+$ and $P_{jj'}^+$ be the sub-paths of $P_{ii'}$ and $P_{jj'}$ comprised between ϕ and the terminals t_i and t_j, respectively. Now we slightly expand the staircase $\mathcal{S}_{ii' | jj'}$ by considering as $\mathcal{S}_{ii' | jj'}$ the region bounded by the paths $P_{ii'}^+, P_{jj'}^+$, and M_{ij} ($P_{ii'}^+$ and $P_{jj'}^+$ are not included in the staircase but M_{ij} and the terminals from the set T_{ij} are). Inside $\mathcal{S}_{ii' | jj'}$, any flow $f^{ki'}$ (or $f^{kj'}$), $k \in T_{ij}$, may be as fractional as possible: it may happen that several l_1-paths between t_k and $t_{i'}$ carry flow $f^{ki'}$. Any such l_1-path intersects one of the paths $P_{ii'}^+$ or $P_{jj'}^+$, therefore the total $f^{ki'}$-flow arriving at $P_{ii'}^+ \cup P_{jj'}^+$ is equal to 1. (This flow can be redirected to ϕ via the paths $P_{ii'}^+$ and $P_{jj'}^+$, and further, along the path $P_{ii'}$, to the terminal $t_{i'}$). Therefore it remains to decide how to round up the flow $f^{ki'}$ inside the expanded staircase $\mathcal{S}_{ii'|jj'}$. For this, notice that either the total $f^{ki'}$-flow carried by the l_1-paths that arrive at $P_{ii'}^+$ is at least $\frac{1}{2}$ or the total $f^{ki'}$-flow on the l_1-paths that arrive at $P_{jj'}^+$ is at least $\frac{1}{2}$.

Phase 2 (procedure RoundStaircase). For a staircase $\mathcal{S}_{ii'|jj'}$ defined by the l_1-paths $P_{ii'}^+$ and $P_{jj'}^+$ and the monotone path M_{ij}, find the lowest terminal $t_m \in T_{ij}$ such that the $f^{mi'}$-flow on l_1-paths between t_m and $t_{i'}$ that arrive first at $P_{ii'}^+$ is $\geq \frac{1}{2}$ (we may suppose without loss of generality that this terminal exists). Let t_s be the terminal of T_{ij} immediately below t_m (this terminal may not exist). By the choice of t_m, the $f^{si'}$-flow on paths which arrive at $P_{jj'}$ is $\geq \frac{1}{2}$. Denote by ϕ' the intersection of the horizontal line passing via the terminal t_m with the path $P_{ii'}^+$. Analogously, let ϕ'' denote the intersection of the vertical line passing via t_s with the path $P_{jj'}^+$. Round up all edges of the horizontal segment $[t_m, \phi']$ and all edges of the vertical segment $[t_s, \phi'']$. If T_{ij} contains terminals located above the horizontal line (t_m, ϕ'), then recursively call RoundStaircase to the expanded staircase defined by $[t_m, \phi']$, the sub-path of $P_{ii'}^+$ comprised between ϕ' and t_i, and the sub-path M_{im} of the monotone path M_{ij} between t_m and α. Analogously, if T_{ij} contains terminals located to the right of the vertical line (t_s, ϕ''), then recursively call RoundStaircase to the expanded staircase defined by $[t_s, \phi'']$, the sub-path of $P_{jj'}^+$ comprised between ϕ'' and t_j, and the sub-path M_{sj} of the monotone path M_{ij} between t_s and β; see Fig. 7 for an illustration.

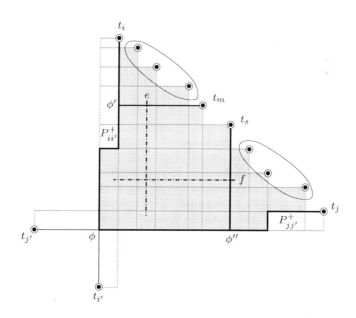

Fig. 7. Procedure RoundStaircase

Let E_0 denote the edges of Γ which belong to the boundary of the Pareto envelope of T. Let E_1 be the set of all edges picked by the procedure RoundStrip and which do not belong to E_0, and let E_2 be the set of all edges picked by the recursive procedure RoundStaircase and which do not belong to $E_0 \cup E_1$. Denote by $N^* = (V^*, E_0 \cup E_1 \cup E_2)$ the resulting rectilinear network. From Lemma 3.2 and the rounding procedures presented above we infer that N^* is a Manhattan network. Let $\mathbf{x}^*$ be the integer solution of (1) associated with N^*, i.e., $x_e^* = 1$ if $e \in E_0 \cup E_1 \cup E_2$ and $x_e^* = 0$ otherwise.

5   Analysis

In this section, we will show that the length of the Manhattan network N^* is at most twice the cost of the optimal fractional solution of (1'), i.e., that

$$\text{cost}(\mathbf{x}^*) = \sum_{e \in E} l_e x_e^* \leq 2 \sum_{e \in E} l_e x_e = 2\text{cost}(\mathbf{x}). \tag{3}$$

Recall that $x_e = x_e^* = 1$ holds for every edge $e \in E_0$. To establish the inequality (3), to every edge $e \in E_1 \cup E_2$ we will assign a set E_e of parallel to e edges such that (i) $\sum_{e' \in E_e} x_{e'} \geq \frac{1}{2}$ and (ii) $E_e \cap E_f = \emptyset$ for any two edges $e, f \in E_1 \cup E_2$.

First pick an edge $e \in E_1$, say $e \in P_{ii'}$ for a strip $R_{ii'}$. If e belongs to a side of this strip, then $x_e \geq \frac{1}{2}$, and in this case we can set $E_e := \{e\}$. Now, if e is the switch of $R_{ii'}$, then E_e consists of anyone of the two edges of ∂P parallel to e. From the definition of strips one conclude that no other switch can be parallel to these edges of ∂P. Therefore each pair of parallel edges of ∂P may appear in at most one set E_e for a switch e.

Finally suppose that $e \in E_2$, say e belongs to the expanded staircase $\mathcal{S}_{ii'|jj'}$. If e belongs to the segment $[t_m, \phi']$, then E_e consists of e and all parallel to e edges of $\mathcal{S}_{ii'|jj'}$ located below e; see Fig. 7. Since every l_1-path between t_m and $t_{i'}$ intersecting the path $P_{ii'}^+$ contains an edge of E_e, we infer that the value of the $f^{mi'}$-flow traversing the set E_e is at least $\frac{1}{2}$, therefore $\sum_{e' \in E_e} x_{e'} \geq \frac{1}{2}$, thus establishing (i). Analogously, if f is an edge of the vertical segment $[t_s, \phi'']$, then E_f consists of f and all parallel to f edges of $\mathcal{S}_{ii'|jj'}$ located to the left of f. Obviously, $E_e \cap E_f = \emptyset$. Since E_e and E_f belong to the region of $\mathcal{S}_{ii'|jj'}$ delimited by the segments $[t_m, \phi']$ and $[t_s, \phi'']$ and the recursive calls of the procedure RoundStaircase concern the staircases disjoint from this region, we deduce that E_e and E_f are disjoint from the sets $E_{e'}$ for all edges e' picked by the recursive calls of RoundStaircase to the staircase $\mathcal{S}_{ii'|jj'}$. Every edge of Γ belongs to at most one staircase, therefore $E_e \cap E_f = \emptyset$ if the edges $e, f \in E$ belong to different staircases. Finally, since there are no terminals of T located below or to the left of the staircase $\mathcal{S}_{ii'|jj'}$, no strip traverses this staircase (a strip intersecting $\mathcal{S}_{ii'|jj'}$ either coincides with $R_{ii'}$ and $R_{jj'}$, or intersects the staircase along segments of the boundary path M_{ij}). Therefore, no edge from E_1 can be assigned to a set E_e for some $e \in E_2 \cap \mathcal{S}_{ii'|jj'}$, thus establishing (ii) and the desired inequality (3). Now, we are in position to formulate the main result of this note:

Theorem 1. *The rounding algorithm described in Section 4 achieves an approximation guarantee of 2 for the minimum Manhattan network problem.*

6 Conclusion

In this paper, we presented a simple rounding algorithm for the minimum Manhattan network problem and we established that the length of the Manhattan network returned by this algorithm is at most twice the cost of the optimal fractional solution of the MMN problem. Nevertheless, experiences show that the ratio between the costs of the solution returned by our algorithm and the optimal solution of the linear programs (1') and (2') is much better than 2. We do not know the worst integrality gap of (1) (the worst gap obtained by computer experiences is about 1.087). Say, is this gap smaller or equal than 1.5? Does there exist a gap in the case when the terminals are the origin and the corners of a staircase?

References

1. M. Benkert, T. Shirabe, and A. Wolff, The minimum Manhattan network problem - approximations and exact solutions, 20th EWCG, March 25-26, Seville (Spain), 2004.
2. G. Chalmet, L. Francis, and A. Kolen, Finding efficient solutions for rectilinear distance location problems efficiently, European J. Operations Research 6 (1981) 117–124.
3. D. Eppstein, Spanning trees and spanners, Handbook of Computational Geometry, J.-R. Sack and J. Urrutia, eds., Elsevier, 1999, pp. 425–461.
4. J. Gudmundsson, C. Levcopoulos, and G. Narasimhan, Approximating a minimum Manhattan network, Nordic J. Computing 8 (2001) 219–232 and Proc. APPROX'99, 1999, pp. 28–37.
5. R. Kato, K. Imai, and T. Asano, An improved algorithm for the minimum Manhattan network problem, ISAAC'02, Lecture Notes Computer Science, vol. 2518, 2002, pp. 344-356.
6. F. Lam, M. Alexanderson, and L. Pachter, Picking alignments from (Steiner) trees, J. Computational Biology 10 (2003) 509–520.
7. K. Nouioua, Une approche primale-duale pour le problème du réseau de Manhattan minimal, RAIRO Operations Research (submitted).
8. E. Tardos, A strongly polynomial algorithm to solve combinatorial linear programs, Operations Research 34 (1986) 250–256.
9. R.E. Wendell, A.P. Hurter, and T.J. Lowe, Efficient points in location theory, AIEE Transactions 9 (1973) 314-321.
10. M. Zachariasen, A catalog of Hanan grid problems, Networks 38 (2001) 76–83.

Packing Element-Disjoint Steiner Trees

Joseph Cheriyan[1,*] and Mohammad R. Salavatipour[2,**]

[1] Department of Combinatorics and Optimization, University of Waterloo
Waterloo, Ontario N2L3G1, Canada
jcheriyan@math.uwaterloo.ca
[2] Department of Computing Science, University of Alberta
Edmonton, Alberta T6G2E8, Canada
mreza@cs.ualberta.ca

Abstract. Given an undirected graph $G(V, E)$ with terminal set $T \subseteq V$ the problem of packing element-disjoint Steiner trees is to find the maximum number of Steiner trees that are disjoint on the nonterminal nodes and on the edges. The problem is known to be NP-hard to approximate within a factor of $\Omega(\log n)$, where n denotes $|V|$. We present a randomized $O(\log n)$-approximation algorithm for this problem, thus matching the hardness lower bound. Moreover, we show a tight upper bound of $O(\log n)$ on the integrality ratio of a natural linear programming relaxation.

1 Introduction

Throughout we assume that $G = (V, E)$, with $n = |V|$, is a simple graph and $T \subseteq V$ is a specified set of nodes (although we do not allow multi-edges, these can be handled by inserting new nodes into the edges). The nodes in T are called *terminal* nodes or *black* nodes, and the nodes in $V - T$ are called *Steiner* nodes or *white* nodes. Following the (now standard) notation on approximation algorithms for graph connectivity problems (e.g. see [16]), by an *element* we mean either an edge or a Steiner node. A *Steiner tree* is a connected, acyclic subgraph that contains all the terminal nodes (Steiner nodes are optional). The problem of packing element-disjoint Steiner trees is to find a maximum-cardinality set of element-disjoint Steiner trees. In other words, the goal is to find the maximum number of Steiner trees such that each edge and each white node is in at most one of these trees. We denote this problem by **IUV**. Here, I denotes identical terminal sets for different trees in the packing, U denotes an undirected graph, and V denotes disjointness for white nodes and edges.

By *bipartite* **IUV** we mean the special case where G is a bipartite graph with node partition $V = T \cup (V - T)$, that is, one of the sets of the vertex bipartition consists of all of the terminal nodes. We will also consider the problem of packing Steiner trees fractionally (or fractional **IUV** for short), with constraints on the

---

* Supported by NSERC grant No. OGP0138432.
** Supported by an NSERC postdoctoral fellowship, the Department of Combinatorics and Optimization at the University of Waterloo, and a university start-up grant at the University of Alberta.

C. Chekuri et al. (Eds.): APPROX and RANDOM 2005, LNCS 3624, pp. 52–61, 2005.

nodes which corresponds to a natural linear programming relaxation of **IUV** (explained later in this section).

IUV captures some of the fundamental problems of combinatorial optimization and graph theory. First, suppose that T consists of just two nodes s and t. Then the problem is to find a maximum-cardinality set of element-disjoint s, t-paths. This problem is addressed by one of the cornerstone theorems in graph theory, namely Menger's theorem [4, Theorem 3.3.1], which states that the maximum number of openly-disjoint s, t-paths equals the minimum number of white nodes whose deletion leaves no s, t-path. The algorithmic problem of finding an optimal set of s, t-paths can be solved efficiently via any efficient maximum s, t-flow algorithm. Another key special case of **IUV** occurs for $T = V$, that is, all the nodes are terminals. Then the problem is to find a maximum-cardinality set of edge-disjoint spanning trees. This problem is addressed by another classical min-max theorem, namely the Tutte/Nash-Williams theorem [4, Theorem 3.5.1]. The algorithmic problem of finding an optimal set of edge-disjoint spanning trees can be solved efficiently via the matroid intersection algorithm. In contrast, the problem **IUV** is known to be NP-hard [3, 7], and the optimal value cannot be approximated within a factor of $\Omega(\log n)$ modulo the P$\neq$NP conjecture [3]. Moreover, this hardness result applies also to bipartite **IUV** and even to the problem of packing Steiner trees fractionally for the bipartite case (see (1) below for more details) via [15, Theorem 4.1]. That is, the optimal value of this linear programming relaxation of bipartite **IUV** cannot be approximated within a factor of $\Omega(\log n)$ modulo the P$\neq$NP conjecture. This is discussed in more detail later. For related results, see [2].

One variant of **IUV** has attracted increasing research interest over the last few years, namely, the problem of packing edge-disjoint Steiner trees (find a maximum-cardinality set of edge-disjoint Steiner trees); we denote this problem by **IUE**. This problem in its full generality has applications in VLSI circuit design (e.g., see [12, 21]). Other applications include multicasting in wireless networks (see [6]) and broadcasting large data streams, such as videos, over the Internet (see [15]). Almost a decade ago, Grötschel et al., motivated by the importance of **IUE** in applications and in theory, studied the problem using methods from mathematical programming, in particular polyhedral theory and cutting-plane algorithms, see [8–12]. Moreover, there is significant motivation from the areas of graph theory and combinatorial optimization, partly based on the relation to the classical results mentioned above, and partly fueled by an exciting conjecture of Kriesell [18] (the conjecture states that the maximum number of edge-disjoint Steiner trees is at least half of an obvious upper bound, namely, the minimum number of edges in a cut that separates some pair of terminals). If this conjecture is settled by a constructive proof, then it may give a 2-approximation algorithm for **IUE**. Recently, Lau [19] made a major advance on this conjecture by presenting a 26-approximation algorithm for **IUE** using new combinatorial ideas. Lau's construction is based on an earlier result of Frank, Kiraly, and Kriesell [7] that gives a 3-approximation for a special case of bipartite **IUV**. (To the best of our knowledge, no other method for **IUE** gives an $O(1)$-approximation guarantee, or even a $o(|T|)$-approximation guarantee).

Here is a summary of the previous results in the area. Frank et al. [7] studied bipartite **IUV**, and focusing on the restricted case where the degree of every white node is $\leq \Delta$ they presented a Δ-approximation algorithm (via the matroid intersection theorem and algorithm). Recently, we [3] showed that (i) **IUV** is hard to approximate within a factor of $\Omega(\log n)$, even for bipartite **IUV** and even for the fractional version of bipartite **IUV**, (ii) **IUV** is APX-hard even if $|T|$ is a small constant, and (iii) we gave an $O(\sqrt{n}\log n)$-approximation algorithm for a generalization of **IUV**. For **IUE**, Jain et al. [15] proved that the problem is APX-hard, and (as mentioned above) Lau [19] presented a 26-approximation algorithm, based on the results of Frank et al. for bipartite **IUV**[1]. Another related topic pertains to the domatic number of a graph and computing near-optimal domatic partitions. Feige et al. [5] presented approximation algorithms and hardness results for these problems. One of our key results is inspired by this work.

Although **IUE** seems to be more natural compared to **IUV**, and although there are many more papers (applied, computational, and theoretical) on **IUE**, the only known $O(1)$-approximation guarantee for **IUE** is based on solving bipartite **IUV**. This shows that **IUV** is a fundamental problem in this area. Our main contribution is to settle (up to constant factors) **IUV** and bipartite **IUV** from the perspective of approximation algorithms. Moreover, our result extends to the capacitated version of **IUV**, where each white (Steiner) node v has a nonnegative integer capacity c_v, and the goal is to find a maximum collection of Steiner trees (allowing multiple copies of any Steiner tree) such that each white node v appears in at most c_v Steiner trees; there is no capacity constraint on the edges, i.e., each edge has infinite capacity. The capacitated version of **IUV** (which contains **IUV** as a special case) may be formulated as an integer program (IP) that has an exponential number of variables. Let $\mathcal{F}$ denote the collection of all Steiner trees in G. We have a binary variable x_F for each Steiner tree $F \in \mathcal{F}$.

$$
\begin{aligned}
\text{maximize} \quad & \sum_{F \in \mathcal{F}} x_F \\
\text{subject to} \quad & \forall v \in V - T : \sum_{F : v \in F} x_F \leq c_v \\
& \forall F \in \mathcal{F} : \quad x_F \in \{0, 1\}
\end{aligned}
\tag{1}
$$

Note that in uncapacitated **IUV** we have $c_v = 1, \forall v \in V - T$. The fractional **IUV** (mentioned earlier) corresponds to the linear programming relaxation of this IP which is obtained by relaxing the integrality condition on x_F's to $0 \leq x_F \leq 1$.

Our main result is the following:

Theorem 1. *(a) There is a polynomial time probabilistic approximation algorithm with a guarantee of $O(\log n)$ and a failure probability of $\frac{O(1)}{\log n}$ for (uncapacitated) **IUV**. The algorithm finds a solution that is within a factor $O(\log n)$ of the optimal solution to fractional **IUV**.*
*(b) The same approximation guarantee holds for capacitated **IUV**.*

---

[1] Although not relevant to this paper, we mention that the directed version of **IUV** has been studied [3], and the known approximation guarantees and hardness lower bounds are within the same "ballpark" according to the classification of Arora and Lund [1].

We call an edge white if both its end-nodes are white, otherwise, the edge is called black (then at least one end-node is a terminal). For our purposes, any edge can be subdivided by inserting a white node. In particular, any edge with both end-nodes black can be subdivided by inserting a white node. Thus, the problem of packing element-disjoint Steiner trees can be transformed into the problem of packing Steiner trees that are disjoint on the set of white nodes. We prefer the formulation in terms of element-disjoint Steiner trees; for example, this formulation immediately shows that **IUV** captures the problem of packing edge-disjoint spanning trees; of course, the two formulations are equivalent.

For two nodes s, t, let $\kappa(s, t)$ denote the maximum number of element-disjoint s, t-paths (an s, t-path means a path with end-nodes s and t); in other words, $\kappa(s, t)$ denotes the maximum number of s, t-paths such that each edge and each white node is in at most one of these paths. The graph is said to be k-*element connected* if $\kappa(s, t) \geq k$, $\forall s, t \in T, s \neq t$, i.e., there are $\geq k$ element-disjoint paths between every pair of terminals. For a graph $G = (V, E)$ and edge $e \in E$, $G - e$ denotes the graph obtained from G by deleting e, and G/e denotes the graph obtained from G by contracting e; see [4, Chapter 1] for more details. As mentioned above, *bipartite* **IUV** means the special case of **IUV** where every edge is black. We call the graph *bipartite* if every edge is black.

Here is a sketch of our algorithm and proof for Theorem 1(a). Let k be the maximum number such that the input graph G is k-element connected. Clearly, the maximum number of element-disjoint Steiner trees is $\leq k$ (informally, each Steiner tree in a family of element-disjoint Steiner trees contributes one to the element connectivity). Note that this upper bound also holds for the optimal *fractional* solution. We delete or contract white edges in G, while preserving the element connectivity, to obtain a bipartite graph G^*; thus, G^* too is k-element connected (details in Section 2). Then we apply our key result (Theorem 3 in Section 3) to G^* to obtain $O(k/\log n)$ element-disjoint Steiner trees; this is achieved via a simple algorithm that assigns a random colour to each Steiner node – it turns out that for each colour, the union of T and the set of nodes with that colour induces a connected subgraph, and hence this subgraph contains a Steiner tree. Finally, we uncontract some of the white nodes to obtain the same number of element-disjoint Steiner trees of G. Note that uncontracting white nodes in a set of element-disjoint Steiner trees preserves the Steiner trees (up to the deletion of redundant edges) and preserves the element-disjointness of the Steiner trees.

2   Reducing IUV to Bipartite IUV

To prove our main result, we first show that the problem can be reduced to bipartite **IUV** while preserving the approximation guarantee. The next result is due to Hind and Oellermann [14, Lemma 4.2]. We had found the result independently (before discovering the earlier works), and have included a proof for the sake of completeness.

Theorem 2. *Given a graph $G = (V, E)$ with terminal set T that is k-element connected (and has no edge with both end-nodes black), there is a poly-time algorithm to obtain a bipartite graph G^* from G such that G^* has the same terminal set and is k-element connected, by repeatedly deleting or contracting white edges.*

Proof. Consider any white edge $e = pq$. We prove that either deleting or contracting e preserves the k-element connectivity of G.

Suppose that $G - e$ is *not* k-element connected. Then by Menger's theorem $G - e$ has a set D of $k - 1$ white nodes whose deletion "separates" two terminals. That is, every terminal is in one of two components of $G - D - e$ and each of these components has at least one terminal; call these two components C_p and C_q. Let s be a terminal in C_p and let t be a terminal in C_q. Let $\mathcal{P}(s, t)$ denote any set of k element-disjoint s, t-paths in G, and observe that one of these s, t-paths, say P_1, contains e (since the k-set $D \cup \{e\}$ "covers" $\mathcal{P}(s, t)$).

By way of contradiction, suppose that the graph $G'' = G/e$, obtained from G by contracting e, is *not* k-element connected. Then focus on G and note that, again by Menger's theorem, it has a set R of k white nodes, $R \supseteq \{p, q\}$, whose deletion "separates" two terminals. That is, there are two terminals that are in different components of $G - R$ (R is obtained by taking a "cut" of $k - 1$ white nodes in G'' and uncontracting one node). This gives a contradiction because: (1) for s, t as above, the s, t-path P_1 in $\mathcal{P}(s, t)$ contains both nodes $p, q \in R$; since $|R| = k$ and $\mathcal{P}(s, t)$ has k element disjoint paths (by the Pigeonhole Principle) another one of the s, t-paths in $\mathcal{P}(s, t)$ say P_k is disjoint from R; hence, $G - R$ has an s, t-path, and (2) for terminals v, w that are both in say C_p (or both in C_q), $G - R$ has a v, t path (arguing as in (1)) and also it has a w, t path (as in (1)), thus $G - R$ has a v, w path.

It is easy to complete the proof: we repeatedly choose any white edge and either delete e or contract e, while preserving the k-element connectivity, until no white edges are left; we take G^* to be the resulting k-element connected bipartite graph.

Clearly, this procedure can be implemented in polynomial time. In more detail, we choose any white edge e (if there exists one) and delete it. Then we compute whether or not the new graph is k-element connected by finding whether $\kappa(s, t) \geq k$ in the new graph for every pair of terminals s, t; this computation takes $O(k|T|^2|E|)$ time. If the new graph is k-element connected, then we proceed to the next white edge, otherwise, we identify the two end nodes of e (this has the effect of contracting e in the old graph). Thus each iteration decreases the number of white edges (which is $O(|E|)$), hence, the overall running time is $O(k|T|^2|E|^2)$.   □

3   Bipartite IUV

This section has the key result of the paper, namely, a randomized $O(\log n)$-approximation algorithm for bipartite **IUV**.

Theorem 3. *Given an instance of bipartite* **IUV** *such that the graph is k-element connected, there is a randomized poly-time algorithm that with probability $1 - \frac{1}{\log n}$ finds a set of $O(\frac{k}{\log n})$ element-disjoint Steiner trees.*

Proof. Without loss of generality, assume that the graph is connected, and there is no edge between any two terminals (if there exists any, then subdivide each such edge by inserting a Steiner node).

For ease of exposition, assume that n is a power of two and k is an integer multiple of $R = 6 \log n$; here, R is a parameter of the algorithm. The algorithm is simple: we color each Steiner node u.r. (uniformly at random) with one of $\frac{k}{R}$ super-colors $i = 1, \ldots, k/R$. For each $i = 1, \ldots, k/R$, let $\mathcal{D}^i$ denote the set of nodes that get the super-color i. We claim that for each i, the subgraph induced by $\mathcal{D}^i \cup T$ is connected with high probability, and hence this subgraph contains a Steiner tree. If the claim holds, then we are done, since we get a set of k/R element-disjoint Steiner trees.

For the purpose of analysis, it is easier to present the algorithm in an equivalent form that has two phases. In phase one, we color every Steiner node u.r. with one of k colors $i = 1, \ldots, k$ and we denote the set of nodes that get the color i by C^i ($i = 1, \ldots, k$). In phase two, we partition the color classes into k/R super-classes where each super-class $\mathcal{D}^j$ ($j = 1, \ldots, k/R$) consists of R consecutive color classes $C_{(j-1)R+1}, C_{(j-1)R+2}, \ldots, C_{jR}$. We do this in R rounds, where in round $1 \leq \ell \leq R$ we have $\mathcal{D}_\ell^j = \bigcup_{i=(j-1)R+1}^{(j-1)R+\ell} C^i$; thus we have $\mathcal{D}^j = \mathcal{D}_R^j$. Consider an arbitrary super-class, say the first one $\mathcal{D}^1$. For an arbitrary $1 \leq \ell < R$, focus on the graph H_ℓ induced by $\mathcal{D}_\ell^1 \cup T$. Let $G_1, \ldots, G_{d_\ell}$ be the connected components of H_ℓ; note that $d_\ell \geq 1$ denotes the number of components of H_ℓ. Suppose that H_ℓ is not connected, i.e. $d_\ell > 1$.

Lemma 1. *Consider any connected component of H_ℓ, say G_1. There is a set $U \subseteq V - T - V(G_1)$ (of white nodes) with $|U| \geq k$ such that each node in U is adjacent to a terminal in G_1 and to a terminal in $G - V(G_1)$.*

Proof. Let $U \subseteq V - V(G_1)$ be a maximum-size set of Steiner nodes such that each node in U has a neighbour in each of G_1 and $G - V(G_1)$; note that none of the nodes in U is in G_1. By way of contradiction, assume that $|U| < k$. Consider $G - U$. An important observation is that every edge of G between G_1 and $G - V(G_1)$ is between a terminal of G_1 and a Steiner node of $G - V(G_1)$; this holds because G is bipartite and G_1 is a subgraph induced by T and some set of white nodes. From this, and by definition of U, there is no edge between G_1 and $G - U - V(G_1)$, i.e., $G - U$ is disconnected (note that there is at least one terminal in G_1 and one terminal in $G - U - V(G_1)$). This contradicts the assumption that G is k element-connected.   ☐

Consider a set U as in the above lemma. If a vertex $s \in U$ has the color $\ell + 1$, then when we add $C^{\ell+1}$ to $\mathcal{D}_\ell^1$, we see that s connects G_1 and another connected component of H_ℓ, because s is adjacent to a terminal in G_1 and to a terminal in $G - V(G_1)$. For every node $s \in U$ we have $\Pr[s \in C^{\ell+1}] = \frac{1}{k}$. Thus, the probability that none of the vertices in U has been colored $\ell + 1$ is at most:

$$\left(1 - \frac{1}{k}\right)^{|U|} \leq \left(1 - \frac{1}{k}\right)^k \leq e^{-1}. \qquad (2)$$

This is an upper bound on the probability that when we add $C^{\ell+1}$ to $\mathcal{D}_\ell^1$, component G_1 *does not* become connected to another connected component G_a, for some $2 \leq a \leq d_\ell$. If every connected component G_i, $1 \leq i \leq d_\ell$, becomes connected to another component, then the number of connected components of H_ℓ decreases to at most $\frac{d_\ell}{2}$ in round $\ell + 1$. If in every round and for every super-class, the number of connected components decreases by a constant factor then, after $O(\log n)$ rounds, every $\mathcal{D}^i \cup T$ forms a connected graph. We show that this happens with sufficiently high probability.

By (2), in round ℓ, any fixed connected component of H_ℓ becomes connected to another component with probability at least $1 - e^{-1}$. So the expected number of connected components of H_ℓ that become connected to another component is $(1 - e^{-1}) \cdot d_\ell$. Thus, if $d_\ell \geq 2$ then defining $\sigma = \frac{1 + e^{-1}}{2}$ we have:

$$E[d_{\ell+1} \mid d_\ell] \leq \sigma \cdot d_\ell. \qquad (3)$$

Define $X_\ell = d_\ell - 1$. Therefore, $X_1, X_2, \ldots, X_\ell, \ldots$, is a sequence of integer random variables that starts with $X_1 = d_1 - 1$. Moreover, for every $\ell \geq 1$, we have $X_\ell \geq 0$, and if $X_\ell = 0$ then $E[X_{\ell+1}] = 0$ and if $X_\ell \geq 1$ then

$$\begin{aligned}
E[X_{\ell+1}|X_\ell] &= E[d_{\ell+1} - 1|d_\ell - 1 \geq 1] \\
&= E[d_{\ell+1}|d_\ell \geq 2] - 1 \\
&\leq \sigma d_\ell - 1 \qquad \text{by (3)} \\
&= \sigma X_\ell + \sigma - 1 \\
&\leq \sigma X_\ell.
\end{aligned}$$

An easy induction shows that $E[X_{\ell+1}] \leq \sigma^\ell X_1$. Since $X_1 \leq n - 1$ and $\sigma < \frac{3}{4}$, we have $E[X_R] \leq \frac{1}{n}$ (recall that $R = 6 \log n$). Therefore, Markov's inequality implies that $\Pr[X_R \geq 1] \leq \frac{1}{n}$. This implies that $\Pr[d_R \geq 2] \leq \frac{1}{n}$, i.e., the probability that $H_R = D^1 \cup T$ is *not* connected is at most $\frac{1}{n}$. As there are $\frac{k}{R}$ super-classes, a simple union-bound shows that the probability that there is at least one D^j ($1 \leq j \leq \frac{k}{R}$) such that $D^j \cup T$ is not connected is at most $\frac{k}{Rn} \leq \frac{1}{\log n}$. Thus, with probability at least $1 - \frac{1}{\log n}$, every super-class D^j (together with T) induces a connected graph, and hence, the randomized algorithm finds $\Omega(k/\log n)$ element-disjoint Steiner trees. $\qquad \square$

4   IUV and Capacitated IUV

Now we complete the proof of Theorem 1 using Theorems 2 and 3.

First, we prove part (a). Let k be the maximum number such that the input graph G is k-element connected. Clearly, the maximum number of element-disjoint Steiner trees is at most k. Apply Theorem 2 to obtain a bipartite graph

G^* that is k-element connected. Apply Theorem 3 to find $\Omega(\frac{k}{\log n})$ element-disjoint Steiner trees in G^*. Then uncontract white nodes to obtain the same number of element-disjoint Steiner trees of G. Moreover, it can be seen that the optimal value of the LP relaxation is at most k (because there exists a set of k white nodes whose deletion leaves no path between some pair of terminals). Thus our integral solution is within a factor $O(\log n)$ of the optimal fractional solution.

Now, we prove part (b) of Theorem 1. Our proof uses ideas from [3, 15, 19]. Consider the IP formulation (1) of capacitated **IUV**. The *fractional packing vertex capacitated Steiner tree* problem is the linear program (LP) obtained by relaxing the integrality condition in the IP to $x_F \geq 0$. As we said earlier, this LP has exponentially many variables, however, we can solve it approximately. Then we show that either rounding the approximate LP solution will result in an $O(\log n)$-approximation or we can reduce the problem to the uncapacitated version of **IUV** and use Theorem 1,(a).

Note that the separation oracle for the dual of the LP is the problem of finding a minimum node-weighted Steiner tree. Using this fact, the proof of Theorem 4.1 in [15] may be adapted to prove the following:

Lemma 2. *There is an α-approximation algorithm for fractional **IUV** if and only if there is an α-approximation algorithm for the minimum node-weighted Steiner tree problem.*

Klein and Ravi [17] (see also Guha and Khuller [13]) give an $O(\log n)$-approximation algorithm for the problem of computing a minimum node-weighted Steiner tree. Their result, together with Lemma 2 implies that:

Lemma 3. *There is a polynomial-time $O(\log n)$-approximation algorithm for fractional **IUV**.*

Define φ and φ_f to be the optimal (objective) values for capacitated **IUV** and for fractional capacitated **IUV**, respectively. Consider an approximately optimal solution to fractional capacitated **IUV** obtained by Lemma 3. Let φ^* denote the approximately optimal (objective) value, and let $Y = \{x_1, \ldots, x_d\}$ denote the set of primal variables that have positive values. One of the features of the algorithm of Lemma 3 (which is also a feature of the algorithm of [15]) is that d (the number of fractional Steiner trees computed) is polynomial in n (even though the LP has an exponential number of variables). If $\sum_{i=1}^{d} \lfloor x_i \rfloor \geq \frac{1}{2} \sum_{i=1}^{d} x_i$ then $Y' = \{\lfloor x_1 \rfloor, \ldots, \lfloor x_d \rfloor\}$ is an integral solution (i.e., a solution for capacitated **IUV**) with value at least $\frac{\varphi^*}{2}$, which is at least $\Omega(\frac{\varphi_f}{\log n})$, and this in turn is at least $\Omega(\frac{\varphi}{\log n})$. In this case the algorithm returns the Steiner trees corresponding to the variables in Y' and stops. This is within an $O(\log n)$ factor of the optimal solution. Otherwise, if $\sum_{i=1}^{d} \lfloor x_i \rfloor < \frac{1}{2} \sum_{i=1}^{d} x_i$ then

$$\varphi^* = \sum_{i=1}^{d} x_i = \sum_{i=1}^{d} \lfloor x_i \rfloor + \sum_{i=1}^{d}(x_i - \lfloor x_i \rfloor) < \frac{\varphi^*}{2} + d.$$

Therefore $\varphi^* < 2d$. This implies that for every Steiner node v, at most a value of $\min\{c_v, O(d\log n)\}$ of the capacity of v is used in any optimal (fractional or integral) solution. So we can decrease the capacity c_v of every Steiner node $v \in V - T$ to $\min\{c_v, O(d\log n)\}$. Note that this value is upper bounded by a polynomial in n. Let this new graph be G'. We are going to modify this graph to another graph G'' which will be an instance of uncapacitated **IUV**. For every Steiner node $v \in G'$ with capacity c_v we replace v with c_v copies of it called $v_1, \ldots, v_{c_v}$ each having unit capacity. The set of terminal nodes stays the same in G' and G''. Then for every edge $uv \in G'$ we create a complete bipartite graph on the copies of v (as one part) and the copies of u (the other part) in G''. This new graph G'' will be the instance of (uncapacitated) **IUV**. It follows that the size of G'' is polynomial in G. Also, it is straightforward to verify that G'' has α element-disjoint Steiner trees if and only if there are α Steiner trees in G satisfying the capacity constraints of the Steiner nodes. Finally, we apply the algorithm of Theorem 1,(a) to graph G''.

5   Concluding Remarks

We presented a simple combinatorial algorithm which finds an integral solution that is within a factor $O(\log n)$ of the optimal integral (and in fact optimal fractional) solution. Recently, Lau [20] has given a combinatorial $O(1)$-approximation algorithm for computing a maximum collection of edge-disjoint Steiner forests in a given graph. His result again relies on the result of Frank et al. [7] for solving (a special case of) bipartite **IUV**. It would be interesting to study the corresponding problem of packing *element-disjoint* Steiner forests.

Acknowledgments

The authors thank David Kempe, whose comments led to an improved analysis in Theorem 3, and Lap chi Lau who brought reference [14] to our attention.

References

1. S.Arora and C.Lund, *Hardness of approximations*, in *Approximation Algorithms for NP-hard Problems*, Dorit Hochbaum Ed., PWS Publishing, 1996.
2. J.Bang-Jensen and S.Thomassé, *Highly connected hypergraphs containing no two edge-disjoint spanning connected subhypergraphs*, Discrete Applied Mathematics 131(2):555-559, 2003.
3. J. Cheriyan and M. Salavatipour, *Hardness and approximation results for packing Steiner trees*, invited to a special issue of Algorithmica. Preliminary version in Proc. ESA 2004, Springer LNCS, Vol 3221, pp 180-191.
4. R.Diestel, *Graph Theory*, Springer, New York, NY, 2000.
5. U. Feige, M. Halldorsson, G. Kortsarz, and A. Srinivasan, *Approximating the domatic number*, SIAM J.Computing 32(1):172-195, 2002. Earlier version in STOC 2000.

6. P. Floréen, P. Kaski, J. Kohonen, and P. Orponen, *Multicast time maximization in energy constrained wireless networks*, in Proc. 2003 Joint Workshop on Foundations of Mobile Computing, DIALM-POMC 2003.
7. A.Frank, T.Király, M.Kriesell, *On decomposing a hypergraph into k connected sub-hypergraphs*, Discrete Applied Mathematics 131(2):373-383, 2003.
8. M.Grötschel, A.Martin, and R.Weismantel, *Packing Steiner trees: polyhedral investigations*, Math. Prog. A 72(2):101-123, 1996.
9. ——, *Packing Steiner trees: a cutting plane algorithm*, Math. Prog. A 72(2):125-145, 1996.
10. ——, *Packing Steiner trees: separation algorithms*, SIAM J. Disc. Math. 9:233-257, 1996.
11. ——, *Packing Steiner trees: further facets*, European J. Combinatorics 17(1):39-52, 1996.
12. ——, *The Steiner tree packing problem in VLSI design*, Mathematical Programming 78:265-281, 1997.
13. S. Guha and S. Khuller, *Improved methods for approximating node weighted Steiner trees and connected dominating sets*, Information and Computation 150:57-74, 1999. Preliminary version in FST&TCS 1998.
14. H. R. Hind and O. Oellermann, *Menger-type results for three or more vertices* Congressus Numerantium 113:179–204, 1996.
15. K. Jain, M. Mahdian, M.R. Salavatipour, *Packing Steiner trees*, in Proc. ACM-SIAM SODA 2003.
16. K. Jain, I. Mandoiu, V. Vazirani and D. Williamson, *A primal-dual schema based approximation algorithm for the element connectivity problem*, in Proc. ACM-SIAM SODA 1999, 99–106.
17. P.Klein and R.Ravi, *A nearly best-possible approximation algorithm for node-weighted Steiner trees*, Journal of Algorithms 19:104-115 (1995).
18. M.Kriesell, *Edge-disjoint trees containing some given vertices in a graph*, J. Combinatorial Theory (B) 88:53-65, 2003.
19. L. Lau, *An approximate max-Steiner-tree-packing min-Steiner-cut theorem*, In Proc. IEEE FOCS 2004.
20. L. Lau, *Packing Steiner forests*, to appear in Proc. IPCO 2005.
21. A.Martin and R.Weismantel, *Packing paths and Steiner trees: Routing of electronic circuits*, CWI Quarterly 6:185-204, 1993.

Approximating the Bandwidth of Caterpillars

Uriel Feige[1] and Kunal Talwar[2]

[1] Weizmann Institute and Microsoft Research
urifeige@microsoft.com
[2] Microsoft Research
kunal@microsoft.com

Abstract. A caterpillar is a tree in which all vertices of degree three or more lie on one path, called the backbone. We present a polynomial time algorithm that produces a linear arrangement of the vertices of a caterpillar with bandwidth at most $O(\log n/\log\log n)$ times the *local density* of the caterpillar, where the local density is a well known lower bound on the bandwidth. This result is best possible in the sense that there are caterpillars whose bandwidth is larger than their local density by a factor of $\Omega(\log n/\log\log n)$. The previous best approximation ratio for the bandwidth of caterpillars was $O(\log n)$. We show that any further improvement in the approximation ratio would require using linear arrangements that do not respect the order of the vertices of the backbone. We also show how to obtain a $(1 + \epsilon)$ approximation for the bandwidth of caterpillars in time $2^{\tilde{O}(\sqrt{n/\epsilon})}$. This result generalizes to trees, planar graphs, and any family of graphs with treewidth $\tilde{O}(\sqrt{n})$.

1 Introduction

To set the ground for presenting our results, let us first define the main terms that we use.

Definition 1. *A* linear arrangement *of a graph is a numbering of its vertices from 1 to n. The* bandwidth *of a linear arrangement is the largest difference between the two numbers given to endpoints of the same edge. The bandwidth of a graph is the minimum bandwidth over all linear arrangements of the graph.*

Definition 2. *A* caterpillar *is a tree composed of a path, called the* backbone, *and other paths, called* strands, *connected to the backbone. The connection point of a strand to the backbone is called the* root *of the strand. Hence all vertices of degree more than two are roots.*

Definition 3. *The* unfolded bandwidth *of a caterpillar is the minimum bandwidth in a linear arrangement that respects the order of the vertices in the backbone.*

Definition 4. *Let $B(v, r)$ denote the set $\{u \in V : d(u, v) \leq r\}$ where $d(\cdot, \cdot)$ denotes the usual shortest path distance in G. The* local density *of a graph ρ_G is defined as*

C. Chekuri et al. (Eds.): APPROX and RANDOM 2005, LNCS 3624, pp. 62–73, 2005.

$$\rho_G = \max_{v \in V} \max_r \frac{|B(v,r)|}{2r}$$

It is easy to check that for any graph G, the local density ρ_G gives a lower bound on the bandwidth of G.

Given an algorithm A for approximating the bandwidth of caterpillars, for every caterpillar G we consider the following four quantities: its local density ρ_G; its bandwidth b_G; its unfolded bandwidth u_G; and the bandwidth of the linear arrangement found by algorithm A, denoted by A_G.

Clearly, $\rho_G \leq b_G \leq u_G \leq A_G$. Previously, it was known [3] that b_G can be as large as $\Omega(\rho_G \log n / \log \log n)$. Also, a simple approximation algorithm [9] obtained $A_G \leq O(\rho_G \log n)$. We present an approximation algorithm for which $A_G = O(\rho_G \log n / \log \log n)$, which is best possible (with respect to ρ_G). We also show that the gap of $\Omega(\log n / \log \log n)$ may be present between every two adjacent quantities in the above list. The gap of $\Omega(\log n / \log \log n)$ between b_G and u_G is especially important, because we show that u_G can in fact be approximated within a constant factor. Without this gap between u_G and b_G, this would improve the approximation ratio for the bandwidth of caterpillars beyond $\log n / \log \log n$.

In a result of a somewhat different nature, we present an algorithm that achieves a $(1 + \epsilon)$ approximation ratio for the bandwidth of trees (and hence also caterpillars) in time $2^{\tilde{O}(\sqrt{n/\epsilon})}$. This result generalizes to families of graphs that have recursive decomposition with separators of size $\tilde{O}(\sqrt{n})$, including for example planar graphs. These results show in particular that in any reduction from 3SAT showing the hardness of approximating the bandwidth of caterpillars or trees, the number of vertices in the resulting graph will be at least quadratic in the size of the 3CNF formula, unless 3SAT has subexponential algorithms.

1.1   Related Work

Chinn *et al.* [2] showed that trees with constant local density can have bandwidth as large as $\Omega(\log n)$. Chung and Seymour [3] exhibited a family of caterpillars with constant local density and bandwidth $\Omega(\frac{\log n}{\log \log n})$.

Monien [11] showed that the bandwidth problem on caterpillars is NP-hard. We note here that the reduction crucially uses a gadget that forces the bandwidth to be folded. Blache, Karpinski and Jürgen [1] showed that the bandwidth of trees is hard to approximate within some constant factor. Unger [14] claimed (without proof) that the bandwidth of caterpillars is hard to approximate within any constant factor.

Haralambides *et al.* [9] showed that for caterpillars, folding strands to one side is an $O(\log n)$-approximation with respect to the local density. For general graphs, Feige [5] gave an $O(\log^{3.5} \sqrt{\log \log n})$ algorithm with respect to the local density lower bound (slightly improved in   [10]). A somewhat improved approximation ratio of $O(\log^3 n \sqrt{\log \log n})$ with respect to a semidefinite programming lower bound was given by Dunagan and Vempala [4]. Gupta [8] gave an $O(\log^{2.5} n)$-approximation algorithm on trees, that on caterpillars gives an $O(\log n)$ approximation. Filmus [7] extended this $O(\log n)$-approximation to graphs formed by many caterpillars sharing a backbone vertex.

An exact algorithm for general graphs, running in time approximately n^B was given by Saxe [13] where B is the optimal bandwidth. Feige and Killian [6] gave a $2^{O(n)}$ time exact algorithm.

2   New Results

2.1   More Definitions

Definition 5. *A* bucket arrangement *of a graph is a placement of its vertices into consecutive buckets, such that the endpoints of an edge are either in the same bucket or in adjacent buckets. The* bucketwidth *is the number of vertices in the most loaded bucket. The* bucketwidth *of a graph is the minimum bucketwidth of all bucket arrangements of the graph.*

The following lemma is well known.

Lemma 1. *For every graph, its bandwidth and bucketwidth differ by at most a factor of 2.*

Proof. Given a linear arrangement of bandwidth b, make every b consecutive vertices into a bucket. Given a bucket arrangement with bucketwidth b, create a linear arrangement bucket by bucket, where vertices in the same bucket are numbered in arbitrary order.   □

As we shall be considering approximation ratios which are much worse than 2, we will consider bandwidth and bucketwidth interchangeably.

Definition 6. *The* unfolded bucketwidth *of a caterpillar is the minimum bucketwidth in a bucket arrangement in which every backbone vertex lies in a different bucket. Hence backbone vertices are placed in order.*

Lemma 2. *For every caterpillar, its unfolded bandwidth and unfolded bucketwidth differ by at most a factor of 2.*

Proof. Given an unfolded linear arrangement of bandwidth b, let every backbone vertex start a new bucket, called a backbone bucket. In regions with no backbone vertex (the leftmost and rightmost regions of the linear arrangement), make every b consecutive vertices into a bucket. It may happen that strands (say, k of them) jump over a backbone bucket in this bucket arrangement, but then this backbone bucket has load at most $b - k$. For each such strand, shift nodes so as to move one vertex from the outermost bucket of the strand to the separating backbone bucket. We omit the details from this extended abstract.

Given an unfolded bucket arrangement with bucketwidth b, create a linear arrangement bucket by bucket, where vertices in the same bucket are numbered in arbitrary order.   □

2.2   Approximating Unfolded Bandwidth

Theorem 1. *The unfolded bucketwidth of a caterpillar can be approximated within a constant factor.*

Proof. Up to a factor of two in the resulting bandwidth, we may assume that a strand is folded either to the right, or to the left (but not partly to the right and partly to the left), i.e. all buckets containing a node from a strand H lie on one side of the bucket containing its root. We call such an unfolded bucket arrangement *locally consistent*. We now formulate the problem of finding a locally consistent unfolded bucket arrangement of minimum bucketwidth as an integer program.

The variables of the integer program are of the form x_{ik}, where i specifies a vertex and k specifies a bucket number. Our intention is that $x_{ik} = 1$ if vertex i is in bucket k, and $x_{ik} = 0$ otherwise. Hence we shall have the *integrality constraints*:

$$x_{ij} \in \{0, 1\} \tag{1}$$

For concreteness, we assume that vertices of the caterpillar are named in the following order. First, the backbone vertices are named from left to right as 1 up to ℓ (where ℓ is the number of backbone vertices), and then the strands of the caterpillar are named one by one, where within a strand, vertices are named in order of increasing distance from the root of the strand. (This naming should not be confused with a linear arrangement. It is just a convention used in formulating the integer program.) We now place the vertices of the backbone in consecutive buckets. This gives the *backbone constraints*:

$$x_{ii} = 1 \quad \text{for all } 1 \le i \le \ell \tag{2}$$

Since vertices of strands might be placed in buckets to the left and to the right of the endpoints of the backbone, the parameter k in x_{ik} specifying the bucket number need not be limited to the range 1 up to ℓ, but will be allowed to range between $-n$ and n.

Now we are ready to present the most important set of constraints. They are more complicated than may appear necessary for the integer program, but this will become useful once we relax the integer program to a polynomial time solvable linear program.

Consider two vertices $i, i+1$ on the same strand, rooted at the backbone vertex r. Then we have the *root constraints*:

$$x_{(i+1)r} \le x_{ir} \tag{3}$$

Moreover, let k be a bucket to the right of bucket r, namely, $k > r$. (Later we will present analogous constraints for $k < r$.) For every $t \ge k$ we have the *right consistency constraints*:

$$\sum_{l=k}^{t} x_{(i+1)l} \le \sum_{l=k-1}^{t} x_{il} \quad \text{and} \quad \sum_{l=k}^{t} x_{il} \le \sum_{l=k}^{t+1} x_{(i+1)l} \tag{4}$$

The consistency constraints indicate that $(i+1)$ appears in a bucket to the right of r only if i preceded it, and that i appears only if $(i+1)$ follows it.

Similar constraints are written for buckets to the left of the root. Namely, for $k < r$ and $t \le k$ we have the *left consistency constraints*:

$$\sum_{l=t}^{k} x_{(i+1)l} \le \sum_{l=t}^{k+1} x_{il} \quad \text{and} \quad \sum_{l=t}^{k} x_{il} \le \sum_{l=t-1}^{k} x_{(i+1)l} \tag{5}$$

Finally, we introduce a constraint specifying that the locally consistent unfolded bucketwidth is at most b. Namely, for every bucket k there is the *bucketwidth constraint*:

$$\sum x_{ik} \le b \tag{6}$$

The above integer program is feasible if and only if there is a locally consistent bucket arrangement of bucketwidth at most b. We now relax the integer program to a linear program by replacing the integrality constraints by nonnegativity constraints $x_{ik} \ge 0$ and *choice constraints*, namely, for every vertex i on a strand of length t rooted at r we have:

$$\sum_{k=r-t}^{r+t} x_{ik} = 1 \tag{7}$$

As the linear program is a relaxation of the integer program, it is feasible whenever the integer program is. As linear programs can be solved in polynomial time, we can obtain a feasible solution to the linear program of bucketwidth at most b^*, where b^* is the minimum locally consistent unfolded bucketwidth. The solution to the linear program is fractional, in the sense the a vertex may fractionally belong to several buckets. We now show how to round the fractional solution to an integer solution, loosing only a constant factor in the bucketwidth.

Consider an arbitrary solution x_{ik} to the linear program. Consider a strand S rooted at r. Consider the smallest $k > r$ such that $\sum_{i \in S} x_{ik} \le 1/4$. We claim that $\sum_{i \in S, k' \ge k} x_{ik'} \le |S|/4$. This claim follows from the following inequality:

$$\sum_{k' > k} x_{ik'} \le \sum_{j \in S; j < i} x_{jk}$$

This inequality can be proved by induction on i. For the first vertex on the strand, this inequality is true because then $\sum_{k' > k} x_{ik} = 0$. Assume now that the inequality was proved for vertex i and then the right consistency constraints and the inductive hypothesis imply for vertex $i+1$ that:

$$\sum_{k' > k} x_{(i+1)k'} \le \sum_{k' \ge k} x_{ik'} \le x_{ik} + \sum_{j \in S; j < i} x_{jk} = \sum_{j \in S; j < i+1} x_{jk}$$

Similar to the above, consider the largest $k < r$ with $\sum_{i \in S} x_{ik} \le 1/4$. It can be shown that $\sum_{i \in S, k' \le k} x_{ik'} \le |S|/4$.

It follows that there is some k (w.l.o.g. we will assume here that $k > r$) such that $\sum_{i \in S; r \le k' < k} x_{ik'} \ge |S|/4$, and $\sum_{i \in S} x_{ik'} \ge 1/4$ for all $r \le k' < k$. Now place the vertices of S one by one in the buckets starting at r and continuing to the right, putting $\lceil 4 \sum_{i \in S} x_{ik'} \rceil$ vertices in bucket k'. The whole strand can be put in these buckets before reaching bucket k. Each bucket suffered a multiplicative factor of at most 8 (rounding up to the nearest integer contributes a factor of at most 2, because $4 \sum_{i \in S} x_{ik'} \ge 1$).

Finally, there is the following issue that we ignored in the proof above, and this is the fact that for the root bucket r, it might not be the case that $\sum_{i \in S} x_{ir} \ge 1/4$. In that case, the integer solution might put a vertex of S in bucket r that we cannot charge against the fractional values of S that were in bucket r. However, this cannot harm the approximation ratio by more than a constant factor, because the number of vertices added by this effect to bucket r cannot be more than twice the local density at r.   □

2.3 Upper Bound

In this section, we present an algorithm for the bandwidth problem on caterpillars and show that it is an $O(\frac{\log n}{\log \log n})$ approximation.

We shall give a bucket arrangement of caterpillar T. More precisely, we shall define maps $x : V \to [n]$ and $y : V \to [n]$ such that

- The function $f : V \to [n] \times [n]$ defined as $f(v) = (x(v), y(v))$ is one-one.
- For every edge $(u, v) \in E$, $|x(u) - x(v)| \le 1$.

In other words, we shall place the vertices of the graph on the integer grid so that adjacent vertices land either in the same column or in adjacent columns. The goal would be to minimize the maximum height of any column.

Let T be a caterpillar with backbone $B = \{1, \dots, p\}$ and strands $\mathcal{H} = \{H_1, \dots, H_q\}$, where each H_i can be specified by its root $r_i \in B$ and length l_i. Let $k = 2 \frac{\log n}{\log \log n}$ so that $k^k > n$. For an assignment f and for a strand H_i let the *height* α_i of H_i be $\max_{v \in H_i} y(v)$ and let the *range* β_i of H_i be $\max_{v \in H_i} |x(v) - x(r_i)|$.

The algorithm proceeds as follows. Let $\mathcal{H}_s = \{H_i \in \mathcal{H} : k^{s-1} \le l_i < k^s\}$ (note that $\mathcal{H}_k$ is empty). We start out by setting $f(j) = (j, 1)$ for $j \in B$. The algorithm has k phases. In phase s, we consider the strand $H_i \in \mathcal{H}_s$ in increasing order of r_i (breaking ties arbitrarily). We fold each strand to its right so as to minimize the maximum height of any column, and subject to that, minimize its range. Amongst the arrangements minimizing the height and the range, we break ties to the left, i.e. fill up columns from left to right.

We now show that the mapping f satisfies $\mathsf{bw}_f \le O(\frac{\log n}{\log \log n} \rho_G)$. We denote by $f^{-1}(X, Y) = \{v : f(v) = (x, y), x \in X, y \in Y\}$ the set of nodes assigned to columns X and rows Y.

Let I_t denote the interval $[4(k+t-1)\rho_G + 1, 4(k+t)\rho_G]$. For a column j, let $\mathcal{H}_t^j = \{H_i \in \mathcal{H}_t : H_i \cap f^{-1}(\{j\}, I_t) \neq \phi\}$ be the set of hair in $\mathcal{H}_t$ that contribute to $f^{-1}(\{j\}, I_t)$.

The tie breaking rule ensures the following.

Observation 2. *Just after H_i is assigned, if H_i contributes to columns j and j', $j < j'$, then the height of j' is no larger than that of j.*

Next we show a few simple lemmas

Lemma 3. *Every strand $H_i \in \mathcal{H}_t^j$ has range β_i at most k^{t-1}.*

Proof. For every $H_i \in \mathcal{H}_t^j$, it must be the case that for all $j' : r_i \leq j' \leq j$, the column j' has height at least $4k\rho_G$ after H_i was considered. Consider the set of vertices $f^{-1}([r_i,j],[1,4k\rho_G])$ assigned to rows r_i through j and columns 1 through $4k\rho_G$ in this partial assignment. Because of the order in which we consider strands, each of these vertices comes from a strand of length less than k^t and the hence the root of every such strand must lie in $[r_i - k^t, j]$. Thus every such vertex lies in $B(r_i, 2k^t)$ and there are at least $2k\rho_G|j - r_i| = 2k\rho_G\beta_i$ such vertices. Since $|B(r_i, 2k^t)| \leq 4k^t\rho_G$, it follows that $\beta_i \leq k^{t-1}$ for every $H_i \in \mathcal{H}_t^j$. $\square$

Corollary 1. *For every j, t, $|\mathcal{H}_t^j| < 4\rho_G$.*

Proof. Every strand $H_i \in \mathcal{H}_t^j$ has its root in $[j - k^{t-1}, j]$. Since each has length at least k^{t-1}, $B(j, 2k^{t-1})$ has size at least $(1 + |\mathcal{H}_t^j|k^{t-1})$. Hence, $|\mathcal{H}_t^j| < 4\rho_G$. $\square$

Lemma 4. *At the end of phase s of the algorithm, the height of (partial) assignment is at most $4(k + s)\rho_G$.*

Proof. We show the claim by induction on s. In the base case $s = 0$, each column has one node and since ρ_G is at least 1, the claim holds.

Suppose the claim is true at end of phase $(t - 1)$. We wish to show that it is also true at the end of phase t.

Consider column j and let $\mathcal{H}_t^j$ be as above. We first argue that each strand in $\mathcal{H}_t^j$ contributes at most one vertex to column j. Assume the contrary and let H_i be the first strand in $\mathcal{H}_t^j$ that contributes at least 2 vertices to column j. Further, let j be the first such column. Then there is a column $j' > j$ such that just before H_i was assigned, j' had height at least one more than that of j. Since we look at strand in each class from left to right, every strand contributing to $f^{-1}(j', I_t)$ at this point has its root to the left of j. Thus by observation 2, j' has height no larger than j, contradicting the assumption. Hence the claim.

By corollary 1, $\mathcal{H}_t^j$ has fewer than $4\rho_G$ strands and hence the induction holds. The claim follows. $\square$

We remark that we have actually shown an $O(\log h / \log\log h)$-approximation, where h is the longest strand length. Note that our algorithm outputs an unfolded arrangement. We show in the next section that it is not possible to beat the $O(\frac{\log n}{\log\log n})$ barrier when outputting an unfolded arrangement.

Finally, we note while this algorithm never folds the backbone, it could be far from optimal for the unfolded bandwidth problem.

Proposition 1. *The algorithm of the above theorem may output a linear arrangement with bandwidth $\Omega(\log n / \log\log n)$ times the unfolded bandwidth.*

Proof. The instance has a backbone of length $2^{k+1} - 1$ and a strand of length $i2^i$ at vertex 2^i and vertex $2^{k+1} - 2^i$. Folding (one vertex per bucket) the first half of the strands to the right and the the second half of the strands to the left gives an arrangement with unfolded bucketwidth $O(\log k)$. On the other hand, our algorithm (in fact any algorithm folding all strands to the right) gives a bucketwidth of k. Since k is $O(\log n)$, the claim follows.   □

2.4 Gap Between Bandwidth and Unfolded Bandwidth

We construct a sequence of caterpillars $C_1, C_2, \ldots$ such that C_k has bucketwidth $O(k)$ and unfolded bucketwidth $\Omega(k^2)$.

C_1 is a caterpillar with a backbone of length three and p strands of length 1 attached to the middle node. The first and last nodes on the backbone are designated s_1 and t_1 respectively (see figure 1).

C_k is constructed by joining in series, a path P_1 of length $(l_k + w_k)$, a copy T_1 of C_{k-1}, a path P_2 of length $2w_k$, another copy T_2 of C_{k-1} and finally a path P_3 of length $(l_k + w_k)$. Moreover, p strands each of length l_k are attached at the first and last vertices of P_2 (see figure 1). The first vertex of P_1 and the last vertex of P_3 are named s_k and t_k respectively. We refer to the backbone vertices in T_1, P_2 and T_2 as the *core* of C_k. The strands attached to P_2 are referred to as *central* strands.

Here w_k and l_k are parameters that we define as follows: $l_1 = 1, w_1 = 0$, $w_k = 2l_{k-1} + 2w_{k-1}$, $l_k = 6kl_{k-1}$. Note that the length of the backbone satisfies the relation $B_k = 4w_k + 2l_k + 2B_{k-1}$. It follows that

Observation 3. *The length of the backbone of C_k is at most $B_k \le 4l_k$.*

We first show that there is an arrangement with small bucketwidth.

Lemma 5. *There is a valid bucket arrangement of C_k with bucketwidth $p + 2k$ into at most $w_{k+1} = 2l_k + 2w_k$ buckets with s_k and t_k in the first and last buckets respectively.*

Proof. The proof uses induction. The base case is immediate.

Suppose that there is such a bucket arrangement for C_{k-1}. We use it to construct a bucket arrangement for C_k (see figure 2). We place the first path P_1 into buckets 1 through $(l_k + w_k)$. The recursive arrangement of T_1 is placed in buckets $(l_k + w_k + 1)$ through $(l_k + 2w_k)$. We place the path P_2 in buckets $(l_k + 2w_k)$ to $(l_k + 1)$. T_2 is placed similarly in buckets $(l_k + 1)$ to $(l_k + w_k)$ and we place P_3 starting at $(l_k + w_k + 1)$ and ending at bucket $2(l_k + w_k)$. Each central strand rooted at the endpoint of P_2 in bucket $(l_k + 2w_k)$ spans buckets $(l_k + 2w_k + 1)$ through $2(l_k + w_k)$ and each central strand rooted at the endpoint in bucket $l_k + 1$ goes into buckets l_k through 1. It is easy to verify that this is a legal bucket arrangement. Moreover, the maximum height of any bucket increases by at most 2, and hence the induction holds.   □

We now show that the unfolded bucketwidth of C_k is large. Consider an unfolded bucket arrangement f. Without loss of generality, s_k falls to the left of

t_k. A bucket is called a *core* bucket if it contains a vertex from the core. Other buckets are referred to as *peripheral*. Note that the core buckets fall between the left set and the right set of peripheral buckets.

Lemma 6. *In any unfolded bucket arrangement of C_k, some core bucket has load more than $\frac{pk}{3}$.*

Proof. We prove this by induction on k

In the base case $k = 1$, we simply use a local density argument. There are $p + 3$ nodes and since the diameter is 2, they all fall in at most 3 buckets. The claim follows.

Suppose the claim holds for C_{k-1} and let f be an unfolded bucket arrangement for C_k. Suppose that f has bucketwidth less than $\frac{pk}{3}$. Consider the $2p$ central strands of G_k.

Proposition 2. *In any arrangement with bucketwidth less than $(\frac{pk}{3})$ at least $(\frac{2p}{3})$ central strands extend to peripheral buckets.*

Proof. Suppose not. Then at least $\frac{4p}{3}$ central strands are wholly contained in the core buckets. The number of core bucket is at most $2(w_k + B_{k-1})$ and each has at most $\frac{pk}{3}$ nodes in the arrangement f. However, since each strand has length l_k and $(\frac{4p}{3})l_k > 2(w_k + B_{k-1})(\frac{kp}{3})$, we get a contradiction. $\qquad\square$

Thus $\frac{2p}{3}$ strands extend to peripheral buckets. Without loss of generality, at least $(\frac{p}{3})$ of these strands extend to the left set of peripheral buckets. Each of these strands contributes at least 1 to the buckets containing the backbone of T_1. Since f also induces an unfolded bucket arrangement of T_1, the induction hypothesis implies that one of the core buckets of T_1 must have load at least $\frac{p(k-1)}{3}$ from vertices in T_1. Adding the load due to the $\frac{p}{3}$ crossing strands, we get an overall load of at least $\frac{pk}{3}$. The claim follows. $\qquad\square$

Taking p to be k, we get a gap of $\Omega(k)$. Since $l_k \leq 6^k k!$, the number of vertices in C_k is $2^{O(k \log k)}$. Hence we get a gap of $\Omega(\frac{\log n}{\log \log n})$. Since (unfolded) bucketwidth and (unfolded) bandwidth are within a constant factor of each other, the gap between unfolded bandwidth and bandwidth is also $\Omega(\frac{\log n}{\log \log n})$.

3   An Approximation Scheme

In this section, we give an approximation scheme for the bandwidth problem on trees that gives a $(1 + \epsilon)$-approximation in time $2^{\tilde{O}(\sqrt{\frac{n}{\epsilon}})}$. We note that the algorithm described here is more complicated than needed; the extra work here enables us to extend the algorithm to a larger class of graphs without any modifications.

We first guess the bandwidth B. If $B \leq \sqrt{\frac{n}{\epsilon}}$, the dynamic programming algorithm of Saxe [13] finds the optimal bandwidth arrangement in time $n^{O(B))}$ which is $2^{\tilde{O}(\sqrt{\frac{n}{\epsilon}})}$ for B as above.

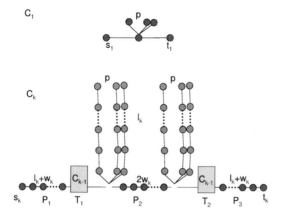

Fig. 1. Construction of caterpillar C_k

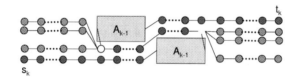

Fig. 2. Low bucketwidth arrangement of caterpillar C_k

In case $B \geq \sqrt{\frac{n}{\epsilon}}$, we run a different algorithm. We shall construct an assignment f of V into K buckets $1, \ldots, K$ such that

- For any bucket i, the number of nodes assigned to it $|f^{-1}(i)|$ is at most ϵB.
- For any $u, v \in V$ such that $(u, v) \in E$, $f(u)$ and $f(v)$ differ by at most $\frac{1}{\epsilon}$.

It is easy to see that if G has bandwidth B, such an assignment always exists for $K = (\frac{n}{\epsilon B})$. Moreover, given such as assignment, we can find a bandwidth $(1 + \epsilon)B$ arrangement of G by picking any ordering that respects the ordering defined by f.

Note that any tree T has a vertex v_T such that deleting v from T breaks it up in components of size at most $\frac{2n}{3}$; we call such a v a *balanced separator* of T. We recurse on each component until each component is a leaf. This sequence of decompositions can be represented in the form of a rooted *decomposition tree* τ with the following properties:

- Each node x of the decomposition tree τ corresponds to a subtree T_x and a balanced separator v_x of T_x.
- The children of an internal node x correspond exactly to the components resulting from deleting v_x from the tree T_x.
- The root node r corresponds to T and the leaves correspond to singletons.
- The depth of the tree is $O(\log n)$.

A partial assignment is a partial function f that maps a subset $D_f \subseteq V$ into K buckets $1, \ldots, K$. We say f is feasible if for every $u, v \in D_f$, $(u, v) \in E$, $f(u)$

and $f(v)$ differ by at most $\frac{1}{\epsilon}$. We say g *extends* f to D' if $D_g = D_f \cup D'$ and f agrees with g on D_f. Given a partial assignment f, its *profile* p_f is defined by the K-tuple $(|f^{-1}(1)|, |f^{-1}(2)|, \ldots, |f^{-1}(K)|)$. For the purposes of our algorithm two partial assignments are considered equivalent if they have the same profile.

Our algorithm does dynamic programming on the decomposition tree τ of T. Let x be an internal node of τ with children $y_1, \ldots, y_k$. Let $f_x^1, \ldots, f_x^{n_x}$ be the set of all feasible partial assignments with domain consisting of all the separator nodes on the r-x path in τ, i.e. f_x^a has domain $P_x = \{v_z : z \text{ is on } r\text{-}x \text{ path in } \tau\}$. Given x, $a \leq n_x$ and $j \leq k$, we compute the list $L_{x,j}^a$ of all possible profiles of a feasible extension g of f_x^a to $\cup_{i \leq j} T_{y_i}$. We use the notation L_x^a for $L_{x,k}^a$ if x has k children in τ.

We populate the dynamic programming table from the leaves moving up the tree as follows. We start with the obvious setting of L_x^a for each leaf x of τ, for each $a \in [n_x]$. For an internal node x, clearly $L_{x,0}^a$ is just the profile of f_x^a. Given $L_{x,(j-1)}^a$ and $L_{y_j}^b$ for every feasible extension $f_{y_j}^b$ of f_x^a to v_{y_j}, we show how to compute $L_{x,j}^a$. The crucial fact here, ensured by our construction of the decomposition tree, is that all the edges leaving T_{y_j} are incident on P_x. Thus given a feasible extension g_1 of f_x^a to $\cup_{i<j} T_{y_i}$ and another feasible extension g_2 of f_x^a to T_{y_j}, they can be combined to give a feasible extension g_3 of f_x^a to $\cup_{i \leq j} T_{y_i}$. Thus for every profile p_1 in $L_{x,(j-1)}^a$ and for every profile p_2 in any of the lists $L_{y_j}^b$ where $f_{y_j}^b$ is a feasible extension of f_x^a, we get a profile p_3 to be placed in $L_{x,j}^a$.

Finally, if T has bandwidth B, at least one of the lists L_r^a contains a profile p_f with $|f^{-1}(i)| \leq \epsilon B$ for each i.

It remains to bound the running time of our algorithm. Since the depth of the tree is $O(\log n)$, for every node x in τ, the number of partial assignments n_x is at most $K^{O(\log n)}$. The number of nodes in τ is clearly at most n and hence the algorithm only need compute $nK^{O(\log n)}$ table entries. The size of each list $L_{x,j}^a$ is bounded by the total number of possible profiles, which is no more than n^K. Each entry of the table can thus be computed in time $O(n^{2K})$. Thus the overall running time of the algorithm is $2^{O(K \log n)}$. Since $K = \frac{n}{\epsilon B}$ and $B \geq \sqrt{\frac{n}{\epsilon}}$, the running time of our algorithm is $2^{\tilde{O}(\sqrt{\frac{n}{\epsilon}})}$. Thus we have shown that

Theorem 4. *The algorithm described above computes a $(1 + \epsilon)$ approximation to the bandwidth of a tree in time $2^{\tilde{O}(\sqrt{\frac{n}{\epsilon}})}$.*

We note that it is easy to modify the algorithm to also give an assignment having such a profile.

Moreover, note that the only property of the tree we have used is the existence of small separators. The algorithm can be naturally modified to work with any family of graphs with (recursive)-small separators—the number of partial assignments to be considered for a table entry now goes up $K^{O(t \log n)}$ where t is an upper bound on the size of the separator. This gives a $2^{\tilde{O}(t + \sqrt{\frac{n}{\epsilon}})}$ time $(1 + \epsilon)$-approximation algorithm for graphs with tree-width t. Recall that planar graphs and other excluded minor families of graphs have separators of size $O(\sqrt{n})$. Thus the running time of our algorithm for such graphs is $2^{\tilde{O}(\sqrt{\frac{n}{\epsilon}})}$.

Acknowledgements

We thank Moses Charikar, Marek Karpinski and Ryan Williams.

References

1. G. Blache, M. Karpinski, and J. Wirtgen. On approximation intractability of the bandwidth problem. Technical report, University of Bonn, 1997.
2. P. Chinn, J. Chvatálová, A. Dewdney, and N. Gibbs. The bandwidth problem for graphs and matrices - survey. *Journal of Graph Theory*, 6(3):223–254, 1982.
3. F. R. Chung and P. D. Seymour. Graphs with small bandwidth and cutwidth. *Discrete Mathematics*, 75:113–119, 1989.
4. J. Dunagan and S. Vempala. On euclidean embeddings and bandwidth minimization. In *APPROX '01/RANDOM '01*, pages 229–240, 2001.
5. U. Feige. Approximating the bandwidth via volume respecting embeddings. *J. Comput. Syst. Sci.*, 60(3):510–539, 2000.
6. U. Feige. Coping with the NP-hardness of the graph bandwidth problem. In *SWAT*, pages 10–19, 2000.
7. Y. Filmus. Master's thesis, Weizmann Institute, 2003.
8. A. Gupta. Improved bandwidth approximation for trees. In *Proceedings of the eleventh annual ACM-SIAM symposium on Discrete algorithms*, pages 788–793, 2000.
9. J. Haralambides, F. Makedon, and B. Monien. Bandwidth minimization: An approximation algorithm for caterpillars. *Mathematical Systems Theory*, pages 169–177, 1991.
10. R. Krauthgamer, J. Lee, M. Mendel, and A. Naor. Measured descent: A new embedding method for finite metrics. In *FOCS*, pages 434–443, 2004.
11. B. Monien. The bandwidth minimization problem for caterpillars with hair length 3 is NP-complete. *SIAM J. Algebraic Discrete Methods*, 7(4):505–512, 1986.
12. C. Papadimitriou. The NP-completeness of the bandwidth minimization problem. *Computing*, 16:263–270, 1976.
13. J. Saxe. Dynamic-programming algorithms for recognizing small-bandwidth graphs in polynomial time. *SIAM J. Algebraic Discrete Methods*, 1:363–369, 1980.
14. W. Unger. The complexity of the approximation of the bandwidth problem. In *FOCS '98: Proceedings of the 39th Annual Symposium on Foundations of Computer Science*, pages 82–91, 1998.

Where's the Winner?
Max-Finding and Sorting with Metric Costs

Anupam Gupta[1,*] and Amit Kumar[2]

[1] Dept. of Computer Science, Carnegie Mellon University, Pittsburgh PA 15213
anupamg@cs.cmu.edu
[2] Dept. of Computer Science & Engineering, Indian Institute of Technology, Hauz Khas
New Delhi, India - 110016
amitk@cse.iitd.ernet.in

Abstract. Traditionally, a fundamental assumption in evaluating the performance of algorithms for sorting and selection has been that comparing any two elements costs one unit (of time, work, etc.); the goal of an algorithm is to minimize the total cost incurred. However, a body of recent work has attempted to find ways to weaken this assumption – in particular, new algorithms have been given for these basic problems of searching, sorting and selection, when comparisons between different pairs of elements have different associated costs.

In this paper, we further these investigations, and address the questions of max-finding and sorting when the comparison costs form a *metric*; i.e., the comparison costs c_{uv} respect the triangle inequality $c_{uv} + c_{vw} \geq c_{uw}$ for all input elements u, v and w. We give the first results for these problems – specifically, we present

- An $O(\log n)$-competitive algorithm for max-finding on general metrics, and we improve on this result to obtain an $O(1)$-competitive algorithm for the max-finding problem in constant dimensional spaces.

- An $O(\log^2 n)$-competitive algorithm for sorting in general metric spaces.

Our main technique for max-finding is to run two copies of a simple natural online algorithm (that costs too much when run by itself) in parallel. By judiciously exchanging information between the two copies, we can bound the cost incurred by the algorithm; we believe that this technique may have other applications to online algorithms.

1 Introduction

The questions of optimal searching, sorting, and selection lie at the very basis of the field of algorithms, with a vast literature on algorithms for these fundamental problems [1]. Traditionally the fundamental assumption in evaluating the performance of these algorithms has been the *unit-cost comparison model* – comparing any two elements costs one unit, and one of the goals is to devise algorithms that minimize the total cost of comparisons performed. Recently, Charikar et al. [2] posed the following problem: given a set V of n elements, where the cost of comparing u and v is c_{uv}, how should we design algorithms for sorting and selection so as to minimize the total cost of

---

* Research partly supported by NSF CAREER award CCF-0448095, and by an Alfred P. Sloan Fellowship.

C. Chekuri et al. (Eds.): APPROX and RANDOM 2005, LNCS 3624, pp. 74–85, 2005.

all the comparisons performed? Note that these costs c_{uv} are known to the algorithm, which can use them to decide on the sequence of comparisons. The case where all comparison costs are identical is just the unit-cost model. To measure the performance of algorithms in this model, Charikar et al. [2] used the framework of *competitive analysis*: they compared the cost incurred by the algorithm to the cost incurred by an optimal algorithm to *prove* the output correct.

The paper of Charikar et al. [2], and subsequent work by the authors [3] and Kannan and Khanna [4] considered sorting, searching, and selection for special cost functions (which are described in the discussion on related work). However, there seems to be no work on the case where the comparison costs form a *metric space*, i.e., where costs respect the triangle inequality $c_{uv} + c_{vw} \geq c_{uw}$ for all $u, v, w \in V$. Such situations may arise if the elements reside at different places in a communication network, and the communication cost of comparing two elements is proportional to the distance between them. An equivalent, and perhaps more natural way of enforcing the metric constraint on the costs is to say that the vertices in V lie in an ambient metric space (X, d) (where $V \subseteq X$), where the cost c_{ij} of comparing two vertices i and j is the distance $d(i, j)$ between them.

Our Results. In this paper, we initiate the study of these problems: in particular, we consider problems of max-finding and sorting with metric comparison costs. For the max-finding problem, our results show that the lower bound of $\Omega(n)$ on the competitive ratio arises only for arguably pathological scenarios, and substantially better guarantees can be given for metric costs.

Theorem 1. *Max-finding with metric costs has an $O(\log n)$-competitive algorithm.*

We improve the result for the special case when the points are located in d-dimensional Euclidean space:

Theorem 2. *There is an $O(d^3)$-competitive randomized algorithm for max-finding when the nodes lie on the d-dimensional grid and the distance between points is given by the ℓ_∞ metric; this yields an $O(d^{3(1+1/p)})$-competitive algorithm for d-dimensional ℓ_p space.*

For the problem of sorting n elements in a metric, we give an $O(\log n)$-competitive algorithm for the case of hierarchically well-separated trees (HSTs). We then use standard results of Bartal [5], and Fakcharoenphol et al. [6] from the theory of metric embeddings to extend our results to general metrics. Our main theorem is the following:

Theorem 3. *There is an $O(\log^2 n)$-competitive randomized algorithm for sorting with metric costs.*

It can be seen that any algorithm for sorting with metric costs must be $\Omega(\log n)$-competitive even when the points lie on a star or a line – indeed, one can model unit-cost sorting and searching sorted lists in these cases. The question of closing the logarithmic gap between the upper and lower bounds remains an intriguing one.

Our Techniques. For the max-finding algorithm for general metrics, we use a simple algorithm that uses $O(\log n)$ rounds, eliminating half the elements in each round while paying at most OPT. Getting better results turns out to be non-trivial even for very simple metrics: an illuminating example is the case where the comparison costs for elements $V = \{v_1, v_2, \ldots, v_n\}$ are given by $c_{v_i v_j} = |i - j|$; i.e., when the metric is generated by a path. (Note that this path has no relationship to the total order on V: it merely specifies the costs.)

Indeed, an $O(1)$-competitive algorithm for the path requires some work: one natural idea is to divide the line into two pieces, recursively find the maximum in each of these pieces, and then compare the two maxima to compute the overall maximum. However, a closer look indicates that this algorithm also gives us a competitive ratio of $\Omega(\log n)$. To fix this problem and reduce the expected cost to $O(OPT)$, we make a simple yet important change: we run *two copies* of the above algorithm in parallel, transferring a small amount of information between the two copies after every round of comparisons. Remarkably, this subtle change in the algorithm gives us the claimed $O(1)$-competitive ratio for the line. In fact, this idea extends to the d-dimensional grid to give us $O(d)$-competitive algorithms – while the algorithm remains virtually unchanged, the proof becomes quite non-trivial for d-dimensions.

For the results on sorting, we first develop an algorithm for the case when the metric is a k-HST (which is a tree where the edges from each vertex to its children are k times shorter than the edge to its parent). For these HST's, we show how to implement a "bottom-up mergesort" to get an (existentially optimal) $O(\log n)$-competitive algorithm; this is then combined with standard techniques to get the $O(\log^2 n)$-competitiveness for general metrics.

Previous Work. The study of the arbitrary cost model for sorting and selection was initiated by Charikar et al. [2]. They showed an $O(n)$-competitive algorithm for finding the maximum for general metrics. Tighter upper and matching lower bounds (up to constants) for finding the maximum were shown by Hartline et al. [7] and independently by the authors [3].

In the latter paper [3], the authors considered the special case of *structured costs*, where each element v_i is assumed to have an inherent *size* s_i, and the cost $c_{v_i v_j}$ of comparing two elements v_i and v_j is of the form $f(s_i, s_j)$ for some function f; as expected, better results could be proved if the function f was "well-behaved". Indeed, for monotone functions f, they gave $O(\log n)$-competitive algorithms for sorting, $O(1)$-competitive algorithms for max-finding, and $O(1)$-competitive algorithms for selection for the special cases of f being addition and multiplication. Subseqently, Kannan and Khanna [4] gave an $O(\log^2 n)$-competitive algorithm for selection with monotone functions f, and an $O(1)$-competitive algorithm when f was the min function.

Formal Problem Definition. The input to our problems is a *complete* graph $G = (V, E)$, with $|V| = n$ vertices. These vertices represent the elements of the total order, and hence each vertex $v \in V$ has a distinct *key value* denoted by key(v). (We use $x \prec y$ to denote that key$(x) \leq$ key(y).) Each edge $e = (u, v)$ has non-negative *length* or *cost* $c_e = c_{uv}$, which is the cost of comparing the elements u and v. We assume that these edge lengths satisfy the triangle inequality, and hence form a metric space.

In this paper, we consider the problems of finding the element in V with the maximum key value, and the problem of sorting the elements in V according to their key values. We work in the framework of competitive analysis, and compare the cost of the comparisons performed by our algorithm to the cost incurred by the optimal algorithm which knows the results of all pairwise comparisons and just has to *prove* that the solution produced by it is correct. We shall denote the optimal solution by $OPT \subseteq E$. Given a set of edges $E' \subseteq E$, we will let $c(E') = \sum_{e \in E'} c_e$, and hence $c(OPT)$ is the optimal cost. Note that a proof for max-finding is a rooted spanning tree of G with the maximum element at the root, and where the key values of vertices monotonically increase when moving from any leaf to the root; for sorting, a proof is the Hamilton path where key values monotonically increase from one end to the other.

2   Max-Finding in Arbitrary Metrics

For arbitrary metrics, we give an algorithm for finding the maximum element $v_{\max}$ of the nodes in V; the algorithm incurs cost at most $O(\log n) \times c(OPT)$. Our algorithm proceeds in stages. In stage i, we have a subgraph $G_i = (V_i, E_i)$ such that V_i contains $v_{\max}$; here $V_i \subseteq V$, and $E_i = V_i \times V_i$ with the same costs as in G. We start with $G_0 = G$; in stage i, if G_i has a single node, then it must be $v_{\max}$, else we do the following steps.

1. Find a minimum cost almost-perfect matching M_i in G_i. (I.e., at most one node remains unmatched.)
2. For every edge $e = (u, v) \in M_i$, compare the end-points of e. (If u is greater than v, then u "wins" and v "loses".)
3. Delete nodes which lost in the comparisons above to get the new set V_{i+1} from V_i.

It is clear that the above algorithm correctly finds $v_{\max}$; there are $O(\log n)$ rounds since $|V_i| = \lceil n/2^i \rceil$, and hence the following lemma immediately implies that the cost incurred is $O(\log n) \times c(OPT)$, this proving Theorem 1.

Lemma 1. *The cost of edges in M_i is at most $c(OPT)$.*

Proof. Given any set of $2k$ vertices in a tree T, one can find a pairing of these vertices into k pairs so that the paths in T between these pairs are edge disjoint (see, e.g., [8, Lemma 2.4]). We use this to pair off the vertices of V_i in the tree OPT; the total cost of the paths between them is an upper bound on a min-cost almost-perfect matching of V_i. Finally, since the paths between the pairs are edge disjoint, their total cost is at most $c(OPT)$, and hence the cost of M_i is at most $c(OPT)$ as well.   □

3   Finding the Maximum on a Line

To improve on the results of the previous section, let us consider the case of the line metric; i.e., where the vertices $V = \{1, 2, \ldots, n\}$ lie on a line, and the cost of comparing two elements in V is just the distance between them in this line. Let us assume that the line is unweighted, and consecutive points in V are at unit distance from each other, and hence $c_{ij} = |i - j|$; we will indicate how to remove this simplifying assumption at the end of this section. We also assume that n is a power of 2. For an element $x \in V$ which is not the maximum, let $g(x)$ be a *nearest* element to x which has a key greater than $\text{key}(x)$, and let $d(x)$ be the distance between x and $g(x)$. Observe that in OPT, the parent of x must be at distance $d(x)$ from x, and hence $c(OPT) = \sum_{x \neq v_{\max}} d(x)$.

Let us first look at a naïve scheme: we start off with a division D_1 of the line into two-node *segments* $\{[1,2],[3,4],\ldots,[n-1,n]\}$. In next division D_2, we pair off segments of D_1 to get $n/4$ segments $\{[1,2,3,4],\ldots,[n-3,n-2,n-1,n]\}$; similarly, D_i has $n/2^i$ segments, each with 2^i nodes. We maintain the invariant that we know the maximum key element in each segment of D_i; when merging two segments, we compute the maximum by comparing the maxima of the two segments. However, this is just the algorithm of Section 2, and if we have $1 \prec 2 \prec \cdots \prec n$, then $c(OPT) = n-1$ but our scheme costs $\Omega(n \log n)$.

An Algorithm Which Almost Works. A natural next attempt is to introduce randomization: to form the division D_2 from D_1, we toss an unbiased coin: if the result is "heads", we merge $[1,2],[3,4]$ into one segment (which we denote using the notation [1-4]), merge $[5,6],[7,8]$ into the segment $[5-8]$, and so on. If the coin comes up "tails", we *shift* over by one: $[1,2]$ forms a segment by itself, and from then on, we merge every two consecutive segments of D_1. Hence, with probability $\frac{1}{2}$, the segments in D_2 are $\{[1-4],[5-8],\ldots\}$, and with probability $\frac{1}{2}$, they are $\{[1-2],[3-6],[7-10],\ldots\}$. To get division D_{i+1} from D_i, we flip an unbiased coin and either merge every pair of consecutive segments of D_i beginning with the *first* segment, or merge every pair of consecutive segments starting at the *second* one. It is easy to see that all segments in D_i, except perhaps the first and last ones, have 2^i nodes. Again, the natural randomized algorithm is to maintain the maximum element in each segment of D_i: when combining segments of D_i to form segments of D_{i+1}, we compare the two maxima to find the maximum of the newly formed segment. (We use *stage i* to refer to the comparisons performed whilst forming D_i; note that stages begin at 1, and there are no comparisons in the first stage.)

The correctness of the procedure is immediate; to calculate the expected cost incurred, we charge the cost of a comparison to the loser – note that each node except $v_{\max}$ pays for exactly one comparison. We would like to show that the expected cost paid by $x \in V$ in our algorithm is $O(d(x))$. Let $S_i(x)$ denote the segment of D_i containing x; the *size* of $|S_i(x)| \leq 2^i$, and the *length* of $S_i(x)$ is $|S_i(x)| - 1$. Note that if $2^k \leq d(x) < 2^{k+1}$, then x definitely wins (and hence does not pay) in stages 1 through k; the following lemma bounds the probability that it loses in any stage $t \geq k+1$. (Recall that depending on the coin tosses, x may nor may lose to $g(x)$.)

Lemma 2. *Let $2^k \leq d(x) < 2^{k+1}$. Then $\mathbf{Pr}[x$ loses in stage $t] \leq 2^{-(t-k-2)}$.*

Proof. Note that the lemma is vacuously true for $t \leq k+2$. Since $d(x) < 2^{k+1}$, nodes x and $g(x)$ must lie in the same or consecutive segments in stage $k+1$. Now for x to lose in stage t, it must not have lost in stages $\{k+2, k+3, \ldots, t-1\}$, and hence the segments containing x and $g(x)$ must not have merged in these stages. Since we make independent decisions at each stage, the probability of this event happening is $(1/2)^{(t-1)-(k+2)+1} = 2^{-(t-k-2)}$. □

Since x may have to pay as much as 2^{t+1} if it loses in stage t, the expected cost for x is $\sum_{t \geq k} \Theta(2^k)$, which may be as large as $\Theta(2^k \cdot (\log n - k))$. Before we indicate how to fix the problem, let us note that our analysis is tight: for the example with $1 \prec 2 \prec \cdots \prec n$, the randomized algorithm incurs a cost $\Omega(n \log n)$.

Two Copies Help: The Double-Random Algorithm. Let us modify the above algorithm to maintain two independent *copies* of the line, which we call L and L'. The partitions in L will be denoted by $D_1, D_2, \ldots$, while those in L' will be called $D'_1, D'_2, \ldots$. These partitions in the two lines are chosen independent of each other. Again, we maintain the maximum element of each segment in D_i and D'_i, but also exchange some information between the lines. Consider the step of merging segments S_1 and S_2 to get a segment S in D_{i+1}, and let x_i be the maximum element of S_i. Before we compare x_1 and x_2, we check if x_1 has lost to an element $y \in S_2$ in some previous stage in L': in this case, we know that $x_1 \prec y \prec x_2$, and hence can avoid comparing x_1 and x_2. Similarly, if x_2 has previously lost in L' to some element $z \in S_1$, we can declare x_1 to be the maximum element of S. Only if neither of these fortuitous events occur, we compare x_1 and x_2. (The process for L' is similar, and uses the information from L's previous rounds.) The correctness of the algorithm follows immediately.

Notice that each element x now loses exactly twice, once in each line, but the second loss may be implicit (without an actual comparison being performed). As before, we say that a node x *loses in stage i of L* (or L') if this is the first stage in which x loses in L (or L'). The node x *pays* in stage i of L (or L') if x loses in stage i of L (or L') *and* an actual comparison was made. While x loses twice, it may possibly pay only once.

Lemma 3. *If $x, y \in V$ are at distance $d(x,y)$, then the probability (in either line) that x and y lie in different segments in D_i is at most $d(x,y)/2^{i-1}$.*

Proof. Let $2^k \le d(x,y) < 2^{k+1}$; the statement is vacuously true for $i - 1 \le k$. In stage $i - 1 > k$, the nodes x and y must lie in either the same or consecutive segments in D_{i-1}. Now, if they were in different segments in D_{i-1} (which inductively happens with probability at most $d(x,y)/2^{i-2}$), the chance that these segments do not merge in stage i is exactly $\frac{1}{2}$, giving us the bound of $d(x,y)/2^{i-2} \times \frac{1}{2}$. □

Let the node $g(x)$ lie to the left of x; the other case is symmetric, and proved identically. Let the distance $d(x) = d(x, g(x))$ satisfy $2^k \le d(x) < 2^{k+1}$. Let $h(x)$ be the nearest point to the right of x such that $x \prec h(x)$, and let $2^m \le d(x, h(x)) < 2^{m+1}$. Note that if x *pays* in stage t of L, then $t \le m+3$. Indeed, if the segment $S_{m+3}(x)$ is the leftmost or the rightmost segment, then it either contains $g(x)$ or $h(x)$, so it must have paid by then. Else, the length of $S_{m+3}(x) = 2^{m+3} - 1$, and since $d(g(x), h(x)) < 2^{m+1} + 2^{k+1} = 2^{m+2}$, the segment $S_{m+3}(x)$ must contain one of $g(x)$ or $h(x)$, so $t \le m+3$. Moreover, since $S_t(x)$ must contain either $g(x)$ or $h(x)$, it follows that $t > k$. The following key lemma shows us that the probability of paying in stage $t \in [k + 1, m]$ is small. (An identical lemma holds for L'.)

Lemma 4. *For $t \in [k + 1, m]$, $\mathbf{Pr}[x$ pays in stage t of $L] \le 2^{-2(t-k)+5}$.*

Proof (Lemma 4). Note that if x pays in stage $t \le m$ of L, then x must have lost to some element to its left in L, since $d(x, h(x)) \ge 2^m$. Depending on whether x loses in L' before stage t or not, there are two cases.

Case I: *x has not lost in D'_{t-1}.* This implies that x and $g(x)$ lie in different segments in L', which by Lemma 3 has probability $\le d(x, g(x))/2^{t-2} \le 2^{-(t-k-3)}$. Now the chance that x loses in L in stage t is $2^{-(t-k-2)}$ (by Lemma 2). Since the partitions are independently chosen, the two events are independent, which proves the lemma.

Case II: *x has lost in stage* $l \leq t - 1$ *in* L'. Since $l \leq m$, x must have lost to some element y to its left in L'; this y is either $g(x)$, or lies to the left of $g(x)$. Consider stage $t - 1$ in L: since the distance $d(y, x) < 2^l \leq 2^{t-1}$, the three elements $x, g(x)$ and y lie in the union of two adjacent segments in D_t. Furthermore, x must lie in a different segment from y and $g(x)$, otherwise x would have already lost in L in stage $t - 1$. Recall that if x loses in stage t in L, it must lose to a node to its left – since $t \leq m$, $h(x)$ is too far to the right. But this implies that $S_{t-1}(x)$ must merge in L with $S_{t-1}(y) = S_{t-1}(g(x))$; in this case, no comparisons would be performed since x had already lost to y in L'. $\square$

Note that this lemma implies that the expected payment of x for stages $k+1$ through m is at most $\sum_{t=k+1}^{m} 2^{-2(t-k)+5} \times 2^t = O(2^k)$. The expected payment in stages $m+1$ to $m + 3$ is at most $3 \cdot O(2^m) = O(2^k)$ by Lemma 2, which proves:

Theorem 4. *The Double-Random algorithm is an $O(1)$-competitive algorithm for max-finding on the line.*

To end, note that the assumption of unit length edges can be removed: by scaling and translation, all distances can be made integers. We can add dummy vertices at all integers i that do not correspond to a vertex in V, where all these vertices have key values less than those of all the non-dummy vertices. Running Double-Random on this augmented line allows us to run the algorithm without increasing the cost. (We can even space and time overhead by maintaining only segments containing at least one non-dummy node.)

4   Max-Finding for Euclidean Metrics

In this section, we extend our algorithm for the line metric to arbirary Euclidean metrics: the basic idea of running two copies of the algorithm and judiciously exchanging information will be used again, but the proof becomes substantially more involved. We give the proof for the 2-d case; the proof for the general case is deferred to the final version of the paper.

The General Double-Random Algorithm. As in the case of the line, we begin with the simplifying assumption that the nodes in V form a subset of the unit-weight $n \times n$ grid; we refer to this underlying grid as $\mathbb{M} = \{1, 2, \ldots, n\} \times \{1, 2, \ldots, n\}$. (This assumption can be easily discharged, as for the case of the line; we omit the details here.) Hence each point $v \in V$ corresponds to a point $(v_x, v_y) \in \mathbb{M}$, with $1 \leq v_x, v_y \leq n$. In fact, if P^x denotes the path along the x-axis from 1 to n, and P^y denotes a similar path along the y-axis, then we can identify the grid $\mathbb{M}$ with the cartesian product $P^x \times P^y$. To construct partitions $D_1, D_2, \ldots$ of the grid, we build stage-i partitions D_i^x and D_i^y for the paths P^x and P^y: the rectangles in $\mathbb{M}$'s partition correspond to the products of the segments in D_i^x and D_i^y, and hence a square in D_{i+1} is formed by merging at most four squares in D_i [1]. The random partitioning schemes for P^x and P^y evolve independently of each other.

---

[1] We abuse notation and say "squares" even though the pieces may be rectangles.

Again, we maintain the invariant that we know the maximum element in each square of the partition D_i, and as in the case of the line, we do not want to perform three comparisons when merging squares in D_i to get D_{i+1}. Hence we maintain two independent copies $\mathbb{M}$ and $\mathbb{M}'$ of the grid, with D_i and D_i' being the partitions in the two grids at stage i. Suppose $\{x_1, x_2, x_3, x_4\}$ are the four maxima of four squares S_i being merged into a new square S in $\mathbb{M}$: for each $i \in [1, 4]$, we check if x_i has lost to some $y \in S$ in a previous stage in $\mathbb{M}'$, and if so, we remove x_i from consideration; we finally compare the x_i's that remain. The correctness of the algorithm is immediate, and we just have to bound the costs incurred.

4.1   Cost of the Double-Random Algorithm in Two Dimensions

We charge the cost of each comparison to the node that loses in that comparison, and wish to upper bound the cost charged to any node $p \in \mathbb{M}$. Let $G(p) = \{q \mid p \prec q\}$ be the set of nodes with keys greater than p; fix a vertex $g(p) \in G(p)$ *closest* to p, and let $d(p)$ be the distance between p and $g(p)$, with $2^\ell \leq d(p) < 2^{\ell+1}$. Since we focus on the node p for the entire argument, we shift our coordinate system to be *centered at p*: we renumber the vertices on the paths P^x and P^y so that the vertex p lies at the "origin" of the 2-d grid. Formally, we label the nodes $p_x \in P^x$ and $p_y \in P^y$ as 0; the other vertices on the paths are labeled accordingly. This naturally defines four quadrants as well. Let D_i^x be the projection of the partition D_i on the line P^x, and D_i^y be its projection of P^y. ($D_i^{x'}$ and $D_i^{y'}$ are defined similarly for the grid $\mathbb{M}'$.)

Let us note an easy lemma, bounding the chance that p and $g(p)$ are separated in the partition D_i in $\mathbb{M}$. (Such a lemma holds for partition D_i' of $\mathbb{M}'$, of course.)

Lemma 5. $\Pr[p \text{ and } g(p) \text{ lie in different squares of } D_i] \leq 2^{-(i-\ell-3)}$.

Proof. Let the distance from p to $g(p)$ along the two axes be $d_x = d(p_x, g(p)_x)$ and $d_y = d(p_y, g(p)_y)$ with $\max\{d_x, d_y\} = d(p)$. By Lemma 3, the projections p_x and $g(p)_x$ do not lie in the same stage-i interval of P^x with probability at most $d_x/2^{i-1}$. Similarly, p_y and $g(p)_y$ do not lie in the same stage-i interval of P^y with probability at most $d_y/2^{i-1}$; a trivial union bound implies that the probability that $2d(p)/2^{i-1} < 2^{-(i-\ell-3)}$. □

We now prove that the expected charge to a node p is $O(2^\ell)$, where the distance between p and a closest point $g(p)$ in the set $G(p) = \{q \in V \mid p \prec q\}$ lies in the interval $[2^\ell, 2^{\ell+1})$. Let $H(p) = G(p) - \{g(p)\}$. Let $S_i(p)$ and $S_i'(p)$ be the squares in D_i and D_i' respectively that contain p. We will consider two events of interest:

1. Let $\mathcal{A}_i$ be the event that p pays in $\mathbb{M}$ in stage i, and the square $S_i(p)$ contains at least one point from $H(p)$, but does not contain $g(p)$.
2. Let $\mathcal{B}_i$ be the event that p pays in $\mathbb{M}$ in stage i, and S_i contains $g(p)$.

Note that $\mathcal{A}_i \cap \mathcal{B}_i = \emptyset$; also, if p pays in stage i in $\mathbb{M}$, then either $\mathcal{A}_i$ or $\mathcal{B}_i$ must occur. Also, $\Pr[\mathcal{A}_i \cup \mathcal{B}_i] > 0$ only when p and some element of $G(p)$ lie in the same square in stage i in $\mathbb{M}$: since any two points in such a square are at ℓ_∞ distance $\leq 2^\ell - 1$ from each other, and each element of $G(p)$ has ℓ_∞ distance at least 2^ℓ from p, it suffices to consider the case $i > \ell$. Theorems 5 and 6 will show that $\sum_i \mathbf{Pr}[\mathcal{A}_i] \times 2^i + \sum_i \mathbf{Pr}[\mathcal{B}_i] \times 2^i = O(2^\ell)$. This shows that p pays only $O(2^\ell)$ in $\mathbb{M}$; a similar bound holds for $\mathbb{M}'$, which proves the claim that Double-Random is $O(1)$-competitive in the case of two-dimensional grids.

Theorem 5. $\sum_i \mathbf{Pr}[\mathcal{A}_i] \times 2^i = O(2^\ell)$.

Proof. Let us define two events $\mathcal{E}_x$ and $\mathcal{E}_y$. Let $\mathcal{E}_x$ be the event that p_x and $g(p)_x$ lie in different segments in D_i^x, and $\mathcal{E}_y$ be the event that p_y and $g(p)_y$ lie in different segments in D_i^y. Note that $\mathcal{A}_i \setminus (\mathcal{E}_x \cup \mathcal{E}_y) = \emptyset$, and hence

$$\mathbf{Pr}[\mathcal{A}_i] \leq \mathbf{Pr}[\mathcal{A}_i \cap \mathcal{E}_x] + \mathbf{Pr}[\mathcal{A}_i \cap \mathcal{E}_y] \tag{4.1}$$

Let us now estimate $\mathbf{Pr}[\mathcal{A}_i \cap \mathcal{E}_x]$, the argument for the other term is similar. Assume (w.l.o.g.) that $g(p)_x$ lies to the left of p_x, and let the points between $g(p)_x$ and p_x (including p_x, but not including $g(p)_x$) in P^x be labeled $p_x^1, p_x^2, \ldots, p_x^k = p_x$ from left to right. Define $\mathcal{F}_j$ as the event that the segment $S_i^x(p)$ in D_i^x containing p_x has p_x^j as its left end-point. Note that the events $\mathcal{F}_j$ are disjoint, and $\mathcal{E}_x = \cup_{j=1}^k \mathcal{F}_j$. Thus it follows that

$$\mathbf{Pr}[\mathcal{A}_i \cap \mathcal{E}_x] = \sum_j \mathbf{Pr}[\mathcal{A}_i \cap \mathcal{F}_j] = \sum_j \mathbf{Pr}[\mathcal{A}_i \mid \mathcal{F}_j] \, \mathbf{Pr}[\mathcal{F}_j]. \tag{4.2}$$

If $\mathcal{F}_j$ occurs then the end-points of the edge connecting p_x^j and p_x^{j-1} (where $p_x^0 = g(p)_x$) lie in different segments of D_i^x. Lemma 3 implies that this can happen with probability at most $\frac{1}{2^{i-1}}$. Thus, we get

$$\mathbf{Pr}[\mathcal{A}_i \cap \mathcal{E}_x] \leq 2^{-(i-1)} \times \sum_j \mathbf{Pr}[\mathcal{A}_i \mid \mathcal{F}_j]. \tag{4.3}$$

Define I_i^j as the segment of length 2^i in P^x containing p_x and having p_x^j as its left end-point. Let $q(i,j) \in H(p)$ be such that $q(i,j)_x \in I_i^j$ and $|q(i,j)_y - p_y|$ is minimum; in other words, the point closest to the x-axis whose projection lies in I_i^j. If no such point exists, then we say that $q(i,j)$ is undefined. Let $\delta(i,j) = |q(i,j)_y - p_y|$ if $q(i,j)$ is defined, and ∞ otherwise. Notice that for a fixed j, $\delta(i,j)$ is a decreasing function of i.

Assume $\mathcal{F}_j$ occurs for some fixed j: for $\mathcal{A}_i \neq \emptyset$, $S_i(p)$ must contain a point in $H(p)$, and hence $\delta(i,j) \leq 2^i$; let $i(j)$ be the smallest value of i for which $\delta(i,j) \leq 2^i$. Due to $\delta(i,j)$ being a decreasing function in i, $\delta(i,j) > 2^i$ for all $i < i(j)$, and for all $i \geq i(j)$, $\delta(i,j) \leq 2^i$. Now suppose $i > i(j)$; note the strict inequality, which ensures that $q(i-1,j)$ exists. Again assume that $\mathcal{F}_j$ occurs: now for $\mathcal{A}_i$ to occur, the square $S_{i-1}(p)$ cannot contain any point of $H(p)$. In particular, it cannot contain $q(i-1,j)$.

Lemma 6. *If $\mathcal{F}_j$ occurs and $i > \ell + 1$, then p_x and $q(i-1,j)_x$ lie in the same segment of D_{i-1}^x.*

Proof. It will suffice to show the claim that segment containing p_x in D_{i-1}^x also has p_x^j as the left end-point; since $q(i-1,j)_x$ also lies in this segment, the lemma follows. To prove the claim, note that the distance $|p_x^j - p_x| \leq |g(p)_x - p_x| - 1 \leq (2^{\ell+1} - 1) - 1 = 2^{\ell+1} - 2$. Since $i > \ell$, it follows that p_x and p_x^j must lie in the same or in adjacent segments of D_{i-1}^x; we claim that the former is true. Indeed, suppose they were in different segments: since the segment of D_{i-1}^x containing p_x^j must have width $2^{i-1} - 1 \geq 2^{\ell+1} - 1$ which is greater than $|p_x^j - p_x|$, it must happen that p_x^j lies in the interior of this segment, and hence $\mathcal{F}_j$ could not occur. $\square$

Note that since the projections of p and $q(i-1,j)$ on the x-axis lie in the same segment implies that the projections p_y and $q(i-1,j)_y$ on the y-axis must lie in different segments of D^y_{i-1}. Since this event is independent of $\mathcal{F}_j$, we can use Lemma 3 to bound the probability: indeed, we get that for $i > i(j)$,

$$\mathbf{Pr}[\mathcal{A}_i \mid \mathcal{F}_j] \le \tfrac{\delta(i-1,j)}{2^{i-1}}. \tag{4.4}$$

We are now ready to prove the theorem.

$$\begin{aligned}
\textstyle\sum_i 2^i \cdot \mathbf{Pr}[\mathcal{A}_i \cap \mathcal{E}_x] &\le 2 \textstyle\sum_i \sum_j \mathbf{Pr}[\mathcal{A}_i \mid \mathcal{F}_j] && \text{(from (4.3))} \\
&= 2 \textstyle\sum_j \sum_{i \ge i(j)} \mathbf{Pr}[\mathcal{A}_i \mid \mathcal{F}_j] \\
&\le 2 \textstyle\sum_j \left(1 + \sum_{i>i(j)} \tfrac{\delta(i-1,j)}{2^{i-1}}\right) && \text{(from (4.4))} \\
&\le 2 \textstyle\sum_j 3 \quad \le 6 \cdot 2^\ell.
\end{aligned}$$

where penultimate inequality follows from the fact that $\delta(i,j)$ is a decreasing function of i, and hence $\sum_{i>i(j)} \tfrac{\delta(i-1,j)}{2^{i-1}}$ is a dominated by a geometric sum. A similar calculation proves that $\sum_i 2^i \cdot \mathbf{Pr}[\mathcal{A}_i \cap \mathcal{E}_y]$ is $O(2^\ell)$, which in turn completes the proof of Theorem 5.   □

Now that we have bounded the charge to p due to the events $\mathcal{A}_i$ by $O(2^\ell)$, we turn our attention to the events $\mathcal{B}_i$, and claim a similar result for this case.

Theorem 6. $\sum_i 2^i \cdot \mathbf{Pr}[\mathcal{B}_i] \le O(2^\ell)$.

Proof (Theorem 6). Recall that if p loses in stage i, then $i > \ell$: hence we define a set of events $\mathcal{E}_{\ell+1}, \ldots, \mathcal{E}_{i-3}$, where $\mathcal{E}_j$ occurs if p loses in stage j of $\mathbb{M}'$. Also, define the event $\mathcal{E}_0$ occur if p does not lose in $\mathbb{M}'$ till stage $i-3$. Note that exactly one of these events can occur, and hence

$$\mathbf{Pr}[\mathcal{B}_i] = \mathbf{Pr}[\mathcal{B}_i \mid \mathcal{E}_0]\mathbf{Pr}[\mathcal{E}_0] + \textstyle\sum_{j=\ell+1}^{i-3} \mathbf{Pr}[\mathcal{B}_i \mid \mathcal{E}_j]\mathbf{Pr}[\mathcal{E}_j]. \tag{4.5}$$

The next two lemmas give us bounds on the probability of each of the terms in the summation.

Lemma 7. *If $i > \ell + 1$, then* $\mathbf{Pr}[\mathcal{B}_i \mid \mathcal{E}_0]\mathbf{Pr}[\mathcal{E}_0] \le 2^{-(2i-2\ell-10)}$.

Proof. Lemma 5 implies that $\mathbf{Pr}[\mathcal{E}_0] \le 2^{-((i-3)-\ell-3)}$. Now given $\mathcal{E}_0$, p must not lose till stage $i-1$ in $\mathbb{M}$ for $\mathcal{B}_i$ to occur. But this event is independent of $\mathcal{E}_0$, and hence Lemma 5 implies that $\mathbf{Pr}[\mathcal{B}_i \mid \mathcal{E}_0]$ is at most $2^{-((i-1)-\ell-3)}$. Multiplying the two completes the proof.   □

Lemma 8. $\mathbf{Pr}[\mathcal{B}_i \mid \mathcal{E}_j]\mathbf{Pr}[\mathcal{E}_j] \le 2^{-(2i-2\ell-9)}$.

Proof. For the event $\mathcal{E}_j$ to occur, p does not lose till stage $j-1$ in $\mathbb{M}'$; now applying Lemma 5 gives us that $\mathbf{Pr}[\mathcal{E}_j] \le 2^{-((j-1)-\ell-3)}$. Also, note that $\ell < j \le i-3$ for us to be in this case.

Now let us condition on $\mathcal{E}_j$ occurring: let p lose to some q in stage j of $\mathbb{M}'$, and hence $|p_x - q_x|, |p_y - q_y| < 2^j$. Now consider stage $i-1$ of $\mathbb{M}$. We claim that p_x, q_x and $g(p)_x$

do not all lie in the same segment of D^x_{i-1}. Indeed, since the distance $|p_y - g(p)_y| < 2^{\ell+1} \leq 2^{i-2}$, the triangle inequality ensures that $|q_y - g(p)_y| \leq |p_y - g(p)_y| + |p_y - q_y| \leq 2^{i-1}$, and hence the distance between any two points in the set $\{p_y, q_y, g(p)_y\}$ is at most 2^{i-1}. Thus two of these points must lie in the same segment in D^y_{i-1} in $\mathbb{M}$. If all three lay in the same segment of D^x_{i-1}, two of these points would lie in the same square in D_{i-1}. Now if p was one of these points, then p would lose before stage i and $\mathcal{B}_i$ would not occur. If $g(p)$ and q would lie in the same square of D_{i-1}, then p and q would be in the same square in D_i, and then p would not pay. Therefore, all three of p_x, q_x and $g(p)_x$ cannot lie in the same segment of D^x_{i-1}; similarly, p_y, q_y and $g(p)_y$ can not lie in the same segment of D^y_{i-1}.

Hence one of the following two events must happen: either (1) $p_x, g(p)_x$ lie in different segments of D^x_{i-1} and p_y, q_y lie in different segments of D^y_{i-1}, or (2) p_x, q_x lie in different segments of D^x_{i-1} and $p_y, g(p)_y$ lie in different segments of D^y_{i-1}. Lemma 3 implies that the probability of either of these events is at most $2^{-(i-\ell-2) \cdot} 2^{-(i-j-2)}$, and hence $\mathbf{Pr}[\mathcal{B}_i | \mathcal{E}_j] \leq 2^{-(2i-\ell-j-5)}$. Finally, multiplying this with $\mathbf{Pr}[\mathcal{E}_j] \leq 2^{-((j-1)-\ell-3)}$ completes the proof.

Now combining (4.5) with Lemmas 7 and 8, we see that if $i > \ell + 1$, then $\mathbf{Pr}[\mathcal{B}_i] \leq O\left(\frac{i-l}{2^{2i-2\ell}}\right)$. Thus,

$$\sum_{i \geq \ell} 2^i \cdot \mathbf{Pr}[\mathcal{B}_i] \leq O\left(2^\ell \cdot \sum_{i > \ell} \left(\frac{i-l}{2^{i-l}}\right)\right) = O(2^\ell). \qquad (4.6)$$

This completes the proof of Theorem 6.

$\square$

5   Sorting with Metric Comparison Costs

We now consider the problem of sorting the points in V according to their key values. Let OPT be the set of $n-1$ edges going between consecutive nodes in sorted order. A rooted tree T is called a 2-HST if the lengths of all edges at any level of T are the same, and the lengths of consecutive edges on any root-leaf path decrease by a factor of exactly 2. We assume that each internal node of T has at least 2 children. Indeed, if a node has exactly one child, we can contract this edge – this will change distances between leaves up to a constant factor only. Let us denote the set of leaves of the 2-HST tree T by V, and let $|V| = n$. The following theorem is the main technical result of this section.

Theorem 7. *Given n elements, and the metric generated by the leaves of a 2-HST, there is an algorithm to sort the elements with a cost of $O(\log n) \times c(OPT)$.*

Using standard results on approximating arbitrary metrics by probability distributions on metrics generated by HSTs [5, 6], the above theorem immediately implies Theorem 3.

Proof (Theorem 7). For any rooted subtree H of T, let $OPT(H)$ denote the optimal set of comparisons to sort the leaves in H, and let $c(OPT(H))$ be their cost. Let h be the root of H, and h's children be $h_1, \ldots, h_r$; let the subtree rooted at h_i be H_i. Consider $OPT(H)$, and let a *segment* of $OPT(H)$ be a maximal sequence of consecutive vertices in $OPT(H)$ belonging to the same sub-tree H_i for some i. Clearly, we can divide $OPT(H)$ uniquely into node-disjoint segments – let segs(H) denote the number of these disjoint segments. Let $d(H)$ denote the cost of an edge joining h to one of its children; recall that all these edges have the same cost. We omit the proof of the following simple lemma.

Lemma 9. $c(OPT(H)) \geq \sum_{i=1}^{r} c(OPT(H_i)) + (\mathsf{segs}(H) - 1) \cdot d(H)$.

Our algorithm sorts the leaves of T in a bottom-up manner, by sorting the leaves of various subtrees, and then merging the results. For subtrees which just consist of a leaf, there is nothing to do. Now let H, h, H_i, h_i be as above, and assume we have sorted the leaves of H_i for all i: we want to merge these sorted lists to get the sorted list for the leaves of H. The following lemma, whose proof we omit, shows that we can do this without paying too much.

Lemma 10. *There is an algorithm to merge the sorted lists for H_i while incurring a cost of $O(\mathsf{segs}(H) \cdot \log n \cdot d(H))$.*

We now complete the proof of Theorem 7. If cost(H) is the cost incurred to sort the subtree H, we claim cost$(H) \leq \alpha \cdot \log n \cdot c(OPT(H))$ for some constant α. The proof is by induction on the height of the tree: the base case is when H is a leaf, and cost$(H) = c(OPT(H)) = 0$. If H, H_i are as above, and if our claim is true for H_i, then Lemma 10 implies that

$$
\begin{aligned}
\mathsf{cost}(H) &\leq \sum_i \mathsf{cost}(H_i) + O(\mathsf{segs}(H) \cdot \log n \cdot d(H)) \\
&\leq \sum_i \alpha \cdot \log n \cdot c(OPT(H_i)) + O(\mathsf{segs}(H) \cdot \log n \cdot d(H)) \\
&\leq \alpha \cdot \log n [\sum_i c(OPT(H_i)) + (\mathsf{segs}(H) - 1)]
\end{aligned}
\tag{5.7}
$$

provided α is large enough. (The last inequality used the fact that since segs$(H) \geq 2$, segs$(H) = O(\mathsf{segs}(H) - 1)$. But (5.7) is at most $\alpha \cdot \log n \cdot c(OPT(H))$, by Lemma 9, which proves Theorem 7. □

References

1. Knuth, D.E.: The art of computer programming. Volume 3: Sorting and searching. Addison-Wesley Publishing Co., Reading, Mass. (1973)
2. Charikar, M., Fagin, R., Guruswami, V., Kleinberg, J., Raghavan, P., Sahai, A.: Query strategies for priced information. In: Proc. 32nd ACM STOC. (2000) 582–591
3. Gupta, A., Kumar, A.: Sorting and selection with structured costs. In: Proc. 42nd IEEE FOCS (2001) 416–425
4. Kannan, S., Khanna, S.: Selection with monotone comparison costs. In: Proc. 14th ACM-SIAM SODA (2003) 10–17
5. Bartal, Y.: Probabilistic approximations of metric spaces and its algorithmic applications. In: Proc. 37th IEEE FOCS. (1996) 184–193
6. Fakcharoenphol, J., Rao, S., Talwar, K.: A tight bound on approximating arbitrary metrics by tree metrics. In: Proc. 35th ACM STOC (2003) 448–455
7. (Hartline, J., Hong, E., Mohr, A., Rocke, E., Yasuhara, K.) As reported in [3].
8. Kleinberg, J.: Detecting a network failure. Internet Math. **1** (2003) 37–55

What About Wednesday?
Approximation Algorithms
for Multistage Stochastic Optimization[*]

Anupam Gupta[1,**], Martin Pál[2,***],
Ramamoorthi Ravi[3,†], and Amitabh Sinha[4]

[1] Dept. of Computer Science, Carnegie Mellon University, Pittsburgh PA 15213
anupamg@cs.cmu.edu
[2] DIMACS Center, Rutgers University, Piscataway, NJ
mpal@acm.org
[3] Tepper School of Business, Carnegie Mellon University, Pittsburgh PA 15213
ravi@cmu.edu
[4] Ross School of Business, University of Michigan, Ann Arbor MI 48109
amitabh@umich.edu

Abstract. We study the problem of multi-stage stochastic optimization
with recourse, and provide approximation algorithms using cost-sharing
functions for such problems. Our algorithms use and extend the Boosted
Sampling framework of [6]. We also show how the framework can be
adapted to give approximation algorithms even when the inflation parameters are correlated with the scenarios.

1 Introduction

Many problems in planning involve making decisions under uncertainty that un-
ravels in several stages. Demand for new services such as cable television and
broadband Internet access originate over time, leading to interesting questions in
the general area of installing infrastructure to support demand that evolves over
time. For instance, a communications company may want to solve the problem
of constructing a network to serve demands as they arise but also keep potential
future growth in hot-spots in mind while making investments in costly optic fiber
cables. While traditional ways to solve such problems involve casting them as
network loading and network expansion problems in each stage and solving for
them sequentially, advances in forecasting methods have made a more integrated
approach feasible: with the availability of forecasts about how future demands
evolve, it is now preferable to use the framework of *multistage stochastic optimization with recourse* to model such problems.

[*] Full paper available at
http://www.tepper.cmu.edu/andrew/ravi/public/pubs.html.
[**] Supported in part by NSF CAREER award CCF-0448095 and an Alfred P. Sloan
Fellowship.
[***] Supported by NSF grant EIA 02-05116.
[†] Supported in part by NSF grant CCR-0105548 and ITR grant CCR-0122581.

C. Chekuri et al. (Eds.): APPROX and RANDOM 2005, LNCS 3624, pp. 86–98, 2005.

Before we talk about the multistage optimization, let us describe the basic ideas via the example of the *two-stage* stochastic Steiner tree problem: Given a metric space with a root node, the Steiner tree problem is to find a minimum length tree that connects the root to a set of specified terminals S. In the two-stage stochastic version considered recently by several authors [6, 7, 9, 16], information about the set of terminals S is revealed only in the second stage while the probability distribution π from which this set is drawn known in the first stage (either explicitly [7, 9, 16], or as a black box from which one can sample efficiently [6]). One can now purchase some edges F_1 in the first stage – and once the set of terminals $S \subseteq V$ is revealed, one can buy some more edges $F_2(S)$ so that $F_1 \cup F_2(S)$ contains a rooted tree spanning S. The goal is to minimize the cost of F_1 plus the expected cost of the edges F_2. Note however that the edges purchased in the second stage F_2 are costlier by a factor of $\sigma > 1$; it is this that motivates the purchase of some anticipatory edges F_1 in an optimal solution. In this paper, we will consider the following k-stage problem, and will give a $2k$-approximation for this problem.

The k-Stage Problem. In the k-*stage* problem, we are allowed to buy a set of edges in each stage to augment our current solution in reaction to the updated information received in that stage in the form of a *signal*; however, the cost of buying any edge increases with each stage. Let us use σ_i to denote the inflation factor of stage i; i.e., how much more expensive each edge is in comparison to stage $i-1$. Assume $\sigma_1 = 1$; hence purchasing an edge e in stage i costs $c_e \prod_{j=1}^{i} \sigma_i$. We assume that costs are non-decreasing, which corresponds to $\sigma_i \geq 1$.

- At the beginning of the i-th stage (where $1 \leq i \leq k - 1$), we receive a *signal* s_i that represents the information gained about future terminals that will arise. After this observation s_i, we know that future signals, as well as the set $\mathbf{S}$ of eventual terminals, will come from a revised distribution conditioned on seeing this signal. After observing the signal s_i, we can purchase some more edges F_i at cost $\prod_{j=1}^{i} \sigma_j c(F_i)$.
- Finally, in the k-th stage we observe the realization of the random variable $\mathbf{S} = S$ of terminals, and have to buy the final set F_k so that $\cup_{i=1}^{k} F_k$ is a Steiner tree spanning S. Our goal is to minimize the expected cost incurred in all stages together, namely $\mathbf{E}[\sum_{i=1}^{k} (\prod_{j \leq i} \sigma_j) c(F_i)]$.

This multistage framework can be naturally extended to model problems like Vertex Cover and Facility Location as well; we give the formal definitions in Section 2. We then extend the *Boosted Sampling* framework from [6] to the multistage situation, and use this to give approximation algorithms for Stochastic Steiner Tree, Facility Location, and Vertex Cover.

1.1   Informal Description of Results

In this paper we extend the *Boosted Sampling* framework for two-stage stochastic optimization problems with recourse. The framework and terminology is defined in Section 2. This framework had been proposed in our earlier paper [6]; given an

---

1. *Boosted Sampling:* Sample σ times from the distribution π to get sets of clients $D_1, \ldots, D_\sigma$.
2. *Building First Stage Solution:* Build an α-approx. solution for the clients $D = \cup_i D_i$.
3. *Building Recourse:* When actual future in the form of a set S of clients appears (with probability $\pi(S)$), augment the sol'n. of Step 2 to a feasible solution for S.

Fig. 1. Algorithm Boost-and-Sample(Π)

inflation parameter σ, and a distribution π over second-stage scenarios, the following procedure was used to translate optimization algorithms for deterministic problems to their two-stage stochastic variants:

In [6], we proved that given an α-approximation algorithm for the deterministic version Det(Π) of any problem Π, boosted sampling would yield an approximation algorithm for the two-stage stochastic version Stoc(Π), as long as the deterministic algorithm satisfied certain technical cost-sharing conditions; moreover, the sampling framework worked even if the probability distribution π over the second-stage client sets was specified using a *black-box* from which one could draw samples efficiently.

Multistage Results. We first show (in Section 3) how to extend the boosted sampling framework to handle k-stage stochastic variants of problems (for all $k \geq 2$), and give the technical conditions under which effective approximations can be obtained. (See Theorems 1 and 2 for the precise statements). As in [6], these technical conditions are phrased in terms of the existence of certain "good" cost-sharing functions related to the approximation algorithms. In particular, we want the cost-shares to satisfy both *strictness* and *cross-monotonicity*; details and definitions appear in Section 2.

Our results, together with the "good" cost-shares for some problems, give us constant-factor approximation algorithms when the number of stages is a constant. In particular, such results are obtainable for the Steiner tree problem (where we can get a $2k$-approximation for the k-stage problem), Facility Location (an approximation of $3 \cdot 2^k$), and Vertex Cover (at most 4^k). A more precise summary of our results for the k-stage versions of these problems is in the last column of Figure 2.

Correlated Inflation. As outlined in Figure 1, boosted sampling assumes a fixed *deterministic* inflation parameter σ. If this inflation parameter $\boldsymbol{\sigma}$ is random

Problem	Approximation ratio α w.r.t. ξ	c-Strictness of ξ	Is ξ X-mono?	k-Stage Stochastic Approx.
Steiner Tree	2	1	Yes	$2k$
Facility Location	3	2	Yes	$3(2^k - 1)$
Vertex Cover	2	2	No	$\frac{2}{3}(4^k - 1)$

Fig. 2. A summary of the results obtainable from this work

but *independent* of the distribution π over scenarios, one can just use $\mathbf{E}[\sigma]$ in the place of σ. This independence assumption is somewhat restrictive, as the equipment prices often correlate with demands. Here, using the expected value can lead to very poor approximations.

In Section 4, we give a simple way to extend the Boosted Sampling framework to the case when σ is arbitrarily correlated with the distribution π without losing anything in the appproximation ratios.

1.2   Related Work

There is a huge body of work in the Operations Research community on multistage stochastic optimization with recourse; the study of stochastic optimization [2, 12] dates back to the work of Dantzig [3] and Beale [1] in 1955. Stochastic linear programming was defined in these papers, and have been very widely studied since, with gradient-based and decomposition-based approaches being known for some versions of stochastic linear programming. On the other hand, only moderate progress has been reported for stochastic integer (and mixed-integer) programming in both theoretical and computational domains; see [13, 17] for details.

The study of stochastic versions of NP-hard problems has received some attention lately in the theoretical computer science community, and approximation algorithms for *two-stage* stochastic programming versions of a variety of combinatorial optimization problems have been devised actively in several recent papers [4, 6, 7, 9, 16, 18].

Shmoys and Swamy [19] have recently shown that for a broad class of multistage stochastic linear programs, a $(1 + \epsilon)$-approximate solution can be found in polynomial time using a Sampled-Average-Approximation approach. They show that for several problems (including Facility Location, Set Cover and Vertex Cover), the LP solution for each stage can be rounded to an integer solution *independently of other stages*. In contrast, our technique requires strict cost shares, but does not depend on the existence of a suitable LP relaxation, or the ability to round each stage independently.

Independently of our work, Hayrapetyan et al. [8] have also devised approximation algorithms for the multistage version of the Stochastic Steiner tree problem that we consider, using a reduction to a variant of an information network gathering problem which they address in their paper. They also provide an $O(k)$-approximation algorithm for multistage Stochastic Steiner tree (our approximation ratio is $2k$). However, their techniques do not seem to extend to the other covering problems that we address in this paper.

2   Basic Model and Notation

Let us define an abstract combinatorial optimization problem Π that we will adapt to a stochastic setting. The optimization problem Π is defined by U, the universe of *clients* (or demands), and the set X of *elements* we can purchase. For a subset $F \subseteq X$ of elements, let $c(F) = \sum_{e \in F} c_e$ denote the *cost* of F. Given a set S of clients, a solution F that *satisfies* each client $j \in S$ is labeled *feasible* for S. The definition of satisfaction naturally depends on the problem; e.g., in the (rooted) Steiner tree problem on a graph $G = (V, E)$, the universe of clients is the set of possible terminals $(U = V)$, the element set X is the set of edges E, and a terminal $j \in S$ is satisfied by $F \subseteq X$ if F contains a path from j to the root vertex r. The cost of a set of edges $F \subseteq X$ is $c(F) = \sum_{e \in F} c_e$.

Given a set $S \subseteq U$ of clients, we let $\mathsf{Sols}(S) \subseteq 2^X$ be the set of *feasible solutions* for S. Given a client set $S \subseteq U$, the *deterministic version* $\mathsf{Det}(\Pi)$ of Π asks us to find a solution $F \in \mathsf{Sols}(S)$ of minimum cost. We denote by $\mathsf{OPT}(S)$ the cost of this minimum cost solution.

Definition 1. *A problem Π is* sub-additive *if for any S and S' being two sets of clients with solutions $F \in \mathsf{Sols}(S)$ and $F' \in \mathsf{Sols}(S')$, we have that (i) $S \cup S'$ is a legal set of clients for Π, and (ii) $F \cup F' \in \mathsf{Sols}(S \cup S')$.*

As in previous papers which give approximation algorithms for two-stage stochastic optimization problems [6, 9, 16], we restrict our attention to sub-additive problems. (Note that the sub-additivity in the rooted Steiner tree problem is ensured by the presence of the root r.)

Given any problem Π, we study the variant when the set of clients (or requirements) is not known in advance, but is revealed gradually. We proceed to build the solution in stages; in each stage, we gain a more precise estimate of the requirements of clients, and then can buy or extend a partial solution (at gradually increasing cost) in response to this updated information. Ultimately, we learn the entire set S of clients or requirements, and then must complete the existing partial solution to a feasible solution $F \in \mathsf{Sols}(S)$.

Multi-stage Stochastic Optimization Problems. We can now define stochastic variants $\mathsf{Stoc}(\Pi)$ of the problem Π. In this model, we obtain increasingly precise forecasts about user demands over several stages as in the Steiner tree example. In the k-stage problem, we are allowed to buy a set of elements in each stage to augment our current solution in reaction to the updated information received in that stage; however, the cost of buying an element $e \in X$ is increasing with each stage. Extending the existing terminology, we use σ_i to denote the inflation factor of stage i; i.e., how much more expensive each element is in comparison to stage $i - 1$. For completeness, we define $\sigma_1 = 1$. Hence purchasing an element $e \in X$ in stage i costs $c_e \prod_{j=1}^{i} \sigma_i$. We assume that costs are non-decreasing, which corresponds to $\sigma_i \geq 1$.

- At the beginning of the i-th stage (where $1 \leq i \leq k - 1$), we receive a *signal* s_i that represents the information gained that we can use to correct our anticipation of the demands. Formally, the signal $\mathbf{s}_i$ is a random variable correlated with $\mathbf{S}$ and in stage i we observe s_i, a realization of $\mathbf{s}_i$. (Note that the signal $\mathbf{s}_1$ is a dummy signal, but we use it to simplify notation.) After this observation s_i, we know that future signals, as well as the set $\mathbf{S}$ of demands will come from the conditional distribution $[\pi | \mathbf{s}_1 = s_1, \mathbf{s}_2 = s_3, \ldots, \mathbf{s}_i = s_i]$. After observing the signal s_i, we can purchase some more elements $F_i \subseteq X$ at cost $\prod_{j=1}^{i} \sigma_j c(F_i)$.

- Finally, in the k-th stage we observe the realization of the random variable $\mathbf{S} = S$, and have to buy the final set F_k so that $\cup_{i=1}^{k} F_k \in \mathsf{Sols}(S)$. Again, our goal is to minimize the expected cost incurred in all stages together, that is, $Z = \mathbf{E}\big[\sum_{i=1}^{k} \big(\prod_{j \leq i} \sigma_j\big) c(F_i) \big]$.
  Note that each of the partial solutions $F_i = F_i(s_1, s_2, \ldots, s_i)$ may depend on signals up to stage i, but not on signals observed in the subsequent stages.

2.1 Cost Sharing Functions

We now define the notion of *cost shares* that we will crucially use to analyze our approximation algorithms. Loosely, a cost-sharing function ξ divides the cost of a solution $F \in \mathsf{Sols}(S)$ among the clients in S. Cost-sharing functions have long been used in game-theory (see, e.g., [11, 14, 15, 20]).

Cost-shares with a closer algorithmic connection were recently defined by Gupta et al. [5] to analyze a randomized algorithm for the Rent-or-Buy network design problem. In this paper, we have to redefine *strict* cost-sharing functions slightly: in contrast to previously used definitions, our cost-shares are defined by, and relative to, an approximation algorithm $\mathcal{A}$ for the problem Π.

A *cost-sharing algorithm* $\mathcal{A}$ for a problem Π takes an instance (X, S) of Π and outputs **(a)** a solution $F \subseteq X$ with $F \in \mathsf{Sols}(S)$, and **(b)** a real value $\xi(X, S, j) \geq 0$ for each client $j \in S$. This value $\xi(X, S, j)$ is called the *cost-share* of client j, and the function $\xi(\cdot, \cdot, \cdot)$ computed by $\mathcal{A}$ is the *cost sharing function* associated with $\mathcal{A}$. A cost-sharing function ξ is *cross-monotone* if for every pair of client sets $S \subseteq T$ and client $j \in S$, we have $\xi(X, T, j) \leq \xi(X, S, j)$.

We also require all cost-sharing functions to provide a lower bound on the cost of the optimal solution, in order to use the cost-sharing function to provide bounds on the cost of our solution. A cost-sharing function ξ is *competitive* if for every client set S, it holds that $\sum_{j \in S} \xi(X, S, j) \leq \mathsf{OPT}(X, S)$. For a subset of clients $S' \subseteq S$, let $\xi(X, S, S')$ denote the sum $\sum_{j \in S'} \xi(X, S, j)$; thus competitiveness is the property that $\xi(X, S, S) \leq \mathsf{OPT}(X, S)$. In this paper, we will focus solely on competitive ξ.

Crucial to our proofs is the notion of *strictness* [6] that relates the cost of extending a solution on S so as to serve more clients T to the cost shares of T. Formally, given a set X of elements and S of clients, let $F = \mathcal{A}(X, S)$ denote the solution found by algorithm $\mathcal{A}$. We create a new *reduced* instance of the problem Π by zeroing out the cost of all elements in F; this instance is denoted by X/F. A cost-sharing algorithm $\mathcal{A}$ is β-*strict* if for any sets of clients S, T there exists a solution $F_T \subseteq X$ constructible in polynomial time such that $\mathcal{A}(X, S) \cup F_T \in \mathsf{Sols}(T)$ and $c(F_T) \leq \beta \times \xi(X, S \cup T, T)$.

A subtly different notion of strictness is *c-strictness*; here the "c" is supposed to emphasize the enhanced role of the cost-shares. Let $S, T \subseteq U$ be sets of clients, and let X be an instance of Π. The cost-sharing function ξ given by an algorithm $\mathcal{A}$ is β-*c-strict* if $\xi(X/\mathcal{A}(X, S), T, T) \leq \beta \times \xi(X, S \cup T, T)$.

In other words, the total cost shares for the set T of clients in the reduced instance $X/\mathcal{A}(X, S)$ is at most β times the cost-shares for T if the clients in S

were present as well. Note that any competitive β-strict cost sharing function ξ is also β-c-strict.

Recall that $\mathcal{A}$ is an α-approximation algorithm if $c(\mathcal{A}(X, S)) \leq \alpha\,\mathsf{OPT}(X, S)$. In this paper, we will need the following stronger guarantee (which goes hand-in-hand with c-strictness): An algorithm $\mathcal{A}$ (with cost-sharing function ξ) is *an α-approximation algorithm with respect to ξ* if $c(\mathcal{A}(X, S)) \leq \alpha\,\xi(X, S, S)$.

If ξ is competitive, then $\xi(X, S, S) \leq c(\mathsf{OPT}(S))$, and thus an α-approximation algorithm w.r.t. ξ is also simply an α-approximation algorithm. This, together with c-strictness, implies that an α-approximation algorithm w.r.t. β-c-strict ξ is $(\alpha\beta)$-strict in the sense of the first definition of strictness.

2.2   New Results on c-Strictness and Multi-stage Approximations

Having laid down the crucial definitions, we can finally state the main theorems for multi-stage stochastic covering problems.

Theorem 1. *There is an $\alpha \cdot \sum_{i=0}^{k-1} \beta^i$-approximation algorithm for the k-stage stochastic problem $\mathsf{Stoc}_k(\Pi)$ if the corresponding problem Π has an α-approximation algorithm $\mathcal{A}$ with respect to a β-c-strict cost-sharing function ξ, and this ξ is cross-monotone.*

A slightly weaker version of the above theorem can be proved without cross-monotone cost-shares: this is useful for problems like Vertex Cover for which good cross-monotone cost-shares do not exist [10].

Theorem 2. *There is an $\alpha \cdot \sum_{i=0}^{k-1} \beta_1 \beta_2{}^i$-approximation algorithm for the k-stage stochastic problem $\mathsf{Stoc}_k(\Pi)$ if the corresponding problem Π has an α-approximation algorithm $\mathcal{A}$ with respect to a β_1-c-strict cost-sharing function ξ, and $\mathcal{A}$ is also β_2-strict with respect to this ξ.*

Finally, we improve the previous results on strictness in [6], and adapt them to the new notion of c-strictness. Due to space constraints, proofs of Theorems 2 and 3 only appear in the full version of the paper. Figure 2 summarizes the results.

Theorem 3.   *1. There is a 2-approximation algorithm for the Minimum Steiner Tree problem w.r.t. a 1-c-strict cost sharing function ξ. Furthermore, this ξ is also cross-monotone.*
   2. The Uncapacitated Facility Location problem admits a 3-approximation algorithm w.r.t. a 2-c-strict ξ. This cost-sharing function ξ is also cross-monotone.
   3. The Vertex Cover problem has 2-approximation algorithm w.r.t. an associated 2-strict (and hence 2-c-strict) cost-sharing function ξ.

3   Multiple Stage Stochastic Optimization

For ease of exposition, let us state the algorithm assuming that the inflation factors σ_i are deterministically known in advance. With some extra work as in

Section 4, randomly varying inflation factors can be handled – the details are deferred to a full version of this paper.

Let us first outline the key idea of the algorithm. In the two-stage Boosted Sampling framework with inflation being σ (see Figure 1), the first stage involved simulating σ independent runs of the second stage and building a solution that satisfied the union of these simulations. We use the same basic idea for the k-stage problem: in each stage i, we simulate σ_{i+1} "copies" of the remaining $(k - i)$-stage stochastic process; each such "copy" provides us with a (random) set of clients to satisfy, and we build a solution that satisfies the union of all these clients. The "base case" of this recursive idea is the final stage, where we get a set S of clients, and just build the required solution for S.

---

1. *(Base case.)* If $i = k$, draw one sample set of clients S_k from the conditional distribution $[\pi | s_1, \ldots, s_k]$. Return the set S_k.
2. *(New samples.)* If $i < k$, draw $\lfloor \sigma_{i+1} \rfloor$ samples of the signal $\mathbf{s}_{i+1}$ from the conditional distribution $[\pi | s_1, \ldots, s_i]$. Let $s^1, \ldots, s^n$ be the sampled signals (where $n = \lfloor \sigma_{i+1} \rfloor$).
3. *(Recursive calls.)* For each sample signal s^j, recursively call Recur-Sample($\Pi, i + 1, s_1, \ldots, s_i, s^j$) to obtain a sample set of clients S^j. Return the set $S_i = \cup_{j=1}^n S^j$.

Fig. 3. Procedure Recur-Sample(Π, stage $i, s_1, \ldots, s_i$)

Our algorithm, in each stage i (except the final, k-th stage), uses a very natural recursive sampling procedure that emulates σ_{i+1} executions of itself on the remaining $(k - i)$ stages. (This sampler is specified in Figure 3.) Having obtained a collection of sampled sets of clients, it then augments the current partial solution to a feasible solution for these sampled sets. Finally, in the k-th stage it performs the ultimate augmentation to obtain a feasible solution for the revealed set of demands. The expected number of calls to the black box required in stage i will be $\prod_{j=i+1}^k \sigma_j$.

The definition of the sampling routine is given in Figure 3, and the procedure Multi-Boost-and-Sample(Π, i) to be executed in round i is specified in Figure 4. Note that if we set $k = 2$, we get back precisely the Boosted Sampling framework from our previous paper [6].

3.1   The Analysis

We will now show that this extended framework can be used to translate an approximation algorithm $\mathcal{A}$ for the deterministic version Det(Π) of a problem Π to its k-stage stochastic version Stoc$_k$(Π). The quality of this translation depends on the approximation guarantee of $\mathcal{A}$ with respect to some c-strict cost-sharing function. The main result of this section is the following:

Theorem 4. *Given a problem Π, if $\mathcal{A}$ is an α-approximation algorithm w.r.t. a β-c-strict cost-sharing function ξ, and if ξ is cross-monotone, then* Multi-Boost-

---

1. *(External signal.)* If the current stage $i < k$, observe the signal s_i. If this is the final stage $i = k$, observe the required set of clients S instead.
2. *(Sample.)* If $i = k$, let $D_k := S$. Else $i < k$, and then use procedure Recur-Sample$(\Pi, i, s_1, \ldots, s_i)$ to obtain a sample set of clients D_i.
3. *(Augment solution.)* Let $B_i = \cup_{j=1}^{i-1} F_j$ be the elements that were bought in earlier rounds. Set the costs of elements $e \in B_i$ to zero. Using algorithm $\mathcal{A}$, find a set of elements $F_i \subseteq X \setminus B_i$ to buy so that $(F_i \cup B_i) \in \mathsf{Sols}(D_i)$.

Fig. 4. Algorithm Multi-Boost-and-Sample(Π, i)

---

and-Sample(Π) *is an* $\alpha \cdot \sum_{i=0}^{k-1} \beta^i$-*approximation algorithm for the k-stage stochastic problem* $\mathsf{Stoc}_k(\Pi)$.

Before we prove Theorem 4, we set the stage for the proof by providing a brief overview of the proof technique, and proving a couple of lemmas which provide useful bounds. A naïve attempt to prove this result along the lines of our previous paper[6] does not succeed, since we have to move between the cost-shares and the cost of the solutions F_i, which causes us to lose factors of $\approx \alpha$ at each step. Instead, we bound all costs incurred in terms of the ξ's: we first argue that the expected sum of cost-shares paid in the first stage is no more than the optimum total expected cost Z^*, and then bound the sum of cost-shares in each consecutive stage in terms of the expected cost shares from the previous stage (with a loss of a factor of β at each stage). Finally, we bound the actual cost of the partial solution constructed at stage i by α times the expected cost shares for that stage, which gives us the geometric sum claimed in the theorem.

Let F^* be an optimal solution to the given instance of $\mathsf{Stoc}_k(\Pi)$. We denote by F_i^* the partial solution built in stage i; recall that $F_i^* = F_i^*(s_1, s_2, \ldots, s_i)$ is a function of the set of all possible i-tuples of signals that could be observed before stage i. The expected cost of this solution can be expressed as $Z^* = \sigma_1 \mathbf{E}[c(F_1^*(\mathbf{s}_1))] + \sigma_1 \sigma_2 \mathbf{E}[c(F_2^*(\mathbf{s}_1, \mathbf{s}_2))] + \cdots + \sigma_1 \ldots \sigma_n \mathbf{E}[c(F_k^*(\mathbf{s}_1, \mathbf{s}_2, \ldots, \mathbf{s}_k))]$.

Lemma 1. *The expected cost share* $\mathbf{E}[\xi(X, D_1, D_1)]$ *is at most the total optimum cost* Z^*.

Proof. Consider D_1, the sample set of clients returned by Recur-Sample$(\Pi, 1, s_1)$. We claim there is a solution $\widehat{F}(D_1)$, such that $\mathbf{E}[c(\widehat{F}(D_1))] \leq Z^*$ (the expectation is over the execution of the procedure Recur-Sample). To construct the solution $\widehat{F}(D_1)$, we consider the tree of recursive calls of the procedure Recur-Sample. For each recursive call Recur-Sample$(\Pi, i, s_1, \ldots, s_i)$, we add the set of elements $F_i^*(s_1, \ldots, s_i)$ to $\widehat{F}(D_1)$. It is relatively straightforward to establish that (1) $\widehat{F}(D_1)$ is a feasible solution for the set D_1 and (2) the expected cost $\mathbf{E}[c(\widehat{F}(D_1))] \leq \mathbf{E}\left[\sum_{i=1}^{k} \left(\prod_{j \leq i} \sigma_j\right) c(F_i^*)\right] = Z^*$.

Since the expected cost of a feasible solution for D_1 is bounded above by Z^*, the competitiveness of ξ implies that this bound must hold for the sum of cost shares as well.

Lemma 2. *Let $\hat{F} = F_1 \cup \cdots \cup F_{i-1}$ be the solution constructed in a particular execution of the first $i - 1$ stages, and let s_i be the signal observed in stage i. Let D_i and D_{i+1} be the random variables denoting the samples returned by the procedure* Recur-Sample *in Stages i and $i+1$, and let F_i be the (random) solution constructed by $\mathcal{A}$ for the set of clients D_i. Then, $\mathbf{E}[\xi(X/(\hat{F} \cup F_i), D_{i+1}, D_{i+1})] \leq \frac{\beta}{\sigma_{i+1}} \cdot \mathbf{E}[\xi(X/\hat{F}, D_i, D_i)]$.*

Proof. Recall that the sampling procedure Recur-Sample$(\Pi, i, s_1, \ldots, s_i)$ gets $n = \lfloor \sigma_{i+1} \rfloor$ independent samples $s^1, s^2, \ldots s^n$ of the signal $\mathbf{s}_{i+1}$ from the distribution π conditioned on $s_1, \ldots, s_i$, and then for each sampled signal calls itself recursively to obtain the n sets $S^1, \ldots, S^n$. Note that the set $D_i = \bigcup_{j=1}^n S^j$ is simply the union of these n sets. On the other hand, the set D_{i+1} is obtained by observing the signal s_{i+1} (which is assumed to come from the same distribution $[\pi|s_1, \ldots, s_i]$), and then calling Recur-Sample with the observed value of s_{i+1}.

We now consider an alternate, probabilistically equivalent view of this process. First take $n+1$ samples $s^1, \ldots, s^{n+1}$ of the signal $\mathbf{s}_{i+1}$ from the distribution $[\pi|s_1, \ldots, s_i]$. Call the procedure SAMPLE$(\pi, i + 1, s_1, \ldots, s_i, s^j)$ for each s^j to obtain sets $S^1, \ldots, S^{n+1}$. Pick an index j uniformly at random from the set of integers $1, \ldots, n+1$. Let $D_{i+1} = S^j$, and let D_i be the union of the remaining n sets. This process of randomly constructing the pair of sets (D_i, D_{i+1}) is clearly equivalent to the original process. Note that $D_i \cup D_{i+1} = \bigcup_{l=1}^{n+1} S^l$.

To simplify notation, let us denote $X/\hat{F}$ by $\hat{X}$. By the definition of β-c-strictness, we first get $\xi(\hat{X}/F_i, D_{i+1}, D_{i+1}) \leq \beta \cdot \xi(\hat{X}, D_i \cup D_{i+1}, D_{i+1})$.

By the relation that D_{i+1} is equivalent to S^j sampled uniformly from $n + 1$ alternates in the equivalent process above and that D_i is the union of the remaining n sets, we have $\mathbf{E}[\xi(\hat{X}, D_i \cup D_{i+1}, S^j)] \leq \mathbf{E}[\frac{1}{n} \times \xi(\hat{X}, D_i \cup D_{i+1}, D_i)]$.

Now we use cross-monotonicity of the cost shares (which says that the cost shares of D_i should not increase when the elements of $D_{i+1} \setminus D_i$ join the fray), and finally get $\mathbf{E}[\xi(\hat{X}, D_i \cup D_{i+1}, D_i)] \leq \mathbf{E}[\xi(\hat{X}, D_i, D_i)]$.

Chaining the above inequalities proves the lemma.

Proof. (of Theorem 4) Recall that the expected cost of the solution given by Algorithm Multi-Boost-and-Sample(Π) is: $E[Z] = E[\sum_{i=1}^k (\prod_{j=1}^i \sigma_j) c(F_i)]$

Using cross-monotonicity and the fact that $\mathcal{A}$ is an α-c-approximation algorithm with respect to the β-c-strict cost-shares ξ, we obtain the following: $E[Z] \leq \alpha E[\sum_{i=1}^k (\prod_{j=1}^i \sigma_j) \xi(X/B_i, F_i, F_i)]$

Using Lemma 2 inductively on $\xi(X/B_i, F_i, F_i)$, we find that $\xi(X/B_i, F_i, F_i) \leq \frac{\beta^i}{\prod_{j=1}^i \sigma_j} \xi(X, D_1, D_1)$. Using this inequality in the bound for $E[Z]$ above: $E[Z] \leq \alpha E[\sum_{i=1}^k \beta^i \xi(X, D_1, D_1)]$

Lemma 1 bounds $\xi(X, D_1, D_1)$ from above by Z^*. Using this bound in the inequality above completes the proof of Theorem 4.

4   Correlated Inflation Factors

In this section, we show how to extend the basic Boosted Sampling framework to work in the case where the inflation factor σ is a random variable arbitrarily correlated with the random scenarios. For brevity, we describe the idea only for the two-stage setting, though the same idea can be used for the multi-stage framework of Section 3.

Formally, let us assume that we have access to a distribution π' over $\mathbb{R}_{\geq 1} \times 2^U$, where $\pi'(\sigma, S)$ is the probability that the set S arrives and the inflation factor is σ. We assume that we know an integer $M \in \mathbb{Z}$ which is an upper bound on the value of the inflation parameter σ; i.e., with probability 1, it should be the case that $\sigma \leq M$ holds. (Note that choosing a pessimistic value of M will only increase the running time, but not degrade the approximation guarantee of the framework.)

---

1. *Boosted Sampling:* Draw M independent samples from the joint distribution π' of $(\boldsymbol{\sigma}, \mathbf{S})$. Let $(\sigma_1, S_1), (\sigma_2, S_2), \ldots, (\sigma_M, S_M)$ denote this collection of samples.
2. *Rejection Stage:* For $i = 1, \ldots, M$, *accept* the sample S_i with probability σ_i / M. Let $S_{i_1}, S_{i_2}, \ldots, S_{i_k}$ be the accepted samples, and let $S = \bigcup_j S_{i_j}$.
3. *First-stage Solution:* Using the algorithm $\mathcal{A}$, construct an α-approximate *first-stage solution* $F^1 \in \mathsf{Sols}(S)$.
4. *Second-stage Recourse:* Let T be the set of clients realized in the second stage. Use an augmenting algorithm to compute the *second-stage* solution F^2 such that $F^1 \cup F^2 \in \mathsf{Sols}(T)$.

Fig. 5. Algorithm General-Boost-and-Sample(Π)

The algorithm General-Boost-and-Sample(Π) is given in Figure 5. Note that if $\boldsymbol{\sigma}$ is a constant, then we would behave identically to the original Boosted Sampling framework. To get some intuition for the new steps, note that if the sampled inflation factor σ_i is large, this indicates that we want to handle the associated S_i in the first stage; on the other hand, if the σ_i is small, we can afford to wait until the second-stage to handle the associated S_i – and this is indeed what the algorithm does, albeit in a probabilistic way. The following is an extension of the main structural theorem proved in our earlier paper [6]:

Theorem 5. *Consider a sub-additive combinatorial optimization problem Π, and let $\mathcal{A}$ be an α-approximation algorithm for its deterministic version $\mathsf{Det}(\Pi)$. If $\mathcal{A}$ admits a β-strict cost sharing function, then* General-Boost-and-Sample(Π) *is an $(\alpha + \beta)$-approximation algorithm for $\mathsf{Stoc}(\Pi)$.*

Proof. Let us transform the "random inflation" stochastic problem instance (X, π') to one with a fixed inflation factor thus: the distribution $\widehat{\pi}(\sigma, S) = \pi'(\sigma, S) \times (\sigma/M)$; note that this ensures that $\sum_{\sigma, S} \widehat{\pi}(\sigma, S) \leq 1$, and hence we can increase the probability $\widehat{\pi}(1, \emptyset)$ so that the sum becomes exactly 1 and $\widehat{\pi}$ is a well-defined probability distribution. The inflation factor for this new instance is set to M, and hence the σ output by $\widehat{\pi}$ is only for expositional ease.

Now the objective for this new problem is to minimize the expected cost under this new distribution, which is $c(F_0) + \sum_{\sigma, S} \widehat{\pi}(\sigma, S) \, M \, c(F_S) = c(F_0) + \sum_{\sigma, S} \pi'(\sigma, S) \, (\sigma/M) \, M \, c(F_S)$, which is the same as the original objective function; hence the two problems are identical, and running Boost-and-Sample on this new distribution $\widehat{\pi}$ with inflation parameter M would give us an $(\alpha + \beta)$-approximation.

Finally, note that one can implement $\widehat{\pi}$ given black-box access to π' by just rejecting any sample (σ, S) with probability σ/M. Including this implementation within Boost-and-Sample gives us precisely the above General-Boost-and-Sample, which completes the proof of the theorem.

This immediately implies a 3.55-approximation for Stochastic Steiner tree, a 4-approximation for Stochastic Vertex Cover, and a 5.45-approximation for Stochastic Facility Location even when the second-stage inflation factors are drawn from a distribution that may be arbitrarily correlated to the set of clients materializing in the second stage.

A naïve attempt to extend the Boosted Sampling algorithm of [6] might proceed by obtaining $E[\sigma]$ samples and invoking the algorithm. Unfortunately, this approach is doomed to fail, as the following example shows. Consider the case where with probability $\frac{1}{2}$, the inflation $\sigma = M \gg 1$ but $S = \emptyset$, and with probability $\frac{1}{2}$, the inflation $\sigma \approx 1$ but $S \neq \emptyset$; the average value of σ is $\approx M/2$, but in fact we should be ignoring the high σ's.

5   Concluding Remarks

While the algorithms described in this paper can provide approximation algorithms with performance guarantee linear in the number of stages for some problems (Stochastic Steiner Tree), this requires 1-c-strict cost-shares, which may not always exist. The question of which k-stage stochastic optimization problems can be approximated within ratios linear in k and which require exponential dependence on k is an intriguing one. We also note that the running time of our algorithm is exponential in k due to the recursive sampling procedure; we leave open the question whether a sub-exponential collection of samples can be used to construct the partial solutions in earlier stages while still resulting in an algorithm with a provably good approximation ratio.

References

1. E. M. L. Beale. On minimizing a convex function subject to linear inequalities. *J. Roy. Statist. Soc. Ser. B.*, 17:173–184; discussion, 194–203, 1955. (Symposium on linear programming.).
2. John R. Birge and François Louveaux. *Introduction to stochastic programming*. Springer Series in Operations Research. Springer-Verlag, New York, 1997.

3. George B. Dantzig. Linear programming under uncertainty. *Management Sci.*, 1:197–206, 1955.
4. Kedar Dhamdhere, R. Ravi, and Mohit Singh. On two-stage stochastic minimum spanning trees. In *Proceedings of the 11th Integer Programming and Combinatorial Optimization Conference*, pages 321–334, 2005.
5. Anupam Gupta, Amit Kumar, Martin Pál, and Tim Roughgarden. Approximations via cost-sharing. In *Proceedings of the 44th Annual IEEE Symposium on Foundations of Computer Science*, pages 606–615, 2003.
6. Anupam Gupta, Martin Pál, R. Ravi, and Amitabh Sinha. Boosted sampling: Approximation algorithms for stochastic optimization problems. In *Proceedings of the 36th ACM Symposium on the Theory of Computing (STOC)*, pages 417–426, 2004.
7. Anupam Gupta, R. Ravi, and Amitabh Sinha. An edge in time saves nine: LP rounding approximation algorithms for stochastic network design. In *Proceedings of the 45th Symposium on the Foundations of Computer Science (FOCS)*, pages 218–227, 2004.
8. A. Hayrapetyan, C. Swamy, and E. Tardos. Network design for information networks. In *Proceedings of the 16th ACM-SIAM Symposium on Discrete Algorithms (SODA)*, pages 933–942, 2005.
9. Nicole Immorlica, David Karger, Maria Minkoff, and Vahab Mirrokni. On the costs and benefits of procrastination: Approximation algorithms for stochastic combinatorial optimization problems. In *Proceedings of the 15th Annual ACM-SIAM Symposium on Discrete Algorithms*, pages 691–700, 2004.
10. Nicole Immorlica, Mohammad Mahdian, and Vahab Mirrokni. Limitations of cross-monotonic cost-sharing schemes. In *Proceedings of the 16th ACM-SIAM Symposium on Discrete Algorithms (SODA)*, 2005.
11. Kamal Jain and Vijay Vazirani. Applications of approximation algorithms to cooperative games. In *Proceedings of the 33rd Annual ACM Symposium on the Theory of Computing (STOC)*, pages 364–372, 2001.
12. Peter Kall and Stein W. Wallace. *Stochastic programming*. Wiley-Interscience Series in Systems and Optimization. John Wiley & Sons Ltd., Chichester, 1994.
13. Willem K. Klein Haneveld and Maarten H. van der Vlerk. *Stochastic Programming*. Department of Econometrics and OR, University of Groningen, Netherlands, 2003.
14. Hervé Moulin and Scott Shenker. Strategyproof sharing of submodular costs: budget balance versus efficiency. *Econom. Theory*, 18(3):511–533, 2001.
15. Martin Pál and Éva Tardos. Group strategyproof mechanisms via primal-dual algorithms. In *Proceedings of the 44th Annual IEEE Symposium on Foundations of Computer Science*, pages 584–593, 2003.
16. R. Ravi and Amitabh Sinha. Hedging uncertainty: Approximation algorithms for stochastic optimization problems. In *Proceedings of the 10th Integer Programming and Combinatorial Optimization Conference*, pages 101–115, 2004.
17. R. Schultz, L. Stougie, and M. H. van der Vlerk. Two-stage stochastic integer programming: a survey. *Statist. Neerlandica*, 50(3):404–416, 1996.
18. David Shmoys and Chaitanya Swamy. Stochastic optimization is (almost) as easy as deterministic optimization. In *Proceedings of the 45th Symposium on the Foundations of Computer Science (FOCS)*, pages 228–237, 2004.
19. David Shmoys and Chaitanya Swamy. Sampling-based approximation algorithms for multi-stage stochastic optimization. Manuscript, 2005.
20. H. P. Young. Cost allocation. In R. J. Aumann and S. Hart, editors, *Handbook of Game Theory*, volume 2, chapter 34, pages 1193–1235. North-Holland, 1994.

The Complexity of Making Unique Choices: Approximating 1-in-k SAT

Venkatesan Guruswami[1,*] and Luca Trevisan[2]

[1] Dept. of Computer Science & Engineering
University of Washington, Seattle, WA 98195
venkat@cs.washington.edu
[2] Computer Science Division
University of California at Berkeley, Berkeley, CA 94720
luca@cs.berkeley.edu

Abstract. We study the approximability of 1-in-kSAT, the variant of Max kSAT where a clause is deemed satisfied when precisely one of its literals is satisfied. We also investigate different special cases of the problem, including those obtained by restricting the literals to be unnegated and/or all clauses to have size exactly k. Our results show that the 1-in-kSAT problem exhibits some rather peculiar phenomena in the realm of constraint satisfaction problems. Specifically, the problem becomes *substantially* easier to approximate with perfect completeness as well as when negations of literals are not allowed.

1 Introduction

Boolean constraint satisfaction problems (CSP) arise in a variety of contexts and their study has generated a lot of algorithmic and complexity-theoretic research. An instance of a Boolean CSP is given by a set of variables and a collection of Boolean constraints, each on a certain subset of variables, and the objective is to find an assignment to the variables that satisfies as many constraints as possible. The most fundamental problem in this framework is of course Max SAT where the constraints are disjunctions of a subset of literals (i.e., variables and their negations), and thus are satisfied if *at least* one of the literals in the subset are set to 1. When the constraints are disjunctions of at most k literals, we get the Max kSAT problem.

In this work we consider constraint satisfaction problems where each constraint requires that *exactly* one literal (from the subset of literals constrained by it) is set to 1. This is a natural variant of SAT which is also NP-hard. We study the version of this problem when all constraints contain at most k literals (for constant $k \geq 3$). We call this problem 1-in-kSAT. For $k = 3$, this problem has often found use as a more convenient starting point compared to 3SAT for NP-completeness reductions. For each $k \geq 3$, determining if all constraints can be satisfied is NP-hard [9]. We study the approximability of this problem and some of its variants.

* Supported in part by NSF Career Award CCF-0343672.

C. Chekuri et al. (Eds.): APPROX and RANDOM 2005, LNCS 3624, pp. 99–110, 2005.
© Springer-Verlag Berlin Heidelberg 2005

In addition to being a natural satisfiability problem that merits study for its own sake, the underlying problem of making unique choices from certain specified subsets is salient to a natural pricing problem. We describe this connection at the end of the introduction.

We now formally define the problems we consider. Let $\{x_1, x_2, \ldots, x_n\}$ be a set of Boolean variables. For a set $S \subseteq \{x_i, \overline{x}_i \mid 1 \leq i \leq n\}$ of literals, the constraint $\mathsf{ONE}(S)$ is satisfied by an assignment to $\{x_1, \ldots, x_n\}$ if exactly one of the literals in S is set to 1 and the rest are set to 0.

Definition 1 (1-in-kSAT **and** 1-in-EkSAT). *Let $k \geq 2$. An instance of 1-in-kSAT consists of a set $\{x_1, x_2, \ldots, x_n\}$ of Boolean variables, and a collection of constraints $\mathsf{ONE}(S_i)$, $1 \leq i \leq m$ for some subsets $S_1, S_2, \ldots, S_m$ of $\{x_i, \overline{x}_i \mid 1 \leq i \leq n\}$, where $|S_i| \leq k$ for each $i = 1, 2, \ldots, m$. When each S_i has size exactly k, we get an instance of Exact 1-in-k Satisfiability, denoted 1-in-EkSAT.*

Definition 2 (1-in-kHS **and** 1-in-EkHS). *Let $k \geq 2$. An instance of 1-in-kHS consists of a set $\{x_1, x_2, \ldots, x_n\}$ of Boolean variables, and a collection of constraints $\mathsf{ONE}(S_i)$, $1 \leq i \leq m$ for some subsets $S_1, S_2, \ldots, S_m$ of $\{x_i \mid 1 \leq i \leq n\}$, where $|S_i| \leq k$ for each $i = 1, 2, \ldots, m$. When each S_i has size exactly k, we get an instance of Exact 1-in-k Hitting Set, denoted 1-in-EkSAT.*

Note that 1-in-kHS (resp. 1-in-EkHS) is a special case of 1-in-kSAT (resp. 1-in-EkSAT) where no negations are allowed. Also, 1-in-kHS is simply the variant of the Hitting Set problem where we are given a family of sets each of size at most k from a universe, and the goal is to pick a subset of the universe which intersects a maximum number of sets from the input family in *exactly* one element. The 1-in-EkHS problem corresponds to the case when each set in the family has exactly k elements. For every $k \geq 3$, 1-in-EkHS is NP-hard [9].

Clearly, the following are both orderings of the problems in decreasing order of their generality: (i) 1-in-kSAT, 1-in-EkSAT, 1-in-EkHS; and (ii) 1-in-kSAT, 1-in-kHS, 1-in-EkHS.

The 1-in-3SAT problem was considered in Schaefer's work on complexity of satisfiability problems [9]. An inapproximability factor of $6/5 - \varepsilon$ was shown for 1-in-E3SAT in [6]. We are unaware of any comprehensive prior investigation into the complexity of approximating 1-in-kSAT and its variants for larger k.

Our Results. For a maximization problem (such as a maximum constraint satisfaction problem), we define an α-approximation algorithm to be one which always delivers solution whose objective value is at least a fraction $1/\alpha$ of the optimum. A random assignment clearly satisfies at least a fraction $k2^{-k}$ fraction of constraints in any 1-in-kSAT instance. This gives a $2^k/k$-approximation algorithm, and no better approximation algorithm appears to be known for general k. We prove that, for sufficiently large k, it is NP-hard to approximate 1-in-EkSAT (and hence also the more general 1-in-kSAT) within a $2^{k-O(\sqrt{k})}$ factor. The result uses a gadget-style reduction from the general Max kCSP problem, but the analysis of the reduction uses a "random perturbation" technique which is a bit different from the standard way of analyzing gadget-based reductions.

The easiest of the problems we consider, namely 1-in-EkHS, has a simple e-approximation algorithm. We prove that this is the best possible – obtaining an $(e-\varepsilon)$-approximation is NP-hard, for every $\varepsilon > 0$ (for large enough k). Using the algorithm for 1-in-EkHS, one can give an $O(\log k)$-approximation algorithm for 1-in-kHS. (Recently, in [4], a hardness result for approximating 1-in-kHS within a factor $\log^{\sigma} n$ has been shown. Here n is the size of the universe, k is allowed to grow with the input, and $\sigma > 0$ is an absolute constant.)

The $2^{k-O(\sqrt{k})}$ inapproximability factor for general 1-in-kSAT says that algorithms that are substantially better than picking a random assignment do not exist, assuming P $\neq$ NP. For satisfiable instances of 1-in-kSAT, however, we are able to give an e-approximation algorithm. Specifically, when given a 1-in-kSAT instance for which an assignment that satisfies *every* clause exists, the algorithm finds an assignment that satisfies at least a fraction $1/e$ of the constraints. This is again the best possible, since our $(e - \varepsilon)$-inapproximability result holds for satisfiable instances of 1-in-EkHS.

Our results highlight the following peculiar behavior of 1-in-kSAT which is unusual for constraint satisfaction problems. The 1-in-kSAT problem becomes much easier to approximate when negations are not allowed (the 1-in-kHS variant, which has a $O(\log k)$-approximation algorithm), or when restricted to satisfiable instances (which has an e-approximation algorithm).

A Pricing Problem. The 1-in-kHS problem is related to a natural optimization problem concerning pricing that was recently considered in [7]. Consider the following pricing question: there is a universe of n items, each in unlimited supply with the seller. There are m customers, and each customer is single-minded and wants to buy a precise subset of at most k items (and this subset is known to the seller). Each customer values his/her subset at one dollar, and thus will buy the subset if and only if it costs at most one dollar in total. The goal is set prices to the items that maximizes the total revenue. If the prices are all either 0 or 1 dollars, then this problem is exactly 1-in-kHS. But the seller could price items in cents, and thereby possibly generating more revenue. However, it can be shown that the optimum with fractional prices is at most e times that with $0, 1$-prices (this follows from Lemma 6). Therefore, this problem admits a constant factor approximation (with approximation ratio independent of k) if and only if 1-in-kHS admits such an algorithm.

The Problem of Constructing "Ad-Hoc Selective Families." Another application of the 1-in-kHS problem is to the computation of *ad-hoc selective families*. An (n, h)-*selective family* is a combinatorial object defined and studied in [2] to deal with a broadcast problem in a radio network of unknown topology. An (n, h)-selective family is a collection $\mathcal{S}$ of subsets of $[n]$ such that for every set $F \subseteq [n]$ such that $|F| \leq h$ there is a set $S \in \mathcal{S}$ such that $|F \cap S| = 1$. More generally, a collection $\mathcal{S}$ of subsets of $[n]$ is *ad-hoc selective* for a collection $\mathcal{F}$ of subsets of $[n]$ if for every set $F \in \mathcal{F}$ there is a set $S \in \mathcal{S}$ such that $|S \cap F| = 1$ (in such a case, we say that S *selects* F). The notion of ad-hoc selective family, introduced in [3], has applications to the broadcast problem in radio networks of *known* topology. Here the computational problem of interest is, given a family

$\mathcal{F}$ (that is related to the topology of the radio network) to find a family $\mathcal{S}$ of smallest size that is ad-hoc selective for $\mathcal{F}$: the family $\mathcal{S}$ determines a schedule for the broadcast on the radio network and the time needed to realize the broadcast depends on the number of sets in the family $\mathcal{S}$. Clementi et al. [3] observe that given an approximation algorithm for 1-in-kHS one can get an approximation algorithm for the ad-hoc selective family problem as follows. Think of $\mathcal{F}$ an instance of 1-in-kHS, where the correspondence is that the instance of 1-in-kHS has a variable for every element of the universe $[n]$ of $\mathcal{F}$, and a constraint for every set of $\mathcal{F}$. An approximation algorithm for 1-in-kHS will find a set S that "selects" a large number of sets in $\mathcal{F}$: then one deletes those sets from $\mathcal{F}$ and repeats the above process. In order to optimize this process, Clementi et al. [3] introduce the idea of dividing $\mathcal{F}$ into sub-families of sets having approximately the same size, and our $O(\log k)$ approximation algorithm for 1-in-kHS is based on a similar idea.

2   Approximation Algorithms

In this section we present a randomized algorithm that delivers solutions within factor $1/e$ of the optimum solution for 1-in-EkHS. We also present a randomized algorithm that given an instance of 1-in-kSAT that is satisfiable, finds an assignment that satisfies an expected fraction $1/e$ of the clauses. Both these algorithms are the best possible in terms of approximation ratio, for large k, as we show in Section 3.

2.1   Approximation Algorithm for 1-in-EkHS

For the simplest variant 1-in-EkHS, there is a trivial e-approximation algorithm.

Theorem 3. *For every integer $k \geq 2$, there is a polynomial time e-approximation algorithm for 1-in-EkHS. The claim holds also when k is not an absolute constant but an arbitrary function of the universe size.*

Proof. Set each variable to 1 independently with probability $1/k$. The probability that a clause of k variables has exactly one variable set to 1 equals $k \cdot \frac{1}{k}(1 - 1/k)^{k-1} = (1-1/k)^{k-1} \geq 1/e$. Therefore the expected fraction of clauses satisfied by such a random assignment is at least $1/e$. The algorithm can be derandomized using the method of conditional expectations. Note that we did not use the exact value of k in the above argument, only that all sets had size k.   ∎

Algorithm for 1-in-kHS. We now consider the case when not all sets have exactly k elements. For this case we do not know any way to approximate within a factor that is independent of k.

Theorem 4. *There exists $c > 0$ such that for every integer $k \geq 2$, there is a polynomial time $c \log k$-approximation algorithm for 1-in-kHS. The claim holds even when k is not a constant but grows with the universe size.*

Proof. Let m be the number of sets in the 1-in-kHS instance. We partition the sets in the 1-in-kHS instance according to their size, placing the sets of size in the range $[2^{i-1}, 2^i)$ in collection $\mathcal{F}_i$ partition, for $i = 1, 2, \ldots, \lceil \log k \rceil$. Pick the partition that has a maximum number of sets, say $\mathcal{F}_j$, breaking ties arbitrarily. Clearly, $|\mathcal{F}_j| \geq \frac{m}{\lceil \log k \rceil}$. Set each element to 1 with probability $1/2^j$. Fix a set set in $\mathcal{F}_j$ that has x elements, $2^{j-1} \leq x < 2^j$. The probability that it has exactly one variable set to 1 equals $\frac{x}{2^j}\left(1 - \frac{1}{2^j}\right)^x$, which is easily seen to be at least $\frac{1}{2e}$. Thus, expected fraction of sets in $\mathcal{F}_j$ that have exactly one element set to 1 is at least $\frac{1}{2e}$. Therefore, we satisfy at least $\frac{m}{2e\lceil \log k \rceil}$ sets. The algorithm can again be derandomized using conditional expectations, and the argument holds for every k in the range $1 \leq k \leq n$, where n is the universe size.   ∎

Remark: Note that in the above algorithm, the upper bound on optimum we used was the total number of sets. With this upper bound, the best approximation factor we can hope for is $O(\log k)$. This is because there are instances of 1-in-kHS with m sets whose optimum is at most $O(\frac{m}{\log k})$. In fact, it can be shown that an instance obtained by picking an equal number of sets in each of the buckets at random will have this property with high probability.

2.2 Approximation Algorithm for 1-in-kSAT with Perfect Completeness

So far we saw algorithms for the case when negations were not allowed. We now give an approximation algorithm for 1-in-kSAT for the case when the instance is in fact satisfiable. Later on, in Section 3, we will show that without this restriction, a strong inapproximability result for 1-in-kSAT holds. We will also show that the factor e is the best possible, even with this restriction.

Theorem 5. *For every $k \geq 2$, there is an e-approximation algorithm for satisfiable instances of 1-in-kSAT. The claim holds even when k is not an absolute constant but an arbitrary function of the number of variables.*

Proof. The approach is to use a linear programming relaxation, and apply randomized rounding to it to obtain a Boolean assignment to the variables. Let $(V, \mathcal{C})$ be an instance of 1-in-kSAT, where $V = \{x_1, x_2, \ldots, x_n\}$ is the set of variables and $\mathcal{C} = \{C_1, C_2, \ldots, C_m\}$ is the set of clauses. For $j = 1, 2, \ldots, m$, define $\mathsf{pos}(C_j) \subseteq V$ to be those variables that appear positively (i.e., unnegated) in C_j, and $\mathsf{neg}(C_j)$ to be those variables that appear negated in C_j. Consider the linear program P with the following constraints in variables $p_1, p_2, \ldots, p_n$:

$$0 \leq p_i \leq 1 \quad \text{for } i = 1, 2, \ldots, n ,$$

$$\sum_{i:x_i \in \mathsf{pos}(C_j)} p_i + \sum_{i:x_i \in \mathsf{neg}(C_j)} (1 - p_i) = 1 \quad \text{for } j = 1, 2, \ldots, m.$$

The above program has a feasible solution. Indeed, let $a : V \to \{0, 1\}$ be an assignment that satisfies all clauses in $\mathcal{C}$. Then clearly $p_i = a(x_i)$ for $i = 1, 2, \ldots, n$ satisfies all the above constraints.

Solve the linear program (P) to find a feasible solution p_i^*, $1 \leq i \leq n$ in polynomial time. We need to convert this solution into an assignment $a^* : V \rightarrow \{0, 1\}$. We do this using randomized rounding. That is, for each x_i independently, we set

$$a(x_i) = \begin{cases} 1 & \text{with probability } p_i^* \\ 0 & \text{with probability } (1 - p_i^*) \end{cases}$$

Now consider the probability that a particular clause, say C_1, is satisfied. Let C_1 depend on r variables, $x_{i_1}, x_{i_2}, \ldots, x_{i_r}$. For $1 \leq \ell \leq r$, define $q_\ell = p_{i_\ell}^*$ if x_{i_ℓ} appears unnegated in C_1, and equal to $1 - p_{i_\ell}^*$ if x_{i_ℓ} appears negated in C_1. Then we have $q_1 + q_2 + \ldots + q_r = 1$. The probability that C_1 is satisfied equals

$$q_1(1 - q_2)(1 - q_3) \cdots (1 - q_r) \; + \; (1 - q_1)q_2(1 - q_3) \cdots (1 - q_r) + \cdots$$

$$\cdots \; + \; (1 - q_1)(1 - q_2) \cdots (1 - q_{r-1})q_r$$

By Lemma 6, this quantity is minimized when $q_1 = q_2 = \cdots = q_r = 1/r$, and thus is at least $(1 - 1/r)^{r-1} \geq 1/e$. Therefore the expected fraction of clauses satisfied by the randomized rounding is at least $1/e$, proving the theorem. ∎

We now state the inequality that was used in the above proof. An elegant proof of this inequality was shown to us by Chris Chang. For reasons of space, we omit the proof here.

Lemma 6. *Let $r \geq 2$ and let $q_1, q_2, \ldots, q_r$ be non-negative integers that sum up to 1. Then the quantity*

$$q_1(1 - q_2) \cdots (1 - q_r) + (1 - q_1)q_2(1 - q_3) \cdots (1 - q_r) + \cdots + (1 - q_1) \cdots (1 - q_{r-1})q_r$$

attains its minimum value when $q_1 = q_2 = \cdots = q_r = 1/r$. In particular, the quantity is at least $(1 - 1/r)^{r-1}$.

3   Inapproximability Results

3.1   Factor $2^{\Omega(k)}$ Hardness for 1-in-EkSAT

We now prove that, if P $\neq$ NP, then, for sufficiently large k, 1-in-EkSAT cannot be approximated within a factor of $2^{k-O(\sqrt{k})}$. The result uses a gadget-style reduction from the constraint satisfaction problem Max EkAND, but the analysis of the reduction uses a "random perturbation" technique which is a bit different from the standard way of analyzing gadget-based reductions.

Preliminaries. We first define the Max EkAND problem. An instance of Max EkAND consists of a set of Boolean variables, and a collection of AND constraints of the form $l_1 \wedge l_2 \wedge \cdots \wedge l_k$ where each l_j is a literal. The goal is to find an assignment that satisfies a maximum number of the AND constraints.

Theorem 7 ([8]). *If* P $\neq$ NP*, for every* $\varepsilon > 0$ *and for every integers* $q \geq 1$ *and* k *such that* $2q+1 \leq k \leq 2q+q^2$*, there is no* $(2^{k-2q} - \varepsilon)$*-approximation algorithm for* Max EkAND*. In particular, for every* $k \geq 7$*,* $\varepsilon > 0$*, there is no* $(2^{k-2\lceil\sqrt{k}\rceil} - \varepsilon)$*-approximate algorithm for Max* kAND*. Furthermore, if* ZPP $\neq$ NP*, then for every* ε *there is a constant* c *such that there is no* $n^{1-\varepsilon}$*-approximate algorithm for Max* $(c \log n)$AND*.*

The furthermore part follows from the result of [8] plus the use of a randomized reduction described in [1].

Inapproximability Result. We describe a reduction from Max EkAND to 1-in-EkSAT.

Lemma 8. *Suppose that there is a polynomial time* β*-approximation algorithm for* 1-in-EkSAT*, for some* $k \geq 3$*. Then there is an* $2ek\beta$*-approximate algorithm for* Max EkAND*.*

Proof. We describe how to map an instance φ_{AND} of Max EkAND into an instance φ_{oik} of 1-in-EkSAT. Let $l_1 \wedge \cdots \wedge l_k$ be a clause of the Max EkAND instance φ_{AND}. We introduce the constraints $\mathsf{ONE}(l_1, \mathrm{neg}l_2, \ldots, \mathrm{neg}l_k)$, $\mathsf{ONE}(\mathrm{neg}l_1, l_2, \ldots, \mathrm{neg}l_k)$, $\ldots$, $\mathsf{ONE}(\mathrm{neg}l_1, \mathrm{neg}l_2, \ldots, l_k)$, where $\mathrm{neg}\ell$ denotes the negation of the literal ℓ. If originally we had m AND constraints, now we have km constraints of 1-in-EkSAT.

Claim 1 *If there is an assignment that satisfies* t *constraints in* φ_{AND}*, then the same assignment satisfies at least* tk *constraints in* φ_{oik}*.*

Proof of Claim. For each clause $l_1 \wedge \cdots \wedge l_k$ that is true, then the k corresponding ONE constraints are satisfied. ∎

Claim 2 *If there is an assignment that satisfies* t' *constraints in* φ_{oik}*, then there is an assignment that satisfies at least* $t'/(2ek^2)$ *constraints in* φ_{AND}*.*

Proof of Claim. Take the assignment a that satisfies t' constraints in φ_{oik} and then randomly perturb it in the following way: for each variable independently, flip its value with probability $1/k$ and leave the value unchanged with probability $1-1/k$; we will estimate the expected number of clauses of φ_{AND} that are satisfied by this random assignment R.

Let us look at a clause $l_1 \wedge \cdots \wedge l_k$ of φ_{AND} and the k corresponding clauses of φ_{oik}. We consider different cases, depending upon how many of the literals $l_1, l_2, \ldots, l_k$ are set to 1 by the assignment a.

- If a satisfies all literals, then it satisfies all k clauses in φ_{oik} and the random assignment satisfies $l_1 \wedge \cdots \wedge l_k$ with probability at least $(1-1/k)^k \geq 8/27$, for $k \geq 3$.
- If a satisfies $k-2$ literals, then it satisfies 2 clauses in φ_{oik}, and the random assignment satisfies $l_1 \wedge \cdots \wedge l_k$ with probability at least $(1-1/k)^{k-2} \cdot k^{-2} \geq 1/(ek^2)$

– In all other cases, a satisfies none of the clauses corresponding to $l_1 \wedge \cdots \wedge l_k$ in φ_{oik}.

In each case, we have that the probability that the assignment satisfies $l_1 \wedge \cdots \wedge l_k$ is at least $1/(2ek^2)$ times the number of constraints corresponding to $l_1 \wedge \cdots \wedge l_k$ in φ_{oik} satisfied by a. If a satisfies t' constraints, it follows that the random assignment satisfies, on average, at least $t'/(2ek^2)$ clauses of φ_{AND}.   ∎

Suppose now that we have a β-approximate algorithm for 1-in-EkSAT and that we are given in input an instance φ_{AND} of Max EkAND. Suppose that the optimum solution for φ_{AND} has cost opt. Then we construct an instance φ_{oik} of 1-in-EkSAT as described above, and give it to the approximation algorithm. By Claim 1, the optimum of φ_{oik} is at least $k \cdot$ opt, and so the algorithm will return a solution of cost at least $k \cdot$ opt$/\beta$. By Claim 2, we get a distribution over assignments for φ_{AND} that, on average, satisfies at least opt$/(2e\beta k)$ constraints. An assignment that satisfies at least as many constraints can be found deterministically using the method of conditional expectations.   ∎

Theorem 9. *If* P $\neq$ NP*, for every sufficiently large k and for every $\varepsilon > 0$, there is no polynomial time $(2^{k-2\lceil \sqrt{k} \rceil}/(2ek) - \varepsilon)$-approximation algorithm for* 1-in-EkSAT*. Furthermore, if* ZPP $\neq$ NP*, for every ε there is a c such that there is no polynomial time $n^{1-\varepsilon}$ approximation algorithm for* 1-in-$(c \log n)$SAT*.*

Proof. Follows from Theorem 7 and from the reduction of Lemma 8. For the "furthermore" part one should observe that the reduction does not increase the size of the input by more that a logarithmic factor.   ∎

3.2   Factor $e - \varepsilon$ Hardness for 1-in-EkHS

In this section, we prove the following hardness result, which shows that the results of Theorems 3 and 5 are tight in terms of the approximation ratio.

Theorem 10. *For every $\varepsilon > 0$, for sufficiently large k, there is no polynomial time $(e - \varepsilon)$-approximation algorithm for* 1-in-EkHS*, unless* P $=$ NP*. Furthermore, the result holds even when the instance of* 1-in-EkHS *is satisfiable.*

Multiprover Systems. Our proof will use the approach behind Feige's hardness result for set cover [5]. We will give a reduction from the multiprover proof system of Feige, which we state in a form convenient to us below. In what follows, we use $[m]$ to denote the set $\{1, 2, \ldots, m\}$.

Definition 11 (p-prover game). *For every integer $p \geq 2$ and a parameter u that is an even integer, an instance $\mathcal{I}$ of the p-prover game of size n is defined as follows.*

– *The instance consists of a p-uniform p-partite hypergraph $\mathcal{H}$ with the following properties:*

- [Vertices] *The vertex set of $\mathcal{H}$ is given by $W = Q_1 \cup Q_2 \cup \cdots \cup Q_p$, where Q_i is the vertices in the i'th part (or prover), and $|Q_i| = Q = n^{u/2}(5n/3)^{u/2}$ for $i = 1, 2, \ldots, p$.*
- [Hyperedges] *There are $R = (5n)^u$ hyperedges in $\mathcal{H}$, labeled by $r \in [R]$. Denote the r'th hyperedge, for $r \in [R]$, by $(q_{r,1}, q_{r,2}, \ldots, q_{r,p})$, where $q_{r,i} \in Q_i$ for $i \in [p]$.*
- [Regularity] *Each vertex in W belongs to precisely R/Q hyperedges.*

– *Define $B = 4^u$ and $A = 2^u$. For each $r \in [R]$, and $i \in [p]$, the instance consists of projections $\pi_{r,i} : [B] \to [A]$ each of which is (B/A)-to-1.*

The goal is to find a labeling $a : W \to [B]$ that "satisfies" as many hyperedges of $\mathcal{H}$ as possible, where we define the notion of when a hyperedge is satisfied below.

– [Strongly satisfied hyperedges] *We say that a labeling $a : W \to [B]$ strongly satisfies a hyperedge $r \in [R]$ if*

$$\pi_{r,1}(a(q_{r,1})) = \pi_{r,2}(a(q_{r,2})) = \cdots = \pi_{r,p}(a(q_{r,p})) \ .$$

– [Weakly satisfied hyperedges] *We say that a labeling $a : W \to [B]$ weakly satisfies a hyperedge $r \in [R]$ if at least two elements of the tuple*

$$\langle \pi_{r,1}(a(q_{r,1})), \ \pi_{r,2}(a(q_{r,2})), \ \cdots, \ \pi_{r,p}(a(q_{r,p})) \rangle$$

are equal.

Feige's result on the above p-prover games can be stated as follows.

Theorem 12. *There exists a constant $0 < c < 1$ such that for every $p \geq 2$ and all large enough u, given an instance $\mathcal{I}$ of the p-prover game with parameter u, it is NP-hard to distinguish between the following two cases, when it is promised that one of them holds:*

– YES INSTANCES: *There is a labeling that strongly satisfies all hyperedges.*
– NO INSTANCES: *No labeling weakly satisfies more than a fraction $p^2 c^u$ of the hyperedges.*

Note that difference between YES and NO instances is not just in the fraction of satisfied hyperedges, but also in how the hyperedge is satisfied (strong vs. weak).

Reduction to 1-in-EkHS. The result of Theorem 10 clearly follows from Theorem 12 and the reduction guaranteed by the following lemma.

Lemma 13. *For every $\varepsilon > 0$, there exists a positive integer p such that for all large enough even u the following holds. Let $k = 4^u$. There is polynomial time reduction from p-prover games with parameter u to 1-in-EkHS that has the following properties:*

– [Completeness]: *If the original instance of the p-prover game is a YES instance (in the sense of Theorem 12), then the instance of 1-in-EkHS produced by the reduction is satisfiable.*

- [Soundness]: If the original instance of the p-prover game is a No instance (in the sense of Theorem 12), then no assignment satisfies more than a fraction $(1/e + \varepsilon)$ of the constraints in the instance of 1-in-EkHS produced by the reduction.

Proof. We begin with describing the reduction. Suppose we are given an instance $\mathcal{I}$ of the p-prover game with parameter u. In the following, we will use the notation and terminology from Definition 11. We define an instance of 1-in-EkHS on the universe

$$U \overset{\text{def}}{=} \{(i, q, a) \mid i \in [p], q \in Q_i, a \in [B]\} .$$

Thus the universe simply corresponds to all possible ⟨vertex, label⟩ pairs. The collection of sets in the instance is given by $\{S_{r,x}\}$ as r ranges over $[R]$ and x over $\{1, 2, \ldots, p\}^A$, where

$$S_{r,x} \overset{\text{def}}{=} \{(i, q_{r,i}, a) \in U \mid x_{\pi_{r,i}(a)} = i\} .$$

Note that the size of each $S_{r,x}$ equals $B = 4^u = k$. This is because $S_{r,x} = \bigcup_{j \in [A]} \{(i, q_{r,i}, a) \mid i = x_j, \ \pi_{r,i}(a) = j\}$, and for each $j \in [A]$, there are precisely B/A elements $a \in [B]$ such that $\pi_{r,i}(a) = j$ (since the projections are (B/A)-to-1).

Let us first argue the completeness (this will also help elucidate the rationale for the choice of the sets $S_{r,x}$).

Claim 3 (Completeness) *Let $a : W \to [B]$ be an assignment that strongly satisfies all hyperedges of $\mathcal{I}$. Consider the subset $C = \{(i, q, a(q)) \mid i \in [p], \ q \in Q_i\}$. Then $|C \cap S_{r,x}| = 1$ for every $r \in [R]$ and $x \in [p]^A$.*

Proof of Claim. Since a strongly satisfies every hyperedge $r \in [R]$, we have $\pi_{r,1}(a(q_{r,1})) = \pi_{r,2}(a(q_{r,2})) = \cdots = \pi_{r,p}(a(q_{r,p}))$, and let $j_r \in [A]$ denote this common value. Also let $i_r = x_{j_r} \in [p]$. Then it is not hard to check that $C \cap S_{r,x} = \{(i_r, q_{r,i_r}, a(q_{r,i_r}))\}$. ∎

Claim 4 (Soundness) *Suppose that some $C \subseteq U$ satisfies $|C \cap S_{r,x}| = 1$ for at least a fraction $1/e + \varepsilon$ of the sets $S_{r,x}$. Then, provided $p \geq 1 + 3/(e\varepsilon)$ and $c^u < \varepsilon/(2p^8)$, $\mathcal{I}$ is not a No instance.*

Proof of Claim. For each $i \in [p]$ and each vertex $q \in Q_i$, define $\mathcal{A}_q = \{a \in [B] \mid (i, q, a) \in C\}$, i.e., $\mathcal{A}_q$ consists of those labels for q that the subset C "picked". We will later use the sets $\mathcal{A}_q$, $q \in W$ to prove that a good labeling exists for $\mathcal{I}$.

Call $r \in [R]$ to be *nice* if at least a fraction $(1/e + \varepsilon/2)$ of the sets $S_{r,x}$, as x ranges over $[p]^A$, satisfy $|C \cap S_{r,x}| = 1$. By an averaging argument, at least a fraction $\varepsilon/2$ of $r \in [R]$ are nice.

Let us now focus on a specific r that is nice. Define

$$D_r = \{(i, b) \mid i \in [p], \ b \in [A], \ (i, q_{r,i}, a) \in C \text{ for some } a \text{ s.t. } \pi_{r,i}(a) = b\} .$$

That is D_r consists of the projections of the assignments in $\mathcal{A}_{q_{r,i}}$ for all vertices $q_{r,i}$ belonging to hyperedge r. Let $|D_r| = M$ and $(i_1, b_1), (i_2, b_2), \ldots, (i_M, b_M)$ be the elements of D_r.

Now if $|C \cap S_{r,x}| = 1$, then *exactly one* of the events $x_{b_j} = i_j$ must hold as j ranges over $[M]$. Since r is nice we know that the fraction of such $x \in [p]^A$ is at least $(1/e + \varepsilon/2)$. We consider the following cases.

Case A: $M > p^3$. Then there are at least $M' \stackrel{\text{def}}{=} \lceil M/p \rceil > p^2$ distinct values among $b_1, b_2, \ldots, b_M$. For definiteness, assume that $b_1, b_2, \ldots, b_{M'}$ are distinct. If exactly one of the events $x_{b_j} = i_j$ holds as j ranges over $[M]$, then certainly at most one j in the range $1 \leq j \leq M'$ satisfies $x_{b_j} = i_j$. The fraction of $x \in [p]^A$ for which $\left| \{j \mid j \in [M'], \ x_{b_j} = i_j\} \right| \leq 1$ is at most

$$\left(1 - \frac{1}{p}\right)^{M'} + M'\left(1 - \frac{1}{p}\right)^{M'-1} \leq e^{-(M'-1)/p}(M' + 1) \leq e^{-p}(p^2 + 2) < 1/e$$

for $p \geq 10$. This contradicts that fact that r is nice. Therefore this case cannot occur and we must have $M \leq p^3$.

Case B: $M \leq p^3$ and all the b_j's are distinct. Clearly, the fraction of $x \in [p]^A$ for $x_{b_j} = i_j$ for exactly one choice of $j \in [M]$ is precisely $\frac{M}{p}(1 - 1/p)^{M-1}$. Now we bound this quantity as follows:

$$
\begin{aligned}
\frac{M}{p}(1 - 1/p)^{M-1} &= \frac{p}{p-1}\frac{M}{p}(1 - 1/p)^M \\
&\leq \frac{p}{p-1}\frac{M}{p}e^{-M/p} \quad \text{(using } 1 - x \leq e^{-x} \text{ for } x \geq 0) \\
&\leq \frac{p}{p-1}\frac{1}{e} \quad \text{(using } xe^{-x} \leq 1/e \text{ for } x \geq 0) \\
&\leq \frac{1}{e} + \frac{\varepsilon}{3}
\end{aligned}
$$

provided $p \geq 1 + 3/(e\varepsilon)$. Again, this contradicts that fact that r is nice. Therefore this case cannot occur either.

Therefore we can conclude the following: if r is nice, then $|D_r| \leq p^3$ and there exist $i_{r,1} \neq i_{r,2} \in [p]$ such that for some $b_r \in [A]$, $\{(i_{r,1}, b_r), (i_{r,2}, b_r)\} \subseteq D_r$.

Consider the following labeling a to W. For each $q \in W$, set $a(q)$ to be a random, uniformly chosen, element of $\mathcal{A}_q$ (if $\mathcal{A}_q$ is empty we set $a(q)$ arbitrarily). Consider a nice r. With probability at least $1/p^6$, we have

$$\pi_{r,i_{r,1}}\big(a(q_{r,i_{r,1}})\big) = \pi_{r,i_{r,2}}\big(a(q_{r,i_{r,2}})\big) = b_r$$

and thus hyperedge r is weakly satisfied by the labeling a.

In particular, there exists a labeling that weakly satisfies at least a fraction $1/p^6$ of the nice r, and hence at least a fraction $\frac{\varepsilon}{2p^6}$ of all $r \in [R]$. If u is large enough so that $p^2 c^u < \frac{\varepsilon}{2p^6}$ (where c is the constant from Theorem 12), we know that $\mathcal{I}$ is not a No instance. ∎

The completeness and soundness claims together yield the lemma. ∎

4   Conclusions

The 1-in-EkSAT problem, while hard to approximate within a $2^{\Omega(k)}$ factor, becomes substantially easier and admits an e-approximation in polynomial time with either one of two restrictions: (i) do not allow negations (which is the 1-in-EkHS problem), (ii) consider satisfiable instances. Such a drastic change in approximability under such restrictions is quite unusual for natural constraint satisfaction problems (discounting problems which become polynomial-time tractable under these restrictions).

We conclude with two open questions:

- Does 1-in-kHS admit a polynomial time $o(\log k)$-approximation algorithm[1]?
- Does the $2^{\Omega(k)}$ hardness for 1-in-EkSAT hold for near-satisfiable instances (for which a fraction $(1 - \varepsilon)$ of the constraints can be satisfied by some assignment)?

Acknowledgments

We would like to thank Chris Chang for his proof of Lemma 6 and Ziv Bar-Yossef for useful discussions.

References

1. M. Bellare, O. Goldreich, and M. Sudan. Free bits, PCP's and non-approximability – towards tight results. *SIAM Journal on Computing*, 27(3):804–915, 1998.
2. B. S. Chlebus, L. Gasieniec, A. Gibbons, A. Pelc, and W. Rytter. Deterministic broadcasting in unknown radio networks. In *Proceedings of the 11th ACM-SIAM Symposium on Discrete Algorithms*, pages 861–870, 2000.
3. A. Clementi, P. Crescenzi, A. Monti, P. Penna, and R. Silvestri. On computing ad-hoc selective families. In *Proceedings of RANDOM'01*, 2001.
4. E. Demaine, U. Feige, M. Hajiaghayi, and M. Salavatipour. Combination can be hard: approximability of the unique coverage problem. Manuscript, April 2005.
5. U. Feige. A threshold of $\ln n$ for approximating set cover. *Journal of the ACM*, 45(4):634–652, 1998.
6. V. Guruswami. Query-efficient checking of proofs and improved PCP characterizations of NP. Master's thesis, MIT, 1999.
7. V. Guruswami, J. Hartline, A. Karlin, D. Kempe, C. Kenyon, and F. McSherry. On profit-maximizing envy-free pricing. In *Proceedings of the 16th ACM-SIAM Symposium on Discrete Algorithms (SODA)*, January 2005.
8. A. Samorodnitsky and L. Trevisan. A PCP characterization of NP with optimal amortized query complexity. In *Proceedings of the 32nd ACM Symposium on Theory of Computing*, 2000.
9. T. J. Schaefer. The complexity of satisfiability problems. In *Proceedings of the 10th ACM Symposium on Theory of Computing*, pages 216–226, 1978.

---

[1] A recent work [4] presents some evidence that perhaps such an algorithm does not exist.

Approximating the Distortion*

Alexander Hall[1] and Christos Papadimitriou[2]

[1] ETH Zürich, Switzerland
alex.hall@gmail.com
[2] UC Berkeley, USA
christos@cs.berkeley.edu

Abstract. Kenyon et al. (STOC 04) compute the distortion between one-dimensional finite point sets when the distortion is small; Papadimitriou and Safra (SODA 05) show that the problem is NP-hard to approximate within a factor of 3, albeit in 3 dimensions. We solve an open problem in these two papers by demonstrating that, when the distortion is large, it is hard to approximate within large factors, even for 1-dimensional point sets. We also introduce additive distortion, and show that it can be easily approximated within a factor of two.

1 Introduction

The *distortion problem* is the following: Given two d-dimensional finite points sets $S, T \subseteq \mathbb{R}^d$ with $|S| = |T|$ and a real number $\delta \in \mathbb{R}$ (the *distortion*) is there a bijection $f : S \to T$ such that we have expansion$(f) \cdot$ expansion$(f^{-1}) \leq \delta$, with expansion$(f) := \max_{x,y \in S} (d(f(x), f(y))/d(x, y))$? Here $d(x, y)$ denotes the Euclidean distance between two points $x, y \in \mathbb{R}^d$.

The distortion problem was introduced by Kenyon et al. [1], who gave an optimal algorithm for 1-dimensional points sets that are known to have distortion less than $3 + 2\sqrt{2}$. Their elaborate dynamic programming algorithm crucially relies on the small distortion guarantee to establish and exploit certain restrictions on the bijection between the two point sets. Papadimitriou and Safra [2] present NP-hardness and inapproximability results, which hold for both small and large distortions – albeit for the 3 dimensional case. In both papers the question of whether the distortion of 1-dimensional point sets can be computed or approximated if it is not known to be small was proposed as an open problem.

In this paper we resolve this question by establishing several strong NP-hardness and inapproximability results. In Section 2.1 we show that the distortion problem is NP-hard in the 1-dimensional case when the distortion is at least $|S|^\varepsilon$, for any $\varepsilon > 0$. The proof is a surprisingly simple reduction from the (strongly NP-complete) *3-partition* problem. By appropriately modifying the proof we show that, in the same range, even the *logarithm* of the distortion cannot be approximated within a factor better than 2 (Theorem 2). This answers

---

* Work partially supported by European Commission – Fet Open project DELIS IST-001907 Dynamically Evolving Large Scale Information Systems.

C. Chekuri et al. (Eds.): APPROX and RANDOM 2005, LNCS 3624, pp. 111–122, 2005.

an open question posed in [1]. For larger distortions (growing faster than $|S|$) we show a further inapproximability result: The distortion cannot be approximated within a ratio better than $L_T^{1-\varepsilon}$, where L_T is the ratio of the largest to the smallest distance in point set T; we point out that an approximation ratio of L_T^2 is always trivial, in any metric space. We make more precise the inapproximability bounds given in [2] for 3 dimensions and arbitrary distortion, by making explicit the dependency of the inapproximability ratio on the magnitude of the distortion. An overview of our inapproximability results:

Dimension	Distortion	Inapproximable within	
	$|S| \geq \delta \geq |S|^{\varepsilon}$	$\delta^{1-\varepsilon'}$	
$d \geq 1$	$\delta \geq |S|$	$\sqrt{\delta} \cdot |S|^{\frac{1}{2}-\varepsilon'}$	Thm. 2
	$\delta \geq |S|^{1+\varepsilon}$	$L_T^{1-\varepsilon'}$	
$d \geq 3$	$\delta > 1$	$\sqrt{9 - 8/\delta^2} - \varepsilon'$	Thm. 3
unbounded d	$\delta > 1$	$\delta - \varepsilon'$	Thm. 4

Motivated by these strong inapproximability results, we introduce a novel variant of the problem that we call the *additive distortion problem*: Given two finite points sets $S, T \subseteq \mathbb{R}^d$ with $|S| = |T|$, find the smallest $\Delta \in \mathbb{R}$ (the *additive distortion*) such that there is a bijection $f : S \to T$ with $d(x,y) - \Delta \leq d(f(x), f(y)) \leq d(x,y) + \Delta$, for all $x, y \in S$. By a modification of the Papadimitriou-Safra construction, it is not hard to see that the additive distortion is NP-hard to approximate by a factor better than 3 in 3 dimensions. In Section 3 we present a 2-approximation algorithm for this problem in the 1-dimensional case and a 5-approximation algorithm for the more general case of embedding an arbitrary metric space onto an 1-dimensional point set. Finally, we conclude by pointing out several open questions raised by this work.

Remark. The first three and the last inapproximability results can be strengthened by a power of 2, if we impose the stronger restriction of expansion$(f) \leq \sqrt{\delta}$ and expansion$(f^{-1}) \leq \sqrt{\delta}$ (which implies a distortion of $\leq \delta$). In particular the third bound becomes $L_T^{2-\varepsilon'}$, which is near optimal since a ratio of L_T^2 is trivial also in this more restricted setting.

Related Work. Especially in view of the drastic increase in the number of publications concerning embeddings of metric spaces, it is astonishing that the low distortion problem was only introduced very recently [1]. Most Computer Science related work in this area focuses on the setting where a given finite metric space is to be embedded into an *infinite* host space, usually a low dimensional Euclidean space. Although methods from this area do not seem to apply to our setting of embedding a finite metric onto another finite metric, we give a brief overview of related work.

From a theoretical point of view there has been a large interest in finding worst case bounds for the distortion of embedding a class of metrics, e.g. see the surveys [3–5]. The problem of finding a good embedding for a given metric ("good" compared to an optimal embedding of this, same metric) is practically more relevant and consequently the vast majority of research in this area – also referred to as multi-dimensional scaling – has been on devising good heuristics. See the web-page of the working group on multi-dimensional scaling [6] for an overview and an extensive list of references. An important theoretical result is Linial et al.'s [7] adaption of Bourgain's construction [8]. They present an approximation algorithm based on semidefinite programming for finding a minimum distortion embedding of a given finite metric. Kleinberg et al. [9] consider approximate embeddings of metrics for which only a small subset of the distances are known. Slivkins [10] recently can improve on the results. Also recently Bădoiu et al. [11] give several approximation algorithms for low distortion embeddings of metrics into $\mathbb{R}^1$ and $\mathbb{R}^2$. The notion of *additive* distortion has also been considered for the case where a finite metric is to be embedded into an infinite host space. Håstad et al. [12] give a 2-approximation for the case of embedding into $\mathbb{R}$ and prove that the problem cannot be approximated within $4/3$, unless P = NP. Later Bădoiu [13] and Bădoiu et al. [14] gave an approximation algorithm and a weakly quasi-polynomial time algorithm, respectively, for 2 dimensions.

Other research loosely related to the distortion problem is on the minimum bandwidth problem (see e.g. [15]) and the maximum similarity problem [16].

2   The Inapproximability of Distortion

2.1   NP-Hardness

Theorem 1. *The distortion problem is NP-hard for 1 dimension and any fixed $\delta \geq |S|^{\varepsilon}$, for any constant $\varepsilon > 0$.*

Proof. We reduce the well known *3-partition* problem to it. In this problem we are given a set A of $3n$ items $A = \{1, \ldots, 3n\}$ with associated sizes $a_1, \ldots, a_{3n} \in \mathbb{N}$, and a bound $B \in \mathbb{N}$, with $B/4 < a_i < B/2$, for each i, and $\sum_{i=1}^{3n} a_i = nB$, and we must decide whether A can be partitioned into n disjoint sets $I_0, \ldots, I_{n-1}$ such that $\sum_{i \in I_j} a_i = B$, for $j = 0, \ldots, n-1$. Note that due to the bounds for the item sizes a_i, all sets I_j must have cardinality 3.

We now describe how to construct the point sets S and T on the line (see the left part of Figure 1). The point set S consists of $3n$ blobs of points $S_1, \ldots, S_{3n}$, where blob S_i has a_i points. Points in a blob are distributed regularly along the line with distance $x := 1/\sqrt{\delta \cdot B}$ from one point to the next. The blobs themselves are also distributed regularly along the line with distance 1 from one to the next, i.e. two neighboring points in different blobs have distance 1. The point set T is very similar: it consists of n blobs $T_1, \ldots, T_n$ of size B. Here the distance of neighboring points within a blob is $1/\sqrt{B}$ and if they are in different blobs it is again 1. Finally, we add two points to both S and T, far away from the blobs. Their distance in T is 1 and their distance in S is $\sqrt{\delta}$ (clearly the points can be added such that they need to be mapped onto each other in order to obtain distortion $\leq \delta$). This ensures that $\text{expansion}(f^{-1}) = \max_{x,y \in S} \left(\frac{d(x,y)}{d(f(x),f(y))} \right) \geq \sqrt{\delta}$.

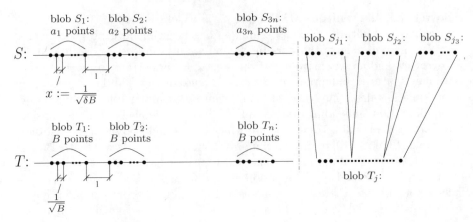

Fig. 1. Left: the point sets S and T on a line constructed from a given instance of 3-partition. Right: the mapping of three blobs in S to one blob in T.

Our aim is to show how to derive a low distortion mapping of points from S to T given a solution of the 3-partition instance and vice versa, assuming that $\delta \geq c \cdot n^2 \cdot B$, for an appropriately chosen constant $c > 1$. We start with the forward direction. The mapping is straight forward: we map blobs $S_{j_1}, S_{j_2}, S_{j_3}$ to blob T_j (simply one blob next to the other, as shown in Figure 1 on the right), if $I_j = \{j_1, j_2, j_3\}$. In order to see that the distortion δ is not violated we check the largest possible changes in distance (relatively speaking) and show that expansion$(f) \leq \sqrt{\delta}$ and expansion$(f^{-1}) \leq \sqrt{\delta}$ hold, leading to three cases:

1. Two neighboring blobs S_a and S_b are spread as far apart as possible. The distance between two points hereby increases from 1 to less than

$$n \cdot \left(B \cdot \frac{1}{\sqrt{B}} + 1 \right) = n \cdot (\sqrt{B} + 1) \leq \sqrt{\delta},$$

   under the assumption $\delta \geq c \cdot n^2 \cdot B$ with some constant c, which is chosen large enough.

2. The leftmost and the rightmost blob in S, i.e. S_1 and S_{3n}, are mapped next to each other into the same blob in T. The distance decreases from less than $n \cdot (B/\sqrt{\delta \cdot B} + 3)$ to $1/\sqrt{B}$. This leads to an upper bound for the relative distance change of

$$\frac{n \cdot \left(\sqrt{\frac{B}{\delta}} + 3 \right)}{\frac{1}{\sqrt{B}}} \leq n\sqrt{B} \left(\frac{1}{\sqrt{c} \cdot n} + 3 \right) \leq \sqrt{\delta},$$

   again with the assumption $\delta \geq c \cdot n^2 \cdot B$ and some appropriately chosen constant c (which clearly exists).

3. What about the distance increase for two points in the same blob? All distances of points within a blob increase by exactly $(1/\sqrt{B})/(1/\sqrt{\delta \cdot B}) = \sqrt{\delta}$.

For the other direction we only need to check that a blob S_i cannot be split up and mapped to two or more different blobs in T.

4. Two neighboring points in a blob in S cannot be mapped to two different blobs in T. The relative increase in distance would be $1/\frac{1}{\sqrt{\delta \cdot B}} > \sqrt{\delta}$. Together with the two "extra" points that force expansion$(f^{-1}) \geq \sqrt{\delta}$ this would yield a distortion $> \delta$.

This ensures that in a low distortion mapping from S to T always exactly three blobs from S will be mapped to one blob from T, as wanted.

In order to obtain the nice lower bound for δ we now add a huge blob to both of them which is far away from all other blobs and whose points are very close to each other. Clearly this can be done in such a manner that all points in this huge blob in S will be mapped directly to the corresponding huge blob in T; without interfering with the mapping of all other points. We started with $|S| = n \cdot B$ points and increase the number of points in the huge blob until $|S|^\varepsilon > c \cdot n^2 \cdot B$. Since ε is a constant the resulting input size will still be bounded by a polynomial. $\diamond$

2.2   Inapproximability

We now describe how to modify the proof in order to obtain the strong inapproximability results for large δ. Let $L_T := \frac{\max_{x,y \in T} d(x,y)}{\min_{x,y \in T} d(x,y)}$ be the ratio of maximum to minimum distance in T.

Theorem 2. *For any $\varepsilon, \varepsilon' > 0$, the 1-dimensional distortion problem for $|S| \geq \delta \geq |S|^\varepsilon$ is inapproximable within a factor of $\delta^{1-\varepsilon'}$, for $\delta \geq |S|$ is inapproximable within a factor of $\sqrt{\delta} \cdot |S|^{\frac{1}{2}-\varepsilon'}$, and for $\delta \geq |S|^{1+\varepsilon}$ is inapproximable within a factor of $L_T^{1-\varepsilon'}$, unless $P = NP$.*

Proof. First of all we introduce a new distance g to be defined presently and replace the inter-blob distances (the distance of 1 from one blob to the next) in T by this distance g.

From the list of cases in the proof of Theorem 1, case 2. and case 3. remain untouched by this change. For case 1. (where the "1" in the expression is now a "g") we choose c in our assumption $\delta \geq c \cdot n^2 \cdot B$ such that $\sqrt{\delta}/2 \geq n\sqrt{B}$ and we choose g such that $\sqrt{\delta}/2 \geq n \cdot g$. Choosing $g := \sqrt{\delta}/2n$ will be fine.

For case 4. the relative increase in distance between two neighboring points in a blob in S would be at least $y := g/\frac{1}{\sqrt{\delta \cdot B}}$ if they were split onto two blobs in T. This would amount to a distortion of $y \cdot \sqrt{\delta}$ (note that expansion$(f^{-1}) \geq \sqrt{\delta}$). The optimum solution is only "allowed" a distortion of δ. Thus, unless $P = NP$, there can be no approximation algorithm with ratio better than

$$\frac{y \cdot \sqrt{\delta}}{\delta} = \frac{g \cdot \sqrt{\delta \cdot B} \cdot \sqrt{\delta}}{\delta} = \sqrt{\delta} \cdot \frac{\sqrt{B}}{2n} \geq \sqrt{\delta} \cdot B^{\frac{1}{2}(1-\varepsilon')}, \qquad (1)$$

with our choice of $g := \sqrt{\delta}/2n$ above. For the inequality we make the assumption of $B \geq n^{4/\varepsilon'}$, which yields $\frac{B}{4n^2} \geq \frac{B}{4B^{\varepsilon'/2}} = \frac{1}{4}B^{1-\varepsilon'/2} \geq B^{1-\varepsilon'}$. Making this assumption poses no problem, since B can easily be increased (i.e. by "blowing it up" similarly to what we do with S), if this should not hold. Let us consider the first statement in the theorem.

$\delta \geq |S|^{\varepsilon}, \delta \leq |S|$. We blow up S and T again, as in the proof of Theorem 1, but we proceed more carefully. Before adding any points to the huge blob, we know that $\delta \leq |S| \leq c \cdot n^2 \cdot B$ (the first by assumption, the second since $|S| = n \cdot B$ holds in the beginning). Now we increase the blob and thereby δ (since by assumption $\delta \geq |S|^{\varepsilon}$) until we have equality $\delta = c'' \cdot n^2 \cdot B$ for some appropriately chosen constant $c'' \geq c$. We obtain $\delta \leq 2 \cdot c'' \cdot B^{\varepsilon'/2} \cdot B \leq B^{1+\varepsilon'}$ and with (1) thus

$$\frac{y \cdot \sqrt{\delta}}{\delta} \geq \sqrt{\delta} \cdot B^{\frac{1}{2}(1-\varepsilon')} \geq \sqrt{\delta} \cdot \delta^{\frac{1}{2} \cdot \frac{1-\varepsilon'}{1+\varepsilon'}} \geq \delta^{1-\varepsilon'}.$$

This gives the first bound for the approximation ratio.

$\delta \geq |S|$. For the second bound in the theorem we lower bound B in terms of $|S|$. To obtain a good bound we will blow up the extra blob of S and T only slightly: we increase the blob until $|S| = B^{1+\varepsilon'}$, which is enough to ensure $\delta \geq B^{1+\varepsilon'} \geq c \cdot B^{1+\varepsilon'/2} \geq c \cdot n^2 \cdot B$ as needed. Note that we assumed $B \geq n^{4/\varepsilon'}$, as before. We insert $|S| = B^{1+\varepsilon'}$ into (1):

$$\frac{y \cdot \sqrt{\delta}}{\delta} \geq \sqrt{\delta} \cdot |S|^{\frac{1}{2} \cdot \frac{1-\varepsilon'}{1+\varepsilon'}} \geq \sqrt{\delta} \cdot |S|^{\frac{1}{2}-\varepsilon'}.$$

$\delta \geq |S|^{1+\varepsilon}$. For the third statement in the theorem we will again search for a lower bound of B in (1), but now by an expression in L_T. In this case we will not blow up S and T with an extra blob, but instead stick to the original construction. The two "extra" points which were added to ensure expansion$(f^{-1}) \geq \sqrt{\delta}$ can be added at distance 1 from the other blobs in S and at distance g from the other blobs in T. Clearly, if a blob S_i is split apart and partly mapped onto these two points, this again yields the distortion given in (1).

For the maximum ratio of distances we have for the set T:

$$L_T \leq \frac{1 + g + n \cdot \left(\frac{B}{\sqrt{B}} + g\right)}{\frac{1}{\sqrt{B}}} = (1 + \sqrt{\delta}/2n)\sqrt{B} + n \cdot B + \sqrt{\delta} \cdot B/2 \leq c' \cdot \sqrt{\delta} \cdot B \cdot n$$

holds, with $g := \sqrt{\delta}/2n$, our assumption $\delta \geq c \cdot n^2 \cdot B$, and an appropriate constant c'. Note that given L_T we need to adjust the minimum size of the input ($|S|$) accordingly.

We replace the assumption made above for B's size by $B \geq n^{\max\{8/\varepsilon', 1/\varepsilon\}}$. The second term in the "max" expression ensures that $c \cdot n^2 \cdot B \leq c \cdot n \cdot B^{\varepsilon} \cdot B \leq (n \cdot B)^{1+\varepsilon} \leq |S|^{1+\varepsilon}$. Since $\delta \geq |S|^{1+\varepsilon}$ holds, we obtain $\delta \geq c \cdot n^2 \cdot B$ as needed. With help of the first term in the "max" expression we obtain for (1): $\frac{y \cdot \sqrt{\delta}}{\delta} \geq \sqrt{\delta} \cdot B^{\frac{1}{2}(1-\varepsilon'/2)}$. We plug in $\frac{L_T}{\sqrt{\delta}} \leq c' \cdot \sqrt{B} \cdot n \leq c' \cdot \sqrt{B} \cdot B^{\varepsilon'/8} \leq B^{\frac{1}{2}(1+\varepsilon'/2)}$:

$$\sqrt{\delta} \cdot B^{\frac{1}{2}(1-\varepsilon'/2)} \geq \sqrt{\delta} \cdot \left(\frac{L_T}{\sqrt{\delta}}\right)^{\frac{1-\varepsilon'/2}{1+\varepsilon'/2}} \geq (L_T)^{\frac{1-\varepsilon'/2}{1+\varepsilon'/2}} \geq L_T^{1-\varepsilon'}.$$

This concludes the proof.   ◇

In connection to the last bound, it is interesting to note that *any* embedding $f : S \to T$ has a distortion of at most $L_T^2 \cdot \delta$, where δ is the optimal distortion, even if (S, d_S) and (T, d_T) are arbitrary metric spaces. A short proof can be found in the full version of the paper.

2.3 Higher Dimensions

For the 3-dimensional problem we can show the following explicit dependence of the inapproximability ratio on the distortion.

Theorem 3. *For any fixed $\delta > 1$ it is* NP-*hard to distinguish whether two given 3-dimensional point sets S and T have distortion $\leq \delta$ or $\geq \sqrt{9\delta^2 - 8}$.*

Proof. By a more detailed analysis of a slight modification of the construction in [2] which we omit here.   ◇

Notice that, in view of the previous theorem, this result is relevant when the distortion is small. Finally, when the dimension is unbounded (that is, for general finite metrics), the reduction of the previous subsection can be adapted to establish that the distortion is even harder to approximate:

Theorem 4. *For any fixed $\delta > 1$ it is* NP-*hard to distinguish whether two given finite metrics S and T have distortion $\leq \delta$ or $\geq \delta^2$.*

Proof. (Sketch.) Repeat the construction with all points in the same blob having distance of 1 from each other, while points belonging to different blobs are at distance δ. This holds for both S and T. If a 3-partition exists, then distances of δ are shrunk to 1, but no distances are dilated, and so the distortion is δ. If a 3-partition does not exist, then certain distances are both shrunk and dilated by δ, and so the distortion is δ^2.   ◇

3 Additive Distortion

For the 1-dimensional additive distortion problem we will show that the simplest possible strategy already yields a 2-approximation. The SWEEP-OR-FLIP algorithm either maps the points in $S := \{s_1, \ldots, s_m\} \subseteq \mathbb{R}$ from left to right onto $T := \{t_1, \ldots, t_m\} \subseteq \mathbb{R}$, or flips the point set and maps them from right to left. In other words, we check the bijections $f(s_i) = t_i$ and $f(s_i) = t_{m-i+1}$ and keep

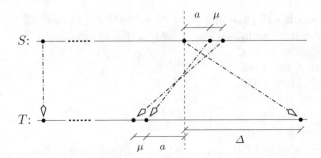

Fig. 2. An example showing that the SWEEP-OR-FLIP strategy does not necessarily yield an optimal distortion. The embedding has additive distortion Δ (note that $2(a + \mu) \leq \Delta$). The "left-to-right" embedding has a larger distortion of $a + \Delta - \mu$. Clearly the other direction has a larger distortion as well.

the better one. It is easy to see (Figure 2) that this is not optimal. In fact, by choosing $a + \mu = \Delta/2$ in the figure we get a gap of $\frac{a + \Delta - \mu}{\Delta} = 1.5 - 2\mu/\Delta$ when comparing the optimal to the SWEEP-OR-FLIP embedding, for arbitrarily small $\mu > 0$.

For the setting where we are given an arbitrary metric space (S, d_S) and want to embed onto points $T := \{t_1, \ldots, t_m\} \subseteq \mathbb{R}$ we will present a straightforward $5 + \varepsilon$-approximation algorithm.

3.1   A 2-Approximation

Before proving that SWEEP-OR-FLIP is a 2-approximation, we give two definitions:

Crossing Points. Consider two points $x, y \in S$ with $x < y$ and for which their counterparts are $f(x) > f(y)$. We say x and y *cross* in the mapping f.

Relative Movement. For fixed S and T, define the *relative movement* of the i-th point to be $\mu_i := t_i - s_i$.

Theorem 5. *The* SWEEP-OR-FLIP *algorithm yields an additive distortion at most* $2 \cdot \Delta$, *where* Δ *is the optimum additive distortion.*

Proof. The proof idea is to fix an optimal embedding f^* and to consider different cases for the relative movements μ_i. We then either show that a distortion of $2 \cdot \Delta$ can be obtained by a "left-to-right" or a "right-to-left" embedding, or arrive at a contradiction by showing that f^* has distortion $> \Delta$. With each of the four steps of the following case analysis we narrow down the situations we need to consider, in terms of the actual relative movements, and also in terms of the mapping of the first and last point in f^*. We start with the latter:

$f^*(s_1) > f^*(s_m)$, **i.e.** s_1 **and** s_m **cross:** If this is the case, we can completely flip T, e.g. by negating all elements of the set without affecting the performance of SWEEP-OR-FLIP. Thus we can assume $f^*(s_1) < f^*(s_m)$.

$|\mu_i| \leq \Delta$, for all $i \in \{1, \ldots, m\}$: Clearly, if we embed left-to-right $f(s_i) = t_i$, the obtained additive distortion is bounded by $2 \cdot \Delta$. To see this take any two points $s_i, s_j \in S$ and note that due to the bounded movement the distance of t_i and t_j can differ by at most $2 \cdot \Delta$ from the distance of s_i and s_j.

$\forall i, j \in \{1, \ldots, m\} : |\mu_i - \mu_j| \leq 2 \cdot \Delta$: Let μ_i be the largest relative movement, and by the previous case assume $\mu_i > \Delta$. Then translate all points in T to the left, until $\mu_i = \Delta$. For all $j \neq i$ and the new relative movements we still have $|\mu_i - \mu_j| \leq 2 \cdot \Delta$. Thus we have $\mu_j \geq -\Delta$ and $\mu_j \leq \mu_i = \Delta$, the former since $\mu_i = \Delta$ and the latter since μ_i is the largest relative movement. In other words, we modified the instance such that the previous case holds. If in the beginning the smallest relative movement is less than $-\Delta$, we proceed analogously translating all points in T to the right.

$\exists i, j \in \{1, \ldots, m\} : |\mu_i - \mu_j| > 2 \cdot \Delta$: Assume $i < j$ and $\mu_i > \mu_j$, otherwise exchange the roles of S and T (the relative movements are negated). Translate all points in T in order to have $\mu_i = \Delta$ and $\mu_j < -\Delta$. Due to a simple counting argument, there must be $k \leq i$ with $f^*(s_k) \geq t_i$ and thus $f^*(s_k) - s_k \geq \mu_i$. Analogously there must be a $l > j$ with $f^*(s_l) \leq t_j$ and thus $f^*(s_l) - s_l \leq \mu_j$. We distinguish the following cases:

s_k and s_l do not cross: See the top picture of Figure 3 for an example. We have $d(s_k, s_l) = s_l - s_k \geq s_j - s_i$ and $d(f^*(s_k), f^*(s_l)) \leq t_j - t_i$. This gives a contradiction: $d(s_k, s_l) - d(f^*(s_k), f^*(s_l)) \geq \mu_i - \mu_j > 2 \cdot \Delta$. See the top picture of Figure 3 for an example.

s_k and s_l cross while s_m and s_k do not: See the middle part of Figure 3. Since f^* projects s_k by at least $\mu_i = \Delta$ to the right, we must have $f^*(s_m) \geq s_m$, otherwise the distance $d(f^*(s_k), f^*(s_m))$ would be less than $d(s_k, s_m) - \Delta$. Similarly since f^* projects l by at least $\mu_j < -\Delta$ to the left, we must have $f^*(s_m) < s_m$, which gives the contradiction.

s_k and s_l cross while s_1 and s_l do not: Same as the previous case.

$k > 1$ and $l < m$, s_k and s_l cross, s_1 crosses s_l, s_m crosses s_k: See the bottom part of Figure 3. Let us start by making sure that this is the only case left. Due to the very first case we know that s_1 and s_m do not cross. Since s_k and s_l do cross, either $k > 1$ or $l < m$ must hold. Assume the former, then since s_1 crosses s_l (which it must due to the previous case) we also have the latter (and again due to the last but one case, s_m must cross s_k).

Now we consider the distance a, b, c, d, e, and f as given in Figure 3, bottom. Since we have an additive distortion of Δ,

$$d + e + f \leq b + \Delta \qquad (2)$$

must hold. Similarly we have $f \geq b + c - \Delta$, $d \geq a + b - \Delta$, and $e \geq a + b + c - \Delta$, which together gives

$$d + e + f \geq 2a + 3b + 2c - 3\Delta.$$

Subtracting (2) we obtain the contradiction

$$0 \geq 2s + 2b + 2c - 4\Delta \geq 2b - 4\Delta > 0.$$

The last inequality holds since $b \geq \mu_i - \mu_j > 2 \cdot \Delta$. We conclude that there cannot be $i, j \in \{1, \dots, m\}$ with $|\mu_i - \mu_j| > 2 \cdot \Delta$.

This gives the stated result.

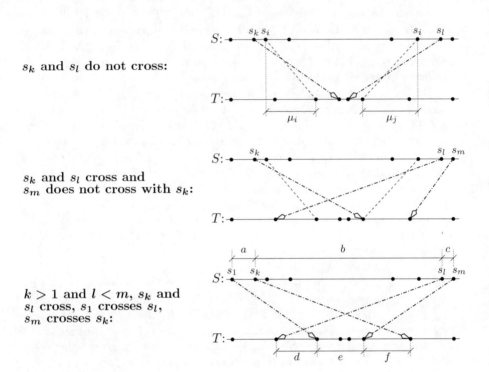

s_k and s_l do not cross:

s_k and s_l cross and s_m does not cross with s_k:

$k > 1$ and $l < m$, s_k and s_l cross, s_1 crosses s_l, s_m crosses s_k:

Fig. 3. Three examples for the cases concerning that $i, j \in \{1, \dots, m\}$ exist such that $\mu_i = \Delta$ and $\mu_j < -\Delta$.

3.2   Embedding an Arbitrary Metric Space to 1D

We are given an arbitrary finite metric space (S, d_S) and $T \subseteq \mathbb{R}$. The algorithm INTERVALS below finds a mapping of the points in S to the points in T within a factor of 5 of the optimum additive distortion. For ease of exposition we start by assuming that we know the optimum distortion Δ. Below we note why the algorithm also works for the case where the distortion is not given. We also assume that we know the point $x \in S$ mapped to t_1 – in fact, we iterate over all $x \in S$.

Feasible Intervals. For $y \in S \setminus \{x\}$ we then define its feasible interval as: $I_y := [t_1 + d_S(x, y) - \Delta, t_1 + d_S(x, y) + \Delta]$. The additive distortion for the pair x, y is $\leq \Delta$ if and only if $f(y) \in I_y$.

ALGORITHM: INTERVALS

---

1. Given $\Delta > 0$ and $x \in S$, compute the feasible intervals I_y for $y \neq x$, and map the remaining nodes of S as follows:
2. Process the feasible intervals by increasing left boundary. For each I_y map y greedily to the leftmost point z in I_y that has not yet been mapped to.

Theorem 6. *The* INTERVALS *algorithm yields an additive distortion of* $5 \cdot \Delta$, *where* Δ *is the optimum additive distortion.*

Proof. Consider the mapping f^* that achieves Δ, and assume $f^*(x) = t_1$. Since there is a bijection that maps each $y \neq x$ to a point $z \in I_y$ (f^* is an example of such a bijection), and since all intervals have the same length, it is clear that the algorithm will find such a bijection, call it f. f can increase the distance between any point pair by at most $4 \cdot \Delta$, since the intervals have a width of $2 \cdot \Delta$. Therefore, the additive distortion of f is within a factor of 5 of the additive distortion. $\diamond$

If INTERVALS succeeds in finding a mapping, it will simply do a left to right mapping of the remaining points $S \setminus \{x\}$ in step 2. Therefore, by iterating over all $x \in S$ and checking for each x the left to right mapping of the rest, we find the same mapping as INTERVALS without knowing Δ in advance.

4   Open Problems

We have made significant progress towards understanding the complexity of computing the distortion of bijections between point sets. But many open problems remain:

- What is the complexity of the distortion problem on the line for large constant values of distortion? The dynamic programming approach seems to exhaust itself after $3 + 2\sqrt{2}$, yet NP-completeness also seems very difficult.
- In view of our results, it seems that, for large distortions, the right quantity to approximate is not δ but $\log \delta$. By Theorem 2 we know that it cannot be approximated by a factor better than 2. Is a constant factor possible? Or is there a generalization of our proof (by some kind of hierarchical 3-partition problem) that shows impossibility?
- In connection to the last open problem, we may want to define the following relaxation of the distortion problem: We are given, besides S, T, and δ, an $\varepsilon > 0$, and we are asked whether there is a bijection between *all but an ε fraction* of S and T such that the distortion of this partial map is δ or less. We conjecture that for any ε there is a polynomial algorithm that approximates $\log \delta$ by a factor of 2.
- Are there better approximation algorithms for the 1-dimensional additive distortion problem? And can one prove the 1-dimensional problem to be NP-complete?

References

1. Kenyon, C., Rabani, Y., Sinclair, A.: Low distortion maps between point sets. In: Proceedings of the 36th STOC. (2004) 272–280
2. Papadimitriou, C., Safra, S.: The complexity of low-distortion embeddings between point sets. In: Proceedings of the 16th SODA. (2005) 112–118
3. Indyk, P.: Algorithmic applications of low-distortion geometric embeddings. In: Tutorial at the 42nd FOCS. (2001) 10–33
4. Linial, N.: Finite metric spaces – combinatorics, geometry and algorithms. In: Proceedings of the International Congress of Mathematicians III, Beijing (2002) 573–586
5. Matousek, J.: Lectures on Discrete Geometry. Springer-Verlag, Graduate Texts in Mathematics, Vol. 212 (2002)
6. Web-page of the working group on multi-dimensional scaling. http://dimacs.rutgers.edu/SpecialYears/2001_Data/Algorithms /AlgorithmsMS.htm (2005)
7. Linial, N., London, E., Rabinovich, Y.: The geometry of graphs and some of its algorithmic applications. Combinatorica 15 (1995) 215–245
8. Bourgain, J.: On lipschitz embedding of finite metric spaces into hilbert space. Isreal Journal of Mathematics 52 (1985) 46–52
9. Kleinberg, J., Slivkins, A., Wexler, T.: Triangulation and embedding using small sets of beacons. In: Proceedings of the 45th FOCS. (2004) 444–453
10. Slivkins, A.: Distributed approaches to triangulation and embedding. In: Proceedings of the 16th SODA. (2005) 640–649
11. Bădoiu, M., Dhamdhere, K., Gupta, A., Rabinovich, Y., Räcke, H., Ravi, R., Sidiropoulos, A.: Approximation algorithms for low-distortion embeddings into low-dimensional spaces. In: Proceedings of the 16th SODA. (2005) 119–128
12. Håstad, J., Ivansson, L., Lagergren, J.: Fitting points on the real line and its application to RH mapping. Lecture Notes in Computer Science 1461 (1998) 465–467
13. Bădoiu, M.: Approximation algorithm for embedding metrics into a two-dimensional space. In: Proceedings of the 14th SODA. (2003) 434–443
14. Bădoiu, M., Indyk, P., Rabinovich, Y.: Approximate algorithms for embedding metrics into lowdimensional spaces. Manuscript (2003)
15. Feige, U.: Approximating the bandwidth via volume respecting embeddings. Journal of Computer and System Sciences 60 (2000) 510–539
16. Akutsu, T., Kanaya, K., Ohyama, A., Fujiyama, A.: Point matching under non-uniform distortions. Discrete Applied Mathematics 127 (2003) 5–21

Approximating the Best-Fit Tree
Under L_p Norms

Boulos Harb*, Sampath Kannan**, and Andrew McGregor***

Department of Computer and Information Science, University of Pennsylvania
Philadelphia, PA 19104, USA
{boulos,andrewm,kannan}@cis.upenn.edu

Abstract. We consider the problem of fitting an $n \times n$ distance matrix M by a tree metric T. We give a factor $O(\min\{n^{1/p}, (k \log n)^{1/p}\})$ approximation algorithm for finding the closest ultrametric T under the L_p norm, i.e. T minimizes $\|T, M\|_p$. Here, k is the number of distinct distances in M. Combined with the results of [1], our algorithms imply the same factor approximation for finding the closest *tree metric* under the same norm. In [1], Agarwala *et al.* present the first approximation algorithm for this problem under L_∞. Ma *et al.* [2] present approximation algorithms under the L_p norm when the original distances are not allowed to contract and the output is an ultrametric. This paper presents the first algorithms with performance guarantees under L_p ($p < \infty$) in the general setting.

We also consider the problem of finding an ultrametric T that minimizes L_{relative}: the sum of the factors by which each input distance is stretched. For the latter problem, we give a factor $O(\log^2 n)$ approximation.

1 Introduction

An *evolutionary tree* for a species set $\mathcal{S}$ is a rooted tree in which the leaves represent the species in $\mathcal{S}$, and the internal nodes represent ancestors. The goal of reconstructing the evolutionary tree is of fundamental scientific importance. Given the increasing availability of molecular sequence data for a diverse set of organisms and our understanding of evolution as a stochastic process, the natural formulation of the tree reconstruction problem is as a maximum likelihood problem – estimate parameters of the evolutionary process that are most likely to have generated the observed sequence data. Here, the parameters include not only rates of mutation on each branch of the tree, but also the topology of the tree itself. It is assumed (although this assumption is not always easy to meet) that the sequences observed at the leaves have been multiply aligned so that each position in a sequence has corresponding positions in the other sequences. It is also assumed for tractability, that each position evolves according to an independent identically distributed process. Even with these assumptions, estimating the most likely tree is a computationally difficult problem.

* This work was supported by NIH Training Grant T32HG00 46.
** This work was supported by NSF CCR98-20885 and NSF CCR01-05337.
*** This work was supported by NSF ITR 0205456.

Recently, *approximately* most likely trees have been found for simple stochastic processes using *distance-based methods* as subroutines [3, 4].

For a distance-based method the input is an $n \times n$ distance matrix M where $M[i, j]$ is the observed distance between species i and j. Given such a matrix, the objective is to find an edge-weighted tree T with leaves labeled 1 through n which minimizes the L_p distance from M where various choices of p correspond to various norms. The tree T is said to *fit* M. When it is possible to define T so that $\|T, M\|_p = 0$, then the distance matrix is said to be *additive*. An $O(n^2)$ time algorithm for reconstructing trees from additive distances was given by Waterman *et al.* [5], who proved in addition that at most one tree can exist. However, real data is rarely additive and we need to solve the norm minimization problem above to find the best tree. Day [6] showed that the problem is NP-hard for $p = 1, 2$.

For the case of $p = \infty$, referred to as the L_∞ norm, [7] showed how the optimal ultrametric tree could be found efficiently and [1] showed how this could be used to find a tree T (not necessarily ultrametric) such that $\|T, M\|_p \leq 3\|T_{\mathrm{OPT}}, M\|_p$ where T_{OPT} is the optimal tree. The algorithm of [1] is the one that is used in [3] and [4] for approximate maximum likelihood reconstruction.

In this paper we explore approximation algorithms under other norms such as L_1 and L_2. We also consider a variant, L_{relative}, of the best-fit objective mentioned above where we seek to minimize the sum of the factors by which each input distance is stretched. The study of L_1 and L_2 norms is motivated by the fact that these are often better measures of fit than L_∞ and the idea that using these methods as subroutines may yield better maximum likelihood algorithms.

1.1   Our Results

We prove the following results:

- We can find an ultrametric tree whose L_p-error is within a factor of $O(\min\{n^{1/p}, (k \log n)^{1/p}\})$ of the optimum, where k is the number of distinct distances in the input matrix.
- We can find an ultrametric tree T whose L_{relative}-error is within a factor of $O(\log^2 n)$ of the optimum.

Our algorithms also solve the problem of finding *non-contracting* ultrametrics, i.e. when $T[i, j]$ is required to be at least $M[i, j]$ for all i, j. More generally, we can require that each output distance is lower bounded by some arbitrary positive value. This generalization allows us to also find *additive* metrics whose L_p-error is within a factor of $O(\min\{n^{1/p}, (k \log n)^{1/p}\})$ of the optimum by appealing to work in [1].

1.2   Related Work

Aside from the aforementioned L_∞ result given in [1], Ma *et al.* [2] present an $O(n^{1/p})$ approximation algorithm for finding non-contracting ultrametrics under

$L_{p<\infty}$. Prior to our results, however, no algorithms with provable approximation guarantees existed for fitting distances by additive metrics under $L_{p<\infty}$ in the general setting.

Some of our results rely on the recent approximation algorithms for the problem of correlation clustering and related problems [8–11]. One of our algorithms can be viewed as performing a hierarchical version of correlation clustering.

Finally, we should mention some recent work that address special cases of our problem. In [12] an algorithm is given that finds a line-embedding of a metric whose L_1-error is $O(\log n)$ away from optimal. If the embedding is further restricted to be a non-contracting line-embedding, then [13] presents an algorithm whose approximation factor is constant.

2   Preliminaries

An *ultrametric* T on a set $[n]$ is a metric that satisfies the following *three-point condition*:

$$\forall x, y, z \in [n] \quad T[x, y] \leq \max\{T[x, z], T[z, y]\} \ .$$

That is, in an ultrametric, triangles are isosceles with the equal sides being longest. An ultrametric is a special kind of tree metric where the distance from the root to all points in $[n]$ (the leaves) is the same. Recall that a *tree metric* (equivalently an *additive* metric) A on $[n]$ is a metric that satisfies the *four-point condition*:

$$\forall w, x, y, z \in [n] \quad A[w, x] + A[y, z] \leq \max\{A[w, y] + A[x, z], A[w, z] + A[x, y]\} \ .$$

Given an $n \times n$ distance matrix M where $M[i, j]$ is the observed distance between objects i and j, our initial objective is to find an edge-weighted ultrametric T with leaves labeled 1 through n which minimizes the L_p distance from M, i.e. T minimizes

$$\|T, M\|_p = \sqrt[p]{\sum_{i,j} |T[i, j] - M[i, j]|^p} \ . \tag{1}$$

We will also look at finding an edge-weighted ultrametric T which minimizes the average stretch of the distances in M, i.e. T minimizes

$$\|T, M\|_{\text{relative}} = \sum_{i,j} \max\left\{ \frac{T[i, j]}{M[i, j]}, \frac{M[i, j]}{T[i, j]} \right\} \tag{2}$$

The entry $T[i, j]$ is the distance between the leaves i and j, which is defined to be the sum of the edge weights on the path between i and j in T. We will also refer to the *splitting distance* of an internal node v of T as the distance between two leaves whose least common ancestor is v. Because T is an ultrametric, the splitting distance of v is simply twice the height of v.

We will assume that the input distances in M are non-negative integers such that

-   $M[x, y] = M[y, x]$; and,
-   $M[x, y] = 0 \iff x = y$.

That is, we will not assume that the distances in M satisfy the triangle inequality. We denote the *distinct* distances in M by,

$$d_k > d_{k-1} > ... > d_2 > d_1 \; .$$

Relationship to Correlation Clustering. The problem of finding an optimal ultrametric T minimizing $\|T, M\|_1$ is closely related to the problem of correlation clustering introduced in [10]. We are interested in the minimization version of correlation clustering which is defined as follows: given a graph G whose edges are labeled "+" (similar) or "–" (dissimilar), cluster the vertices so as to minimize the number of pairs incorrectly classified with respect to the input labeling. That is, minimize the number of "–" edges within clusters plus the number of "+" edges between clusters. We will simply refer to this problem as *correlation clustering*. Note that the number of clusters is not specified in the input.

In fact, when G is complete, correlation clustering is equivalent to the problem of finding an optimal ultrametric under the L_1 norm when the input distances in M are restricted to 1 and 2. An edge (i, j) in the graph labeled "+" (resp. "–") is equivalent to the entry $M[i, j]$ being 1 (resp. 2). It is clear that an optimal ultrametric is an optimal clustering, and vice versa. Hence, the APX-hardness of finding an optimal ultrametric under the L_1 norm follows directly from [11, Theorem 11].

In [11], Charikar, Guruswami and Wirth give a factor $O(\log n)$ approximation to correlation clustering on general *weighted* graphs using linear programming. In an instance of correlation clustering that is weighted, each edge e has a weight w_e which can be either positive or negative. The objective is then to minimize

$$\sum_{e:w_e>0} (|w_e| \text{ if } e \text{ is split}) + \sum_{e:w_e<0} (|w_e| \text{ if } e \text{ is not split}) \; .$$

The bound for the LP relaxation is established via an application of the region growing procedure of Garg, Vazirani and Yannakakis [14]. We will state their theorem below for reference as our algorithm in section 3.1 uses their algorithm as a sub-procedure.

Theorem 1 ([11, Theorem 1]). *There is a polynomial time algorithm that achieves an $O(\log n)$ approximation for correlation clustering on general weighted graphs.*

3   Main Results

Both our algorithms take as input a set of splitting distances we call S that depends on the error norm. The distances in the constructed ultrametrics will be a subset of the given set S. The following lemma quantifies the affect of restricting the output distances to certain sets.

Lemma 1. *(a) There exists an ultrametric T with $T[i,j] \in \{d_1, d_2, \ldots, d_k\}$ for all i, j that is optimal under the L_1 norm.*

(b) There exists an ultrametric T with $T[i,j] \in \{d_1, d_2, \ldots, d_k\}$ for all i, j such that

$$\|T, M\|_p \leq 2\|T_{\mathrm{OPT}}, M\|_p \ ,$$

for $p \geq 2$.

(c) Assuming $d_k = O(poly(n))$, there exists an ultrametric T that uses $O(\log_{1+\epsilon} n)$ distances such that

$$\|T, M\|_{\mathrm{relative}} \ \leq \ (1+\epsilon)\|T_{\mathrm{OPT}}, M\|_{\mathrm{relative}} \ ,$$

where $\epsilon > 0$.

Proof. (a) Say an internal node v is *undesirable* if its distance $h(v)$ to any of its leaves satisfies $2h(v) \notin \{d_1, d_2, \ldots, d_k\}$. Suppose T_{OPT} is an optimal ultrametric with undesirable nodes. We will modify T_{OPT} so that it has one less undesirable node. Let v be the lowest undesirable node in T_{OPT} and let $d = 2h(v) \in (d_\ell, d_{\ell+1})$ for some $1 \leq \ell \leq k-1$. Define the following two multisets:

$$D_\ell = \{M[a,b] : a, b \text{ are in different subtrees of } v \text{ and } M[a,b] \leq d_\ell\} \ ,$$
$$D_{\ell+1} = \{M[a,b] : a, b \text{ are in different subtrees of } v \text{ and } M[a,b] \geq d_{\ell+1}\}.$$

Then the contribution of the distances in $D_\ell \cup D_{\ell+1}$ to $\|T_{\mathrm{OPT}}, M\|_1$ is

$$\sum_{\alpha \in D_\ell} (d - \alpha) + \sum_{\beta \in D_{\ell+1}} (\beta - d) \ .$$

The expression above is linear in d. If its slope ≥ 0 then set $h(v) = d_\ell/2$, and if the slope < 0 then set $h(v) = \min\{d_{\ell+1}/2, h(v')\}$ where v' is the parent of v. Such a change can only improve the cost of the tree.

(b) For $p \geq 2$, let T_{OPT} be an optimal ultrametric with undesirable nodes. We will transform T_{OPT} to an ultrametric T with no undesirable nodes such that $\|T, M\|_p \leq 2\|T_{\mathrm{OPT}}, M\|_p$. Let,

$$\|T_{\mathrm{OPT}}, M\|_p^p = \sum_u g_u(2h(u)) \ ,$$

where the sum is over the internal nodes of T_{OPT} and $g_u(x)$ is the cost of setting the splitting distance of node u to x. Again, let v be the lowest undesirable node and define D_ℓ and $D_{\ell+1}$ as above. Fix $d = 2h(v) \in (d_\ell, d_{\ell+1})$. We claim that $\min\{g_v(d_\ell), g_v(d_{\ell+1})\} \leq 2^p g_v(d)$.

If $d \leq (d_\ell + d_{\ell+1})/2$, then we can set $h(v) = d_\ell/2$ since for all $\alpha \in D_\ell$, $d_\ell - \alpha \leq d - \alpha$ and for all $\beta \in D_{\ell+1}$, $\beta - d_\ell \leq 2(\beta - d)$. Otherwise, we set $h(v) = d_{\ell+1}/2$. We are assuming w.l.o.g. that v has no parent in the region $(d_\ell, d_{\ell+1})$ since if such a parent v' exists, $h(v')$ will also be set to $d_{\ell+1}/2$.

(c) Let $D(T_{\text{OPT}})$ be the set of distances in an optimal ultrametric that minimizes $\|T, M\|_{\text{relative}}$. Group the distances in $D(T_{\text{OPT}})$ geometrically, i.e. for some $\epsilon > 0$, group the distances into the following buckets:

$$[1, 1+\epsilon], \ \left(1+\epsilon, (1+\epsilon)^2\right], \ldots, \ \left((1+\epsilon)^{s-1}, (1+\epsilon)^s\right] \ .$$

Let t be the largest distance in $D(T_{\text{OPT}})$. Clearly, $t \leq d_k = O(\text{poly}(n))$. Hence, the number of buckets $s = \log_{1+\epsilon} t = O(\log_{1+\epsilon} n)$. Now consider an ultrametric T' that sets $T'[i,j] = (1+\epsilon)^\ell$ if the optimal $T[i,j] \in ((1+\epsilon)^{\ell-1}, (1+\epsilon)^\ell]$.

$$\|T', M\|_{\text{relative}} = \sum_{i,j} \max\left\{\frac{T'[i,j]}{M[i,j]}, \frac{M[i,j]}{T'[i,j]}\right\}$$

$$\leq \sum_{i,j} \max\left\{\frac{(1+\epsilon)T[i,j]}{M[i,j]}, \frac{M[i,j]}{T[i,j]}\right\}$$

$$\leq (1+\epsilon)\|T_{\text{OPT}}, M\|_{\text{relative}} \ .$$

For ease of notation, we adopt the following conventions. Let $G = (V, E)$ be the graph representing M in the natural way. For an edge $e = (i,j)$ denote its input distance $M[i,j]$ by m_e and its output distance $T[i,j]$ by t_e. As described in section 2, w_e will code for the label and the weight $|w_e|$ on the edge passed to the correlation clustering algorithm. The lower bound on e, λ_e, is the minimum value e can contract, i.e. $t_e \geq \lambda_e$.

Supplying our algorithm with an edge lower bounds matrix Λ allows us, for example, to solve non-contracting versions of the objective functions we seek to minimize where for all e, $t_e \geq m_e$ by simply setting $\Lambda = M$. We will also use these lower bounds in section 4 when constructing general additive metrics under L_p norms.

In the following two subsections we present algorithms for our problem. The first algorithm is suitable if the number of distinct distances, k, in M is small. Otherwise, the second algorithm is more suitable.

3.1   Algorithm 1

Our algorithm takes as input a set of *splitting distances* S. Each distance in the constructed tree will belong to this set. Let $|S| = \kappa$ and number the splitting distances in ascending order $s_1 < s_2 < \ldots < s_\kappa$. The algorithm considers the splitting distances in descending order, and when considering s_l it may set some distances $T[i,j] = s_l$. If a distance of the tree is not set at this point, it will later be set to $\leq s_{l-1}$. The decision of which distances to set to s_l and which distances to set to $\leq s_{l-1}$ will be made using correlation clustering. See Fig. 1 for the description of the algorithm.

Theorem 2. *Algorithm 1 can be used to find an ultrametric T such that any one of the following holds:*

Algorithm *Correlation-Clustering-Splitting*(G, S, Λ)
($*$ Uses correlation clustering to decide how to split $*$)
1.   Let all edges be "unset"
2.   **for** $l = \kappa$ to 1:
3.   **do** Do correlation clustering on the graph induced by the unset edges with
                weights:
                    -If $m_e \geq s_l$ and $\lambda_e < s_l$ then,
                    $w_e = -(f(m_e, s_{l-1}) - f(m_e, s_l))$
                    -If $\lambda_e = s_l$ then $w_e = -\infty$
                    -If $m_e = s_i < s_l$ then $w_e = f(s_i, s_l)$
4.   **for** For each unset edge e split between different clusters:
5.   **do** $t_e \leftarrow s_l$ and mark e as "set"

Fig. 1. Algorithm 1 (The function f is defined in Thm. 2)

1. $\|T, M\|_p \leq O((k \log n)^{1/p}) \|T_{\text{OPT}}, M\|_p$ *if* $S = \{d_1, \ldots, d_k\}$ *and* $f(m_e, t_e) = |m_e - t_e|^p$.
2. $\|T, M\|_{\text{relative}} \leq O(\log^2 n) \|T_{\text{OPT}}, M\|_{\text{relative}}$ *if* $S = \{(1 + \epsilon)^i : 0 \leq i \leq \log_{1+\epsilon} d_k\}$ *and* $f(m_e, t_e) = \max\{\frac{t_e}{m_e}, \frac{m_e}{t_e}\}$.

Proof. Our algorithm produces an ultrametric T where the splitting distance of each node is restricted to be from the set S, i.e. $t_e \in S$ for all e. The proof below shows that the algorithm gives a $O(|S| \log n)$–approximation to $\sum_e f(m_e, t'_e)$ where T' is the optimal ultrametric satisfying $t'_e \in S$ for all e. The results in the theorem will then follow by appealing to Lemma 1.

Consider the correlation clustering instance performed in iteration l of the algorithm. Let $\text{cost}_{\text{OPT}}(l)$ be the optimal value for this instance and let $\text{cost}(l)$ be the cost of our solution.

Claim 1: $\sum_{1 \leq l \leq \kappa} \text{cost}(l) = \sum_e f(m_e, t_e)$.
Consider each edge e in turn. Let $t_e = s_l$. If $s_l > m_e$, then in the lth iteration we pay $f(m_e, s_l)$ for this edge. If $s_l < m_e = s_{l'}$, then in each iteration i, $l' \geq i > l$, we pay $f(s_{l'}, s_{i-1}) - f(s_{l'}, s_i)$; hence, in total we pay $f(s_{l'}, s_l) = f(m_e, t_e)$.

Claim 2: $\text{cost}_{\text{OPT}}(l) \leq \sum_e f(m_e, t'_e)$
Consider the following solution to the correlation clustering problem at iteration l induced by T': for all unset edges e if $t'_e \geq s_l$ we split e and if $t'_e < s_l$ we don't split e. We claim that the cost of this solution for the correlation clustering problem is less that $\sum_e f(m_e, t'_e)$. Consider each edge e in turn.

- $t'_e < s_l$ and $m_e < s_l$: Not splitting this edge contributes nothing to the correlation clustering objective.
- $t'_e \geq s_l$ and $m_e < s_l$: Splitting this edge contributes $f(s_l, m_e)$ to the correlation clustering objective but contributes $f(t'_e, m_e) \geq f(s_l, m_e)$ to $\sum_e f(m_e, t'_e)$.
- $t'_e < s_l$ and $m_e \geq s_l$: Not splitting this edge contributes $f(m_e, s_{l-1}) - f(m_e, s_l)$ to the correlation clustering objective but contributes $f(m_e, t'_e) \geq f(m_e, s_{l-1}) \geq f(m_e, s_{l-1}) - f(m_e, s_l)$ to $\sum_e f(m_e, t'_e)$.
- $t'_e \geq s_l$ and $m_e \geq s_l$: Splitting this edge contributes nothing to the correlation clustering objective.

Algorithm *Min-Cut-Splitting*(G, S, s_{l^*}, Λ)
(∗ Uses min cuts to work out splits ∗)
1. $l \leftarrow l^* + 1$
2.   Min-Split-Cost$\leftarrow \infty$
3.   **repeat**
4. $l \leftarrow l - 1$
5.   Push-Down-Cost $\leftarrow \sum_e (\max\{0, m_e - s_l\})^p - (\max\{0, m_e - s_{l^*}\})^p$
6.   **if** there exists an edge $e = (s, t)$ such that $\lambda_e = s_l$
7.   **then** Find min-(s, t) cut C in G with edge weights
                $w_e = (\max\{0, s_l - m_e\})^p$
8.   **else** Find min-cut C in G with edge weights
                $w_e = (\max\{0, s_l - m_e\})^p$
9.   Cut-Cost $\leftarrow$ the cost of the cut
10.   **if** Cut-Cost+Push-Down-Cost $\leq$ Min-Split-Cost
11.   **then** Best-Cut$\leftarrow C$
12.   Best-Splitting-Point$\leftarrow s_l$
13.   Min-Split-Cost $\leftarrow$ Cut-Cost+Push-Down-Cost
14.   **until** $l = 0$ or there exists an edge e with $\lambda_e = s_l$
15.   **for** all edges e in Best-Cut:
16.   **do** $t_e \leftarrow$ Best-Splitting-Point
17.   **for** each connected component of $G' \in (V, E \setminus$ Best-Cut$)$:
18.   **do** *Min-Cut-Splitting*$(G', S,$Best-Splitting-Point$, \Lambda)$

Fig. 2. Algorithm 2

Summing over all edges, the contributions to both objective functions gives the second claim.

Combining the above claims with Thm. 1, the tree we construct has the following property,

$$\sum_e f(t_e, m_e) = \sum_{1 \leq l \leq \kappa} \text{cost}(l) \leq O(\kappa \log n) \sum_e f(m_e, t'_e) \ .$$

The theorem follows.

3.2   Algorithm 2

Our second algorithm also takes as input a set of *splitting distances* S and, as before, each distance in the constructed tree belongs to this set. However while the approximation guarantee of the first algorithm depended on $|S|$, the approximation guarantee of the second algorithm depends only on n. At each step the first algorithm decided whether or not to place internal nodes at height s_l, and, if it did, how to partition the nodes below. In our second algorithm, at each step we instead decide the height at which we should place the next internal node and its partition. See Fig. 2 for the description of the algorithm. The first call to the algorithm sets $s_{l^*} = s_\kappa$.

Theorem 3. *Algorithm 2 can be used to find an ultrametric T such that any one of the following holds:*

1. $\|T, M\|_1 \leq n\|T_{\text{OPT}}, M\|_1$ if $S = \{d_1, \ldots, d_k\}$.
2. For $p \geq 2$, $\|T, M\|_p \leq 2n^{1/p}\|T_{\text{OPT}}, M\|_p$ if $S = \{d_1, \ldots, d_k\}$.

Proof. Our algorithm produces a ultrametric T where the splitting distance of each node is restricted to be from the set S, i.e. $t_e \in S$ for all e. The proof below shows that the algorithm gives an n–approximation to $\|T', M\|_p^p$ where T' is the optimal ultrametric satisfying $t'_e \in S$ for all e. The results in the theorem will then follow by appealing to Lemma 1.

Claim 1: The sum of Min-Split-Cost over all recursive calls of *Min-Cut-Splitting* equals $\|T, M\|_p^p$.

  Consider an edge $e = (i, j)$ and let v be the lowest common ancestor of i and j in T. If $m_e \leq t_e$ then we paid $(t_e - m_e)^p$ for this edge in the Cut-Cost when splitting at v. If $m_e > t_e$, consider the internal nodes on the path from root to v that have splitting distances $\leq m_e$, $m_e \geq s_{i_1} > s_{i_2} > \ldots s_{i_j} = t_e$. We paid a total of

$$(m_e - s_{i_2})^p + [(m_e - s_{i_3})^p - (m_e - s_{i_2})^p] + \ldots + \left[(m_e - s_{i_j})^p - (m_e - s_{i_{j-1}})^p\right]$$
$$= (m_e - t_e)^p$$

for this edge as Push-Down-Costs.

Claim 2: The Min-Split-Cost of each call is at most $\|T', M\|_p^p$

  Consider a call $Min\text{-}Cut\text{-}Splitting(\widehat{G} = (\widehat{V}, \widehat{E}), \cdot, s_l, \cdot)$. If there exists an $e \in \widehat{E}$ such that $t'_e \geq s_l$, then $\{e \in \widehat{E} : t'_e \geq s_l\}$ contains at least one cut of which let C be the cut of minimum weight. For edges $e \in C$ the cost of cutting e is $(\max\{0, s_l - m_e\})^p \leq |t'_e - m_e|^p$. Hence the Cut-Cost is $\leq \|T', M\|_p^p$. The Push-Down-Cost is 0 since we are cutting in the first iteration of the loop; therefore,

$$\text{Min-Split-Cost} \leq \|T', M\|_p^p .$$

If all $e \in \widehat{E}$ satisfy $t'_e < s_l$ then let the splitting point be $s_{l'} = \max_{e \in \widehat{E}}\{t'_e\}$. The Push-Down-Cost is then at most

$$\sum_{e \in \widehat{E}}(\max\{0, m_e - s_{l'}\})^p \leq \sum_{e \in \widehat{E}:m_e > t'_e} (m_e - t'_e)^p .$$

  Now the set of edges $\{e \in \widehat{E} : t'_e = s_{l'}\}$ contains at least one cut and, as before, choosing the minimum weight cut, call it C, results in the Cut-Cost being equal to $\sum_{e \in C}(\max\{0, s_{l'} - m_e\})^p = \sum_{e \in C:t'_e > m_e}(t'_e - m_e)^p$. Hence,

$$\text{Min-Spilt-Cost} \leq \sum_{e \in \widehat{E}:m_e > t'_e} (m_e - t'_e)^p + \sum_{e \in C:t'_e > m_e}(t'_e - m_e)^p \leq \|T', M\|_p^p .$$

  The number of recursive calls of *Min-Cut-Splitting* is $n - 1$ because each call fixes an internal node of the tree being constructed and the tree has n leaves.

  Therefore, $\|T, M\|_p^p \leq (n - 1)\|T', M\|_p^p$ and the theorem follows. Note that while a slightly better analysis gives that $\|T, M\|_p^p \leq D\|T', M\|_p^p$ where D is the depth of the recursion tree, D can be as much as $n - 1$.

4   Extension to Additive Trees

In this section, we will generalize our results to approximating the input matrix M by general additive metrics under any L_p norm. Our generalization depends on the following theorem from [1],

Theorem 4 (see [1, Theorem 6.2]). *If $\mathcal{G}(M)$ is an algorithm which achieves an α-approximation to the optimal a-restricted ultrametric under the L_p norm, then there is an algorithm $\mathcal{F}(M)$ which achieves a 3α-approximation to the optimal additive metric under the same norm.*

We will show how our algorithms from section 3 can be used to produce a-restricted ultrametrics. We start with the definition of an a-restricted ultrametric from [1].

Definition 1. *For a point a, an ultrametric T^a is a-restricted with respect to a distance matrix M if*
(1) $T^a[a,i] = 2\mu_a$ *for all $i \neq a$,*
(2) $2\mu_a \geq T^a[i,j] \geq 2(\mu_a - \min\{M[a,i], M[a,j]\})$ for all i,j
where $\mu_a = \max_i M[a,i]$.

The definition of an a-restricted ultrametric immediately implies a procedure for approximating the distance $\|T^a_{\text{OPT}}, M\|_p$ between an optimal T^a_{OPT} and M. For a point a, let M^a be the matrix M with row a and column a deleted. And let Λ^a be the $n-1 \times n-1$ edge lower bounds matrix where

$$\Lambda^a[i,j] = 2(\mu_a - \min\{M[a,i], M[a,j]\}) \ ,$$

for all $i,j \in [n] \setminus \{a\}$, $i \neq j$. Given G^a, the graph representing M^a, and Λ^a our algorithms now find an a-restricted ultrametric T^a such that

$$\|T^a, M\|_p \leq O(\min\{n^{1/p}, (k \log n)^{1/p}\}) \|T^a_{\text{OPT}}, M\|_p \ .$$

Appealing to Thm. 4, we have a $O(\min\{n^{1/p}, (k \log n)^{1/p}\})$-approximation to the optimal additive metric under L_p.

5   Conclusions and Further Work

In this paper we have looked at embedding metrics into additive trees and ultrametrics. We have presented two algorithms, one suitable when the number of distinct distances in the metric is small, and one suitable when the number of distinct distances is large. Both algorithms are intrinsically greedy; they construct trees in a top-down fashion, establishing each internal node in turn by considering the immediate cost of the split it defines. Using these algorithms we provide the first approximation guarantees for this problem; however, there is scope for improving those guarantees.

Addendum: We recently learned that, independent of our work, Ailon and Charikar [15] have obtained improved results. They use ideas similar to those in our work.

References

1. Agarwala, R., Bafna, V., Farach, M., Paterson, M., Thorup, M.: On the approximability of numerical taxonomy (fitting distances by tree metrics). SIAM J. Comput. **28** (1999) 1073–1085
2. Ma, B., Wang, L., Zhang, L.: Fitting distances by tree metrics with increment error. J. Comb. Optim. **3** (1999) 213–225
3. Farach, M., Kannan, S.: Efficient algorithms for inverting evolution. Journal of the ACM **46** (1999) 437–450
4. Cryan, M., Goldberg, L., Goldberg, P.: Evolutionary trees can be learned in polynomial time in the two state general markov model. SIAM J. Comput **31** (2001) 375 – 397
5. Waterman, M., Smith, T., Singh, M., Beyer, W.: Additive evolutionary trees. J. Theoretical Biology **64** (1977) 199–213
6. Day, W.: Computational complexity of inferring phylogenies from dissimilarity matrices. Bulletin of Mathematical Biology **49** (1987) 461–467
7. Farach, M., Kannan, S., Warnow, T.: A robust model for finding optimal evolutionary trees. Algorithmica **13** (1995) 155–179
8. Emanuel, D., Fiat, A.: Correlation clustering - minimizing disagreements on arbitrary weighted graphs. In Battista, G.D., Zwick, U., eds.: ESA. Volume 2832 of Lecture Notes in Computer Science., Springer (2003) 208–220
9. Demaine, E.D., Immorlica, N.: Correlation clustering with partial information. In Arora, S., Jansen, K., Rolim, J.D.P., Sahai, A., eds.: RANDOM-APPROX. Volume 2764 of Lecture Notes in Computer Science., Springer (2003) 1–13
10. Bansal, N., Blum, A., Chawla, S.: Correlation clustering. In: Proc. of the 43rd IEEE Annual Symposium on Foundations of Computer Science. (2002) 238
11. Charikar, M., Guruswami, V., Wirth, A.: Clustering with qualitative information. In: Proc. of the 44th IEEE Annual Symposium on Foundations of Computer Science. (2003) 524
12. Dhamdhere, K.: Approximating additive distortion of embeddings into line metrics. In Jansen, K., Khanna, S., Rolim, J.D.P., Ron, D., eds.: APPROX-RANDOM. Volume 3122 of Lecture Notes in Computer Science., Springer (2004) 96–104
13. Dhamdhere, K., Gupta, A., Ravi, R.: Approximation algorithms for minimizing average distortion. In Diekert, V., Habib, M., eds.: STACS. Volume 2996 of Lecture Notes in Computer Science., Springer (2004) 234–245
14. Garg, N., Vazirani, V.V., Yannakakis, M.: Approximate max-flow min-(multi)cut theorems and their applications. SIAM J. Comput. **25** (1996) 235–251
15. Ailon, N., Charikar, M.: Personal comunication (2005)

Beating a Random Assignment

Gustav Hast

Department of Numerical Analysis and Computer Science
Royal Institute of Technology, 100 44 Stockholm, Sweden
ghast@nada.kth.se

Abstract. MAX CSP(P) is the problem of maximizing the weight of
satisfied constraints, where each constraint acts over a k-tuple of literals
and is evaluated using the predicate P. The approximation ratio of a
random assignment is equal to the fraction of satisfying inputs to P. If
it is NP-hard to achieve a better approximation ratio for MAX CSP(P),
then we say that P is *approximation resistant*. Our goal is to characterize
which predicates that have this property.
A general approximation algorithm for MAX CSP(P) is introduced. For
a multitude of different P, it is shown that the algorithm beats the random assignment algorithm, thus implying that P is not approximation
resistant. In particular, over 2/3 of the predicates on four binary inputs
are proved not to be approximation resistant, as well as all predicates on
$2s$ binary inputs, that have at most $2s + 1$ accepting inputs.
We also prove a large number of predicates to be approximation resistant.
In particular, all predicates of arity $2s + s^2$ with less than 2^{s^2} non-
accepting inputs are proved to be approximation resistant, as well as
almost 1/5 of the predicates on four binary inputs.

1 Introduction

For a constraint satisfaction problem, CSP, the objective is to satisfy a collection
of constraints by finding a good assignment. In the maximization version, MAX
CSP, each constraint has a weight and the objective is to maximize the weight
of satisfied constraints. By restricting what types of constraints that are allowed,
different types of problems can be defined. Two well-known examples are MAX
CUT and MAX E3SAT.

A constraint in a MAX CUT instance consists of two binary variables, x_i and
x_j, and it is satisfied iff $x_i \neq x_j$. Constraints of MAX E3SAT are disjunctions
over three literals, where a literal is a variable or a negated variable. Both these
problems are NP-complete. A naive algorithm for these problems, as well as all
MAX CSPs, is simply assigning all variables a random value, without looking
at the constraints. It is easy to see that such a random assignment on average
satisfies half of the constraints in a MAX CUT instance and 7/8 of the constraints
in a MAX E3SAT instance.

An algorithm is said to α-approximate a problem if the value of a produced
solution is at least $\alpha \cdot OPT$, where OPT is the value of an optimal solution.
For randomized algorithms, we allow this value to be an expected value over the

C. Chekuri et al. (Eds.): APPROX and RANDOM 2005, LNCS 3624, pp. 134–145, 2005.

random choices made by the algorithm. A major breakthrough in approximating various MAX CSPs was made when Goemans and Williamson used a semidefinite relaxation approach in order to 0.878-approximate MAX CUT. Before this, there was no known polynomial time $1/2 + \epsilon$-approximation for any constant $\epsilon > 0$, thus no known efficient algorithms outperformed a random assignment in approximating MAX CUT. Since then, the semidefinite relaxation technique has been used extensively in order to achieve better approximation ratios for a large number of MAX CSP problems, e.g. in [16, 18, 22, 23].

But the use of semidefinite relaxation cannot be applied to all CSPs with success. Based on PCP techniques, a different line of research produced results showing NP-hardness of approximating various problems [1, 2, 6, 8, 19, 20]. Håstad [13] introduced a technique that enabled very efficient analysis of PCPs and could show that there is no hope to efficiently approximate MAX E3SAT better than 7/8. Thus, in short of proving P = NP, the best possible efficient approximation algorithm for MAX E3SAT, up to low order terms, is simply to pick a random assignment.

MAX CSP(P) is the problem of maximizing the weight of satisfied constraints, where each constraint acts over a k-tuple of literals and is evaluated using the predicate P. As Creignou [4] and Khanna et al. [17] showed, MAX CSP(P) is MAX SNP-hard for all predicates depending on at least two variables. Thus, only for the unary predicate the problem can be solved optimally. An interesting distinction is whether a predicate allows a nontrivial approximation, i.e., can be approximated better than using a random assignment. If it is NP-hard to outperform a random assignment on MAX CSP(P), then we say that P is *approximation resistant*. There are many reasons for studying approximation resistance. Approximation resistance is a fundamental property of a predicate which determines if anything nontrivial can be done in approximating the MAX CSP. It is thus an important structural question. We believe that understanding this concept is a key to comprehending what it is that makes some problems hard to approximate. Approximation resistant predicates also play a fundamental role in the design of efficient PCPs.

There are no predicates of arity two that are approximation resistant [10], even if the input variables are non-binary [14]. Håstad [13] showed that if P depends on three binary inputs and accepts all odd parity inputs or all even parity inputs, then P is approximation resistant. Zwick [22] completed the characterization for predicates of arity three by producing approximation algorithms for MAX CSP(P), for all other predicates P, that outperform a random assignment. Samorodnitsky and Trevisan [21] constructed approximation resistant predicates of arity $2s + s^2$ with only 2^{2s} accepting assignments. A predicate P implies P' iff $P(y) \Rightarrow P'(y)$ for all y. In this paper we show that all predicates implied by a Samorodnitsky-Trevisan predicate are approximation resistant as well. As a consequence, all predicates of arity $2s + s^2$ with at most 2^{s^2} non-accepting inputs are also approximation resistant. On the other hand, we also show that if a predicate of arity $2s$ has at most $2s + 1$ accepting inputs, then the predicate is not approximation resistant. Thus, there is a tendency that the more inputs a predicate accepts, the harder it is to beat a random assignment for the corresponding MAX CSP. It could be tempting to assume that if a predicate P is approximation resistant and P implies P', then P' is approximation resistant as well. As Zwick showed, this is true for all predicates of arity three and it is also true if P is a Samorodnitsky-Trevisan predicate. However, this is not true in general. While studying predicates of arity four, we observed that the predicate $(\overline{x}_1 \wedge (x_2 \equiv x_3)) \vee (x_1 \wedge (x_2 \equiv x_4))$, that has been shown to be approximation resistant [11], implies that $\overline{x}_2$, x_3 and x_4 cannot all be equal. Zwick [22] showed that this latter predicate is not approximation resistant.

1.1   Our Contribution

Among other contributions, we introduce a general technique for approximating MAX CSP instances. It works by finding a solution that gives a positive contribution on the small degree terms of the Fourier spectra of the objective function, while seeing to that the higher order terms give a smaller positive or negative contribution. By using a global analysis of the instance, we prove that the method beats a random assignment on MAX $CSP(P)$ for a multitude of different predicates P, thus showing them not to be approximation resistant. In addition, we give non-trivial conditions that optimal solutions to hard MAX CSP instances have to adhere to. We also show a large number of predicates to be approximation resistant.

Method. The objective function of a MAX $CSP(P)$ instance, I, is the weighted sum of the indicator variables of the constraints. In Section 2 we show that each such term can be written as a multilinear expression. The whole sum can thus also be written as a multilinear expression

$$I(x_1, \ldots x_n) = \sum_{i=1}^{m} w_i P(z_{i1}, \ldots z_{ik}) = \sum_{S \subseteq [n], |S| \leq k} c_S \prod_{i \in S} x_i . \tag{1}$$

In fact this is a Fourier spectra of the objective value function. Given a random assignment, where each binary variable x_i is assigned -1 or 1 randomly and independently, all terms of the above expression that includes at least one variable x_i have an expected value of zero. The only term that gives a contribution to the expectation is then the constant term, $c_\emptyset$. Thus, the expected value of a random assignment is equal to $c_\emptyset$. In order to beat a random assignment, we have to produce an assignment with an expected weight of at least $c_\emptyset + \gamma W$, for a constant $\gamma > 0$ and where W is the total weight of the instance.

If there is large weight on the linear terms of the Fourier spectra, i.e., $\sum_{i=1}^{n} |c_{\{i\}}| = \delta W$ for a constant $\delta > 0$, then it is easy to assign values to the variables such that linear terms give a contribution of δW. The problem is, that there is no control on the higher order terms which possibly could give a negative contribution that cancel out the contribution from the linear terms. In order to reduce the contribution of the higher order terms, we therefore only give a small bias towards the solution that maximize the linear terms. With probability ε, depending on k and δ, we assign x_i to $\mathrm{sgn}(c_{\{i\}})$, and otherwise assign x_i a random value. The linear terms then give a positive contribution of $\varepsilon\delta W$ at the same time as the absolute contribution of each term of order i is decreased with a factor of ε^i. The total absolute contribution of the higher order terms can thereby be made to be at most $1/3$ of the contribution of the linear terms by choosing an ε small enough. Thus, the algorithm finds an assignment with expected value of at least $c_\emptyset + 2\varepsilon\delta W/3$ and thereby it beats a random assignment.

We can apply the same method on the bi-linear terms, if there exists an assignment that gives a non-negligible value on the those terms. In that case, we use Charikar and Wirth's approximation algorithm for quadratic programs [3], which is based on semidefinite programming, in order to find an assignment that gives a non-negligible value on the bi-linear terms. We use that algorithm, because even if the optimal value is small it is guaranteed to outperform the expected value of a random assignment. We note that we also could use one of the MAX CUT algorithms in [9, 23].

In order to show that a predicate is not approximation resistant, it is enough to create an approximation algorithm that beats a random assignment on satisfiable and almost satisfiable instances. This is because a random assignment achieves the same expected objective value irrespective of the satisfiability of the instance. Thus, satisfiable and almost satisfiable instances are the ones where a random assignment achieves the worst approximation ratio. By combining the algorithm for almost satisfiable instances with a random assignment, we then get a better approximation ratio than using a random assignment alone. For some predicates P it is possible to show that an assignment that almost satisfies an instance of MAX CSP(P) has to give a non-negligible positive value on either the linear or bi-linear terms of the Fourier spectra. Thus satisfying the condition for our algorithm to outperform a random assignment, which in turn shows that P is not approximation resistant.

Application 1. The predicates P and P' are of the same *type* if MAX CSP(P) and MAX CSP(P') are essentially the same problem. There are exactly 400 different non-trivial predicate types of arity four. By using the above technique we show that 270 of these are not approximation resistant. In particular, all predicates with at most six accepting inputs are not approximation resistant. Extending methods in [13] we derive that 70 predicate types are approximation resistant, thus all but 60 predicate types of arity four are characterized.

Application 2. We show that if P is a predicate of arity $2s$ and has at most $2s+1$ accepting inputs, then it is not approximation resistant. This shows that the approximation resistant predicate of Engebretsen and Holmerin [5], with arity six and eight accepting inputs, is optimal in the sense of minimizing the number of accepting inputs. The method also gives lower bounds on the soundness for a natural class of probabilistically checkable proofs. These PCPs express NP and have perfect or almost perfect completeness and soundness negligibly higher than the acceptance probability of a random proof.

Application 3. Our method enables us to characterize optimal solutions to hard instances of some MAX CSPs. Similar insights, in particular for constraints of arity three, were achieved in [7].

Approximation Resistant Predicates. A Samorodnitsky-Trevisan predicate $P_{ST^{s,t}}$ consists of st different parity checks, each acting on three bits of the input,

$$P_{ST^{s,t}}(x_1, \ldots x_s, x'_1, \ldots x'_t, x''_1, \ldots x''_{st}) = \bigwedge_{\substack{1 \le i \le s \\ 1 \le j \le t}} x_i \oplus x'_j \oplus x''_{(i-1)t+j} \ .$$

Samorodnitsky and Trevisan [21] showed that such a predicate is approximation resistant. In this paper we revisit the proof and show that all predicates implied by a Samorodnitsky-Trevisan predicate are approximation resistant as well. A consequence of this is that all predicates of arity $s + t + st$ with less than 2st non-accepting inputs are approximation resistant. We also show that predicates of arity four and five, having at most two and five non-accepting inputs respectively, are approximation resistant. This is optimal in the sense that there are predicates with one more non-accepting input that we show are not approximation resistant.

1.2   Organization of the Paper

Section 2 gives some notation and definitions used in the paper. In Section 3 we describe the method of exploiting the existence of a good solution on low-degree terms in order to beat a random assignment. In the following section we use this method in order to show that predicates with few accepting inputs are not approximation resistant. We also show that predicates with few non-accepting inputs are approximation resistant. In Section 5 we take a closer look at predicates of arity four. The subsequent section shows that approximation resistance is a non-monotone property. Section 7 is devoted to characterizing optimal solutions to hard instances of MAX CSP(P). All proofs are omitted due to space constraints. They can be found in [12].

2   Preliminaries

A predicate P maps elements from $\{\pm 1\}^k$ to $\{0, 1\}$. For notational convenience we let input bits have value -1, denoting true, and 1, denoting false. If P accepts an input y, then $P(y) = 1$, otherwise $P(y) = 0$. Thus, the set of accepting inputs to P is denoted by $P^{-1}(1)$.

MAX CSP(P) is the MAX CSP problem where all constraints are evaluated using a predicate $P : \{\pm 1\}^k \to \{0, 1\}$. More formally, an instance of MAX CSP(P) consists of m weighted constraints, each one acting over a k-tuple of literals, $(z_{i1}, \ldots z_{ik})$, taken from the set $\{x_1, \ldots x_n, \bar{x}_1, \ldots \bar{x}_n\}$. All variables in such a tuple are assumed to be distinct. A constraint is satisfied if P maps its k-tuple of literals onto one. The objective is to maximize $\sum_{i=1}^{m} w_i P(z_{i1}, \ldots z_{ik})$, where w_i is the (non-negative) weight of the i:th constraint. The weight of an input $(y_1, \ldots y_k)$ for an assignment a to an instance of MAX CSP(P) is $\sum_{i:(z_{i1},\ldots z_{ik})=y} w_i$, where each constraint tuple is evaluated on a. Two predicates, P and P', are of the same *type* iff there is a permutation π on $[k]$ and $a \in \{\pm 1\}^k$ such that $P(x_1, \ldots x_k) = P'(a_1 x_{\pi(1)}, \ldots a_k x_{\pi(k)})$ for all $x \in \{\pm 1\}^k$. A MAX CSP(P) instance can be expressed as a MAX CSP(P') instance by permuting and applying a bitmask to the constraint tuples, so clearly they are equivalent problems.

The following definition formally describes an approximation resistant predicate.

Definition 1. *A predicate P is* approximation resistant *if, for every constant $\epsilon > 0$, it is* NP-*hard to approximate* MAX CSP(P) *within* $2^{-k}|P^{-1}(1)| + \epsilon$.

A predicate can be seen as a sum of conjunctions, each conjunction corresponding to an accepting input to P. If $x, y \in \{\pm 1\}^k$ then $\sum_{S \subseteq [k]} \prod_{i \in S} x_i y_i$ equals 2^k if $x = y$ and otherwise the sum is zero. Thus, a conjunction $\alpha_1 x_1 \wedge \ldots \wedge \alpha_k x_k$, where $\alpha \in \{\pm 1\}^k$, can be arithmetized as

$$\psi_\alpha(x_1, \ldots x_k) = 2^{-k} \sum_{S \subseteq [k]} \prod_{i \in S} \alpha_i x_i = \begin{cases} 1 \text{ if } \alpha = (x_1, \ldots x_k) \\ 0 \text{ otherwise} \end{cases} .$$

A predicate can thus be expressed as a multilinear expression

$$P(x_1, \ldots x_k) = \sum_{\alpha \in P^{-1}(1)} \psi_\alpha(x_1, \ldots x_k) . \tag{2}$$

We let $P^{(i)}$ denote the sum of the i-degree terms of the multilinear expression P, and $P^{(\geq i)}$ denotes the sum of the terms of at least degree i in P. Thus,

$$P^{(i)}(x_1, \ldots x_k) = 2^{-k} \sum_{\alpha \in P^{-1}(1)} \sum_{S \subseteq [k], |S|=i} \prod_{i \in S} \alpha_i x_i .$$

3   General Technique

As mentioned in the introduction, the objective function of a MAX CSP(P) instance can be described as a multilinear expression, see (1). In this section we show that if there is an assignment that gives a non-negligible positive value to the linear or bi-linear terms of this expression, then we can do better than just picking a random assignment.

Consider the linear terms of the objective value function, $I^{(1)}(x_1, \ldots x_n) = \sum_{i=1}^n c_{\{i\}} x_i$. It is easy to see that the linear terms are maximized by assigning x_i to $\mathrm{sgn}(c_{\{i\}})$, giving the value $\sum_i |c_{\{i\}}|$. However, in order to control the expected value of higher order terms Algorithm LIN picks an assignment that is ε-biased towards this assignment. The expected value of the i'th order terms decrease their value with a factor of ε^i. By choosing a small enough bias we ensure that the expected value of the non-constant terms is dominated by the linear terms. The following theorem quantifies the performance of Algorithm LIN.

For each $i := 1, \ldots n$ do:

- with probability ε: assign $x_i := \begin{cases} 1 \text{ if } c_{\{i\}} \geq 0 \\ -1 \text{ otherwise} \end{cases}$,
- or else assign x_i according to an unbiased coin flip.

Fig. 1. Algorithm LIN - for approximating MAX CSP(P).

Theorem 2. *Let* $I(x_1, \ldots x_n) = \sum_{S \subseteq [n], |S| \leq k} c_S \prod_{i \in S} x_i$ *be the objective value function for an instance of* MAX CSP(P), $P : \{\pm 1\}^k \to \{0, 1\}$. *Denote the sum of all constraints' weights* W. *If* $\sum_i |c_{\{i\}}| \geq \delta W$ *for a constant* $\delta > 0$, *then Algorithm* LIN, *with* $\varepsilon = \delta/2k$, *produces a solution of expected weight at least* $c_\emptyset + \frac{\delta^2}{3k} W$.

The method for the linear terms can also be made to work for the bi-linear terms. If there exists an assignment that makes the sum of the bi-linear terms non-negligibly positive, then Algorithm BILIN outperforms a random assignment.

Input: A MAX CSP(P) instance I

1. Approximate the quadratic program that contains of the bi-linear terms, $I^{(2)}$, using the Charikar-Wirth algorithm [3]. Let $a_1, \ldots a_n$ be the produced solution.
2. If $I^{(1)}(a_1, \ldots a_n) < 0$ then do: $a_i := \overline{a}_i$ for $i := 1, \ldots n$.
3. Set $\alpha = I^{(2)}(a_1, \ldots a_n)/W$, where W is the total weight in I. Set $\varepsilon = \max(\alpha/2k^{3/2}, 0)$, where k is the arity of P.
4. For $i := 1, \ldots n$ do:
   - With probability ε: assign $x_i := a_i$,
   - or else assign x_i according to an unbiased coin flip.

Fig. 2. Algorithm BILIN - for approximating MAX CSP(P).

Theorem 3. *Let* $I(x_1, \ldots x_n) = \sum_{S \subseteq [n], |S| \leq k} c_S \prod_{i \in S} x_i$ *be the objective value function for an instance of* MAX CSP(P), $P : \{\pm 1\}^k \to \{0, 1\}$. *Denote the sum of all constraints' weights* W. *If there exists, for a constant* $\delta > 0$, *an assignment such that* $\sum_{1 \leq i < j \leq n} c_{\{i,j\}} x_i x_j \geq \delta W$, *then Algorithm* BILIN *produces a solution of expected weight at least* $c_\emptyset + \kappa W$, *where* $\kappa > 0$ *and only depends on constants* δ *and* k.

From Theorems 2 and 3 we derive Lemma 4 and its generalization, Lemma 5. These lemmas are used in order to show that predicates are not approximation resistant. A crucial observation in their proofs is that in order to achieve a better approximation algorithm than a random assignment, we only have to outperform a random assignment on satisfiable or almost satisfiable instances. Those instances are the ones where a random assignment achieves the worst approximation ratio.

Lemma 4. *Let $P : \{\pm 1\}^k \to \{0, 1\}$ be a predicate. If $P^{(1)}(y) + P^{(2)}(y) > 0$ for all $y \in P^{-1}(1)$, then P is not approximation resistant.*

Lemma 5. *Let $P : \{\pm 1\}^k \to \{0, 1\}$ be a predicate. An instance of MAX CSP(P) has the objective value function $I(x_1, \ldots x_n) = \sum_{i=1}^m w_i P(z_{i1}, \ldots z_{ik})$. For an assignment to $x_1, \ldots x_n$, define the weight on each input y, $u_y = \sum_{i:(z_{i1},\ldots z_{ik})=y} w_i$. For constants C and $\delta > 0$, there exist constants $\varepsilon > 0$ and $\kappa > 0$ such that if $\sum_{y \in P^{-1}(1)} u_y (C \cdot P^{(1)}(y) + P^{(2)}(y)) \geq \delta W$, where $W = \sum_{i=1}^m w_i$, then either the value of the assignment is less than $(1 - \varepsilon)W$, or Algorithm LIN or Algorithm BILIN will achieve an approximation ratio of at least $2^{-k}|P^{-1}(1)| + \kappa$.*

By using a gadget construction to MAX 2SAT and applying Zwick's algorithm for almost satisfiable MAX 2SAT instances, we show the following theorem.

Theorem 6. *Let $P : \{\pm 1\}^k \to \{0, 1\}$ be a predicate. If for some constant C, $C \cdot P^{(1)}(y) + P^{(2)}(y)$ is zero for at most two $y \in P^{-1}(1)$ and strictly positive for all other $y \in P^{-1}(1)$, then P is not approximation resistant.*

4 Predicates on k Variables

If a predicate has few accepting inputs, then we can show by using Lemma 4 that the predicate is not approximation resistant.

Theorem 7. *Let $P : \{\pm 1\}^k \to \{0, 1\}$, $k \geq 3$. If P has at most $2\lfloor k/2 \rfloor + 1$ accepting inputs, then P is not approximation resistant.*

Remark 8. Engebretsen and Holmerin [5] designed an approximation resistant predicate $P : \{\pm 1\}^6 \to \{0, 1\}$, with only eight accepting inputs. The above theorem is thus tight for the case $k = 6$ as it stipulates that if a predicate has at most seven accepting inputs, then it is not approximation resistant.

In the remaining part of this section we describe sufficient conditions on predicates to be approximation resistant. The following theorem shows that all predicates implied by a Samorodnitsky-Trevisan predicate is approximation resistant. It turns out that a large part of the proof is essentially a reiteration of the proof of Håstad and Wigderson [15], which is a simpler proof of the Samorodnitsky-Trevisan PCP than the original one.

Theorem 9. *Let $P : \{\pm 1\}^{s+t+st} \to \{0, 1\}$, be a predicate that is implied by $P_{ST^{s,t}}$, i.e., $P_{ST^{s,t}}^{-1}(1) \subseteq P^{-1}(1)$, then P is approximation resistant.*

If a predicate P has enough accepting inputs, then there is at least one predicate of the same type as P, such that it is implied by a Samorodnitsky-Trevisan predicate. Using Theorem 9, we then conclude that P is approximation resistant.

Theorem 10. *Let $k \geq s + t + st$ and $P : \{\pm 1\}^k \to \{0, 1\}$ be a predicate with at most $2^{st} - 1$ non-accepting inputs, then P is approximation resistant.*

Now we consider predicates of arity five. Theorem 10, with $s = 1$ and $t = 2$, lets us conclude that if $|P^{-1}(0)| \leq 3$ then P is approximation resistant. By taking more care in the analysis we are able to show the following theorem.

Theorem 11. *Let P be a predicate on 5 binary variables with at most 5 non-accepting inputs, then P is approximation resistant.*

Remark 12. The next section shows that there are predicates of arity four, having exactly three non-accepting inputs, which are not approximation resistant. This implies that there are also predicates of arity five, having exactly six non-accepting inputs, which are not approximation resistant. Thus, Theorem 11 is optimal in this sense.

5   Predicates on Four Variables

In this section we take a closer look at predicates of arity four. Theorem 7 stipulates that if a predicate $P : \{\pm 1\}^4 \rightarrow \{0, 1\}$ has at most five accepting inputs then it is not approximation resistant. That was already known as a result of the MAX 4CSP 0.33-approximation algorithm of Guruswami et al. [11], because $5/2^4 < 0.33$. By applying Theorem 6 we show that no predicate that accepts exactly six inputs is approximation resistant. We also show many predicates on four variables to be approximation resistant by extending methods from [13]. Out of exactly 400 different non-trivial predicate types of arity four we show that 79 are approximation resistant and 275 are not approximation resistant. For every number of accepting inputs, the number of predicate types that have been characterized as non-trivially approximable and approximation resistant, respectively, are shown in Table 1. For a detailed description on how these results were obtained, see [12].

Table 1. Approximability of MAX CSP(P) for predicates of arity four.

# accepting inputs	1	2	3	4	5	6	7	8	9	10	11	12	13	14	15	all
# non-trivially approx	1	4	6	19	27	50	50	52	27	26	9	3	1	0	0	275
# approx resistant	0	0	0	0	0	0	0	16	6	22	11	15	4	4	1	79
# unknown	0	0	0	0	0	0	6	6	23	2	7	1	1	0	0	46

6   Approximation Resistance Is Non-monotone

There is one predicate of arity four that we know to be approximation resistant but is not implied by parity,

$$\text{GLST}(x_1, x_2, x_3, x_4) = (\overline{x}_1 \wedge (x_2 \equiv x_3)) \vee (x_1 \wedge (x_2 \equiv x_4)) . \qquad (3)$$

It was shown to be approximation resistant in [11]. Zwick [22] showed that the not all equal predicate of arity three, $3\text{NAE}(x_1, x_2, x_3) = (x_1 \oplus x_2) \vee (x_1 \oplus x_3)$, is not approximation resistant. For all approximation resistant predicates P of arity three, it is true that predicates implied by P is approximation resistant as well. This is true for all Samorodnisky-Trevisan predicates as well. A natural guess could be that it is true in general. However, the observation that $\text{GLST}(x_1, x_2, x_3, x_4)$ implies $3\text{NAE}(\bar{x}_2, x_3, x_4)$ gives the following theorem which falsifies that guess.

Theorem 13. *It is not generally true that if P is approximation resistant and P' is implied by P, then P' is approximation resistant as well.*

7   Characterization of Hard Instances

A hard MAX CSP instance is an instance that is satisfiable or almost satisfiable and for which it is NP-hard to approximate it significantly better than picking a random assignment. In this section we characterize optimal and near-optimal solutions to such instances by bounding the weight of particular inputs for such solutions. Our main tool for this task is Lemma 5.

Zwick characterized all predicates of arity three [22]. From this work we know that approximation resistant predicates of arity three are of the same type as either 3XOR which accepts if and only if the parity of the input bits is odd, 3NTW which accepts if and only if not exactly two of the input bits are true, $3\text{OXR}(x_1, x_2, x_3) = x_1 \vee (x_2 \oplus x_3)$, or $3\text{OR}(x_1, x_2, x_3) = x_1 \vee x_2 \vee x_3$.

Theorem 14. *Let P be 3NTW, 3OXR or 3OR, and let I be an instance of MAX CSP(P) of total weight W. For every $\gamma > 0$, there exist $\varepsilon > 0$ and $\kappa > 0$ such that if there exists a $(1 - \varepsilon)$-satisfying assignment with at least weight γW on even parity inputs, then I can be $(s2^{-3} + \kappa)$-approximated in probabilistical polynomial time, where s is the number of accepting inputs to P.*

Remark 15. In [7], Feige studied the problem of distinguishing between random MAX CSP(3OR) instances and almost satisfiable instances. He shows that the only type of almost satisfiable instances that cannot be distinguished from a random instance is if an optimal assignment makes almost all constraints have either one or three literals true. For all other almost satisfiable instances a certificate of non-randomness can be acquired. By applying Algorithms LIN and BILIN, it can be shown that the certificate implies that we can find an assignment that beats a random assignment. From this Theorem 14 follows.

Using the same method on the GLST predicate, we show that optimal solutions to hard instances of MAX CSP(GLST) have almost all weight on only four of the eight accepting inputs to GLST.

Theorem 16. *Let I be an instance of MAX CSP(GLST). For every $\gamma > 0$, there exist $\varepsilon > 0$ and $\kappa > 0$ such that if there exists a $(1-\varepsilon)$-satisfying assignment with total weight $\gamma w_{tot}(I)$ on inputs*

$$\{(1, 1, 1, 1), (1, -1, -1, -1), (-1, 1, 1, 1), (-1, -1, -1, -1)\} ,$$

then I can be $(1/2 + \kappa)$-approximated in probabilistical polynomial time.

Acknowledgments

I am very grateful to Johan Håstad for much appreciated help. I also thank a referee for bringing [7] to my attention.

References

1. Sanjeev Arora, Carsten Lund, Rajeev Motwani, Madhu Sudan, and Mario Szegedy. Proof verification and hardness of approximation problems. *Journal of the ACM*, 45(3):501–555, 1998. Preliminary version appeared in FOCS 1992.
2. Mihir Bellare, Oded Goldreich, and Madhu Sudan. Free bits, PCPs, and nonapproximability - towards tight results. *SIAM Journal on Computing*, 27(3):804–915, 1998.
3. Moses Charikar and Anthony Wirth. Maximizing quadratic programs: Extending Grothendieck's inequality. In *Proceedings of the 45th Annual IEEE Symposium on Foundations of Computer Science*, pages 54–60, 2004.
4. Nadia Creignou. A dichotomy theorem for maximum generalized satisability problems. *Journal of Computer and System Sciences*, 51(3):511–522, 1995.
5. Lars Engebretsen and Jonas Holmerin. More efficient queries in PCPs for NP and improved approximation hardness of maximum CSP. In *Proceedings of STACS 2005, Lecture Notes in Computer Science 3404*, pages 194–205, 2005.
6. Uriel Feige. A threshold of ln n for approximating set cover. *Journal of the ACM*, 45(4):634–652, 1998.
7. Uriel Feige. Relations between average case complexity and approximation complexity. In *Proceedings of the 34th Annual ACM Symposium on Theory of Computing*, pages 534–543, 2002.
8. Uriel Feige, Shafi Goldwasser, László Lovász, Shmuel Safra, and Mario Szegedy. Interactive proofs and the hardness of approximating cliques. *Journal of the ACM*, 43(2):268–292, 1996. Preliminary version appeared in FOCS 1991.
9. Uriel Feige and Michael Langberg. The RPR^2 rounding technique for semidefinite programs. In *Proceedings of ICALP*, pages 213–224, 2001.
10. Michel X. Goemans and David P. Williamson. Improved Approximation Algorithms for Maximum Cut and Satisfiability Problems Using Semidefinite Programming. *Journal of the ACM*, 42:1115–1145, 1995.
11. Venkatesan Guruswami, Daniel Lewin, Madhu Sudan, and Luca Trevisan. A tight characterization of NP with 3 query PCPs. In *Proceedings of the 39th Annual IEEE Symposium on Foundations of Computer Science*, pages 8–17, 1998.
12. Gustav Hast. *Beating a Random Assignment*. PhD thesis, Royal Institute of Technology, 2005.
13. Johan Håstad. Some optimal inapproximability results. *Journal of the ACM*, 48(4):798–859, 2001. Preliminary version appeared in STOC 1997.
14. Johan Håstad. Every 2-CSP allows nontrivial approximation. In *Proceedings of the 37th Annual ACM Symposium on Theory of Computing*, pages 740–746, 2005.
15. Johan Håstad and Avi Wigderson. Simple analysis of graph tests for linearity and PCP. *Random Structures and Algorithms*, 22(2):139–160, 2003.
16. Howard Karloff and Uri Zwick. A 7/8-approximation algorithm for MAX 3SAT? In *Proceedings of the 38th Annual IEEE Symposium on Foundations of Computer Science*, pages 406–415, 1997.

17. Sanjeev Khanna, Madhu Sudan, Luca Trevisan, and David P. Williamson. The approximability of constraint satisfaction problems. *SIAM Journal on Computing*, 30(6):1863–1920, 2000.
18. Michael Lewin, Dror Livnat, and Uri Zwick. Improved rounding techniques for the MAX 2-SAT and MAX DI-CUT problems. In *Proceedings of 9th IPCO, Lecture Notes in Computer Science 2337*, pages 67–82, 2002.
19. Carsten Lund and Mihalis Yannakakis. On the hardness of approximating minimization problems. *Journal of the ACM*, 41(5):960–981, 1994. Preliminary version appeared in STOC 1993.
20. Christos H. Papadimitriou and Mihalis Yannakakis. Optimization, approximation, and complexity classes. *Journal of Computer and System Sciences*, 43(3):425–440, 1991.
21. Alex Samorodnitsky and Luca Trevisan. A PCP characterization of NP with optimal amortized query complexity. In *Proceedings of the 32nd Annual ACM Symposium on Theory of Computing*, pages 191–199, 2000.
22. Uri Zwick. Approximation algorithms for constraint satisfaction problems involving at most three variables per constraint. In *Proceedings of the 9th Annual ACM-SIAM Symposium on Discrete Algorithms*, pages 201–210, 1998.
23. Uri Zwick. Outward rotations: a tool for rounding solutions of semidefinite programming relaxations, with applications to MAX CUT and other problems. In *Proceedings of the 31st Annual ACM Symposium on Theory of Computing*, pages 679–687, 1999.

Scheduling on Unrelated Machines
Under Tree-Like Precedence Constraints

V.S. Anil Kumar[1,*], Madhav V. Marathe[2],
Srinivasan Parthasarathy[3,**], and Aravind Srinivasan[3,**]

[1] Basic and Applied Simulation Science (CCS-DSS)
Los Alamos National Laboratory, MS M997
P.O. Box 1663, Los Alamos, NM 87545
anil@lanl.gov
[2] Virginia Bio-informatics Institute
and Department of Computer Science Virginia Tech, Blacksburg 24061
mmarathe@vbi.vt.edu
[3] Department of Computer Science, University of Maryland
College Park, MD 20742
{sri,srin}@cs.umd.edu

Abstract. We present polylogarithmic approximations for the $R|prec|C_{max}$ and $R|prec|\sum_j w_j C_j$ problems, when the precedence constraints are "treelike" – i.e., when the undirected graph underlying the precedences is a forest. We also obtain improved bounds for the weighted completion time and flow time for the case of chains with restricted assignment – this generalizes the job shop problem to these objective functions. We use the same lower bound of "congestion+dilation", as in other job shop scheduling approaches. The first step in our algorithm for the $R|prec|C_{max}$ problem with treelike precedences involves using the algorithm of Lenstra, Shmoys and Tardos to obtain a processor assignment with the congestion + dilation value within a constant factor of the optimal. We then show how to generalize the random delays technique of Leighton, Maggs and Rao to the case of trees. For the weighted completion time, we show a certain type of reduction to the makespan problem, which dovetails well with the lower bound we employ for the makespan problem. For the special case of chains, we show a dependent rounding technique which leads to improved bounds on the weighted completion time and new bicriteria bounds for the flow time.

1 Introduction

The most general scheduling problem involves unrelated parallel machines and precedence constraints, i.e., we are given: (i) a set of n jobs with precedence constraints that induce a partial order on the jobs; (ii) a set of m machines, each of which can process at most one job at any time, and (iii) an arbitrary set of

* Research supported by the Department of Energy under Contract W-7405-ENG-36.
** Research supported in part by NSF Award CCR-0208005 and NSF ITR Award CNS-0426683.

C. Chekuri et al. (Eds.): APPROX and RANDOM 2005, LNCS 3624, pp. 146–157, 2005.

integer values $\{p_{i,j}\}$, where $p_{i,j}$ denotes the time to process job j on machine i. Let C_j denote the completion time of job j. Subject to the above constraints, two commonly studied versions are (i) minimize the *makespan*, or the maximum time any job takes, i.e. $\max_j\{C_j\}$ – this is denoted by $R|prec|C_{max}$, and (ii) minimize the weighted completion time – this is denoted by $R|prec|\sum_j w_j C_j$. Numerous other variants, involving release dates or other objectives have been studied (see e.g. Hall [7]).

Almost optimal upper and lower bounds are known for the versions of the above problems without precedence constraints (i.e., the $R||C_{max}$ and $R||\sum_j w_j C_j$ problems) [3, 13, 23], but very little is known in the presence of precedence constraints. The only case of the general $R|prec|C_{max}$ problem for which non-trivial approximations are known is the case where the precedence constraints are chains – this is the job shop scheduling problem (see Shmoys *et al.* [22]), which itself has a long history. The first result for job shop scheduling was the breakthrough work of Leighton et al. [14, 15] for packet scheduling, which implied an $O(\log n)$ approximation for the case of unit processing costs. Leighton et al. [14, 15] introduced the "random delays" technique, and almost all the results on the job shop scheduling problem [5, 6, 22] are based on variants of this technique. Shmoys et al. [22] also generalize job-shop scheduling to DAG-shop scheduling, where the operations of each job form a DAG, instead of a chain, with the additional constraint that *the operations within a job can be done only one at a time.*

The only results known for the case of arbitrary number of processors with more general precedence constraints are for identical parallel machines (denoted by $P|prec|C_{max}$; Hall [7]), or for related parallel machines (denoted by $Q|prec|C_{max}$) [2, 4]. The weighted completion time objective has also been studied for these variants [3, 8]. When the number of machines is constant, polynomial time approximation schemes are known [9, 11]. Note that all of the above discussion relates to *non-preemptive* schedules, i.e., once the processing of a job is started, it cannot be stopped until it is completely processed; preemptive variants of these problems have also been well studied (see e.g. Schulz and Skutella [20]).

Far less is known for the weighted completion time objective in the same setting, instead of the makespan. The known approximations are either for the case of no precedence constraints [23], or for precedence constraints with identical/related processors [8, 19]. To the best of our knowledge, no non-trivial bound is known on the weighted completion time on unrelated machines, in the presence of precedence constraints of *any kind.*

Here, motivated by applications such as evaluating large expression-trees and tree-shaped parallel processes, we consider the special case of the $R|prec|C_{max}$ and $R|prec|\sum_j w_j C_j$ problems, where the precedences form a forest, i.e., the undirected graph underlying the precedences is a forest. Thus, this naturally generalizes the job shop scheduling problem, where the precedence constraints form a collection of disjoint directed chains.

Summary of Results. We present the first polynomial time approximation algorithms for the $R|prec|C_{max}$ and $R|prec|\sum_j w_j C_j$ problems, under "treelike" precedences. As mentioned earlier, these are the first non-trivial generalizations of the job shop scheduling problems to precedence constraints which are not chains. Since most of our results hold in the cases where the precedences form a forest (i.e., the undirected graph underlying the DAG is a forest), we will denote the problems by $R|forest|C_{max}$, and $R|forest|\sum_j w_j C_j$, respectively, to simplify the description – this generalizes the notation used by Jansen and Solis-oba [10] for the case of chains.

1. The $R|forest|C_{max}$ Problem. We obtain a polylogarithmic approximation for this problem in Section 2. We employ the same lower bound used in [5, 6, 14, 22]: $LB \doteq \max\{P_{max}, \Pi_{max}\}$, where P_{max} is the maximum processing time along any directed path and Π_{max} is the maximum processing time needed by any machine, for a fixed assignment of jobs to machines. Let $p_{max} = \max_{i,j} p_{i,j}$ be the maximum processing time of any job on any machine. We obtain an $O(\frac{\log^2 n}{\log \log n} \lceil \frac{\log \min(p_{max}, n)}{\log \log n} \rceil)$ approximation to the $R|forest|C_{max}$ problem. When the forests are out-trees or in-trees, we show that this factor can be improved to $O(\log n \cdot \lceil \log(\min\{p_{max}, n\})/\log \log n \rceil)$; for the special case of unit processing times, this actually becomes $O(\log n)$. We also show that the lower-bound LB cannot be put to much better use: even in the case of trees – for unit processing costs, we show instances whose optimal schedule is $\Omega(LB \cdot \log n)$.

Our algorithm for solving $R|forest|C_{max}$ follows the overall approach used to solve the job shop scheduling problem (see, e.g. Shmoys *et al.* [22]) and involves two steps: (1) We show how to compute a processor assignment within a $(\frac{3+\sqrt{5}}{2})$-factor of LB, by extending the approach of Lenstra *et al.* [13], and, (2) We design a poly-logarithmic approximation algorithm for the resulting variant of the $R|prec|C_{max}$ problem with pre-specified processor assignment, and forest shaped precedences.

2. The $R|forest|\sum_j w_j C_j$ Problem. In Section 3, We show a reduction from $R|prec|\sum_j w_j C_j$ to $R|prec|C_{max}$ of the following form: if there is a schedule of makespan $(P_{max} + \Pi_{max}) \cdot \rho$ for the latter, then there is an $O(\rho)$-approximation algorithm for the former. We exploit this, along with the fact that our approximation guarantee for $R|forest|C_{max}$ is of the form "$(P_{max} + \Pi_{max})$ times polylog", to get a polylogarithmic approximation for the $R|forest|\sum_j w_j C_j$ problem. Our reduction is similar in spirit to that of Queyranne and Sviridenko [19]: both reductions employ geometric time windows and linear constraints on completition times for bounding congestion and dilation. However, the reduction in [19] is meant for *identical* parallel machines while our reduction works for *unrelated* machines. Further, [19] works for job-shop and dag-shop problems under the assumption that *no two operations from the same DAG can be executed concurrently, although no precedence relation might exist between the two operations*; in contrast, we do not impose this restriction and allow concurrent processing subject to precedence and assignment constraints being satisfied.

3. Minimizing Weighted Completion Time and Flow Time on Chains.

For a variant of the $R|forest|\sum_j w_j C_j$ problem where (i) the forest is a collection of chains (i.e., the weighted completion time variant of the job shop scheduling problem), and (ii) for each machine i and operation v, $p_{i,v} \in \{p_v, \infty\}$ (i.e., the *restricted-assignment variant*), we show a better approximation of $O(\log n/\log\log n)$ to the weighted completion time in Section 4. Our result ensures that (i) the precedence constraints are satisfied with *probability* 1, and (ii) for any (v, t), the probability of scheduling v at time t equals its fractional (LP) value $x_{v,t}$. This result also leads to a bicriteria $(1 + o(1))$–approximation for weighted flow time variant of this problem, using $O(\log n/\log\log n)$ copies of each machine.

Due to space limitations, several proofs and algorithm details are omitted here and are deferred to the full version this paper.

2   The $R|forest|C_{max}$ Problem

Consider a (fractional) assignment x of jobs to machines, where $x_{i,j}$ is the fraction of job j assigned to machine i. In the description below, we will use the terms "node" and "job" interchangably; we will not use the term "operation" to refer to nodes of a DAG, because we do not have the job shop or dag shop constraints that at most one node in a DAG can be processed at a time. As before, P_{max} denotes the maximum processing time along any directed path, i,e., $P_{max} = \max_{\text{path } P}\{\sum_{j\in P}\sum_i x_{i,j}p_{i,j}\}$. Also, Π_{max} denotes the maximum load on any machine, i.e., $\Pi_{max} = \max_i\{\sum_j x_{i,j}p_{i,j}\}$. Our algorithm for the $R|forest|C_{max}$ problem involves the following two steps:

Step 1:   We first construct a processor assignment for which the value of $\max\{P_{\max}, \Pi_{\max}\}$ is within a constant factor $((3 + \sqrt{5})/2)$ of the smallest-possible. This is described in Section 2.1.

Step 2:   Solve the GDSS problem we get from the previous step to get a schedule of length polylogarithmically more than $\max\{P_{\max}, \Pi_{\max}\}$. This is described in Section 2.2.

2.1   Step 1: A Processor Assignment Within a Constant Factor of $\max\{P_{\max}, \Pi_{\max}\}$

We now describe the algorithm for processor assignment, using some of the ideas from Lenstra *et al.* [13]. Let T be our "guess" for the optimal value of $LB = \max\{P_{\max}, \Pi_{\max}\}$. Define $S_T = \{(i, j) \mid p_{ij} \leq T\}$. Let J and M denote the set of jobs and machines, respectively. We now define a family of linear programs $LP(T)$, one for each value of $T \in \mathbf{Z}^+$, as follows: **(A1)** $\forall j \in J \sum_i x_{ij} = 1$, **(A2)** $\forall i \in M \sum_j x_{ij}p_{ij} \leq T$, **(A3)** $\forall j \in J z_j = \sum_i p_{ij}x_{ij}$, **(A4)** $\forall (j' \prec j)c_j \geq c_{j'} + z_j$, **(A5)** $\forall j \in J c_j \leq T$. The constraints (A1) ensure that each job is assigned a machine, constraints (A2) ensure that the maximum fractional load on any machine ($\Pi_{\max}$) is at most T. Constraints (A3) define the fractional processing time z_j for a job j and (A4) capture the precedence constraints amongst jobs

(c_j denotes the fractional completion of time of job j). We note that $\max_j c_j$ is the fractional P_{max}. Constraints (A5) state that the fractional $P_{\max}$ value is at most T.

Let T^* be the smallest value of T for which $LP(T)$ has a feasible solution. It is easy to see that T^* is a lower bound on LB. We now present a rounding scheme which rounds a feasible fractional solution $LP(T^*)$ to an integral solution. Let X_{ij} denote the indicator variable which denotes if job j was assigned to machine i in the integral solution, and let C_j be the integer analog of c_j and z_j. We first modify the x_{ij} values using *filtering* (Lin and Vitter [16]). Let $\mu = \frac{3+\sqrt{5}}{2}$. For any (i, j), if $p_{ij} > \mu z_j$, then set x_{ij} to zero. This step could result in a situation where, for a job j, the fractional assignment $\sum_i x_{ij}$ drops to a value r such that $r \in [1 - \frac{1}{\mu}, 1)$. So, we scale the (modified) values of x_{ij} by a factor of at most $\gamma = \frac{\mu}{\mu-1}$. Let $\mathcal{A}$ denote this fractional solution. Crucially, we note that any rounding of $\mathcal{A}$, which ensures that only non-zero variables in $\mathcal{A}$ are set to non-zero values in the integral solution, has an integral $P_{\max}$ value which is at most μT^*. This follows from the fact that if $X_{ij} = 1$ in the rounded solution, then $p_{ij} \leq \mu z_j$. Hence, it is easy to see that by induction, for any job j, C_j is at most $\mu c_j \leq \mu T^*$.

We now show how to round $\mathcal{A}$. Recall that Lenstra *et al.* [13] present a rounding algorithm for unrelated parallel machines scheduling *without* precedence constraints with the following guarantee: if the input fractional solution has a fractional $\Pi_{\max}$ value of x, then the output integral solution has an integral $\Pi_{\max}$ value of at most $x + \max_{x_{ij}>0} p_{ij}$. We use $\mathcal{A}$ as the input instance for the rounding algorithm of Lenstra *et al.* [13]. Note that $\mathcal{A}$ has a fractional $\Pi_{\max}$ value of at most γT^*. Further, $\max_{x_{ij}>0} p_{ij} \leq T^*$. This results in an integral solution I whose $P_{\max}$ value is at most μT^*, and whose $\Pi_{\max}$ value is at most $(\gamma+1)T^*$. Observe that, setting $\mu = \frac{3+\sqrt{5}}{2}$ results in $\mu = \gamma + 1$. Finally, we note that the optimal value of T can be arrived at by a bisection search in the range $[0, np_{\max}]$, where $n = |J|$ and $p_{\max} = \max_{i,j} p_{ij}$. Since T^* is a lower bound on LB, we have the following result.

Theorem 1. *The above algorithm computes a processor assignment for each job such that the value of* $\max\{P_{\max}, \Pi_{\max}\}$ *for the resulting assignment is within a* $(\frac{3+\sqrt{5}}{2})$-*factor of the optimal.*

2.2   Step 2: Solving the GDSS Problem Under Treelike Precedences

We first consider the case when the precedences are a collection of directed in-trees or out-trees. We then extend this to the case where the precedences form an arbitrary forest (i.e., the underlying undirected graph is a forest). Since the processor assignment is already specified in the GDSS problem, we will use the notation $m(v)$ to denote the machine to which node v is assigned. Also, since the machine is already fixed, the processing time for node v is also fixed, and is denoted by p_v.

GDSS on Out-/In-Arborescences. An out-tree is a tree rooted at some node, say r, with all edges directed away from r; an in-tree is a tree obtained by reversing all the directions in an out-tree. In the discussion below, we only focus on out-trees; the same results can be obtained for in-trees. The algorithm for out-trees requires a careful partitioning of the tree into blocks of chains, and giving random delays at the start of each chain in each of the blocks – thus the delays are spread all over the tree. The head of the chain waits for all its ancestors to finish running, after which it waits for an amount of time equal to its random delay. After this, the entire chain is allowed to run without interruption. Of course, this may result in an infeasible schedule where multiple jobs simultaneously contend for the same machine (at the same time). We show that this contention is low and can be resolved by expanding the infeasible schedule produced above.

Chain Decomposition. We define the notions of *chain decomposition* of a graph and its *chain width*; the decomposition for an out-directed arborescence is illustrated in Figure 1. Given a DAG $G(V, E)$, let $d_{in}(u)$ and $d_{out}(u)$ denote the in-degree and out-degree, respectively, of u in G. A *chain decomposition* of $G(V, E)$ is a partition of its vertex set into subsets $B_1, \ldots, B_\lambda$ (called *blocks*) such that the following properties hold: (i) The subgraph induced by each block B_i is a collection of vertex-disjoint directed chains, (ii) For any $u, v \in V$, let $u \in B_i$ be an ancestor of $v \in B_j$. Then, either $i < j$, or $i = j$ and u and v belong to the same directed chain of B_i, (iii) If $d_{out}(u) > 1$, then none of u's out-neighbors are in the same block as u. The *chain-width* of a DAG is the minimum value λ such that there is a chain decomposition of the DAG into λ blocks.

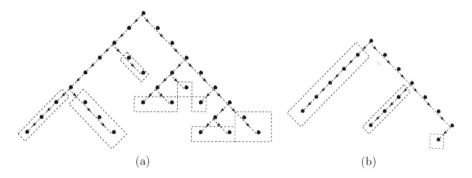

(a)   (b)

Fig. 1. Chain decomposition of an out-directed arborescence. The vertices enclosed within the boxes in figures 1(a) and 1(b) are in blocks B_3 and B_2 respectively while the remaining vertices are in block B_1.

Well Structured Schedules. We now state some definitions motivated by those in Goldberg *et al.* [6]. Given a GDSS instance with a DAG $G(V, E)$ and given a chain decomposition of G into λ blocks, we construct a *B-delayed schedule* for it as follows; B is an integer that will be chosen later. Each job v which is the head of a chain in a block is assigned a delay $d(v)$ in $\{0, 1, \ldots, B-1\}$. Let v be the head of chain C_i. Job v waits for $d(v)$ amount of time after all its predecessors have finished running, after which the jobs of C_i are scheduled consecutively (of course, the resulting schedule might be infeasible). A *random B-delayed schedule* is a B-delayed schedule in which all the delays have been chosen independently and uniformly at random from $\{0, 1, \ldots, B-1\}$. For a B-delayed schedule S, the *contention* $C(M_i, t)$ is the number of jobs scheduled on machine M_i in the time interval $[t, t+1)$. As in [6, 22], we assume w.l.o.g. that all job lengths are powers of two. This can be achieved by multiplying each job length by at most a factor of two (which affects our approximation ratios only by a constant factor). A delayed scheduled S is *well-structured* if for each k, all jobs with length 2^k begin in S at a time instant that is an integral multiple of 2^k. Such schedules can be constructed from randomly delayed schedules as follows. First create a new GDSS instance by replacing each job $v = (m(v), p_v)$ by the job $\hat{v} = (m(v), 2p_v)$. Let S be a random B-delayed schedule for this modified instance, for some B; we call S a *padded random B-delayed schedule*. From S, we can construct a well-structured delayed schedule, S', for the original GDSS instance as follows: insert v with the correct boundary in the slot assigned to $\hat{v}$ by S. S' will be called a *well-structured random B-delayed schedule* for the original GDSS instance.

Our Algorithm. We now describe our algorithm; for the sake of clarity, we occasionally omit floor and ceiling symbols (e.g., "$B = \lceil 2\Pi_{\max}/\log(np_{\max})\rceil$" is written as "$B = 2\Pi_{\max}/\log(np_{\max})$"). As before let $p_{\max} = \max_v p_v$.

1. Construct a chain decomposition of the DAG $G(V, E)$ and let λ be its chain width.
2. Let $B = 2\Pi_{\max}/\log(np_{\max})$. Construct a padded random B-delayed schedule S by first increasing the processing time of each job v by a factor of 2 (as described above), and then choosing a delay $d(v) \in \{0, \ldots, B-1\}$ independently and uniformly at random for each v.
3. Construct a well-structured random B-delayed schedule S' as described above.
4. Construct a valid schedule S'' using the technique from Goldberg *et al.* [6] as follows:
   (a) Let the makespan of S' be L.
   (b) Partition the schedule S' into *frames* of length $p_{\max}$; i.e., into the set of time-intervals $\{[ip_{\max}, (i+1)p_{\max}), \ i = 0, 1, \ldots, \lceil L/p_{\max}\rceil - 1\}$.
   (c) For each frame, use the frame-scheduling technique from [6] to produce a feasible schedule for that frame. Concatenate the schedules of all frames to obtain the final schedule.

The following theorem shows the performance guarantee of the above algorithm, when given a chain decomposition.

Theorem 2. *Let $p_{\max} = \max_{i,j} p_{ij}$. Given an instance of treelike GDSS and a chain decomposition of its DAG $G(V, E)$ into λ blocks, the schedule S'' produced by the above algorithm has makespan $O(\rho \cdot (P_{\max} + \Pi_{\max}))$ with high probability, where $\rho = \max\{\lambda, \log n\} \cdot \lceil \log(\min\{p_{\max}, n\})/\log\log n\rceil$. Furthermore, the algorithm can be derandomized.*

The proof of Theorem 3 demonstrates a chain decomposition of width $O(\log n)$ for any out-tree: this completes the algorithm for an out-tree. An identical argument would work for the case of a directed in-tree. We note that the notions of chain decomposition and chain-width for the out-directed arborescences are similar to those of caterpillar decomposition and caterpillar dimension for trees (see Linial *et al.* [17]). However, in general, a caterpillar decomposition for an arborescence need not be a chain-decomposition and vice-versa.

Theorem 3. *There is a deterministic polynomial-time approximation algorithm for solving the GDSS problem when the underlying DAG is restricted to be an in/out tree. The algorithm computes a schedule with makespan $O((P_{\max} + \Pi_{max}) \cdot \rho)$, where $\rho = \log n \cdot \lceil \log(\min\{p_{\max}, n\}) / \log \log n \rceil$. In particular, we get an $O(\log n)$–approximation in the case of unit-length jobs.*

GDSS on Arbitrary Forest-Shaped DAGs. We now consider the case where the undirected graph underlying the DAG is a forest. The chain decomposition algorithm described for in/out-trees does not work for arbitrary forests: instead of following the approach for in/out-trees, we observe that once we have a chain decomposition, the problem restricted to a block of chains is precisely the job shop scheduling problem. This allows us to reduce the $R|forest|C_{max}$ problem to a set of job shop problems, for which we use the algorithm of Goldberg *et al.* [6]. While this is simpler than the algorithm for in/out-trees, we incur another logarithmic factor in the approximation guarantee.

The following lemma and theorem show that a good decomposition can be computed for forests which can be exploited to yield a good approximation ratio.

Lemma 1. *Every DAG T whose underlying undirected graph is a forest, has a chain decomposition into γ blocks, where $\gamma \leq 2(\lceil \lg n \rceil + 1)$.*

Theorem 4. *Given a GDSS instance and a chain decomposition of its DAG $G(V, E)$ into γ blocks, there is a deterministic polynomial-time algorithm which delivers a schedule of makespan $O((P_{\max} + \Pi_{\max}) \cdot \rho)$, where $\rho = \frac{\gamma \log n}{\log \log n} \lceil \frac{\log \min(p_{\max}, n)}{\log \log n} \rceil$. Thus, Lemma 1 implies that $\rho = O(\frac{\log^2 n}{\log \log n} \lceil \frac{\log \min(p_{\max}, n)}{\log \log n} \rceil)$ is achievable.*

3   The $R|forest|\sum_j w_j C_j$ Problem

We now consider the objective of minimizing weighted completion time, where the given weight for each job j is $w_j \geq 0$. Given an instance of $R|prec|\sum_j w_j C_j$ where the jobs have not been assigned their processors, we now reduce it to instances of $R|prec|C_{\max}$ with processor assignment. More precisely, we show the following: let $P_{\max}$ and $\Pi_{\max}$ denote the "dilation" and "congestion" as usual; if there exists a schedule of makespan $\rho \cdot (P_{\max} + \Pi_{\max})$ for the latter, then there is a $O(\rho)$-approximation algorithm for the former. Let the machines and jobs be indexed by i and j; $p_{i,j}$ is the (integral) time for processing job j on machine i, if we choose to process j on i. We now present an LP-formulation for $R|prec|\sum_j w_j C_j$ which has the following variables: for $\ell = 0, 1, \ldots$, variable $x_{i,j,\ell}$ is the indicator variable which denotes if "job j is processed on machine i, and completes in the time interval $(2^{\ell-1}, 2^\ell]$"; for job j, C_j is its completion time, and z_j is the time spent on processing it. The objective is to minimize $\sum_j w_j C_j$ subject to: **(1)** $\forall j$, $\sum_{i,\ell} x_{i,j,\ell} = 1$, **(2)** $\forall j$, $z_j = \sum_i p_{i,j} \sum_\ell x_{i,j,\ell}$, **(3)** $\forall (j \prec k)$, $C_k \geq C_j + z_j$, **(4)** $\forall j$, $\sum_{i,\ell} 2^{\ell-1} x_{i,j,\ell} < C_j \leq \sum_{i,\ell} 2^\ell x_{i,j,\ell}$, **(5)** $\forall (i, \ell)$, $\sum_j p_{i,j} \sum_{t \leq \ell} x_{i,j,t} \leq 2^\ell$, **(6)** $\forall \ell$ $\forall$maximal chains $\mathcal{P}$, $\sum_{j \in \mathcal{P}} \sum_i p_{i,j} \sum_{t \leq \ell} x_{i,j,t} \leq 2^\ell$, **(7)** $\forall (i, j, \ell)$, $(p_{i,j} > 2^\ell) \Rightarrow x_{i,j,\ell} = 0$, **(8)** $\forall (i, j, \ell)$, $x_{i,j,\ell} \geq 0$

Note that (5) and (6) are "congestion" and "dilation" constraints respectively. Our reduction proceeds as follows. Solve the LP, and let the optimal fractional solution be denoted by variables $x_{i,j,\ell}^*$, C_j^*, and z_j^*. We do the following filtering, followed by an assignment of jobs to (machine, time-frame) pairs.

Filtering: For each job j, note from the first inequality in (4) that the total "mass" (sum of $x_{i,j,\ell}$ values) for the values $2^\ell \geq 4C_j^*$, is at most $1/2$. We first set $x_{i,j,\ell} = 0$ if $2^\ell \geq 4C_j^*$, and scale each $x_{i,j,\ell}$ to $x_{i,j,\ell}/(1 - \sum_{\ell' \geq 4C_j^*} \sum_i x_{i,j,\ell'})$, if ℓ is such that $2^\ell < 4C_j^*$ – this ensures that equation (1) still holds. After the filtering, each non-zero variable increases by at most a factor if 2. Additionally, for any fixed j, the following property is satisfied: consider the largest value of ℓ such that $x_{i,j,\ell}$ is non-zero; let this value be ℓ'; then, $2^{\ell'} = O(C_j^*)$. The right-hand-sides of (5) and (6) become at most $2^{\ell+1}$ in the process and the C_j values increase by at most a factor of two.

Assigning Jobs to Machines and Frames. For each j, set $F(j)$ to be the frame $(2^{\ell-1}, 2^\ell]$, where ℓ is the index such that $4C_j^* \in F(j)$. Let $G[\ell]$ denote the sub-problem which is restricted to the jobs in this frame. Let $P_{\max}(\ell)$ and $\Pi_{\max}(\ell)$ be the fractional congestion and dilation, respectively, for the sub-problem restricted to $G[\ell]$. From constraints (5) and (6), and due to our filtering step, which at most doubles any non-zero variable, it follows that both $P_{\max}(\ell)$ and $\Pi_{\max}(\ell)$ are $O(2^\ell)$. We now perform a processor assignment as follows: for each $G[\ell]$, we use the processor assignment scheme in Section 2.1 to assign processors to jobs. This ensures that the integral $P_{\max}(\ell)$ and $\Pi_{\max}(\ell)$ values are at most a constant times their fractional values.

Scheduling: First schedule all jobs in $G[1]$; then schedule all jobs in $G[2]$, and so on. We can use any approximation algorithm for makespan-minimization, for each of these scheduling steps. It is easy to see that we get a feasible solution: for any two jobs j_1, j_2, if $j_1 \prec j_2$, then $C_{j_1}^* \leq C_{j_2}^*$ and frame $F(j_1)$ occurs before $F(j_2)$ and hence gets scheduled first.

Theorem 5. *If there exists an approximation algorithm which yields a schedule whose makespan is $O((P_{\max} + \Pi_{\max}) \cdot \rho)$, then there is also an $O(\rho)$–approximation algorithm for minimizing weighted completion time. Thus, Theorem 4 implies that $\rho = O(\frac{\log^2 n}{\log \log n} \lceil \frac{\log \min(p_{\max}, n)}{\log \log n} \rceil)$ is achievable.*

Proof. Consider any job j which belongs to $G[\ell]$; both $P_{\max}(\ell)$ and $\Pi_{\max}(\ell)$ are $O(2^\ell)$, the final completion time of job j is $O(\rho 2^\ell)$. Since $2^\ell = O(C_j^*)$, the theorem follows.

4   Weighted Completion Time and Flow Time on Chains

We now consider the case of $R|prec|\sum_j w_j C_j$, where the processor-assignment is *not* prespecified. We consider the *restricted-assignment variant* of this problem [13, 21], where for every job v, there is a value p_v such that for all machines i, $p_{i,v} \in \{p_v, \infty\}$. Let $S(v)$ denote the set of machines such that $p_{i,v} = p_v$. We focus on the case where the precedence DAG is a disjoint union of chains with all p_u being polynomially-bounded positive integers in the input-size N.

We now present our approximation algorithms for weighted completion time. The algorithm and proof techniques for flow time is similar to weighted completition time and is omitted from this version.

Weighted Completion Time. Recall that $N = \max_v\{n, m, p_v\}$ denote the "input size". In this section, we obtain an $O(\log N/\log\log N)$-approximation algorithm for the minimum weighted completion time problem. We first describe an LP relaxation. Let $T = \sum_v p_v$. In the following LP, for ease of exposition, we assume that all the p_v values are equal to one. Our algorithm easily generalizes to the case where the p_v values are arbitrary positive integers, and the LP is polynomial-sized if all p_v values are polynomial in N. Let $\prec$ denote the immediate predecessor relation, i.e., if $u \prec v$, then they both belong to the same chain and u is an immediate predecessor of v in this chain. Note that if v is the first job in its chain, then it has no predecessor. In the time-indexed LP formulation below, the variable $x_{v,i,t}$ denotes the fractional amount of job v that is processed on machine i at time t. The objective is $\min\sum_v w(v)C(v)$, subject to: **(B1)** $\forall v, \sum_{i\in S(v)}\sum_{t\in[1,\ldots,T]} x_{v,i,t} = 1$, **(B2)** $\forall\, i \in [1,\ldots m], \forall\, t \in [1,\ldots T]$, $\sum_v x_{v,i,t} \leq 1$, **(B3)** $\forall\, v, \forall\, t \in [1,\ldots T]$, $z_{v,t} = \sum_{i\in[1,\ldots m]} x_{v,i,t}$, **(B4)** $\forall\, u \prec v, \forall\, t \in [1,\ldots T]$, $\sum_{t'\in[1,\ldots,t]} z_{v,t'} \leq \sum_{t'\in[1,\ldots,t-1]} z_{u,t'}$, **(B5)** $\forall v, C(v) = \sum_{t\in[1,\ldots T]} t\cdot z_{v,t}$, **(B6)** $\forall\, v, \forall\, i \in S(v)$, $\forall t \in [1,\ldots,T]$, $x_{v,i,t} \geq 0$.

The constraints (B1) ensure that all jobs are processed completely, and (B2) ensure that at most one job is (fractionally) assigned to any machine at any time. The variable $z_{v,t}$ denotes the fractional amount of job v that has been processed on all machines at time t. Constraints (B3) and (B4) are the precedence constraints and (B6) define the completion time $C(v)$ for job v.

Our algorithm proceeds as follows. We first solve the above LP optimally. Let OPT be the optimal value of the LP and let x and z denote the optimal solution values in the rest of the discussion. We define a rounding procedure for each chain such that the following hold:

1. Let $Z_{v,t}$ be the indicator random variable which denotes if v is executed at time t in the rounded solution. Let $X_{v,i,t}$ be the indicator random variable which denotes if v is executed at time t on machine i in the rounded solution. Then $\mathbf{E}[Z_{v,t}] = z_{v,t}$ and $\mathbf{E}[X_{v,i,t}] = x_{v,i,t}$.

2. All precedence constraints are satisfied in the rounded solution.
3. Jobs in different chains are rounded independently.

After the $Z_{v,t}$ values have been determined, we do the machine assignment as follows: if $Z_{v,t} = 1$, then job v is assigned to machine i with probability $(x_{v,i,t}/z_{v,t})$. In general, this assignment strategy might result in jobs from *different* chains executing on the same machine at the same time, and hence an *infeasible* schedule. Let C_1 denote the cost of this infeasible solution. Property **1** above ensures that $\mathbf{E}[C_1] = OPT$. Let Y be the random variable which denotes the maximum contention of any machine at any time. We obtain a feasible solution by "expanding" each time slot by a factor of Y.

We now show our rounding procedure for the jobs of a specific chain such that properties **1** and **2** hold; different chains are handled independently as follows: for each chain Γ, we choose a value $r(\Gamma) \in [0, 1]$ uniformly and independently at random. For each job v belonging to chain Γ, $Z_{v,t'} = 1$ iff $\sum_{t=1}^{t'-1} z_{v,t} < r(\Gamma) \leq \sum_{t=1}^{t'} z_{v,t}$. Bertsimas *et al.* [1] show other applications for such rounding techniques. A moment's reflection shows that property **1** holds due to the randomized rounding and property **2** holds due to Equation (B4). A straight forward application of the Chernoff-type bound from [18] yields the following lemma:

Lemma 2. *Let $\mathcal{E}$ denote the event that $Y \leq (\alpha \log N / \log \log N)$, where $\alpha > 0$ is a suitably large constant. Event $\mathcal{E}$ occurs after the randomized machine assignment with high probability: this probability can be made at least $1 - 1/N^{\beta}$ for any desired constant $\beta > 0$, by letting the constant α be suitably large.*

Finally, we note that we expand an infeasible schedule only if the event $\mathcal{E}$ occurs. Otherwise, we can repeat the randomized machine assignment until event $\mathcal{E}$ occurs and expand the resultant infeasible schedule. Let the final cost of our solution be C. We now have an $O(\log N / \log \log N)$-approximation as follows:

$$\mathbf{E}[C \mid \mathcal{E}] \leq \mathbf{E}[O\left(\frac{\log N}{\log \log N}\right) \cdot C_1 \mid \mathcal{E}] \leq O\left(\frac{\log N}{\log \log N}\right) \cdot \frac{\mathbf{E}[C_1]}{\Pr[\mathcal{E}]} \leq O\left(\frac{\log N}{\log \log N}\right) \cdot OPT.$$

Acknowledgments

We are thankful to David Shmoys and the anonymous APPROX 2005 referees for valuable comments.

References

1. D. Bertsimas, C.-P. Teo and R. Vohra. *On Dependent Randomized Rounding Algorithms*. Operations Research Letters, 24(3):105–114, 1999.
2. C. Chekuri and M. Bender. *An Efficient Approximation Algorithm for Minimizing Makespan on Uniformly Related Machines*. Journal of Algorithms, 41:212–224, 2001.
3. C. Chekuri and S. Khanna. *Approximation algorithms for minimizing weighted completion time*. Handbook of Scheduling, 2004.

4. F. A. Chudak and D. B. Shmoys. *Approximation algorithms for precedence-constrained scheduling problems on parallel machines that run at different speeds.* Journal of Algorithms, 30(2):323–343, 1999.
5. U. Feige and C. Scheideler. *Improved bounds for acyclic job shop scheduling.* Combinatorica, 22:361–399, 2002.
6. L.A. Goldberg, M. Paterson, A. Srinivasan and E. Sweedyk. *Better approximation guarantees for job-shop scheduling.* SIAM Journal on Discrete Mathematics, Vol. 14, 67–92, 2001.
7. L. Hall. *Approximation Algorithms for Scheduling.* in *Approximation Algorithms for NP-Hard Problems*, Edited by D. S. Hochbaum. PWS Press, 1997.
8. L. Hall, A. Schulz, D.B. Shmoys, and J. Wein. *Scheduling to minimize average completion time: Offline and online algorithms.* Mathematics of Operations Research, 22:513–544, 1997.
9. K. Jansen and L. Porkolab, *Improved Approximation Schemes for Scheduling Unrelated Parallel Machines.* Proc. ACM Symposium on Theory of Computing (STOC), pp. 408–417, 1999.
10. K. Jansen and R. Solis-oba. *Scheduling jobs with chain precedence constraints.* Parallel Processing and Applied Mathematics, PPAM, LNCS 3019, pp. 105–112, 2003.
11. K. Jansen, R. Solis-Oba and M. Sviridenko. *Makespan Minimization in Job Shops: A Polynomial Time Approximation Scheme.* Proc. ACM Symposium on Theory of Computing (STOC), pp. 394–399, 1999.
12. S. Leonardi and D. Raz. In *Approximating total flow time on parallel machines.* In Proc. ACM Symposium on Theory of Computing, 110–119, 1997.
13. J. K. Lenstra, D. B. Shmoys and É. Tardos. *Approximation algorithms for scheduling unrelated parallel machines.* Mathematical Programming, Vol. 46, 259–271, 1990.
14. F.T. Leighton, B. Maggs and S. Rao. *Packet routing and jobshop scheduling in O(congestion + dilation) Steps,* Combinatorica, Vol. 14, 167–186, 1994.
15. F. T. Leighton, B. Maggs, and A. Richa, *Fast algorithms for finding O(congestion + dilation) packet routing schedules.* Combinatorica, Vol. 19, 375–401, 1999.
16. J. H. Lin and J. S. Vitter. *ε-approximations with minimum packing constraint violation.* In Proceedings of the ACM Symposium on Theory of Computing, 1992, pp. 771–782.
17. N. Linial, A. Magen, and M.E. Saks. *Trees and Euclidean Metrics.* In Proceedings of the ACM Symposium on Theory of Computing, 169–175, 1998.
18. A. Panconesi and A. Srinivasan. *Randomized distributed edge coloring via an extension of the Chernoff-Hoeffding bounds,* SIAM Journal on Computing, Vol. 26, 350–368, 1997.
19. M. Queyranne and M. Sviridenko. *Approximation algorithms for shop scheduling problems with minsum objective,* Journal of Scheduling, Vol. 5, 287–305, 2002.
20. A. Schulz and M. Skutella. *The power of α-points in preemptive single machine scheduling.* Journal of Scheduling 5(2): 121–133, 2002.
21. P. Schuurman and G. J. Woeginger. *Polynomial time approximation algorithms for machine scheduling: Ten open problems.* Journal of Scheduling 2:203–213, 1999.
22. D.B. Shmoys, C. Stein and J. Wein. *Improved approximation algorithms for shop scheduling problems,* SIAM Journal on Computing, Vol. 23, 617–632, 1994.
23. M. Skutella. *Convex quadratic and semidefinite relaxations in scheduling.* Journal of the ACM, 46(2):206–242, 2001.

Approximation Algorithms for Network Design and Facility Location with Service Capacities

Jens Maßberg and Jens Vygen

Research Institute for Discrete Mathematics, University of Bonn
Lennéstr. 2, 53113 Bonn, Germany

Abstract. We present the first constant-factor approximation algorithms for the following problem: Given a metric space (V, c), a set $D \subseteq V$ of terminals/ customers with demands $d : D \to \mathbb{R}_+$, a facility opening cost $f \in \mathbb{R}_+$ and a capacity $u \in \mathbb{R}_+$, find a partition $D = D_1 \dot{\cup} \cdots \dot{\cup} D_k$ and Steiner trees T_i for D_i $(i = 1, \ldots, k)$ with $c(E(T_i)) + d(D_i) \leq u$ for $i = 1, \ldots, k$ such that $\sum_{i=1}^{k} c(E(T_i)) + kf$ is minimum.

This problem arises in VLSI design. It generalizes the bin-packing problem and the Steiner tree problem. In contrast to other network design and facility location problems, it has the additional feature of upper bounds on the service cost that each facility can handle.

Among other results, we obtain a 4.1-approximation in polynomial time, a 4.5-approximation in cubic time and a 5-approximation as fast as computing a minimum spanning tree on (D, c).

1 Introduction

Facility location and network design problems have received much attention in the last decade. Approximation algorithms have been designed for many variants. In most cases, a set of terminals/customers, possibly with demands, has to be served by one or more facilites. A facility serves its customers either by direct links (a star centered at the facility) or by an arbitary connected network (a Steiner tree). Facilities (and sometimes edges) often have capacities, i.e. they can serve customers of limited total demand only.

In contrast to all previously considered models, our facilites have a capacity bounding total customer demand plus service cost. This is an essential constraint in VLSI design, but it may well occur in other applications.

1.1 Problem Statement

We use standard notation (see, e.g., [13]) and consider the following problem: Given a metric space (V, c), a set $D \subseteq V$ of terminals/customers with demands $d : D \to \mathbb{R}_+$, a facility opening cost $f \in \mathbb{R}_+$ and a capacity $u \in \mathbb{R}_+$, find a partition $D = D_1 \dot{\cup} \cdots \dot{\cup} D_k$ and Steiner trees T_i for D_i $(i = 1, \ldots, k)$ with

$$c(E(T_i)) + d(D_i) \leq u \qquad \text{for } i = 1, \ldots, k \qquad (1)$$

such that

C. Chekuri et al. (Eds.): APPROX and RANDOM 2005, LNCS 3624, pp. 158–169, 2005.

$$\sum_{i=1}^{k} c(E(T_i)) + kf \tag{2}$$

is minimum.

So the objective function is the sum of service cost (total length of Steiner trees) and facility cost (f times the number of Steiner trees).

1.2 Results

We describe three approximation algorithms for this problem. Their running times are dominated by the first step, in which algorithm A computes a minimum spanning tree on (D, c), algorithm B computes an approximate Steiner tree for D in (V, c'), and algorithm C computes an approximate tour in (D, c''). Here c' and c'' are new metric cost functions.

Let α be the Steiner ratio (supremum of length of a minimum spanning tree over length of a optimum Steiner tree). Let β and γ be the performance ratios of Steiner tree and TSP approximation algorithms, respectively.

Then algorithm A has performance guarantee $1 + 2\alpha$, algorithm B has performance guarantee $\max\{3.705, 3\beta\}$, and algorithm C has performance guarantee $\max\{4, 3\gamma\}$.

This means that we have a 4.5-approximation for general metrics, using Christofides' algorithm [2], where the most time-consuming step is computing a minimum weight perfect matching on a subset of customers.

Moreover, we have a 5-approximation for general metrics, whose running time is dominated by finding a minimum spanning tree in (D, c). Finally, if we run algorithm B with the Robins-Zelikovsky algorithm, we even get a 4.099-approximation algorithm, however only with an enourmous (though polynomial) running time.

1.3 Motivation: VLSI Design

The problem arises in VLSI design when designing a clock network. The storage elements (flip-flops, latches) of a chip (in the following: customers) need to get a periodic clock signal. Often they have to receive it directly from special modules (clock splitters, local clock buffers), in the following: facilities.

The task is to place a set of facilites on the chip and connect each customer to a facility via a Steiner tree consisting of horizontal and vertical wires. As each facility can drive only a limited electrical capacitance, this means that there is an upper bound on the weighted sum of the wire length and the total input capacitance of the customers served by each facility.

At this stage the main goal is to minimize power consumption. Wires (proportional to their length) as well as facilities (proportional to their number) contribute to the power consumption. As we can always place a facility somewhere on a Steiner tree connecting the customers, we arrive at our problem formulation. In this case, the metric is $(\mathbb{R}^2, \ell_1)$.

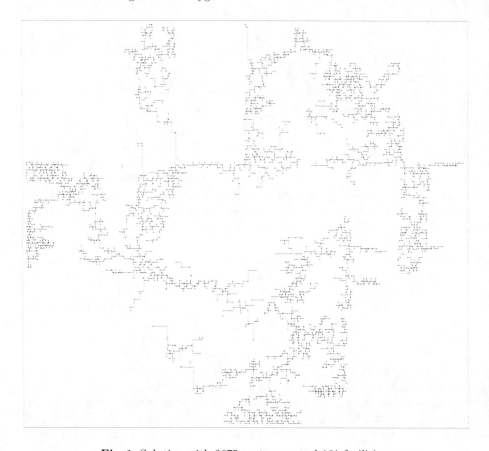

Fig. 1. Solution with 3675 customers and 161 facilities.

A state-of-the-art industrial clocktree design tool is described in [9]. Of course, this has many other components. For example, the facilities have to be served by a clock signal, too. More importantly, each customer is associated with a time interval, and the time intervals of customers served by the same facility must have nonzero intersection. We do not know whether there is an approximation algorithm for the resulting generalization. However, for many clock networks we can allow zero skew, i.e. the intersection of all time intervals is nonempty. In this case the problem considered in this paper is an excellent model of the practical problem. Experimental results are shown in section 3.5.

Real life clocktree instances can have more than 100 000 terminals. Figure 1 shows a solution for a practical instance with 3675 customers (inst1 of section 3.5).

1.4 Related Work

To our knowledge, our problem has not been considered yet in the literature. However, it generalizes several classical NP-hard combinatorial optimization problems.

It includes the Steiner tree problem: simply set u and f large enough. The currently best known approximation algorithm for Steiner trees, by Robins and Zelikovsky [16], achieves an approximation ratio of 1.55.

It also includes the soft-capacitated facility location problem in the case where all elements of V are identical potential facilities. The best known approximation ratio for the soft-capacitated facility location problem is 3 [3], and 2 in the case of uniform demands [14].

Our problem reduces to the bin-packing problem if there are no service costs, i.e. $c \equiv 0$. The best approximation ratio for bin-packing is 1.5 (unless P=NP) [17], although there is a fully polynomial asymptotic approximation scheme [11, 13].

Finally, our problem is loosely related to several clustering problems like the k-tree cover problem (e.g. [5]), the vehicle routing problem (e.g. [18]), and other network design problems such as connected facility location or the rent-or-buy problem.

1.5 Complexity

It is obvious that the problem is strongly NP-hard and MAXSNP-hard as it contains the Steiner tree problem and the bin-packing problem. Moreover we can show the following:

Proposition 1. *There is no* $(2 - \epsilon)-approximation$ *algorithm (for any $\epsilon > 0$) for any class of metrics where the Steiner tree problem cannot be solved exactly in polynomial time.*

Proof. Otherwise the Steiner tree decision problem which is NP-complete (see [7]) can be solved in polynomial time:

Assume we have a $(2 - \epsilon)-approximation$ algorithm for some $\epsilon > 0$, and $S = \{s_1, \ldots, s_n\}$, $k \in \mathbb{R}_+$, is an instance for the Steiner tree decision problem "Is there a Steiner tree with terminals S and length $\leq k$?". We construct an instance for our problem by taking S as the set of terminals, setting $l(s) = 0 \ \forall s \in S$, $u = k$ and $f = k\frac{2-\epsilon}{\epsilon}$. Then the $(2-\epsilon)-approximation$ algorithm computes a solution consisting of one facility if and only if there is a Steiner tree of length $\leq k$. □

For example, this applies to the class of finite metric spaces [12] as well as the Euclidean plane $(\mathbb{R}^2, \ell_2)$ [6] and the rectilinear plane $(\mathbb{R}^2, \ell_1)$ [7].

Regardless of the metric, there is no $\frac{3}{2}$-approximation algorithm unless P = NP, as follows by transformation from the NP-complete PARTITION problem.

1.6 Geometric Instances

For geometric instances (in particular $(V, c) = (\mathbb{R}^2, \ell_p)$ with $p \in \{1, 2\}$) of the Steiner tree problem, the TSP and other problems, Arora [1] developped an approximation scheme. Similar techniques can be applied also to our problem to get a 2-approximation. In fact, we get $(1+\epsilon)$ times the service cost of an optimum solution (for any fixed positive ϵ), but due violated capacity constraints up to twice its facility cost.

By Proposition 1, this 2-approximation is best possible. However, Arora's algorithm is too slow for practical purposes, and we do not even know how to obtain a 3-approximation with a practically fast algorithm. Note that in our VLSI application we have the rectilinear plane $(\mathbb{R}^2, \ell_1)$, and better algorithms for this case (than the factor 4 that we will present) would be particularly interesting.

2 A Lower Bound

Recall that the Steiner ratio α for a given metric is the minimum $\alpha \in \mathbb{R}$ so that for all finite sets of points S the length of a minimum spanning tree on S is at most α times the length of a minimum Steiner tree on S. For general metrics we have $\alpha = 2$. In the metric space $(\mathbb{R}^2, \ell_1)$ it has been shown that $\alpha = \frac{3}{2}$ ([10]).

Let (V, c, D, u, f) be an instance of our problem. We define a *k-spanning (k-Steiner) forest* to be a forest F with $V(F) = D$ ($D \subseteq V(F) \subseteq V$, respectively) containing exactly k connected components.

Proposition 2. *Let α be the Steiner ratio of the given metric. For $k \in \{1, \ldots, n\}$ let $F^k_{Steiner}$ be the edge set of a minimum cost k-Steiner forest and F^k_{spann} the edge set of a minimum cost k-spanning forest. Then*

$$c(F^k_{Steiner}) \leq c(F^k_{spann}) \leq \alpha c(F^k_{Steiner}).$$

Proof. Every k-spanning forest is a k-Steiner forest, so the first inequality holds. If we replace every component T of $F^k_{Steiner}$ by a minimal spanning tree on $T \cap D$, the result is a k-spanning forest of cost at most $\alpha \cdot c(F_{Steiner})$. The cost of a minimal k-spanning forest cannot be greater. □

Now we construct a sequence $F_1, \ldots, F_n$, where (D, F_i) is an i-spanning forest of minimum cost. We start by choosing (D, F_1) as a minimum spanning tree. Let $e_1, \ldots, e_{n-1}$ be the edges of F_1 so that $c(e_1) \geq \ldots \geq c(e_{n-1})$. For $i = 2$ to n set $F_i = F_{i-1} \setminus \{e_{i-1}\}$.

Proposition 3. *(D, F_i) is an i-spanning forest of minimum cost for each i.*

Proof. By induction on i. (V, F_1) is a minimum cost 1-spanning forest by construction. Removing an edge increases the number of components by one, so by induction F_i is an i-spanning forest.

In order to show the optimality of F_i we recall the fact that for two forests $(D, W_1), (D, W_2)$ with $|W_1| < |W_2|$ there is an edge $e \in W_2 \setminus W_1$ so that $(D, W_1 \cup \{e\})$ is still a forest (matroid property, see e.g. [13, p. 281]).

Assume that F_{i-1} is an $i-1$-spanning forest of minimum cost, and consider F_i. Let F be an i-spanning forest of minimum cost and $e \in F_{i-1} \setminus F$ so that

$F \cup \{e\}$ is still a forest. Recalling that e_{i-1} is the most expensive edge of F_{i-1} and $F \cup \{e\}$ is an $(i-1)$-forest we get

$$
\begin{aligned}
c(F_i) + c(e_{i-1}) = c(F_{i-1}) \\
\leq c(F \cup \{e\}) \\
= c(F) + c(e) \\
\leq c(F) + c(e_{i-1})
\end{aligned}
$$

so (D, F_i) has minimum cost. □

We have shown:

Corollary 1. *The cost of any k-Steiner forest is at least $\frac{1}{\alpha}c(F_k)$.*

Proof. This follows from Proposition 2 and Proposition 3. □

The next step is to compute a lower bound on the number of facilities (Steiner trees) of a feasible solution. A *feasible* k-Steiner forest is a k-Steiner forest where inequality (1) holds for each of the components $T_1, \ldots, T_k$.

Proposition 4. *If $\frac{1}{\alpha}c(F_t) + d(D) > t \cdot u$, then there is no feasible t-Steiner forest.*

Proof. Let F be a feasible t-Steiner forest with components $T_1, \ldots, T_t$. Summing up all t inequalities (1) we get

$$
c(E(F)) + d(D) = \sum_{i=1}^{t} (c(E(T_i)) + d(D \cap V(T_i))) \leq t \cdot u.
$$

By Corollary 1 we know that $c(E(F)) \geq \frac{1}{\alpha}c(F_t)$, so $\frac{1}{\alpha}c(F_t) + d(D) \leq t \cdot u$. □

Let t' be the smallest positive integer so that

$$
\frac{1}{\alpha}c(F_{t'}) + d(D) \leq t' \cdot u.
$$

Thus t' is a lower bound for the number of facilities (Steiner trees) of a feasible solution. We conclude:

Theorem 1. $\min_{t \geq t'} \left(\frac{1}{\alpha}c(F_t) + t \cdot f \right)$ *is a lower bound for the cost of an optimal solution.*

We denote by t'' the smallest t for which the minimum is obtained, $L_r := \frac{1}{\alpha}c(F_{t''})$, and $L_f := t'' \cdot f$. Then $L_r + L_f$ is a lower bound on the cost of an optimum solution, and

$$
L_r + d(D) \leq L_f \frac{u}{f}. \tag{3}
$$

3   The Algorithms

We propose three algorithms. All of them start by constructing a forest with relatively small total cost. Then they split connected components violating (1).

For $D' \subseteq D$ and a tree T' with $D' \subseteq V(T') \subseteq V$ we define $\mathrm{load}(T', D') := c(E(T')) + d(D')$. The pair (D', T') is called *overloaded* if $\mathrm{load}(T', D') > u$. Overloaded components will be split.

3.1   Algorithm A

Algorithm A first computes $F_{t''}$. If $(T', D \cap V(T'))$ is not overloaded for any component T of $F_{t''}$, we have a feasible solution costing at most α times the optimum.

Otherwise let T' be a connected component of $F_{t''}$ where $(T', D \cap V(T'))$ is overloaded. We will split off subtrees and reduce the load by at least $\frac{u}{2}$ for each additional component.

Choose an arbitrary vertex $r \in V$ and orient T' as an arborescence rooted at r. For a vertex $v \in V$ we denote by T_v the subtree of T' rooted at v.

Now let $v \in V$ be a vertex of maximum distance (number of edges) from r with $\mathrm{load}(T_v, D \cap V(T_v)) > u$. If there is a successor w of v with $\mathrm{load}(T_w, D \cap V(T_w)) + c(v, w) \geq \frac{u}{2}$, then we split off T_w by removing the edge (v, w).

Let $v_1, \ldots, v_q$ be the successors of v with $x_i := \mathrm{load}(T_{v_i}, D \cap V(T_{v_i})) + c(v, v_i) \leq u$ $(i = 1, \ldots, q)$. Regard $x_1, \ldots, x_q$ and $x_{q+1} := d(v)$ as an instance of the bin-packing problem with bin capacity u, and apply the next-fit algorithm for bin packing. The result is a partition $\{1, \ldots, q+1\} = b_1 \dot{\cup} \cdots \dot{\cup} b_p$ with $\sum_{i \in b_j} x_i \leq u$ for $j = 1, \ldots, p$ and $\sum_{i \in b_j \cup b_{j+1}} x_i > u$ for $j = 1, \ldots, p-1$.

We split off the subtrees corresponding to the first $p' := 2\lfloor \frac{p}{2} \rfloor$ bins, and thus reduce the load by more than $\frac{u}{2} p'$.

More precisely, let $V_i := V(T_{v_i})$ $(i = 1, \ldots, q)$ and $V_{q+1} := \{v\}$. We replace T' by $p'+1$ trees: For $j = 1, \ldots, p'$ we have the pair $(T'[\bigcup_{i \in b_j} V_i \cup \{v\}], D \cap \bigcup_{i \in b_j} V_i)$ (which is not overloaded; note that v just serves as a Steiner point and does not contribute to the load unless $q + 1 \in b_j = b_p$), and finally we have $T'' := T' - \bigcup_{i \in b_1 \cup \cdots \cup b_{p'}} V_i$. Note that $\mathrm{load}(T'', D \cap V(T'')) \leq \mathrm{load}(T', D \cap V(T')) - \frac{u}{2} p'$.

Splitting continues until no overloaded components are left.

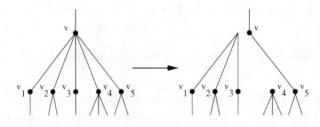

Fig. 2. Splitting an overloaded component. In this example two new trees have been split off. The new component including v_1, v_2 and v_3 has got an additional Steiner point at the position of v.

The number of new components generated for tree T is at most $\frac{2}{u}load(T)$. Thus the cost of the solution computed by algorithm A is at most

$$c(F_{t''}) + t''f + \frac{2}{u}(c(F_{t''}) + d(D))f = \alpha L_r + L_f + \frac{2f}{u}(\alpha L_r + d(D)).$$

Using (3) we get:

Lemma 1. *Algorithm A computes a solution of cost at most* $\alpha L_r + 3L_f + 2\frac{f}{u}(\alpha - 1)L_r$. $\qquad\square$

As $\frac{f}{u}L_r \leq L_f$ by (3), we get:

Corollary 2. *Algorithm A computes a solution of cost at most* $(2\alpha + 1)$ *the optimum.* $\qquad\square$

For small values of $\frac{f}{u}$, the performance ratio is better, as can be seen by analysing Lemma 1:

Corollary 3. *For instances with* $\frac{f}{u} \leq \phi$, *algorithm A computes a solution of cost at most* $\max\{3, \frac{\alpha+2\alpha\phi+\phi}{1+\phi}\}$ *times the optimum.*

3.2   Algorithm B

Let L_r^* and L_f^* be facility and service cost of an optimum solution, respectively. Clearly,

$$L_r^* + d(D) \leq L_f^* \frac{u}{f}. \qquad (4)$$

Let c' be the metric defined by $c'(v,w) := \min\{c(v,w), \frac{u}{2}\}$ for all $v, w \in V$. Algorithm B computes a Steiner tree F for D in (V, c') with some approximation algorithm. Then it deletes the edges of length $\frac{u}{2}$ from F. If the resulting forest contains overloaded components, they are split as in algorithm A (but with c' instead of c). Then the total number of components is $1 + \frac{2}{u}(c'(F) + d(D))$.

We have a total cost of

$$c'(F) + f + \frac{2f}{u}(c'(F) + d(D)) \qquad (5)$$

Note that an optimum solution can be extended to a Steiner tree for D by adding $\frac{L_f^*}{f} - 1$ edges. Hence there is a Steiner tree for D in (V, c') of length $L_r^* + (\frac{L_f^*}{f} - 1)\frac{u}{2}$. F is at most β times more expensive. Hence we can bound the total cost by

$$c'(F)(1 + \frac{2f}{u}) + f + \frac{2f}{u}d(D)$$

$$\leq \beta(L_r^* + \frac{uL_f^*}{2f} - \frac{u}{2})(1 + \frac{2f}{u}) + f + \frac{2f}{u}d(D)$$

$$\leq \beta L_r^* + \frac{2\beta f L_r^*}{u} + \frac{\beta L_f^* u}{2f} + \beta L_f^* + \frac{2f}{u}d(D)$$

$$\leq \beta L_r^* + \frac{2\beta f L_r^*}{u} + \frac{\beta L_f^* u}{2f} + \beta L_f^* + 2L_f^* - 2\frac{f}{u}L_r^*,$$

where we used (4) in the last inequality.

The right-hand side is a convex function in $\frac{f}{u}$, and it is equal to $\frac{3}{2}\beta L_r^* + 3\beta L_f^*$ for $\frac{f}{u} \in \{\frac{L_f^*}{L_r^*}, \frac{\beta}{4(\beta-1)}\}$, so it is at most $\frac{3}{2}\beta L_r^* + 3\beta L_f^*$ for all values in between. As $\frac{f}{u} \leq \frac{L_f^*}{L_r^*}$ by (4), we have:

Lemma 2. *Algorithm B yields a 3β-approximation unless $\frac{f}{u} < \frac{\beta}{4(\beta-1)}$*

A simple calculation shows:

Theorem 2. *Running algorithm A and algorithm B and taking the better solution yields a performance ratio of* $\max\{3\beta, \frac{13+6\alpha+\sqrt{169-84\alpha+36\alpha^2}}{10}\}$. *In particular we have a performance ratio of* $\max\{3\beta, 3.705\}$ *for any metric.*

In particular, using the Robins-Zelikovsky [16] algorithm we get a 4.648-approximation in polynomial time.

However, we can do better by a closer look at this algorithm. Recall that the Robins-Zelikovsky algorithm works with a parameter k and analyzes all k-restricted full components; its running time is $O(n^{4k})$ and the length of the Steiner tree that it computes is at most

$$c'(E(Y^*)) + c'(L^*)\ln\left(1 + \frac{c'(F_1) - c'(E(Y^*))}{c'(L^*)}\right),$$

where (D, F_1) is a minimum spanning tree, Y^* is any k-restricted Steiner tree for D in (V, c'), and L^* is a *loss* of Y^*, i.e. a minimum cost subset of $E(Y^*)$ connecting each Steiner point of degree at least three in Y^* to a terminal.

Let $\epsilon > 0$ fixed. For $k \geq 2^{\lceil \frac{1}{\epsilon} \rceil}$, the optimum k-restricted Steiner tree is at most a factor $1 + \epsilon$ longer than an optimum Steiner tree [4]. Thus, taking an optimum k-restricting Steiner tree for each component of our optimum solution and adding edges to make the graph connected, we get a k-restricted Steiner tree Y^* with $c(E(Y^*)) \leq (1+\epsilon)L_r^* + (\frac{L_f}{f} - 1)\frac{u}{2}$. and with loss L^* of length $c(L^*) \leq \frac{1+\epsilon}{2}L_r^*$. Moreover, $c'(F_1) \leq \alpha L_r^* + (\frac{L_f}{f} - 1)\frac{u}{2}$.

We conclude

$$c'(F) \leq c'(E(Y^*)) + c'(L^*)\ln\left(1 + \frac{(\alpha - 1 - \epsilon)L_r^*}{c'(L^*)}\right)$$

$$\leq c'(E(Y^*)) + (\alpha - 1)L_r^* \frac{c'(L^*)}{(\alpha-1)L_r^*}\ln\left(1 + \frac{(\alpha-1)L_r^*}{c'(L^*)}\right)$$

$$\leq c'(E(Y^*)) + L_r^* \frac{1+\epsilon}{2}\ln(1 + 2(\alpha - 1)),$$

as $\max\{x\ln(1 + \frac{1}{x}) \mid 0 < x \leq \frac{1+\epsilon}{2(\alpha-1)}\} = \frac{1+\epsilon}{2(\alpha-1)}\ln(1 + \frac{2(\alpha-1)}{1+\epsilon})$.

For $\alpha = 2$ we get

$$c'(F) \leq (1+\epsilon)\left(1 + \frac{\ln 3}{2}\right)L_r^* + \left(\frac{L_f}{f} - 1\right)\frac{u}{2}.$$

Writing $\beta := (1+\epsilon)\left(1 + \frac{\ln 3}{2}\right)$ (note that $\beta < 1.5495$ for $\epsilon \leq 10^{-4}$), and plugging this into (5) we get that algorithm B computes a solution of total cost at most

$$c'(F)(1 + \frac{2f}{u}) + f + \frac{2f}{u}d(D)$$

$$\leq \beta L_r^* + \frac{2(\beta-1)fL_r^*}{u} + \frac{L_f^* u}{2f} + 3L_f^*,$$

which is at most $(\beta + \frac{1}{2})L_r^* + (2\beta+1)L_f^*$ unless $\frac{f}{u} < \frac{1}{4(\beta-1)} < \frac{1}{2}$. In this case we can run algorithm A to get a 3-approximation:

Theorem 3. *Running algorithm A if $2f < u$, and otherwise running algorithm B with the Robins-Zelikovsky algorithm with parameter $k = 2^{10000}$, is a 4.099-approximation algorithm.* $\square$

3.3   Algorithm C

Let c'' be the metric defined by $c''(v, w) := \min\{c(v, w), u\}$ for all $v, w \in V$. Algorithm C computes a tour F for D in (V, c'') using some approximation algorithm for the TSP.

We achieve the final Steiner forest by deleting one of the longest edges of F and splitting the resulting path into paths $P_1, \ldots, P_{k'}$ such that $\mathrm{load}(P_i, D \cap V(P_i)) \leq u$ and $\mathrm{load}(P_i', D \cap V(P_i')) > u$ for $i = 1, \ldots, k' - 1$, where P_i' is P_i plus the edge connecting P_i and P_{i+1}.

Due to the page limit, we only claim without proofs that

Theorem 4. *Algorithm C yields a 3γ-approximation unless $\frac{f}{u} < \frac{\gamma}{2(\gamma-1)}$. Running algorithm A and algorithm C and taking the better solution yields a performance ratio of $\max\{3\gamma, 4\}$ for any metric.*

In particular, using Christofides' algorithm [2] we have a 4.5-approximation algorithm. Details will be included in the full paper.

3.4   Running Times and Tight Examples

In algorithm A, the time to determine the initial minimum spanning tree (D, F_1) dominates the running time.

In algorithm B and C, the running times of the Steiner tree and TSP approximation algorithms, respectively, dominate the overall running time.

We can construct examples showing that the performance guarantee of algorithm A is not better than five, but we don't know whether algorithms B and C are tight.

3.5   Experimental Results

As mentioned in Section 1.3 the problem considered in this paper arises in VLSI design. We implemented algorithm A for the $(\mathbb{R}^2, \ell_1)$ metric and ran it on several instances from industrial chips.

168   Jens Maßberg and Jens Vygen

Table 1. Experimental results.

	inst1	inst2	inst3	inst4	inst5	inst6
number of terminals	3 675	17 140	45 606	54 831	109 224	119 461
total demand of terminals	9.09	38.54	97.62	121.87	250.24	258.80
min demand of terminals	0.0025	0.0022	0.0021	0.0021	0.0021	0.0021
max demand of terminals	0.0025	0.0087	0.0022	0.0086	0.0093	0.0023
facility opening cost f	0.127	0.127	0.127	0.127	0.127	0.127
capacity u	0.148	0.110	0.110	0.110	0.124	0.110
length min. spanning tree	13.72	60.35	134.24	183.37	260.36	314.48
lower bound cost	23.07	112.70	251.06	363.28	531.05	689.19
lower bound facilities	117	638	1475	2051	3116	3998
lower bound service cost	8.21	31.68	63.73	102.80	135.32	181.45
final solution cost	32.52	174.50	377.29	531.03	762.15	981.61
final number of facilities	161	947	2171	2922	4156	5525
final service cost	12.08	54.23	101.57	159.93	234.34	279.93
lower bound factor	1.41	1.55	1.59	1.46	1.44	1.42

To improve the results and to consider the special metric we changed the algorithm in three points.

By Lemma 1 the cost of the final solution is at most $\alpha L_r + 3L_f + 2\frac{f}{u}(\alpha - 1)L_f$. If we don't remove one of the edges e when constructing the initial forest, the number of facilities of the forest is reduced by one and by this the cost is decreased by f while the service cost is increased by $c(e)$. Moreover the additional service cost has to be driven by facilities which increases the upper bound by $\frac{2f}{u}c(e)$. So the approximation factor is not increased if we don't remove edges e with $c(e) + \frac{2f}{u}c(e) \leq f$. As the inital forest then has less components we probably get a partition with less facilities.

After constructing the initial forest, we recompute each Steiner tree using special ℓ_1 Steiner tree heuristics (Hanan's algorithm [8] for small instances, Matsuyama and Takahashi's algorithm [15] for all other instances) and take this tree if it is shorter than the spanning tree. So we reduce the service cost.

When splitting a component at a vertex we compute all feasible bin-packings and choose one with minimum number of bins that minimizes the load of the remaining tree. As w.l.o.g. each vertex of an ℓ_1-Steiner tree has at most degree 4, this doesn't increase the computation time signifiantly.

Table 1 shows the computational results of six instances coming from real clocktrees. The solution of inst1 is shown in figure 1. The computation time for each instance was less than 3 minutes.

By Corollary 3 each computed solution costs at most 3 times the optimum. In practice we get a factor of about 1.5 to the lower bound.

References

1. S. Arora. Polynomial time approximation schemes for euclidean tsp and other geometric problems. In *Journal of the ACM 45*, pages 753–782, 1998.

2. N. Christofides. Worst-case analysis of a new heuristic for the traveling salesman problem. Technical report, CS-93-13, G.S.I.A., Carnegie Mellon University, Pittsburgh, 1976.
3. F. A. Chudak and D. B. Shmoys. Improved approximation algorithms for a capacitated facility location problem. In *Proceedings of the 10th annual ACM-SIAM symposium on Discrete algorithms (SODA '99)*, pages 875–876, 1999.
4. D.-Z. Du, Y. Zhang, and Q. Feng. On better heuristic for Euclidean Steiner minimum trees. In *32nd Annual IEEE Symposium on Foundations of Computer Science*, pages 431–439, 1991.
5. G. Even, N. Garg, J. Könemann, R. Ravi, and A. Sinha. Covering graphs using trees and stars. In *Proceddings of 6th International Workshop on Approximation Algorithms for Combinatorial Optimization Problems and 7th International Workshop on Randomization and Approximation Techniques in Computer Science (APPROX-RANDOM)*, pages 24–35, 2003.
6. M. R. Garey, R. L. Graham, and D. S. Johnson. The complexity of computing steiner minimal trees. In *SIAM Journal on Applied Mathematics 32*, pages 835–859, 1977.
7. M.R. Garey and D.S. Johnson. The rectilinear steiner problem is np-complete. In *SIAM Journal on Applied Mathematics 32*, pages 826–834, 1977.
8. M. Hanan. On steiner's problem with rectilinear distance. *SIAM Journal on Applied Mathematics 14(2)*, pages 255–265, 1966.
9. S. Held, B. Korte, J. Maßberg, M. Ringe, and J. Vygen. Clock scheduling and clocktree construction for high performance asics. In *Proceedings of the 2003 IEEE/ACM international conference on Computer-aided design (ICCAD '03)*, pages 232–240. IEEE Computer Society, 2003.
10. F.K. Hwang. On steiner minimal trees with rectilinear distance. In *SIAM Journal on Applied Mathematics 30*, pages 104–114, 1976.
11. N. Karmarkar and R. M. Karp. An efficient approximation scheme for the one-dimensional bin packing problem. In *23rd Annual IEEE Symposium on Foundations of Computer Science*, pages 312–320, 1982.
12. R. M. Karp. Reducibility among combinatorial problems. In R. E. Miller and J. W. Thatcher, editors, *Complexity of Computer Computations*, pages 85–103. Plenum Press, New York, 1972.
13. B. Korte and J. Vygen. *Combinatorial Optimization, Theory and Algorithms, third edition*. Springer, 2005.
14. M. Mahdian, Y. Ye, and J. Zhang. A 2-approximation algorithm for the soft-capacitated facility location problem. In *Proceedings of 6th International Workshop on Approximation Algorithms for Combinatorial Optimization (APPROX)*, pages 129–140, 2003.
15. A. Matsuyama and H. Takahashi. An approximate solution for the steiner problem in graphs. *Mathematica Japonica 24(6)*, pages 573–577, 1980.
16. Gabriel Robins and Alexander Zelikovsky. Improved steiner tree approximation in graphs. In *Symposium on Discrete Algorithms*, pages 770–779, 2000.
17. D. Simchi-Levi. New worst-case results for the bin-packing problem. In *Naval Research Logistics, 41*, pages 579–585, 1994.
18. P. Toth and D. Vigo, editors. *The Vehicle Routing Problem*. SIAM monographs on discrete mathematics and applications. 2002.

Finding Graph Matchings in Data Streams

Andrew McGregor[*]

Department of Computer and Information Science, University of Pennsylvania
Philadelphia, PA 19104, USA
andrewm@cis.upenn.edu

Abstract. We present algorithms for finding large graph matchings in
the streaming model. In this model, applicable when dealing with massive graphs, edges are streamed-in in some arbitrary order rather than
residing in randomly accessible memory. For $\epsilon > 0$, we achieve a $\frac{1}{1+\epsilon}$
approximation for maximum cardinality matching and a $\frac{1}{2+\epsilon}$ approximation to maximum weighted matching. Both algorithms use a constant
number of passes and $\tilde{O}(|V|)$ space.

1 Introduction

Given a graph $G = (V, E)$, the *Maximum Cardinality Matching* (MCM) problem
is to find the largest set of edges such that no two adjacent edges are selected.
More generally, for an edge-weighted graph, the *Maximum Weighted Matching*
(MWM) problem is to find the set of edges whose total weight is maximized
subject to the condition that no two adjacent edges are selected. Both problems
are well studied and exact polynomial solutions are known [1–4]. The fastest
of these algorithms solves MWM with running time $O(nm + n^2 \log n)$ where
$n = |V|$ and $m = |E|$.

However, for massive graphs in real world applications, the above algorithms
can still be prohibitively computationally expensive. Examples include the virtual screening of protein databases. (See [5] for other examples.) Consequently
there has been much interest in faster algorithms, typically of $O(m)$ complexity,
that find good approximate solutions to the above problems. For MCM, a linear
time approximation scheme was given by Kalantari and Shokoufandeh [6]. The
first linear time approximation algorithm for MWM was introduced by Preis [7].
This algorithms achieved a 1/2 approximation ratio. This was later improved
upon by the $(2/3 - \epsilon)$ linear time[1] approximation algorithm given by Drake and
Hougardy [5].

In addition to concerns about time complexity, when computing with massive graphs it is no longer reasonable to assume that we can store the entire
input graph in *random access memory*. In this case the above algorithms are not
applicable as they require random access to the input. With this in mind, we

---

[*] This work was supported by NSF ITR 0205456.
[1] Note that here, and throughout this paper, we assume that ϵ is an arbitrarily small
constant.

C. Chekuri et al. (Eds.): APPROX and RANDOM 2005, LNCS 3624, pp. 170–181, 2005.
© Springer-Verlag Berlin Heidelberg 2005

consider the *graph stream model* discussed in [8–10]. This is a particular formulation of the *data stream model* introduced in [11–14]. In this model the edges of the graph stream-in in some arbitrary order. That is, for a graph $G = (V, E)$ with vertex set $V = \{v_1, v_2, \ldots, v_n\}$ and edge set $E = \{e_1, e_2, \ldots, e_m\}$, a *graph stream* is the sequence $e_{i_1}, e_{i_2}, \ldots, e_{i_m}$, where $e_{i_j} \in E$ and $i_1, i_2, \ldots, i_m$ is an arbitrary permutation of $\{1, 2, \ldots, m\}$.

The main computational restriction of the model is that we have limited space and therefore we can not store the entire graph (this would require $\tilde{O}(m)$ space.) In this paper we restrict our attention to algorithms that use $\tilde{O}(n)$ [2] space. This restriction was identified as a "sweet-spot" for graph streaming in a summary article by Muthukrishnan [8] and subsequently shown to be necessary for the verification of even the most primitive of graph properties such as connectivity [10]. We may however have multiple passes of the graph stream (as was assumed in [10, 11]). To motivate this assumption one can consider external memory systems in which seek times are typically the bottleneck when accessing data. In this paper, we will assume that we can only have constant passes. Lastly it is important to note that while most streaming algorithms use $\text{polylog}(m)$ space on a stream of length m, this is not always the case. Examples include the streaming-clustering algorithm of Guha *et al.* [15] that uses m^ϵ space and the "streaming" [3] algorithm of Drineas and Kannan [16] that uses space $\sqrt{m}$.

Our Results: In this paper we present algorithms that achieve the following approximation ratios:

1. For $\epsilon > 0$, a $\frac{1}{1+\epsilon}$ approximation to maximum cardinality matching.
2. For $\epsilon > 0$, a $\frac{1}{2+\epsilon}$ approximation to maximum weighted matching.

MCM and MWM have previously been studied under similar assumptions by Feigenbaum *et al.* [9]. The best previously attained results were a $\frac{1}{6}$ approximation to MWM and for $\epsilon > 0$ and a $\frac{2}{3+\epsilon}$ approximation to MCM on the assumption that the graph is a bipartite graph. Also in the course of this paper we tweak the $\frac{1}{6}$ approximation to MWM to give a $\frac{1}{3+2\sqrt{2}}$ approximation to MWM that uses only one pass of the graph stream.

2   Unweighted Matchings

2.1   Preliminaries

In this section we describe a streaming algorithm that, for $\epsilon > 0$, computes a $1/(1 + \epsilon)$ approximation to the MCM of the streamed graph. The algorithm will use a constant number of passes. We start by giving some basic definitions common to many matching algorithms.

---

[2]   Sometimes known as the *semi-streaming* space restriction.
[3]   Instead of "streaming," the authors of [16] use the term "pass-efficient" algorithms.

Definition 1 (Basic Matching Theory Definitions). *Given a matching M in a graph $G = (V, E)$, we call a vertex free if it does not appear as the end point of any edge in M. A length $2i+1$ augmenting path is a path $u_1 u_2 \ldots u_{2i+2}$ where u_1 and u_{2i+2} are free vertices and $(u_j, u_{j+1}) \in M$ for even j and $(u_j, u_{j+1}) \in E \setminus M$ for odd j.*

Note that if M is a matching and P is an augmenting path then $M \triangle P$ (the symmetric difference of M and P) is a matching of size strictly greater than M. Our algorithm will start by finding a maximal matching and then, by finding short augmenting paths, increase the size of the matching by making local changes. Note that finding a maximal matching is easily achieved in one pass – we select an edge iff we have not already selected an adjacent edge. Finding maximal matchings in this way will be at the core of our algorithm and we will make repeated use of the fact that the maximum matching has cardinality at most twice that of any maximal matching.

The following lemma establishes that, when there are few short augmenting paths, the size of the matching found can be lower-bound in terms of the size of the maximum cardinality matching OPT.

Lemma 1. *Let M be a maximal matching and OPT be a matching of maximum cardinality. Consider the connected components in $\text{OPT} \triangle M$. Ignoring connected components with the same number of edges from M as from OPT, let $\alpha_i M$ be the number of connected components with i edges from M. Then*

$$\max_{1 \leq i \leq k} \alpha_i \leq \frac{1}{2k^2(k+1)} \Rightarrow M \geq \frac{\text{OPT}}{1 + 1/k}$$

Proof. In each connected component with i edges from M there is either i or $i+1$ edges from OPT. Therefore, $\text{OPT} \leq \sum_{1 \leq i \leq k} \alpha_i \frac{i+1}{i} |M| + \frac{k+2}{k+1}(1 - \sum_{1 \leq i \leq k} \alpha_i)|M|$. By assumption

$$\sum_{1 \leq i \leq k} \alpha_i \frac{i+1}{i} + \frac{k+2}{k+1}(1 - \sum_{1 \leq i \leq k} \alpha_i) \leq \frac{1}{k(k+1)} + \frac{k+2}{k+1} = (1 + 1/k)$$

The result follows.

So, if there are $\alpha_i M$ components in $\text{OPT} \triangle M$ with $i+1$ edges from OPT and i edges from M, then there are at least $\alpha_i M$ length $2i + 1$ augmenting paths for M. Finding an augmenting path allows us to increase the size of M. Hence, if $\max_{1 \leq i \leq k} \alpha_i$ is small we already have a good approximation to OPT whereas, if $\max_{1 \leq i \leq k} \alpha_i$ is large then there exists $1 \leq i \leq k$ such that there are many length $2i + 1$ augmenting paths.

2.2   Description of the Algorithm

Now we have defined the basic notion of augmenting paths, we are in a position to give an overview of our algorithm. We have just reasoned that, if our matching

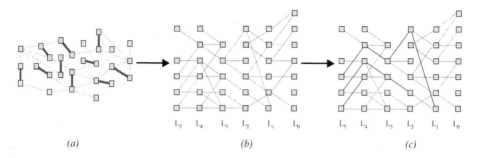

(a)   *(b)*   *(c)*

Fig. 1. A schematic of the procedure for finding length 9 augmenting paths. Explained in the text.

is not already large, then there exists augmenting paths of some length no greater than $2k+1$. Our algorithm looks for augmenting paths of each of the k different lengths separately. Consider searching for augmenting paths of length $2i+1$. See Fig. 1 for a schematic of this process when $i = 4$. In this figure, (a) depicts the graph G with heavy solid lines denoting edges in the current matching. To find length $2i + 1$ we randomly "project" the graph (and current matching) into a set of graphs, $\mathcal{L}_i$, which we now define.

Definition 2. *Consider a graph whose n nodes are partitioned into $i+2$ layers $L_{i+1}, \ldots L_0$ and whose edgeset is a subset of $\cup_{1 \leq j \leq i+1}\{(u,v) : u \in L_j, v \in L_{j-1}\}$. We call the family of such graphs $\mathcal{L}_i$. We call a path $u_l, u_{l-1}, \ldots u_0$ such that $u_j \in L_j$ an l-path.*

The random projection is performed as follows. The image of a free node in $G = (V, E)$ is a node in either L_{i+1} or L_0. The image of a matched edge $e = (v, v')$ is a node, either $u_{(v,v')}$ or $u_{(v',v)}$, in one of $L_i, \ldots L_1$ chosen at random. The edges in the projected graph G' are those in G that are "consistent" with the mapping of the free nodes and the matched edges, ie. there is an edge between a node $u_{(v_1,v_2)} \in L_j$ and $u_{(v_3,v_4)} \in L_{j-1}$ if there is an edge $(v_2, v_3) \in E$. Now note that an $i+1$-path in G' corresponds to a $2i+1$ augmenting path in G. Unfortunately the converse is not true, there may be $2i + 1$ augmenting paths in G that do not correspond to $i + 1$-paths in G' because we only consider consistent edges. However, we will show later that a constant fraction of augmenting paths exist (with high probability) as $i + 1$-paths in G'. In Figure 1, (b) depicts G', the layered graph formed by randomly projecting G into $\mathcal{L}_i$.

We now concern ourselves with finding a nearly maximal set of node disjoint $i + 1$-paths in a graph G'. See algorithm *Find-Layer-Paths* in Figure 2. The algorithm finds node disjoint $i+1$-paths by doing something akin to a depth first search. Finding a maximal set of node disjoint $i + 1$-paths can easily be achieved in the RAM model by actually doing a DFS, deleting nodes of found $i + 1$-paths and deleting edges when backtracking. Unfortunately this would necessitative too many passes in the streaming model as each backtrack potentially requires another pass of the data. Our algorithm in essence blends a DFS and BFS in such a way that we can substantially reduce the number of backtracks required. This will come at the price of possibly stopping prematurely, ie. when there may still exist some $i+1$-paths that we have not located.

The algorithm first finds a maximal matching between L_{i+1} and L_i. Let S' be the subset of nodes L_i involved in this first matching. It then finds a maximal matching between S' and L_{i-1}. We continue in this fashion, finding a matching between $S'' = \{u \in L_{i-1} : u \text{ matched to some } u' \in L_i\}$ and L_{i-2}. One can think of the algorithm as growing node disjoint paths from left to right. (Fig. 1 (c) tries to capture this idea. Here, the solid lines represent matchings between the layers.) If the size of the maximal matching between some subset S of a level L_j and L_{j-1} falls below a threshold we declare all vertices in S to be dead-ends and conceptually remove them from the graph (in the sense that we never again use these nodes while try to find $i+1$-paths.) At this point we start back-tracking. It is the use of this threshold that ensures a limit on the amount of back-tracking performed by the algorithm. However, because of the threshold, it is possible that a vertex may be falsely declared to be a dead-end, ie. there may still be a node disjoint path that uses this vertex. With this in mind we want the threshold to be low such that this does not happen often and we can hope to find all but a few of a maximal set of node disjoint $i+1$-paths. When we grow some node disjoint paths all the way to L_0, we remove these paths and recurse on the remaining graph. For each node v, the algorithm maintains a tag indicating if it is a "Dead End" or, if we have found a $i+1$ path involving v, the next node in the path.

It is worth reiterating that in each pass of the stream we simply find a maximal matching between some set of nodes. The above algorithm simply determines within which set of nodes we find a maximal matching.

Our algorithm is presented in detail in Fig. 2. Here we use the notation $s \in_R S$ to denote choosing an element s uniformly at random from a set S. Also, for a matching M, $\Gamma_M(u) = v$ if $(u, v) \in M$ and $\emptyset$ otherwise.

2.3   Correctness and Running Time Analysis

We first argue that the use of thresholds in *Find-Layer-Paths* ensures that we find all but a small number of a maximal set of $i+1$-paths.

Lemma 2 (Running Time and Correctness of *Find-Layer-Paths*). *Given $G' \in \mathcal{L}_i$, Find-Layer-Paths algorithm finds at least $(\gamma - \delta)|M|$ of the $i+1$-paths where $\gamma|M|$ is the size of some maximal set of $i+1$-paths. Furthermore the algorithm takes a constant number of passes.*

Proof. First note that *Find-Layer-Paths*$(\cdot, \cdot, \cdot, l)$ is called with argument $\delta^{2^{i+1-l}}$. During the running of *Find-Layer-Paths*$(\cdot, \cdot, \cdot, l)$ when we run line 15, the number of $i+1$-paths we rule out is at most $2\delta^{2^{i+1-l}}|L_{l-1}|$ (the factor 2 comes from the fact that a maximal matching is at least half the size of a maximum matching.) Let E_l be the number of times *Find-Layer-Paths*$(\cdot, \cdot, \cdot, l)$ is called: $E_{i+1} = 1$, $E_l \le E_{l+1}/\delta^{2^{i+1-l}}$ and therefore $E_l \le \delta^{-\sum_{0 \le j \le i-l} 2^j} = \delta^{-2^{i-l+1}+1}$. Hence, we

Algorithm *Find-Matching*(G, ϵ)
(∗ Finds a matching ∗)
Output: A matching
1.   Find a maximal matching M
2. $k \leftarrow \lceil \frac{1}{\epsilon} + 1 \rceil$
3. $r \leftarrow 4k^2(8k + 10)(k - 1)(2k)^k$
4.   **for** $j = 1$ to r:
5.   **for** $i = 1$ to k:
6.   **do** $M_i \leftarrow$ *Find-Aug-Paths*(G, M, i)
7. $M \leftarrow \mathrm{argmax}_{M_i} |M_i|$
8.   **return** M

Algorithm *Find-Aug-Paths*(G, M, i)
(∗ Finds length $2i + 1$ augmenting paths for a matching M in G ∗)
1. $G' \leftarrow$ *Create-Layer-Graph*(G, M, i)
2. $\mathcal{P} =$ *Find-Layer-Paths*$(G', L_{i+1}, \frac{1}{r(2k+2)}, i + 1)$
3.   **return** $M \triangle \mathcal{P}$

Algorithm *Create-Layer-Graph*(G, M, i)
(∗ Randomly constructs $G' \in \mathcal{L}_i$ from a graph G and matching M ∗)
1.   **if** v is a free vertex **then** $l(v) \in_R \{0, i + 1\}$
2.   **if** $e = (u, v) \in M$ **then** $j \in_R [i]$ and $l(e) \leftarrow j, l(u) \leftarrow ja$ and $l(v) \leftarrow jb$ or vice versa.
3. $E_i \leftarrow \{(u, v) \in E : l(u) = i + 1, l(v) = ia\}, E_0 \leftarrow \{(u, v) \in E : l(u) = 1b, l(v) = 0\}$
4.   **for** $j = 0$ to $i + 1$
5.   **do** $L_j \leftarrow l^{-1}(j)$
6.   **for** $j = 1$ to $i - 1$
7.   **do** $E_j \leftarrow \{(u, v) \in E : l(u) = (j + 1)b, l(v) = ja\}$
8.   **return** $G' = (L_{i+1} \cup L_i \cup \ldots \cup L_0, E_i \cup E_{i-1} \cup \ldots \cup E_0)$

Algorithm *Find-Layer-Paths*(G', S, δ, j)
(∗ Finds many j-paths from $S \subset L_j$ ∗)
1.   Find maximal matching M' between S and untagged vertices in L_{j-1}
2. $S' \leftarrow \{v \in L_{j-1} : \exists u, (u, v) \in M'\}$
3.   **if** $j = 1$
4.   **then if** $u \in \Gamma_{M'}(L_{j-1})$ **then** $t(u) \leftarrow \Gamma_{M'}(u), t(\Gamma_{M'}(u)) \leftarrow \Gamma_{M'}(u)$
5.   **if** $u \in S \setminus \Gamma_{M'}(L_{j-1})$ **then** $t(u) \leftarrow$ "Dead End "
6.   **return**
7.   **repeat**
8.   *Find-Layer-Paths*$(G', S', \delta^2, j - 1)$
9.   **for** $v \in S'$ such that $t(v) \neq$ "Dead End"
10.   **do** $t(\Gamma_{M'}(v)) \leftarrow v$
11.   Find maximal matching M' between untagged vertices in S and L_{j-1}.
12. $S' \leftarrow \{v \in L_{j-1} : \exists u, (u, v) \in M'\}$
13.   **until** $|S'| \leq \delta |L_{j-1}|$
14.   **for** $v \in S$ untagged
15.   **do** $t(b) \leftarrow$ "Dead End".
16.   **return**

Fig. 2. An Algorithm for Finding Large Cardinality Matchings. (See text for an informal description.)

remove at most $2E_l\delta^{2^{i+1-l}}|L_l| \le 2\delta|L_l|$. Note that when nodes are labeled as dead-ends in a call to *Find-Layer-Paths*$(\cdot,\cdot,\cdot,1)$, they really are dead-ends and declaring them such rules out no remaining $i+1$-paths. Hence the total number of paths not found is at most $2\delta\sum_{1\le j\le i}|L_j| \le 2\delta|M|$. The number of invocations of the recursive algorithm is

$$\sum_{1\le l\le i+1} E_l \le \sum_{1\le l\le i+1} \delta^{-2^{i+1-l}+1} \le \delta^{-2^{i+1}}$$

i.e. $O(1)$ and each invocation requires one pass of the data stream to find a maximal matching.

When looking for length $2i+1$ augmenting paths for a matching M in graph G, we randomly create a layered graph $G' \in \mathcal{L}_{i+1}$ using *Create-Layer-Graph* such that $i+1$-paths in G' correspond to length $2i+1$ augmenting paths. We now need to argue that a) many of the $2i+1$ augmenting paths in G exist in G' as $i+1$-paths and b) that finding a maximal, rather that a maximum, set of $i+1$-paths in G' is sufficient for our purposes.

Theorem 1. *If G has $\alpha_i M$ length $2i+1$ augmenting paths, then the number of length $i+1$-paths found in G' is at least*

$$(b_i\beta_i - \delta)|M| \ ,$$

where $b_i = \frac{1}{2i+2}$ and β_i is a random variables distributed as $\mathbf{Bin}(\alpha_i|M|, \frac{1}{2(2i)^i})$.

Proof. Consider a length $2i+1$ augmenting path $P = u_0u_1\ldots u_{2i+1}$ in G. The probability that P appears as an $i+1$-path in G' is at least,

$$2\mathbb{P}\left(l(u_0)=0\right)\mathbb{P}\left(l(u_{2i+1})=i+1\right)\prod_{j\in[i]}\mathbb{P}\left(l(u_{2j})=ja \text{ and } l(u_{2j-1})=jb\right)=\frac{1}{2(2i)^i}.$$

Given that the probability of each augmenting path existing as a $i+1$-path in G' is independent, the number of length $i+1$-paths in G' is distributed as $\mathbf{Bin}(\alpha_i|M|, \frac{1}{2(2i)^i})$. The size of a maximal set of node disjoint $i+1$-paths is at least a $\frac{1}{2i+2}$ fraction of the maximum size node-disjoint set $i+1$-paths. Combining this with Lemma 2 gives the result.

Finally, we argue that we only need to try to augment our initial matching a constant number of times.

Theorem 2 (Correctness). *With probability $1-f$ by running $O(\log\frac{1}{f})$ copies of the algorithm Find-Matching in parallel we find a $1-\epsilon$ approximation to the matching of maximum cardinality.*

Proof. We show that the probability that a given run of *Find-Matching* does not find a $(1+\epsilon)$ approximation is bounded above by e^{-1}.

Define a *phase* of the algorithm to be one iteration of the loop started at line 4 of *Find-Matching*. At the start of phase p of the algorithm, let M_p be the current matching. In the course of phase p of the algorithm we augment M_p by at least $|M_p|(\max_{1\leq i\leq k}(b_i\beta_{i,p})-\delta)$ edges where $\beta_{i,p}|M_p| \sim \mathbf{Bin}(\alpha_{i,p}|M_p|, \frac{1}{2(2i)^i})$. Let A_p be the value of $|M_p|\max_i(b_i\beta_{i,p})$ in the pth phase of the algorithm. Assume that for each of the r phases of the algorithm $\max \alpha_{i,p} \geq \alpha^* := \frac{1}{2k^2(k-1)}$. (By Lemma 1, if this is ever not the case, we already have a sufficiently sized matching.) Therefore, A_p dominates $b_k\mathbf{Bin}(\alpha^*|M_1|, \frac{1}{2(2k)^k})$. Let $(X_p)_{1\leq p\leq r}$ be independent random variables, each distributed as $b_k\mathbf{Bin}(\alpha^*|M_1|, \frac{1}{2(2k)^k})$. Therefore,

$$\mathbb{P}\left(|M_1| \prod_{1\leq p\leq r} (1 + \max\{0, \max_{1\leq i\leq k}(b_i\beta_{i,p}) - \delta\}) \geq 2|M_1|\right)$$

$$\geq \mathbb{P}\left(\sum_{1\leq p\leq r} \max_{1\leq i\leq k} b_i\beta_{i,p} \geq 2 + r\delta\right)$$

$$\geq \mathbb{P}\left(\sum_{1\leq p\leq r} X_p \geq |M_1|\frac{2+r\delta}{b_k}\right)$$

$$= \mathbb{P}\left(Z \geq |M_1|(4k+5)\right) \; ,$$

for $\delta = b_k/r$ where $Z = \mathbf{Bin}(\alpha^*|M_1|r, \frac{1}{2(2k)^k})$. Finally, by an application of the Chernoff bound,

$$\mathbb{P}\left(Z \geq |M_1|(4k+5)\right) = 1 - \mathbb{P}\left(Z < \mathbb{E}(Z)/2\right) > 1 - e^{-2(8k+10)|M_1|} \geq 1 - e^{-1} \; ,$$

for $r = 2(2k)^k(8k+10)/\alpha^*$. Of course, since M_1 is already at least half the size of the maximal matching. This implies that with high probability, at some point during the r phases our assumption that $\max \alpha_{i,p} \geq \alpha^*$, became invalid and at this point we had a sufficiently large matching.

3   Weighted Matching

We now turn our attention to finding maximum weighted matchings. Here each edge $e \in E$ of our graph G has a weight $w(e)$ (wlog. $w(e) > 0$). For a set of edges S let $w(S) = \sum_{e\in S} w(e)$. We seek to maximize $w(S)$ subject to the constraint that S contains no two adjacent edges.

Consider the algorithms given in Fig. 3. The algorithm *Find-Weighted-Matching* can be viewed as a parameterization of the one pass algorithm given in [9] in which γ was implicitly equal to 1. The algorithm greedily collects edges as they stream past. The algorithm maintains a matching M at all points. On seeing an edge e, if $w(e) > (1 + \gamma)w(\{e'|e' \in M, e'$ and e share an end point$\})$ then the algorithm removes any edges in M sharing an end point with e and adds e to M. The algorithm *Find-Weighted-Matching-Multipass* generalizes this to a multi-pass algorithm that in effect, repeats the one pass algorithm until the improvement yielded falls below some threshold. We start by recapping some notation introduced in [9]. While unfortunately macabre, this notation is nevertheless helpful for developing intuition.

Definition 3. *In a given pass of the graph stream, we say that an edge e is* born *if $e \in M$ at some point during the execution of the algorithm. We say that an edge is* killed *if it was born but subsequently removed from M by a newer heavier edge. This new edge* murdered *the killed edge. We say an edge is a* survivor *if it is born and never killed. For each survivor e, let the* Trail of the Dead *be the set of edges $T(e) = C_1 \cup C_2 \cup \ldots$, where $C_0 = \{e\}$, $C_1 = \{$the edges murdered by $e\}$, and $C_i = \cup_{e' \in C_{i-1}}\{$the edges murdered by $e'\}$.*

Lemma 3. *For a given pass let the set of survivors be S. The weight of the matching found at the end of that pass is therefore $w(S)$.*

1. $w(T(S)) \leq w(S)/\gamma$
2. $\textsc{Opt} \leq (1+\gamma)\,(w(T(S)) + 2w(S))$

Proof. 1. For each murdering edge e, $w(e)$ is at least $(1+\gamma)$ the cost of murdered edges, and an edge has at most one murderer. Hence, for all i, $w(C_i) \geq (1+\gamma)w(C_{i+1})$ and therefore $(1+\gamma)w(T(e)) = \sum_{i \geq 1}(1+\gamma)w(C_i) \leq \sum_{i \geq 0} w(C_i) = w(T(e)) + w(e)$. The first point follows.
2. We can charge the costs of edges in $\textsc{Opt}$ to the $S \cup T(S)$ such that each edge $e \in T(S)$ is charged at most $(1+\gamma)w(e)$ and each edge $e \in S$ is charged at most $2(1+\gamma)w(e)$. See [9] for details.

Hence in the one pass algorithm we get an $\frac{1}{\frac{1}{\gamma}+3+2\gamma}$ approximation ratio since

$$\textsc{Opt} \leq (1+\gamma)(w(T(S)) + 2w(S)) \leq (3 + \frac{1}{\gamma} + 2\gamma)w(S)$$

The maximum of this function is achieved for $\gamma = \frac{1}{\sqrt{2}}$ giving approximation ratio $\frac{1}{3+2\sqrt{2}}$. This represents only a slight improvement over the $1/6$ ratio attained previously. However, a much more significant improvement is realized in the multi-pass algorithm *Find-Weighted-Matching-Multipass*.

Theorem 3. *The algorithm Find-Weighted-Matching-Multipass finds a $\frac{1}{2(1+\epsilon)}$ approximation to the maximum weighted matching. Furthermore, the number of passes required is at most,*

$$\frac{\log(3/2 + \sqrt{2})}{\log \frac{(2\epsilon/3)^3}{(1+2\epsilon/3)^2 - (2\epsilon/3)^3}} + 1 \ .$$

Proof. First we prove that the number of passes is as claimed. We increase the weight of our solution by a factor $1+\kappa$ each time we do a pass and we start with a $1/(3+2\sqrt{2})$ approximation. Hence, if we take, $\log_{1+\kappa}(3/2 + \sqrt{2})$ passes we have already found a maximum weighted matching. Substituting in $\kappa = \frac{\gamma^3}{(1+\gamma)^2 - \gamma^3}$ establishes the bound on the number of passes.

Algorithm *Find-Weighted-Matching(G, γ)*
($*$ Finds Large Weighted Matchings in One Pass $*$)
1. $M \leftarrow \emptyset$
2.   **for** each edge $e \in G$
3.   **do if** $w(e) > (1 + \gamma)w(\{e'|e' \in M, e' \text{ and } e \text{ share an end point}\})$
4.   **then** $M \leftarrow M \cup \{e\} \setminus \{e'|e' \in M, e' \text{ and } e \text{ share an end point}\}$
5.   **return** M

Algorithm *Find-Weighted-Matching-Multipass(G, ϵ)*
($*$ Finds Large Weighted Matchings $*$)
1. $\gamma \leftarrow \frac{2\epsilon}{3}$
2. $\kappa \leftarrow \frac{\gamma^3}{(1+\gamma)^2 - \gamma^3}$
3.   Find a $\frac{1}{3+2\sqrt{2}}$ weighted matching, M
4.   **repeat**
5. $S \leftarrow w(M)$
6.   **for** each edge $e \in G$
7.   **do if** $w(e) > (1 + \gamma)w(\{e'|e' \in M, e' \text{ and } e \text{ share an end point}\})$
8.   **then** $M \leftarrow M \cup \{e\} \setminus \{e'|e' \in M, e' \text{ and } e \text{ share an end point}\}$
9.   **until** $\frac{w(M)}{S} \leq 1 + \kappa$
10. **return** M

Fig. 3. Algorithms for Finding Large Weighted Matchings.

Let M_i be the matching constructed after the i-th pass. Let $B_i = M_i \cap M_{i-1}$. Now, $(1 + \gamma)(w(M_{i-1}) - w(B_i)) \leq w(M_i) - w(B_i)$ and so,

$$\frac{w(M_i)}{w(M_{i-1})} = \frac{w(M_i)}{w(M_{i-1}) - w(B_i) + w(B_i)} \geq \frac{(1+\gamma)w(M_i)}{w(M_i) + \gamma w(B_i)} \ .$$

If $\frac{w(M_i)}{w(M_{i-1})} < (1 + \kappa)$, then we deduce that $w(B_i) \geq \frac{\gamma - \kappa}{\gamma + \gamma\kappa}w(M_i)$. Appealing to Lemma 3, this means that, for all i,

$$OPT \leq (1/\gamma + 3 + 2\gamma)(w(M_i) - w(B_i)) + 2(1 + \gamma)w(B_i) \ ,$$

since edges in B_i have empty trails of the dead. So if $w(B_i) \geq \frac{\gamma - \kappa}{\gamma + \gamma\kappa}w(M_i)$ we get that,

$$OPT \leq (1/\gamma + 3 + 2\gamma)(w(M_i) - w(B_i)) + 2(1 + \gamma)w(B_i)$$
$$\leq (1/\gamma + 3 + 2\gamma - (1/\gamma + 1)\frac{\gamma - \kappa}{\gamma + \gamma\kappa})w(M_i)$$
$$\leq (2 + 3\gamma)w(M_i) \ .$$

Since $\gamma = \frac{2\epsilon}{3}$ the claimed approximation ratio follows.

4 Conclusions and Open Questions

New constant pass streaming algorithms, using $\tilde{O}(n)$ space, have been presented for the MCM and MWM problems. The MCM algorithms uses a novel randomized technique that allows us to find augmenting paths and thereby find a matching of size $\text{OPT}/(1 + \epsilon)$. The MWM algorithm builds upon previous work to find a matching whose weight is at least $\text{OPT}/(2 + \epsilon)$.

It is worth asking if there exists a streaming algorithm that achieves a $1/(1+\epsilon)$ approximation for the MWM problem. It is possible, although non-trivial, to extend some of the ideas for the MCM problem to deal with the case when edges are weighted. The main problem lies in the fact that, in the weighted case, there are "augmenting cycles" in addition to (weight-)augmenting paths (naturally defined). Unfortunately, finding cycles in the streaming model seems to be inherently difficult. In particular, this author fears that lower bounds concerning finding the girth of a graph [10] and finding common neighborhoods [17] may suggest a negative result.

Finally, the careful reader may have noticed that the number of passes required by the MCM algorithm has a rather strong dependence on ϵ in the sense that, as ϵ becomes small, the number of passes necessary, grows very quickly. This paper was rather cavalier with the dependence on ϵ as we were primarily concerned with ensuring that the number of passes was independent of n. However, for the sake of practical applications, a weaker dependence would be desirable.

References

1. Edmonds, J.: Maximum matching and a polyhedron with 0,1-vertices. J. Res. Nat. Bur. Standards **69** (1965) 125–130
2. Gabow, H.N.: Data structures for weighted matching and nearest common ancestors with linking. In: Proc. ACM-SIAM Symposium on Discrete Algorithms. (1990) 434–443
3. Hopcroft, J.E., Karp, R.M.: An $n^{5/2}$ algorithm for maximum matchings in bipartite graphs. SIAM J. Comput. **2** (1973) 225–231
4. Micali, S., Vazirani, V.: An $O(\sqrt{V}E)$ algorithm for finding maximum matching in general graphs. In: Proc. 21st Annual IEEE Symposium on Foundations of Computer Science. (1980)
5. Drake, D.E., Hougardy, S.: Improved linear time approximation algorithms for weighted matchings. In Arora, S., Jansen, K., Rolim, J.D.P., Sahai, A., eds.: RANDOM-APPROX. Volume 2764 of Lecture Notes in Computer Science., Springer (2003) 14–23
6. Kalantari, B., Shokoufandeh, A.: Approximation schemes for maximum cardinality matching. Technical Report LCSR–TR–248, Laboratory for Computer Science Research, Department of Computer Science. Rutgers University (1995)
7. Preis, R.: Linear time 1/2-approximation algorithm for maximum weighted matching in general graphs. In Meinel, C., Tison, S., eds.: STACS. Volume 1563 of Lecture Notes in Computer Science., Springer (1999) 259–269
8. Muthukrishnan, S.: Data streams: Algorithms and applications. (2003) Available at http://athos.rutgers.edu/~muthu/stream-1-1.ps.

9. Feigenbaum, J., Kannan, S., McGregor, A., Suri, S., Zhang, J.: On graph problems in a semi-streaming model. In: Proc. 31st International Colloquium on Automata, Languages and Programming. (2004) 531–543
10. Feigenbaum, J., Kannan, S., McGregor, A., Suri, S., Zhang, J.: Graph distances in the streaming model: The value of space. Proc. 16th ACM-SIAM Symposium on Discrete Algorithms (2005)
11. Munro, J., Paterson, M.: Selection and sorting with limited storage. Theoretical Computer Science **12** (1980) 315–323
12. Henzinger, M.R., Raghavan, P., Rajagopalan, S.: Computing on data streams. Technical Report 1998-001, DEC Systems Research Center (1998)
13. Alon, N., Matias, Y., Szegedy, M.: The space complexity of approximating the frequency moments. Journal of Computer and System Sciences **58** (1999) 137–147
14. Feigenbaum, J., Kannan, S., Strauss, M., Viswanathan, M.: An approximate L^1 difference algorithm for massive data streams. SIAM Journal on Computing **32** (2002) 131–151
15. Guha, S., Mishra, N., Motwani, R., O'Callaghan, L.: Clustering data streams. In: Proc. 41th IEEE Symposium on Foundations of Computer Science. (2000) 359–366
16. Drineas, P., Kannan, R.: Pass efficient algorithms for approximating large matrices. In: Proc. 14th ACM-SIAM Symposium on Discrete Algorithms. (2003) 223–232
17. Buchsbaum, A.L., Giancarlo, R., Westbrook, J.: On finding common neighborhoods in massive graphs. Theor. Comput. Sci. **1-3** (2003) 707–718

A Primal-Dual Approximation Algorithm for Partial Vertex Cover: Making Educated Guesses[*]

Julián Mestre

Department of Computer Science
University of Maryland, College Park, MD 20742
jmestre@cs.umd.edu

Abstract. We study the PARTIAL VERTEX COVER problem, a generalization of the well-known VERTEX COVER problem. Given a graph
$G = (V, E)$ and an integer s, the goal is to cover all but s edges, by
picking a set of vertices with minimum weight. The problem is clearly
NP-hard as it generalizes the vertex cover problem. We provide a primal-
dual 2-approximation algorithm which runs in $O(V \log V + E)$ time. This
represents an improvement in running time from the previously known
fastest algorithm.

Our technique can also be applied to a more general version of the problem. In the PARTIAL CAPACITATED VERTEX COVER problem each vertex
u comes with a capacity k_u and a weight w_u. A solution consists of a
function $x : V \to \mathbb{N}_0$ and an orientation of all but s edges, such that
the number edges oriented toward any vertex u is at most $x_u k_u$. The
cost of the cover is given by $\sum_{v \in V} x_v w_v$. Our objective is to find a cover
with minimum cost. We provide an algorithm with the same performance
guarantee as for regular partial vertex cover. In this case no algorithm
for the problem was known.

1 Introduction

The VERTEX COVER problem has been widely studied [12, 16]. In the simplest
version, we are given a graph $G = (V, E)$ and are required to pick a minimum
subset of vertices to cover all the edges of the graph. An edge is said to be covered
if at least one of its incident vertices is picked. The problem is NP-hard and was
one of the first problems shown to be NP-hard in Karp's seminal paper [15].
However, several different approximation algorithms have been developed for
it [12]. These algorithms develop solutions within twice the optimal in polynomial
time.

Recently several generalizations of the vertex cover problem have been considered. Bshouty and Burroughs [4] were the first to study the PARTIAL VERTEX
COVER problem. In this generalization we are not required to cover all edges of
the graph, any s edges may be left uncovered. By allowing some fixed number
of edges to be left uncovered we hope to obtain a cover with substantially lower
cost. It can be regarded as a trade off between cost and coverage.

[*] Research supported by NSF Awards CCR 0113192 and CCF 0430650

C. Chekuri et al. (Eds.): APPROX and RANDOM 2005, LNCS 3624, pp. 182–191, 2005.

The paper by Bshouty and Burroughs [4] was the first to give a factor 2 approximation using LP-rounding. Subsequently, combinatorial algorithms with the same approximation guarantee were proposed. Houchbaum [13] presented an $O(EV \log \frac{V^2}{E} \log V)$ time algorithm; Bar-Yehuda [1] developed an $O(V^2)$ time algorithm using the local-ratio technique; finally, a primal-dual algorithm which runs in $O(V(E + V \log V))$ time was given by Gandhi et al. [8].

The main result of this paper is a primal-dual 2-approximation algorithm for partial vertex cover which runs in $O(V \log V + E)$. This represents an improvement in running time over the previously known best solution for the problem. One additional benefit of our method is that it can be used to solve a more general version of the problem.

Motivated by a problem in computational biology, the CAPACITATED VERTEX COVER problem was proposed by Guha et al. [9]. For each vertex $v \in V$ we are given a capacity k_v and a weight w_v. A *capacitated vertex cover* is a function $x : V \to \mathbb{N}_0$ such that there exists an orientation of all edges in which the number of edges directed into $v \in V$ is at most $k_v x_v$. When e is oriented toward v we say e is *covered by* or *assigned to* v. The weight of the cover is $\sum_{v \in V} w_v x_v$, we want to find a cover with minimum weight. Factor 2 approximation algorithms have been developed [7, 9] for this problem.

By allowing s edges to be left unassigned we get a new problem: the PARTIAL CAPACITATED VERTEX COVER problem. Thus we bring together two generalization of the original vertex cover problem that have only been studied separately. Using the primal-dual algorithm of [8] for regular capacitated vertex cover coupled with our method we show a 2-approximation for the more general problem.

For vertex cover, the best (known) approximation factor for general graphs is $2 - o(1)$ [3, 10] but the best constant factor remains 2. Some algorithms [2, 6, 11] that achieve this later bound rank among the first approximation algorithms and can be interpreted using the primal-dual method, where a maximal dual solution is found and the cost of the cover is charged to the dual variables. It is interesting to see how far the same primal-dual method can be taken to tackle our much more general version of vertex cover.

Finally we mention that the techniques used in this paper are not restricted to vertex-cover-like problems. For instance, the primal-dual algorithm of Jain and Vazirani [14] for the FACILITY LOCATION problem can be used in conjunction with our method to design a faster approximation algorithm for the the partial version of the problem which was studied by Charikar et at. [5].

2   LP Formulation

Let us describe an integer linear program formulation for the partial vertex cover problem. Each vertex $v \in V$ has associated a variable $x_v \in \{0, 1\}$ which indicates if v is picked for the cover. For every edge $e \in E$ either one of its endpoints is picked or the edge is left uncovered, the variable $p_e \in \{0, 1\}$ indicates the later. The number of uncovered edges cannot exceed s. Figure 1 show the LP relaxation of the IP just described.

$$\min \sum_{v \in V} w_v x_v$$

subject to:
$$p_e + x_u + x_v \geq 1 \qquad \forall e = \{u, v\} \in E$$
$$\sum_{e \in E} p_e \leq s$$
$$x_u, p_e \geq 0$$

Fig. 1. LP for Partial Vertex Cover

As it stands, the LP exhibits an unbounded integrality gap. Consider the following example: a star where the center node has degree d, all leaves have unit weight while the center node has weight 10. Suppose we must cover two edges, that is, $s = d - 2$. The optimal integral solution (OPT) must pick two leaves, the cost of the solution is therefore 2. On the other hand, a fractional solution can pick $2/d$ of the center node, every edge is covered by this amount which adds up to 2. The cost is $20/d$, choose d big enough to make the cost as small as desired.

Suppose we knew that $h \in V$ is the most expensive vertex in OPT. We modify the problem slightly by forcing the solution to pick h and disallowing picking more expensive vertices. This modification does not increase the cost of the integral optimal solution, but it does narrow the integrality gap. In the star example above if we knew that any leaf can be the most expensive vertex in OPT, then the modification effectively closes the integrality gap. Provided we know which is the most expensive vertex in OPT, Gandi *et al.* [8] made use of the above modification to develop a primal-dual 2-approximation algorithm for partial vertex cover. Unfortunately we don't know, so we must guess. Their procedure is run on each possible choice of h, and the best cover produced is returned. Each run takes $O(V \log V + E)$ time and therefore the whole procedure runs in $O(V(V \log V + E))$ time.

Our approach is along these lines, but instead of doing exhaustive guessing we run our algorithm only once and make different guesses along the way. One may say that the algorithm *makes educated guesses*, this allows us to bring down the running time to $O(V \log V + E)$.

3 Primal-Dual Algorithm

The dual program for partial vertex cover is given in Figure 2. By $\delta(v)$ we refer to the edges incident to vertex v.

Initially all dual variables are set to 0, and all edges are unassigned. The algorithm works in iterations, each consisting of a *pruning step* followed by a *dual update step*. As the algorithm progresses we build a cover $C \subseteq V$, which initially is empty. Along the way we may disallow picking some vertices, we keep track of these with the set R, which also starts off empty.

In the pruning step we check every vertex $v \in V \setminus \{C \cup R\}$. If $C + v$ is a feasible cover, we guess $C + v$ as a possible cover and then we disallow picking

$$\max \sum_{e \in E} y_e - sz$$

subject to:
$$\sum_{e \in \delta(v)} y_e \leq w_v \qquad \forall v \in V$$
$$y_e \leq z \qquad \forall e \in E$$
$$y_e, z \geq 0$$

Fig. 2. The dual for Partial Vertex Cover

v, i.e., we set $w_v \leftarrow \infty$ and add it to R. Notice that many vertices may be guessed/disallowed in a single pruning step.

If $|E(R)| > s$ we must stop as there is no hope of finding a feasible cover anymore, even if all vertices in $V \setminus R$ are picked. We thus return, among the guessed covers, the one with the least weight.

In the dual update step we uniformly raise z and the y_e variables of unassigned edges until some vertex u becomes tight, notice that $u \in V \setminus R$. If multiple vertices become tight then arbitrarily pick one to process. Vertex u is added to C, free edges in $\delta(u)$ are assigned to u and their dual variables are frozen. After this is done we proceed to the next iteration. Realize that after adding u the cover is still not feasible, otherwise u would have been guessed and disallowed in the pruning step.

The algorithm terminates when $|E(R)| > s$, at this point we know OPT must use at least one vertex in R. Let h be the first such vertex to be added to R, and C the cover at the moment h was guessed. By definition $C + h$ is a feasible solution, suppose its cost is at most twice that of OPT. The algorithm returns the guessed cover with minimum weight, thus it constitutes a 2-approximation for partial vertex cover. Now it all boils down to proving the following lemma.

Lemma 1. *Let h be the first vertex in OPT to be added to R, and C the cover constructed by the algorithm when h was guessed. Then $w(C + h) \leq 2\,\mathrm{OPT}$.*

As a warm-up we first show that $w(C + h) \leq 2\,\mathrm{OPT} + w_h$, this implies a 3-factor approximation. To do so we modify the problem slightly by disallowing picking vertices added to R before h, this does not change the cost of the integral optimal solution. In fact the algorithm does exactly this by changing the weight of those vertices to ∞. We denote by OPT', LP' and DL' the cost of the optimal integral, fractional primal and dual solutions for the new problem. Note that $\mathrm{DL}' = \mathrm{LP}' \leq \mathrm{OPT}' = \mathrm{OPT}$.

The dual solution (y, z) constructed by the algorithm at the moment h was guessed constitutes a feasible solution for the dual of the modified problem. Every vertex $v \in C$ is tight ($w_v = \sum_{e \in \delta(v)} y_e$), thus charging once the dual variables of edges incident to v is enough to pay for v. Assigned edges are charged at most twice, therefore $w(C) \leq 2 \sum_{e \text{ assigned}} y_e$. Since the set C is not a feasible solution more than s edges remain unassigned. These edges have $y_e = z$ and are not charged, therefore $w(C) \leq 2(\sum_{e \in E} y_e - sz) \leq 2\,\mathrm{OPT}$, and the bound follows.

Proof (of Lemma 1). To obtain the desired bound we must modify the problem even further: In addition to disallowing picking vertices added to R before h we force every solution to pick h. Like before this should not change the cost of the integral optimal solution and so the cost of a feasible dual solution for the new problem is a lower bound on the cost of OPT.

These modifications to the problem bring some changes to the LP formulation. The weight of the vertices that came to R before h are set to ∞. Since we are forced to pick h, we set $x_h = 1$. As a result a constant w_h term appears in the objective function of both the primal and the dual and the dual constraint for x_h disappears. Every edge $e \in \delta(h)$ is covered by h, so the primal constraint for e and the variable p_e go away; in the dual this means variable y_e and the constraint $y_e \leq z$ disappear. The dual solution (y, z) constructed by the algorithm is not feasible as variables y_e for $e \in \delta(h)$ have non-zero value. We construct another solution y' by letting $y'_e = 0$ for $e \in \delta(h)$ and $y'_e = y_e$ otherwise[1]. The cost of the new solution (y', z) is $\sum_{e \in E} y'_e - sz + w_h$.

If we proceed as before to account for the cost of vertices $v \in C$ we run into two problems. The first comes from neighbors of h for which $w_v = \sum_{e \in \delta(v)} y'_e + y_{vh}$. This means we cannot fully pay for v just using $\sum_{e \in \delta(v)} y'_e$, and so we consider the term y_{vh} to be overcharge. Thus the total overcharge from neighbors of h is at most $\sum_{e \text{ assigned } \in \delta(h)} y_e$.

The second problem arises when trying to pay for l, the last vertex added to C. Suppose that just before l became tight there were f free edges, note that these edges have $y_e = z$. When l is added to C, t_l of these edges are assigned to it, since C is not feasible we have $f - t_l > s$. Assume that of the remaining free edges, t_h are incident to h. Because $C + h$ is feasible we have $f - (t_l + t_h) \leq s$. After we switch from y to y' only $f - t_h$ edges remain with $y'_e = z$, notice that $f - t_h > s$, otherwise h would have been guessed earlier. Among these edges, s should be left uncharged as the term $-sz$ in the objective function cancels their contribution. Vertex l needs to charge t_l of these edges, but only $f - t_h - s \leq t_l$ are available. The shortfall is $t_l - (f - t_h - s) < t_h$, again we consider this as overcharge which is bounded by $t_h z = \sum_{e \text{ unassigned } \in \delta(h)} y_e$.

The total overcharge is therefore $\sum_{e \in \delta(h)} y_e \leq w_h$ and so the lemma follows:

$$w(C + h) \leq 2(\sum_{e \in E} y'_e - sz) + w_h + w_h \leq 2(\sum_{e \in E} y'_e - sz + w_h) \leq 2\text{OPT}$$

□

For the implementation we keep two heaps for the vertices. One heap tells us when a vertex will become tight, which is easily determined by looking at its weight, the current value of the dual variables of edges incident to it and the number of those edges that are free. The second heap tells us how many free edges are incident to the vertex.

---

[1] Although dual variables for $e \in \delta(h)$ really disappear, the analysis is cleaner if we keep them but force them to be 0

In the dual update step we take out the vertex which will become tight the soonest and update the heap key for its neighbors. In the pruning step we look at the second heap and fetch vertices with the largest number of free incident edges. Checking if its addition to C make the solution feasible can be done in constant time by keeping track of how many edges C covers.

All in all at most $2|V|$ remove and at most $2|E|$ increase/decrease key operations are performed. Using Fibonacci heaps it takes $O(V \log V + E)$ time. This finishes the proof of the main result of the paper.

Theorem 1. *There is a 2-approximation algorithm for Partial Vertex Cover which runs in $O(V \log V + E)$.*

4   Generalizations

Our technique can be used to solve a more general version of the vertex cover problem. In the partial capacitated vertex cover problem every vertex $u \in V$ has a capacity k_u and a weight w_u. A single copy of u can cover at most k_u edges incident to it, the number of copies picked is given by $x_u \in \mathbb{N}_0$. Every edge $e = \{u, v\} \in E$ is either left uncovered ($p_e = 1$) or is assigned to one of its endpoints, variables y_{eu} and y_{ev} indicate this. The number of edges assigned to a vertex u cannot exceed $k_u x_u$, while the ones left unassigned cannot be more than s. Figure 3 shows the LP relaxation of the program just described. An additional constraint $x_v \geq y_{ev}$ is needed to fix the integrality gap.

$$\min \sum_{v \in V} w_v x_v$$

subject to:

$$
\begin{array}{ll}
p_e + y_{eu} + y_{ev} \geq 1 & \forall e = \{u, v\} \in E \\
k_v x_v \geq \sum_{e \in \delta(v)} y_{ev} & \forall v \in V \\
x_v \geq y_{ev} & \forall v \in e \in E \\
\sum_{e \in E} p_e \leq s & \\
p_e, y_{ev}, x_v \geq 0 &
\end{array}
$$

Fig. 3. LP for partial capacitated vertex cover

Like before our algorithm works in iterations, each having a pruning and a dual update step. For the later we follow the algorithm in [9] which is described here for completeness. We will use the dual program of Figure 4.

Initially all variables are set to 0 and all edges are unassigned. A vertex u is said to be high degree if more than k_u unassigned edges are incident to u, otherwise u is low degree. As we will see a vertex may start off as high degree and later become low degree as edges get assigned to its neighbors. We define L_u to be the set of unassigned edges incident to u at the moment u becomes low degree. For vertices that start off as low degree we define $L_u = \delta(u)$.

In the dual update step we raise α_e for all unassigned edges, because of the constraint $\alpha_e \leq q_v + l_{ev}$ one of the terms on the right hand side must be raised by the same amount. Which variable is increased depends on the nature of v: If v is

$$\max \sum_{e \in E} \alpha_e - sz$$
subject to:
$$k_v q_v + \sum_{e \in \delta(v)} l_{ev} \le w_v \qquad \forall\, v \in V$$
$$\alpha_e \le q_v + l_{ev} \qquad \forall\, v \in e \in E$$
$$\alpha_e \le z \qquad \forall\, e \in E$$
$$q_v, l_{ev}, \alpha_e, z \ge 0$$

Fig. 4. The dual

high degree we rise q_v otherwise we rise l_{ev}. When some vertex u becomes tight we *open* it: If u is high degree we assign to it all the free edges in $\delta(u)$, otherwise u is low degree and all edges in L_u are assigned to it. Notice that in the later case some of the edges in L_u might have already been assigned. In particular this means that some of the edges originally assigned to a high degree vertex when it was opened may later be taken away by the opening of low degree neighbors. Therefore the decision of how many copies to pick of a vertex is differed until the end of the algorithm. In what follows when we talk about current solution we mean the current assignment of edges.

In the pruning step when processing a vertex u, although in principle we are allowed to pick multiple copies of a vertex, we check if adding just one copy of u to the current solution makes it feasible, if so guess the solution plus u and then disallow picking u, i.e.: add it to R and set $w_u \leftarrow \infty$.

A consequence of the pruning step is that if in the dual update step the vertex u to become tight is low degree then opening u cannot make the solution feasible. On the other hand when u is high degree opening u may make the solution feasible. In which case the algorithm *ends prematurely*: we assign just enough edges to make the solution feasible, i.e., s edges are left unassigned, and we return the best solution among the previously guessed covers and the current solution. If on the other hand this does not happen and R grows to the point where $|E(R)| > s$ we stop and return the best guessed solution.

Let us first consider the case where the algorithm ends prematurely, we will show that we can construct a solution with cost $2(\sum_{e \in E} \alpha_e - sz)$. Constructing the solution is simple: we pick enough copies of every vertex to cover the edges assigned to it.

Let u be one of the opened vertices, furthermore suppose u was low degree when opened, this means only one copy is needed. If u started off as low degree then $w_u = \sum_{e \in L_u} l_{eu} = \sum_{e \in L_u} \alpha_e$ as $q_u = 0$. On the other hand if u became low degree afterwards then $|L_u| = k_u$ and only edges in L_u have nonzero l_{eu}, thus:

$$w_u = q_u k_u + \sum_{e \in L_u} l_{eu} = \sum_{e \in L_u} q_u + l_{eu} = \sum_{e \in L_u} \alpha_e$$

In either cases we can pay for u by charging once the edges in L_u.

Now let us consider the case when u is high degree. By definition more than k_u edges were assigned to u when it became tight, note that some of these edges may be later taken away by low degree neighbors. Let R_u be the edges assigned to u at the end of the algorithm. There are two cases to consider. If $|R_u| < k_u$ we only need one copy of u, charging once the edges originally assigned to u (which are more than k_u and have $\alpha_e = q_u$) is enough to pay for $w_u = k_u q_u$. Otherwise $\lceil \frac{|R_u|}{k_u} \rceil$ copies of u are needed. Unfortunately we cannot pay for them using $\sum_{e \in R_u} \alpha_e = |R_u| q_u = \frac{|R_u|}{k_u} w_u$ which is enough to pay for the first $\lfloor \frac{|R_u|}{k_u} \rfloor$ copies but may be less than $\lceil \frac{|R_u|}{k_u} \rceil w_u$. The pathological case happens when $|R_u| = k_u + 1$, two copies are needed but edges in R_u only have a dual cost of $w_u(1 + \frac{1}{k_u})$. The key observation is that edges in R_u will not be charged from the other side, therefore we can charge any k_u edges twice to pay for the extra copy.

How many times can a single edge be charged? At most twice, either from a single endpoint or once from each endpoint. We are leaving s edges uncharged with $\alpha_e = z$, therefore the solution can be paid with $2(\sum_{e \in E} \alpha_e - sz)$.

Before ending prematurely the algorithm may have disallowed some vertices. If none of them is used by OPT, modifying the problem by disallowing picking them does not increase the cost of the integral optimal solution and therefore the cost of the dual solution at the end is a lower bound on OPT. It follows from the above analysis that the cover found has cost at most 2OPT. Otherwise at least one of the guesses was correct. Likewise if the algorithm terminates because $|E(R)| > s$ there must be at least one correct guess. In either case let h be the first vertex added to R which is used by OPT.

We define a new problem where we are not allowed to pick vertices added to R before h and we are forced to pick at least one copy of h. Notice that we place no restriction on which edges h covers and that extra copies may be picked, therefore the cost of the integral optimal solution for the new problem does not change. The primal changes consist in replacing x_h with $x_h + 1$. As a result the objective function becomes $\sum_{e \in E} \alpha_e - sz - k_h q_h - \sum_{e \in \delta(h)} l_{eh} + w_h$.

The dual solution (α, q, l, z) constructed by the algorithm when h was guessed is a feasible solution for the dual of the new problem. As usual the cost of this solution is a lower bound on OPT.

Consider the edge assignment when h was guessed, before constructing the cover we need to modify the assignment. If h is low degree then assign edges in L_h to it, otherwise assign any k_h free edges incident to h. After this is done less than s edges may be left unassigned, we need to free some of the edges assigned to the last vertex to become tight as to leave exactly s edges unassigned with $\alpha_e = z$. Now build the cover by picking as many copies of each vertex to cover the edges assigned to it.

By switching to the modified problem we gain a w_h term in the objective function of the dual, while some cost is lost, namely $k_h q_h + \sum_{e \in \delta(h)} l_{eh} = \sum_{e \in D} \alpha_e$ where D is the set of edges assigned to h. We can regard the contribution of edges in D to the dual cost as disappearing. In turn, this causes some overcharge when paying for the rest of the solution. A similar argument as the one used for regular partial vertex cover shows that the overcharge is at most $\sum_{e \in D} \alpha_e \leq w_h$. Add the cost of h and the whole solution can be paid with twice the cost of the dual which is ≤ 2 OPT.

Returning the solution with minimum weight among the ones guessed guarantees producing a cover with cost at most twice that of the optimal solution.

The algorithm can be implemented using Fibonacci heaps to find out which vertex will become tight next and to do the pruning, the total running time is $O(V \log V + E)$. The section is thus summarized in the following theorem.

Theorem 2. *There exists a 2-approximation algorithm for Partial Capacitated Vertex Cover which runs in $O(V \log V + E)$*

5   Conclusion

We have developed two algorithms for partial vertex cover problems. Our pruning/dual update technique is quite general and may be applied to other covering problems where a primal-dual already exist for the special case of $s = 0$. For instance a similar algorithm may be developed for facility location using Jain and Vazirani's algorithm [14] for the dual update step.

Acknowledgment

The author would like to thank Samir Khuller for suggesting developing an algorithm for partial vertex cover that does not use exhaustive guessing.

References

1. R. Bar-Yehuda. Using homogenous weights for approximating the partial cover problem. In *Proceedings of the 10th Annual ACM-SIAM Symposium on Discrete Algorithms (SODA'99)*, pages 71–75, 1999.
2. R. Bar-Yehuda and S. Even. A linear time approximation algorithm for approximating the weighted vertex cover. *Journal of Algorithms*, 2:198–203, 1981.
3. R. Bar-Yehuda and S. Even. A local-ratio theorem for approximating the weighted vertex cover problem. *Annals of Discrete Mathematics*, 25:27–46, 1985.
4. N. Bshouty and L. Burroughs. Massaging a linear programming solution to give a 2-approximation for a generalization of the vertex cover problem. In *Proceedings of the 15th Annual Symposium on the Theoretical Aspects of Computer Science (STACS'98)*, pages 298–308, 1998.
5. M. Charikar, S. Khuller, D. M. Mount, and G. Narasimhan. Algorithms for facility location problems with outliers. In *soda01*, pages 642–651, 2001.
6. K. L. Clarkson. A modification of the greedy algorithm for vertex cover. *Information Processing Letters*, 16(1):23–25, 1983.
7. R. Gandhi, S. Khuller, S. Parthasarathy, and A. Srinivasan. Dependent rounding in bipartite graphs. In *Proceedings of the 43rd Annual IEEE Symposium on Foundations of Computer Science (FOCS'02)*, pages 323–332, 2002.
8. R. Gandhi, S. Khuller, and A. Srinivasan. Approximation algorithms for partial covering problems. In *Proceedings of the 11th International Colloquium on Automata, Languages, and Programming (ICALP'01)*, pages 225–236, 2001.

9. S. Guha, R. Hassin, S. Khuller, and E. Or. Capacitated vertex covering with applications. In *Proceedings of the 13th Annual ACM-SIAM Symposium on Discrete Algorithms (SODA'02)*, pages 858–865, 2002.
10. E. Halperin. Improved approximation algorithms for the vertex cover problem in graphs and hypergraphs. In *Proceedings of the 11th Annual ACM-SIAM Symposium on Discrete Algorithms (SODA'00)*, pages 329–337, 2000.
11. D. S. Hochbaum. Approximation algorithms for the set covering and vertex cover problems. *SIAM Journal on Computing*, 11:385–393, 1982.
12. D. S. Hochbaum, editor. *Approximation Algorithms for NP–hard Problems*. PWS Publishing Company, 1997.
13. D. S. Hochbaum. The *t*-vertex cover problem: Extending the half integrality framework with budget constraints. In *Proceedings of the 1st International Workshop on Approximation Algorithms for Combinatorial Optimization Problems (APPROX'98)*, pages 111–122, 1998.
14. K. Jain and V. V. Vazirani. Approximation algorithms for metric facility location and k-median problems using the primal-dual schema and lagrangian relaxation. *Journal of the ACM*, 48(2):274–296, 2001.
15. R. M. Karp. Reducibility among combinatorial problems. In *Complexity of Computer Computations*, pages 85–103. Plenum Press, 1972.
16. V. V. Vazirani. *Approximation Algorithms*. Springer-Verlag, 2001.

Efficient Approximation of Convex Recolorings[*]

Shlomo Moran[1,**] and Sagi Snir[2]

[1] Computer Science dept., Technion, Haifa 32000, Israel
moran@cs.technion.ac.il
[2] Mathematics dept. University of California, Berkeley, CA 94720, USA
ssagi@math.berkeley.edu

Abstract. A coloring of a tree is convex if the vertices that pertain to any color induce a connected subtree; a partial coloring (which assigns colors to some of the vertices) is convex if it can be completed to a convex (total) coloring. Convex coloring of trees arises in areas such as phylogenetics, linguistics, etc. e.g., a perfect phylogenetic tree is one in which the states of each character induce a convex coloring of the tree. Research on perfect phylogeny is usually focused on finding a tree so that few predetermined partial colorings of its vertices are convex.

When a coloring of a tree is not convex, it is desirable to know "how far" it is from a convex one. In [MS05], a natural measure for this distance, called *the recoloring distance* was defined: the minimal number of color changes at the vertices needed to make the coloring convex. This can be viewed as minimizing the number of "exceptional vertices" w.r.t. to a closest convex coloring. The problem was proved to be NP-hard even for colored strings.

In this paper we continue the work of [MS05], and present a 2-approximation algorithm of convex recoloring of strings whose running time $O(cn)$, where c is the number of colors and n is the size of the input, and an $O(cn^2)$ 3-approximation algorithm for convex recoloring of trees.

1 Introduction

A phylogenetic tree is a tree which represents the course of evolution for a given set of species. The leaves of the tree are labeled with the given species. Internal vertices correspond to hypothesized, extinct species. A *character* is a biological attribute shared among all the species under consideration, although every species may exhibit a different *character state*. Mathematically, if X is the set of species under consideration, a character on X is a function C from X into a set $\mathcal{C}$ of character states. A character on a set of species can be viewed as a *coloring* of the species, where each color represents one of the character's states. A natural biological constraint is that the reconstructed phylogeny have the property that each of the characters could have evolved without reverse or convergent transitions: In a reverse transition some species regains a character state of some old

---

[*] A preliminary version of the results in this paper appeared in [MS03].
[**] This research was supported by the Technion VPR-fund and by the Bernard Elkin Chair in Computer Science.

C. Chekuri et al. (Eds.): APPROX and RANDOM 2005, LNCS 3624, pp. 192–208, 2005.

ancestor whilst its direct ancestor has lost this state. A convergent transition occurs if two species possess the same character state, while their least common ancestor possesses a different state.

In graph theoretic terms, the lack of reverse and convergent transitions means that the character is *convex* on the tree: for each state of this character, all species (extant and extinct) possessing that state induce a single *block*, which is a maximal monochromatic subtree. Thus, the above discussion implies that in a phylogenetic tree, each character is likely to be convex or "almost convex". This makes convexity a fundamental property in the context of phylogenetic trees to which a lot of research has been dedicated throughout the years. The *Perfect Phylogeny* (PP) problem, whose complexity was extensively studied (e.g. [Gus91, KW94, AFB96, KW97, BFW92, Ste92]), seeks for a phylogenetic tree that is simultaneously convex on each of the input characters. *Maximum parsimony* (MP) [Fit81, San75] is a very popular tree reconstruction method that seeks for a tree which minimizes the parsimony score defined as the number of mutated edges summed over all characters (therefore, PP is a special case of MP). [GGP+96] introduce another criterion to estimate the distance of a phylogeny from convexity. They define the *phylogenetic number* as the maximum number of connected components a single state induces on the given phylogeny (obviously, phylogenetic number one corresponds to a perfect phylogeny). Convexity is a desired property in other areas of classification, beside phylogenetics. For instance, in [Be00, BDFY01] a method called *TNoM* is used to classify genes, based on data from gene expression extracted from two types of tumor tissues. The method finds a separator on a binary vector, which minimizes the number of "1" in one side and "0" in the other, and thus defines a convex vector of minimum Hamming distance to the given binary vector. In [HTD+04], distance from convexity is used (although not explicitly) to show strong connection between strains of Tuberculosis and their human carriers.

In a previous work [MS05], we defined and studied a natural distance from a given coloring to a convex one: the *recoloring distance*. In the simplest, unweighted model, this distance is the minimum number of color changes at the vertices needed to make the given coloring convex (for strings this reduces to Hamming distance from a closest convex coloring). This model was extended to a weighted model, where changing the color of a vertex v costs a nonnegative weight $w(v)$. The most general model studied in [MS05] is the *non-uniform* model, where the cost of coloring vertex v by a color d is an arbitrary nonnegative number $cost(v, d)$.

It was shown in [MS05] that finding the recoloring distance in the unweighted model is NP-hard even for strings (trees with two leaves), and few dynamic programming algorithms for exact solutions of few variants of the problem were presented.

In this work we present two polynomial time, constant ratio approximation algorithms, one for strings and one for trees. Both algorithms are for the weighted (uniform) model. The algorithm for strings is based on a lower bound technique which assigns penalties to colored trees. The penalties can be computed in linear

time, and once a penalty is computed, a recoloring whose cost is smaller than
the penalty is computed in linear time. The 2-approximation follows by showing
that for a string, the penalty is at most twice the cost of an optimal convex
recoloring. This last result does not hold for trees, where a different technique
is used. The algorithm for trees is based on a recursive construction that uses a
variant of the local ratio technique [BY00, BYE85], which allows adjustments of
the underlying tree topology during the recursive process.

The rest of the paper is organized as follows. In the next section we present
the notations and define the models used. In Section 3 we define the notion of
penalty which provides lower bounds on the optimal cost of convex recoloring
of any tree. In Section 4, we present the 2-approximation algorithm for the
string. In Section 5 we briefly explain the local ratio technique, and present
the 3-approximation algorithm for the tree. We conclude and point out future
research directions in Section 6.

2   Preliminaries

A colored tree is a pair (T, C) where $T = (V, E)$ is a tree with vertex set $V =
\{v_1, \ldots, v_n\}$, and C is a *coloring* of T, i.e. - a function from V onto a set of colors
$\mathcal{C}$. For a set $U \subseteq V$, $C|_U$ denotes the restriction of C to the vertices of U, and
$C(U)$ denotes the set $\{C(u) : u \in U\}$. For a subtree $T' = (V(T'), E(T'))$ of T,
$C(T')$ denotes the set $C(V(T'))$. A *block* in a colored tree is a maximal set of
vertices which induces a monochromatic subtree. A *d-block* is a block of color d.
The number of d-blocks is denoted by $n_b(C, d)$, or $n_b(d)$ when C is clear from
the context. A coloring C is said to be *convex* if $n_b(C, d) = 1$ for every color
$d \in \mathcal{C}$. The number of d-*violations* in the coloring C is $n_b(C, d) - 1$, and the total
number of *violations* of C is $\sum_{c \in \mathcal{C}}(n_b(C, d) - 1)$. Thus a coloring C is convex
iff the total number of violations of C is zero (in [FBL03] the above sum, taken
over all characters, is used as a measure of the distance of a given phylogenetic
tree from perfect phylogeny).

The definition of convex coloring is extended to *partially colored* trees, in
which the coloring C assigns colors to some subset of vertices $U \subseteq V$, which
is denoted by $Domain(C)$. A partial coloring is said to be convex if it can be
extended to a total convex coloring (see [SS03]). Convexity of partial and total
coloring have simple characterization by the concept of *carriers*: For a subset
U of V, $carrier(U)$ is the minimal subtree that contains U. For a colored tree
(T, C) and a color $d \in C$, $carrier_T(C, d)$ (or $carrier(C, d)$ when T is clear) is
the carrier of $C^{-1}(d)$. We say that C has the *disjointness property* if for each
pair of colors $\{d, d'\}$ it holds that $carrier(C, d) \cap carrier(C, d') = \emptyset$. It is easy to
see that a total or partial coloring C is convex iff it has the disjointness property
(in [DS92] convexity is actually defined by the disjointness property).

When some (total or partial) input coloring (C, T) is given, any other coloring
C' of T is viewed as a *recoloring* of the input coloring C. We say that a recoloring
C' of C *retains* (the color of) a vertex v if $C(v) = C'(v)$, otherwise C' *overwrites*
v. Specifically, a recoloring C' of C overwrites a vertex v either by changing the

color of v, or just by *uncoloring* v. We say that C' retains (overwrites) a set of vertices U if it retains (overwrites resp.) every vertex in U. For a recoloring C' of an input coloring C, $\mathcal{X}_C(C')$ (or just $\mathcal{X}(C')$) is the set of the vertices overwritten by C', i.e.

$$\mathcal{X}_C(C') = \{v \in V : [v \in Domain(C)] \bigwedge [(v \notin Domain(C')) \vee (C(v) \neq C'(v))]\}.$$

With each recoloring C' of C we associate a *cost*, denoted as $cost_C(C')$ (or $cost(C')$ when C is understood), which is the number of vertices overwritten by C', i.e. $cost_C(C') = |\mathcal{X}_C(C')|$. A coloring C^* is an *optimal convex recoloring of C*, or in short an *optimal recoloring of C*, and $cost_C(C^*)$ is denoted by $OPT(T, C)$, if C^* is a convex coloring of T, and $cost_C(C^*) \leq cost_C(C')$ for any other convex coloring C' of T.

The above cost function naturally generalizes to the *weighted* version: the input is a triplet (T, C, w), where $w : V \to \mathbb{R}^+ \cup \{0\}$ is a weight function which assigns to each vertex v a nonnegative weight $w(v)$. For a set of vertices X, $w(X) = \sum_{v \in X} w(v)$. The cost of a convex recoloring C' of C is $cost_C(C') = w(\mathcal{X}(C'))$, and C' is an optimal convex recoloring if it minimizes this cost.

The above unweighted and weighted cost models are *uniform*, in the sense that the cost of a recoloring is determined by the set of overwritten vertices, regardless the specific colors involved. [MS05] defines also a more subtle *non uniform model*, which is not studied in this paper.

Let AL be an algorithm which receives as an input a weighted colored tree (T, C, w) and outputs a convex recoloring of (T, C, w), and let $AL(T, C, w)$ be the cost of the convex recoloring output by AL. We say that AL is an r-*approximation* algorithm for the convex tree recoloring problem if for all inputs (T, C, w) it holds that $AL(T, C, w)/OPT(T, C, w) \leq r$ [GJ79, Hoc97, Vaz01].

We complete this section with a definition and a simple observation which will be useful in the sequel. Let (T, C) be a colored tree. A coloring C^* is an *expanding* recoloring of C if in each block of C^* at least one vertex v is retained (i.e., $C(v) = C^*(v)$).

Observation 1 *let $(T = (V, E), C, w)$ be a weighted colored tree, where $w(V) > 0$. Then there exists an expanding optimal convex recoloring of C.*

Proof. Let C' be an optimal recoloring of C which uses a minimum number of colors (i.e. $|C'(V)|$ is minimized). We shall prove that C' is an expanding recoloring of C.

Since $w(V) > 0$, the claim is trivial if C' uses just one color. So assume for contradiction that C' uses at least two colors, and that for some color d used by C', there is no vertex v s.t. $C(v) = C'(v) = d$. Then there must be an edge (u, v) such that $C'(u) = d$ but $C'(v) = d' \neq d$. Therefore, in the uniform cost model, the coloring C'' which is identical to C' except that all vertices colored d are now colored by d' is an optimal recoloring of C which uses a smaller number of colors - a contradiction.

In view of Observation 1 above, we assume in the sequel (sometimes implicitly) that the given optimal convex recolorings are expanding.

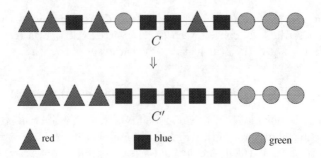

Fig. 1. C' is a convex recoloring for C which defines the following penalties: $p_{green}(C') = 1$, $p_{red}(C') = 2$, $p_{blue}(C') = 3$.

3   Lower Bounds via Penalties

In this section we present a general lower bound on the recoloring distance of weighted colored trees. Although for a general tree this bound can be fairly poor, in the next section we present an algorithm for convex recoloring of strings, which always finds a convex recoloring whose cost is at most twice this lower bound, and hence it is a 2-approximation algorithm for strings.

Let (T, C, w) be a weighted colored tree. For a color d and $U \subseteq V(T)$ let:

$$penalty_{C,d}(U) = w(U \cap \overline{C^{-1}(d)}) + w(\overline{U} \cap C^{-1}(d))$$

Informally, when the vertices in U induce a subtree, $penalty_{C,d}(U)$ is the total weight of the vertices which must be overwritten to make U the unique d-block in the coloring: a vertex v must be overwritten either if $v \in U$ and $C(v) \neq d$, or if $v \notin U$ and $C(v) = d$.

$penalty_C(C')$, the penalty of a convex recoloring C' of C, is the sum of the penalties of all the colors, with respect to the color blocks of C':

$$penalty_C(C') = \sum_{d \in C} penalty_{C,d}(C'^{-1}(d))$$

Figure 1 depicts the calculation of a penalty associated with a convex recoloring C' of C.

In the sequel we assume that the input colored tree (T, C) is fixed, and omit it from the notations.

Claim. $penalty(C') = 2cost(C')$

Proof. From the definitions we have

$$penalty(C') = \sum_{d \in C} w \left(\{v \in V : C'(v) = d \text{ and } C(v) \neq d\} \cup \right.$$

$$\left. \{v \in V : C'(v) \neq d \text{ and } C(v) = d\}\right)$$

$$= 2w(\{v \in V : C'(v) \neq C(v)\}) = 2cost(C')$$

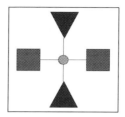

Fig. 2. A convex recoloring must overwrite at least one of the large lateral blocks (a triangle or a rectangle).

As can be seen in Figure 1, $penalty(C') = 6$ while $cost(C') = 3$.

For each color d, p_d^* is the penalty of a block which minimizes the penalty for d:

$$p_d^* = \min\{penalty_d(V(T')) : T' \text{ is a subtree of } T\}$$

Corollary 1. *For any recoloring C' of C,*

$$\sum_{d \in C} p_d^* \leq \sum_{d \in C} penalty_d(C') = 2cost(C').$$

Proof. The inequality follows from the definition of p_d^*, and the equality from Claim 3.

Corollary 1 above provides a lower bound on the cost of convex recoloring of trees. It can be shown that this lower bound can be quite poor for trees, that is: $OPT(T, C)$ can be considerably larger than $(\sum_{d \in C} p_d^*)/2$. For example, any convex recoloring of the tree in Figure 2, must overwrite at least one of the (large) lateral blocks in the tree, while $(\sum_{d \in C} p_d^*)/2$ in that tree is the weight of the (small) central vertex (the circle). However in the next section we show that this bound can be used to obtain a polynomial time 2-approximation for convex recoloring of strings.

4   A 2-Approximation Algorithm for Strings

Let a weighted colored string (S, C, w), where $S = (v_1, \ldots, v_n)$, be given. For $1 \leq i \leq j \leq n$, $S[i, j]$ is the substring $(v_i, v_{i+1}, \ldots, v_j)$ of S. The algorithm starts by finding for each d a substring $B_d = S[i_d, j_d]$ for which $penalty_d(S[i_d, j_d]) = p_d^*$. It is not hard to verify that B_d consists of a subsequence of consecutive vertices in which the difference between the total weight of d-vertices and the total weight of other vertices (i.e. $w(B_d \cap C^{-1}(d)) - w(B_d \setminus C^{-1}(d))$) is maximized, and thus B_d can be found in linear time. We say that a vertex v is *covered by color* d if it belongs to B_d. v is *covered* if it is covered by some color d, and it is *free* otherwise.

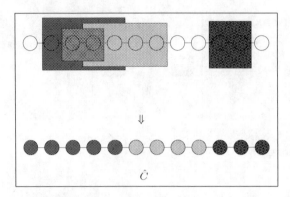

Fig. 3. The upper part of the figure shows the optimal blocks on the string and the lower part shows the coloring returned by the algorithm.

We describe below a linear time algorithm which, given the blocks B_d, defines a convex coloring $\hat{C}$ so that $cost(\hat{C}) < \sum_d p_d^*$, which by Corollary 1 is a 2-approximation to a minimal convex recoloring of C.

$\hat{C}$ is constructed by performing one scan of S from left to right. The scan consists of at most c stages, where stage j defines the $j-th$ block of $\hat{C}$, to be denoted F_j, and its color, d_j, as follows.

Let d_1 be the color of the leftmost covered vertex (note that v_1 is either free or covered by d_1). d_1 is taken to be the color of the first (leftmost) block of $\hat{C}$, F_1, and $\hat{C}(v_1)$ is set to d_1. For $i > 1$, $\hat{C}(v_i)$ is determined as follows: Let $\hat{C}(v_{i-1}) = d_j$. Then if $v_i \in B_{d_j}$ or v_i is free, then $\hat{C}(v_i)$ is also set to d_j. Else, v_i must be a covered vertex. Let d_{j+1} be one of the colors that cover v_i. $\hat{C}(v_i)$ is set to d_{j+1} (and v_i is the first vertex in F_{j+1}).

Observation 2 $\hat{C}$ is a convex coloring of S.

Proof. Let d_j be the color of the $j-th$ block of $\hat{C}$, F_j, as described above. The convexity of $\hat{C}$ follows from the the following invariant, which is easily proved by induction: For all $j \geq 1$, $\cup_{k=1}^{j} F_k \supseteq \cup_{k=1}^{j} B_{d_k}$. This means that, for all j, no vertex to the right of F_j is covered by d_j, and hence no such vertex is colored by d_j.

Thus it remains to prove

Lemma 1. $cost(\hat{C}) \leq \sum_{d \in C} p_d^*$.

Proof. $cost(\hat{C}) = \sum \{w(v_i) : C(v_i) \neq \hat{C}(v_i)\}$. Thus $w(v_i)$ is added to $cost(\hat{C})$ only if $C(v_i) = d$ and $\hat{C}(v_i) = d'$ for some distinct d', d. By the algorithm, $\hat{C}(v_i) = d'$ only if $v_i \in B_{d'}$ or v_i is free. In the first case $w(v_i)$ is accounted for in $p_{d'}^*$. In the second case it is accounted for in p_d^*. In both cases $w(v_i)$ is added to sum on the righthand side, which proves the inequality.

5 A 3-Approximation Algorithm for Tree

In this section we present a polynomial time algorithm which approximates the minimal convex coloring of a weighted tree by factor three. The input is a triplet (T, C, w), where w is a nonnegative weight function and C is a (possibly partial) coloring whose domain is the set $support(w) = \{v \in V : w(v) > 0\}$.

We first introduce the notion of covers w.r.t. colored trees. A set of vertices X is a *convex cover* (or just a *cover*) for a colored tree (T, C) if the (partial) coloring $C_X = C|_{[V \setminus X]}$ is convex (i.e., C can be transformed to a convex coloring by overwriting only vertices in X). Thus, if C' is a convex recoloring of (T, C), then $\mathcal{X}_C(C')$, the set of vertices overwritten by C', is a cover for (T, C), and cost of C' is $w(\mathcal{X}(C'))$. Moreover, deciding whether a subset $X \subseteq V$ is a cover for (T, C), and constructing a total convex recoloring C' of C such that $\mathcal{X}(C') \subseteq X$ in case it is, can be done in $O(n \cdot n_c)$ time. Therefore, finding an optimal convex total recoloring of C is polynomially equivalent to finding an optimal cover X, or equivalently a partial convex recoloring C' of C so that $w(\mathcal{X}(C')) = w(X)$ is minimized.

Our approximation algorithm makes use of the local ratio technique, which is useful for approximating optimization covering problems such as vertex cover, dominating set, minimum spanning tree, feedback vertex set and more [BYE85, BBF99, BY00]. We hereafter describe it briefly:

The input to the problem is a triplet $(V, \Sigma \subseteq 2^V, w : V \to \mathbb{R}^+)$, and the goal is to find a subset $X \in \Sigma$ such that $w(X)$ is minimized, i.e. $w(X) = OPT(V, \Sigma, w) = \min_{Y \in \Sigma} w(Y)$ (in our context V is the set of vertices, and Σ is the set of covers). The local ratio principle is based on the following observation (see e.g. [BY00]):

Observation 3 *For every two weight functions w_1, w_2:*

$$OPT(V, \Sigma, w_1) + OPT(V, \Sigma, w_2) \leq OPT(V, \Sigma, w_1 + w_2)$$

Now, given our initial weight function w, we select w_1, w_2 s.t. $w_1 + w_2 = w$ and $|supprt(w_1)| < |support(w)|$. We first apply the algorithm to find an r-approximation to (V, Σ, w_1) (in particular, if $V \setminus support(w_1)$ is a cover, then it is an optimal cover to (V, Σ, w_1)). Let X be the solution returned for (V, Σ, w_1), and assume that $w_1(X) \leq r \cdot OPT(V, \Sigma, w_1)$. If we could also guarantee that $w_2(X) \leq r \cdot OPT(V, \Sigma, w_2)$ then by Observation 3 we are guaranteed that X is also an r-approximation for $(V, \Sigma, w_1 + w_2 = w)$. The original property, introduced in [BYE85], which was used to guarantee that $w_2(X) \leq r \cdot OPT(V, \Sigma, w_2)$ is that w_2 is r-*effective*, that is: for every $X \in \Sigma$ it holds that $w_2(X) \leq r \cdot OPT(V, \Sigma, w_2)$ (note that if $V \in \Sigma$, the above is equivalent to requiring that $w_2(V) \leq r \cdot OPT(V, \Sigma, w_2)$).

Theorem 4. *[BYE85] Given $X \in \Sigma$ s.t. $w_1(X) \leq r \cdot OPT(V, \Sigma, w_1)$. If w_2 is r-effective, then $w(X) = w_1(X) + w_2(X) \leq r \cdot OPT(V, \Sigma, w)$.*

We start by presenting two applications of Theorem 4 to obtain a 3-approximation algorithm for convex recoloring of strings and a 4-approximation algorithm for convex recoloring of trees.

3-string-APPROX:

Given an instance of the convex weighted string problem (S, C, w):

1. If $V \setminus support(w)$ is a cover then $X \leftarrow V \setminus support(w)$. Else:
2. Find 3 vertices $x, y, z \in support(w)$ s.t. $C(x) = C(z) \neq C(y)$ and y lies between x and z.
   (a) $\varepsilon \leftarrow \min\{w(x), w(y), w(z)\}$
   (b) $w_2(v) = \begin{cases} \varepsilon & \text{if } v \in \{x, y, z\} \\ 0 & \text{otherwise.} \end{cases}$
   (c) $w_1 \leftarrow w - w_2$
   (d) $X \leftarrow$ 3-string-APPROX$(S, C|_{support(w_1)}, w_1)$

Note that a (partial) coloring of a string is not convex iff the condition in 2 holds. It is also easy to see that w_2 is 3-effective, since any cover Y must contain at least one vertex from any triplet described in condition 2, hence $w_2(Y) \geq \varepsilon$ while $w_2(V) = 3\varepsilon$.

The above algorithm cannot serve for approximating convex tree coloring since in a tree the condition in 2 might not hold even if $V \setminus support(w)$ is not a cover. In the following algorithm we generalize this condition to one which must hold in any non-convex coloring of a tree, in the price of increasing the approximation ratio from 3 to 4.

4-tree-APPROX:

Given an instance of the convex weighted tree problem (T, C, w):

1. If $V \setminus support(w)$ is a cover then $X \leftarrow V \setminus support(w)$. Else:
2. Find two pairs of (not necessarily distinct) vertices (x_1, x_2) and (y_1, y_2) in $support(w)$ s.t. $C(x_1) = C(x_2) \neq C(y_1) = C(y_2)$, and $carrier(\{x_1, x_2\}) \cap carrier(\{y_1, y_2\}) \neq \emptyset$:
   (a) $\varepsilon \leftarrow \min\{w(x_i), w(y_i)\}, i = \{1, 2\}$
   (b) $w_2(v) = \begin{cases} \varepsilon & \text{if } v \in \{x_1, x_2, y_1, y_2\} \\ 0 & \text{otherwise.} \end{cases}$
   (c) $w_1 \leftarrow w - w_2$
   (d) $X \leftarrow$ 4-tree-APPROX$(S, C|_{support(w_1)}, w_1)$

The algorithm is correct since if there are no two pairs as described in step 2, then $V \setminus support(w)$ is a cover. Also, it is easy to see that w_2 is 4-effective. Hence the above algorithm returns a cover with weight at most $4 \cdot OPT(T, C, w)$.

We now describe algorithm 3-tree-APPROX. Informally, the algorithm uses an iterative method, in the spirit of the local ratio technique, which approximates the solution of the input (T, C, w) by reducing it to (T', C', w_1) where $|support(w_1)| < |support(w)|$. Depending on the given input, this reduction is either of the local ratio type (via an appropriate 3-effective weight function) or, the input graph is replaced by a smaller one which preserves the optimal solutions.

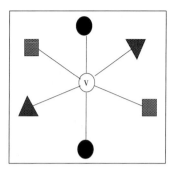

Fig. 4. Case 2: a vertex v is contained in 3 different carriers.

3-tree-APPROX(T, C, w)
On input (T, C, w) of a weighted colored tree, do the following:

1. If $V \setminus support(w)$ is a cover then $X \leftarrow V \setminus support(w)$. Else:
2. $(T', C', w_1) \leftarrow REDUCE(T, C, w)$. \The function $REDUCE$ guarantees that $|support(w_1)| < |support(w)|$
   (a) $X' \leftarrow$ 3-tree-APPROX(T', C', w_1).
   (b) $X \leftarrow UPDATE(X', T)$. \The function $UPDATE$ guarantees that if X' is a 3-approximation to (T', C', w_1), then X is a 3-approximation to (T, C, w).

Next we describe the functions $REDUCE$ and $UPDATE$, by considering few cases. In the first two cases we employ the local ratio technique.

Case 1: $support(w)$ contains three vertices x, y, z such that y lies on the path from x to z and $C(x) = C(z) \neq C(y)$.
In this case we use the same reduction of 3-string-APPROX:
Let

$$\varepsilon = \min\{w(x), w(y), w(z)\} > 0.$$

Then

$$REDUCE(T, C, w) = (T, C|_{support(w_1)}, w_1),$$

where $w_1(v) = w(v)$ if $v \notin \{x, y, z\}$, else $w_1(v) = w(v) - \varepsilon$. The same arguments which imply the correctness of 3-string-APPROX imply that if X' is a 3-approximation for (T', C', w_1), then it is also a 3-approximation for (T, C, w), thus we set

$$UPDATE(X', T) = X'.$$

Case 2: Not Case 1, and T contains a vertex v such that $v \in \cap_{i=1}^{3} carrier(d_i, C)$ for three distinct colors d_1, d_2 and d_3 (see Figure 4).
In this case we must have that $w(v) = 0$ (else Case 1 would hold), and there are three *designated pairs* of vertices $\{x_1, x_2\}, \{y_1, y_2\}$ and $\{z_1, z_2\}$ such that $C(x_i) = d_1$, $C(y_i) = d_2, C(z_i) = d_3 (i = 1, 2)$, and v lies on each of the three paths connecting these three pairs (see Figure 4). We set

$$REDUCE(T, C, w) = (T, C|_{support(w_1)}, w_1),$$

where w_1 is defined as follows.
Let

$$\varepsilon = \min\{w(x_i), w(y_i), w(z_i) : i = 1, 2\}.$$

Then $w_1(v) = w(v)$ if v is not in one of the designated pairs, else $w_1(v) = w(v) - \varepsilon$. Finally, any cover for (T, C) must contain at least two vertices from the set $\{x_i, y_i, z_i : i = 1, 2\}$, hence $w - w_1 = w_2$ is 3-effective, and by the local ratio theorem we can set

$$UPDATE(X', T) = X'.$$

Case 3: Not Cases 1 and 2.
Root T at some vertex r and for each color d let r_d be the root of the subtree $carrier(d, C)$. Let d_0 be a color for which the root r_{d_0} is farthest from r. Let $\bar{T}$ be the subtree of T rooted at r_{d_0}, and let $\hat{T} = T \setminus \bar{T}$ (see Figure 5). By the definition of r_{d_0}, no vertex in $\hat{T}$ is colored by d_0, and since Case 2 does not hold, there is a color d' so that $\{d_0\} \subseteq C(V(\bar{T})) \subseteq \{d_0, d'\}$.

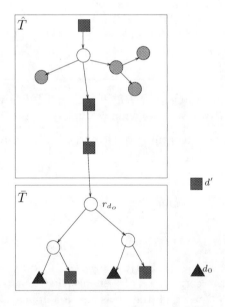

Fig. 5. Case 3: Not case 1 nor 2. $\bar{T}$ is the subtree rooted at r_{d_0} and $\hat{T} = T \setminus \bar{T}$.

Subcase 3a: $C(V(\bar{T})) = \{d_0\}$ (see Figure 6).
In this case, $carrier(d_0, C) \cap carrier(d, C) = \emptyset$ for each color $d \neq d_0$, and for each optimal solution X it holds that $X \cap V(\bar{T}) = \emptyset$. We set

$$REDUCE(T, C, w) \leftarrow (\hat{T}, C|_{V(\hat{T})}, w|_{V(\hat{T})}).$$

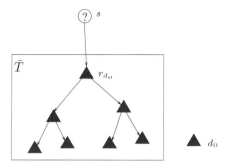

Fig. 6. Case 3a: No vertices of $\hat{T}$ are colored by d'.

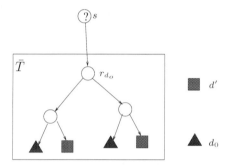

Fig. 7. Case 3b: $r_{d_0} \in T_{d_0} \cap carrier(d')$.

The 3-approximation X' to (T', C', w_1) is also a 3-approximation to (X, C, w), thus

$$UPDATE(X', T) = X'.$$

We are left with the last case, depicted in Figure 7.

Subcase 3b: $C(V(\bar{T})) = \{d_0, d'\}$.
In this case we have that $r_{d_0} \in carrier(d_0, C) \cap carrier(d', C)$ and $w(r_{d_0}) = 0$ (since Case 1 does not hold).

Informally, $REDUCE(T, C, w)$ modifies the tree T by replacing the subtree $\bar{T}$ by a smaller subtree $\bar{T}_0$, which contains only two vertices, and which encodes three possible recolorings of $\bar{T}$, one of which must be used in an optimal recoloring of T. The tree T', resulted from replacing $\bar{T}$ by $\bar{T}_0$ in the tree T, has smaller support than T, since $|support(w) \cap V(\bar{T})| \geq 3$. This last inequality holds since, by the fact that r_{d_0} lies between two vertices colored d_0 and between two vertices colored d', $V(\bar{T})$ must contain at least two vertices colored d_0 and at least one vertex colored d'.

Observation 5 *There is an optimal convex coloring C' which satisfies the following: $C'(v) \neq d_0$ for any $v \in V(\hat{T})$, and $C'(v) \in \{d_0, d'\}$ for any $v \in V(\bar{T})$.*

Proof. Let $\hat{C}$ be an expanding optimal convex recoloring of (T, C). We will show that there is an optimal coloring C' satisfying the lemma such that $cost(C') \leq cost(\hat{C})$. Since $\hat{C}$ is expanding and optimal, at least one vertex in $\bar{T}$ is colored either by d_0 or by d'. Let U be a set of vertices in $\bar{T}$ so that $carrier(U)$ is a maximal subtree all of whose vertices are colored by colors not in $\{d_0, d'\}$. Then $carrier(U)$ must have a neighbor u in $\bar{T}$ s.t. $\hat{C}(u) \in \{d_0, d'\}$. Changing the colors of the vertices in U to $\hat{C}(u)$ does not increase the cost of the recoloring. This procedure can be repeated until all the vertices of $\bar{T}$ are colored by d_0 or by d'. A similar procedure can be used to change the color of all the vertices in $\hat{T}$ to be different from d_0. It is easy to see that the resulting coloring C' is convex and $cost(C') \leq cost(\hat{C})$.

The function $REDUCE$ in Subcase 3b is based on the following observation: Let C' be any optimal recoloring of T satisfying Observation 5, and let s be the parent of r_{d_0} in T. Then $C'|_{V(\bar{T})}$, the restriction of the coloring C' to the vertices of $\bar{T}$, depends only on whether $carrier(d', C')$ intersects $V(\hat{T})$, and in this case if it contains the vertex s. Specifically, $C'_{V(\bar{T})}$ must be one of the three colorings of $V(\bar{T})$, C_{high}, C_{medium} and C_{min}, according to the following three scenarios:

1. $carrier(d', C') \cap V(\hat{T}) \neq \emptyset$ and $s \notin carrier(d', C')$. Then it must be the case that C' colors all the vertices in $V(\bar{T})$ by d_0. This coloring of $\bar{T}$ is denoted as C_{high}.
2. $carrier(d', C') \cap V(\hat{T}) \neq \emptyset$ and $s \in carrier(d', C')$. Then $C'|_{\bar{T}}$ is a coloring of minimal possible cost of $\bar{T}$ which either equals C_{high} (i.e. colors all vertices by d_0), or otherwise colors r_{d_0} by d'. This coloring of $\bar{T}$ is called C_{medium}.
3. $carrier(d', C') \cap V(\hat{T}) = \emptyset$. Then $C'|_{\bar{T}}$ must be an optimal convex recoloring of $\bar{T}$ by the two colors d_0, d'. This coloring of $\bar{T}$ is called C_{min}.

We will show soon that the colorings C_{high}, C_{medium} and C_{min} above can be computed in linear time. The function $REDUCE$ in Subcase 3b modifies the tree T by replacing $\bar{T}$ by a subtree $\bar{T}_0$ with only 2 vertices, r_{d_0} and v_0, which encodes the three colorings $C_{high}, C_{medium}, C_{min}$. Specifically,

$$REDUCE(T, C, w) = (T', C', w_1)$$

where (see Figure 8):

- T' is obtained from T by replacing the subtree $\bar{T}$ by the subtree $\bar{T}_0$ which contains two vertices: a root r_{d_0} with a single descendant v_0.
- $w_1(v) = w(v)$ for each $v \in V(\hat{T})$. For r_{d_0} and v_0, w_1 is defined as follows: $w_1(r_{d_0}) = cost(C_{medium}) - cost(C_{min})$ and $w_1(v_0) = cost(C_{high}) - cost(C_{min})$.
- $C'(v) = C(v)$ for each $v \in V(\hat{T})$; if $w_1(r_{d_0}) > 0$ then $C'(r_{d_0}) = d_0$ and if $w_1(v_0) > 0$ then $C'(v_0) = d'$. (If $w_1(u) = 0$ for $u \in \{r_{d_0}, v_0\}$, then $C'(u)$ is undefined).

Figure 8 illustrates $REDUCE$ for case 3b. In the figure, C_{high} requires overwriting all d' vertices and therefore costs 3, C_{medium} requires overwriting one

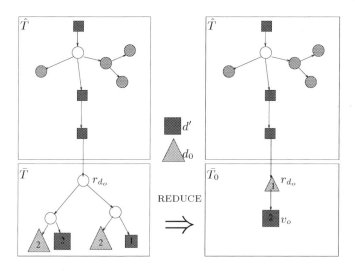

Fig. 8. $REDUCE$ of case 3b: $\bar{T}$ is replaced with $\bar{T}_0$ where $w_1(r_{d_0}) = C_{medium} - C_{min} = 1$ and $w_1(v_0) = C_{high} - C_{min} = 2$.

d_0 vertex and costs 2 and C_{min} is the optimal coloring for $\bar{T}$ with cost 1. The new subtree $\bar{T}_0$ reflects these weight with $w_1(r_{d_0}) = C_{medium} - C_{min} = 1$ and $w_1(v_0) = C_{high} - C_{min} = 2$.

Claim. $OPT(T', C', w_1) = OPT(T, C, w) - cost(C_{min})$.

Proof. We first show that $OPT(T', C', w_1) \leq OPT(T, C, w) - cost(C_{min})$. Let C^* be an optimal recoloring of C satisfying Observation 5, and let $X^* = \mathcal{X}(C^*)$. By the discussion above, we may assume that $C^*|_{V(\bar{T})}$ has one of the forms C_{high}, C_{medium} or C_{min}. Thus, $X^* \cap V(\bar{T})$ is either $\mathcal{X}(C_{high}), \mathcal{X}(C_{medium})$ or $\mathcal{X}(C_{min})$. We map C^* to a coloring C' of T' as follows: for $v \in V(\hat{T})$, $C'(v) = C^*(v)$. C' on r_{d_0} and v_0 is defined as follows:

- If $C^*|_{V(\bar{T})} = C_{high}$ then $C'(r_{d_0}) = C'(v_0) = d_0$, and $cost(C'|_{V(\bar{T})}) = w_1(v_0)$;
- If $C^*|_{V(\bar{T})} = C_{medium}$ then $C'(r_{d_0}) = C'(v_0) = d'$, and $cost(C'|_{V(\bar{T})}) = w_1(r_{d_0})$;
- If $C^*|_{V(\bar{T})} = C_{min}$ then $C'(r_{d_0}) = d_0, C'(v_0) = d'$, and $cost(C'|_{V(\bar{T})}) = 0$.

Note that in all three cases, $cost(C') = cost(C^*) - cost(C_{min})$.

The proof of the opposite inequality

$$OPT(T, C, w) - cost(C_{min}) \leq OPT(T', C', w_1)$$

is similar.

Corollary 2. C^* *is optimal recoloring of* (T, C, w) *iff* C' *is an optimal recoloring of* (T', C', w_1).

We now can define the $UPDATE$ function for Subcase 3b: Let $X' = 3-tree-$ $APPROX(T', C', w_1)$. Then X' is a disjoint union of the sets $\hat{X}' = X' \cap V(\hat{T})$ and $\bar{X}'_0 = X' \cap V(\bar{T}_0)$. Moreover, $\bar{X}'_0 \in \{\{r_{d_0}\}, \{v_0\}, \emptyset\}$. Then $X \leftarrow UPDATE(X') = \hat{X}' \cup \bar{X}'$, where $\bar{X}'$ is $\mathcal{X}(C_{high})$ if $\bar{X}'_0 = \{r_{d_0}\}$, is $\mathcal{X}(C_{medium})$ if $\bar{X}'_0 = \{v_0\}$, and is $\mathcal{X}(C_{min})$ if $\bar{X}'_0 = \emptyset$. Note that $w(X) = w(X') + cost(C_{min})$. The following inequalities show that if $w_1(X')$ is a 3-approximation to $OPT(T', C', w_1)$, then $w(X)$ is a 3-approximation to $OPT(T, C, w)$:

$$w(X) = w_1(X') + cost(C_{min}) \le 3OPT(T', C', w_1) + cost(C_{min})$$
$$< 3(OPT(T', C', w_1) + cost(C_{min})) = 3OPT(T, C, w)$$

5.1   A Linear Time Algorithm for Subcase 3b

In Subcase 3b we need to compute C_{high}, C_{medium} and C_{min}. The computation of C_{high} is immediate. C_{medium} and C_{min} can be computed by the following simple, linear time algorithm that finds a minimal cost convex recoloring of a bi-colored tree, under the constraint that the color of a given vertex r is predetermined to one of the two colors.

Let the weighted colored tree (T, C, w) and the vertex r be given, and let $\{d_1, d_2\} = C(T)$. For $i \in \{1, 2\}$, let C_i the minimal cost convex recoloring which sets the color of r to d_i (note that a coloring with minimum cost in $\{C_1, C_2\}$ is an optimal convex recoloring of (T, C)). We illustrate the computation of C_1 (the computation of C_2 is similar):
Compute for every edge $e = (u \rightarrow v)$ a cost defined by

$$cost(e) = w(\{v' : v' \in T(v) \text{ and } C(v') = d_1\}) +$$
$$w(\{v' : v' \in [T \setminus T(v)] \text{ and } C(v') = d_2\})$$

where $T(v)$ is the subtree rooted at v. This can be done by one post order traversal of the tree. Then, select the edge $e^* = (u_0 \rightarrow v_0)$ which minimizes this cost, and set $C_1(w) = d_2$ for each $w \in T(v_0)$, and $C_1(w) = d_1$ otherwise.

5.2   Correctness and Complexity

We now summarize the discussion of the previous section to show that the algorithm terminates and return a cover X which is a 3-approximation for (T, C, w).
Let $(T = (V, E), C, w)$ be an input to 3-tree-APPROX. if $V \setminus support(w)$ is a cover then the returned solution is optimal. Else, in each of the cases, $REDUCE(T, C, w)$ reduces the input to (T', C', w_1) such that $|support(w_1)| < |support(w)|$, hence the algorithm terminates within at most $n = |V|$ iterations. Also, as detailed in the previous subsections, the function $UPDATE$ guarantees that if X' is a 3-approximation for (T', C', w_1) then X is a 3-approximation to (T, C, w). Thus after at most n iterations the algorithm provides a 3-approximation to the original input.
Checking whether Case 1, Case 2, Subcase 3a or Subcase 3b holds at each stage requires $O(cn)$ time for each of the cases, and computing the function

REDUCE after the relevant case is identified requires linear time in all cases. Since there are at most n iterations, the overall complexity is $O(cn^2)$. Thus we have

Theorem 6. *Algorithm 3-tree-APPROX is a polynomial time 3-approximation algorithm for the minimum convex recoloring problem.*

6 Discussion and Future Work

In this work we showed two approximation algorithms for colored strings and trees, respectively. The 2-approximation algorithm relies on the technique of penalizing a colored string and the 3-approximation algorithm for the tree extends the local ratio technique by allowing dynamic changes in the underlying graph.

Few interesting research directions which suggest themselves are:

- Can our approximation ratios for strings or trees be improved.
- This is a more focused variant of the previous item. A problem has a *polynomial approximation scheme* [GJ79, Hoc97], or is *fully approximable* [PM81], if for each ε it can be ε-approximated in $p_\varepsilon(n)$ time for some polynomial p_ε. Are the problems of optimal convex recoloring of trees or strings fully approximable, (or equivalently have a polynomial approximation scheme)?
- Alternatively, can any of the variant be shown to be APX-hard [Vaz01]?
- The algorithms presented here apply only to uniform models. The *non uniform model*, motivated by weighted maximum parsimony [San75], assumes that the cost of assigning color d to vertex v is given by an arbitrary nonnegative number $cost(v,d)$ (note that, formally, no initial coloring C is assumed in this cost model). In this model $cost(C')$ is defined only for a total recoloring C', and is given by the sum $\sum_{v \in V} cost(v, C'(v))$. Finding non-trivial approximation results for this model is challenging.

Acknowledgments

We would like to thank Reuven Bar Yehuda, Arie Freund and Dror Rawitz for very helpful discussions.

References

[AFB96]   R. Agrawala and D. Fernandez-Baca. Simple algorithms for perfect phylogeny and triangulating colored graphs. *International Journal of Foundations of Computer Science*, 7(1):11–21, 1996.
[BBF99]   V. Bafna, P. Berman, and T. Fujito. A 2-approximation algorithm for the undirected feedback vertex set problem. *SIAM J. on Discrete Mathematics*, 12:289–297, 1999.
[BDFY01] A. Ben-Dor, N. Friedman, and Z. Yakhini. Class discovery in gene expression data. In *RECOMB*, pages 31–38, 2001.

[Be00]   M. Bittner and et.al. Molecular classification of cutaneous malignant melanoma by gene expression profiling. *Nature*, 406(6795):536–40, 2000.

[BFW92]   H.L. Bodlaender, M.R. Fellows, and T. Warnow. Two strikes against perfect phylogeny. In *ICALP*, pages 273–283, 1992.

[BY00]   R. Bar-Yehuda. One for the price of two: A unified approach for approximating covering problems. *Algorithmica*, 27:131–144, 2000.

[BYE85]   R. Bar-Yehuda and S. Even. A local-ratio theorem for approximating the weighted vertex cover problem. *Annals of Discrete Mathematics*, 25:27–46, 1985.

[DS92]   A. Dress and M.A. Steel. Convex tree realizations of partitions. *Applied Mathematics Letters,*, 5(3):3–6, 1992.

[FBL03]   D. Fernández-Baca and J. Lagergren. A polynomial-time algorithm for near-perfect phylogeny. *SIAM Journal on Computing*, 32(5):1115–1127, 2003.

[Fit81]   W. M. Fitch. A non-sequential method for constructing trees and hierarchical classifications. *Journal of Molecular Evolution*, 18(1):30–37, 1981.

[GGP⁺96]   L.A. Goldberg, P.W. Goldberg, C.A. Phillips, Z Sweedyk, and T. Warnow. Minimizing phylogenetic number to find good evolutionary trees. *Discrete Applied Mathematics*, 71:111–136, 1996.

[GJ79]   M. R. Garey and D. S. Johnson. *Computers and Intractability; A Guide to the Theory of NP-Completeness*. W.H. Freeman and Company, 1979.

[Gus91]   D. Gusfield. Efficient algorithms for inferring evolutionary history. *Networks*, 21:19–28, 1991.

[Hoc97]   D. S. Hochbaum, editor. *Approximation Algorithms for NP-Hard Problem*. PWS Publishing Company, 1997.

[HTD⁺04]   A. Hirsh, A. Tsolaki, K. DeRiemer, M. Feldman, and P. Small. From the cover: Stable association between strains of mycobacterium tuberculosis and their human host populations. *PNAS*, 101:4871–4876, 2004.

[KW94]   S. Kannan and T. Warnow. Inferring evolutionary history from DNA sequences. *SIAM J. Computing*, 23(3):713–737, 1994.

[KW97]   S. Kannan and T. Warnow. A fast algorithm for the computation and enumeration of perfect phylogenies when the number of character states is fixed. *SIAM J. Computing*, 26(6):1749–1763, 1997.

[MS03]   S. Moran and S. Snir. Convex recoloring of strings and trees. Technical Report CS-2003-13, Technion, November 2003.

[MS05]   S. Moran and S. Snir. Convex recoloring of strings and trees: Definitions, hardness results and algorithms. In *WADS*, 2005.

[PM81]   A. Paz and S. Moran. Non deterministic polynomial optimization probems and their approximabilty. *Theoretical Computer Science*, 15:251–277, 1981. Abridged version: Proc. of the 4th ICALP conference, 1977.

[San75]   D. Sankoff. Minimal mutation trees of sequences. *SIAM Journal on Applied Mathematics*, 28:35–42, 1975.

[SS03]   C. Semple and M.A. Steel. *Phylogenetics*. Oxford University Press, 2003.

[Ste92]   M. Steel. The complexity of reconstructing trees from qualitative characters and subtrees. *Journal of Classification*, 9(1):91–116, 1992.

[Vaz01]   V. Vazirani. *Approximation Algorithms*. Springer, Berlin, germany, 2001.

Approximation Algorithms
for Requirement Cut on Graphs

Viswanath Nagarajan[*] and Ramamoorthi Ravi[**]

Tepper School of Business, Carnegie Mellon University, Pittsburgh PA 15213
{viswa,ravi}@cmu.edu

Abstract. In this paper, we unify several graph partitioning problems including multicut, multiway cut, and k-cut, into a single problem. The input to a requirement cut problem is an undirected edge-weighted graph $G = (V, E)$, and g groups of vertices $X_1, \cdots, X_g \subseteq V$, each with a requirement r_i between 0 and $|X_i|$. The goal is to find a minimum cost set of edges whose removal separates each group X_i into at least r_i disconnected components.

We give an $O(\log n \log(gR))$ approximation algorithm for the requirement cut problem, where n is the total number of vertices, g is the number of groups, and R is the maximum requirement. We also show that the integrality gap of a natural LP relaxation for this problem is bounded by $O(\log n \log(gR))$. On trees, we obtain an improved guarantee of $O(\log(gR))$. There is a natural $\Omega(\log g)$ hardness of approximation for the requirement cut problem.

1 Introduction

Graph partitioning problems form a fundamental area of the study of approximation algorithms. The simplest graph partitioning problem is probably the well known s-t minimum cut problem [1]. Here we have an edge weighted graph containing two specified terminals s and t, and the goal is to find a minimum weight set of edges to disconnect s and t. The classical result of Ford and Fulkerson [1] proved a max-flow min-cut duality which related the maximum flow and minimum cut problems.

Multicuts: In the seminal work of Leighton and Rao [2], the authors gave an approximate max-flow min-cut theorem for a generalization of $s - t$ maximum flow to multicommodity flow. The flow problem considered here is *concurrent multicommodity flow*, and the corresponding cut is the *sparsest cut*. Here we have an undirected graph G with edge capacities. There are several commodities, each having a source, sink, and a specified demand. A maximum concurrent flow is one that maximizes the fraction of all demands that can be feasibly routed. A sparsest cut is one that minimizes the ratio of edge capacity of the cut to

---

[*] Supported by NSF ITR grant CCR-0122581 (The ALADDIN project)
[**] Supported in part by NSF grants CCR-0105548, CCF-0430751 and ITR grant CCR-0122581 (The ALADDIN project).

C. Chekuri et al. (Eds.): APPROX and RANDOM 2005, LNCS 3624, pp. 209–220, 2005.
© Springer-Verlag Berlin Heidelberg 2005

the demand separated by the cut. The original paper [2] considered uniform instances where there is a commodity for every pair of vertices, and one unit of demand for each commodity. The results in [2] showed a max-flow min-cut ratio of $O(\log n)$ where n is the number of vertices in the graph. This was extended to the case of arbitrary demands in a series of papers [3–7]. The best max-flow min-cut ratio is $\Theta(\log k)$ (k is the number of commodities). Garg, Vazirani, and Yannakakis [4] also considered the related *maximum multicommodity flow*. Here the objective is to maximize the total flow routed over all commodities. The corresponding cut problem is *multicut* - given a set of source-sink pairs $\{(s_1, t_1), \cdots, (s_k, t_k)\}$, a multicut is a set of edges whose removal separates each s_i from t_i. In [4], the authors proved that the ratio of the minimum multicut to the maximum multicommodity flow is $O(\log k)$, and they gave an $O(\log k)$ approximation algorithm for the minimum multicut problem.

Multiway Cut: In this problem, we have a set X of terminals, and the goal is to remove a minimum cost set of edges so that no two terminals are in the same connected component. Karger et al. [8] gave an approximation algorithm with guarantee 1.3438.

Multi-multiway Cut: Recently, Avidor and Langberg [9] extended multiway cut and multicut to a *multi-multiway cut* problem. Here we are given g sets of vertices $X_1, \cdots, X_g$, and we wish to find a minimum cost set of edges whose removal completely disconnects each of the sets $X_1, \cdots, X_g$. For this problem, the authors [9] presented an $O(\log g)$ approximation algorithm.

Steiner Multicut: Another interesting graph partitioning problem is the *Steiner multicut* problem [10]. In this problem, we are given g sets of vertices $X_1, \cdots, X_g$. The goal is to find a minimum cost set of edges that *separates* each of $X_1, \cdots, X_g$. We say that a set of vertices S is separated, if S is not contained in a single connected component. Klein et al. [10] presented an $O(\log^3 gt)$ approximation algorithm, where $t = \max_{i=1}^{g} |X_i|$ is the maximum size of a set of vertices.

k-Cuts: Another well studied graph partitioning problem is the *k-cut* problem [11]. Given an undirected graph with edge costs, the goal is to find a minimum cost set of edges that separates the graph into at least k connected components. Saran and Vazirani [11] gave the first approximation algorithm for this problem which achieves a guarantee of 2. More recently, Chekuri et al. [12] considered the *Steiner k-cut* problem. This is a generalization of the k-cut problem, where in addition to the graph, a subset X of vertices is specified as terminals. The objective is to find a minimum cost set of edges whose removal results in at least k disconnected components, each containing a terminal. Chekuri et al. [12] showed that the greedy algorithm of [11] can be modified to get a 2-approximation for this problem. They also showed how to round a natural LP relaxation to achieve the same bound.

Our goal in this work is to unify all the graph partitioning problems mentioned above into a single problem that contains all these as subproblems (See Figure 1). The input to a *requirement cut* problem is an undirected graph $G = (V, E)$ with non-negative costs $c(e)$ on its edges. There are g groups of

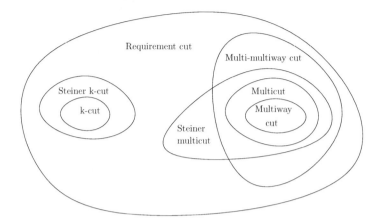

Fig. 1. Containment of cut problems

vertices $X_1, X_2, \cdots X_g \subseteq V$; for each group i we are given a requirement r_i between 0 and $|X_i|$. The aim is to find a minimum cost set of edges whose removal separates each group i into at least r_i disconnected components (each of which contains at least one member from the group). Below, we summarize how the requirement cut problem generalizes many of the graph partitioning problems.

Multicut : $|X_i| = r_i = 2$ for all $i = 1, \cdots, g$.
Steiner k-cut : $g = 1$, $X_1 = X$ the set of terminals, and $r_1 = k$.
Multi-multiway cut : $r_i = |X_i|$ for all i.
Steiner multicut : $r_i = 2$ for all i.

1.1 Our Results

We obtain an $O(\log n \log(gR))$ approximation algorithm for the requirement cut problem on general graphs, where n is the number of vertices in the graph, g is the number of groups and R is the maximum requirement of any group. We present two algorithms achieving this guarantee. The first (and more interesting) algorithm is via rounding a suitable LP relaxation. We also show that when restricted to trees, the approximation ratio can be improved to $O(\log(gR))$. On the other hand, we know that this problem is at least $\Omega(\log g)$ hard to approximate, even on a star (Section 2.2). The LP rounding procedure is described in Section 2. The second algorithm is based on the greedy heuristic for set cover. Here we compute a minimum cost effective cut (defined later) at each stage and add it to the solution. This greedy step is an interesting problem in itself, and has been studied in Klein et al. [10] as the Steiner ratio cut problem. We provide an improved approximation algorithm for this problem, and show that it yields an $O(\log n \log(gR))$ approximation algorithm for the requirement cut problem. The greedy algorithm is presented in Section 3.

The LP rounding algorithm on trees generalizes the randomized rounding for set cover, whereas the second method extends the greedy algorithm. While the first algorithm relies on the tree structure for rounding and hence incurs a log-squared overhead, the second method leaves some hope for improvement. The running time of the first algorithm is better, as it involves solving only one linear program.

As noted in the introduction, the Steiner multicut problem of Klein et al. is a special case of the requirement cut problem; so our algorithm implies an approximation of $O(\log n \log g)$ for Steiner multicut. However, we note that using a similar idea even the algorithm in [10] achieves this improved guarantee.

2   LP Based Algorithm for Requirement Cut

In this section, we present an $O(\log n \cdot \log(gR))$ approximation algorithm for requirement cut via LP rounding. We first formulate requirement cut as an integer program, and obtain its linear relaxation. Then we consider the case when the input graph is a tree, and show that randomized rounding gives an $O(\log(gR))$ approximation. Finally we show how requirement cut on a graph can be reduced to requirement cut on a tree, to obtain an approximation algorithm for the general case.

2.1   IP Formulation and a Linear Relaxation

Here we look at an integer program for the requirement cut problem, and a linear relaxation for it. This formulation is a generalization of that used in Chekuri et al. [12] for the Steiner k-cut problem. We also prove a property of the optimal LP solution that is useful in the rounding process. We may assume that the input graph $G = (V, E)$ is complete, by adding edges of zero cost. Our IP has a 0-1 variable d_e for each edge $e \in E$, which represents whether or not this edge is cut.

$$\min \sum_{e \in E} c_e d_e$$
$$s.t.$$
$$(Steiner - IP) \sum_{e \in T_i} d_e \geq r_i - 1 \; \forall T_i : \text{Steiner tree on } X_i, \; \forall i = 1 \cdots g$$
$$d_e \in \{0, 1\} \qquad \forall e \in E$$

Note that this IP is an exact formulation of the requirement cut problem. Unfortunately, the LP relaxation of this IP does not have a polynomial time separation oracle, and we don't know how to solve it efficiently. So we consider a relaxation of the Steiner tree constraints, by requiring that all *spanning trees* on the induced graph $G[X_i]$ have length at least $r_i - 1$. Since the minimum spanning tree can be efficiently computed, this LP has a polynomial time separation oracle and can be solved in polynomial time using the ellipsoid algorithm.

$$\min \sum_{e \in E} c_e d_e$$
$$s.t.$$
$$(LP) \sum_{e \in T_i} d_e \geq r_i - 1 \; \forall T_i : \text{spanning tree in } G[X_i], \; \forall i = 1 \cdots g$$
$$d : \text{metric}$$
$$d_e \in [0, 1] \qquad \forall e \in E$$

Let d^* be an optimal solution to this LP, and OPT_f its cost. Clearly the cost of the optimum requirement cut $OPT \geq OPT_f$. Let d be defined by $d_e = \min\{2 \cdot d_e^*, 1\}$. It is easy to check that d is a metric, and all edge lengths are in $[0, 1]$. The following claim is easy to see using the MST heuristic for Steiner tree.

Claim 1 *For any group X_i ($i = 1, \cdots g$), the minimum Steiner tree on X_i w.r.t. d has length at least $r_i - 1$.*

2.2   LP Rounding for Tree Requirement Cut

We consider here the special case that the input graph is a tree $T = (V, E)$. We show how to round the above mentioned LP relaxation within a factor of $O(\log gR)$ to obtain an approximation algorithm for requirement cut on trees. We note that even this restriction is at least as hard as set-cover. Consider a star with edges corresponding to sets in the set-cover instance. For each element j, we create a group that contains the root, and all the leaves of edges that correspond to sets containing j. Further, we set all requirements to 2, and all edge costs equal to 1. Clearly any feasible solution in this requirement cut instance is equivalent to a solution in the set cover instance. This shows that requirement cut on trees is at least $\Omega(\log g)$ hard to approximate [13].

The algorithm begins by solving (LP) optimally and obtaining the solution d defined in the previous section. We now describe how d is rounded to an integral solution. Our algorithm proceeds in phases, and augments the (partial) solution in each phase. Let $C^k \subseteq E$ denote the partial solution at the start of the phase k ($C^1 = \phi$). $F^k = T \setminus C^k$ denotes the forest in round k, with the edges in C^k removed. The rounding in phase k flips a coin with probability of success d_e for each edge $e \in F^k$, independently, and adds all edges that were successes, to the partial solution to get C^{k+1}. It is clear that the expected cost in this phase is exactly $\sum_{e \in F} c_e d_e \leq 2 \sum_{e \in E} c_e d_e^* = 2 \cdot OPT_f$.

We define the *residual requirement* r_i^k of a group X_i at the start of phase k to be the difference between r_i (requirement of group i) and the number of connected components containing X_i in F^k. The rounding procedure ends when the residual requirement of each group is 0. We will bound the expected number of phases by $O(\log gR)$ which gives us the desired approximation guarantee. For this, we show that in each phase, the total residual requirement (summed over all groups) goes down by a constant factor in expectation. This technique was used in the paper of Konjevod et al [14] on the covering Steiner problem.

Rounding in a Single Phase. The analysis here is for a single phase, and we drop the superscript k for ease of notation. For a group i, define F_i to be the sub-forest induced by X_i in F. Let H_i be the forest obtained from F_i by short cutting over all degree 2 Steiner (non X_i) vertices. So all Steiner vertices in H_i have degree at least 3. Such a forest is useful because of the following lemma, which is used in bounding the residual requirement at the end of this phase. The proof of this lemma is a simple counting argument, and we omit it.

Lemma 1 *In any Steiner forest with each non terminal having degree at least 3, removal of any m edges results in at least $\lceil \frac{m+1}{2} \rceil$ more components containing terminals.*

Let r'_i denote the residual requirement of group i at the start of the current phase. For a subgraph F' of F, by length $d(F')$ of F' we mean the total length of edges in F', $d(F') = \sum_{e \in F'} d_e$. Let $p_{u,v}$ denote the probability that vertices u and v are disconnected in this phase of rounding. Then we have the following lemma.

Lemma 2 *The total probability weight on H_i, $\sum_{e \in H_i} p_e$, is at least $(1 - \frac{1}{e})d(H_i)$.*

Proof: For those edges e of H_i which are also edges in F, it is clear that $p_e = d_e$. Now consider an edge $(u,v) \in H_i$ that is obtained by short cutting a path P in F_i. We claim that $p_{u,v} \geq (1 - \frac{1}{e})d_{u,v}$. Since each edge is rounded independently, and u and v are separated if any of the edges in P is removed, $p_{u,v} = 1 - \Pi_{e \in P}(1 - d_e) \geq 1 - e^{-d(P)}$. We consider the following 2 cases:

- $d(P) \leq 1$. Note that $1 - e^{-y} \geq (1 - 1/e)y$ for $y \in [0,1]$. So $p_{u,v} \geq (1 - 1/e)d(P) \geq (1 - 1/e)d_{u,v}$, since d is a metric.
- $d(P) \geq 1$. In this case $p_{u,v} \geq 1 - 1/e \geq (1 - 1/e)d_{u,v}$, since $d_{u,v} \in [0,1]$.

Thus, summing $p_{u,v}$ over all edges $(u,v) \in H_i$ we get the lemma. ∎

Now, consider adding edges to forest H_i to make it a Steiner tree on X_i. If X_i appears in c_i connected components in F, we need to add $c_i - 1$ edges. Since every edge has d-length at most 1, adding $c_i - 1$ edges increases the length of H_i by at most $c_i - 1$. But from Claim 1, every Steiner tree on X_i has length at least $r_i - 1$. So we get $d(H_i) \geq (r_i - 1) - (c_i - 1) = r'_i$. Lemma 2 then gives $\sum_{e \in H_i} p_e \geq (1 - 1/e)r'_i \geq \frac{r'_i}{2}$.

Consider a $0/1$ random variable Z^i_e for each edge $e = (u,v) \in H_i$ which is 1 iff the connection between u and v is broken. Note that $Pr[Z^i_e = 1] = p_e$. Let $Y_i = \sum_{e \in H_i} Z^i_e$ which is the number of edges cut in forest H_i. Note that although H_i is not a subgraph of F, separating vertices in H_i is equivalent to separating them in F. Now, $E[Y_i] = \sum_{e \in H_i} E[Z^i_e] = \sum_{e \in H_i} p_e \geq \frac{r'_i}{2}$. Let N_i be the increase in the number of components of group i in this phase. Since H_i is a forest that satisfies the conditions of Lemma 1, we have $N_i \geq Y_i/2$. Thus $E[N_i] \geq \frac{r'_i}{4}$.

Bounding the Number of Phases. Let random variable R^k_i denote the residual requirement of group i at the *start* of phase k, and N^k_i the increase in the number of components of group i in phase k. So, $N^k_i = R^k_i - R^{k+1}_i$. Our analysis so far holds for any round k. So we have

$$
\begin{aligned}
E[R^{k+1}_i | R^k_i = r'_i] &= E[R^k_i - N^k_i | R^k_i = r'_i] \\
&= r'_i - E[N^k_i | R^k_i = r'_i] \\
&\leq r'_i - \frac{r'_i}{4} \\
&= \frac{3r'_i}{4}
\end{aligned}
$$

So unconditionally, $E[R_i^{k+1}] \leq \frac{3E[R_i^k]}{4}$. Now let $R^k = \sum_{i=1}^{g} R_i^k$ be the total residual requirement at the start of round k. By linearity of expectation,
$E[R^{k+1}] \leq \frac{3}{4}\sum_{i=1}^{g} E[R_i^k] = \frac{3}{4}E[R^k]$. Thus, we have by induction that after k rounds, $E[R^k] \leq (3/4)^k E[R^0] \leq (3/4)^k gR$. So choosing $k = 4\ln(gR)$,
$E[R^k] \leq 1/2$, and by Markov inequality, $Pr[R^k \geq 1] \leq 1/2$. Thus, we have
proved the following.

Theorem 1 *There is a randomized rounding procedure which computes a solution to the requirement cut on trees, of cost $O(\log(gR)) \cdot OPT_f$ and succeeds with
probability at least half.*

2.3   LP Rounding for General Graphs

In this section we show how the LP relaxation (LP) yields an $O(\log n \log(gR))$
approximation algorithm for requirement cut on general graphs. This will also
show that the integrality gap of (LP) is at most $O(\log n \log(gR))$. The integrality
gap of the linear program for set cover gives an $\Omega(\log g)$ integrality gap for LP.
Let LP_G denote the LP relaxation of the requirement cut on $G = (V(G), E(G))$,
and OPT_f its optimal cost. Our rounding procedure uses the optimal solution
d^* of LP_G, and embeds it into a distribution of tree metrics. Then we use the
rounding procedure for trees to solve requirement cut on the resulting tree.

We use the embedding results [15] of F,R&T, to embed metric (d^*, G) into
a distribution of dominating trees $\mathcal{T}$, with a logarithmic overhead. In [15], it is
shown that this can be done in polynomial time. Let $(\kappa, T) \in_R \mathcal{T}$ be a random
instance from this distribution, where $T = (V(T), E(T))$ is a Steiner tree on
$V(G)$, and κ is the tree metric obtained by assigning some distances to $E(T)$.
For an edge $e \in E(T)$, we denote by sep_e the set of edges of G that are separated
by e in the tree T, *i.e.* $sep_e = \{(i, j)|i, j \in V(G), e$ lies on $i - j$ path in T$\}$. We
define costs on the edges of T as $c'_e = \sum_{(i,j) \in sep_e} c_{i,j}$, and on non tree edges as
0. Consider a tree requirement cut instance on T with the cost function c'. The
groups $X_1, \cdots, X_g$ and corresponding requirements $r_1, \cdots, r_g$ are the same as
in the original requirement cut on G. It is easy to see that a feasible (integral)
solution to the requirement cut on T of cost C, corresponds to a feasible integral
solution on G, of cost at most C.

Now consider the linear relaxation LP_T of this requirement cut on tree T.
Define $\kappa'_{u,v} = \min\{\kappa_{u,v}, 1\}$ for all $u, v \in V(T)$. Note that κ' is a metric in $[0, 1]$.
Since d^* is a metric with values only in $[0, 1]$, and κ restricted to $V(G)$ dominates
d^*, so does κ'. So for any spanning tree T_i on a group X_i, its length under κ',
$\kappa'(T_i) \geq d^*(T_i)$ its length under d^*. Since d^* is a feasible solution for LP_G, we
get $\kappa'(T_i) \geq r_i - 1$. Thus κ' is a feasible solution for LP_T on tree T.

The cost of this fractional solution is

$$C = \sum_{u,v \in V(T)} c'_{u,v} \cdot \kappa'_{u,v} = \sum_{e \in E(T)} \kappa'_e \sum_{(i,j) \in sep_e} c_{i,j} = \sum_{i,j \in V(G)} c_{i,j} \cdot \kappa'_{i,j}$$

Since κ dominates κ', we get $C \leq \sum_{i,j \in V(G)} c_{i,j} \cdot \kappa_{i,j}$. Now from [15] we have
the following theorem

Theorem 2 *([15]) : For all $i, j \in V(G)$, $\kappa_{i,j} \geq d_{i,j}^*$ and the expected value $E[\kappa_{i,j}] \leq \rho \cdot d_{i,j}^*$, where $\rho = O(\log n)$.*

Using this, and linearity of expectation, we get $E[C] \leq \rho \cdot \sum_{i,j \in V(G)} c_{i,j} d_{i,j}^* = \rho \cdot OPT_f$. Now using the rounding for trees (Theorem 1), we get the following approximation for general graphs.

Theorem 3 *There is a randomized rounding procedure which computes a solution to the requirement cut on graphs, of cost $O(\log n \cdot \log(gR)) \cdot OPT_f$ and succeeds with probability at least half.*

Remark: Improved approximation for small sized groups. We note that there is an alternate algorithm that gives an $O(t \log g)$ approximation ratio, where t is the size of the largest group. This follows from the algorithm for multi-multiway cut [9]. We first solve the LP relaxation (LP) to get a metric d^*. The following argument holds for any group i. The MST on X_i, T_i, has length at least $r_i - 1$. Since each edge has length at most 1, T_i has at least $r_i - 1$ edges of length $\geq \frac{1}{t}$. Consider removing the $r_i - 1$ longest edges in T_i to get r_i connected components. Pick one vertex from each component to obtain set $S_i = \{s_1, s_2, \cdots, s_{r_i}\} \subseteq X_i$. Since T_i is an MST on X_i, each pairwise distance in S_i is at least $\frac{1}{t}$.

We define metric $d' = \min\{t \cdot d^*, 1\}$. In metric d', all pairwise distances in S_i are 1 ($i = 1, \cdots, g$). We define a multi-multiway cut instance on the sets $S_1, S_2, \cdots, S_g$. Now d' is a feasible fractional solution to this instance. So using the region growing procedure described in [9], we get a cut that separates each group X_i into the required number of pieces and has cost at most $O(t \cdot \log g) \cdot OPT_f$. This implies an improvement when the largest group size is a constant.

3   Deterministic Greedy Algorithm

In this section we present an alternate greedy algorithm that achieves the same guarantee of $O(\log(gR) \log n)$ for the requirement cut problem. As before, we assume that the graph G is complete. Our algorithm works in phases, and we maintain a partial solution in every phase of the algorithm. Each phase is a greedy step and augments this partial solution. The algorithm ends when all requirements have been satisfied. A partial solution is just a set of edges $A \subseteq E$. Recall the definitions of *residual graph*, and *residual requirement*, w.r.t. a partial solution A (Section 2.2). At the start of our algorithm, $A = \phi$ and the residual requirement of group i ($i = 1, \cdots, g$) is $r_i - 1$. We call a group i *active* if it has positive residual requirement in the current residual graph.

Consider a residual graph $G' = (V, E')$. A *bond* in G' is an edge cut of the form $\partial S = \{(i, j) \in E' | i \in S, j \notin S\}$, where $S \subseteq V$ is a set of vertices. We define $cov(\partial S)$, the *coverage* of ∂S to be the number of active groups separated by C in G'. The *cost-effectiveness* of bond ∂S in G' is $eff(\partial S) = \frac{c(\partial S)}{cov(\partial S)}$, the ratio of the cost of the bond to its coverage.

The greedy approach for the requirement cut problem is to remove a bond of minimum cost effectiveness in each phase. But the problem of finding the minimum cost effective bond generalizes sparsest cut (See next section), and we do not know of an exact algorithm. However, we can obtain an approximate solution for this problem, and work with this as the greedy choice. For a phase k, let G_k denote the residual graph at the start of this phase, ϕ_k the total residual requirement in G_k, and M_k the minimum cost effective bond in G_k. Below OPT^* is the cost of the optimal requirement cut on G.

Lemma 3 $eff(M_k) \leq \frac{2OPT^*}{\phi_k}$. *Without loss of generality, M_k is contained in some connected component of G_k.*

Proof: Consider the connected components $S_1, S_2, \cdots S_l$ in the graph G after removing the edges of the optimal solution. It is clear that each edge that is cut is incident on exactly 2 of these connected components. So we have $\sum_{j=1}^{l} c(\partial S_j) = 2OPT^*$. Let $S_1', S_2', \cdots, S_p'$ be the restrictions of $S_1, S_2, \cdots S_l$ to the connected components of G_k. Note that the total additional requirement satisfied on removing $\partial S_j'$ in G_k is just $cov(\partial S_j')$, for any $j = 1, \cdots, p$. Now since the optimal solution must satisfy all the residual requirements in G_k, we have $\sum_{j=1}^{p} cov(\partial S_j') \geq \phi_k$. Let $c'(\partial S)$ denote the cost of bond S in graph G_k (this may only be smaller than the cost of bond S in G). It is clear that $\sum_{j=1}^{p} c'(\partial S_j') = \sum_{j=1}^{l} c'(\partial S_j) \leq 2OPT^*$. We now have

$$\min_{j=1}^{p} \frac{c'(\partial S_j')}{cov(\partial S_j')} \leq \frac{\sum_{j=1}^{p} c'(\partial S_j')}{\sum_{j=1}^{p} cov(\partial S_j')} \leq 2\frac{OPT^*}{\phi_k}$$

So the minimum cost effective bond M_k has $eff(M_k) \leq \frac{2OPT^*}{\phi_k}$.

By a similar averaging argument, we may assume that M_k is in some connected component of G_k.

3.1   Approximating the Minimum Cost Effective Bond

In making the greedy choice, we wish to compute a bond of minimum cost effectiveness. Formally, we are given a connected graph $H = (V_H, E_H)$ with costs on edges, and g groups of vertices $X_1, \cdots, X_g \subseteq V_H$. We want a bond that minimizes the ratio of the cost of the cut to the number of groups it separates. We observe that this is the *minimum ratio Steiner cut* problem that is solved in Klein et al [10]. In [10] they give an $O(\log^2 gt)$ approximation algorithm for this problem, where t is the maximum size of a group. Here, using the results of Fakcharoenphol et al. [15], we improve the approximation ratio to $O(\log n)$. Note that when all group sizes and requirements are 2, the Steiner ratio problem reduces to sparsest cut [2].

The following is an LP relaxation for the Steiner ratio problem. The MST algorithm gives us a polynomial time separation oracle for this LP. We may assume that the input graph H is complete, by adding edges of zero cost.

$$\min \sum_{e \in H} c_e l_e$$
$$s.t.$$
$$(LP_H) \qquad \begin{aligned} \sum_{i=1}^{g} \sum_{e \in T_i} l_e &\geq 1 \; \forall (T_1, \cdots, T_g) \text{ where} \\ T_i &: \text{Spanning tree on } H[X_i], i = 1, \cdots, g \end{aligned}$$
$$l_e \in [0,1]$$
$$l : \text{metric}$$

To observe that this is indeed a relaxation of the Steiner ratio problem, consider the 0-1 metric l' corresponding to the optimal solution B (which is a bond) : $l'(u,v) = 0$ iff u and v are in the same connected component in $H \setminus \partial B$, and 1 otherwise. Let σ be the number of groups separated by ∂B. Clearly, the sum of the MSTs in $H[X_i]$ ($i = 1, \cdots g$) is also σ. Now $l = \frac{1}{\sigma} l'$ is a feasible solution to (LP_H), and has cost $\frac{c(\partial B)}{\sigma} = eff(\partial B)$.

Let l be the optimal solution of (LP_H), and $OPT' = \sum_{e \in H} c_e \cdot l_e$ its cost. We now describe how to round l to get an approximate integral cut. We use the FRT-embedding to embed metric (l, H) to a distribution $\mathcal{T}$ of dominating tree metrics. Let $(T, \kappa) \in_R \mathcal{T}$ be a random sample from $\mathcal{T}$. Then restating Theorem 2, we have the following : For all $i, j \in V_H$, $\kappa_{i,j} \geq l_{i,j}$ and the expected value $E[\kappa_{i,j}] \leq O(\log n) \cdot l_{i,j}$.

Each edge f in T corresponds to a bond sep_f (defined in Section 2.3) in the original graph H. We say edge $f \in T$ separates group i (denoted $f|X_i$) if removing edge f from T disconnects X_i. We denote by $T[X_i]$, the smallest subtree of T containing X_i. We have,

$$\min_{f \in T} \frac{c(sep_f)}{cov(sep_f)} = \min_{f \in T} \frac{\kappa_f c(sep_f)}{\kappa_f \sum_{i=1}^{g} 1_{f|X_i}} \leq \frac{\sum_{f \in T} \kappa_f \sum_{(i,j) \in sep_f} c_{i,j}}{\sum_{f \in T} \kappa_f \sum_{i=1}^{g} 1_{f|X_i}}$$

$$= \frac{\sum_{i,j} c_{i,j} \sum_{f:(i,j) \in sep_f} \kappa_f}{\sum_{f \in T} \kappa_f \sum_{i=1}^{g} 1_{f|X_i}} = \frac{\sum_{i,j} c_{i,j} \kappa_{i,j}}{\sum_{i=1}^{g} \sum_{f:f|X_i} \kappa_f}$$

$$= \frac{\sum_{i,j} c_{i,j} \kappa_{i,j}}{\sum_{i=1}^{g} \kappa(T[X_i])}$$

$$\leq 2 \cdot \frac{\sum_{i,j} c_{i,j} \kappa_{i,j}}{\sum_{i=1}^{g} \kappa(MST_i)} \qquad (1)$$

$$\leq 2 \cdot \frac{\sum_{i,j} c_{i,j} \kappa_{i,j}}{\sum_{i=1}^{g} l(MST_i)} \qquad (2)$$

$$\leq 2 \cdot \sum_{i,j} c_{i,j} \kappa_{i,j} \qquad (3)$$

where the only non trivial inequalities are the last three. Since $T[X_i]$ is a Steiner tree on X_i, the length of an MST on X_i is at most 2 times the length of $T[X_i]$ - so (1) follows. For (2) note that κ (restricted to V_H) dominates l, and (3) follows from the feasibility of l in (LP_H). In [15], they describe how their tree embedding algorithm can be derandomized to find a *single tree* with small weighted average stretch. Here if we think of $c_{i,j}$ as the weights, we can find a single tree metric (T, κ) s.t. $\sum_{i,j} c_{i,j} \kappa_{i,j} \leq \rho \cdot \sum_{i,j} c_{i,j} l_{i,j}$, where $\rho = O(\log n)$. The minimum cost effective bond C corresponding to an edge of T is output as the approximate solution. From the above argument, $eff(C) \leq 2\rho \cdot OPT'$, and we get an $O(\log n)$ approximation.

The greedy step in phase k of the requirement cut algorithm is as follows. Run the Steiner ratio algorithm for each connected component of the residual graph G_k. Pick the bond C_k of minimum cost effectiveness among those obtained. Let h_k be the number of active groups separated by C_k (in G_k). Then lemma 3, with the Steiner ratio approximation gives the following.

Lemma 4 *The cost effectiveness of the cut picked in the k-th iteration of the greedy algorithm $eff(C_k) = \frac{c(C_k)}{h_k} \leq 2\rho \cdot eff(M_k) \leq 4\rho \cdot \frac{OPT^*}{\phi_k}$.*

Our analysis to bound the cost incurred by the greedy algorithm is similar to that used in other set cover based algorithms [16]. The decrease in residual requirement is $\phi_k - \phi_{k+1} = h_k$, the coverage of the greedy step. Using lemma 4, we get $\phi_{k+1} \leq (1 - \frac{c(C_k)}{4\rho OPT^*})\phi_k$. Using this recursively, we get

$$\phi_m \leq \Pi_{k=1}^{m-1}(1 - \frac{c(C_k)}{4\rho OPT^*}) \cdot \phi_1$$
$$\leq e^{-\frac{\sum_{k=1}^{m-1} c(C_k)}{4\rho OPT^*}} \cdot \phi_1$$

where m is the number of phases in the greedy algorithm. So we get $\sum_{k=1}^{m-1} c(C_k)$ $\leq 4\rho \cdot \ln(\frac{\phi_1}{\phi_m})OPT^* \leq 4\rho \ln(gR)OPT^*$, since $\phi_m \geq 1$ and $\phi_1 \leq gR$. Observe that the coverage in the last phase, $h_m = \phi_m$. Now from lemma 4, $c(C_m) \leq 4\rho OPT^*$. So the cost of the greedy solution is at most $4\rho(\ln(gR) + 1)OPT^*$. Also, the number of iterations is at most gR, which gives a polynomial time algorithm.

Theorem 4 *The (approximate) greedy algorithm is an $O(\log n \log(gR))$ approximation algorithm for the requirement cut problem on graphs.*

As a consequence of this theorem, we see that when restricted to trees, the greedy step is trivial : bonds are just edges, and we can enumerate all of them. So the greedy algorithm is an $O(\log(gR))$ approximation for trees.

4   Conclusions and Open Problems

In this paper, we introduced the requirement cut problem, which contains several graph partitioning problems as special cases. For general graphs, we have two $O(\log n \log(gR))$ approximation algorithms : one based on LP rounding, and the other a greedy algorithm. When restricted to trees, both these algorithm have a better performance guarantee of $O(\log(gR))$.

One obvious open question is whether the approximation ratio can be improved. One approach is to try improving the approximation guarantee for the Steiner ratio problem. The recent techniques of Arora, Rao, Vazirani [17], do not seem to apply directly here. The rounding procedure in Section 2 shows that the integrality gap of (LP) is at most $O(\log n \log(gR))$. There is a lower bound of $\Omega(\log g)$ due to set cover. It will be interesting to know if there is a worse example. Even in the case of a tree there is a gap between the upper bound of $O(\log(gR))$ and lower bound of $\Omega(\log g)$.

References

1. L.R. Ford, J., Fulkerson, D.: Maximal flow through a network. Canadian Journal of Math. **8** (1956)
2. Leighton, T., Rao, S.: An approximate max-flow min-cut theorem for uniform multicommodity flow problems, with applications to approximation algorithms. In Proc. 29th IEEE Annual Symposium on Foundations of Computer Science (1988) 422–431
3. Klein, P., Rao, S., Agarwal, A., Ravi, R.: An approximate max-flow min-cut relation for multicommodity flow, with applications. Proc. 31st IEEE Annnual Symposium on Foundations of Computer Science (1990) 726–737
4. Garg, N., Vazirani, V.V., Yannakakis, M.: Approximate max-flow min-(multi)cut theorems and their applications. ACM Symposium on Theory of Computing (1993) 698–707
5. Plotkin, S.A., Tardos, E.: Improved bounds on the max-flow min-cut ratio for multicommodity flows. ACM Symposium on Theory of Computing (1993) 691–697
6. Aumann, Y., Rabani, Y.: An $\log k$ approximate min-cut max-flow theorem and approximation algorithm. SIAM J. Comput. **27** (1998) 291–301
7. Linial, N., London, E., Rabinovich, Y.: The geometry of graphs and some of its algorithmic applications. FOCS **35** (1994) 577–591
8. D.R.Karger, P.N.Klein, C.Stein, M.Thorup, N.E.Young: Rounding algorithms for a geometric embedding for minimum multiway cut. ACM Symposium on Theory of Computing (1999) 668–678
9. Avidor, A., Langberg, M.: The multi-multiway cut problem. Proceedings of SWAT (2004) 273–284
10. Klein, P.N., Plotkin, S.A., Rao, S., Tardos, E.: Approximation algorithms for steiner and directed multicuts. J. Algorithms **22** (1997) 241–269
11. Saran, H., Vazirani, V.V.: Finding k cuts within twice the optimal. SIAM J. Comput. **24** (1995)
12. Chekuri, C., Guha, S., Naor, J.S.: The steiner k-cut problem. Proc. of ICALP (2003)
13. Feige, U.: A threshold of $\ln n$ for approximating set cover. J. ACM **45** (1998) 634–652
14. Konjevod, G., Ravi, R., Srinivasan, A.: Approximation algorithms for the covering steiner problem. Random Struct. Algorithms **20** (2002) 465–482
15. Fakcharoenphol, J., Rao, S., Talwar, K.: A tight bound on approximating arbitrary metrics by tree metrics. Proceedings of the 35th Annual ACM Symposium on Theory of Computing. (2003)
16. Klein, P.N., Ravi, R.: A nearly best-possible approximation algorithm for node-weighted steiner trees. J. Algorithms **19** (1995) 104–115
17. Arora, S., Rao, S., Vazirani, U.: Expander flows, geometric embeddings and graph partitioning. STOC (2004) 222–231

Approximation Schemes for Node-Weighted Geometric Steiner Tree Problems

Jan Remy and Angelika Steger

Institut für Theoretische Informatik, ETH Zürich, Switzerland
{jremy,steger}@inf.ethz.ch

Abstract. In this paper, we consider the following variant of the geometric Steiner tree problem. Every point u which is not included in the tree costs a penalty of $\pi(u)$ units. Furthermore, every Steiner point we use costs c_S units. The goal is to minimize the total length of the tree plus the penalties. We prove that the problem admits a polynomial time approximation scheme, if the points lie in the plane. Our PTAS uses a new technique which allows us to bypass major requirements of Arora's framework for approximation schemes for geometric optimization problems [1]. It may thus open new possibilities to find approximation schemes for geometric optimization problems that have a complicated topology. Furthermore the techniques we use provide a more general framework which can be applied to geometric optimization problems with more complex objective functions.

1   Introduction

The Model. In this paper we consider a geometric optimization problem which we call NODE-WEIGHTED GEOMETRIC STEINER TREE or NWGST for short. An instance of this problem consists of a set of points $\mathscr{P}$ in the plane, a penalty function $\pi : \mathscr{P} \to \mathbb{Q}_+$ and a $c_S \in \mathbb{Q}_+$. A solution $\mathcal{ST}$ is a (geometric) spanning tree on $V(\mathcal{ST}) \cup S(\mathcal{ST})$, where $V(\mathcal{ST}) \subseteq \mathscr{P}$ and $S(\mathcal{ST}) \subset \mathbb{R}^d$. The points in $S(\mathcal{ST})$ are called *Steiner points*. Let $E(\mathcal{ST})$ denote the set of line segments or edges used by $\mathcal{ST}$. We minimize

$$\text{val}\,(\mathcal{ST}) = \sum_{\{u,v\}\in E(\mathcal{ST})} \ell_q\,(\{u,v\}) + \sum_{u\in\mathscr{P}-V(\mathcal{ST})} \pi(u) + |S(\mathcal{ST})|c_S, \quad (1)$$

where $\ell_q\,(\{u,v\}) = d_q\,(u,v)$ measures the length of the edge $\{u,v\}$ in the $\mathscr{L}_q$-metric. Since it is easy to compute a spanning tree of minimum length, the main problem is in fact to choose $V(\mathcal{ST})$ and $S(\mathcal{ST})$ optimally.

The objective function can be intuitively understood as follows. We pay for the total length of the tree plus a penalty of c_S units for every Steiner point we use. Furthermore, we are charged $\pi(u)$ units for every point $u \in \mathscr{P}$ which is not included in the tree. Throughout the paper $\mathcal{ST}^*$ denotes the optimal solution for the input set $\mathscr{P}$.

If we chose $c_S = 0$ and $\pi(u) = \infty$ then we obtain the well-known *geometric Steiner tree problem* [7, 10]. Furthermore, by choosing just $c_S = 0$, we obtain the

C. Chekuri et al. (Eds.): APPROX and RANDOM 2005, LNCS 3624, pp. 221–232, 2005.

prize-collecting variant of the geometric Steiner tree problem. In this paper we
will prove

Theorem 1. *Let $d = 2$, that is, $\mathscr{P} \in \mathbb{R}^2$. Then for every fixed ε, there is a polynomial time algorithm which computes a $(1+\varepsilon)$-approximation to* NWGST *.*

This means that NWGST and thus also the PRIZE-COLLECTING GEOMET-
RIC STEINER TREE admit a polynomial time approximation scheme (PTAS).
We would like to mention that our PTAS can also be applied to further variants
of node-weighted Steiner problems and other tree and tour problems. We will
give examples in Section 5. There, we will also show that the problem admits
a quasi-polynomial time approximation scheme (QPTAS) in higher dimensions.
It is well-known, that the existence of a QPTAS implies that the problem is not
$\mathcal{APX}$-hard, provided SAT $\notin$ DTIME$[n^{\text{polylog}(n)}]$.

For the sake of brevity we will use the following notations in this paper.
Unless stated otherwise, $\log n$ always denotes the logarithm of n to base 2. For
every solution $\mathcal{ST}$, $\ell_q(\mathcal{ST})$ denotes the total length of the $\mathcal{ST}$. Furthermore,
$\pi(\mathcal{ST})$ and $c_S(\mathcal{ST})$ denote the total amount of penalties we pay for unconnected
points and the use of Steiner points, respectively.

Previous Work. The input of the *Steiner tree problem in networks* is a weighted
graph $\mathcal{G} = (\mathcal{V}, \mathcal{E})$ and a set $K \subseteq V$. The goal is to find a tree of minimum
weight in $\mathcal{G}$ which includes all vertices in K. This problem is known to be $\mathcal{APX}$-
complete [6] and it does therefore not admit a PTAS unless $\mathcal{P} = \mathcal{NP}$. However,
there are many constant-factor approximation algorithms for this problem. The
currently best algorithm is by Robins and Zelikovsky [15]. It yields a $(1+\ln 3/2)$-
approximation. There is also a node-weighted version of this problem [10] which
is similar to our model. It was shown by Klein and Ravi [11] that there is
an approximation algorithm with performance ratio $2\ln|K|$. Furthermore, they
proved that it is not possible to achieve a ratio better than logarithmic, unless
SAT $\in$ DTIME$[n^{\text{polylog}(n)}]$. A similar setup was considered by moss and Rabani
[13]. In essence they combined the unweighted Steiner tree problem with packing
problems. A 2-approximation algorithm for the price collecting variant of the
Steiner tree problem in networks is due to Goemans and Williamson [9].

It was shown by Arora [1, 2] and independently by Mitchell [12] that the
geometric Steiner tree problem admits a PTAS. This is also the best we can
hope for, since the geometric variant is strongly $\mathcal{NP}$-hard [8]. Furthermore, Arora
claimed in [2] that there is a QPTAS for the price collecting variant. Talwar [16]
shows that Arora's method extends to metrics that satisfy certain properties but
the complexity of his approximation schemes is quasi-polynomial. The Steiner
tree problem is extensively discussed in the textbooks of Hwang, Richards and
Winter [10] and of Prömel and Steger [14].

Motivation. NWGST is a very natural generalization of the geometric Steiner
tree problem and it has interesting applications. In practice it usually costs to
include an additional node into a network. For example, additional switches
could reduce the cable length of the computer network but they are expensive
to install.

We can not simply adapt Arora's PTAS for the geometric Steiner tree problem. One reason is that the so-called *patching lemma* seems to be no longer true if one induces costs for the Steiner points. The patching lemma states that one can reduce the number of crossing of some Steiner tree $\mathcal{ST}$ with a line segment S of length x to 1 and that the required modifications to $\mathcal{ST}$ increase its length only slightly. Arora's method extends to many geometric optimization problems provided one can find an equivalent for the patching lemma. In very few cases this lemma is not required, like for the k-median problem [4]. On the other hand, there are some open geometrical problems which include complicated topology for which no equivalent of the patching lemma is known. Examples are triangulation problems or the degree-restricted spanning tree problem [3]. Thus, for the design of approximation schemes for geometrical problems, new techniques avoiding the patching lemma are of interest of their own.

Outline of the Algorithm. We briefly review the main ideas of Arora's approximation scheme for the geometric Steiner tree problem. Arora subdivides the smallest rectangle enclosing all points recursively by using a quadtree such that the rectangles of the leafs of the quadtree contain at most one point. The structure of a Steiner tree $\mathcal{ST}$ inside a rectangle of this quadtree can be described by specifying *i)* the locations where edges of $\mathcal{ST}$ cross the boundary of the rectangle and *ii)* how these locations are connected inside the rectangle. Both parameters define the configuration of this rectangle. Given a rectangle and a configuration, Arora's algorithm optimizes locally as follows. It computes the best solution for this configuration by enumerating all configurations of the children of the rectangle and combining their optimum solutions. As the optimum solutions for rectangles containing at most one point are easy to compute, the optimum solutions of all rectangles can be computed bottom-up by dynamic programming. In other words a look-up table is maintained which contains for each configuration-rectangle pair the corresponding optimum. The complexity of this dynamic program depends on the size of this table, i.e., the size of the quadtree and the number of different configurations per rectangle. We shall see later that the quadtree has logarithmic height and contains therefore polynomial many rectangles. In order to reduce the number of configurations per rectangle, one has to do some rounding. This is achieved as follows. Firstly, edges are only allowed to cross the boundary of a rectangle at one out of $\mathcal{O}(\log n)$ prespecified locations. Secondly, the tree may cross the boundary of a rectangle at most constantly many times. In this way one can show that there exist only polynomially many different configurations for each rectangle and the dynamic program therefore terminates in polynomial time. On the other hand, by reducing the number of configurations the dynamic program optimizes only over a subclass of all trees. One therefore also has to show that this subclass contains a tree whose length differs only slightly from those of an optimum tree. By using the patching lemma and introducing randomness into the quadtree one can show that this is indeed the case.

Unfortunately, our problem NWGST seems not to fit in this framework. Although one can use a similar setup, the fact that additional Steiner points add additional costs to the tree makes it hard to imagine that an optimum tree can always be changed in such a way that it crosses the boundary of all rectangles only at constantly many positions. We therefore use a completely different approach to define configurations. Unlike Arora we do not specify the locations where the tree edges cross the boundary of the a rectangle. Instead we aim at specifying the structure of the tree within a rectangle and at specifying at which locations it should be connected to points outside of the rectangle. Basically, this is achieved by subdividing the rectangle into $\mathcal{O}\left(\log n\right)$ many cells. A configuration specifies which cells contain an endpoint of an edge crossing the boundary and specifies to which component inside of the rectangle those points belong. In this way we restrict the number of different locations of end points of edges crossing the boundary of a rectangle but not the number of such edges.

In the remainder of this paper we explain the details of our algorithm. Firstly, we need a slightly different definition of a quadtree; this is described in Section 2. Secondly, we need an appropriate subdivision of a rectangle into $\mathcal{O}\left(\log n\right)$ many cells. The structure we use is defined in Section 3.1. Finally, we have to define a class of standardized trees that can be represented by a configuration. Such trees are discussed in Section 3.2. The dynamic program is described in Section 4. It is not possible to compute the optimal standardized tree but we can show that our algorithm computes a $(1 + \varepsilon)$-approximation to it. Due to space limitations all proofs are omitted. For details the reader is referred to a preprint of the full paper which is available at http://www.ti.inf.ethz.ch/as/people/remy.

2   Preliminaries

An instance of the NWGST problem is *well-rounded* if all points and Steiner points have odd integral coordinates, and the side length of the bounding box is $L = \mathcal{O}\left(n^2\right)$, i.e., $\mathscr{P} \subset \{1, 3, 5, ..., L - 1\}^2$. Note that this definition slightly differs from the one Arora uses for his approximation scheme [1].

Lemma 1. *If there is a PTAS for well-rounded instances of* NWGST *then there is also a PTAS for arbitrary instances.*

Furthermore, we adapt the concept of *shifted quadtrees* or *shifted dissections* [1, 2] to our purposes. We require that L is a power of 2. If this is not the case, we simply enlarge the bounding box appropriately. We choose two integers a and b with $a, b \in \{0, 2, ..., L - 2\}$. The vertical line with x-coordinate a and the horizontal line with y-coordinate b split the bounding box into four smaller rectangles. We enlarge those rectangles such that their side length is L. By $\mathfrak{S}_0$ we denote the area covered by the four enlarged rectangles. Note that $\mathfrak{S}_0$ is a square of side length $L_0 = 2L$. We now subdivide $\mathfrak{S}_0$ using an ordinary quad tree until we obtain squares of size 2×2. We denote the quad tree generated in this fashion by $QT_{a,b}$ to emphasize that its structure depends on the choice of a and b.

Throughout we denote vertical dissection lines with *even* x-coordinates by V and horizontal ones with *even* y-coordinates by H. A vertical line V with x-coordinate x has *level l with respect to a* if there is an *odd* (potentially negative) integer i, such that $x = a + i \cdot \frac{L}{2^t}$. We write: $\mathrm{lev}(V, a) = l$. Similarly we define $\mathrm{lev}(H, b)$. The *level* of a square $\mathfrak{S}$ in the quadtree $QT_{a,b}$ is defined as follows: The squares in the very first subdivision have level 0 and a square $\mathfrak{S}$ at level l is subdivided by a horizontal line H and a vertical line V with $\mathrm{lev}(H, b) = \mathrm{lev}(V, a) = l+1$ into four squares at level $l+1$. The level of a square $\mathfrak{S}$ is denoted by $\mathrm{lev}(\mathfrak{S})$. Let us shortly summarize salient properties of shifted quadtrees.

Observation 2 *For a shifted quadtree $QT_{a,b}$ the following is true:*

1. *The subdivision uses only horizontal and vertical lines that have even coordinates.*
2. *The depth of $QT_{a,b}$ is $\log L =: t$.*
3. *The side length of a square $\mathfrak{S}$ at level l is $S = L/2^l$.*
4. *The number of vertical (horizontal) lines in $QT_{a,b}$ at level l is at most 2^l.*

Note that t is integral by choice of L. This property will turn out to be useful in Section 3.1.

3 (s, t)-Maps and Standardized Solutions

3.1 (s, t)-Maps

From now on, we only consider well-rounded instances of NWGST. As mentioned in the introduction we need some subdivision of $\mathfrak{S} \in QT_{a,b}$ into cells. A natural idea would be to use a regular $(d \times d)$-grid. Unfortunately, it will turn out that we have to choose $d = \Omega(st) = \Omega(\log n)$ to obtain a $(1 + 1/s)$-approximation, where $s = \mathcal{O}(1/\varepsilon)$. Therefore this grid has $\Omega(\log^2 n)$ many cells. This is to much, since we later store some bits per rectangle and since we require that we have polynomial many states per rectangle. Indeed, such grids already appear in quasi-polynomial approximation schemes of [4, 16].

This problem is solved by using so-called (s, t)-maps which have only $\mathcal{O}(\log n)$ cells. The intuition behind (s, t)-maps is very simple. In the PTAS special care has to be taken for edges that cross boundaries of squares. Due to complexity, we do not explicitly specify the endpoints of such edges. Instead, we will argue that it suffices to describe them as "edges" between cells. This has the effect that our PTAS can only reconstruct that the endpoints are somewhere within the cells. Thus, it may choose any point inside the cell. However, the length of the edge varies by at most twice the side lengths ($\mathscr{L}_1$-metric) of the cells. The definition of (s, t)-maps comes directly from the following simple observation. We can estimate the endpoints of long edges more roughly than those of short edges, since we have only to assure that the absolute error is at most $1/s$ of the edge's optimal length. Edges that reach deep into $\mathfrak{S}$ are long and thus we can make the cells inside $\mathfrak{S}$ larger than those which are close to the boundary. The definition of an (s, t)-*map* follows exactly this idea. We will assure that the side length of a cell is at most

$$\max \left\{ \frac{1}{st} \langle \text{side length of } \mathfrak{S} \rangle, \frac{1}{s} \langle \text{distance between } \mathfrak{C} \text{ and boundary of } \mathfrak{S} \rangle \right\}$$

and that the map has $\mathcal{O}\left(s^2t\right)$ many cells. In essence, an (s,t)-map can be regarded as a $(st \times st)$-grid with cell sizes growing to the interior.

Now we define (s,t)-maps more formally. For arbitrary $s \in \mathbb{N}$, we define a function $\beta : \mathbb{N}_0 \to \mathbb{N}_0$ as

$$\beta(i) = \begin{cases} \lfloor \frac{j}{4s} \rfloor & ,j < 8s \\ 2 & ,8s \leq j \leq 9s - 1 \\ \lfloor \frac{j-s}{4s} \rfloor & ,j \geq 9s \end{cases}.$$

Let S be the side length of $\mathfrak{S}$ and let $d = d(s) = 8s2^\tau$. Here τ is the smallest integer such that $2^\tau \geq t$, where t is the depth of the shifted quadtree. Note that this implies $\tau < 1 + \log t$. Therefore, we have $8st \leq d < 16st$.

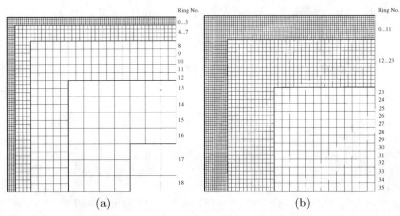

Fig. 1. The upper-left part of an (s,t)-map with $s = 1$ (a). The cell size doubles every four rings - except the 12th ring. In (b) the same region is covered by an (s,t)-map with multiplicity $s = 3$

We subdivide $\mathfrak{S}$ in *rings* as illustrated in Figure 1 for $s = 1$ and $s = 3$. The jth ring has width $2^{\beta(j)}S/d$ and consists of congruent axis parallel squares of the same size. We call this subdivision the (s,t)-*map* and write $\mathcal{M}[\mathfrak{S}]$. We have groups of $4s$ rings of same size. The only exception is the third group which contains $5s$ rings. Note that the boundaries of the cells overlap. Since we later require that the cells partition the rectangle we assume that the upper and left boundary always belong to the corresponding neighboring cell if there is any.

Let γ count the number of groups. Since the width of a ring in the ith group is $2^{i-1}S/d$, γ must satisfy $(5s - 4s)2^2\frac{S}{d} + 4s\sum_{i=1}^{\gamma} \frac{2^{i-1}S}{d} \overset{!}{=} \frac{S}{2}$, and thus $\gamma = \log(d/8s) \leq 1 + \log t$. Note that γ is integral by definition of d. In the remainder of this section we state some properties of (s,t)-maps. In some sense, (s,t)-maps and grids share salient properties. However, the number of cells in a $d \times d$ grid is $d^2 = s^2t^2$ while an (s,t)-map has $\mathcal{O}\left(s \cdot d\right)$ cells.

Lemma 3. $\mathcal{M}[\mathfrak{S}]$ *contains* $\mathcal{O}\left(s^2t\right)$ *cells.*

Lemma 4. *Let* $\mathfrak{C} \in \mathcal{M}[\mathfrak{S}]$ *have side length* $C \geq 2S/d$ *and let* $p \in \mathscr{P}$ *be a point which is contained in* $\mathfrak{C}$. *Then the distance of* p *to the boundary of* $\mathfrak{S}$ *in any* $\mathscr{L}_q$*-metric is at least* $2sC$.

Let $\mathfrak{S}$ be a square at level l and let $\mathcal{M}[\mathfrak{S}]$ be an (s,t)-map on $\mathfrak{S}$. Furthermore, let $\mathfrak{S}'$ be a child of $\mathfrak{S}$ in $QT_{a,b}$ and let $\mathfrak{C}'$ be a cell of $\mathcal{M}[\mathfrak{S}']$. A cell $\mathfrak{C}$ in $\mathcal{M}[\mathfrak{S}]$ is a *parent of* $\mathfrak{C}'$ if $\mathfrak{C}' \cap \mathfrak{C} \neq \emptyset$, i.e. if the cells overlap. Analogously, we write child $[\mathfrak{C}]$ for the child set of $\mathfrak{C}$.

Lemma 5. *Every cell has a unique parent.*

3.2   Standardized Solutions

In this section, we will consider trees that enjoy certain structural properties. Roughly speaking, our PTAS optimizes over all such trees. The main result of this section is Theorem 2 which states that there exists an almost optimal tree having this structure. Let a and b be fixed and let $QT_{a,b}$ denote the corresponding shifted quadtree with (s,t)-maps on all its squares. Furthermore, let ST be an arbitrary Steiner tree. An *internal component* of $\mathfrak{S} \in QT_{a,b}$ is a connected component in the induced Steiner tree $ST[\mathscr{X} \cap \mathfrak{S}]$, where $\mathscr{X}$ is the set of points spanned by ST. Let $k(\mathfrak{S})$ count the number of internal components of $\mathfrak{S}$. An edge $\{u,v\} \in ST$ is an *external edge of* $\mathfrak{S}$, if exactly one endpoint of $\{u,v\}$ is contained in $\mathfrak{S}$. Let $\mathscr{E}(\mathfrak{S}, ST)$ denote the set of endpoints (within $\mathfrak{S}$) of external edges. An edge $\{u,v\} \in ST$ has *level* l or *appears at level* l, if l is the least level, where $\{u,v\}$ is external. The level of $\{u,v\}$ is denoted by $\mathrm{lev}(\{u,v\})$. By removing all external edges, we obtain a set of connected components within $\mathfrak{S}$. These are exactly the internal components of $\mathfrak{S}$.

Definition 1. *Let* $r \in \mathbb{N}$, $r \geq 2$. *A Steiner tree* ST *is* (r,s)*-standardized with respect to* a *and* b *if every square* $\mathfrak{S} \in QT_{a,b}$ *is* (r,s)*-standardized, i.e.,* $\mathfrak{S}$ *satisfies*

(S1) $k(\mathfrak{S}) \leq r$.
(S2) *For all* $\mathfrak{C} \in \mathcal{M}[\mathfrak{S}]$, *all points in* $\mathscr{E}(\mathfrak{S}, ST) \cap \mathfrak{C}$ *belong to the same internal component of* $\mathfrak{S}$.

The following definition is essentially important for the proof of Theorem Theorem 3. Since it is omitted in this paper, the reader can skip this paragraph. Let $\mathbf{S}^*$ denote the set of all Steiner trees which have the same point set as ST^*, i.e., $ST \in \mathbf{S}^*$ if and only if $V(ST) = V(ST^*)$ and $S(ST) = S(ST^*)$. For fixed a and b, one can define a transitive relation $\lhd$ on $\mathbf{S}^*$. For $ST, ST' \in \mathbf{S}^*$ we have $ST \lhd ST'$ if and only if there exists a bijection $\gamma : E(ST) \to E(ST')$ such that $\mathrm{lev}(e) \leq \mathrm{lev}(\gamma(e))$.

Theorem 2. *For all* $a, b \in \{0, 2, 4, \ldots, L-2\}$ *there exists an* (r,s)*-standardized Steiner tree* $ST^*_{a,b} \in \mathbf{S}^*$ *with* $ST^* \lhd ST^*_{a,b}$ *such that if* a *and* b *are chosen u.a.r. then* $\mathbb{E}\left[\mathrm{val}\left(ST^*_{a,b}\right) - \mathrm{val}(ST^*)\right] \leq \mathcal{O}(1/s + 1/\sqrt{r})\,\mathrm{val}(ST^*)$.

As mentioned earlier, our PTAS tries to find the optimal standardized tree. Unfortunately it is not possible to compute the optimal one, thus we will compute a (good) approximation. We later bound the difference between the cost of our solution and val $(\mathcal{ST}_{a,b}^*)$. There we need the relation $\lhd$ to compare this difference with the optimum (cf. Theorem 3).

4   The Approximation Scheme

Although (r, s)-standardized trees have nice properties, we do not see an approach to compute the optimal one in polynomial time. Instead, our PTAS computes an almost optimal (r, s)-standardized tree. For fixed a and b, consider the optimal standardized Steiner tree $\mathcal{ST}_{a,b}^*$ and two squares $\mathfrak{S}_u$ and $\mathfrak{S}_v$ at level l. Assume that the two squares have a common parent $\mathfrak{S}$. Let $e = \{u, v\}$ be an edge that is contained within $\mathfrak{S}$ but has one endpoint in $\mathfrak{S}_u$ an the other in $\mathfrak{S}_v$, i.e, e appears at level l. There are cells $\mathfrak{C}_u$ and $\mathfrak{C}_v$ containing the endpoints of e in $\mathfrak{S}_u$ and $\mathfrak{S}_v$, respectively. The tree $\mathcal{ST}_{a,b}^*$ has the following structural property. All edges which appear at level l and have an endpoint in $\mathfrak{C}_u$ connect to the same internal component of $\mathfrak{S}_u$. The points in $\mathfrak{C}_u$ which belong to this component are from a structural point of view equivalent, since instead of connecting v to u we can connect v to any other of those points without adding a cycle or loosing connectivity. Although the length of such an edge may be greater than $\ell_q(e)$, the absolute error is at most twice the side length of $\mathfrak{C}_u$. Of course, the same is true for $\mathfrak{C}_v$. Therefore, the distance between the centers of $\mathfrak{C}_u$ and $\mathfrak{C}_v$ is a good estimation to the length of e.

If we are willing to accept this error, it is obvious that the following information is sufficient to describe a standardized Steiner tree within $\mathfrak{S}$. Firstly, we have to specify in which cells we have endpoints of external edges. Intuitively speaking, such cells, later called portals, represent the component to which all external edges should be connected. Secondly, it is necessary to encode which cells represent the same internal component of $\mathfrak{S}$. This motivates the following strategy for our PTAS. In a first phase we compute the "optimal" structure of our solution. This structure is described by edges between centers of cells and we minimize the total length of the cell-to-cell edges we use. In a second phase, those cell-to-cell edges are replaced by "real" edges.

4.1   Square Configuration

We describe the encoding in more detail. Let $\mathfrak{S}$ be a square in $QT_{a,b}$. The *configuration* $\mathcal{C}(\mathfrak{S})$ *of* $\mathfrak{S}$ is with respect to the irregular map $\mathcal{M}[\mathfrak{S}]$ and is given by the following parameters.

1. For every cell $\mathfrak{C}$ we store a bit $\rho(\mathfrak{C})$ which indicates whether $\mathfrak{C}$ is a *portal cell*.
2. For every cell $\mathfrak{C}$ we store a bit $\alpha(\mathfrak{C})$ which indicates whether $\mathfrak{C}$ is an *anchor cell*. We require, that every anchor is also a portal, i.e., $\alpha(\mathfrak{C}) \Rightarrow \rho(\mathfrak{C})$.
3. A partition Z of the portals of $\mathfrak{S}$ into at most r sets.

The anchor bit is not required to encode the interior of a square but necessary to guarantee that our PTAS returns a connected tree. This has the following reason. Intuitively speaking, $\rho(\mathfrak{C})$ indicates an obligation to connect to $\mathfrak{C}$ on some larger square. In this context the anchor bit indicates that this obligation is handed over to the next larger square. This means that we have to connect at the current level to every portal which is not an anchor. Since $d = 8st = \mathcal{O}(\log n)$, we obtain

Lemma 6. *The number of configurations for $\mathfrak{S}$ is at most $2^{\mathcal{O}(s^2 t \log r)}$.*

The configuration, where no bit is set is called the *empty configuration*. We write $\mathcal{C}(\mathfrak{S}) = \bigcirc$. The *configuration of* $QT_{a,b}$, denoted by $\mathcal{C}(QT_{a,b})$ is a set which contains tuples $\langle \mathfrak{S}, \mathcal{C}(\mathfrak{S}) \rangle$ representing the configurations of all the squares in $QT_{a,b}$ including $\mathfrak{S}_0$.

4.2 The Algorithm: First Phase

For every square $\mathfrak{S}$ and every configuration $\mathcal{C}(\mathfrak{S})$, we compute a value $T[\mathfrak{S}, \mathcal{C}(\mathfrak{S})]$ which can be seen as an almost sharp upper bound to the cost of the optimal subtree within $\mathfrak{S}$ which has the structural properties specified by $\mathcal{C}(\mathfrak{S})$. Then, $T[\mathfrak{S}_0, \bigcirc]$ is an upper bound to val $\left(\mathcal{ST}^*_{a,b}\right)$. Since val $\left(\mathcal{ST}^*_{a,b}\right)$ is a good upper bound to val $(\mathcal{ST}^*)$, it suffices to quantify the gap between val $\left(\mathcal{ST}^*_{a,b}\right)$ and $T[\mathfrak{S}_0, \bigcirc]$. This will be done in Section 4.3.

If $\mathfrak{S}$ is a leaf of $QT_{a,b}$ then we have exactly one cell $\mathfrak{C}_0$ which contains a point with odd integral coordinates. This point need not be contained in $\mathscr{P}$. First we check whether $\rho(\mathfrak{C}) = 0$ for all $\mathfrak{C} \in \mathcal{M}[\mathfrak{S}] - \{\mathfrak{C}_0\}$. If this is not the case, we store $T[\mathfrak{S}, \mathcal{C}(\mathfrak{S})] = \infty$. Otherwise, we have to determine the penalty we have to pay in $\mathfrak{S}$. We have three cases. If $\rho(\mathfrak{C}_0) = 1$ and $\mathscr{P} \cap \mathfrak{C}_0 = \emptyset$ then we place a Steiner point into $\mathfrak{C}_0$ and set $T[\mathfrak{S}, \mathcal{C}(\mathfrak{S})] = c_S$. If $\rho(\mathfrak{C}_0) = 0$ and $\mathscr{P} \cap \mathfrak{C}_0 = \{u\}$ then we store $T[\mathfrak{S}, \mathcal{C}(\mathfrak{S})] = \pi(u)$. In all other cases, $T[\mathfrak{S}, \mathcal{C}(\mathfrak{S})] = 0$.

If $\mathfrak{S}$ is an internal node of $QT_{a,b}$ we proceed as follows. Assume that $\mathfrak{S}$ has level l and let $\mathfrak{S}_1, ..., \mathfrak{S}_4$ denote its children. We enumerate all combinations of configurations for $\mathfrak{S}_1, ..., \mathfrak{S}_4$. Let $\mathcal{C}(\mathfrak{S}_1), ..., \mathcal{C}(\mathfrak{S}_4)$ be such configurations. For every square $\mathfrak{S}_i$, $\mathcal{C}(\mathfrak{S}_i)$ defines a partition $Z(\mathfrak{S}_i)$ of the portals in $\mathfrak{S}_i$ into at most r classes. We now consider a (not necessarily unique) forest $\mathcal{Z}(\mathfrak{S}_i)$ with the portals of $\mathfrak{S}_i$ as vertices. $\mathcal{Z}(\mathfrak{S}_i)$ must have the property that two portals $\mathfrak{C}_1$ and $\mathfrak{C}_2$ belong to the same connected component of $\mathcal{Z}(\mathfrak{S}_i)$ if and only if they are in the same partition of $Z(\mathfrak{S}_i)$. For the sake of brevity, we use $\mathcal{Z}' = \bigcup_{i=1}^{4} \mathcal{Z}(\mathfrak{S}_i)$. Let $\mathfrak{C}$ be a portal and let $\mathfrak{C}' \in$ child $[\mathfrak{C}]$. If $\alpha(\mathfrak{C}') = 1$ then we say that $\mathfrak{C}'$ is an *anchor hook* of $\mathfrak{C}$. Recall, that $\mathcal{Z}'$ is a graph on the portals of the children of $\mathfrak{S}$. A choice of configurations $\mathcal{C}(\mathfrak{S}_1), ..., \mathcal{C}(\mathfrak{S}_4)$ is *admissible* if each of the following conditions holds.

(C1) Every portal in $\mathfrak{S}$ has at least one anchor hook.
(C2) The parent in $\mathfrak{S}$ of every anchor in some $\mathfrak{S}_i$ is a portal, unless $\mathfrak{S} = \mathfrak{S}_0$.
(C3) All anchors of a portal in $\mathfrak{S}$ belong to the same connected component of $\mathcal{Z}'$.

Furthermore, there exists a graph $\mathcal{D}$ on the portals of $\bigcup_i \mathfrak{S}_i$ which has the following properties.

(C4) Every portal in $\mathfrak{S}_i$, $1 \leq i \leq 4$ which is not an anchor is incident to some edge of $\mathcal{D}$.

(C5) Every connected component of $\mathcal{D} \cup \mathcal{Z}'$ contains at least one anchor, unless $\mathfrak{S} = \mathfrak{S}_0$.

(C6) Two portals $\mathfrak{C}_1$ and $\mathfrak{C}_2$ of $\mathfrak{S}$ belong to the same partition class of $Z(\mathfrak{S})$ if and only if their anchors belong to the same connected component of $\mathcal{D} \cup \mathcal{Z}'$.

(C7) $\mathcal{D}$ uses only edges which cross the dissection lines that divide $\mathfrak{S}$ into its children.

The relaxations of (C2) and (C5) are necessary to compute $T[\mathfrak{S}_0, \bigcirc]$. Since the configuration $\bigcirc$ does not contain any portal at all, there would be no admissible configuration of the children.

The length of an edge in $\mathcal{D}$ is the distance of the centers of the portals which it connects in the $\mathscr{L}_q$-metric. If there is no combination which is admissible, we store ∞ in $T[\mathfrak{S}, \mathcal{C}(\mathfrak{S})]$. Otherwise, we choose admissible configurations $\mathcal{C}(\mathfrak{S}_1), \ldots, \mathcal{C}(\mathfrak{S}_4)$ and a graph $\mathcal{D}$ such that

$$\sum_{i=1}^{4} T[\mathfrak{S}_i, \mathcal{C}(\mathfrak{S}_i)] + \ell_q(\mathcal{D}) + \sum_{\{c_1, c_2\} \in \mathcal{D}} [C(c_1) + C(c_2)] \tag{2}$$

is minimal. Here $C(c_1)$ and $C(c_2)$ denote the side length of the portal which has center c_1 and c_2, respectively. In essence, the third part (the sum over all edges in $\mathcal{D}$) of the formula above is necessary to bound the deviation in length from the optimal standardized solution. In the sequel, this will be explained in more detail. For each choice of configurations we have to find the optimal graph $\mathcal{D}$. This can be achieved using some generalization of Kruskal's algorithm. Details are omitted due to space constraints. Then it is clear that we can find the configurations $\mathcal{C}^*(\mathfrak{S}_1), \ldots, \mathcal{C}^*(\mathfrak{S}_4)$ that minimize (2) by exhaustive search. Let $\mathcal{D}_{\mathfrak{S}}^*$ denote the optimal graph $\mathcal{D}$ corresponding to $\mathcal{C}^*(\cdot)$.

We compute $T[\mathfrak{S}_0, \bigcirc]$ by using a dynamic programming approach. If we proceed bottom-up, it suffices to read the $T[\mathfrak{S}_i, \mathcal{C}(\mathfrak{S}_i)]$ from the lookup table. We obtain $T[\mathfrak{S}_0, \bigcirc]$ which corresponds to some configuration $\mathcal{C}(QT_{a,b})$ that we have computed implicitly. One can easily check, that the running time of this dynamic program is $n^{\mathcal{O}(s^2 \cdot \log r)}$ by using $d = \mathcal{O}(s^2 t) = \mathcal{O}(s^2 \log n)$.

4.3   The Algorithm: Second Phase

In the first phase we obtain a configuration for $QT_{a,b}$ corresponding to $T[\mathfrak{S}_0, \bigcirc]$. This configuration contains a collection of graphs $D(\mathscr{P}) := \{\mathcal{D}_{\mathfrak{S}}^* : \mathfrak{S} \in QT_{a,b}\}$. In some sense, this collection specifies which of the components are connected in which square. Even more, the edges to be used are roughly described by edges between centers of cells. During the second phase of our approximation scheme, we construct a Steiner tree $\mathcal{ST}$ from $D(\mathcal{P})$. This time, we proceed in top-down fashion. Let $\mathcal{ST} = \emptyset$. For some square $\mathfrak{S}$ at level l, we consider the graph $\mathcal{D}_{\mathfrak{S}}^*$. Let c be some edge in $\mathcal{D}_{\mathfrak{S}}^*$ which connects (the centers) of two cells $\mathfrak{C}_1$ and $\mathfrak{C}_2$. By recursively following the anchors of both $\mathfrak{C}_1$ and $\mathfrak{C}_2$ we determine points u_1 and u_2 which serve as endpoints. In other words, starting at portal $\mathfrak{C}_1$ we choose one of its anchors, then we choose an anchor hook of this anchor, and so on. Thus, we go to lower and lower levels in $QT_{a,b}$ until we reach a leaf. In a leaf, we find by construction a point in $\mathcal{P}$ or Steiner point which we choose as u_1. We proceed similarly for $\mathfrak{C}_2$, add the edge $\{u_1, u_2\}$ to $\mathcal{ST}$ and mark c as done. As u_1 and u_2 are within $\mathfrak{C}_1$ and $\mathfrak{C}_2$, (2) implies

Lemma 7. $\ell_q (\mathcal{ST}) \leq T[\mathfrak{S}_0, \bigcirc]$. $\qquad\qquad\qquad\qquad\qquad\qquad\qquad\qquad\qquad\square$

In addition, we have to prove that $\mathcal{ST}$ is a Steiner tree on $V(\mathcal{ST}) \cup S(\mathcal{ST})$, i.e., $\mathcal{ST}$ is cycle free and connected and spans all points in $V(\mathcal{ST}) \cup S(\mathcal{ST})$. This can be checked using (C1) - (C7). Details are omitted due to space constraints. Now, one can show for a, b uniformly chosen at random that

$$\mathbb{E}\left[T[\mathfrak{S}, \bigcirc] - \mathrm{val}\left(\mathcal{ST}_{a,b}^*\right)\right] = \mathcal{O}\left(1/s + 1/\sqrt{r}\right) \mathrm{val}\left(\mathcal{ST}^*\right).$$

We can even show that there are a, b such that both $T[\mathfrak{S}, \bigcirc]$ and $\mathrm{val}\left(\mathcal{ST}_{a,b}^*\right)$ are good upper bounds. Then one can prove using Theorem 2 and Lemma 7 the following statement.

Theorem 3. *There are $a, b \in \{0, 2, 4 \ldots, L - 2\}$ such that*

$$\ell_q (\mathcal{ST}) \leq \mathcal{O}\left(1 + \frac{1}{s} + \frac{1}{\sqrt{r}}\right) \mathrm{val}\left(\mathcal{ST}^*\right).$$

Choosing $s = \mathcal{O}\left(1/\varepsilon\right)$ and $r = \mathcal{O}\left(1/\varepsilon^2\right)$ proves Theorem 1.

5   Generalizations

A natural question is, whether there is also PTAS for $\mathcal{P} \in \mathbb{R}^d$, for fixed $d \geq 3$. In this case the number of cells in a (s, t)-map is $\mathcal{O}\left(t^{d-1}\right)$ and thus the running time of our algorithm is no longer polynomial. However, we can construct a quasi-polynomial time approximation scheme. Instead of using (s, t)-maps it is more convenient to place a d-dimensional grid of granularity $\mathcal{O}\left(R\varepsilon/t\right)$ on a square of side length S. The remaining parts of our proofs have straight-forward equivalents.

Our approach also extends to other tree problems. For example, assume that the input contains an additional set S_0 which restricts locations for Steiner points, i.e., every solution should satisfy $S(\mathcal{ST}) \subseteq S_0$. This problem has a simple PTAS, if Steiner points have zero cost. It is then sufficient to add S_0 to $\mathcal{P}$, to set the corresponding penalties to 0 and to chose $c_S = \infty$. If we assign (potential different) costs to the locations in S_0, the problem still admits a PTAS which can be obtained by slightly modifying our algorithm. This is insofar surprising, as the corresponding network problem is not likely to be in $\mathcal{APX}$ [11].

A third example is the Vehicle Routing-Allocation Problem (VRAP) [5] for which our method also yields a PTAS. In VRAP not all customers need be visited by the vehicle on its salesman tour. However customers not visited either have to be allocated to some customer on one of the vehicle tour or left isolated. As in the NWGST there are penalties for allocated or isolated customers.

References

1. S. Arora. Polynomial time approximation schemes for euclidean traveling salesman and other geometric problems. *Journal of the ACM*, 45(5):753–782, 1998.
2. S. Arora. Approximation schemes for NP-hard geometric optimization problems: A survey. *Mathematical Programming, Series B*, 97:43–69, 2003.
3. S. Arora and K. Chang. Approximation schemes for degree-restricted MST and red-blue separation problems. *Algorithmica*, 40(3):189–210, 2004.
4. S. Arora, P. Raghavan, and S. Rao. Approximation schemes for Euclidean k-medians and related problems. In *Proceedings of the 30th Annual ACM Symposium on Theory of Computing*, pages 106–113, 1998.
5. J. E. Beasley and E. Nascimento. The vehicle routing allocation problem: a unifying framework. *TOP*, 4:65–86, 1996.
6. M. Bern and P. Plassmann. The Steiner tree problem with edge length 1 and 2. *Information Processing Letters*, 32:171–176, 1989.
7. R. Courant and H. Robbins. *What is Mathematics?* Oxford University Press, 1941.
8. M. Garey, R. Graham, and D. Johnson. The complexity of computing Steiner minimal trees. *SIAM Journal on Applied Mathematics*, 32:835–859, 1977.
9. M. X. Goemans and D. P. Williamson. A general approximation technique for constrained forest problems. *SIAM Journal on Computing*, 24(2):296–317, 1995.
10. F. K. Hwang, D. Richards, and P. Winter. *The Steiner Tree Problem*, volume 53 of *Annals of Discrete Mathematics*. North-Holland, 1995.
11. P. Klein and R. Ravi. A nearly best-possible approximation algorithm for node-weighted Steiner trees. *Journal of Algorithms*, 19:104–115, 1995.
12. J. S. B. Mitchell. Guillotine subdivisions approximate polygonal subdivisions: A simple polynomial-time approximation scheme for geometric TSP, k-MST and related problems. *SIAM Journal on Computing*, 28(4):1298–1309, 1999.
13. A. Moss and Y. Rabani. Approximation algorithms for constrained node weighted Steiner tree problems. In *Proceedings of the 33rd Annual ACM Symposium on Theory of Computing*, pages 373–382, 2001.
14. H. J. Prömel and A. Steger. *The Steiner Tree Problem — A Tour through Graphs, Algorithms, and Complexity*. Advanced Lectures in Mathematics. Vieweg Verlag, 2002.
15. G. Robins and A. Zelikovsky. Improved Steiner tree approximation in graphs. In *Proceedings of the 11th annual ACM-SIAM symposium on Discrete algorithms*, pages 770–779, 2000.
16. K. Talwar. Bypassing the embedding: algorithms for low dimensional metrics. In *Proceedings of the 36th Annual ACM Symposium on Theory of Computing*, 2004.

Towards Optimal Integrality Gaps for Hypergraph Vertex Cover in the Lovász-Schrijver Hierarchy

Iannis Tourlakis

Princeton University, Department of Computer Science, Princeton, NJ 08544
itourlak@cs.princeton.edu

Abstract. "Lift-and-project" procedures, which tighten linear relaxations over many rounds, yield many of the celebrated approximation algorithms of the past decade or so, even after only a constant number of rounds (e.g., for MAX-CUT, MAX-3SAT and SPARSEST-CUT). Thus proving super-constant round lowerbounds on such procedures may provide evidence about the inapproximability of a problem.

We prove an integrality gap of $k - \epsilon$ for linear relaxations obtained from the trivial linear relaxation for k-uniform hypergraph VERTEX COVER by applying even $\Omega(\log \log n)$ rounds of Lovász and Schrijver's LS lift-and-project procedure. In contrast, known PCP-based results only rule out $k - 1 - \epsilon$ approximations. Our gaps are tight since the trivial linear relaxation gives a k-approximation.

1 Introduction

Several papers [1, 2, 5, 9] have appeared over the past few years studying the quality of linear programming (LP) and semidefinite programming (SDP) relaxations derived by so-called "lift-and-project" procedures. These procedures (for examples, see Lovász and Schrijver [12] and Sherali and Adams [13]) start with an initial relaxation and add, in a controlled manner, more and more inequalities satisfied by all integral solutions in a bid to obtain tighter and hence, hopefully, better relaxations.

One motivation for studying lift-and-project procedures is the large gap that remains for some problems (such as VERTEX COVER) between approximation ratios achieved by known algorithms and ratios ruled out by PCPs. On the one hand, lift-and-project methods may lead to improved approximation algorithms for such problems. For example, the celebrated SDP-based algorithms of [11] for MAX-CUT and of [4] for SPARSEST CUT are efficiently derived by lift-and-project methods. On the other hand, ruling out good approximation algorithms based on lift-and-project methods might give evidence about the problem's true inapproximability.

In this paper we concentrate on proving integrality gaps for relaxations obtained by the LS lift-and-project technique defined by Lovász and Schrijver [12] (definitions in Section 2). The problem we study is VERTEX COVER on k-uniform

C. Chekuri et al. (Eds.): APPROX and RANDOM 2005, LNCS 3624, pp. 233–244, 2005.

hypergraphs. Progress on graph-related problems is sometimes made by considering hypergraph analogues, motivating our study of hypergraph VERTEX COVER. As with graph VERTEX COVER, there is a gap between the factors achieved by known approximation algorithms and those ruled out by PCPs: While PCP-based hardness results for k-uniform hypergraph VERTEX COVER rule out $k-1-\epsilon$ polynomial-time approximations [7], only $k - o(1)$ approximation algorithms are known. Our main result is that for all $\epsilon > 0$ and all sufficiently large n there exist k-uniform hypergraphs on n vertices for which $\Omega(\log \log n)$ rounds of LS are necessary to obtain a relaxation for VERTEX COVER with integrality gap less than $k - \epsilon$.

1.1   Related Work

Alekhnovich et al. [1] also considered integrality gaps of LS tightenings for k-uniform hypergraph VERTEX COVER proving that a gap of $k - 1 - \epsilon$ remains after a linear number of even LS_+ tightenings (LS_+, defined in Section 2, is a stronger version of LS where positive semidefinite constraints are also added). However, the work most closely related to ours is the paper of Arora et al. [2]. They show that an integrality gap of $2 - \epsilon$ remains for graph VERTEX COVER even after $\Omega(\sqrt{\log n})$ rounds of LS. However, obtaining an analogous integrality gap of $k - \epsilon$ for k-uniform hypergraph VERTEX COVER for a non-constant number of rounds remained open since their analysis was especially tailored for graphs. The main technical contribution of this paper is adapting the proof framework of [2] to reason about hypergraphs. We discuss this further after some necessary technical definitions in Section 2 below.

   Other recent papers studying Lovász-Schrijver tightenings are [1, 5, 6, 9, 10, 14]. Stephen and Tunçel [14] show that $\Omega(\sqrt{n})$ rounds of even the stronger LS_+ procedure are required to derive some simple inequalities for the MATCHING polytope. Cook and Dash [6] and Goemans and Tunçel [10] both show how for some relaxations n rounds of LS_+ are required to derive some simple inequalities. Feige and Krauthgamer [9] show that a large gap remains for INDEPENDENT SET after $\Omega(\log n)$ rounds of LS_+, while Buresh-Oppenheim et al. [5] show that the gap for MAX-kSAT, $k \geq 5$ remains $(2^k - 1)/2^k - \epsilon$ even after $\Omega(n)$ rounds of LS_+. In addition to their results for hypergraph VERTEX COVER mentioned above, Alekhnovich et al. [1] extend the results of [5] to MAX-3SAT as well as proving an integrality gap of $(1 - \epsilon) \ln n$ after $\Omega(n)$ rounds of LS_+ for SET COVER.

2   Lovász-Schrijver Liftings and Our Methodology

We will use the VERTEX COVER problem to explain relaxations and how to tighten them using LS liftings. The integer program (IP) characterization for a k-uniform hypergraph $G = (V, E)$ is: minimize $\sum_{i \in V} v_i$ such that $v_{j_1} + \ldots + v_{j_k} \geq 1$ for all hyperedges $\{j_1, \ldots, j_k\}$, where $v_i \in \{0, 1\}$. The integer hull I is the convex hull of all solutions to this program. The standard LP relaxation is to allow $0 \leq v_i \leq 1$. Clearly the value of the LP is at most that of the IP. Let P be the convex hull of all solutions to the LP in $[0,1]^n$. A linear relaxation is *tightened* by adding constraints that also hold for the integral hull; in general this gives some polytope P' such that $I \subseteq P' \subseteq P$. The quality of a linear relaxation is measured by the ratio $\frac{\text{optimum value over } I}{\text{optimum value over } P}$, called its *integrality gap*.

Lovász and Schrijver [12] present several "lift-and-project" techniques for deriving tighter and tighter relaxations of a 0-1 integer program. These procedures take an n dimensional relaxed polytope, "lift" it to n^2 dimensions, add new constraints to the high-dimensional polytope, and then project back to the original space. Three progressively stronger such techniques, LS_0, LS and LS_+ are given in [12]; we will focus on LS, but also define LS_+ below for completeness.

The notation uses homogenized inequalities. Let G be a k-uniform hypergraph and let $\text{VC}(G)$ be the cone in $\Re^{n+1}$ that contains $(x_0, x_1, \ldots, x_n)$ iff it satisfies the edge constraints $x_{i_1} + \ldots + x_{i_k} \geq x_0$ for each edge $\{i_1, \ldots, i_k\} \in G$. All cones in what follows will be in $\Re^{n+1}$ and we will be interested in the slice cut by the hyperplane $x_0 = 1$. Denote by $N^r(\text{VC}(G))$ and $N_+^r(\text{VC}(G))$ the feasible cones of all inequalities obtained from r rounds of the LS and LS_+ lifting procedures, respectively. The following lemma characterizes the effect of one round of the N and N_+ operators. (In the remainder of the paper we index columns and rows starting from 0 rather than 1, and e_i denotes the ith unit vector.)

Lemma 1 ([12]). *Let Q be a cone in $\Re^{n+1}$. Then $y \in \Re^{n+1}$ is in $N(Q)$ iff there exists a symmetric matrix $Y \in \Re^{(n+1) \times (n+1)}$ such that:*

1. $Y e_0 = diag(Y) = y$,
2. $Y e_i, Y(e_0 - e_i) \in Q$, for all $i = 1, \ldots, n$.

Moreover, $y \in N_+(Q)$ iff Y is in addition positive semidefinite.

To prove that $y \in N^{r+1}(Q)$, we have to construct a specific matrix Y and prove that the $2n$ vectors defined in Lemma 1 are in $N^r(Q)$. Such a matrix Y is called a *protection matrix* for y since it "protects" y for one round. The points y we protect will always have $y_0 = 1$ in which case condition 2 above is equivalent to the following condition (below P is the projection of Q along $x_0 = 1$):

2′. For each i such that $x_i = 0$, $Y e_i = 0$; for each i such that $x_i = 1$, $Y e_i = y$; Otherwise $Y e_i / x_i$ and $Y(e_0 - e_i)/(1 - x_i)$ are both in P.

The simplest possible Y has $Y_{ij} = y_i y_j$ except along the diagonal where $Y_{ii} = y_i$. Then $Y e_i / x_i$ and $Y(e_0 - e_i)/(1 - x_i)$ are identical to y except in the ith coordinate which are now 1 and 0, respectively. This matrix was used in [5, 10].

However, such a protection matrix does not work for us: To prove gaps of $k - \epsilon$ on k-uniform hypergraphs we will protect the point $y_\gamma = (1, \frac{1}{k} + \gamma, \ldots, \frac{1}{k} + \gamma)$ where $\gamma > 0$ is an arbitrarily small constant. For y_γ, the vector $Y(e_0 - e_i)/(1 - x_i)$ given by the simple protection matrix is not guaranteed to be in $\text{VC}(G)$: if coordinate i is changed to 0 in y_γ, we violate all edge constraints involving the ith vertex. So more sophisticated protection matrices are needed.

For the case when $k = 2$ (i.e., for graphs), Arora et al. [2] used LP duality to show that appropriate protection matrices exist for at least $\Omega(\sqrt{\log n})$ rounds. In particular, they used a special combinatorial form of Farkas's lemma specifically applicable to the constraints involved in defining protection matrices for graphs.

We will also reduce the problem of showing our protection matrices exists to the feasibility of linear programs; however, we will need a much more complicated form of Farkas's lemma than in the graph case. As such, we will only be able to carry out our arguments for $O(\log \log n)$ rounds.

3   Lowerbounds for Hypergraph Vertex Cover

Theorem 1. *For all $k \geq 2$, $\epsilon > 0$ there exist constants $n_0(k, \epsilon), \delta(k, \epsilon) > 0$ s.t. for every $n \geq n_0(k, \epsilon)$ there exists a k-uniform hypergraph G on n vertices for which the integrality gap of $N^r(\mathrm{VC}(G))$ is at least $k-\epsilon$ for all $r \leq \delta(k, \epsilon) \log \log n$.*

Theorem 1 will follow from the following two theorems.

Theorem 2. *For all $k \geq 2$ and any $\epsilon > 0$ there exists an $n_0(k, \epsilon)$ such that for every $n \geq n_0(k, \epsilon)$ there exist k-uniform hypergraphs with n vertices and $O(n)$ hyperedges having $\Omega(\log n)$ girth but no independent set of size greater than ϵn.*

Theorem 3. *Fix $\gamma > 0$ and let $y_\gamma = (1, \frac{1}{k} + \gamma, \frac{1}{k} + \gamma, \ldots, \frac{1}{k} + \gamma)$. Let G be a k-uniform hypergraph such that $\mathrm{girth}(G) \geq \frac{20}{\gamma} r 5^r$. Then $y_\gamma \in N^r(\mathrm{VC}(G))$.*

Theorem 2 is an easy extension to hypergraphs of a result proved by Erdős [8]. The remainder of this section will be devoted to proving Theorem 3.

3.1   Proof of Theorem 3

As in previous LS lowerbounds, the theorem will be proved by induction where the inductive hypothesis will require some set of vectors to be in $N^m(\mathrm{VC}(G))$ for $m \leq r$. These vectors will be essentially all-$(\frac{1}{k} + \gamma)$ except possibly for a few small neighbourhoods where they can take arbitrary nonnegative values so long as the edge constraints for G are satisfied. The exact characterization is given by the following definition similar to an analogous definition in [2]:

Definition 1. *Let $S \subseteq [n]$, R be a positive integer, and $\gamma > 0$. A nonnegative vector $(\alpha_0, \alpha_1, \ldots, \alpha_n)$ with $\alpha_0 = 1$ is an (S, R, γ)-vector if the entries satisfy the edge constraints and if $\alpha_j = \frac{1}{k} + \gamma$ for each $j \notin \cup_{w \in S} Ball(w, R)$. Here $Ball(w, R)$ denotes the set of vertices within distance R of w in the graph.*

Let $R_r = 0$ and let $R_m = 5R_{m+1} + 1/\gamma$ for $0 \leq m < r$. Note $R_m \leq \mathrm{girth}(G)/20$.

To prove Theorem 3 we prove the following inductive claim. The theorem is a subcase of $m = r$:

Inductive Claim for $N^m(\mathrm{VC}(G))$: *For every set S of $r - m$ vertices, every (S, R_m, γ)-vector is in $N^m(\mathrm{VC}(G))$.*

The base case $m = 0$ is trivial since (S, R_m, γ)-vectors satisfy the edge constraints for G. In the remainder of this section we prove the Inductive Claim for $m + 1$ assuming truth for m.

To that end, let α be an (S, R_{m+1}, γ)-vector where $|S| = r - m - 1$. By induction, every $(S \cup \{i\}, R_m, \gamma)$-vector is in $N^m(\text{VC}(G))$. So to prove that $\alpha \in N^{m+1}(\text{VC}(G))$ it suffices by Lemma 1 to exhibit an $(n+1) \times (n+1)$ symmetric matrix Y such that:

A. $Y e_0 = diag(Y) = \alpha$,

B. For each i such that $\alpha_i = 0$, $Y e_i = 0$; for each i such that $\alpha_1 = 1$, $Y e_0 = Y e_i$; otherwise, $Y e_i / \alpha_i$ and $Y(e_0 - e_i)/(1 - \alpha_i)$ are $(S \cup \{i\}, R_m, \gamma)$-vectors.

As was done in [2] and [12], we will write these conditions as a linear program and show that the program is feasible, proving the existence of Y. Our notation will assume symmetry, namely Y_{ij} represents $Y_{\{i,j\}}$.

Condition **A** requires:

$$Y_{kk} = \alpha_k \qquad \forall k \in \{1, \ldots, n\}. \tag{1}$$

Condition **B** requires first of all that $Y e_j / \alpha_j$ and $Y(e_0 - e_j)/(1 - \alpha_j)$ satisfy the edge constaints: For all $i \in \{1, \ldots, n\}$ and all $\{j_1, \ldots, j_k\} \in E$:

$$\alpha_i \leq Y_{ij_1} + \ldots + Y_{ij_\ell} \leq \alpha_i + (\alpha_{j_1} + \ldots + \alpha_{j_k} - 1). \tag{2}$$

Vertices i, t are called a *distant pair* if $t \notin \cup_{w \in S \cup \{i\}} Ball(w, 5R_{m+1} + 1/\gamma)$. (Note then that $\alpha_t = \frac{1}{k} + \gamma$.) Condition **B** requires that for such a pair, the tth coordinates of $Y e_i / \alpha_i$ and $Y(e_0 - e_i)/(1 - \alpha_i)$ are $\frac{1}{k} + \gamma$. In particular,

$$Y_{it} = \alpha_i \alpha_t = \alpha_i \left(\frac{1}{k} + \gamma \right). \tag{3}$$

Note that distant pairs have the property that every path in G that connects them contains at least $4R_m + 1/\gamma - 1$ hyperedges each of which α oversatisfies by $k\gamma$, and at most R_m hyperedges which α does *not* oversatisfy by $k\gamma$.

Finally, condition **B** requires that $Y e_i / \alpha_i$, $Y(e_0 - e_i)/(1 - \alpha_i)$ are in $[0, 1]^{n+1}$:

$$0 \leq Y_{ij} \leq \alpha_i, \qquad \forall i, j \in \{1, \ldots, n\}, i \neq j \tag{4}$$

$$-Y_{ij} \leq 1 - \alpha_i - \alpha_j, \qquad \forall i, j \in \{1, \ldots, n\}, i \neq j \tag{5}$$

The above constraints are equivalent to the following four constraint families:

$$Y_{ij} \leq \beta(i, j), \qquad \forall i, j \in \{1, \ldots, n\} \tag{6}$$

$$-Y_{ij} \leq \delta(i, j), \qquad \forall i, j \in \{1, \ldots, n\} \tag{7}$$

$$Y_{ij_1} + \ldots + Y_{ij_k} \leq a(i, j_1, \ldots, j_k), \qquad \forall \{j_1, \ldots, j_k\} \in E \tag{8}$$

$$-Y_{ij_1} - \ldots - Y_{ij_k} \leq b(i, j_1, \ldots, j_k), \qquad \forall \{j_1, \ldots, j_k\} \in E \tag{9}$$

Here (1) $\beta(i, j) = \alpha_i \alpha_j$ if i, j is a distant pair and $\beta(i, j) = \min(\alpha_i, \alpha_j)$ otherwise; (2) $\delta(i, j) = -\alpha_i$ if $i = j$; $\delta(i, j) = -\alpha_i \alpha_j$ if i, j is a distant pair; and $\delta(i, j) = 1 - \alpha_i - \alpha_j$ otherwise; (3) $a(i, j_1, \ldots, j_k) = \alpha_i + (\alpha_{j_1} + \ldots + \alpha_{j_k} - 1)$; and (4) $b(i, j_1, \ldots, j_k) = -\alpha_i$. Note that $\beta(i, j) + \delta(i, j) \geq 0$ always, since $\alpha \in [0, 1]^{n+1}$.

To prove the consistency of constraints (6)–(9), we will use a special combinatorial version of Farkas's Lemma similar in spirit to that used in [12] and [2]. Before giving the exact form, we require some definitions.

Definition 2. *A tiling* (W, P, N) *for* G *is a connected* k-*uniform hypergraph* $H = (W, P, N)$ *on vertices* W *and two disjoint* k-*edge sets* P *and* N *such that:*

1. *Each vertex in* W *is labelled by* Y_{ij}, $i, j \in \{1, \ldots, n\}$. *Note that distinct vertices need not have different labels.*
2. *Each vertex belongs to at most one edge in* P *and at most one edge in* N *(in particular, all edges in* P *are mutually disjoint, as are all edges in* N*).*
3. *All edges in* $P \cup N$ *are of the form* $\{Y_{ij_1}, \ldots, Y_{ij_k}\}$ *where* $\{j_1, \ldots, j_k\} \in E$.

The edges of a tiling are called tiles. *Vertices in* W *not incident to any tile in* P *are called* unmatched negative vertices; *vertices in* W *not incident to any tile in* N *are called* unmatched positive vertices. *Let* U_N *and* U_P *denote the sets of unmatched negative and unmatched positive vertices, respectively. A vertex labelled* Y_{ii} *is called* diagonal. *Denote the set of unmatched positive diagonal vertices in* W *by* U_{PD}. *A vertex labelled* Y_{ij} *is called a* distant pair *if* $\{i, j\}$ *are a distant pair. Given a tile* $\{Y_{ij_1}, \ldots, Y_{ij_k}\}$ *in* P *or* N, *call* $\{j_1, \ldots, j_k\} \in E$ *its* bracing edge *and* i *its* bracing node. *An edge* $\{j_1, \ldots, j_k\}$ *in* G *is called* overloaded *if* $\alpha_{j_1} = \ldots = \alpha_{j_k} = \frac{1}{k} + \gamma$. *A tile* $\{Y_{ij_1}, \ldots, Y_{ij_k}\}$ *in* H *is overloaded if its bracing edge is overloaded.*

Given a tiling $H = (W, P, N)$ for G, define the following sums:

$$S_1^H = \sum_{\{Y_{ij_1}, \ldots, Y_{ij_k}\} \in P} a(i, j_1, \ldots, j_k) + \sum_{\{Y_{ij_1}, \ldots, Y_{ij_k}\} \in N} b(i, j_1, \ldots, j_k), \qquad (10)$$

$$S_2^H = \sum_{Y_{ij} \in U_P} \delta(i, j) + \sum_{Y_{ij} \in U_N} \beta(i, j). \qquad (11)$$

Finally, let $S^H = S_1^H + S_2^H$.

Lemma 2 (Special Case of Farkas's Lemma). *Constraints (6)–(9) are unsatisfiable iff there exists a tiling* $H = (W, P, N)$ *for* G *such that* $S^H < 0$.

Proof. Note first that by the general form of Farkas's lemma constraints (6)–(9) are unsatisfiable iff there exists a positive rational linear combination of them where the LHS is 0 and the RHS is negative.

Now suppose that there exists a tiling $H = (W, P, N)$ such that $S^H < 0$. Consider the following linear integer combination of the constraints: (1) For each tile $e = \{Y_{ij_1}, \ldots, Y_{ij_k}\} \in H$, if $e \in P$, add the constraint $Y_{ij_1} + \ldots + Y_{ij_k} \leq a(i, j_1, \ldots, j_k)$; if $e \in N$, add the constraint $-Y_{ij_1} - \ldots - Y_{ij_k} \leq b(i, j_1, \ldots, j_k)$; (2) For each $v \in U_N$ labelled Y_{ij}, add the constraint $Y_{ij} \leq \beta(i, j)$; (3) For each $v \in U_P$ labelled Y_{ij}, add the constraint $-Y_{ij} \leq \delta(i, j)$. But then, for this combination of constraints the LHS equals 0 while the RHS equals $S^H < 0$. So by Farkas's lemma the constraints are unsatisfiable.

Now assume on the other hand that the constraints are unsatisfiable. So there exists a positive rational linear combination of the constraints such that the LHS is 0 and the RHS is negative. By clearing out denominators, we can assume that the linear combination has *integer* coefficients. Hence, as $\beta(i, j) + \delta(i, j) \geq 0$

always, our combination must contain constraints of type (8) and (9). Moreover, since the LHS is 0, for each Y_{ij} appearing in the integer combination there must be a corresponding occurrence of $-Y_{ij}$. But then, it is easy to see that the constraints in the integer linear combination can be grouped into a set of tilings $\{H_i = (W_i, P_i, N_i)\}$ such that the RHS of the linear combination equals $\sum_i S^{H_i}$. Since the RHS is negative, it must be that at least one of the tilings H in the set is such that $S^H < 0$. The lemma follows.

So to show that the constraints for the matrix Y are consistent and thus complete the proof of the Inductive Claim for $m + 1$ (and complete the proof of Theorem 3), we will show that $S^H \geq 0$ for any tiling $H = (W, P, N)$ for G. To that end, fix a tiling $H = (W, P, N)$ for G. Our analysis divides into three cases depending on the size of U_{PD}, the set of unmatched positive diagonal vertices in H. In the first (and easiest) case, $|U_{PD}| = 0$; in the second, $|U_{PD}| \geq 2$; and in the final, $|U_{PD}| = 1$. We will show that $S^H \geq 0$ in all these cases . To reduce clutter, we drop the superscript H from S_1^H, S_2^H and S^H. In what follows, let C be the subgraph of G induced by the bracing edges of all tiles in H.

We first note two easy facts about H used below:

Proposition 1.   *1. Suppose H contains a diagonal vertex. Then C is connected. Moreover, for any vertex labelled Y_{ij} in H, there exists a path p in H such that the bracing edges corresponding to the tiles in p form a path p' from i to j in C.*
2. The distance between any two diagonal vertices in H is at least $\mathrm{girth}(G)/2$.

Proof. Part (2) is sketched in the Appendix. We leave part (1) as an easy exercise.

Case 1: No Unmatched Positive Diagonal Vertices

Consider the following sum:

$$S_2' = \sum_{Y_{ij} \in U_N} \alpha_i \alpha_j - \sum_{Y_{ij} \in U_P} \alpha_i \alpha_j. \qquad (12)$$

Note that since $\alpha \in [0, 1]^{n+1}$, it follows that $-\alpha_i \alpha_j \leq 1 - \alpha_i - \alpha_j$ and $\alpha_i \alpha_j \leq \alpha_i$. So since there are no unmatched positive diagonal vertices in the tiling, $\alpha_i \alpha_j \leq \beta(i, j)$ and $-\alpha_i \alpha_j \leq \delta(i, j)$ for all unmatched vertices labelled Y_{ij} in the tiling, and hence, $S_2 \geq S_2'$. So to show $S \geq 0$ in this case, it suffices to show $S_1 + S_2' \geq 0$.

To that end, consider the following sum:

$$\sum_{\{Y_{ij_1}, \ldots, Y_{ij_k}\} \in P} (-\alpha_i \alpha_{j_1} - \ldots - \alpha_i \alpha_{j_k}) + \sum_{\{Y_{ij_1}, \ldots, Y_{ij_k}\} \in N} (\alpha_i \alpha_{j_1} + \ldots + \alpha_i \alpha_{j_k}). \quad (13)$$

By properties (2) and (3) of a tiling, it follows that (13) telescopes and equals S_2'. But then, to show that $S \geq 0$ in this case it suffices to show that for each $\{Y_{ij_1}, \ldots, Y_{ij_k}\} \in P$,

$$a(i, j_1, \ldots, j_k) - [\alpha_i(\alpha_{j_1} + \ldots + \alpha_{j_k})] = (1 - \alpha_i)(\alpha_{j_1} + \ldots + \alpha_{j_k} - 1), \quad (14)$$

is nonnegative (the first term comes from the term in S_1 for the tile, and the second from the term for the tile in (13)), and for each $\{Y_{ij_1}, \ldots, Y_{ij_k}\} \in N$,

$$b(i, j_1, \ldots, j_k) + [\alpha_i(\alpha_{j_1} + \ldots + \alpha_{j_k})] = \alpha_i(\alpha_{j_1} + \ldots + \alpha_{j_k} - 1), \qquad (15)$$

is nonnegative (again, the first term comes from S_1 and the second from (13)). But (14) and (15) are both nonnegative since the bracing edges for all tiles are in G and α satisfies the edge constraints for G. Hence, $S \geq 0$ in this case.

Case 2: At Least 2 Unmatched Positive Diagonal Vertices

We first define some notation to refer to quantities (14) and (15) which will be used again in this case: For a tile $e = \{Y_{ij_1}, \ldots, Y_{ij_k}\}$, define $\zeta(e)$ to be $(1 - \alpha_i)(\alpha_{j_1} + \ldots + \alpha_{j_k} - 1)$ if $e \in P$ and be $\alpha_i(\alpha_{j_1} + \ldots + \alpha_{j_k} - 1)$ if $e \in N$.

In Case 1 we showed that $S \geq 0$ by (1) defining a sum S'_2 such that $S \geq S_1 + S'_2$, then (2) noting that $S_1 + S'_2 = \sum_{e \in H} \zeta(e)$, and finally (3) showing that $\zeta(e) \geq 0$ for all tiles e in H. Unfortunately, in the current case, since H contains unmatched positive diagonal vertices, it is no longer true that $S \geq S_1 + S'_2$. However, it is easy to see that $S \geq (S_1 + S'_2) - \sum_{Y_{ii} \in U_{PD}} (\alpha_i - \alpha_i^2)$. In particular, $S \geq \sum_{e \in H} \zeta(e) - \sum_{Y_{ii} \in U_{PD}} (\alpha_i - \alpha_i^2)$. So since $\zeta(e) \geq 0$ for all tiles, to show that $S \geq 0$ for the current case, it suffices to show that for "many" tiles e in H, $\zeta(e)$ is sufficiently large so that $\sum_{e \in H} \zeta(e) \geq \sum_{Y_{ii} \in U_{PD}} (\alpha_i - \alpha_i^2)$.

We require the following lemma:

Lemma 3. *If there are $\ell \geq 2$ diagonal vertices in H, then there exist ℓ disjoint paths of length* $\text{girth}(G)/4 - 2$ *in H which do not involve any diagonal vertices.*

Proof. Let Q be a tree in H that spans all the diagonal vertices in H (such a tree exists since H is connected). For each diagonal vertex $u \in H$ let $B(u)$ be the tiles in Q with distance at most $\text{girth}(G)/4 - 1$ from u. Then for all diagonal vertices $u, v \in H$, the balls $B(u)$ and $B(v)$ must be disjoint: otherwise, the distance between u and v would be less than $\text{girth}(G)/2$, contradicting part (2) of Proposition 1. Since Q is a spanning tree, for each diagonal vertex $u \in H$, there exists a path of length at least $\text{girth}(G)/4 - 2$ in $B(u)$.

By the lemma, for each vertex $v \in U_{PD}$ there exists a path p_v in H of length $\text{girth}(G)/4 - 2$ such that for distinct $u, v \in U_{PD}$ the paths p_u and p_v share no tiles. We will show (provided that n is sufficiently large) that for each $v \in U_{PD}$, $\sum_{e \in p_v} \zeta(e) \geq 1$. Since $\alpha_i - \alpha_i^2 \leq \frac{1}{4}$, it will follow that $S \geq 0$, completing the proof for this case.

So fix $v \in U_{PD}$. Since $R^m \leq \text{girth}(G)/20$, p_v contains at least $\text{girth}(G)/20$ disjoint pairs of adjacent overloaded tiles (i.e., $\alpha_j = \frac{1}{k} + \gamma$ for all vertices j in the bracing edges of the two tiles in the pair). Let $e_1 = \{Y_{qr_1}, \ldots, Y_{qr_k}\} \in P$ and $e_2 = \{Y_{st_1}, \ldots, Y_{st_k}\} \in N$ be such a pair. Note that $\zeta(e_1) = (1 - \alpha_i)k\gamma$ and $\zeta(e_2) = \alpha_s k\gamma$. Now, either $s = r_\ell$ for some $\ell \in [k]$, or $s = q$. If $s = r_\ell$, then $\alpha_s = \frac{1}{k} + \gamma$ and $\zeta(e_1) + \zeta(e_2) \geq k\gamma(\frac{1}{k} + \gamma)$. If instead $s = q$, then $\zeta(e_1) + \zeta(e_2) \geq k\gamma$. In either case, $\zeta(e_1) + \zeta(e_2) \geq k\gamma(\frac{1}{k} + \gamma)$. Hence, summing over all girth$(G)/20$ disjoint pairs of adjacent overloaded tiles, it follows that $\sum_{e\in p_v} \zeta(e) \geq k\gamma(\frac{1}{k} + \gamma)\,\text{girth}(G)/20$. As desired, the latter is indeed greater than 1 for large n since girth$(G) = \Omega(\log n)$.

Case 3: Exactly 1 Unmatched Positive Diagonal Vertex

The argument in Case 2 actually rules out the possibility that the tiling contains two or more diagonal vertices, unmatched or not. Hence, we will assume that our tiling contains exactly one diagonal vertex labelled without loss of generality by Y_{11}. We consider two subcases.

Subcase 1: At least one of the unmatched vertices is a distant pair:
Arguing as in Case 2, we have in this subcase that $S = [\sum_{e\in H} \zeta(e)] - (\alpha_1 - \alpha_1^2)$ where, moreover, $\zeta(e) \geq 0$ for all tiles $e \in H$. Hence, to show that $S \geq 0$ in this subcase, it suffices to show that there exists a subset $H' \subseteq H$ such that $\sum_{e\in H'} \zeta(e) \geq \frac{1}{4} \geq \alpha_1 - \alpha_1^2$.

To that end, let v be an unmatched vertex in H labelled Y_{ij} where $\{i,j\}$ is a distant pair. By part (1) of Proposition 1 there is a path p in H such that the bracing edges corresponding to the tiles in p form a path q in C connecting vertices i and j. Since i, j are distant, q must contain, by definition, at least $4R_{m+1} + 1/\gamma$ overloaded edges. Moreover, it is not hard to see that there exists a sub-path p' of p of length at most $5R_{m+1} + 1/\gamma$ such that the bracing edges of the tiles in p' include all these overloaded edges. Hence p' must contain at least $1/\gamma$ disjoint pairs of overloaded tiles. But then, arguing as in Case 2, it follows that $\sum_{e\in p} \zeta(e) \geq (1/\gamma)(k\gamma(\frac{1}{k} + \gamma)) \geq \frac{1}{4}$, and hence that $S \geq 0$ in this subcase.

Subcase 2: None of the unmatched vertices is a distant pair:
Note first that we can assume that H contains no cycles: If it does, then it is easy to see that C must also contain a cycle, and hence, we can use the ideas from Case 2 to show that $S \geq 0$.

So assume H has no cycles and define a tree T as follows: There is a node in T for each tile in H. The root $root(T)$ corresponds to the tile containing Y_{11}. There is an edge between two nodes in T iff the tiles corresponding to the nodes share a vertex. Note that T is a tree since H is acyclic. For a node $v \in T$, let $Tile(v)$ denote its corresponding tile in H. Finally, for a node $v \in T$ we abuse notation and say $v \in P$ (resp., $v \in N$) if $Tile(v) \in P$ (resp., $Tile(v) \in N$).

Recursively define a function t on the nodes of T as follows: Let v be a node in T and suppose $Tile(v) = \{Y_{ij_1}, \dots, Y_{ij_k}\}$. If $v \in P$, then

$$t(v) = a(i, j_1, \dots, j_k) + \sum_{v' \in Child(v)} t(v') + \sum_{\text{unmatched } Y_{ij_\ell} \text{ in } Tile(v)} \delta(i, j_\ell). \quad (16)$$

If instead $v \in N$, then

$$t(v) = b(i, j_1, \dots, j_k) + \sum_{v' \in Child(v)} t(v') + \sum_{\text{unmatched } Y_{ij_\ell} \text{ in } Tile(v)} \beta(i, j_\ell). \quad (17)$$

A simple induction now shows that $t(root(T)) = S$ (recall that $root(T)$ corresponds to the tile containing Y_{11}). Hence, to show that $S \geq 0$ in this subcase it suffices to show that $t(root(T)) \geq 0$. This will follow from the following lemma:

Lemma 4. *Given* $v \in T$, *let* $e = Tile(v) = \{Y_{ij_1}, \ldots, Y_{ij_k}\}$. *Assume, without loss of generality, that* $Y_{ij_1}, \ldots, Y_{ij_\ell}$ *(where* $0 \leq \ell \leq k$*) are the labels of the vertices in* e *shared with the tile corresponding to* v*'s parent (where* $\ell = 0$ *iff* $v = root(T)$*). Then,*

1. *If* $v = root(T)$, *then* $t(v) = t(root(T)) \geq 0$; *otherwise,*
2. *If* $v \in P$, *then* $t(v) \geq \min(\alpha_i, \sum_{r=1}^{\ell} \alpha_{j_r})$;
3. *If instead* $v \in N$, *then* $t(v) \geq \min(0, 1 - \alpha_i - \sum_{r=1}^{\ell} \alpha_{j_r})$.

Proof. The proof is by induction on the size of the subtree at v. For the base case, let v be a leaf. Hence, the vertices labelled by $Y_{ij_{\ell+1}}, \ldots, Y_{ij_k}$ in e are all unmatched non-distant pair vertices (since H contains no unmatched distant pair vertices). Suppose $v \in P$. By equation (16),

$$t(v) = \alpha_i + \sum_{r=1}^{k} \alpha_r - 1 + \sum_{r=\ell+1}^{k} \delta(i, j_r). \tag{18}$$

There are two subcases to consider. In the first subcase, $j_r \neq i$ for all $r = \ell + 1, \ldots, k$. Then $\delta(i, j_r) = 1 - \alpha_i - \alpha_r$ for all $r = \ell + 1, \ldots, k$ and it follows that $t(v) = \sum_{r=1}^{\ell} \alpha_r + (k - \ell - 1)(1 - \alpha_i) \geq \sum_{r=1}^{\ell} \alpha_r$. In the second subcase, $j_r = i$ for some r, say $r = k$. This can only happen for one r since H only contains one diagonal vertex, namely Y_{11}. In particular, v must be $root(T)$. Then $\delta(i, j_k) = -\alpha_i$ and $\delta(i, j_r) = 1 - \alpha_i - \alpha_r$ for all $r = \ell + 1, \ldots, k - 1$. Since α satisfies the edge constraints for G, it follows that $t(v) = t(root(T)) \geq 0$.

Suppose instead that $v \in N$. By equation (17), $t(v) = -\alpha_i + \sum_{r=\ell+1}^{k} \beta(i, j_r)$. Now, $\beta(i, j_r) \geq \min(\alpha_i, \alpha_{j_r})$ for all $r = \ell + 1, \ldots, k$. If $\beta(i, j_r) = \alpha_i$ for some j_r, then $t(v) \geq 0$ as desired. So assume $\beta(i, j_r) = \alpha_{j_r}$ for all $r = \ell + 1, \ldots, k$. Since α satisfies the edge constraints for G, $\sum_{r=1}^{k} \alpha_{j_r} \geq 1$. But then, $t(v) \geq 1 - \alpha_i - \sum_{r=1}^{\ell} \alpha_{j_r}$, completing the proof for the base case.

For the inductive step, let v be any node in T and assume the lemma holds for all children of v. Assume first that $v \in P$ and hence, all children of v are in N (by definition of a tiling). By the induction hypothesis, there are two possibilities: either (a) $t(v') \geq 0$ for all children v' of v, and there are no unmatched vertices in $Tile(v)$, or (b) there exists some unmatched vertex in $Tile(v)$ or $t(v') \geq 1 - \alpha_i - \sum_{s=1}^{t} \alpha_{j_{r_s}}$ for some child v' of v. In case (a), it follows from (16) and from from the fact that α satisfies the edge constraints for G that $t(v) \geq \alpha_i$. In case (b), using the arguments from the base case, it follows that $t(v) \geq 0$ when $v = root(T)$, and $t(v) \geq \sum_{r=1}^{\ell} \alpha_{j_r}$ when $v \neq root(T)$.

Now assume that $v \in N$. The arguments from the above case when $v \in P$, as well as the arguments from the base case, can now be adapted to show that $t(v) \geq \min(0, 1 - \alpha_i - \sum_{r=1}^{\ell} \alpha_{j_r})$. Note that in in this case $v \neq root(T)$ since $root(T) \in P$. The inductive step, and hence the lemma, now follow.

So $S \geq 0$ in Case 3 also, and the Inductive Claim now follows for $m + 1$.

4 Discussion

The integrality gap of $2 - \epsilon$ obtained in [2] for graph VERTEX COVER holds for $\Omega(\sqrt{\log n})$ rounds of LS (improved to $\Omega(\log n)$ rounds in [3]). We conjecture that our integrality gaps in the hypergraph case should also hold for at least $\Omega(\log n)$ rounds. Indeed, examining the proof of Theorem 3, it can be shown that by redefining the recursive definition of R_m to be $R_m = R_{m+1} + 1/\gamma$, then all cases considered in the proof except Subcase 1 of Case 3 can be argued for $\Omega(\log n)$ rounds. While it can be argued that $S \geq 0$ for $\Omega(\sqrt{\log n})$ rounds for graphs in this subcase (and in fact for $\Omega(\log n)$ rounds if the definition of (S, R, γ)-vectors is further refined as in [3]), a proof for hypergraphs eludes us.

The integrality gaps of $k - 1 - \epsilon$ given in [1] for k-uniform hypergraph VERTEX COVER held not only for LS but also for LS_+ liftings, the stronger semidefinite version of Lovász-Schrijver liftings. As mentioned in the introduction, obtaining optimal integrality gaps for both graph and hypergraph VERTEX COVER for LS_+ remains a difficult open question.

Finally, can our techniques be applied to other problems? For instance, could our techniques be used to show that, say, even after $\log \log n$ rounds of LS the integrality gap of the linear relaxation for MAX-CUT remains larger than the approximation factor attained by the celebrated SDP-based Goemans-Williamson algorithm [11]?

Acknowledgements

I would like to thank Sanjeev Arora and Elad Hazan for many helpful discussions. I would also like to thank the anonymous referees for helpful comments.

References

1. M. Alekhnovich, S. Arora, and I. Tourlakis. Towards strong nonapproximability results in the Lovász-Schrijver hierarchy. In *Proceedings of the 37th Annual Symposium on the Theory of Computing*, New York, May 2005. ACM Press.
2. S. Arora, B. Bollobás, and L. Lovász. Proving integrality gaps without knowing the linear program. In *Proceedings of the 43rd Symposium on Foundations of Computer Science (FOCS-02)*, pages 313–322, Los Alamitos, Nov. 16–19 2002.
3. S. Arora, B. Bollobás, L. Lovász, and I. Tourlakis. Proving integrality gaps without knowing the linear program. Manuscript, 2005.
4. S. Arora, S. Rao, and U. Vazirani. Expander flows, geometric embeddings and graph partitioning. In *Proceedings of the thirty-sixth annual ACM Symposium on Theory of Computing (STOC-04)*, pages 222–231, New York, June 13–15 2004. ACM Press.
5. J. Buresh-Oppenheim, N. Galesi, S. Hoory, A. Magen, and T. Pitassi. Rank bounds and integrality gaps for cutting planes procedures. In *FOCS: IEEE Symposium on Foundations of Computer Science (FOCS)*, 2003.
6. W. Cook and S. Dash. On the matrix-cut rank of polyhedra. *Mathematics of Operations Research*, 26(1):19–30, 2001.

7. I. Dinur, V. Guruswami, S. Khot, and O. Regev. A new multilayered PCP and the hardness of hypergraph vertex cover. In ACM, editor, *Proceedings of the Thirty-Fifth ACM Symposium on Theory of Computing, San Diego, CA, USA, June 9–11, 2003*, pages 595–601, New York, NY, USA, 2003. ACM Press.
8. P. Erdős. Graph theory and probability. *Canadian Journal of Mathematics*, 11:34–38, 1959.
9. U. Feige and R. Krauthgamer. The probable value of the Lovász–Schrijver relaxations for maximum independent set. *SIAM Journal on Computing*, 32(2):345–370, Apr. 2003.
10. M. X. Goemans and L. Tunçel. When does the positive semidefiniteness constraint help in lifting procedures. *Mathematics of Operations Research*, 26:796–815, 2001.
11. M. X. Goemans and D. P. Williamson. .878-approximation algorithms for MAX CUT and MAX 2SAT. In *Proceedings of the Twenty-Sixth Annual ACM Symposium on the Theory of Computing*, pages 422–431, Montréal, Québec, Canada, 23–25 May 1994.
12. L. Lovász and A. Schrijver. Cones of matrices and set-functions and 0-1 optimization. *SIAM Journal on Optimization*, 1(2):166–190, May 1991.
13. H. D. Sherali and W. P. Adams. A hierarchy of relaxations between the continuous and convex hull representations for zero-one programming problems. *SIAM Journal on Discrete Mathematics*, 3(3):411–430, Aug. 1990.
14. T. Stephen and L. Tunçel. On a representation of the matching polytope via semidefinite liftings. *Mathematics of Operations Research*, 24(1):1–7, 1999.

A   Proof Sketch of Part (2) of Proposition 1

Since H is connected, there exists a path between any two diagonal vertices in H. We will show that the subgraph of G induced by the bracing edges of the tiles in such path must contain a cycle. Part (2) will then follow.

To that end, let q be an arbitrary path in H comprised, in order, of tiles $e_1, \ldots, e_r$. Consider these tiles in order beginning with e_1. As long as the bracing node in successive tiles does not change, then the bracing edges of these tiles form a path q' in G. If the bracing node changes at some tile, say e_i, then the bracing edge for e_i starts a new path q'' in G. Moreover, some vertex $w \in G$ from the last edge visited in q' becomes the (new) bracing node for e_i. The bracing edges of the tiles following e_i now extend q'' until an edge e_j is encountered with yet a different bracing node. But then, the bracing edge for e_j must contain w. Hence the bracing edge for e_j must extend q'. Continuing this argument, we see that each time the bracing node switches, the tiles switch back and forth from contributing to q' or q''.

Each time the current tile e_ℓ contains some diagonal vertex labelled Y_{ii}, the bracing node i for e_ℓ is also contained in the bracing edge for e_ℓ. But then, since the bracing node belongs to, say, q' while the bracing edge belongs to q'', it follows that q' and q'' must intersect at vertex i in G. Hence, if a tile path q contains *two* diagonal vertices, q' and q'' must intersect twice. Part (2) follows.

Bounds for Error Reduction with Few Quantum Queries

Sourav Chakraborty[1], Jaikumar Radhakrishnan[2,3], and Nandakumar Raghunathan[1]

[1] Department of Computer Science
University of Chicago, Chicago
IL 60637, USA
{sourav,nanda}@cs.uchicago.edu
[2] Toyota Technological Institute at Chicago
IL 60637, USA
jaikumar@tti-c.org
[3] School of Technology and Computer Science,
Tata Institute of Fundamental Research, Mumbai 400005, India

Abstract. We consider the quantum database search problem, where we are given a function $f : [N] \rightarrow \{0, 1\}$, and are required to return an $x \in [N]$ (a target address) such that $f(x) = 1$. Recently, Grover [G05] showed that there is an algorithm that after making one quantum query to the database, returns an $X \in [N]$ (a random variable) such that

$$\Pr[f(X) = 0] = \epsilon^3,$$

where $\epsilon = |f^{-1}(0)|/N$. Using the same idea, Grover derived a t-query quantum algorithm (for infinitely many t) that errs with probability only ϵ^{2t+1}. Subsequently, Tulsi, Grover and Patel [TGP05] showed, using a different algorithm, that such a reduction can be achieved for all t. This method can be placed in a more general framework, where given any algorithm that produces a target state for some database f with probability of error ϵ, one can obtain another that makes t queries to f, and errs with probability ϵ^{2t+1}. For this method to work, we do not require prior knowledge of ϵ. Note that no classical randomized algorithm can reduce the error probability to significantly below ϵ^{t+1}, even if ϵ is known. In this paper, we obtain *lower bounds* that show that the amplification achieved by these quantum algorithms is essentially optimal. We also present simple alternative algorithms that achieve the same bound as those in Grover [G05], and have some other desirable properties. We then study the best reduction in error that can be achieved by a t-query quantum algorithm, when the initial error ϵ is known to lie in an interval of the form $[\ell, u]$. We generalize our basic algorithms and lower bounds, and obtain nearly tight bounds in this setting.

1   Introduction

In this paper, we consider the problem of reducing the error in quantum search algorithms by making a small number of queries to the database. Error reduction in the form of amplitude amplification is one of the central tools in the design of efficient quantum search algorithms [G98a, G98b, BH+02]. In fact, Grover's database search algorithm [G96, G97] can be thought of as amplitude amplification applied to the trivial

C. Chekuri et al. (Eds.): APPROX and RANDOM 2005, LNCS 3624, pp. 245–256, 2005.

algorithm that queries the database at a random location and succeeds with probability at least $\frac{1}{N}$. The key feature of quantum amplitude amplification is that it can boost the success probability from a small quantity δ to a constant in $O(1/\sqrt{\delta})$ steps, whereas, in general a classical algorithm for this would require $\Omega(1/\delta)$ steps. This basic algorithm has been refined, taking into account the number of solutions and the desired final success probability $1 - \epsilon$. For example, Buhrman, Cleve, de Wolf and Zalka [BC+99] obtained the following:

Theorem [BC+99]: Fix $\eta \in (0, 1)$, and let $N > 0$, $\epsilon \geq 2^{-N}$, and $t \leq \eta N$. Let T be the optimal number of queries a quantum computer needs to search with error $\leq \epsilon$ through an unordered list of N items containing at least t solutions. Then $\log 1/\epsilon \in \Theta(T^2/N + T\sqrt{t/N})$ (Note that the constant implicit in the Θ notation can depend on η).

Recently, Grover [G05] considered error reduction for algorithms that err with small probability. The results were subsequently refined and extended by Tulsi, Grover and Patel [TGP05]. Let us describe their results in the setting of the database search problem, where, given a database $f : [N] \rightarrow \{0, 1\}$, we are asked to determine an $x \in f^{-1}(1)$. If $|f^{-1}(0)| = \epsilon N$, then choosing x uniformly at random will meet the requirements with probability at least $1 - \epsilon$. This method makes no queries to the database. If one is allowed one classical query, the error can be reduced to ϵ^2 and, in general, with t classical queries one can reduce the probability of error to ϵ^{t+1}. It can be shown that no classical t-query randomized algorithm for the problem can reduce the probability of error significantly below ϵ^{t+1}, even if the value of ϵ is known in advance. Grover [G05] presented an interesting algorithm that makes one quantum query and returns an x that is in $f^{-1}(1)$ with probability $1 - \epsilon^3$. Tulsi, Grover and Patel [TGP05] showed an iteration where one makes just one query to the database and performs a measurement, so that after t iterations of this operator the error is reduced to ϵ^{2t+1}. This algorithm works for all ϵ and is not based on knowing ϵ in advance. Thus this iteration can be said to exhibit a "fixed point" behavior [G05, TGP05], in that the state approaches the target state (or subspace) closer with each iteration, just as it does in randomized classical search. The iteration used in the usual Grover search algorithm [G98a, G98b] does not have this property. Note, however, that if the initial success probability is $\frac{1}{N}$, these new algorithms make $\Omega(N)$ queries to the database, whereas the original algorithm makes just $O(\sqrt{N})$ queries.

In [G05], the database is assumed to be presented by means of an oracle of the form $|x\rangle \rightarrow \exp(f(x)\pi i/3))|x\rangle$. The standard oracle for a function f used in earlier works on quantum search is $|x\rangle|b\rangle \mapsto |x\rangle|b \oplus f(x)\rangle$, where $\oplus$ is addition modulo two. It can be shown that the oracle assumed in [G05] cannot be implemented by using just one query to the standard oracle. In Tulsi, Grover and Patel [TGP05] the basic iteration uses the controlled version of the oracle, namely $|x\rangle|b\rangle|c\rangle \mapsto |x\rangle|b \oplus c \cdot f(x)\rangle|c\rangle$.

In this paper, we present a version of the algorithm that achieves the same reduction in error as in [G05], but uses the standard oracle. In fact, our basic one-query algorithm has the following natural interpretation. First, we note that for $\delta \leq \frac{3}{4}$, there is a one-query quantum algorithm $\mathcal{A}_\delta$ that makes no error if $|f^{-1}(0)| = \delta N$. Then, using a simple computation, one can show that the one-query algorithm corresponding to $\delta = \frac{1}{N}$ errs with probability less than ϵ^3 when $|f^{-1}(0)| = \epsilon N$. One can place these algorithms in a more general framework, just as later works due to Grover [G98a, G98b] and Brassard, Hoyer, Mosca and Tapp [BH+02] placed Grover's original database search algorithm [G96, G97] in the general amplitude amplification framework. The framework is as follows: Suppose there is an algorithm G that guesses a solution to a problem along with a witness, which can be checked by another algorithm T. If the guess returned by G is correct with probability $1 - \epsilon$, then there is another algorithm that uses G, G^{-1}, makes t queries to T, and guesses correctly with probability $1 - \epsilon^{2t+1}$.

These algorithms show that, in general, a t-query quantum algorithm can match the error reduction obtainable by any $2t$-query randomized algorithm. Can one do even better? The main contribution of this paper are the *lower bounds* on the error probability of t-query algorithms. We show that the amplification achieved by these algorithms is essentially optimal (see Section 2.1 for the precise statement). Our result does not follow immediately from the result of Buhrman, Cleve, de Wolf and Zalka [BC+99] cited above because of the constants implicit in the θ notation, but with a slight modification of their proofs one can derive a result similar to ours (see Section 2.1). Our lower bound result uses the polynomial method of Beals, Cleve, Buhrman, Mosca and de Wolf [BB+95] combined with an elementary analysis based on the roots of low degree polynomials, but unlike previous proofs using this method, we do not rely on any special tools for bounding the rate of growth of low degree polynomials.

2   Background, Definitions and Results

We first review the standard framework for quantum search. We assume that the reader is familiar with the basics of quantum circuits, especially the quantum database search algorithm of Grover [G96, G97] (see, for example, Nielsen and Chuang [NC, Chapter 6]). The database is modelled as a function $f : [N] \to S$, where $[N] \overset{\Delta}{=} \{0, 1, 2, \ldots, N-1\}$ is the set of addresses and S is the set of possible items to be stored in the database. For our purposes, we can take S to be $\{0, 1\}$. When thinking of bits we identify $[N]$ with $\{0, 1\}^n$. Elements of $[N]$ will be called addresses, and addresses in $f^{-1}(1)$ will be referred to as targets. In the quantum model, the database is provided to us by means of an oracle unitary transformation T_f, which acts on an $(n + 1)$-qubit space by sending the basis vector $|x\rangle|b\rangle$ to $|x\rangle|b \oplus f(x)\rangle$. For a quantum circuit $\mathcal{A}$ that makes queries to a database oracle in order to determine a target, we denote by $\mathcal{A}(f)$ the random variable (taking values in $[N]$) returned by $\mathcal{A}$ when the database oracle is T_f.

Definition 1. *Let $\mathcal{A}$ be a quantum circuit for searching databases of size N. For a database f of size N, let $\mathrm{err}_{\mathcal{A}}(f) = \Pr[\mathcal{A}(f)$ is not a target state]. When ϵN is an integer in $\{0, 1, 2, \ldots, N\}$, let $\mathrm{err}_{\mathcal{A}}(\epsilon) = \max\limits_{f:|f^{-1}(0)|=\epsilon N} \mathrm{err}_{\mathcal{A}}(f)$.*

Using this notation, we can state Grover's result as follows.

Theorem 1 (Grover [G05]). *For all N, there is a one-query algorithm $\mathcal{A}$, such that for all ϵ (assuming ϵN is an integer), $\mathrm{err}_{\mathcal{A}}(\epsilon) = \epsilon^3$.*

This error reduction works in a more general setting. Let $[N]$ represent the set of possible solutions to some problem, and let $f : [N] \to \{0, 1\}$ be the function that checks that the solution is correct; as before we will assume that we are provided access to this function via the oracle T_f. Let G be a unitary transform that guesses a solution in $[N]$ that is correct with probability $1 - \epsilon$. Our goal is to devise another guessing algorithm $\mathcal{B}$ that using T_f, G and G^{-1} produces a guess that is correct with significantly better probability. Let $\mathcal{B}(T_f, G)$ be the answer returned by $\mathcal{B}$ when the checker is T_f and the guesser is G.

Theorem 2 (Grover [G05]). *There is an algorithm $\mathcal{B}$ that uses T_f once, G twice and G^{-1} once, such that $\Pr[f(\mathcal{B}(T_f, G)) = 0] = \epsilon^3$, where ϵ is the probability of error of the guessing algorithm G.*

Note that Theorem 1 follows from Theorem 2 by taking G to be the Hadamard transformation, which produces the uniform superposition on all N states when applied to the state $|0\rangle$.

Theorem 3 ([G05, TGP05]). *For all $t \geq 0$ and all N, there is a t-query quantum database search algorithm such that, for all ϵ (ϵN is an integer), $\mathrm{err}_A(\epsilon) = \epsilon^{2t+1}$.*

In Grover [G05], this result was obtained by recursive application of Theorem 2, and worked only for infinitely many t. Tulsi, Grover and Patel [TGP05] rederived Theorems 1 and 2 using a different one-query algorithm, which could be applied iteratively to get Theorem 3.

From now on when we consider error reduction for searching a database f and use the notation $\mathrm{err}_A(\epsilon)$, ϵ will refer to $|f^{-1}(0)|/N$; in particular, we assume that $\epsilon N \in \{0, 1, \ldots, N - 1\}$. However, for the general framework, ϵ can be any real number in $[0, 1]$.

2.1 Our Contributions

As stated earlier, in order to derive the above results, Grover [G05] and Tulsi, Grover and Patel [TGP05] assume that access to the database is available using certain special types of oracles. In the next section, we describe alternative algorithms that establish Theorem 1 while using only the standard oracle $T_f : |x\rangle|b\rangle \rightarrow |x\rangle|b \oplus f(x)\rangle$. The same idea can be used to obtain results analogous to Theorems 2. By recursively applying this algorithm we can derive a version of Theorem 3 for t of the form $\frac{3^i-1}{2}$ where i is the number of recursive applications. Our algorithms and those in Tulsi, Grover and Patel [TGP05] use similar ideas, but were obtained independently of each other.

We also consider error reduction when we are given a lower bound on the error probability ϵ, and obtain analogs of Theorems 1 and 2 in this setting.

Theorem 4 (Upper bound result).

(a) For all N and $\delta \in [0, \frac{3}{4}]$, there is a one-query algorithm A_δ such that for all $\epsilon \geq \delta$,

$$\mathrm{err}_{A_\delta}(f) \leq \epsilon \left[\frac{\epsilon - \delta}{1 - \delta}\right]^2.$$

(b) For all $\delta \in [0, \frac{3}{4}]$, there is an algorithm $\mathcal{B}_\delta$ that uses T_f once and G twice and G^{-1} once, such that

$$\mathrm{err}_{\mathcal{B}_\delta}(T_f, G) \leq \epsilon \left[\frac{\epsilon - \delta}{1 - \delta} \right]^2 .$$

The case $\epsilon = \delta$ corresponds to the fact that one can determine the target state with certainty if $|f^{-1}(0)|$ is known exactly and is at most $\frac{3N}{4}$. Furthermore, Theorems 1 and 2 can be thought of as special cases of the above Proposition corresponding to $\delta = 0$. In fact, by taking $\delta = \frac{1}{N}$ in the above proposition, we obtain the following slight improvement over Theorem 1.

Corollary 1. *For all N, there is a one-query database search algorithm $\mathcal{A}$ such that for all ϵ (where $\epsilon N \in \{0, 1, \ldots, N\}$), we have $\mathrm{err}_{\mathcal{A}}(\epsilon) \leq \epsilon \left[\frac{\epsilon - \frac{1}{N}}{1 - \frac{1}{N}} \right]^2 .$*

Lower bounds: The main contribution of this work is our lower bound results. We show that the reduction in error obtained in Theorem 1 and 2 are essentially optimal.

Theorem 5 (Lower bound result). *Let $0 < \ell \leq u < 1$ be such that ℓN and uN are integers.*

(a) For all one-query database search algorithms $\mathcal{A}$, for either $\epsilon = \ell$ or $\epsilon = u$,

$$\mathrm{err}_{\mathcal{A}}(\epsilon) \geq \epsilon^3 \left(\frac{u - \ell}{u + \ell - 2\ell u} \right)^2 .$$

(b) For all t-query database search algorithms $\mathcal{A}$, there is an $\epsilon \in [\ell, u]$ such that ϵN is an integer, and

$$\mathrm{err}_{\mathcal{A}}(\epsilon) \geq \epsilon^{2t+1} \left(\frac{b-1}{b+1} - \frac{1}{N\ell(b+1)} \right)^{2t} ,$$

where $b = \left(\frac{u}{\ell} \right)^{\frac{1}{t+1}}$, and we assume that $N\ell(b-1) > 1$.

In particular, this result shows that to achieve the same reduction in error, a quantum algorithm needs to make roughly at least half as many queries as a classical randomized algorithm. A similar result can be obtained by modifying the proof in Buhrman, Cleve, de Wolf and Zaka [BC+99]: there is a constant $c > 0$, such that for all $u > 0$, all large enough N and all t-query quantum search algorithms for databases of size N, there is an $\epsilon \in (0, u]$ (ϵN is an integer) such that $\mathrm{err}_{\mathcal{A}}(\epsilon) \geq (cu)^{2t+1}$.

3   Upper Bounds: Quantum Algorithms

In this section, we present algorithms that justify Theorems 1, 2 and 3, but by using the standard database oracle. We then modify these algorithms to generalize and slightly improve these theorems.

3.1   Alternative Algorithms Using the Standard Oracle

We first describe an alternative algorithm $\mathcal{A}_0$ to justify Theorem 1. This simple algorithm (see Figure 1) illustrates the main idea used in all our upper bounds. We will work with n qubits corresponding to addresses in $[N]$ and one ancilla qubit. Although we do not simulate Grover's oracle directly, using the ancilla, we can reproduce the effect the complex amplitude used there by real amplitudes. As we will see, the resulting algorithm has an intuitive explanation and also some additional properties not enjoyed by the original algorithm.

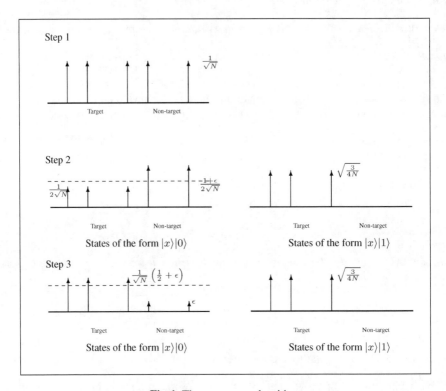

Fig. 1. The one-query algorithm

Step 1: We start with the uniform superposition on $[N]$ with the ancilla bit in the state $|0\rangle$.

Step 2: For targets x, transform $|x\rangle|0\rangle$ to $\frac{1}{2}|x\rangle|0\rangle + \sqrt{\frac{3}{4}}|x\rangle|1\rangle$. The basis states $|x\rangle|0\rangle$ for $x \in f^{-1}(0)$, are not affected by this transformation.

Step 3: Perform an inversion about the average controlled on the ancilla bit being $|0\rangle$, and then measure the address registers.

Step 1 is straightforward to implement using the n-qubit Hadamard transform H_n. For Step 2, using one-query to T_f, we implement a unitary transformation U_f, which maps

$|x\rangle|0\rangle$ to $|x\rangle|0\rangle$ if $f(x) = 0$ and to $|x\rangle\left(\frac{1}{2}|0\rangle + \sqrt{\frac{3}{4}}|1\rangle\right)$, if $f(x) = 1$. One such transformation is $U_f = (I_n \otimes R^{-1})T_f(I_n \otimes R)$, where I_n is the n-qubit identity operator and R is the one-qubit gate for rotation by $\frac{\pi}{12}$ (that is, $|0\rangle \xrightarrow{R} \cos(\frac{\pi}{12})|0\rangle + \sin(\frac{\pi}{12})|1\rangle$ and $|1\rangle \xrightarrow{R} \cos(\frac{\pi}{12})|1\rangle - \sin(\frac{\pi}{12})|0\rangle$). The inversion about the average is usually implemented as $A_n = -H_n(I_n - 2|0\rangle\langle0|)H_n$. The controlled version we need is then given by

$$A_{n,0} = A_n \otimes |0\rangle\langle0| + I_n \otimes |1\rangle\langle1|.$$

Let $H' = H_n \otimes I$. The final state is $|\phi_f\rangle = A_{n,0}U_fH'|0\rangle|0\rangle$.

To see that the algorithm works as claimed, consider the state just before the operator $A_{n,0}$ is applied. This state is

$$\frac{1}{\sqrt{N}}\left[\sum_{x\in f^{-1}(1)}\frac{1}{2}|x\rangle|0\rangle \sum_{x\in f^{-1}(0)}|x\rangle|0\rangle\right] + \frac{1}{\sqrt{N}}\sum_{x\in f^{-1}(1)}\sqrt{\frac{3}{4}}|x\rangle|1\rangle.$$

Suppose $|f^{-1}(0)| = \epsilon N$. The "inversion about the average" is performed only on the first term, so the non-target states receive no amplitude from the second term. The average amplitude of the states in the first term is $\frac{1}{2\sqrt{N}}(1+\epsilon)$ and the amplitude of the states $|x\rangle|0\rangle$ for $x \in f^{-1}(0)$ is $\frac{1}{\sqrt{N}}$. Thus, after the inversion about the average the amplitude of $|x\rangle|0\rangle$ for $x \in f^{-1}(0)$ is $\frac{\epsilon}{\sqrt{N}}$. It follows that if we measure the address registers in the state $|\phi_f\rangle$, the probability of observing a non-target state is exactly

$$|f^{-1}(0)| \cdot \frac{\epsilon^2}{N} = \epsilon^3.$$

Remark: Note that this algorithm actually achieves more. Suppose we measure the ancilla bit in $|\phi_f\rangle$, and find a 1. Then, we are assured that we will find a target on measuring the address registers. Furthermore, the probability of the ancilla bit being 1 is exactly $\frac{3}{4}(1 - \epsilon)$. One should compare this with the randomized one-query algorithm that with probability $1 - \epsilon$ provides a guarantee that the solution it returns is correct. The algorithm in [G05] has no such guarantee associated with its solution. However, the algorithm obtained by Tulsi, Grover and Patel [TGP05] gives a guarantee with probability $\frac{1}{2}(1 - \epsilon)$.

The general algorithm $\mathcal{B}_0$ needed to justify Theorem 2 is similar. We use G instead of H_n; that is, we let $H' = G \otimes I$, $A_n = G(2|0\rangle\langle0| - I_n)G^{-1}$, and, as before,

$$A_{n,0} = A_n \otimes |0\rangle\langle0| + I_n \otimes |1\rangle\langle1|.$$

The final state is obtained in the same way as before $|\phi_f\rangle = A_{n,0}U_fH'|0\rangle|0\rangle$.

Remark: As stated, we require the controlled version of G and G^{-1} to implement $A_{n,0}$. However, we can implement G with the uncontrolled versions themselves from the following alternative expression for $A_{n,0}$:

$$A_{n,0} = (G \otimes I)[(2|0\rangle\langle0| - I_n) \otimes |0\rangle\langle0| + I_n \otimes |1\rangle\langle1|](G^{-1} \otimes I).$$

We can estimate the error probability of this algorithm using the following standard calculation. Suppose the probability of obtaining a non-target state on measuring the address registers in the state $G|0\rangle$ is ϵ. Let us formally verify that the probability of obtaining a non-target state on measuring the address registers in the state $|\phi_f\rangle$ is ϵ^3. This follows using the following routine calculation. We write

$$G|\mathbf{0}\rangle = \alpha|t\rangle + \beta|t'\rangle,$$

where $|t\rangle$ is a unit vector in the "target space" spanned by $\{|x\rangle : f(x) = 1\}$, and $|t'\rangle$ is a unit vector in the orthogonal complement of the target space. By scaling $|t\rangle$ and $|t'\rangle$ by suitable phase factors, we can assume that α and β are real numbers. Furthermore $\beta^2 = \epsilon$. The state after the application of U_f is then given by

$$\left(\frac{\alpha}{2}|t\rangle + \beta|t'\rangle\right)|0\rangle + \sqrt{\frac{3}{4}}\alpha|t\rangle|1\rangle. \tag{1}$$

Now, the second term is not affected by $A_{n,0}$, so the amplitude of states in the subspace of non-target states is derived entirely from the first term, which we denote by $|u\rangle$. To analyze this contribution we write $|u\rangle$, using the basis

$$|v\rangle = \alpha|t\rangle + \beta|t'\rangle; \tag{2}$$
$$|v'\rangle = \beta|t\rangle - \alpha|t'\rangle. \tag{3}$$

That is, $|u\rangle = \left(\frac{\alpha^2}{2} + \beta^2\right)|v\rangle|0\rangle - \frac{\alpha\beta}{2}|v'\rangle|0\rangle$.

Since $A_{n,0}|v\rangle = |v\rangle$ and $A_{n,0}|v'\rangle = -|v'\rangle$, we have

$$A_{n,0}|u\rangle = \left(\frac{\alpha^2}{2} + \beta^2\right)|v\rangle|0\rangle + \frac{\alpha\beta}{2}|v'\rangle|0\rangle.$$

Returning to the basis $|t\rangle$ and $|t'\rangle$ (using (2) and (3)), we see that the amplitude associated with $|t'\rangle$ in this state is β^3. Thus, the probability that the final measurement fails to deliver a target address is exactly $\beta^6 = \epsilon^3$.

Remark: The algorithm $\mathcal{B}_0$ can be used recursively to get a t-query algorithm that achieves the bound Theorem 3. Just as in the one-query algorithm, by measuring the ancilla bits we can obtain a guarantee; this time the solution is accompanied with guarantee with probability at least $(1 - \frac{1}{t} - \frac{6\log t}{t(\log\frac{1}{\epsilon})^{\log_3 4}})$. The t-query algorithm obtained by Tulsi, Grover and Patel [TGP05] has significantly better guarantees: it certifies that its answer is correct with probability at least $1 - \epsilon^{2t}$.

3.2   Algorithms with Restrictions on ϵ

As stated above, for each $\delta \in [0, 1]$, there is a one-query quantum algorithm $\mathcal{A}_\delta$ that makes no error if the $|f^{-1}(0)| = \delta N$ (or, in the general setting, if G is known to err with probability at most $\frac{3}{4}$). Let us explicitly obtain such an algorithm $\mathcal{A}_\delta$ by slightly modifying the algorithm above. The idea is to ensure that the inversion about the average performed in Step 3 reduces the amplitude of the non-target states to zero. For this, we only need to replace U_f by $U_{f,\delta}$, which maps $|x\rangle|0\rangle$ to $|x\rangle|0\rangle$ if $f(x) = 0$ and to $|x\rangle(\alpha|0\rangle + \beta|1\rangle)$, if $f(x) = 1$, where $\alpha = \dfrac{1 - 2\delta}{2(1 - \delta)}$ and $\beta = \sqrt{1 - \alpha^2}$.

Also, one can modify the implementation of U_f above, replacing $\frac{\pi}{12}$ by $\frac{\sin^{-1}(\alpha)}{2}$ (note that $\delta \leq \frac{3}{4}$ implies that $|\alpha| \leq 1$), and implement $U_{f,\delta}$ using just one-query to T_f.

Proposition 1. *Let* $|f^{-1}(0)| = \delta \leq \frac{3}{4}$. *Then,* $\mathrm{err}_{\mathcal{A}_\delta}(f) = 0$.

An analogous modification for the general search gives us an algorithm $\mathcal{B}_\delta(T, G)$ that has no error when G produces a target state for T with probability exactly $1 - \delta$. We next observe that the algorithms $\mathcal{A}_\delta$ and $\mathcal{B}_\delta$ perform well not only when the original probability is known to be δ but also if the original probability is $\epsilon \geq \delta$. This justifies Theorem 4 claimed above.

Proof of Theorem 4: We will only sketch the calculations for part (a). The average amplitude of all the states of the form $|x\rangle|0\rangle$ is $(\frac{1}{\sqrt{N}})(1 - 2\delta + \epsilon)/(2(1 - \delta))$. From this it follows that the amplitude of a non-target state after the inversion about the average is $(\frac{1}{\sqrt{N}})(\epsilon - \delta)/(1 - \delta)$. Our claim follows from this by observing that there are exactly ϵN non-target states.   □

4   Lower Bounds

In this section, we show that the algorithms in the previous section are essentially optimal. For the rest of this section, we fix a t-query quantum search algorithm to search a database of size N. Using the polynomial method we will show that no such algorithm can have error probability significantly less than ϵ^{t+1}, for a large range of ϵ.

The proof has two parts. First, using standard arguments we observe that $\mathrm{err}_{\mathcal{A}}(\epsilon)$ is a polynomial of degree at most $2t + 1$ in ϵ.

Lemma 1. *Let $\mathcal{A}$ be a t-query quantum search algorithm for databases of size N. Then, there is a univariate polynomial $r(Z)$ with real coefficients and degree at most $2t + 1$, such that for all ϵ*

$$\mathrm{err}_{\mathcal{A}}(\epsilon) \geq r(\epsilon).$$

Furthermore, $r(x) \geq 0$ for all $x \in [0, 1]$.

In the second part, we analyze such low degree polynomials to obtain our lower bounds. We present this analysis first, and return to the proof of Lemma 1 after that.

4.1   Analysis of Low Degree Polynomials

Definition 2 (Error polynomial). *We say that a univariate polynomial $r(Z)$ is an error polynomial if (a) $r(z) \geq 0$ for all $z \in [0, 1]$, (b) $r(0) = 0$, and (c) $r(1) = 1$.*

Our goal is to show that an error polynomial of degree at most $2t+1$ cannot evaluate to significantly less than ϵ^{2t+1} for many values of ϵ. For our calculations, it will be convenient to ensure that all the roots of such a polynomial are in the interval $[0, 1)$.

Lemma 2. *Let $r(Z)$ an error polynomial of degree $2t + 1$ with $k < 2t + 1$ roots in the interval $[0, 1)$. Then, there is another error polynomial $q(Z)$ of degree at most $2t + 1$ such that $q(z) \leq r(z)$ for all $z \in [0, 1]$, and $q(Z)$ has at least $k + 1$ roots in the interval $[0, 1)$.*

Proof. Let $\alpha_1, \alpha_2, \ldots, \alpha_k$ be the roots of $r(x)$ in the interval $[0, 1)$. Hence we can write

$$r(Z) = \prod_{i=1}^{k}(Z - \alpha_i)r'(Z),$$

where $r'(Z)$ does not have any roots in $[0, 1)$. Now, by substituting $Z = 1$, we conclude that $r'(1) \geq 1$. Since $r'(Z)$ does not have any roots in $[0, 1)$, it follows that $r'(z) > 0$ for all $z \in [0, 1)$.

The idea now, is to subtract a suitable multiple of the polynomial $1 - Z$ from $r'(Z)$ and obtain another polynomial $r''(Z)$ which has a root in $[0, 1)$. Since $1 - Z$ is positive in $[0, 1)$, $r''(Z)$ is at most $r'(Z)$ in this interval. The polynomial $q(Z)$ will be defined by $q(Z) = \prod_{\alpha \in R}(Z - \alpha)r''(Z)$. To determine the multiple of $1 - Z$ we need to subtract, consider $\lambda(c) = \min_{z \in [0,1)} r'(Z) - c(1 - Z)$. Since $\lambda(c)$ is continuous, $\lambda(0) > 0$ and $\lambda(c) < 0$ for large enough c, it follows that $\lambda(c_0) = 0$ for some $c_0 > 0$. Now, let $r''(Z) = r(Z) - c_0(1 - Z)$. □

By repeatedly applying Lemma 2 we obtain the following.

Lemma 3. *Let $r(Z)$ be an error polynomial of degree at most $2t + 1$. Then, there is an error polynomial $q(Z)$ of degree exactly $2t + 1$ such that $q(z) \leq r(z)$ for all $z \in [0, 1]$, and $q(Z)$ has $2t + 1$ roots in the interval $[0, 1)$.*

We can now present the proof of Theorem 5, our main lower bound result.

Proof of Theorem 5: Consider the case $t = 1$. By Lemma 1, it is enough to show that an error polynomial $r(Z)$ of degree at most three is bounded below as claimed. By Lemma 3, we may assume that all three roots of $r(Z)$ lie in $[0, 1)$. Since $r(0) = 0$ and $r(z) \geq 0$ in $[0, 1)$, we may write $r(Z) = aZ(Z - \alpha)^2$ for some $\alpha \in [0, 1)$ and some positive a; since $r(1) = 1$, we conclude that $a = \frac{1}{(1-\alpha)^2}$. Thus, we need to determine the value of α so that $t(\alpha) = \max_{x \in \{\ell, u\}} \frac{r(x)}{x^3}$ is as small as possible. Consider the function $t_x(\alpha) = \frac{r(x)}{x^3} = \left(\frac{x - \alpha}{(1-\alpha)x}\right)^2$. Note that for all x, $t_x(\alpha)$ is monotonically increasing in $|x - \alpha|$. It follows that $t(\alpha)$ is minimum for some $\alpha \in [\ell, u]$. For α in this interval $t_\ell(\alpha)$ is an increasing function of α and $t_u(\alpha)$ is a decreasing function of α. So $t(\alpha)$ is minimum when $t_\ell(\alpha) = t_u(\alpha)$. It can be checked by direct computation that when $\alpha = \frac{2\ell u}{\ell + u}$,

$$t_\ell(\alpha) = t_u(\alpha) = \left(\frac{u - \ell}{u + \ell - 2\ell u}\right)^2.$$

This establishes part (a) of Theorem 5.

To establish part (b), we show that an error polynomial of degree at most $2t + 1$ satisfies the claim. As before, by Lemma 3, we may assume that $r(Z)$ has all its roots in $[0, 1)$. Furthermore, since $r(Z) \geq 0$, we conclude that all roots in $(0, 1)$ have even multiplicity. Thus we may write

$$r(Z) = \frac{Z(Z - \alpha_1)^2(Z - \alpha_2)^2 \cdots (Z - \alpha_t)^2}{(1 - \alpha_1)^2(1 - \alpha_2)^2 \cdots (1 - \alpha_t)^2}.$$

Now, let $b = (\frac{u}{\ell})^{\frac{1}{t+1}}$. Consider subintervals $\{(\ell b^j, \ell b^{j+1}] : j = 0, 1, \ldots, t\}$. One of these intervals say $\ell b^{j_0}, \ell b^{j_0+1}$ has no roots at all. Let ϵ be the mid point of the interval, that is, $\epsilon = (\ell b^{j_0} + \ell b^{j_0+1})/2$. Then, we have $(\epsilon - \alpha_j)^2 \geq \left(\frac{\ell b^{j_0+1} - \ell b^{j_0}}{2}\right)^2$, and since $(1 - \alpha_j)^2 \leq 1$, we have $\frac{r(\epsilon)}{\epsilon^{2t+1}} \geq \left(\frac{b-1}{b+1}\right)^{2t}$.

This establishes part (b). The term $-\frac{1}{N\ell(b+1)}$ appears in the statement of Theorem 5 because we need to ensure that ϵN is an integer. □

4.2   Proof of Lemma 1

We will use the following notation. Let $p(X_1, X_2, \ldots, X_N)$ be a polynomial in N variables $X_1, X_2, \ldots, X_N$ with real coefficients. For a database $f : \{0, 1\}^N \rightarrow \{0, 1\}$, let

$$p(f) \overset{\Delta}{=} p(f(1), f(2), \ldots, f(N)).$$

Also, in the following $\mathbf{X}$ denotes the sequence of variables $X_1, X_2, \ldots, X_N$.
    The key fact we need is the following.

Theorem 6 ([BB+95]). *Let $\mathcal{A}$ be a t-query quantum database search algorithm. Then, for $i = 1, 2, \ldots, N$, there is a multilinear polynomial $p_i(\mathbf{X})$ of degree at most $2t$, such that for all f.*

$$\Pr[\mathcal{A}(f) = i] = p_i(f).$$

Furthermore, $p_i(\mathbf{x}) \geq 0$ for all $\mathbf{x} \in [0, 1]^N$.

Lemma 4. *Let $\mathcal{A}$ be a t-query quantum database search algorithm. Then, there is a multilinear polynomial $p(\mathbf{X})$ of degree at most $2t + 1$ such that for all f,*

$$\mathrm{err}_{\mathcal{A}}(f) = p_{\mathcal{A}}(f).$$

Proof. Using the polynomials $p_i(X)$ from Theorem 6, define

$$p_{\mathcal{A}}(\mathbf{X}) = \sum_{i=1}^{n}(1 - X_i)p_i(\mathbf{X}).$$

Clearly, $p(f) = \sum_{i=1}^{n}(1 - f(i))p_i(f) = \sum_{i \in f^{-1}(0)} \Pr[\mathcal{A}(f) = i] = \mathrm{err}_{\mathcal{A}}(f).$ □

We can now prove Lemma 1. For a permutation σ of N and $f : [N] \rightarrow \{0, 1\}$, let σf be the function defined by $\sigma f(i) = f(\sigma(i))$.

Note that $|f^{-1}(0)| = |(\sigma f)^{-1}(0)|$. Now,

$$\frac{1}{N!}\sum_{\sigma} p_{\mathcal{A}}(\sigma f) = \mathrm{E}_{\sigma}[\mathrm{err}_{\mathcal{A}}(\sigma f)] \leq \max_{\sigma} \mathrm{err}_{\mathcal{A}}(\sigma f) \leq \mathrm{err}_{\mathcal{A}}(\epsilon), \tag{4}$$

where $|f^{-1}(0)| = \epsilon N$.

Let $\sigma \mathbf{X}$ be the sequence $X_{\sigma(1)}, X_{\sigma(2)}, \ldots, X_{\sigma(N)}$, and let

$$p_{\mathcal{A}}^{\mathrm{sym}}(\mathbf{X}) = \frac{1}{N!}\sum_{\sigma} p_{\mathcal{A}}(\sigma \mathbf{X}).$$

Then, by (4), we have $p_{\mathcal{A}}^{\mathrm{sym}}(f) = \frac{1}{N!}\sum_{\sigma} p_{\mathcal{A}}(\sigma f) \leq \mathrm{err}_{\mathcal{A}}(\epsilon)$.

Now, $p_{\mathcal{A}}^{\mathrm{sym}}(\mathbf{X})$ is a symmetric multilinear polynomial in N variables of degree at most $2t + 1$. For any such polynomial, there is a univariate polynomial $q(Z)$ of degree at most $2t + 1$ such that if we let $\hat{p}(\mathbf{X}) = q(\sum_{i=1}^{N} X_i)/N)$, then for all f,

$$\hat{p}(f) = p_{\mathcal{A}}^{\mathrm{sym}}(f) \leq \mathrm{err}_{\mathcal{A}}(\epsilon).$$

(See Minsky and Papert [MP].) Now, $\hat{p}(f) = q((f(1) + f(2) + \ldots + f(N))/N) = q(1 - \epsilon)$. To complete the proof, we take $r(Z) = q(1 - Z)$. □

Acknowledgements

We thank the referees for their helpful comments.

References

[BB+95] R. Beals, H. Buhrman, R. Cleve, M. Mosca, R. de Wolf. Quantum lower bounds by polynomials, FOCS (1998): 352–361. quant-ph/9802049.

[BC+99] H. Buhrman, R. Cleve, R. de Wolf, C. Zalka. Bounds for Small-Error and Zero-Error Quantum Algorithms, FOCS (1999): 358–368.

[BH+02] G. Brassard, P. Hoyer, M. Mosca, A. Tapp: Quantum amplitude amplification and estimation. In S.J. Lomonaco and H.E. Brandt, editors, Quantum Computation and Information, AMS Contemporary mathematics Series (305), pp. 53–74, 2002. quant-ph/0005055.

[G96] L.K. Grover: A fast quantum mechanical algorithm for database search. STOC (1996): 212-219. quant-ph/9605043.

[G97] L.K. Grover: Quantum Mechanics helps in searching for a needle in a haystack. Physical Review Letters 79(2), pp. 325-328, July 14, 1997. quant-ph/9706033.

[G98a] L.K. Grover: A framework for fast quantum mechanical algorithms, Proc. 30th ACM Symposium on Theory of Computing (STOC), 1998, 53–63.

[G98b] L.K. Grover. Quantum computers can search rapidly by using almost any transformation, Phy. Rev. Letters, 80(19), 1998, 4329–4332. quant-ph/9711043.

[G05] L.K. Grover. A different kind of quantum search. March 2005. quant-ph/0503205.

[MP] M.L. Minsky and S.A. Papert. Perceptrons: An Introduction to Computational Geometry. MIT Press (1968).

[NC] M.A. Nielsen and I.L. Chuang: Quantum Computation and Quantum Information. Cambridge University Press (2000).

[TGP05] T. Tulsi, L.K. Grover, A. Patel. A new algorithm for directed quantum search. May 2005. quant-ph/0505007.

Sampling Bounds for Stochastic Optimization

Moses Charikar[1,*], Chandra Chekuri[2,**], and Martin Pál[2,***]

[1] Computer Science Dept., Princeton University, Princeton, NJ 08544
moses@cs.princeton.edu
[2] Lucent Bell Labs, 600 Mountain Avenue, Murray Hill, NJ 07974
chekuri@reserch.bell-labs.com, mpal@acm.org

Abstract. A large class of stochastic optimization problems can be modeled as minimizing an objective function f that depends on a choice of a vector $x \in X$, as well as on a random external parameter $\omega \in \Omega$ given by a probability distribution π. The value of the objective function is a random variable and often the goal is to find an $x \in X$ to minimize the expected cost $E_\omega[f_\omega(x)]$. Each ω is referred to as a *scenario*. We consider the case when Ω is large or infinite and we are allowed to sample from π in a black-box fashion. A common method, known as the SAA method (sample average approximation), is to pick sufficiently many independent samples from π and use them to approximate π and correspondingly $E_\omega[f_\omega(x)]$. This is one of several scenario reduction methods used in practice.

There has been substantial recent interest in two-stage stochastic versions of combinatorial optimization problems which can be modeled by the framework described above. In particular, we are interested in the model where a parameter λ bounds the relative factor by which costs increase if decisions are delayed to the second stage. Although the SAA method has been widely analyzed, the known bounds on the number of samples required for a $(1 + \varepsilon)$ approximation depend on the variance of π even when λ is assumed to be a fixed constant. Shmoys and Swamy [13, 14] proved that a polynomial number of samples suffice when f can be modeled as a linear or convex program. They used modifications to the ellipsoid method to prove this.

In this paper we give a different proof, based on earlier methods of Kleywegt, Shapiro, Homem-De-Mello [6] and others, that a polynomial number of samples suffice for the SAA method. Our proof is not based on computational properties of f and hence also applies to integer programs. We further show that small variations of the SAA method suffice to obtain a bound on the sample size even when we have only an approximation algorithm to solve the sampled problem. We are thus able to extend a number of algorithms designed for the case when π is given explicitly to the case when π is given as a black-box sampling oracle.

* Supported by NSF ITR grant CCR-0205594, DOE Early Career Principal Investigator award DE-FG02-02ER25540, NSF CAREER award CCR-0237113, an Alfred P. Sloan Fellowship and a Howard B. Wentz Jr. Junior Faculty Award.
** Supported in part by an ONR basic research grant MA14681000 to Lucent Bell Labs.
*** Work done while at DIMACS, supported by NSF grant EIA 02-05116.

C. Chekuri et al. (Eds.): APPROX and RANDOM 2005, LNCS 3624, pp. 257–269, 2005.
© Springer-Verlag Berlin Heidelberg 2005

1   Introduction

Uncertainty in data is a common feature in a number of real world problems. Stochastic optimization models uncertain data using probability distributions. In this paper we consider problems that are modeled by the two-stage stochastic minimization program

$$\min_{x \in X} f(x) = c(x) + \mathbf{E}_\omega[q(x, \omega)]. \tag{1}$$

An important context in which the problem (1) arises is *two-stage stochastic optimization with recourse*. In this model, a *first-stage decision* $x \in X$ has to be made while having only probabilistic information about the future, represented by the probability distribution π on Ω. Then, after a particular future *scenario* $\omega \in \Omega$ is realized, a *recourse action* $r \in R$ may be taken to ensure that the requirements of the scenario ω are satisfied. In the two-stage model, $c(x)$ denotes the cost of taking the first-stage action x. The cost of the second stage in a particular scenario ω, given a first-stage action x, is usually given as the optimum of the second-stage minimization problem

$$q(x, \omega) = \min_{r \in R}\{\text{cost}_\omega(x, r) \mid (x, r) \text{ is a feasible solution for scenario } \omega\}.$$

We give an example to illustrate some of the concepts. Consider the following facility location problem. We are given a finite metric space in the form of a graph that represents the distances in some underlying transportation network. A company wants to build service centers at a number of locations to best serve the demand for its goods. The objective is to minimize the cost of building the service centers subject to the constraint that each demand point is within a distance B from its nearest center. However, at the time that the company plans to build the service centers, there could be uncertainty in the demand locations. One way to deal with this uncertainty is to make decisions in two or more stages. In the first stage certain service centers are built, and in the second stage, when there is a clearer picture of the demand, additional service centers might be built, and so on. How should the company minimize the overall cost of building service centers? Clearly, if building centers in the second stage is no more expensive than building them in the first stage, then the company will build all its centers in the second stage. However, very often companies cannot wait to make decisions. It takes time to build centers and there are other costs such as inflation which make it advantageous to build some first stage centers. We can assume that building a center in the second stage costs at most some $\lambda \geq 1$ times more than building it in the first stage. This tradeoff is captured by (1) as follows. X is the set of all n-dimensional binary vectors where n is the number of potential locations for the service centers. A binary vector x indicates which centers are built in the first stage and $c(x)$ is the cost of building them. The uncertainty in demand locations is modeled by the probability distribution on Ω. A scenario $\omega \in \Omega$ is characterized by set of demand locations in ω. Thus, given ω and the first stage decision x, the recourse action is to build additional centers so that all demands in ω are within a distance B of some center. The cost of building these additional centers is given by $q(x, \omega)$. Note that $q(x, \omega)$ is itself an optimization problem very closely related to the original k-center problem.

How does one solve problems modeled by (1)? One key issue is how the probability distribution π is specified. In some cases the number of scenarios in Ω is small and π is explicitly known. In such cases the problem can be solved as a deterministic problem using whatever mathematical programming method (linear, integer, non-linear) is applicable for minimizing c and q. There are however situations in which Ω is too large or infinite and it is infeasible to solve the problem by explicitly listing all the scenarios. In such cases, a natural approach is to take some number, N, of independent samples $\omega_1, \ldots, \omega_N$ from the distribution π, and approximate the function f by the *sample average function*

$$\hat{f}(x) = c(x) + \frac{1}{N} \sum_{i=1}^{N} q(x, \omega_i). \tag{2}$$

If the number of samples N is not too large, finding an $\hat{x}$ that minimizes $\hat{f}(\hat{x})$ may be easier than the task of minimizing f. One might then hope that for a suitably chosen sample size N, a good solution $\hat{x}$ to the sample average problem would be a good solution to the problem (1). This approach is called the *sample average approximation* (SAA) method. The SAA method is an example of a *scenario reduction* technique, in that it replaces a complex distribution π over a large (or even infinite) number of scenarios by a simpler, empirical distribution π' over N observed scenarios. Since the function (2) is now deterministic, we can use tools and algorithms from deterministic optimization to attempt to find its exact or approximate optimum.

The SAA method is well known and falls under the broader area of Monte Carlo sampling. It is used in practice and has been extensively studied and analyzed in the stochastic programming literature. See [9, 10] for numerous pointers. In a number of settings, in particular for convex and integer programs, it is known that the SAA method converges to the true optimum as $N \to \infty$. The number of samples required to obtain an *additive* ε approximation with a probability $(1 - \delta)$ has been analyzed [6]; it is known to be polynomial in the dimension of X, $1/\varepsilon$, $\log 1/\delta$ and the quantity $V = \max_{x \in X} V(x)$ where $V(x)$ is the variance of the random variable $q(x, w)$. This factor V need not be polynomial in the input size even when π is given explicitly.

The two-stage model with recourse has gained recent interest in the theoretical computer science community following the work of Immorlica et al. [5] and Ravi and Sinha [11]. Several subsequent works have explored this topic [1–4, 7, 13, 14]. The emphasis in these papers is primarily on *combinatorial optimization* problems such as shortest paths, spanning trees, Steiner trees, set cover, facility location, and so on. Most of these problems are NP-hard even when the underlying single stage problem is polynomial time solvable, for example the spanning tree problem. Thus the focus has been on approximation algorithms and in particular on *relative* error guarantees. Further, for technical and pragmatic reasons, an additional parameter, the *inflation factor* has been introduced. Roughly speaking, the inflation factor, denoted by λ, upper bounds the relative factor by which the second stage decisions are more expensive when compared to the first stage. It is reasonable to expect that λ will be a small constant, say under 10, in many practical situations.

In this new model, for a large and interesting class of problems modeled by linear programs, Shmoys and Swamy [13] showed that a relative $(1 + \varepsilon)$ approximation can be obtained with probability at least $(1 - \delta)$ using a number of samples that is polynomial in the input size, λ, $\log 1/\delta$ and $1/\varepsilon$. They also established that a polynomial dependence on λ is necessary. Thus the dependence on V is eliminated. Their first result [13] does not establish the guarantee for the SAA method but in subsequent work [14], they established similar bounds for the SAA method. Their proof is based on the ellipsoid method where the samples are used to compute approximate sub-gradients for the separation oracle. We note two important differences between these results when compared to earlier results of [6]. The first is that the new bounds obtained in [13, 14] guarantee that the optimum solution $\bar{x}$ to the sampled problem $\hat{f}$ satisfies the property that $f(\bar{x}) \leq (1 + \varepsilon)f(x^*)$ with sufficiently high probability where x^* is an optimum solution to f. However they do not guarantee that the value $f(x^*)$ can be estimated to within a $(1 + \varepsilon)$ factor. It can be shown that estimating the expected value $E_\omega[q(x, \omega)]$ for any x requires the sample size to depend on V. Second, the new bounds, by relying on the ellipsoid method, limit the applicability to when X is a continuous space while the earlier methods applied even when X is a discrete set and hence could capture integer programming problems. In fact, in [6], the SAA method is analyzed for the discrete case, and the continuous case is analyzed by discretizing the space X using a fine grid. The discrete case is of particular interest in approximation algorithms for combinatorial optimization. In this context we mention that the boosted sampling algorithm of [3] uses $O(\lambda)$ samples to obtain approximation algorithms for a class of network design problems. Several recent results [4, 5, 7] have obtained algorithms that have provably good approximation ratios when π is given explicitly. An important and useful question is whether these results can be carried over to the case when π can only be sampled from. We answer this question in the positive. Details of our results follow.

1.1   Results

In this paper we show that the results of Shmoys and Swamy [13, 14] can also be derived using a modification to the basic analysis framework in the methods of [6, 10]: this yields a simple proof relying only on Chernoff bounds. Similar to earlier work [6, 10], the proof shows that the SAA method works because of *statistical* properties of X and f, and not on *computational* properties of optimizing over X. This allows us to prove a result for the discrete case and hence we obtain bounds for integer programs. The sample size that we guarantee is strongly polynomial in the input size: in the discrete setting it depends on $\log |X|$ and in the continuous setting it depends on the dimension of the space containing X. We also extend the ideas to approximation algorithms. In this case, the plain SAA method achieves guarantee of $(1+\varepsilon)$ times the guarantee of the underlying approximation algorithm only with $O(\epsilon)$ probability. To obtain a probability of $1-\delta$ for any given δ, we analyze two minor variants of SAA. The first variant repeats SAA $O(\frac{1}{\varepsilon}\log\frac{1}{\delta})$ times independently and picks the solution with the smallest sample value. The second variant rejects a small fraction of high cost samples before running the approximation algorithm on the samples. This latter result was also obtained independently by Ravi and Singh [12].

2   Preliminaries

In the following, we consider the stochastic optimization problem in its general form (1). We consider stochastic two stage problems that satisfy the following properties.

(A1) **Non-negativity.** The functions $c(x)$ and $q(x,\omega)$ are non-negative for every first stage action x and every scenario ω.

(A2) **Empty First Stage.** We assume that there is an empty first-stage action, $0 \in X$. The empty action incurs no first-stage cost, i.e. $c(0) = 0$, but is least helpful in the second stage. That is, for every $x \in X$ and every scenario ω, $q(x,\omega) \le q(0,\omega)$.

(A3) **Bounded Inflation Factor.** The inflation factor λ determines the relative cost of information. It compares the difference in cost of the "wait and see" solution $q(0,\omega)$ and the cost of the best solution with hindsight, $q(x,\omega)$, relative to the first stage cost of x. Formally, the inflation factor $\lambda \ge 1$ is the least number such that for every scenario $\omega \in \Omega$ and every $x \in X$, we have

$$q(0,\omega) - q(x,\omega) \le \lambda c(x). \qquad (3)$$

We note that the above assumptions capture both the discrete and continuous case problems considered in recent work [1, 3, 5, 11, 13, 14].

We work with exact and approximate minimizers of the sampled problem. We make the notion precise.

Definition 1. *An $x^* \in X$ is said to be an* exact minimizer *of the function $f(\cdot)$ if for all $x \in X$ it holds that $f(x^*) \le f(x)$. An $\bar{x} \in X$ is an α-approximate minimizer of the function $f(\cdot)$, if for all $x \in X$ it holds that $f(\bar{x}) \le \alpha f(x)$.*

The main tool we will be using is the Chernoff bound. We will be using the following version of the bound (see e.g. [8, page 98]).

Lemma 1 (Chernoff bound). *Let $X_1,\ldots,X_N$ be independent random variables with $X_i \in [0,1]$ and let $X = \sum_{i=1}^{N} X_i$. Then, for any $\varepsilon \ge 0$, we have $\Pr[\,|X - \mathbf{E}[X]|\, > \varepsilon N\,] \le 2\exp(-\varepsilon^2 N)$.*

Throughout the rest of the paper when we refer to an event happening with probability β, it is with respect to the randomness in sampling from the distribution π over Ω.

3   Discrete Case

We start by discussing the case when the first stage decision x ranges over a finite set of choices X. In the following, let x^* denote an optimal solution to the true problem (1), and Z^* its value $f(x^*)$. In the following we assume that ε is small, say $\varepsilon < 0.1$.

Theorem 1. *Any exact minimizer $\bar{x}$ of the function $\hat{f}(\cdot)$ constructed with $\Theta(\lambda^2 \frac{1}{\varepsilon^4} \log |X| \log \frac{1}{\delta})$ samples is, with probability $1 - 2\delta$, a $(1 + O(\varepsilon))$-approximate minimizer of the function $f(\cdot)$.*

In a natural attempt to prove Theorem 1, one might want to show that if N is large enough, the functions $\hat{f}$ will be close to f, in that with high probability $|f(x) - \hat{f}(x)| \leq \varepsilon f(x)$. Unfortunately this may not be the case, as for any particular x, the random variable $q(x, \omega)$ may have very high variance. However, intuitively, the high variance of $q(x, \omega)$ can only be caused by a few "disaster" scenarios of very high cost but low probability, whose cost is not very sensitive to the particular choice of the first stage action x. Hence these high cost scenarios do not affect the choice of the optimum $\bar{x}$ significantly. We formalize this intuition below.

For the purposes of the analysis, we divide the scenarios into two classes. We call a scenario ω *high*, if its second stage "wait and see" cost $q(0, \omega)$ exceeds a threshold M, and *low* otherwise. We set the threshold M to be $\lambda Z^*/\varepsilon$.

We approximate the function f by taking N independent samples $\omega_1, \ldots, \omega_N$. We define the following two functions to account for the contributions of low and high scenarios respectively.

$$\hat{f}_l(x) = \frac{1}{N} \sum_{i:\omega_i \text{ low}} q(x, \omega_i) \quad \text{and} \quad \hat{f}_h(x) = \frac{1}{N} \sum_{i:\omega_i \text{ high}} q(x, \omega_i).$$

Note that $\hat{f}(x) = c(x) + \hat{f}_l(x) + \hat{f}_h(x)$. We make a similar definition for the function $f(\cdot)$. Let $p = \Pr_\omega[\omega \text{ is a high scenario}]$.

$$f_l(x) = \mathbf{E}[q(x, \omega) | \omega \text{ is low}] \cdot (1 - p) \quad \text{and} \quad f_h(x) = \mathbf{E}[q(x, \omega) | \omega \text{ is high}] \cdot p$$

so that $f(x) = c(x) + f_l(x) + f_h(x)$.

We need the following bound on p.

Lemma 2. *The probability mass p of high scenarios is at most $\frac{\varepsilon}{(1-\varepsilon)\lambda}$.*

Proof. Recall that x^* is a minimizer of f, and hence $Z^* = f(x^*)$. We have

$$Z^* \geq f_h(x^*) = p \cdot \mathbf{E}[q(x^*, \omega) \mid \omega \text{ is high}] \geq p \cdot [M - \lambda c(x^*)],$$

where the inequality follows from the fact that for a high scenario ω, $q(0, \omega) \geq M$ and by Axiom (A3), $q(x, \omega) \geq q(0, \omega) - \lambda c(x)$. Substituting $M = \lambda Z^*/\varepsilon$, and using that $c(x^*) \leq Z^*$ we obtain

$$Z^* \geq Z^* \left(\frac{\lambda}{\varepsilon} - \lambda \right) p.$$

Solving for p proves the claim.

To prove Theorem 1, we show that each of the following properties hold with probability at least $1 - \delta$ (in fact, the last property holds with probability 1).

(P1) For every $x \in X$ it holds that $|f_l(x) - \hat{f}_l(x)| \leq \varepsilon Z^*$.
(P2) For every $x \in X$ it holds that $\hat{f}_h(0) - \hat{f}_h(x) \leq 2\varepsilon c(x)$.
(P3) For every $x \in X$ it holds that $f_h(0) - f_h(x) \leq 2\varepsilon c(x)$.

Proving that these properties hold is not difficult, and we will get to it shortly; but let us first show how they imply Theorem 1.

Proof of Theorem 1. With probability $1 - 2\delta$, we can assume that all three properties (P1–P3) hold. For any $x \in X$ we have

$$
\begin{aligned}
f_l(x) &\leq \hat{f}_l(x) + \varepsilon Z^* && \text{by (P1)} \\
f_h(x) &\leq f_h(0) && \text{by (A2)} \\
0 &\leq \hat{f}_h(x) + 2\varepsilon c(x) - \hat{f}_h(0) && \text{by (P2)}
\end{aligned}
$$

Adding the above inequalities and using the definitions of functions f and $\hat{f}$ we obtain

$$f(x) - \hat{f}(x) \leq \varepsilon Z^* + 2\varepsilon c(\bar{x}) + f_h(0) - \hat{f}_h(0). \tag{4}$$

By a similar reasoning we get the opposite inequality

$$\hat{f}(x) - f(x) \leq \varepsilon Z^* + 2\varepsilon c(x) + \hat{f}_h(0) - f_h(0). \tag{5}$$

Now, let x^* and $\bar{x}$ be minimizers of the functions $f(\cdot)$ and $\hat{f}(\cdot)$ respectively. Hence we have $\hat{f}(\bar{x}) \leq \hat{f}(x^*)$. We now use (4) with $x = \bar{x}$ and (5) with $x = x^*$. Adding them up, together with the fact that $\hat{f}(\bar{x}) \leq \hat{f}(x^*)$ we get

$$f(\bar{x}) - 2\varepsilon c(\bar{x}) \leq f(x^*) + 2\varepsilon c(x^*) + 2\varepsilon Z^*.$$

Noting that $c(x) \leq f(x)$ holds for any x and that $Z^* = f(x^*)$, we get that $(1 - 2\varepsilon)f(\bar{x}) \leq (1 + 4\varepsilon)f(x^*)$. Hence, with probability $(1 - 2\delta)$, we have that $f(\bar{x}) \leq (1 + O(\varepsilon))f(x^*)$.

Now we are ready to prove properties (P1–P3). We will make repeated use of the Chernoff bound stated in Lemma 1. Properties (P2) and (P3) are an easy corrolary of Axiom A3 once we realize that the probability of drawing a high sample from the distribution π is small; and that the fraction of high samples we draw will be small as well with high probability. Let N_h denote the number of high samples in $\omega_1, \ldots, \omega_N$.

Lemma 3. *With probability $1 - \delta$, $N_h/N \leq 2\varepsilon/\lambda$.*

Proof. Let X_i be an indicator variable that is equal to 1 if the sample ω_i is high and 0 otherwise. Then $N_h = \sum_{i=1}^{N} X_i$ is a sum of i.i.d. 0-1 variables, and $\mathbf{E}[N_h] = pN$. From Lemma 2, $p \leq \frac{\varepsilon}{(1-\varepsilon)\lambda}$.

Using Chernoff bounds,

$$\Pr\left[N_h - Np > \frac{\varepsilon}{\lambda} N \left(2 - \frac{1}{1 - \varepsilon}\right)\right] \leq \exp\left(-\frac{\varepsilon^2}{\lambda^2} \frac{(1 - 2\varepsilon)^2}{(1 - \varepsilon)^2} N\right).$$

With $\varepsilon < 1/3$ and N chosen as in Theorem 1, this probability is at most δ.

Corollary 1. *Property (P2) holds with probability* $1-\delta$, *and property (P3) holds with probability* 1.

Proof. By Lemma 3, we can assume that the number of high samples $N_h \leq 2N\varepsilon/\lambda$ for $\varepsilon < 1/3$. Then,

$$\hat{f}_h(0) - \hat{f}_h(x) = \frac{1}{N} \sum_{i:\omega_i \text{ high}} q(0,\omega_i) - q(x,\omega_i) \leq \frac{N_h}{N}\lambda c(x).$$

The inequality comes from Axiom A3. Since $N_h/N \leq 2\varepsilon/\lambda$, the right hand side is at most $2\varepsilon c(x)$, and thus Property (P2) holds with probability $1 - \delta$.

Using Lemma 2, by following the same reasoning (replacing sum by expectation) we obtain that Property (P3) holds with probability 1.

Now we are ready to prove that property (P1) holds with probability $1 - \delta$. This is done in Lemma 4. The proof of this lemma is the only place where we use the fact that X is finite. In Section 5 we show property (P1) to hold when $X \subseteq \mathbb{R}^n$, under an assumption that the function $c(\cdot)$ is linear and that $q(\cdot,\cdot)$ satisfies certain Lipschitz-type property.

Lemma 4. *With probability* $1 - \delta$ *it holds that for all* $x \in X$,

$$|f_l(x) - \hat{f}_l(x)| \leq \varepsilon Z^*.$$

Proof. First, consider a fixed first stage action $x \in X$. Note that we can view $\hat{f}_l(x)$ as the arithmetic mean of N independent copies $Q_1, \ldots, Q_N$ of the random variable Q distributed as

$$Q = \begin{cases} q(x,\omega) & \text{if } \omega \text{ is low} \\ 0 & \text{if } \omega \text{ is high} \end{cases}$$

Observe that $f_l(x) = \mathbf{E}[Q]$. Let Y_i be the variable Q_i/M and let $Y = \sum_{i=1}^N Y_i$. Note that $Y_i \in [0,1]$ and $\mathbf{E}[Y] = \frac{N}{M}f_l(x)$. We apply the Chernoff bound from Lemma 1 to obtain the following.

$$\Pr\left[\left|Y - \frac{N}{M}f_l(x)\right| > \frac{\varepsilon^2}{\lambda}N\right] \leq 2\exp\left(-\frac{\varepsilon^4}{\lambda^2}N\right).$$

With N as in Theorem 1, this probability is at most $\delta/|X|$. Now, taking the union bound over all $x \in X$, we obtain the desired claim.

4   Approximation Algorithms and SAA

In many cases of our interest, finding an exact minimizer of the function $\hat{f}$ is computationally hard. However, we may have an algorithm that can find approximate minimizers of functions $\hat{f}$.

First, we explore the performance of the plain SAA method used with an α-approximation algorithm for minimizing the function $\hat{f}$. The following lemma is an adaptation of Theorem 1 to approximation algorithms.

Lemma 5. *Let $\bar{x}$ be a α-approximate minimizer for $\hat{f}$. Then, with probability $(1 - 2\delta)$,*

$$f(\bar{x})(1 - 2\varepsilon) \leq (1 + 6\varepsilon)\alpha f(x^*) + (\alpha - 1)(\hat{f}_h(0) - f_h(0)).$$

Proof. Again, with probability $1 - 2\delta$, we can assume that Properties (P1–P3) hold. Following the proof of Theorem 1, the Lemma follows from Inequalities (4-5) and the fact that $\hat{f}(\bar{x}) \leq \alpha\hat{f}(x^*)$.

From the above lemma, we see that $\bar{x}$ is a good approximation to x^* if $\hat{f}_h(0) - f_h(0)$ is small. Since $\hat{f}_h(x^*)$ is an unbiased estimator of $f_h(x^*)$, by Markov's inequality we have that $\hat{f}_h(x^*) \leq (1 + 1/k)f_h(x^*)$ holds with probability $\frac{1}{k+1}$. Thus, if we want to achieve multiplicative error $(1+\varepsilon)$, we must be content with probability of success only proportional to $1/\varepsilon$. It is not difficult to construct distributions π where the Markov bound is tight.

There are various ways to improve the success probability of the SAA method used in conjunction with an approximation algorithm. We propose two of them in the following two sections. Our first option is to boost the success probability by repetition: in Section 4.1 we show that by repeating the SAA method $\varepsilon^{-1} \log \delta^{-1}$ times independently, we can achieve success probability $1 - \delta$. An alternate and perhaps more elegant method is to reject the high cost samples; this does not significantly affect the quality of any solution, while significantly reducing the variance in evaluating the objective function. We discuss this in Section 4.2.

4.1   Approximation Algorithms and Repeating SAA

As we saw in Lemma 5, there is a reasonable probability of success with SAA even with an approximation algorithm for the sampled problem. To boost the probability of success we independently repeat the SAA some number of times and we pick the solution of lowest cost. The precise parameters are formalized in the theorem below.

Theorem 2. *Consider a collection of k functions $\hat{f}^1, \hat{f}^2, \ldots, \hat{f}^k$, such that $k = \Theta(\varepsilon^{-1} \log \delta^{-1})$ and the $\hat{f}^i$ are independent sample average approximations of the function f, using $N = \Theta(\lambda^2 \varepsilon^{-4} \cdot k \cdot \log |X| \log \delta^{-1})$ samples each. For $i = 1, \ldots, k$, let $\bar{x}^i$ be an α-approximate minimizer of the function $\hat{f}^i$. Let $i = \arg\min_j \hat{f}^j(\bar{x}^j)$. Then, with probability $1 - 3\delta$, $\bar{x}^i$ is an $(1 + O(\varepsilon))\alpha$-approximate minimizer of the function $f(\cdot)$.*

Proof. We call the i-th function $\hat{f}^i$ good if it satisfies Properties (P1) and (P2). The number of samples N has been picked so that each $\hat{f}^i$ is good with probability at least $1 - 2\delta/k$ and hence all samples are good with probability $1 - 2\delta$.

By Markov inequality, the probability that $\hat{f}_h^i(0) < (1+\varepsilon)f_h(0)$ is at least $1/\varepsilon$. Hence the probability that none of these events happens is at most $(1 - \varepsilon)^k < \delta$. Thus, with probability $1 - \delta$ we can assume that there is an index j for which

$\hat{f}_h^j(0) \le (1+\varepsilon)f_h(0)$. As $f_h(0) \le (1+2\varepsilon)Z^*$ easily follows from Property (P2), with probability $1-\delta$ we have

$$\hat{f}_h^j(0) \le (1+4\varepsilon)Z^*. \tag{6}$$

Let $i = \operatorname{argmin}_\ell \hat{f}^\ell(\bar{x}^\ell)$. Using the fact that $\hat{f}^i(\bar{x}^i) \le \hat{f}^j(\bar{x}^j)$ and that x^i and x^j are both α-approximate minimizers of their respective functions, we get

$$\hat{f}^i(\bar{x}^i) \le \frac{1}{\alpha}\hat{f}^i(\bar{x}^i) + \frac{\alpha-1}{\alpha}\hat{f}^j(\bar{x}^j) \le \hat{f}^i(x^*) + (\alpha-1)\hat{f}^j(x^*). \tag{7}$$

Substituting $\bar{x}^i$ and x^* for x in Inequalities (4) and (5) respectively, we obtain

$$f(\bar{x}^i) \le \hat{f}^i(\bar{x}^i) + 2\varepsilon Z^* + 2\varepsilon c(\bar{x}^i) + f_h(0) - \hat{f}_h(0) \tag{8}$$

$$\hat{f}(x^*) \le f(x^*) + 2\varepsilon Z^* + 2\varepsilon c(x^*) + \hat{f}_h(0) - f_h(0). \tag{9}$$

Adding inequalities (7), (8), and (9), we get

$$f(\bar{x}^i) - 2\varepsilon c(\bar{x}^i) \le (\alpha-1)\hat{f}^j(x^*) + f(x^*) + 4\varepsilon Z^* + 2\varepsilon c(x^*). \tag{10}$$

Using Equation 6 with Lemma 5 to bound $\hat{f}^j(x^*)$ finishes the proof.

4.2   Approximation Algorithms and Sampling with Rejection

Instead of repeating the SAA method multiple times to get a good approximation algorithm, we can use it only once, but ignore the high cost samples. The following lemma makes this statement precise.

Lemma 6. *Let $g: X \mapsto \mathbb{R}$ be a function satisfying $|f_l(x)+c(x)-g(x)| = O(\varepsilon)Z^*$ for every $x \in X$. Then any α-approximate minimizer $\bar{x}$ of the function $g(\cdot)$ is also an $\alpha(1+O(\varepsilon))$-approximate minimizer of the function $f(\cdot)$.*

Proof. Let $\bar{x}$ be an α-approximate minimizer of g. We have

$$f(\bar{x}) \le g(\bar{x}) + O(\varepsilon Z^*) + f_h(\bar{x})$$
$$g(\bar{x}) \le \alpha(c(x^*) + f_l(x^*) + O(\varepsilon Z^*))$$

By Axiom (A2) and Property (P3) we can replace $f_h(\bar{x})$ in the first inequality by $f_h(x^*) + \varepsilon c(x^*)$. Adding up, we obtain

$$f(\bar{x}) \le \alpha(c(x^*) + f_l(x^*)) + \alpha O(\varepsilon Z^*) + \varepsilon c(x^*) + f_h(x^*) \le (1+O(\varepsilon))\alpha Z^*.$$

According to Lemma 6, a good candidate for the function g would be $g(x) = c(x) + \hat{f}_l(x)$, since by Lemma 4 we know that $|f_l(x) - \hat{f}_l(x)| \le \varepsilon Z^*$ holds with probability $1-\delta$. However, in order to evaluate $\hat{f}_l(x)$, we need to know the value Z^* to be able to classify samples as high or low. If Z^* is not known, we can approximate f_l by the function

$$\bar{f}_l(x) = \frac{1}{N}\sum_{i=1}^{N-2\varepsilon N/\lambda} q(x, \omega_i)$$

where we assume that the samples were reordered so that $q(0, \omega_1) \leq q(0, \omega_2) \leq \cdots \leq q(0, \omega_N)$. In other words, we throw out $2\varepsilon N/\lambda$ samples with highest recourse cost.

Lemma 7. *With probability $1 - \delta$, for all $x \in X$ it holds that $|\bar{f}_l(x) - f_l(x)| \leq 3\varepsilon Z^*$.*

Proof. By Lemma 3, with probability $1 - \delta$ we can assume that $\bar{f}_l$ does not contain any high samples, and hence $\bar{f}_l(x) \leq \hat{f}_l(x)$. Since there can be at most $2\varepsilon N/\lambda$ samples that contribute to $\hat{f}_l$ but not to $\bar{f}_l$, and all of them are low, we get $\hat{f}_l(x) - \bar{f}_l(x) \leq M \cdot 2\varepsilon/\lambda = 2\varepsilon Z^*$, and hence $|\bar{f}_l(x) - \hat{f}_l(x)| \leq 2\varepsilon Z^*$. Finally, by Lemma 4 we have that $|\hat{f}_l(x) - f_l(x)| \leq \varepsilon Z^*$.

Theorem 3. *Let $\omega_1, \omega_2, \ldots, \omega_N$ be independent samples from the distribution π on Ω with $N = \Theta(\lambda^2 \epsilon^{-4} \cdot \log|X| \log \delta^{-1})$. Let $\omega'_1, \omega'_2, \ldots, \omega'_N$ be a reordering of the samples such that $q(0, \omega'_1) \leq q(0, \omega'_2) \ldots \leq q(0, \omega'_N)$. Then any α-approximate minimizer $\bar{x}$ of the function $\bar{f}(x) = c(x) + \frac{1}{N} \sum_{i=1}^{N'} q(x, \omega'_i)$ with $N' = (1 - 2\varepsilon/\lambda)N$ is a $(1 + O(\varepsilon))\alpha$-approximate minimizer of $f(\cdot)$.*

In many situations, computing $q(x, \omega)$ (or even $q(0, \omega)$) requires us to solve an NP-hard problem. Hence we cannot order the samples as we require in the above theorem. However, if we have an approximation algorithm with ratio β for computing $q(\cdot, \cdot)$, we can use it to order the samples instead. For the above theorem to be applicable with such an approximation algorithm, the number of samples, N, needs to increase by a factor of β^2 and N' needs to be $(1 - 2\varepsilon/(\beta\lambda))N$.

5   From the Discrete to the Continuous

So far we have assumed that X, the set of first-stage decisions, is a finite set. In this section we demonstrate that this assumption is not crucial, and extend Theorems 1, 2 and 3 to the case when $X \subseteq \mathbb{R}^n$, under reasonable Lipschitz type assumptions on the functions $c(\cdot)$ and $q(\cdot, \cdot)$.

Since all our theorems depend only on the validity of the three properties (P1–P3), they continue to hold in all settings where (P1–P3) can be shown to hold. Properties (P2) and (P3) are a simple consequence of Axiom (A3) and hold irrespective of the underlying action space X. Hence, to extend our results to a continuous setting, we only need to check the validity of Property (P1). In the rest of this section, we show (P1) to hold for $X \subseteq \mathbb{R}^n_+$, assuming that the first stage costs is linear, i.e. $c(x) = c^t \cdot x$ for some real vector $c \geq 0$, and that the recourse function satisfies the following property.

Definition 2. We say that the recourse function $q(\cdot, \cdot)$ is (λ, c)-*Lipschitz*, if the following inequality holds for every scenario ω:

$$|q(x, \omega) - q(x', \omega)| \leq \lambda \sum_{i=1}^{n} c_i |x_i - x'_i|.$$

Note that the Lipschitz property implies Axiom (A3) in that any (λ, c)-Lipschitz recourse function q satisfies $q(0, \omega) - q(x, \omega) \leq \lambda c(x)$.[1]

We use a standard meshing argument: if two functions $\hat{f}$ and f do not differ by much on a dense enough finite mesh, because of bounded gradient, they must approximately agree in the whole region covered by the mesh. This idea is by no means original; it has been used in the context of stochastic optimization by various authors (among others, [6, 14]). We give the argument for the sake of completeness.

Our mesh is an n-dimensional grid of points with $\varepsilon/(n\alpha\lambda c_i)$ spacing in each dimension $1 \leq i \leq n$.

Since the i-th coordinate $\bar{x}_i$ of any α-approximate minimizer $\bar{x}$ cannot be larger than $\alpha Z^*/c_i$ (as otherwise the first stage cost would exceed αZ^*), we can assume that the feasible region lies within the bounding box $0 \leq x_i \leq \alpha Z^*/c_i$ for $1 \leq i \leq n$. Thus, the set X' of mesh points can be written as

$$X' = \left\{ \left(i_1 \frac{\varepsilon Z^*}{n\lambda c_1}, i_2 \frac{\varepsilon Z^*}{n\lambda c_2}, \ldots, i_n \frac{\varepsilon Z^*}{n\lambda c_n} \right) \middle| (i_1, i_2, \ldots, i_n) \in \{0, 1, \ldots, \lceil n\alpha\lambda/\varepsilon \rceil\}^n \right\}.$$

We claim the following analog of Lemma 4.

Lemma 8. *If $N \geq \theta(\lambda^2 \frac{1}{\varepsilon^4} n \log(n\lambda/\varepsilon) \log \delta)$, then with probability $1 - \delta$ we have that $|\hat{f}_l(x) - f_l(x)| \leq 3\varepsilon Z^*$ holds for every $x \in X$.*

Proof. The size of X' is $(1 + n\alpha\lambda/\varepsilon)^n$, and hence $\log|X'| = O(n \log(n\lambda/\varepsilon))$. Hence Lemma 4 guarantees that with probability $1 - \delta$, $|\hat{f}_l(x') - f_l(x')| \leq \varepsilon Z^*$ holds for every $x' \in X'$.

For a general point $x \in X$, there must be a nearby mesh point $x' \in X'$ such that $\sum_{i=1}^n c_i |x_i - x_i'| \leq \varepsilon Z^*/\lambda$. By Lipschitz continuity of q we have that $|f_l(x) - f_l(x')| \leq \varepsilon Z^*$ and $|\hat{f}_l(x) - \hat{f}_l(x')| \leq \varepsilon Z^*$. By triangle inequality,

$$|\hat{f}_l(x) - f_l(x)| \leq |\hat{f}_l(x) - \hat{f}_l(x')| + |\hat{f}_l(x') - f_l(x')| + |f_l(x') - f_l(x)| \leq 3\varepsilon Z^*.$$

6   Concluding Remarks

In some recent work, Shmoys and Swamy [15] extended their work on two-stage problems to multi-stage problems. Their new work is not based on the ellipsoid method but still relies on the notion of a sub-gradient and thus requires X to be continuous set. We believe that our analysis in this paper can be extended to the multi-stage setting even when X is a discrete set and we plan to explore this in future work.

Acknowledgments

We thank Retsef Levi, R. Ravi, David Shmoys, Mohit Singh and Chaitanya Swamy for useful discussions.

---

[1] This not necessarily true for non-linear $c(\cdot)$.

References

1. K. Dhamhere, R. Ravi and M. Singh. On two-stage Stochastic Minimum Spanning Trees. *Proc. of IPCO*, 2005.
2. A. Flaxman, A. Frieze, M. Krivelevich. On the random 2-stage minimum spanning tree. *Proc. of SODA*, 2005.
3. A. Gupta, M. Pál, R. Ravi, and A. Sinha. Boosted sampling: Approximation algorithms for stochastic optimization. In *Proceedings of the 36th Annual ACM Symposium on Theory of Computing*, 2004.
4. A. Gupta, R. Ravi, and A. Sinha. An edge in time saves nine: Lp rounding approximation algorithms. In *Proceedings of the 45th Annual IEEE Symposium on Foundations of Computer Science*, 2004.
5. N. Immorlica, D. Karger, M. Minkoff, and V. Mirrokni. On the costs and benefits of procrastination: Approximation algorithms for stochastic combinatorial optimization problems. In *Proceedings of the 15th Annual ACM-SIAM Symposium on Discrete Algorithms*, 2004.
6. A. J. Kleywegt, A. Shapiro, and T. Homem-De-Mello. The sample average approximation method for stochastic discrete optimization. *SIAM J. on Optimization*, 12:479-502, 2001.
7. M. Mahdian. Facility Location and the Analysis of Algorithms through Factor-Revealing Programs. *Ph.D. Thesis*, MIT, June 2004.
8. R. Motwani and P. Raghavan. *Randomized Algorithms*. Cambridge University Press, 1995.
9. *Stochastic Programming*. A. Ruszczynski, and A. Shapiro editors. Vol 10 of *Handbook in Operations Research and Management Science*, Elsevier 2003.
10. A. Shapiro. Montecarlo sampling methods. In A. Ruszczynski, and A. Shapiro editors, *Stochastic Programming*, Vol 10 of *Handbook in Operations Research and Management Science*, Elsevier 2003.
11. R. Ravi and A. Sinha. Hedging uncertainty: Approximation algorithms for stochastic optimization problems. In *Proceedings of the 10th International Conference on Integer Programming and Combinatorial Optimization (IPCO)*, 2004.
12. R. Ravi and M. Singh. Personal communication, February 2005.
13. D. Shmoys and C. Swamy. Stochastic optimization is (almost) as easy as deterministic optimization. *Proc. of FOCS*, 2004.
14. D. Shmoys and C. Swamy. The Sample Average Approximation Method for 2-stage Stochastic Optimization. Manuscript, November 2004.
15. D. Shmoys and C. Swamy. Sampling-based Approximation Algorithms for Multi-stage Stochastic Optimization. Manuscript, 2005.

An Improved Analysis of Mergers

Zeev Dvir[*] and Amir Shpilka[**]

Department of Computer Science
Weizmann institute of science
Rehovot, Israel
{zeev.dvir,amir.shpilka}@weizmann.ac.il

Abstract. Mergers are functions that transform k (possibly dependent) random sources into a single random source, in a way that ensures that if one of the input sources has min-entropy rate δ then the output has min-entropy rate close to δ. Mergers have proven to be a very useful tool in explicit constructions of *extractors* and *condensers*, and are also interesting objects in their own right. In this work we present a new analysis of the merger construction of [6]. Our analysis shows that the min-entropy rate of this merger's output is actually $0.52 \cdot \delta$ instead of $0.5 \cdot \delta$, where δ is the min-entropy rate of one of the inputs. To obtain this result we deviate from the usual linear algebra methods that were used by [6] and introduce a new technique that involves results from additive number theory.

1 Introduction

Mergers are functions that take as input k samples, taken from k (possibly dependent) random sources, each of length n-bits. It is assumed that one of these random sources, who's index is unknown, is sufficiently random, in the sense that it has min-entropy[1] $\geq \delta n$. We want the merger to output an n'-bit string (n' could be smaller than n) that will be close to having min-entropy at least $\delta' n'$, where δ' is not considerably smaller than δ. To achieve this, the merger is allowed to use an additional small number of truly random bits called a *seed*. The goals in merger constructions are a) to minimize the seed length, b) to maximize the min-entropy of the output and c) to minimize the error (that is, the statistical distance between the merger's output and some high min-entropy source).

The notion of *merger* was first introduced by Ta-Shma [11], in the context of explicit constructions of *extractors*[2]. Recently, Lu, Reingold, Vadhan and

---

[*] Research supported by Israel Science Foundation (ISF) grant.
[**] Research supported by the Koshland fellowship.

[1] A source has min-entropy $\geq b$ if none of its values is obtained with probability larger than 2^{-b}.

[2] An extractor is a function that transforms a source with min-entropy b into a source which is close to uniform, with the aid of an additional random seed. For a more detailed discussion of extractors see [9].

C. Chekuri et al. (Eds.): APPROX and RANDOM 2005, LNCS 3624, pp. 270–281, 2005.
© Springer-Verlag Berlin Heidelberg 2005

Wigderson [6] gave a very simple and beautiful construction of mergers based on Locally-Decodable-Codes. This construction was used in [6] as a building block in an explicit construction of extractors with nearly optimal parameters. More recently, [7] generalized the construction of [6], and showed how this construction (when combined with other techniques) can be used to construct *condensers*[3] with constant seed length. The analysis of the merger constructed in [7] was subsequently refined in [4].

The merger constructed by [6] takes as input k strings of length n, one of which has min-entropy b, and outputs a string of length n that is close to having min-entropy at least $0.5 \cdot b$. Loosely speaking, the output of the merger is computed as follows: treat each input block as a vector in the vector space F^m, where F is some small finite field, and output a uniformly chosen linear combination of these k vectors. The analysis of this construction is based on the following simple idea: In every set of linear combinations, who's density is at least γ (where γ is determined by the size of F), there exist two linear combinations that, when put together, determine the 'good' source (that is, the 'good' source can be computed from both of them deterministically). Therefore, one of these linear combinations must have at least half the entropy of the 'good' source (this reasoning extends also to min-entropy). As a results we get that for most seed values (linear combinations) the output has high min-entropy, and the result follows. This is of course an over-simplified explanation, but it gives the general idea behind the proof.

In this paper we present an alternative analysis to the one just described. Our analysis relies on two results from the field of additive-number theory. The first is Szemeredi's theorem ([10],[5]) on arithmetic progressions of length three (also known as Roth's theorem [8]). This theorem states that in every subset of $\{1, \ldots, N\}$, who's density is at least $\delta(N)$, there exits an arithmetic progression of length three. For our purposes we use a quantitative version of this theorem as proven by Bourgain [3]. The second result that we rely on is a lemma of Bourgain [2] that deals with sum-sets and difference-sets of integers. Roughly speaking, the lemma says that if the sum-set of two sets of integers is very small, then their difference-set cannot be very large (for a precise formulation see Section 3). It is interesting to note that this is not the first time that results from additive number-theory are used in the context of randomness extraction. A recent result of Barak, Imagliazzo and Wigderson [1] uses results from this field to construct multi-source extractors.

Using these two results we are able to show that the min-entropy outputted by the aforementioned merger is $0.52 \cdot b$ and not $0.5 \cdot b$ as was previously known. One drawback of our analysis is that the length of the seed is required to be $O(k \cdot \gamma^{-2})$ in order for the output error to be γ, where in the conventional analysis the seed length can be as short as $O(k \cdot \log(\gamma^{-1}))$. This however does not present a problem in many of the current applications of mergers, where the

---

[3] A condenser is a function that transforms a source with min-entropy rate δ into a source which is close to having min-entropy rate $\delta' > \delta$, with the aid of an additional random seed.

error parameter and the number of input sources are both constants and the seed length is also required to be a constant. One place where our analysis can be used in order to simplify an existing construction is in the extractor construction of [7]. There, the output of the merger is used as an input to an extractor that requires the min-entropy rate of its input to be larger than one-half. In [7] this problem is addressed by a more complicated merger construction who's output length is shorter than n. our analysis shows that the more simple construction of [6] could be used instead, since its output min-entropy rate is larger than one-half.

1.1   Somewhere-Random-Sources

An *n-bit random source* is a random variable X that takes values in $\{0,1\}^n$. We denote by supp(X) $\subset \{0,1\}^n$ the support of X (i.e. the set of values on which X has non-zero probability). For two n-bit sources X and Y, we define the statistical distance between X and Y to be

$$\Delta(X,Y) \stackrel{\triangle}{=} \frac{1}{2} \sum_{a \in \{0,1\}^n} |\mathbf{Pr}[X=a] - \mathbf{Pr}[Y=a]| \,.$$

We say that a random source X (of length n bits) has min-entropy $\geq b$ if for every $x \in \{0,1\}^n$ the probability for $X = x$ is at most 2^{-b}.

Definition 1.1 (Min-entropy) *Let X be a random variable distributed over $\{0,1\}^n$. The min-entropy of X is defined as*[4]

$$\mathrm{H}^\infty(\mathrm{X}) \stackrel{\triangle}{=} \min_{x \in \mathrm{supp}(\mathrm{X})} \log\left(\frac{1}{\mathbf{Pr}[X=x]}\right).$$

Definition 1.2 ((n,b)-Source) *We say that X is an (n,b)-source, if X is an n-bit random source, and $\mathrm{H}^\infty(\mathrm{X}) \geq b$.*

A *somewhere-(n,b)-source* is a source comprised of several blocks, such that at least one of the blocks is an (n,b)-source. Note that we allow the other source blocks to depend arbitrarily on the (n,b)-source, and on each other.

Definition 1.3 ($(n,b)^{1:k}$-Source)) *A k-places-somewhere-(n,b)-source, or an $(n,b)^{1:k}$-source, is a random variable $X = (X_1, \ldots, X_k)$, such that every X_i is of length n bits, and at least one X_i is of min-entropy $\geq b$. Note that $X_1, \ldots, X_k$ are not necessarily independent*[5].

---

[4] All logarithms in this paper are taken base 2.

[5] It is possible to define somewhere-(n,b)-sources in a more general way to also include convex combinations of sources of the type described by Definition 1.3. However, it suffices to consider sources of this simpler type for the task of merger constructions.

1.2 Mergers

A merger is a function transforming an $(n, b)^{1:k}$-source into a source which is γ-close (i.e. it has statistical distance $\leq \gamma$) to an (m, b')-source. Naturally, we want b'/m to be as large as possible, and γ to be as small as possible. We allow the merger to use an additional small number of truly random bits, called a *seed*. A Merger is *strong* if for almost all possible assignments to the seed, the output is close to be an (m, b')-source. A merger is *explicit* if it can be computed in polynomial time.

Definition 1.4 (Merger) *A function* $M : \{0,1\}^d \times \{0,1\}^{n \cdot k} \to \{0,1\}^m$ *is a* $[d, (n, b)^{1:k} \mapsto (m, b') \sim \gamma]$-*merger if for every* $(n, b)^{1:k}$-*source* X, *and for an independent random variable* Z *uniformly distributed over* $\{0,1\}^d$, *the distribution* $M(Z, X)$ *is* γ-*close to a distribution of an* (m, b')-*source. We say that* M *is* **strong** *if the average over* $z \in \{0,1\}^d$ *of the minimal distance between the distribution of* $M(z, X)$ *and a distribution of an* (m, b')-*source is* $\leq \gamma$.

We now give a formal definition of the merger we wish to analyze. To simplify the analysis we will assume in several places that certain quantities are integers. This, however, will not affect our results in any significant way.

Construction 1.5 ([6]) *Let* n, k *be integers,* p *a prime number, and let* $l = \frac{n}{\log(p)}$. *We define a function* $M : \{0,1\}^d \times \{0,1\}^{n \cdot k} \to \{0,1\}^n$, *with* $d = \log(p) \cdot k$, *in the following way: Let* F *denote the field* GF(p). *Given* $z \in \{0,1\}^d$, *we think of* z *as a vector* $(z_1, \ldots, z_k) \in F^k$. *Given* $x = (x_1, \ldots, x_k) \in \{0,1\}^{n \cdot k}$, *we think of each* $x_i \in \{0,1\}^n$ *as a vector in* F^l. *The function* M *is now defined as*

$$M(z, x) = \sum_{i=1}^{k} z_i \cdot x_i \in F^l,$$

where the operations are preformed in the vector space F^l. *Intuitively, one can think of* M *as* $M : F^k \times \left(F^l\right)^k \to F^l$.

1.3 Our Results

We prove the following theorem:

Theorem 1 *Let* $0 < \gamma < 1$ *be any constant,* $k > 0$ *a constant integer, and let* p *be a prime larger than[6]* $\exp(\gamma^{-2})$. *Let*

$$M : \{0,1\}^d \times \{0,1\}^{n \cdot k} \to \{0,1\}^n,$$

be as in Construction 1.5, where $d = \log(p) \cdot k$ *(and underlying field* GF(p)*). Then for any constant* $\alpha > 0$ *there exists a constant* b_0 *such that for all* $n \geq b \geq b_0$, M *is a* $[d, (n, b)^{1:k} \mapsto (n, b') \sim \gamma]$-*strong merger with*

$$b' = (0.52 - \alpha) \cdot b.$$

---

[6] In writing $\exp(f)$ we mean $2^{O(f)}$.

From Theorem 1 we see that in order to get a merger with error γ we need to choose the underlying field to be of size at least $\exp(\gamma^{-2})$. It is well known that for every integer n, there is a prime between n and $2n$. Therefore we can take p to be $O\left(\exp(\gamma^{-2})\right)$ and have that the length of the random seed is $d = \log(p) \cdot k = O\left(k \cdot \gamma^{-2}\right)$ bits long. Hence, for constant γ and k the length of the random seed used by the merger is constant. Notice that finding the prime p that answers our demands can be done by testing all integers in the range of interest for primality (if γ is constant then this will take a constant amount of time).

1.4   Organization

In Section 2 we give our analysis of Construction 1.5 and prove Theorem 1. The analysis presented in Section 2 relies on two central claims that we prove in Section 3.

2   Analysis of Construction 1.5

In this section we present our improved analysis of Construction 1.5, and prove Theorem 1. The analysis will go along the same lines as in [4] and will differ from it in two claims that we will prove in Section 3. We begin with some notations that will be used throughout the paper.

Let $X = (X_1, \ldots, X_k) \in \{0,1\}^{n \cdot k}$ be a somewhere (n,b)-source, and let us assume w.l.o.g. that $\mathrm{H}^\infty(X_1) \geq b$. Let $0 < \gamma < 1$ be any constant, and let $p \geq \exp(\gamma^{-2})$ be a prime number. Let $M : F^k \times \left(F^l\right)^k \to F^l$, be as in Construction 1.5, where $F = \mathrm{GF}(p)$, $d = \log(p) \cdot k$ and $n = \log(p) \cdot l$. Our goal is to analyze the min-entropy of $M(Z, X)$ where Z will denote a random variable uniformly distributed over F^k. In particular, we would like to show that the random variable $M(Z, X)$ is γ-close to having min-entropy $\geq (0.52 - \alpha) \cdot b$ for all constant α.

For every $z \in F^k$ we denote by $Y_z \triangleq M(z, X)$ the random variable given by the output of M on the fixed seed value z (recall that, in Construction 1.5, every seed value corresponds to a specific linear combination of the source blocks). Let $u \triangleq 2^d = p^k$ be the number of different seed values, so we can treat the set $\{0,1\}^d$ as the set[7] $[u]$. We can now define $Y \triangleq (Y_1, \ldots, Y_u) \in (F^l)^u$. The random variable Y is a function of X, and is comprised of u blocks, each one of length $n = \log(p) \cdot l$ bits, representing the output of the merger on all possible seed values. We will first analyze the distribution of Y as a whole, and then use this analysis to describe the output of M on a uniformly chosen seed.

Definition 2.1 *Let $D(\Omega)$ denote the set of all probability distributions over a finite set Ω. Let $\mathcal{P} \subset D(\Omega)$ be some property. We say that $\mu \in D(\Omega)$ is γ-close to a convex combination of distributions with property $\mathcal{P}$, if there exists*

---

[7] For an integer n, we write $[n] \triangleq \{1, 2, \ldots, n\}$.

constants $\alpha_1, \ldots, \alpha_t, \gamma > 0$, and distributions $\mu_1, \ldots, \mu_t, \mu' \in D(\Omega)$ such that the following three conditions hold[8]:(1) $\mu = \sum_{i=1}^{t} \alpha_i \mu_i + \gamma \mu'$. (2) $\sum_{i=1}^{t} \alpha_i + \gamma = 1$. (3)$\forall i \in [t]$, $\mu_i \in \mathcal{P}$.

Let Y be the random variable defined above, and let $\mu : (F^l)^u \to [0,1]$ be the probability distribution of Y (i.e. $\mu(y) = \mathbf{Pr}[Y = y]$). We would like to show that μ is exponentially (in b) close to a convex combination of distributions, each having a certain property which will be defined shortly.

Given a probability distribution μ on $(F^l)^u$ we define for each $z \in [u]$ the distribution $\mu_z : F^l \to [0,1]$ to be the restriction of μ to the z's block. More formally, we define

$$\mu_z(y) \stackrel{\triangle}{=} \sum_{y_1, \ldots, y_{z-1}, y_{z+1}, \ldots, y_u \in F^l} \mu(y_1, \ldots, y_{z-1}, y, y_{z+1}, \ldots, y_u).$$

Next, let $\alpha > 0$. We say that a distribution $\mu : (F^l)^u \to [0,1]$ is α-good if for at least $(1 - \gamma/2) \cdot u$ values of $z \in [u]$, μ_z has min-entropy at least $(0.52 - \alpha) \cdot b$. The statement that we would like to prove is that the distribution of Y is close to a convex combination of α-good distributions (see Definition 2.1). As we will see later, this is good enough for us to be able to prove Theorem 1. The following lemma states this claim in a more precise form.

Lemma 2.2 (Main Lemma) Let $Y = (Y_1, \ldots, Y_u)$ be the random variable defined above, and let μ be its probability distribution. Then, for any constant $\alpha > 0$, μ is $2^{-\Omega(b)}$-close to a convex combination of α-good distributions.

We prove Lemma 2.2 in subsection 2.1. The proof of Theorem 1, which follows quite easily from Lemma 2.2, is essentially the same as in [4] (with modified parameters). Due to space constraints we omit the proof of Theorem 1 (the proof appears in the full version of the paper).

2.1 Proof of Lemma 2.2

In order to prove Lemma 2.2 we prove the following slightly stronger lemma.

Lemma 2.3 Let $X = (X_1, \ldots, X_k)$ be an $(n,b)^{1:k}$-source, and define $Y = (Y_1, \ldots, Y_u)$ and μ be as in Lemma 2.2. Then for any constant $\alpha > 0$ there exists an integer $t \geq 1$, and a partition of $\{0,1\}^{n \cdot k}$ into $t + 1$ sets $W_1, \ldots, W_t, W'$, such that:

1. $\mathbf{Pr}_X[X \in W'] \leq 2^{-\Omega(b)}$.

---

[8] In condition 1, we require that the convex combination of the μ_i's will be strictly smaller than μ. This is not the most general case, but it will be convenient for us to use this definition.

2. *For every $i \in [t]$ the probability distribution of $Y \mid X \in W_i$ (that is - of Y conditioned on the event $X \in W_i$) is α-good. In other words: for every $i \in [t]$ there exist at least $(1 - \gamma/2) \cdot u$ values of $z \in [u]$ for which*

$$H^\infty(Y_z | X \in W_i) \geq (0.52 - \alpha) \cdot b.$$

Before proving Lemma 2.3 we give the proof of Lemma 2.2 which follows easily from Lemma 2.3.

Proof of Lemma 2.2: The lemma follows immediately from Lemma 2.3 and from the following equality, which holds for every partition $W_1, \ldots, W_t, W'$, and for every y:

$$\mathbf{Pr}[Y = y] = \sum_{i=1}^{t} \mathbf{Pr}[X \in W_i] \cdot \mathbf{Pr}[Y = y \mid X \in W_i]$$
$$+ \mathbf{Pr}[X \in W'] \cdot \mathbf{Pr}[Y = y \mid X \in W'].$$

If the partition $W_1, \ldots, W_t, W'$ satisfies the two conditions of Lemma 2.3 then from Definition 2.1 it is clear that Y is exponentially (in b) close to a convex combination of α-good distributions. □

Proof of Lemma 2.3: Every random variable Y_z is a function of X, and so it partitions $\{0,1\}^{n \cdot k}$ in the following way:

$$\{0,1\}^{n \cdot k} = \bigcup_{y \in \{0,1\}^n} (Y_z)^{-1}(y),$$

where $(Y_z)^{-1}(y) \overset{\triangle}{=} \{x \in \{0,1\}^{n \cdot k} \mid Y_z(x) = y\}$. For each $z \in [u]$ we define the set

$$B_z \overset{\triangle}{=} \bigcup_{\{y \mid \mathbf{Pr}[Y_z = y] > 2^{-(0.52 - \alpha/2) \cdot b}\}} (Y_z)^{-1}(y)$$
$$= \left\{ x' \in \{0,1\}^{n \cdot k} \;\middle|\; \mathbf{Pr}_X[Y_z(X) = Y_z(x')] > 2^{-(0.52 - \alpha/2) \cdot b} \right\}.$$

Intuitively, B_z contains all values of x that are "bad" for Y_z, where in "bad" we mean that $Y_z(x)$ is obtained with relatively high probability in the distribution $Y_z(X)$.

Definition 2.4 (Good Triplets) *Let $(z_1, z_2, z_3) \in [u]^3$ be a triplet of seed values. Since each seed value is actually a vector in F^k we can write each z_i ($i = 1, 2, 3$) as a vector $(z_{i1}, \ldots, z_{ik})$, where each z_{ij} is an integer in the range $0 \ldots (p - 1)$. We say that the triplet (z_1, z_2, z_3) is **good** if the following two conditions hold:*

1. *For all $2 \leq j \leq k$, $z_{1j} = z_{2j} = z_{3j}$.*
2. *There exists a positive integer $0 < a < p$ such that $z_{21} = z_{11} + a$ and $z_{31} = z_{11} + 2a$, where the equalities are over $F = \text{GF}(p)$.*

That is, the vectors z_1, z_2, z_3 are identical in all coordinates different from one, and their first coordinates form an arithmetic progression of length three in $F = GF(p)$.

The next two claims are the place where our analysis differs from that of [6] and [4]. We devote Section 3 to the proofs of these two claims. The first claim shows that the intersection of the "bad" sets $B_{z_1}, B_{z_2}, B_{z_3}$ for a good triplet (z_1, z_2, z_3) is small:

Claim 2.5 *For every good triplet (z_1, z_2, z_3) it holds that*

$$\mathbf{Pr}_X[X \in B_{z_1} \cap B_{z_2} \cap B_{z_3}] \leq C_\alpha \cdot 2^{-(\alpha/4)\cdot b},$$

where C_α is a constant depending only on α.

The second claim shows that every set of seed values who's density is larger than $\gamma/2$ contains a good triplet.

Claim 2.6 *Let $T \subset [u]$ be such that $|T| > (\gamma/2) \cdot u$. Then T contains a good triplet.*

The proof of Lemma 2.3 continues along the same lines as in [4]. We define for each $x \in \{0,1\}^{n \cdot k}$ a vector $\pi(x) \in \{0,1\}^u$ in the following way :

$$\forall z \in [u] \quad , \quad \pi(x)_z = 1 \iff x \in B_z.$$

For a vector $\pi \in \{0,1\}^u$, let $w(\pi)$ denote the weight of π (i.e. the number of 1's in π). Since the weight of $\pi(x)$ denotes the number of seed values for which x is "bad", we would like to somehow show that for most x's $w(\pi(x))$ is small. This can be proven by combining Claim 2.5 with Claim 2.6, as shown by the following claim.

Claim 2.7

$$\mathbf{Pr}_X[w(\pi(X)) > (\gamma/2) \cdot u] \leq u^3 \cdot C_\alpha \cdot 2^{-(\alpha/4)\cdot b}.$$

Proof. If x is such that $w(\pi(x)) > (\gamma/2) \cdot u$ then, by Claim 2.6, we know that there exists a good triplet (z_1, z_2, z_3) such that $x \in B_{z_1} \cap B_{z_2} \cap B_{z_3}$. Therefore we have $\mathbf{Pr}_X[w(\pi(X)) > (\gamma/2) \cdot u] \leq \mathbf{Pr}_X[\exists \text{ a good triplet } (z_1, z_2, z_3) \text{ s.t } x \in B_{z_1} \cap B_{z_2} \cap B_{z_3}]$. Now, using the union bound and Claim 2.5 we can bound this probability by $u^3 \cdot C_\alpha \cdot 2^{-(\alpha/4)\cdot b}$.

From Claim 2.7 we see that every x (except for an exponentially small set) is contained in at most $(\gamma/2) \cdot u$ sets B_z. The idea is now to partition the space $\{0,1\}^{n \cdot k}$ into sets of x's that have the same $\pi(x)$. If we condition the random variable Y on the event $\pi(X) = \pi_0$, where π_0 is of small weight, we will get an α-good distribution. We now explain this idea in more details. We define the following sets

$$BAD_1 \overset{\triangle}{=} \{\pi' \in \{0,1\}^u \mid w(\pi') > (\gamma/2) \cdot u\},$$

$$BAD_2 \triangleq \left\{ \pi' \in \{0,1\}^u \mid \mathbf{Pr}_X[\pi(X) = \pi'] < 2^{-(\alpha/2)\cdot b} \right\},$$

$$BAD \triangleq BAD_1 \cup BAD_2.$$

The set $BAD \subset \{0,1\}^u$ contains values $\pi' \in \{0,1\}^u$ that cannot be used in the partitioning process described in the last paragraph. There are two reasons why a specific value $\pi' \in \{0,1\}^u$ is included in BAD. The first reason is that the weight of π' is too large (i.e. larger than $(\gamma/2) \cdot u$), these values of π' are included in the set BAD_1. The second less obvious reason for π' to be excluded from the partitioning is that the set of x's for which $\pi(x) = \pi'$ is of extremely small probability. These values of π' are bad because we can say nothing about the min-entropy of Y when conditioned on the event[9] $\pi(X) = \pi'$.

Having defined the set BAD, we are now ready to define the partition required by Lemma 2.3. Let $\{\pi^1, \ldots, \pi^t\} = \{0,1\}^u \backslash BAD$. We define the sets $W_1, \ldots, W_t, W' \subset \{0,1\}^{n\cdot k}$ as follows:

- $W' = \{x \mid \pi(x) \in BAD\}$.
- $\forall i \in [t]$, $W_i = \{x \mid \pi(x) = \pi^i\}$.

Clearly, the sets $W_1, \ldots, W_t, W'$ form a partition of $\{0,1\}^{n\cdot k}$. We will now show that this partition satisfies the two conditions required by Lemma 2.3. To prove the first part of the lemma note that the probability of W' can be bounded by (using Claim 2.7 and the union-bound)

$$\mathbf{Pr}_X[X \in W'] \leq \mathbf{Pr}_X[\pi(X) \in BAD_1] + \mathbf{Pr}_X[\pi(X) \in BAD_2]$$
$$\leq u^3 \cdot C_\alpha \cdot 2^{-(\alpha/4)\cdot b} + 2^u \cdot 2^{-(\alpha/2)\cdot b} = 2^{-\Omega(b)}.$$

We now prove that $W_1, \ldots, W_t$ satisfy the second part of the lemma. Let $i \in [t]$, and let $z \in [u]$ be such that $(\pi^i)_z = 0$ (there are at least $(1 - \gamma/2) \cdot u$ such values of z). Let $y \in \{0,1\}^n$ be any value. If $\mathbf{Pr}[Y_z = y] > 2^{-(0.52-\alpha/2)\cdot b}$ then $\mathbf{Pr}[Y_z = y \mid X \in W_i] = 0$ (this follows from the way we defined the sets B_z and W_i). If on the other hand $\mathbf{Pr}[Y_z = y] \leq 2^{-(0.52-\alpha/2)\cdot b}$ then $\mathbf{Pr}[Y_z = y \mid X \in W_i] \leq \frac{\mathbf{Pr}[Y_z=y]}{\mathbf{Pr}[X\in W_i]} \leq 2^{-(0.52-\alpha/2)\cdot b}/2^{-(\alpha/2)\cdot b} = 2^{-(0.52-\alpha)\cdot b}$. Hence, for all values of y we have $\mathbf{Pr}[Y_z = y \mid X \in W_i] \leq 2^{-(0.52-\alpha)\cdot b}$. We can therefore conclude that for all $i \in [t]$, $H^\infty(Y_z|X \in W_i) \geq (0.52 - \alpha) \cdot b$. This completes the proof of Lemma 2.3. □

3   Proving Claim 2.5 and Claim 2.6 Using Results from Additive Number Theory

In this section we prove Claim 2.5 and Claim 2.6. These two claims are the only place in which our analysis differs from that of [6] and [4]. In the proofs we use

---

[9] Consider the extreme case where there is only one $x_0 \in \{0,1\}^{n\cdot k}$ with $\pi(x_0) = \pi'$. In this case the min-entropy of Y, when conditioned on the event $X \in \{x_0\}$, is zero, even if the weight of $\pi(x_0)$ is small.

two results from additive number theory. The first is a quantitative version of Roth's theorem [8] given by Bourgain [2]. The second is a Lemma of Bourgain that deals with sum-sets and difference-sets.

3.1   Proof of Claim 2.5

The proof of the claim relies on the following result from additive number theory due to Bourgain [2][10].

Lemma 3.1 ([2]) *For every $\epsilon > 0$ there exists a constant C_ϵ such that the following holds: Let A, B be subsets of an abelian group G. Let $\Gamma \subset A \times B$, and define*

$$S \overset{\triangle}{=} \{a + b \mid (a,b) \in \Gamma\},$$

$$D \overset{\triangle}{=} \{a - b \mid (a,b) \in \Gamma\}.$$

Suppose that there exists $K > 0$ such that $|A|, |B|, |S| \leq K$, then

$$|D| < C_\epsilon \cdot K^{(1/0.52)+\epsilon}.$$

Before we can apply Lemma 3.1 we need some notations. Let $U \overset{\triangle}{=} B_{z_1} \cap B_{z_2} \cap B_{z_3}$. We define for every $i = 1, 2, 3$ the set $V_i \overset{\triangle}{=} \{Y_{z_i}(x) \mid x \in U\}$. Next, we define a subset $\Gamma \subset V_1 \times V_3$ as follows

$$\Gamma \overset{\triangle}{=} \{(v_1, v_3) \mid \exists x \in U \ \ s.t \ \ Y_{z_1}(x) = v_1 \ \ and \ \ Y_{z_3}(x) = v_3\}.$$

We now define the sets S and D as in Lemma 3.1, where the roles of A and B are taken by V_1 and V_3.

$$S \overset{\triangle}{=} \{v_1 + v_3 \mid (v_1, v_3) \in \Gamma\},$$

$$D \overset{\triangle}{=} \{v_1 - v_3 \mid (v_1, v_3) \in \Gamma\}.$$

We also define

$$K \overset{\triangle}{=} 2^{(0.52-\alpha/2)\cdot b},$$

and

$$\hat{U} \overset{\triangle}{=} \{x_1 \in \{0,1\}^n \mid \exists x_2, \ldots, x_k \in \{0,1\}^n \ \ s.t \ \ (x_1, \ldots, x_k) \in U\}.$$

the following claim states several facts that, when combined, will enable us to use Lemma 3.1 on the sets we have defined.

Claim 3.2 *the following is true:*

1. $|V_1|, |V_2|, |V_3| \leq K$.
2. $|S| \leq |V_2| \leq K$.
3. $|\hat{U}| \leq |D|$.

---

[10] This result appears in [2] with respect to subsets of a torsion-free group. However, it is easily seen from the proof that it holds for subsets of any abelian group.

Proof. 1. Follows directly from the definition of the sets B_{z_i} and V_i. Each value $v \in V_i$ is a "heavy element" of the random variable Y_{z_i}. That is, the probability that $Y_{z_i} = v$ is at least $2^{-(0.52-\alpha/2)\cdot b} = K^{-1}$, and so there can be at most K such values.

2. What we will show is that the set S is contained in the set $2V_2 \overset{\triangle}{=} \{2 \cdot v \mid v \in V_2\}$ (these two sets are actually equal, but we will not need this fact). To see this, recall that from the definition of a good triplet we have that for every $x \in \{0,1\}^{n \cdot k}$

$$Y_{z_1}(x) + Y_{z_3}(x) = 2 \cdot Y_{z_2}(x). \tag{1}$$

Let $v \in S$. From the definition of S (and of Γ) we know that there exists $x \in U$ and $v_1 \in V_1, v_3 \in V_3$ such that $Y_{z_1}(x) = v_1, Y_{z_3}(x) = v_3$ and $v = v_1 + v_3$. From Eq.1 we now see that $v = 2 \cdot Y_{z_2}(x)$, and therefore $v \in 2V_2$. The inequality now follows from the fact that $|V_2| = |2V_2|$.

3. This follows in a similar manner to 2. We will show that the set $\hat{U}$ is contained in the set $c \cdot D \overset{\triangle}{=} \{c \cdot v \mid v \in D\}$, for some $0 < c < p$ (again, the two sets are actually equal, but we will not use this fact). From the definition of a good triplet we know that there exists $0 < c < p$ such that for every $x = (x_1, \ldots, x_k) \in \{0,1\}^{n \cdot k}$

$$c \cdot (Y_{z_1}(x) - Y_{z_3}(x)) = x_1. \tag{2}$$

Let $x_1 \in \hat{U}$. From the definition of $\hat{U}$ we know that there exist $x_2, \ldots, x_k \in \{0,1\}^n$ such that $x = (x_1, \ldots, x_k) \in U$. From Eq.2 it follows that $x_1 \in c \cdot D$, since $Y_{z_1}(x) - Y_{z_3}(x) \in D$ by definition.

Let $\epsilon \overset{\triangle}{=} \frac{\alpha}{4} \cdot \frac{1}{(0.52-\alpha/2)}$. From the first two parts of Claim 3.2 we see that we can apply Lemma 3.1 with $A = V_1$ and $B = V_3$ to get that $|D| < C_\epsilon \cdot K^{(1/0.52)+\epsilon}$, (where C_ϵ depends only on ϵ, which in turn depends only on α). Substituting the values of ϵ and K we see that (we can write C_α instead of C_ϵ)

$$|D| < C_\alpha \cdot 2^{b \cdot (0.52-\alpha/2) \cdot (1/0.52+\epsilon)} = C_\alpha \cdot 2^{b \cdot \left(1 + \epsilon(0.52-\alpha/2) - \frac{\alpha}{2} \cdot \frac{1}{0.52}\right)}$$

$$= C_\alpha \cdot 2^{b \cdot \left(1 + \frac{\alpha}{4} - \frac{\alpha}{2} \cdot \frac{1}{0.52}\right)} \leq C_\alpha \cdot 2^{b \cdot \left(1 - \frac{\alpha}{4}\right)},$$

where C_α depends only on α.

Using the third part of Claim 3.2 and Eq. 3 we conclude that $|\hat{U}| \leq |D| \leq C_\alpha \cdot 2^{b \cdot \left(1 - \frac{\alpha}{4}\right)}$. We can therefore bound the probability of U by $\mathbf{Pr}_x[X \in U] \leq \mathbf{Pr}_{X_1}[X_1 \in \hat{U}] \leq 2^{-b} \cdot |\hat{U}| \leq 2^{-b} \cdot \left(C_\alpha \cdot 2^{b \cdot \left(1 - \frac{\alpha}{4}\right)}\right) = C_\alpha \cdot 2^{-(\alpha/4) \cdot b}$, (the second inequality follows from the fact that the min-entropy of X_1 is at least b). This completes the proof of Claim 2.5. □

3.2 Proof of Claim 2.6

The claim follows from Roth's theorem [8] on arithmetic progressions of length three. For our purposes we require the quantitative version of this theorem as proven by Bourgain [3].

Theorem 3.3 ([3]) *Let $\delta > 0$, let $N \geq \exp(\delta^{-2})$ and let $A \subset \{1, \ldots, N\}$ be a set of size at least δN. Then A contains an arithmetic progression of length three.*

Each element in T is a vector in F^k (recall that $F = \mathrm{GF}(\mathrm{p})$). A simple counting argument shows that T must contain a subset T' such that (a) $|T'| > (\gamma/2) \cdot p$. (b) All vectors in T' are identical in all coordinates different than one. Using Theorem 3.3 and using the fact that p was chosen to be greater than $\exp(\gamma^{-2})$, we conclude that there exists a triplet in T' such that the first coordinates of this triplet form an arithmetic progression. This is a good triplet, since in T' the vectors are identical in all coordinates different than one. $\square$

Acknowledgements

The authors would like to thank Ran Raz, Omer Reingold and Avi Wigderson for helpful conversations. A.S. would also like to thank Oded Goldreich for helpful discussions on related problems.

References

1. Boaz Barak, Russell Impagliazzo, and Avi Wigderson. Extracting randomness using few independent sources. In *45th Symposium on Foundations of Computer Science (FOCS 2004)*, pages 384–393, 2004.
2. Jean Bourgain. On the dimension of kakeya sets and related maximal inequalities. *Geom. Funct. Anal.*, (9):256–282, 1999.
3. Jean Bourgain. On triples in arithmetic progression. *Geom. Funct. Anal.*, (9):968–984, 1999.
4. Zeev Dvir and Ran Raz. Analyzing linear mergers. *Electronic Colloquium on Computational Complexity (ECCC)*, (025), 2005.
5. Timothy Gowers. A new proof of szemeredi's theorem. *Geom. Funct. Anal.*, (11):465–588, 2001.
6. Chi-Jen Lu, Omer Reingold, Salil Vadhan, and Avi Wigderson. Extractors: optimal up to constant factors. In *STOC '03: Proceedings of the thirty-fifth annual ACM symposium on Theory of computing*, pages 602–611. ACM Press, 2003.
7. Ran Raz. Extractors with weak random seeds. *STOC 2005 (to appear)*, 2005.
8. Klaus F Roth. On certain sets of integers. *J. Lond. Math. Soc.*, (28):104–109, 1953.
9. Ronen Shaltiel. Recent developments in extractors. *Bulletin of the European Association for Theoretical Computer Science*, 77:67–95, 2002.
10. Endre Szemeredi. On sets of integers containing no k elements in arithmetic progression. *Acta. Arith.*, (27):299–345, 1975.
11. Amnon Ta-Shma. On extracting randomness from weak random sources (extended abstract). In *STOC '96: Proceedings of the twenty-eighth annual ACM symposium on Theory of computing*, pages 276–285. ACM Press, 1996.

Finding a Maximum Independent Set
in a Sparse Random Graph*

Uriel Feige and Eran Ofek

Weizmann Institute of Science,
Department of Computer Science and Applied Mathematics,
Rehovot 76100, Israel
{uriel.feige,eran.ofek}@weizmann.ac.il

Abstract. We consider the problem of finding a maximum independent set in a random graph. The random graph G is modelled as follows. Every edge is included independently with probability $\frac{d}{n}$, where d is some sufficiently large constant. Thereafter, for some constant α, a subset I of αn vertices is chosen at random, and all edges within this subset are removed. In this model, the planted independent set I is a good approximation for the maximum independent set I_{max}, but both $I \setminus I_{max}$ and $I_{max} \setminus I$ are likely to be nonempty. We present a polynomial time algorithms that with high probability (over the random choice of random graph G, and without being given the planted independent set I) finds a maximum independent set in G when $\alpha \geq \sqrt{c_0 \log d/d}$, where c_0 is some sufficiently large constant independent of d.

1 Introduction

Let $G = (V, E)$ be a graph. An independent set I is a subset of vertices which contains no edges. The problem of finding a maximum size independent set in a graph is NP-hard. Moreover, for any $\epsilon > 0$ there is no $n^{1-\epsilon}$ (polynomial time) approximation algorithm for it unless NP=ZPP [12]. The best approximation ratio currently known for maximum independent set [6] is $O(n(\log \log n)^2/(\log n)^3)$.

In light of the above mentioned negative results, one may try to design a heuristic which performs well on typical instances. Karp [14] proposed trying to find a maximum independent set in a random graph. However, even this problem appears to be beyond the capabilities of current algorithms. For example let $G_{n,1/2}$ denote the random graph on n vertices obtained by choosing randomly and independently each possible edge with probability $1/2$. A random $G_{n,1/2}$ graph has almost surely maximum independent set of size $2(1 + o(1)) \log_2 n$. A simple greedy algorithm almost surely finds an independent set of size $\log_2 n$ [11]. However, there is no known polynomial time algorithm that almost surely finds an independent set of size $(1 + \epsilon) \log_2 n$ (for any $\epsilon > 0$).

To further simplify the problem, Jerrum [13] and Kucera [15] proposed a *planted model* in which a random graph $G_{n,1/2}$ is chosen and then a clique of size k is randomly

---

* This work was supported in part by a grant from the G.I.F., the German-Israeli Foundation for Scientific Research and Development. Part of this work was done while the authors were visiting Microsoft Research in Redmond, Washington.

C. Chekuri et al. (Eds.): APPROX and RANDOM 2005, LNCS 3624, pp. 282–293, 2005.

placed in the graph. (A clique in a graph G is an independent set in the edge complement of G, hence all algorithmic results that apply to one of the problems apply to the other.) Alon Krivelevich and Sudakov [2] gave an algorithm based on spectral techniques that almost surely finds the planted clique for $k = \Omega(\sqrt{n})$. More generally, one may extend the range of parameters of the above model by planting an independent set in $G_{n,p}$, where p need not be $1/2$, and may also depend on n. The $G_{n,p,\alpha}$ model is as follows: n vertices are partitioned at random into two sets of vertices, I of size αn and C of size $(1 - \alpha)n$. No edges are placed within the set I, thus making it an independent set. Every other possible edge (with at least one endpoint not in I) is added independently at random with probability p. The goal of the algorithm, given the input G (without being given the partition into I and C) is to find a maximum independent set. Intuitively, as α becomes smaller the size of the planted independent is closer to the probable size of the maximum independent set in $G_{n,p}$ and the problem becomes harder.

We consider values of p as small as d/n where d is a large enough constant. A difficulty which arises in this sparse regime (e.g. when d is constant) is that the planted independent set I is not likely to be a maximum independent set. Moreover, with high probability I is not contained in any maximum independent set of G. For example, there are expected to be $e^{-d}n$ vertices in C of degree one. It is very likely that two (or more) such vertices $v, w \in C$ will have the same neighbor, and that it will be some vertex $u \in I$. This implies that every maximum independent set will contain v, w and not u, and thus I contains vertices that are not contained in any maximum independent set.

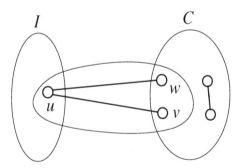

Fig. 1. The vertex $u \in I$ is not contained in any maximum independent set because no other edges touch v, w

A similar argument shows that there are expected to be $e^{-\Omega(d)}n$ isolated edges. This implies that there will be an exponential number of maximum independent sets.

1.1 Our Result

We give a polynomial time algorithm that searches for a maximum independent set of G. Given a random instance of $G_{n,\frac{d}{n},\alpha}$, the algorithm almost surely succeeds, when $d > d_0$ and $\alpha \geq \sqrt{c_0 \log d/d}$ (d_0, c_0 are some universal constants). The parameter d can be also an arbitrary increasing function of n.

Remark: A significantly more complicated version of our algorithm works for a wider range of parameters, namely, for $\alpha \geq \sqrt{c_0/d}$ rather than $\alpha \geq \sqrt{c_0 \log d/d}$. The improved algorithm will appear in the full version of this paper [1].

1.2   Related Work

For $p = 1/2$, Alon Krivelevich and Sudakov [2] gave an efficient spectral algorithm which almost surely finds the planted independent set when $\alpha = \Omega(1/\sqrt{n})$. For the above mentioned parameters, the planted independent set is almost surely the unique maximum independent set.

A few papers deal with *semi-random* models which extend the planted model by enabling a mixture of random and adversarial decisions. Feige and Kilian [7] considered the following model: a random $G_{n,p,\alpha}$ graph is chosen, then an adversary may add arbitrarily many edges between I and C, and make arbitrary changes (adding or removing edges) inside C. For any constant $\alpha > 0$ they give a heuristic that almost surely outputs a list of independent sets containing the planted independent set, whenever $p > (1 + \epsilon) \ln n/\alpha n$ (for any $\epsilon > 0$). The planted independent set may not be the only independent set of size αn since the adversary has full control on the edges inside C. Possibly, this makes the task of finding the planted independent set harder.

In [8] Feige and Krauthgamer considered a less adversarial semi-random model in which an adversary is allowed to add edges to a random $G_{n,\frac{1}{2},\frac{1}{\sqrt{n}}}$ graph. Their algorithm almost surely extracts the planted independent set and certifies its optimality.

Heuristics for optimization problems different than max independent set will be discussed in the following section.

Technique and Outline of the Algorithm.   Our algorithm builds on ideas from the algorithm of Alon and Kahale [1], which was used for recovering a planted 3-coloring in a random graph. The algorithm we propose has the following 3 phases:

1. Get an approximation of I, C denoted by I', C', OUT, where I' is an independent set. The error term $|C \triangle C'| + |I \triangle I'|$ should be at most $e^{-c \log d} n$ where c is a large enough universal constant (this phase is analogous to the first two phases of [1]).
2. Move to OUT vertices of I', C' which have non typical degrees.
   We stop when I', C' become *promising*: every vertex of C' has at least 4 edges to I' and no vertex of I' has edges to OUT. At this point we have a promising partial solution I', C' and the error term (with respect to I, C) is still small. Using the fact that (almost surely) random graphs have no small dense sets, it can be shown that I' is *extendable*: $I' \subseteq I_{max}$ for some optimal solution I_{max}.
3. Extend the independent set I' optimally using the vertices of OUT. This is done by finding a maximum independent in graph induced on OUT and adding it to I'. With high probability the structure of OUT will be easy enough so that a maximum independent set can be efficiently found. OUT is a random graph of size $n/poly(d)$. Notice however, that the set OUT depends on the graph itself thus we can not argue that it is a random $G_{\frac{n}{poly(d)},\frac{d}{n}}$ graph.

---

[1] Can be found in http://wisdom.weizmann.ac.il/~erano

The technique of [1] was implemented successfully on various problems in the planted model: hypergraph coloring, 3-SAT, 4-NAE, min-bisection (by Chen and Frieze [3] , Flaxman [9] , Goerdt and Lanka [10], Coja-Oghlan [4] respectively).

Perhaps the work closest in nature to the work in the current paper is that of Amin Coja-Oghlan [4] on finding a bisection in a sparse random graph. Both in our work and in that of [4], one is dealing with an optimization problem, and the density of the input graph is such that the planted solution is not an optimal solution. The algorithm for bisection in [4] is based on spectral techniques, and has the advantage that it provides a certificate showing that the solution that it finds is indeed optimal. Our algorithm for maximum independent set does not use spectral techniques and does not provide a certificate for optimality.

An important difference between planted models for independent set and those for other problems such as 3-coloring and min-bisection is that in our case the planted classes I, C are not symmetric. The lack of symmetry between I and C makes some of the ideas used for the more symmetric problems insufficient. In the approach of [1], a vertex is removed from its current color class and placed in OUT if its degree into some other current color class is very different than what one would typically expect to see between the two color classes. This procedure is shown to "clean" every color class C from all vertices that should have been from a different color class, but were wrongly assigned to class C in phase 1 of the algorithm. (The argument proving this goes as follows. Every vertex remaining in the wrong color class by the end of phase 2 must have many neighbors that are wrongly assigned themselves. Thus the set of wrongly assigned vertices induces a small subgraph with large edge density. But G itself does not have any such subgraphs, and hence by the end of phase 2 it must be the case that all wrongly assigned vertices were moved into OUT.) It turns out that this approach works well when classes are of similar nature (such as color classes, or two sides of a bisection), but does not seem to suffice in our case where I' is supposed to be an independent set whereas C' is not. Specifically, the set I' might still contain wrongly assigned vertices, and might not be a subset of a maximum independent set in the graph. Under these circumstances, phase 3 will not result in a maximum independent set. Our solution to this problem involves the following aspects, not present in previous work. In phase 2 we remove from I' every vertex that has even one edge connecting it to OUT. This adds more vertices to OUT and may possibly create large connected components in OUT. Indeed, we do not show that OUT has no large connected components, which is a key ingredient in previous approaches. Instead, we analyze the 2-core of OUT and show that the 2-core has no large components. Then, in phase 3, we use dynamic programming to find a maximum independent set in OUT, and use the special structure of OUT to show that the algorithm runs in polynomial time.

1.3 Notation and Terminology

Let $G = (V, E)$ and let $U \subset V$. The subgraph of G induced by the vertices of U is denoted by $G[U]$. We denote by $\deg^E(v)_U$ the number of edges from E that connect a vertex v to $U \subset V$; when E is clear from the context we will use $\deg(v)_U$. We use $\Gamma(U)$ to denote the vertex neighborhood of $U \subset V$ (excluding U). The parameter d (specifying the expected degree in the random graph G) is assumed to be sufficiently large, and some of the inequalities that we shall derive implicitly use this assumption, without stating it explicitly. The term *with high probability* (w.h.p.) is used to denote a sequence of probabilities that tends to 1 as n tends to infinity.

2   The Algorithm

FindIS(G)

1. (a) Set: $I_1 = \{v : \deg(v) < d - \alpha d/2\}$,
   $\quad\quad\quad C_1 = \{v : \deg(v) \geq d - \alpha d/2\}$,
   $\quad\quad\quad OUT_1 = \emptyset$.
   (b) For every edge (u, v) such that both u, v are in I_1, move u, v to OUT_1.
2. Set $I_2 = I_1$, $C_2 = C_1$, $OUT_2 = OUT_1$.
   A vertex $v \in C_2$ is *removable* if $\deg(v)_{I_2} < 4$.
   Iteratively: find a removable vertex v. Move v and $\Gamma(v) \cap I_2$ to OUT_2.
3. Output the union of I_2 and a maximum independent set of $G[OUT_2]$. We will explain later how this is done efficiently.

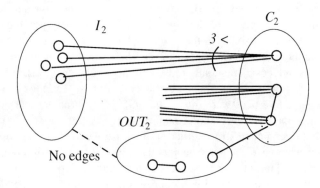

Fig. 2. After step 2 of the algorithm, I_2 is an independent set, there are no edges between I_2 and OUT_2, and every vertex $v \in C_2$ has at least four neighbors in I_2

3   Correctness

When $d > n^{32/c_0}$ a simple argument (using the union and the Chernoff bounds), which we omit, shows that after step 1a of the algorithm it holds that $I_1 = I, C_1 = C$. In fact, when $d > n^{32/c_0}$ the planted independent set is the unique maximum independent set and this case was already covered in [5]. From now we will assume that $d < n^{32/c_0}$.

Let I_{max} be a maximum independent set of G. We establish two theorems. Theorem 1 guarantees the algorithm correctness and Theorem 2 guarantees its efficient running time. Here we present these two theorems, and their proofs are deferred to later sections.

Theorem 1. *W.h.p. there exists I_{max} such that $I_2 \subseteq I_{max}, C_2 \cap I_{max} = \emptyset$.*

Definition 1. *The 2-core of a graph G is the maximal subgraph in which the minimal degree is at least 2.*

Clearly the 2-core is unique and can be found by iteratively removing vertices whose degree is smaller than 2.

Theorem 2. *W.h.p. the largest connected component in the 2-core of $G[OUT_2]$ has cardinality of at most $2 \log n$.*

Let G be any graph. Those vertices of G that do not belong to the 2-core form trees. Each such tree is either disconnected from the 2-core or it is connected by exactly one edge to the 2-core. To find a maximum independent set of $G[OUT_2]$ we need to find a maximum independent set in each connected component of $G[OUT_2]$ separately. For each connected component D_i of $G[OUT_2]$ we find the maximum independent set as follows: let C_i be the intersection of D_i with the 2-core of G. We enumerate all possible independent sets in C_i (there are at most $2^{|C_i|}$ possibilities), each one of them can be optimally extended to an independent set of D_i by solving (separately) a maximum independent set problem on each of the trees connected to C_i. For some trees we may have to exclude the tree vertex which is connected to C_i if it is connected to a vertex of the independent set that we try to extend. On each tree the problem can be solved by dynamic programming.

Corollary 1. *A maximum independent set of $G[OUT_2]$ can be found efficiently.*

3.1 Dense Sets and Degree Deviations

In proving the correctness of the algorithm, we will use structural properties of the random graph G. In particular, such a random graph most likely has no small dense sets (small sets of vertices that induce many edges). This fact will be used on several occasions to derive a proof by contradiction. Namely, certain undesirable outcomes of the algorithm cannot occur, as otherwise they lead to a discovery of a small dense set. The lemmas relating to these properties are rather standard and their proofs are omitted due to lack of space.

Lemma 1. *Let G be a random graph taken from $G_{n,p,\alpha}$ ($p = \frac{d}{n}$, $d < n^{32/c_0}$). The following holds:*

1. *W.h.p. for every set $U \subset V$ of cardinality smaller than $2n/d^5$ the number of edges inside U is bounded by $\frac{4}{3}|U|$.*
2. *Let $c \geq 3$. With probability of at least $n^{-0.9(c-1)}$ for every set of vertices U of size smaller than n/d^2 the number of edges inside U is less than $c|U|$.*
3. *W.h.p. there is no $C' \subseteq C$ such that $\frac{n}{2d^5} \leq |C'| \leq \frac{2n \log d}{d}$ and $|\Gamma(C') \cap I| \leq |C'|$.*

Corollary 2. *Let G be a graph which has the property from Lemma 1 part 1. Let A, B be any two disjoint sets of vertices each of size smaller than n/d^5. If every vertex of B has at least 2 edges going into A, then $|A| \geq |B|/2$.*

Lemma 2. *Let $d < n^{32/c_0}$. The following hold with probability $> 1 - e^{-n^{0.1}}$:*

1. *The number of vertices from I which are not I_1 is at most $e^{-\alpha^2 d/64}n$.*
2. *The number of vertices from C whose degree into I is $< \alpha d/2$ is at most $e^{-\alpha^2 d/64}n$.*
3. *The number of vertices from C which are not in C_1 is most $e^{-\alpha^2 d/64}n$.*
4. *The number of edges that contain a vertex with degree at least $3d$ is at most $3e^{-d}dn$.*

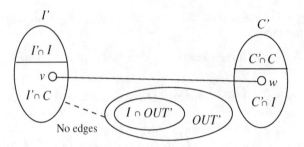

Fig. 3. A vertex $v \in (I' \cap C)$ which has exactly one edge into $I \cap C'$

3.2   Proof of Theorem 1

We would have liked to prove that with high probability $I_2 \subseteq I \subseteq I_2 \cup OUT_2$. However, this is incorrect when d is a constant.

Lemma 3. *Let I be any independent set of G and let $C \triangleq V \setminus I$. Let I', C', OUT' be an arbitrary partition of V for which I' is an independent set. If the following hold:*

1. *$|(I' \cap C) \cup (I \cap C')| < n/d^5$.*
2. *Every vertex of C' has 4 neighbors in I'. There are no edges between I' and OUT'.*
3. *The graph G has no small dense subsets as described in Lemma 1 part 1.*

then there exists an independent set I_{new} (and $C_{new} \triangleq V \setminus I_{new}$) such that $I' \subseteq I_{new}, C' \subseteq C_{new}$ and $|I_{new}| \geq |I|$.

Proof. If we could show that on average a vertex of $U = (I' \cap C) \cup (I \cap C')$ contributes at least $4/3$ internal edges to U, then U would form a small dense set that contradicts Lemma 1. This would imply that $U = (I' \cap C) \cup (I \cap C')$ is the empty set, and we could take $I_{new} = I$ in the proof of Lemma 3. The proof below extends this approach to cases where we cannot take $I_{new} = I$.

Every vertex $v \in C'$ has at least 4 edges into vertices of I'. Since I is an independent set it follows that every vertex of $I \cap C'$ has at least 4 edges into $I' \cap C$. To complete the argument we would like to show that every vertex of $I' \cap C$ has at least 2 edges into $I \cap C'$. However, some vertices $v \in I' \cap C$ might have less than two neighbors in $I \cap C'$. In this case, we will modify I to get an independent set I_{new} (and $C_{new} \triangleq V \setminus I_{new}$) at least as large as I, for which every vertex of $I' \cap C_{new}$ has 2 neighbors in $I_{new} \cap C'$. This is done iteratively; after each iteration we set $I = I_{new}, C = C_{new}$. Consider a vertex $v \in (I' \cap C)$ which has strictly less than 2 edges into $I \cap C'$:

- If v has no neighbors in $I \cap C'$, then define $I_{new} = I \cup \{v\}$. I_{new} is an independent set because v (being in I') has no neighbors in I' nor in OUT'.
- If v has only one edge into $w \in (I \cap C')$ then define $I_{new} = (I \setminus \{w\}) \cup \{v\}$. I_{new} is an independent set because v (being in I') has no neighbors in I' nor in OUT'. The only neighbor of v in $I \cap C'$ is w.

The three properties (from the statement of this lemma) are maintained also with respect to I_{new}, C_{new} (replacing I, C): properties 2, 3 are independent on the sets I, C and property 1 is maintained since after each iteration it holds that $|(I' \cap C_{new}) \cup (I_{new} \cap C')| < |(I' \cap C) \cup (I \cap C')|$.

When the process ends, let U denote $(I' \cap C) \cup (I \cap C')$. Each vertex of $I' \cap C$ has at least 2 edges into $I \cap C'$, thus $|I \cap C'| \geq \frac{1}{2}|I' \cap C|$ (see Corollary 2). Each vertex of $I \cap C'$ has 4 edges into $I' \cap C$ so the number of edges in U is at least $4|I \cap C'| \geq 4|U|/3$ and also $|U| < n/d^5$, which implies that U is empty (by Lemma 1 part 1).   □

To use Lemma 3 with $I = I_{max}, I' = I_2$, we need to show that condition 1 is satisfied, i.e. $|I_{max} \triangle I_2| < n/d^5$.

Lemma 4. *With high probability* $|I_{max} \triangle I_2| < n/d^5$.

Proof. $|I_{max} \triangle I_2| \leq |I_{max} \triangle I| + |I \triangle I_2|$. By Lemma 2 the number of wrongly assigned vertices (in step 1a) is $e^{-\alpha^2 d/64}n$. By Lemma 5 $|OUT_2| < e^{-\alpha^2 d/70}n$. It follows that $|I \triangle I_2| << n/d^5$. It remains to bound $|I_{max} \triangle I|$:

$$|I_{max} \triangle I| = |I_{max} \setminus I| + |I \setminus I_{max}| \leq 2|I_{max} \setminus I| = 2|I_{max} \cap C|.$$

$I_{max} = (I_{max} \cap I) \cup (I_{max} \cap C)$. One can always replace $I_{max} \cap C$ with $\Gamma(I_{max} \cap C) \cap I$ to get an independent set $(I_{max} \cap I) \cup (\Gamma(I_{max} \cap C) \cap I)$. The maximum independent set in $G[C]$ is w.h.p. of cardinality at most $\frac{2n \log d}{d}$ (the proof is standard using first moment), this upper bounds $|I_{max} \cap C|$. From Lemma 1 part 3 if $|I_{max} \cap C| > n/(2d^5)$ then $|\Gamma(C \cap I_{max}) \cap I| > |I_{max} \cap C|$ which contradicts the maximality of I_{max}.   □

It follows that I_2 is contained in some maximum independent set I_{max}. It remains to show that the 2-core of OUT_2 has no large connected components.

3.3 Proof of Theorem 2

Lemma 5. *With probability of at least* $1 - e^{-c_1\alpha^2 d \log n}$ *the cardinality of* OUT_2 *is at most* $e^{-\alpha^2 d/70}n$ *(where c_1 is a universal constant independent of c_0).*

Proof. We will use the assumption $d < n^{-32/c_0}$. If u, v were moved to OUT_1 in step 1b then at least one of them belongs to $C \cap I_1$. Thus:

$$|OUT_1| \leq |C \cap I_1| + |\Gamma(C \cap I_1)| <$$
$$e^{-\alpha^2 d/64}n + 3de^{-\alpha^2 d/64}n + 3e^{-d}dn < e^{-\alpha^2 d/64 + \log d + O(1)}n$$

with probability $> 1 - e^{-n^{0.1}}$ (we use Lemma 2 parts 3, 4).

Let $C' = \{v \in C : \deg(v)_I < \alpha d/2\}$. The cardinality of C' is at most $e^{-\alpha^2 d/64}$ with probability at least $1 - e^{-n^{0.1}}$ (using Lemma 2 part 2). We will now show that $|OUT_2| \leq 2(|OUT_1| + |C'|)$. Let $U = OUT_1 \cup C'$. Start adding to U vertices of $OUT_2 \setminus U$ by the order in which the algorithm moved them to OUT_2. In each step we add at most 4 vertices to U. Assume that at some point $|U|$ becomes larger than $2(|OUT_1| + |C'|)$. The number of edges inside U is at least:

$$\frac{1}{4}\frac{1}{2}|U|\alpha d/2 = \frac{\alpha d}{16}|U|. \tag{1}$$

At this point $|U| \leq e^{-\alpha^2 d/64 + \log d + O(1)} n + 4 << n/d^5$ and U contains $\frac{\alpha d}{16}|U|$ edges. By Lemma 1 part 2 the probability that G contains such a dense set U is at most $e^{-0.9 \log n(\alpha d/16 - 1)} \leq e^{-c_1 \alpha^2 d \log n}$ (we use here $\alpha d >> \alpha^2 d$). □

Having established that OUT_2 is small, we would now like to establish that its structure is simple enough to allow one to find a maximum independent set of $G[OUT_2]$ in polynomial time. Establishing such a structure would have been easy if the vertices of OUT_2 were chosen independently at random, because a small random subgraph of a random graph G is likely to decompose into connected components no larger than $O(\log n)$. However, OUT_2 is extracted from G using some deterministic algorithm, and hence might have more complicated structure. For this reason, we shall now consider the 2-core of $G[OUT_2]$ and bound the size of its connected components.

Let A denote the 2-core of $G[OUT_2]$. In order to show that A has no large connected component, it is enough to show that A has no large tree. We were unable to show such a result for a general tree. Instead, we prove that A has no large *balanced* trees, that is trees in which at least $1/3$ fraction of the vertices belong to C. Fortunately, this turns out to be enough. Any set of vertices $U \subset V$ is called *balanced* if it contains at least $\frac{|U|}{3}$ vertices from C. We use the following reasoning: any maximal connected component of A is balanced - see Lemma 6 below. Furthermore, any balanced connected component of size at least $2 \log n$ (in vertices) must contain a balanced tree of size is in $[\log n, 2 \log n - 1]$ – see Lemma 7. We then complete the argument by showing that OUT_2 does not contain a balanced tree with size in $[\log n, 2 \log n]$.

Lemma 6. *Every maximal connected component of the 2-core of OUT_2 is balanced.*

Proof. Let A_i be such a maximal connected component. Every vertex of A_i has degree of at least 2 in A_i because A_i is a maximal connected component of a 2-core. $|A_i| \leq |OUT_2| < \frac{n}{d^5}$. If $\frac{|A_i \cap I|}{|A_i|}$ is more than $\frac{2}{3}$, then the number of internal edges in A_i is $> 2 \cdot \frac{2}{3}|A_i| > \frac{4}{3}|A_i|$ which contradicts Lemma 1. □

Lemma 7. *Let G be a connected graph whose vertices are partitioned into two sets: C and I. Let $\frac{1}{k}$ be a lower bound on the fraction of C vertices, where k is an integer. For any $1 \leq t \leq |V(G)|/2$ there exists a tree whose size is in $[t, 2t - 1]$ and at least $\frac{1}{k}$ fraction of its vertices are C.*

Proof. We use the following well know fact: any tree T contains a *center* vertex v such that each subtree hanged on v contains strictly less than half of the vertices of T.

Let T be an arbitrary spanning tree of G, with center v. We proceed by induction on the size of T. Consider the subtrees $T_1, ..., T_k$ hanged on v. If there exists a subtree T_j with at least t vertices then also $T \setminus T_j$ has at least t vertices. In at least one of $T_j, T \setminus T_j$ the fraction of C vertices is at least $\frac{1}{k}$ and the lemma follows by induction on it. Consider now the case in which all the trees have less than t vertices. If in some subtree T_j the fraction of C vertices is at most $\frac{1}{k}$, then we remove it and apply induction to $T \setminus T_j$. The remaining case is that in all the subtrees the fraction of C vertices is strictly more than $\frac{1}{k}$. In this case we start adding subtrees to the root v until for the first time the number of vertices is at least t. At this point we have a tree with at most $2t - 1$ vertices and the fraction of C vertices is at least $\frac{1}{k}$. To see that the fraction of C vertices is at least $\frac{1}{k}$, we only need to prove that the tree formed by v and the first subtree has $\frac{1}{k}$ fraction of C vertices. Let r be the number of C vertices in the first subtree and let b be the number of vertices in it. Since k is integer we have: $\frac{r}{b} > \frac{1}{k} \implies \frac{r}{b+1} \geq \frac{1}{k}$. □

We shall now prove that OUT_2 contains no balanced tree of size in $[\log n, 2\log n]$. To simplify the following computations, we will assume that $\alpha < 1/2$ (the proof can be modified to work for any α, we omit the details). Fix t to be some value in $[\log n, 2\log n]$. We will use the fact that the number of spanning trees of a t-clique is t^{t-2} (Cayley's formula). The probability that OUT_2 contains a balanced tree of size t is at most:

$$\sum_{\substack{T \text{ is balanced,} \\ |T|=t}} \Pr[T \subseteq E] \cdot \Pr[V(T) \subseteq OUT_2 \mid T \subseteq E] \leq \qquad (2)$$

$$\binom{n}{t} t^{t-1} \left(\frac{d}{n}\right)^{t-1} \max_{\substack{T \text{ balanced,} \\ |T|=t}} \{\Pr[V(T) \subseteq OUT_2 | T \subseteq E]\} \leq \qquad (3)$$

$$n\,(ed)^t \max_{\substack{T \text{ is balanced,} \\ |T|=t}} \{\Pr[V(T) \subseteq OUT_2 \mid T \subseteq E]\} \leq \qquad (4)$$

$$e^{t(\log d+1)+\log n} \max_{\substack{T \text{ is balanced,} \\ |T|=t}} \{\Pr[V(T) \subseteq OUT_2 \mid T \subseteq E]\} \qquad (5)$$

To upper bound the above expression by $o(1/\log n)$ (so we can use union bound over all choices of t), it is enough to prove that for some universal constant c_1 and any fixed balanced tree T of size t it holds that:

$$\Pr[V(T) \subseteq OUT_2 \mid T \subseteq E] \leq e^{-c_1(\alpha^2 dt)}.$$

The last term is bounded by $\leq e^{-c_1 c_0 t \log d}$ because $\alpha^2 d \geq c_0 \log d$. By choosing $c_0 > 3/c_1$ we derive the required bound. We will use the following equality

$$\Pr_E[V(T) \subseteq OUT_2 \mid T \subseteq E] = \Pr_E[V(T) \subseteq OUT_2(E \cup T)], \qquad (6)$$

which is true because the distribution of E given that $T \subseteq E$ is exactly the distribution of $E \cup T$. We have to show that for any balanced tree T of size t, $\Pr[V(T) \subseteq OUT_2(E \cup T)] \leq e^{-c_1 \alpha^2 dt}$. We use the technique introduced at [1] and modify it to our setting. To simplify the notation we will use F to denote the algorithm FindIS defined at Section 2. Given a fixed balanced tree T of size $\leq 2\log n$, we define an intermediate algorithm F' that knows the partition I, C and also $I(T)$ (which is $I \cap V(T)$). F' has no knowledge of which vertices are in $C(T)$ (in fact they can be chosen after the algorithm is run). F' is identical to F except for the following difference: after step 1(a), it uses a step 1'(a) that puts all vertices of $I(T)$ and all vertices of $C \cap I_1$ in OUT_1, it then continues to step 1(b) after which, it also throws from I_1 (to OUT_1) all the vertices connected to OUT_1 (step 1(c)).

We use $OUT_2'(E)$ to denote OUT_2 in the outcome of $F'(E)$. As we shall see, F' dominates F in the following sense: $OUT_2'(E)$ contains $OUT_2(E \cup T)$. Nevertheless, since F' has no knowledge of $C(T)$, the set $C(T)$ is likely to be in $OUT_2'(E)$ as any other subset of C with the same size (where the random variable is E). An argument similar to the one in Lemma 5 (which we omit) shows that $OUT_2'(E)$ is likely to be small: $|OUT_2'(E)| < e^{-\alpha^2 d/70} n$ with probability $> 1 - e^{c_1 \alpha^2 d \log n}$. We get:

$$\Pr_E[C(T) \subseteq OUT_2(E \cup T) \leq \Pr_E[C(T) \subseteq OUT_2'(E)]$$

$$\leq \Pr_E[C(T) \subseteq OUT_2'(E) \mid \#OUT_2'(E) \cap C < e^{-\alpha^2 d/70}] + e^{-c_1 \alpha^2 d \log n}.$$

Given that $|OUT_2'(E) \cap C| = m$, the set $OUT_2'(E) \cap C$ is just a random subsets of C of size m. It then follows that $\Pr[C(T) \subseteq OUT_2'(E) \mid \#OUT_2'(E) \cap C < e^{-\alpha^2 d/70}]$ is bounded by the probability that a binary random variable $X \sim Bin(m, p = \frac{|C(T)|}{|C|-m})$ has $|C(T)|$ successes. Since $m \leq e^{-\alpha^2 d/70} n$ and $|C(T)| \geq t/3$ this probability is bounded by:

$$\binom{m}{t/3} p^{t/6} \leq \left(\frac{me}{t/3} \cdot \frac{t/3}{|C| - m} \right)^{t/3} \leq \left(\frac{e^{-\alpha^2 d/70+1} n}{n/3} \right)^{t/3} \leq e^{-c_1 \alpha^2 dt}.$$

In the second inequality we use $(1 - \alpha)n - m > n/3$ which is true for $\alpha < 1/2$. (If we let $(1 - \alpha)$ to be too small we get a small factor in the denominator which becomes significant. Nevertheless, a more careful estimation in equation (3) yields a factor that cancels it out.) It follows that $\Pr_E[V(T) \subseteq OUT_2(E \cup T)] < e^{-c_1 \alpha^2 dt}$ as needed.

It remains to show that $OUT_2(E \cup T) \subseteq OUT_2'(E)$. This is done in Lemma 8.

Lemma 8. $OUT_2(E \cup T) \subseteq OUT_2'(E \cup T) \subseteq OUT_2'(E)$.

Proof. There are three executions that we consider:

(i) $F(E \cup T)$ which produces $I_1^{(i)}, C_1^{(i)}$ in step 1 and $I_2^{(i)}, C_2^{(i)}$ in step 2,

(ii) $F'(E \cup T)$ which produces $I_1^{(ii)}, C_1^{(ii)}$ in step 1 and $I_2^{(ii)}, C_2^{(ii)}$ in step 2,

(iii) $F'(E)$ which produces $I_1^{(iii)}, C_1^{(iii)}$ in step 1 and $I_2^{(iii)}, C_2^{(iii)}$ in step 2.

First we analyse step 1 and show that: $I_1^{(i)} \supseteq I_1^{(ii)} \supseteq I_1^{(iii)}$ and $C_1^{(i)} \supseteq C_1^{(ii)} \supseteq C_1^{(iii)}$. The inclusions $I_1^{(i)} \supseteq I_1^{(ii)}, C_1^{(i)} \supseteq C_1^{(ii)}$ are easy as after step 1(a) they are in fact equalities and in steps 1'(a),1(b),1(c) F' removes more vertices then what F removes in 1(b) to OUT_1. We now prove the inclusions $I_1^{(ii)} \supseteq I_1^{(iii)}, C_1^{(ii)} \supseteq C_1^{(iii)}$. The only difference (due to T edges) in step 1(a) is that some vertices of $I_1^{(iii)} \cap V(T)$ will be put in $C_1^{(ii)}$ (instead of being in $I_1^{(ii)}$). This does not pose a problem since anyway all the vertices of $I_1^{(iii)} \cap V(T)$ are moved to OUT_1 by F' at step 1'(a) (because $I_1^{(iii)} \cap V(T)$ is contained in $I(T) \cup (C \cap I_1^{(iii)})$). Any two vertices which are removed in step 1(b) from $I_1^{(ii)}$ will be removed from $I_2^{(iii)}$ either in step 1(b) or in step 1(c).

Given the inclusions of step 1 we are ready to prove the inclusions of step 2: $I_2^{(i)} \supseteq I_2^{(ii)} \supseteq I_2^{(iii)}$ and $C_2^{(i)} \supseteq C_2^{(ii)} \supseteq C_2^{(iii)}$. Notice that these inclusions imply the Lemma. We begin with $I_2^{(i)} \supseteq I_2^{(ii)}, C_2^{(i)} \supseteq C_2^{(ii)}$. Consider the execution of step 2 of $F(E \cup T)$. We show a parallel execution of step 2 of $F'(E \cup T)$, for which the invariant $I_2^{(i)} \supseteq$

$I_2^{(ii)}, C_2^{(i)} \supseteq C_2^{(ii)}$ is kept. It is enough to show one possible execution of step 2 of $F'(E \cup T)$, because the order by which vertices are removed does not affect the final outcome (once a vertex becomes removable it will be removed). Initially, the required inclusions are true due to the inclusions after step 1. Whenever a vertex $v \in C_2^{(i)}$ is removed, it becomes removable from $C_2^{(ii)}$ and we remove it (if it had already been removed then also its neighbors from $I_2^{(ii)}$ had been removed because F' ensures there are no edges between I_2 and OUT_2). Proving the inclusions $I_2^{(ii)} \supseteq I_2^{(iii)}, C_2^{(ii)} \supseteq C_2^{(iii)}$ is done in a similar way, only that now we have the edges of T which influence the execution of $F'(E \cup T)$, but do not exist in the execution of $F'(E)$. Again, whenever a vertex $v \in C_2^{(ii)}$ is removed, it becomes removable from $C_2^{(iii)}$ and we remove it (if it is was not already moved to OUT_2). Let u be a neighbor of v from $I_2^{(ii)}$ that is moved together with v to OUT_2. If $(u, v) \in E$ then u can not stay in $I_2^{(iii)}$ as there are no edges between I_2 and OUT_2 in F'. If $(u, v) \in T$ then u is either in $I_2^{(iii)} \cap C(T)$ or in $I_2^{(iii)} \cap I(T)$. In both cases u belonged to either $I_1^{(iii)} \cap C$ or $I(t)$ and was already removed in step 1'(a).   □

References

1. N. Alon and N. Kahale. A spectral technique for coloring random 3-colorable graphs. *SIAM Journal on Computing*, 26(6):1733–1748, 1997.
2. N. Alon, M. Krivelevich, and B. Sudakov. Finding a large hidden clique in a random graph. *Random Structures and Algorithms*, 13(3-4):457–466, 1988.
3. H. Chen and A. Frieze. Coloring bipartite hypergraphs. IPCO 1996, 345–358.
4. A. Coja-Oghlan. A spectral heuristic for bisecting random graphs. SODA 2005, 850–859.
5. A. Coja-Oghlan. Finding Large Independent Sets in Polynomial Expected Time. STACS 2003, 511–522.
6. U. Feige. Approximating maximum clique by removing subgraphs. *Siam J. on Discrete Math.*, 18(2):219–225, 2004.
7. U. Feige and J. Kilian. Heuristics for semirandom graph problems. *Journal of Computing and System Sciences*, 63(4):639–671, 2001.
8. U. Feige and R. Krauthgamer. Finding and certifying a large hidden clique in a semirandom graph. *Random Structures and Algorithms*, 16(2):195–208, 2000.
9. A. Flaxman. A spectral technique for random satisfiable 3cnf formulas. SODA 2003, 357–363.
10. A. Goerdt and A. Lanka. On the hardness and easiness of random 4-sat formulas. ISAAC 2004, 470–483.
11. G. Grimmet and C. McDiarmid. On colouring random graphs. *Math. Proc. Cam. Phil. Soc.*, 77:313–324, 1975.
12. J. Håstad. Clique is hard to approximate within $n^{1-\epsilon}$. *Acta Mathematica*, 182(1):105–142, 1999.
13. M. Jerrum. Large clique elude the metropolis process. *Random Structures and Algorithms*, 3(4):347–359, 1992.
14. R. M. Karp. The probabilistic analysis of some combinatorial search algorithms. In J. F. Traub, editor, *Algorithms and Complexity: New Directions and Recent Results*, pages 1–19. Academic Press, New York, 1976.
15. L. Kučera. Expected complexity of graph partitioning problems. *Discrete Appl. Math.*, 57(2-3):193–212, 1995.

On the Error Parameter of Dispersers

Ronen Gradwohl[1], Guy Kindler[2], Omer Reingold[1,*], and Amnon Ta-Shma[3]

[1] Department of Computer Science and Applied Math, Weizmann Institute of Science
{ronen.gradwohl,omer.reingold}@weizmann.ac.il
[2] Institute for Advanced Study, Princeton
gkindler@ias.edu
[3] School of Computer Science, Tel-Aviv University
amnon@tau.ac.il

Abstract. Optimal dispersers have better dependence on the error than optimal extractors. In this paper we give *explicit* disperser constructions that beat the best possible extractors in some parameters. Our constructions are not strong, but we show that having such explicit *strong* constructions implies a solution to the Ramsey graph construction problem.

1 Introduction

Extractors [8, 18] and dispersers [13] are combinatorial structures with many random-like properties[1]. Extractors are functions that take two inputs – a string that is not uniformly distributed, but has some randomness, and a shorter string that is completely random – and output a string that is close to being uniformly distributed. Dispersers can be seen as a weakening of extractors: they take the same input, but output a string whose distribution is only guaranteed to have a large support. Both objects have found many applications, including simulation with weak sources, deterministic amplification, construction of depth-two super-concentrators, hardness of approximating clique, and much more [7]. For nearly all applications, *explicit* constructions are required.

Extractors and dispersers have several parameters: the longer string is called the input string, and its length is denoted by n, whereas the shorter random string is called the seed, and its length is denoted by d. Additional parameters are the the output length m, the amount of required entropy in the source k and the error parameter ϵ. The two parameters that are of central interest for this paper are the seed length d and the entropy loss $\Lambda = k + d - m$, a measure of how much randomness is lost by an application of the disperser/extractor.

The best extractor construction (which is known to exist by nonconstructive probabilistic arguments) has seed length $d = \log(n - k) + 2\log\frac{1}{\epsilon} + \Theta(1)$ and entropy loss $\Lambda = \log\frac{1}{\epsilon} + \Theta(1)$, and this was shown to be tight by [10]. However, the situation for optimal dispersers is different. It can be shown non-explicitly that there exist dispersers with seed length $d = \log(n-k) + \log\frac{1}{\epsilon} + \Theta(1)$ and entropy loss $\Lambda = \log\log\frac{1}{\epsilon} + \Theta(1)$, and, again, this was shown to be tight by [10]. In particular we see that optimal dispersers can have shorter seed length and significantly smaller entropy loss than the best possible extractors. Up to this work, however, all such explicit constructions also incur a significant cost in other parameters.

---

* Research supported by US-Israel Binational Science Foundation Grant 2002246.

[1] For formal definitions, see Sect. 2.

C. Chekuri et al. (Eds.): APPROX and RANDOM 2005, LNCS 3624, pp. 294–305, 2005.
© Springer-Verlag Berlin Heidelberg 2005

1.1 Our Results

Ta-Shma and Zuckerman [17], and Reingold, Vadhan, and Wigderson [12] gave a way to construct dispersers with small dependence on ϵ via an error reduction technique[2]. Their approach requires a disperser for high min-entropy sources, and therefore our first result is a good disperser for such sources. Next, we show a connection between dispersers with optimal dependence on ϵ and constructions of bipartite Ramsey graphs[3]. This connection holds for dispersers that work well for low min-entropies, and so our second construction focuses on this range.

A Good Disperser for High Entropies. Our goal is to construct a disperser for very high min-entropies ($k = n-1$) but with very small error ϵ, as well as the correct entropy loss of $\log \log \frac{1}{\epsilon}$ and seed length of $\log \frac{1}{\epsilon}$.

One high min-entropy disperser associates the input string with a vertex of an expander graph (see Definition 2.8), and the output with the vertex reached after taking one step on the graph [4]. This disperser has the optimal $d = \log \frac{1}{\epsilon}$, but its entropy loss $\Lambda = d = \log \frac{1}{\epsilon}$ is exponentially larger than the required $\log \log \frac{1}{\epsilon}$.

A better disperser construction, in terms of the entropy loss, was given by Reingold, Vadhan, and Wigderson [12], who constructed high min-entropy dispersers in which the degree and entropy loss are nearly optimal. Their constructions are based on the extractors obtained by the Zig-Zag graph product, but the evaluation time includes factors of $\text{poly}(\frac{1}{\epsilon})$ or even $2^{1/\epsilon}$, so they are inefficient for super-polynomially small error. The reason these constructions incur this cost is that they view their extractors as bipartite graphs, and then define their dispersers as the same graph, but with the roles of the left-hand-side and right-hand-side reversed. This inversion of an extractor is nontrivial, and thus requires much computation.

Our constructions are a new twist on the old theme of using random walks on expander graphs to reduce the error. Formally, we get,

Theorem 1.1. *For every $\epsilon = \epsilon(n) > 0$, there exists an efficiently constructible $(n - 1, \epsilon)$-disperser DSP : $\{0,1\}^n \times \{0,1\}^d \mapsto \{0,1\}^m$ with $d = (2 + o(1)) \log \frac{1}{\epsilon}$ and entropy loss $\Lambda = (1 + o(1)) \log \log \frac{1}{\epsilon}$.*

As a corollary of one of the lemmas used in this construction, Lemma 3.1, we also get the following construction:

Corollary 1.1. *There exists an efficiently constructible $(n - 1, \epsilon)$-disperser DSP : $\{0,1\}^n \times \{0,1\}^d \mapsto \{0,1\}^m$ with $d = O(\log \frac{1}{\epsilon})$ and entropy loss $\Lambda = \log \log \frac{1}{\epsilon} + O(1)$.*

Error Reduction for Dispersers. An error reduction takes a disperser with, say, constant error, and converts it to a disperser with the desired (small) error ϵ. One way for achieving error reduction for dispersers was suggested by Ta-Shma and Zuckerman [17], and Reingold, Vadhan, and Wigderson [12], and is obtained by first applying the

---

[2] See Sect. 1.1 for details.
[3] For a formal definition, see Sect. 4.

constant error disperser, and then applying a disperser with only error ϵ that works well for sources of very high min-entropy (such as $k = n - 1$). Using the disperser of Theorem 1.1 we get:

Theorem 1.2 (error reduction for dispersers). *Suppose there exists an efficiently constructible $(k, \frac{1}{2})$-disperser $\mathrm{DSP}_1 : \{0,1\}^n \times \{0,1\}^{d_1} \mapsto \{0,1\}^{m_1}$ with entropy loss Λ_1. Then for every $\epsilon = \epsilon(n) > 0$ there exists an efficiently constructible (k, ϵ)-disperser $\mathrm{DSP} : \{0,1\}^n \times \{0,1\}^d \mapsto \{0,1\}^m$ with entropy loss $\Lambda = \Lambda_1 + (1 + o(1)) \log \log \frac{1}{\epsilon}$ and $d = d_1 + (2 + o(1)) \log \frac{1}{\epsilon}$.*

If we take the best explicit disperser construction for constant error (stated in Theorem 2.2), and apply an error reduction based on Corrolary 1.1, we get:

Corollary 1.2. *There exists an efficiently constructible (k, ϵ)-disperser $\mathrm{DSP} : \{0,1\}^n \times \{0,1\}^d \mapsto \{0,1\}^m$ with seed length $d = O(\log \frac{n}{\epsilon})$ and entropy loss $\Lambda = O(\log n) + \log \log \frac{1}{\epsilon}$.*

A Disperser for Low Entropies. Next, we construct a disperser for low min-entropies. The seed length of this disperser is $d = O(k) + (1 + o(1)) \log \frac{1}{\epsilon}$ and the entropy loss is $\Lambda = O(k + \log \log \frac{1}{\epsilon})$. Formally,

Theorem 1.3. *There exists an efficiently constructible (k, ϵ)-disperser $\mathrm{DSP} : \{0,1\}^n \times \{0,1\}^d \mapsto \{0,1\}^m$ with $d = O(k) + (1 + o(1)) \log \frac{1}{\epsilon}$ and $\Lambda = O(k + \log \log \frac{1}{\epsilon})$ entropy loss.*

Notice that when ϵ is small (say $2^{-n^{2/3}}$) and k is small (say $O(\log(n))$) the entropy loss is optimal up to constant factors, and the seed length is optimal with respect to ϵ. At first glance this may seem like only a small improvement over the previous construction, which was obtained using the error reduction. However, we now shortly argue that the reduction of the seed length from $O(\log(n)) + 2 \log \frac{1}{\epsilon}$ to $O(\log(n)) + 1 \cdot \log \frac{1}{\epsilon}$ is significant.

First we note that for some applications, such as the Ramsey graph construction we discuss below, the constant coefficient is a crucial parameter. A seed length of $O(\log(n)) + 2 \log \frac{1}{\epsilon}$ does not yield any improvement over known constructions of Ramsey graphs, whereas a seed length of $O(\log(n)) + 1 \cdot \log \frac{1}{\epsilon}$ implies a vast improvement.

Second, we argue that in some sense the improvement in the seed length from $2 \log \frac{1}{\epsilon}$ to $1 \cdot \log \frac{1}{\epsilon}$ is equivalent to the exponential improvement of $\log \frac{1}{\epsilon}$ to $O(\log \log \frac{1}{\epsilon})$ in the entropy loss. Say $\mathrm{DSP} : \{0,1\}^{n_1} \times \{0,1\}^{d_1} \to \{0,1\}^{n_2}$ is a (k_1, ϵ_1) disperser. We can view DSP as a bipartite graph with $N_1 = 2^{n_1}$ vertices on the left, $N_2 = 2^{n_2}$ vertices on the right, with an edge (v_1, v_2) in the bipartite graph if and only if $\mathrm{DSP}(v_1, y) = v_2$ for some $y \in \{0,1\}^{d_1}$. The disperser property then translates to the property that any set $A_1 \subseteq \{0,1\}^{n_1}$ of size $K_1 = 2^{k_1}$, and any set $A_2 \subseteq \{0,1\}^{n_2}$ of size $\epsilon_1 N_2$ have an edge between them. This property is symmetric, and so every disperser DSP from $[N_1]$ to $[N_2]$ can equivalently be thought of as a disperser from $[N_2]$ to $[N_1]$, which we call the *inverse* disperser.

It then turns out that for certain settings of the paraments, if a disperser has seed length dependence of $1 \cdot \log \frac{1}{\epsilon}$ on the error, then the inverse disperser has entropy loss

$O(\log \log \frac{1}{\epsilon})^4$. However, as noted earlier, inverting a disperser is often computationally difficult, which is the reason we require explicit constructions of dispersers in both directions.

The Connection to Ramsey Graphs. Our final result is a connection between dispersers and bipartite Ramsey graphs. Ramsey graphs have the property that for every subset of vertices on the left-hand-side and every subset of vertices on the right-hand-side, there exists both an edge and a non-edge between the subsets. We show that bipartite Ramsey graphs can be attained by constructing a disperser with $O(\log \log \frac{1}{\epsilon})$ entropy loss and $1 \cdot \log \frac{1}{\epsilon}$ seed length, as well as the additional property of being *strong*[5]. This connection is formalized in Theorem 4.1.

2   Preliminaries

2.1   Dispersers and Strong Dispersers

Dispersers are formally defined as follows:

Definition 2.1 (disperser). *A bipartite graph $G = (L, R, E)$ is a (k, ϵ)-disperser if for every $S \subset L$ with $|S| \geq 2^k$, $|\Gamma(S)| \geq (1 - \epsilon)|R|$, where $\Gamma(S)$ denotes the set of neighbors of the vertices in S.*

As noted earlier, we will denote $|L| = 2^n$ and $|R| = M = 2^m$. Also, the left-degree of the disperser will be denoted by $D = 2^d$.

We can describe dispersers as functions $\text{DSP} : \{0,1\}^n \times \{0,1\}^d \mapsto \{0,1\}^m$. DSP is a disperser if, when choosing x uniformly at random from S and r uniformly at random from $\{0,1\}^d$, the distribution of $\text{DSP}(x, r)$ has support of size $(1 - \epsilon)M$.

We will also be interested in strong dispersers. Loosely speaking, this means that they have a large support for most seeds. Equivalently, if we were to concatenate the seed to the output of the disperser, then this extended output would have a large support. Here we actually give a slightly more general definition, to include the case of almost-strong dispersers, which have a large support when "most" of the seed is concatenated to the output.

Definition 2.2 (strong disperser). *Denote by $x_{[0..t-1]}$ the first t bits of a string x. A (k, ϵ)-disperser $\text{DSP} : \{0,1\}^n \times \{0,1\}^d \mapsto \{0,1\}^m$ is strong in t bits if for every $S \subseteq L$ with $|S| \geq 2^k$, $|\{(\text{DSP}(x,r), r_{[0..t-1]}) : x \in S, r \in \{0,1\}^d\}| \geq (1 - \epsilon)2^{m+t}$. DSP is a strong disperser if it is strong in all d bits.*

We also need the following proposition, which is implicit in [12].

---

[4] We remark that the entropy loss lower bound in [10] is obtained by proving a seed length lower bound, and then using this equivalence.

[5] See Definition 2.2.

Proposition 2.1 ([12]). *Let*

- $\mathrm{DSP}_1 : \{0,1\}^{n_1} \times \{0,1\}^{d_1} \mapsto \{0,1\}^{m_1}$ *be a* (k, ϵ_1)-*disperser with entropy loss* Λ_1, *and let*
- $\mathrm{DSP}_2 : \{0,1\}^{m_1} \times \{0,1\}^{d_2} \mapsto \{0,1\}^{m_2}$ *be an* $(m_1 - \log \frac{1}{1-\epsilon_1}, \epsilon_2)$-*disperser with entropy loss* Λ_2.

Then $\mathrm{DSP} : \{0,1\}^{n_1} \times \{0,1\}^{d_1+d_2} \mapsto \{0,1\}^{m_2}$, *where*

$$\mathrm{DSP}(x, r_1, r_2) = \mathrm{DSP}_2(\mathrm{DSP}_1(x, r_1), r_2) \ ,$$

is a (k, ϵ_2)-*disperser with entropy loss* $\Lambda = \Lambda_1 + \Lambda_2$.

2.2   Extractors

To formally describe extractors, we first need a couple of other definitions:

Definition 2.3 (statistical difference). *For two distributions* X *and* Y *over some finite domain, denote the* statistical difference *between them by* $\Delta(X, Y)$, *where:*

$$\Delta(X, Y) = \frac{1}{2} \sum_{i \in \mathrm{supp}(X \cup Y)} |\Pr[X = i] - \Pr[Y = i]| \ .$$

X *and* Y *are* ε-close *if* $\Delta(X, Y) \leq \varepsilon$.

We also need a measure of the randomness of a distribution.

Definition 2.4 (min-entropy). *The* min-entropy *of a distribution* X, *denoted by* $H_\infty(X)$, *is defined as*

$$H_\infty(X) = \min_{i \in \mathrm{supp}(X)} \log \frac{1}{\Pr[X = i]} \ .$$

Definition 2.5 (extractor). $\mathrm{EXT} : \{0,1\}^n \times \{0,1\}^d \mapsto \{0,1\}^m$ *is a* (k, ϵ)-*extractor if, for any distribution* X *with* $H_\infty(X) \geq k$, *when choosing* x *according to* X *and* r *uniformly at random from* $\{0,1\}^d$, *the distribution of* $\mathrm{EXT}(x, r)$ *is* ϵ-*close to uniform.*

As with dispersers, we also consider strong extractors:

Definition 2.6 (strong extractor). $\mathrm{EXT} : \{0,1\}^n \times \{0,1\}^d \mapsto \{0,1\}^m$ *is a* (k, ϵ)-*extractor if, for any distribution* X *with* $H_\infty(X) \geq k$, *when choosing* x *according to* X *and* r *uniformly at random from* $\{0,1\}^d$, *the distribution of* $(\mathrm{EXT}(x, r), r)$ *is* ϵ-*close to uniform.*

Finally, for both dispersers and extractors, and $S \subseteq (n)$ and $T \subseteq (d)$, denote by $\mathrm{EXT}(S, T) = \{\mathrm{EXT}(s, t) : s \in S, t \in T\}$ and $\mathrm{DSP}(S, T) = \{\mathrm{DSP}(s, t) : s \in S, t \in T\}$.

2.3   Previous Explicit Constructions

In this paper we are interested in explicit constructions, which we now define formally.

Definition 2.7 (explicit family). *A family of functions* $f_n : \{0,1\}^n \times \{0,1\}^{d_n} \mapsto \{0,1\}^{m_n}$ *is an explicit family of extractors/dispersers if for every* n, f_n *is an extractor/disperser, and if there exists an algorithm* A *such that, given* $x \in \{0,1\}^n$ *and* $r \in \{0,1\}^{d_n}$, A *computes* $f_n(x,r)$ *in time polynomial in* n.

We will refer to extractors and dispersers as explicit when we mean that there exist explicit families of such functions.

In our construction, we will use the following extractor of Srinivasan and Zuckerman [14] and disperser of Ta-Shma, Umans and Zuckerman [16]:

Theorem 2.1 ([14]). *There exists an efficiently constructible strong* (k,ϵ)-*extractor* $\mathrm{EXT}_{SZ} : \{0,1\}^n \times \{0,1\}^d \mapsto \{0,1\}^m$ *with* $d = O(m + \log \frac{n}{\epsilon})$ *and* $\Lambda = 2\log\frac{1}{\epsilon} + O(1)$.

Theorem 2.2 ([16]). *For every* $k < n$ *and any constant* $\epsilon > 0$, *there exists an efficiently constructible disperser* $\mathrm{DSP}_{TUZ} : \{0,1\}^n \times \{0,1\}^d \mapsto \{0,1\}^m$ *with* $d = O(\log n)$ *and* $m = k - 3\log n - O(1)$.

2.4   Expander Graphs

A main tool we use in our construction is an expander graph.

Definition 2.8 (expander). *Let* $G = (V,E)$ *be a regular graph with normalized adjacency matrix* A *(the adjacency matrix divided by the degree), and denote by* λ *the second largest eigenvalue (in absolute value) of* A. *Then* G *is an* (N,D,α)-*expander if* G *is* D-*regular,* $|V| = N$, *and* $\lambda \le \alpha$.

An expander G is explicit if there exists an algorithm that, given any vertex $v \in V$ and index $i \in \{0,1,\ldots,D-1\}$, computes the i'th neighbor of v in time $\mathrm{poly}(\log|V|)$. Intuitively, the smaller the value of α, the better the expander, implying better parameters for our construction. Thus, we wish to use graphs with the optimal $\alpha \le \frac{2}{\sqrt{D}}$, called Ramanujan Graphs.

A minor technicality involving Ramanujan Graphs is that there are known explicit constructions only for certain values of N and D. However, for any given N and D, it is possible to find a description of the Ramanujan Graphs of Lubotzky, Phillips and Sarnak [6] that are sufficient for our needs. These graphs satisfy all the requirements, with $|V| \in [N, (1+\delta) \cdot N]$ and degree D, and can be found in time $\mathrm{poly}(n, \frac{1}{\delta})$ [1, 4].

Since the Ramanujan Graphs of [6] are Cayley graphs, they also have the useful property of being consistently labelled: in any graph $G = (V,E)$, the label of an edge $(u,v) \in E$ is i if following edge i leaving u leads to v. The graph is consistently labelled if for all vertices $u, v, w \in V$, if $(u,v) \in E$ with label i and $(w,v) \in E$ with label j, then $i \ne j$. Loosely speaking, if we take two walks on a consistently labelled graph following the same list of labels but starting at different vertices, then the walks will not converge into one walk.

The reason expander graphs will be useful is Kahale's Expander Path Lemma [5]:

Lemma 2.1 ([5]). *Let G be an (N, D, α)-expander, and let $B \subset V$ with density $\beta = \frac{|B|}{|V|}$. Choose $X_0 \in V$ uniformly at random, and let $X_0, \ldots, X_t$ be a random walk on G starting at X_0. Then*

$$\Pr[\forall i, X_i \in B] \leq (\beta + \alpha)^t \ .$$

3 A High Min-Entropy Disperser

In this section we prove Theorem 1.1. Our technique is motivated by the construction of Cohen and Wigderson [3]. In the context of deterministic amplification, Cohen and Wigderson [3] used a random walk on an expander graph to reduce the error from polynomially small to exponentially small. Translating their work into the current notion of a disperser, their construction has the following parameters: $d = (2 + \alpha) \log(1/\epsilon)$ and $\Lambda = \log \log(1/\epsilon)$, for some constant α. Cohen and Wigderson used this disperser as a sampling procedure, and they were only interested in the case $k \geq n - \frac{1}{\text{poly}(n)}$. In fact, their construction does not attain the above parameters for smaller entropy thresholds, such as $k = n - 1$.

3.1 The Basic Construction

Lemma 3.1. *For all positive w, t, and δ there exists an $(n - c, \epsilon)$-disperser DSP : $\{0,1\}^n \times \{0,1\}^d \mapsto \{0,1\}^m$ computable in time $\text{poly}(n, \frac{1}{\delta})$ with:*

- $\epsilon \leq \left(1 - \frac{1}{(1+\delta)2^c} + \frac{2}{2^{w/2}}\right)^t$
- $d = wt + \log t$
- $\Lambda \leq \log t$

Proof. Consider a Ramanujan Graph $G = (V, E)$ with $|V| \in [N, (1 + \delta)N]$ and degree 2^w, which can be constructed in time $\text{poly}(\frac{1}{\delta})$ (see Sect. 2.4). Associate with an arbitrary set of N vertices of G the strings $\{0,1\}^n$, which in turn correspond to the vertices on the left side of the disperser. Again, for every string $s \in \{0,1\}^d$, we think of $s = r \circ i$, where $r \in \{0,1\}^{wt}$ and $i \in \{0,1\}^{\log t}$. The disperser is defined as $\text{DSP}(x, r \circ i) = (y, r)$, where $y \in V$ is the vertex of G reached after taking a walk consisting of the last i steps encoded by r, starting at the vertex associated with x. Now define $S \subset \{0,1\}^n$, an arbitrary set of starting vertices, such that $|S| \geq 2^{n-c}$.

For a pair (y, r), denote by $\overleftarrow{r}_y$ the walk r ending at y, backwards. By backwards we mean that if the i'th step of the walk is from v to u, then the $(t - i)$'th step of $\overleftarrow{r}_y$ has the label of the edge from u to v. If (y, r) is bad (i.e. for no $x \in S$ and $i \in \{0,1\}^{\log t}$, $\text{DSP}(x, r \circ i) = (y, r)$), then the walk $\overleftarrow{r}_y$ starting at vertex y in G never hits a vertex $x \in S$. This implies that in all t steps, the walk stays in $\overline{S}$. Note that because the graph is consistently labelled, there is a bijection from pairs (y, r) to pairs $(y, \overleftarrow{r}_y)$. Thus, to bound ϵ, it suffices to bound the probability that a random walk in G stays in $\overline{S}$. We use the Expander Path Lemma (Lemma 2.1) to bound this probability by

$$\epsilon \leq \left(\frac{|\overline{S}|}{|V|} + \lambda\right)^t ,$$

where λ is the second largest eigenvalue (in absolute value) of G. Since $|S| \geq \frac{N}{2^c}$,

$$\frac{|\overline{S}|}{|V|} \leq \frac{|V| - \frac{N}{2^c}}{|V|} \leq \frac{(1+\delta)N - \frac{N}{2^c}}{(1+\delta)N} = 1 - \frac{1}{(1+\delta)2^c} \ .$$

Thus,

$$\epsilon \leq \left(1 - \frac{1}{(1+\delta)2^c} + \lambda\right)^t \leq \left(1 - \frac{1}{(1+\delta)2^c} + \frac{2}{2^{w/2}}\right)^t$$

for a Ramanujan Graph. It is clear that $d = wt + \log t$. Since the length of the output is at least $n + wt$, the entropy loss $\Lambda \leq \log t$. $\square$

3.2   Parameters

By using Lemma 3.1 with $w = 6$, $t = \frac{5}{2}\log\frac{1}{\epsilon}$, and $\delta = \frac{1}{\text{poly}(n)}$ we get Corollary 1.1 of Section 1.1.

For our high min-entropy disperser, we wish to reduce the coefficient in the seed length. Thus, we will compose two different dispersers guaranteed by Lemma 3.1. The effect of composing two dispersers is captured by Proposition 2.1.

The first disperser will have entropy requirement $n - 1$, and error $\epsilon_1 = \frac{2}{2^{w/2}}$. This is chosen in order to satisfy $\epsilon_1 = \lambda$, the second eigenvalue of the expanders used in the disperser constructions, and thus simplify the analysis. The second disperser will take a subset of $\{0,1\}^{n_2}$ with error $\epsilon_1 = \frac{2}{2^{w/2}}$ (or, equivalently, entropy requirement $n_2 - \log\frac{1}{1-\epsilon_1}$), and have error ϵ, as desired. Both dispersers will be as guaranteed by Lemma 3.1, with the same w. This means that both expander graphs will have the same degree, and thus the same bound on λ, but their sizes will be different. Note that for the first expander, we do not care too much about the size of $|V|$, and so we choose δ to be some constant, to yield a graph with $|V| \in [N, \frac{3}{2}N]$. For the second expander, however, this size is relevant, so we must choose a smaller δ.

We now prove Theorem 1.1.

Proof. Our disperser DSP will consist of the composition of two other dispersers, an $(m_1 - 1, \epsilon_1)$-disperser $\text{DSP}_1 : \{0,1\}^n \times \{0,1\}^{d_1} \mapsto \{0,1\}^{m_1}$ and a $(n - \log\frac{1}{1-\epsilon_1}, \epsilon)$-disperser $\text{DSP}_2 : \{0,1\}^{m_1} \times \{0,1\}^{d_2} \mapsto \{0,1\}^m$. These dispersers will be as guaranteed by Lemma 3.1, with parameters $t_1, w_1, \delta_1 = \frac{1}{2}$ and t_2, w_2, δ_2 respectively. Recall that t_i indicates the length of the walk, w_i is the number of bits needed to specify a step in the expander graph, and δ_i is related to the size of the expander graph. We will have

$$\text{DSP}(x, r_1 \circ r_2 \circ i_1 \circ i_2) = \text{DSP}_2(\text{DSP}_1(x, r_1 \circ i_1), r_2 \circ i_2) \ .$$

Let $w = w_1 = w_2$, a parameter we will choose later. For DSP_1, we wish its error to be $\epsilon_1 \leq \frac{2}{2^{w/2}}$. By Lemma 3.1, $\left(1 - \frac{2}{3} \cdot \frac{1}{2} + \frac{2}{2^{w/2}}\right)^{t_1} = \left(\frac{2}{3} + \frac{2}{2^{w/2}}\right)^{t_1} \leq \frac{2}{2^{w/2}}$. If $\lambda < \frac{1}{6}$ (which it will be), then it suffices to satisfy $\left(\frac{3}{4}\right)^{t_1} \leq \frac{2}{2^{w/2}}$. This occurs whenever $t_1 \geq \frac{3}{2}w$, so set $t_1 = \frac{3}{2}w$. Thus, DSP_1 will output m_1 bits, with error $\epsilon_1 \leq \frac{2}{2^{w/2}}$. Furthermore, the required seed length $d_1 = t_1 w = \frac{3}{2}w^2$.

We now apply DSP_2 on these strings, with $\delta_2 = \frac{1}{1-\epsilon_1} - 1$. The construction time is $\text{poly}(\frac{1}{\delta_2}) = 2^{O(w)}$. By Lemma 3.1, the error is

$$\epsilon \leq \left(1 - \frac{1-\epsilon_1}{1+\delta_2} + \lambda\right)^{t_2} < (2\epsilon_1 + \lambda)^{t_2} = \left(\frac{6}{2^{w/2}}\right)^{t_2}.$$

This implies that

$$t_2 = \frac{\log \epsilon}{\log \frac{6}{2^{w/2}}} = \log \frac{1}{\epsilon}\left(\frac{1}{\log \frac{2^{w/2}}{6}}\right)$$

$$< \log \frac{1}{\epsilon}\left(\frac{2}{w-6}\right).$$

The required seed length is

$$d_2 = t_2 w = \frac{2w}{w-6}\log \frac{1}{\epsilon} = (2 + \frac{12}{w-6})\log \frac{1}{\epsilon}.$$

We now analyze the parameters we obtained in the composition of DSP_1 and DSP_2. By Proposition 2.1, the total seed length of DSP is $d = (2 + \frac{12}{w-6})\log \frac{1}{\epsilon} + \frac{3}{2}w^2$. If we pick $w = \omega(1)$, but also $w = o(\log\log \frac{1}{\epsilon})$, then $d = (2 + o(1))\log \frac{1}{\epsilon}$, as claimed. The entropy loss $\Lambda \leq \log[(2 + \frac{12}{w-6})\log \frac{1}{\epsilon}] + \log \frac{3}{2}w^2$, and with the above choice of w, we get $\Lambda = (1 + o(1))\log\log \frac{1}{\epsilon}$. $\qquad\square$

3.3   Error Reduction for Dispersers

The error reduction for dispersers of Theorem 1.2 is attained by composing DSP_1 with our high min-entropy disperser, as in Proposition 2.1.

4   Dispersers and Bipartite Ramsey Graphs

Our construction of low min-entropy dispersers was motivated by the connection of such dispersers to bipartite Ramsey graphs.

Definition 4.1. *A bipartite graph $G = (L, R, E)$ with $|L| = |R| = N$ is (s,t)-Ramsey if for every $S \subseteq L$ with $|S| \geq s$ and every $T \subseteq R$ with $|T| \geq t$, the vertices of S and the vertices of T have at least one edge and at least one non-edge between them in G.*

By the probabilistic method, it can be shown that $(2\log N, 2\log N)$-Ramsey graphs exist. However, the best known explicit constructions are of (N^δ, N^δ)-Ramsey graphs, for any constant $\delta > 0$ [2]. We now argue that constructions of low min-entropy dispersers may provide new ways of constructing bipartite Ramsey graphs.

Suppose we have some strong (k, ϵ)-disperser DSP that outputs 1 bit, and has seed length $d = s_n + \log \frac{1}{\epsilon}$. Consider the bipartite graph $G = (L, R, E)$, such that there is an edge between $x \in L$ and $y \in R$ if and only if $\text{DSP}(x, y) = 1$. Then for every $S \subset L$

with $|S| \geq 2^k$, and every $T \subset R$ with $|T| > 2\epsilon|R|$, there must be an edge and a non-edge between S and T. To see this, suppose there was no non-edge. Then this implies that $\mathrm{DSP}(S,t) = 1 \circ t$, for all $t \in T$, and for no $t \in T$ does $\mathrm{DSP}(S,t) = 0 \circ t$. This means that the disperser misses more than an ϵ-fraction of the outputs, a contradiction.

How good of a Ramsey graph does this yield? If we set $k = O(\log n)$ and $\epsilon = 2^{-n-s_n}$, then we get a $2^n \times 2^n$ graph that is $(\mathrm{poly}(n), 2^{s_n} + 1)$-Ramsey. If $s_n = O(\log n)$, then this would yield a $(\mathrm{poly}(n), \mathrm{poly}(n))$-Ramsey graph, which is significantly better than any current construction, even for the non-bipartite case.

Theorem 4.1 is a generalization of the above:

Theorem 4.1. *Let* $\mathrm{DSP} : \{0,1\}^n \times \{0,1\}^d \mapsto \{0,1\}^1$ *be a strong* (k,ϵ)-disperser *with* $d = s_n + t \log \frac{1}{\epsilon}$, *where* s_n *is only a function of* n. *Let* $\epsilon = 2^{-\frac{n-s_n}{t}}$, *implying that* $d = n$. *Define the bipartite graph* $G = (L, R, E)$ *with* $|L| = |R| = 2^n$, *such that there is an edge between* $x \in L$ *and* $y \in R$ *if and only if* $\mathrm{DSP}(x,y) = 1$. *Then* G *is* $(2^k, 2^{\frac{t-1}{t}n + \frac{s_n}{t} + 1} + 1)$-*Ramsey.*

Note that the coefficient t plays a crucial role in the quality of the Ramsey graph. If $t = 1$, then it is possible to get extremely good Ramsey graphs. On the other hand, if the seed length is $d > 2 \log \frac{1}{\epsilon}$, then the graph is worse than $(2^k, \sqrt{2^n})$-Ramsey.

5   A Low Min-Entropy Disperser

In this section, we construct a disperser with seed length $d = O(\log n) + (1+o(1))\log\frac{1}{\epsilon}$ that is almost strong.

Lemma 5.1. *There exists an efficiently constructible* (k,ϵ)-disperser $\mathrm{DSP} : \{0,1\}^n \times \{0,1\}^d \mapsto \{0,1\}^m$ *with* $d = O(k) + (1+o(1))\log\frac{1}{\epsilon}$ *that is strong in* $d - O(\log k + \log\log\frac{1}{\epsilon})$ *bits. If the strong seed bits are concatenated to the output of the disperser, then the entropy loss of* DSP *is* $\Lambda = O(k + \log\log\frac{1}{\epsilon})$.

Consider the following intuition. Given some source with min-entropy $O(\log n)$, we first apply an extractor with m output bits (and think of m as a constant) and $\frac{1}{2} \cdot 2^{-m}$ error. The error is small enough relative to the output length to guarantee that half of the seeds are "good" in the sense that all possible output strings are hit, and so the error for them is 0. The error over the seeds, however, is large, because only half of the seeds are good. Our next step is then to use a disperser with error half, to sample a good seed. The main advantage of this approach is that we obtain a disperser with a very low error ϵ, by constructing only objects with relatively high error.

We now formalize the above discussion, and prove Theorem 1.3.

Proof. We wish to construct a (k,ϵ)-disperser $\mathrm{DSP} : \{0,1\}^n \times \{0,1\}^d \mapsto \{0,1\}^m$. We use the following two ingredients:

- $\mathrm{EXT} : \{0,1\}^n \times \{0,1\}^{d_E} \mapsto \{0,1\}^m$ that is a $(k, 2^{-m-1})$-extractor, and,
- $\mathrm{DSP}' : \{0,1\}^{k'+\log\frac{1}{\epsilon}} \times \{0,1\}^{d'} \mapsto \{0,1\}^{d_E}$ that is a $(k', \frac{1}{2})$-disperser, with $d = k' + \log\frac{1}{\epsilon} + d'$.

We will specify the parameters later. Now, Given $x \in \{0,1\}^n$, $r_1 \in \{0,1\}^{k'+\log\frac{1}{\epsilon}}$, and $r_2 \in \{0,1\}^{d'}$, define

$$\text{DSP}(x, r_1 \circ r_2) = \text{EXT}(x, \text{DSP}'(r_1, r_2)) \ .$$

Fix $S \subseteq \{0,1\}^n$ with $|S| \geq 2^k$. We say a seed s is "good" for S if $|\text{EXT}(S, s)| = 2^m$, and "bad" otherwise. As EXT is a strong extractor, at least half of all seeds $s \in \{0,1\}^{d_E}$ are good for S.

Claim. For all S as above, and all but an ϵ-fraction of the $r_i \in \{0,1\}^{k'+\log\frac{1}{\epsilon}}$, $\text{DSP}'(r_i, \{0,1\}^{d'})$ hits a seed s that is good for S.

Proof. Suppose towards a contradiction that more than an ϵ-fraction of the r_i do not hit any s that is good for S. This implies that for at least $\epsilon \cdot 2^{k'+\log\frac{1}{\epsilon}} = 2^{k'}$ x's, $\text{DSP}'(x, \{0,1\}^{d'})$ misses more than half of the outputs. Consider the set T of all such x's. Then $|\text{DSP}'(T, \{0,1\}^{d'})| < \frac{1}{2} \cdot 2^{d_E}$, contradicting the fact that DSP' is a $(k', \frac{1}{2})$-disperser. □

Recall our disperser construction $\text{DSP}(x, r_1 \circ r_2) = \text{EXT}(x, \text{DSP}'(r_1, r_2))$, and denote by R the length of r_1. The claim shows that for all but an ϵ-fraction of the r_1's, $\text{DSP}'(r_1, r_2))$ hits an s that is good for S, and $\text{EXT}(S, s)$ hits all of $\{0,1\}^m$. Thus, for all but an ϵ-fraction of the r_1's, $\text{EXT}(S, \text{DSP}'(r_1, r_2))$ hits all of $\{0,1\}^m$. This implies that DSP is a (k, ϵ)-disperser that is strong in R bits.

To complete the proof, let $\text{EXT} = \text{EXT}_{SZ}$ from Theorem 2.1, and let $\text{DSP}' = \text{DSP}_{TUZ}$ from Theorem 2.2, with $k' = O(k)$ and $m = O(k)$. Then $d' = O(\log(k + \log\frac{1}{\epsilon})) \leq O(\log k + \log\log\frac{1}{\epsilon})$, and $d = R + d' = O(k) + (1 + o(1))\log\frac{1}{\epsilon}$. Thus, DSP is strong in $R = d - d'$ bits, as claimed. Finally, the entropy loss is $O(k)$ from the application of EXT_{SZ}, and an additional $O(\log k + \log\log\frac{1}{\epsilon})$ from the seed of DSP_{TUZ}. Thus, the total entropy loss is $\Lambda = O(k + \log\log\frac{1}{\epsilon})$. □

Plugging in $k = O(\log n)$ we get:

Corollary 5.1. *There exists an efficiently constructible $(O(\log n), \epsilon)$-disperser* DSP : $(n) \times (d) \mapsto (m)$ *with* $d = O(\log n) + (1+o(1))\log\frac{1}{\epsilon}$ *that is strong in* $d - O(\log\log n + \log\log\frac{1}{\epsilon})$ *bits. The entropy loss of* DSP *is* $\Lambda = O(\log n + \log\log\frac{1}{\epsilon})$.

6   Discussion

In this work we addressed one aspect of sub-optimality in dispersers, the dependence on ϵ, and constructed min-entropy dispersers in which the entropy loss is optimally dependent on ϵ.

Another aspect has to do with the dependence of the entropy loss on n. Say DSP : $\{0,1\}^n \times \{0,1\}^d \to \{0,1\}^m$ is a (k, ϵ) disperser with small error ϵ and close to optimal entropy-loss Λ. For any source $X \subseteq \{0,1\}^n$ of size 2^k, DSP is close to being one-to-one on X (to be more precise, every image is expected to have 2^Λ pre-images, and the disperser property guarantees we are not far from that). In particular every such disperser is also an extractor with some worse error, and if we fix ϵ to be some constant, this disperser would actually be an extractor with constant error.

It follows that if we could improve Corollary 1.2 to have entropy loss independent of n, we would get an extractor with constant error, optimal seed length and constant entropy loss! No current extractor construction can attain such entropy loss, regardless of the error, without incurring a large increase in seed length (generally, $d = \log^2 n$ is necessary). So such dispersers may provide a new means of obtaining better extractors.

A different issue is the distinction between strong and non-strong dispersers. It seems that a possible conclusion from this work is that the property of being strong becomes very significant when dealing with objects that are close to being optimal.

Acknowledgements

We would like to thank Ronen Shaltiel for very useful discussions.

References

1. N. Alon, J. Bruck, J. Naor, M. Naor and R. Roth. Construction of asymptotically good, low-rate error-correcting codes through pseudo-random graphs. In *Transactions on Information Theory*, 38:509-516, 1992. IEEE.
2. Boaz Barak, Guy Kindler, Ronen Shaltiel, Benny Sudakov, and Avi Wigderson. Simulating independence: New constructions of condensers, Ramsey graphs, dispersers, and extractors. In *37th STOC*, 2005.
3. Aviad Cohen and Avi Wigderson. Dispersers, deterministic amplification, and weak random sources. In *Proc. of 30th FOCS*, pages 14-19, 1989.
4. Oded Goldreich, Avi Wigderson. Tiny families of functions with random properties: a quality-size trade-off for hashing. Random Structures and Algorithms 11:4, 1997.
5. Nabil Kahale. Better expansion for Ramanujan graphs. In *33rd FOCS*, 1992, pages 398-404.
6. A. Lubotzky, R. Phillips, and P. Sarnak. Explicit expanders and the Ramanujan conjectures. In *18th STOC*, 1986, pages 240-246.
7. Noam Nisan and Amnon Ta-Shma. Extracting randomness: A survey and new constructions. *Journal of Computer and System Sciences*, 58(1):148-173, 1999.
8. Noam Nisan and David Zuckerman. More deterministic simulation in logspace. In *25th STOC*, 1993, pages 235-244.
9. Ran Raz. Extractors with weak random seeds. In *37th STOC*, 2005.
10. Jaikumar Radhakrishnan and Amnon Ta-Shma. Tight bounds for depth-two superconcentrators. In *38th FOCS*, pages 585-594, Miami Beach, Florida, 20-22 October 1997. IEEE.
11. Ran Raz, Omer Reingold, and Salil Vadhan. Error reduction for extractors. In *40th FOCS*, 1999. IEEE.
12. Omer Reingold, Salil Vadhan, and Avi Wigderson. Entropy waves, the zig-zag graph product, and new constant-degree expanders and extractors. In *Proc. 41st FOCS*, pages 3-13, 2000.
13. Michael Sipser. Expanders, randomness, or time versus space. *Journal of Computer and System Sciences*, 36(3):379-383, June 1988.
14. Aravind Srinivasan and David Zuckerman. Computing with very weak random sources. In *35th FOCS*, 1994, pages 264-275.
15. Amnon Ta-Shma. Almost optimal dispersers. In *30th STOC*, pages 196-202, Dallas, TX, May 1998. ACM.
16. Amnon Ta-Shma, Christopher Umans, and David Zuckerman. Loss-less condensers, unbalanced expanders, and extractors. In *Proc. 33rd STOC*, pages 143-152, 2001.
17. Amnon Ta-Shma and David Zuckerman. Personal communication.
18. David Zuckerman. General weak random sources. In *31st FOCS*, 1990.

Tolerant Locally Testable Codes

Venkatesan Guruswami* and Atri Rudra**

Department of Computer Science & Engineering
University of Washington
Seattle, WA 98195
{venkat,atri}@cs.washington.edu

Abstract. An error-correcting code is said to be *locally testable* if it has
an efficient spot-checking procedure that can distinguish codewords from
strings that are far from every codeword, looking at very few locations
of the input in doing so. Locally testable codes (LTCs) have generated
a lot of interest over the years, in large part due to their connection to
Probabilistically checkable proofs (PCPs). The ability to correct errors
that occur during transmission is one of the big advantages of using a
code. Hence, from a coding-theoretic angle, local testing is potentially
more useful if in addition to accepting codewords, it also accepts strings
that are close to a codeword (in contrast, local testers can have arbitrary behavior on such strings, which potentially annuls the benefits of
error-correction). This would imply that when the tester accepts, one can
follow-up the testing with a (more expensive) decoding procedure to correct the errors and recover the transmitted codeword, while if the tester
rejects, we can save the effort of running the more expensive decoding
algorithm.
In this work, we define such testers, which we call *tolerant testers* following some recent work in property testing [13]. We revisit some recent
constructions of LTCs and show how one can make them locally testable
in a tolerant sense. While we do not optimize the parameters, the main
message from our work is that there are explicit tolerant LTCs with
similar parameters to LTCs.

1 Introduction

Locally testable codes (LTCs) have been the subject of much research over the
years and there has been heightened activity and progress on them recently [4–
6, 10–12]. LTCs are error-correcting codes which have a testing procedure with
the following property– given oracle access to a string which is a purported
codeword, these testers "spot check" the string at very few locations, accepting if
the string is indeed a codeword, and rejecting with high probability if the string is
"far-enough" from every codeword. Such spot-checkers arise in the construction
of Probabilistically checkable proofs (PCPs) [1, 2] (see the recent survey by
Goldreich [10] for more details on the interplay between LTCs and PCPs). Note

---

* Supported in part by NSF Career Award CCF-0343672.
** Supported in part by NSF Career Award CCF-0343672.

C. Chekuri et al. (Eds.): APPROX and RANDOM 2005, LNCS 3624, pp. 306–317, 2005.

that in the definition of LTCs, there is no requirement on the tester for input strings that are very close to a codeword. This "asymmetry" in the way the tester accepts and rejects an input reflects the way PCPs are defined, where we only care about accepting perfectly correct proofs with high probability. However, the crux of error-correcting codes is to tolerate and *correct* a few errors that could occur during transmission of the codeword (and not just be able to detect errors). In this context, the fact that a tester can reject received words with few errors is not satisfactory. A more desirable (and stronger) requirement in this scenario would be the following– we would like the tester to make a quick decision on whether or not the purported codeword is close to any codeword. If the tester declares that there is probably a close-by codeword, we then use a decoding algorithm to decode the received word. If on the other hand, we can say with high confidence that the received word is far away from all codewords then we do not run our expensive decoding algorithm.

In this work we introduce the concept of *tolerant testers*. These are codeword testers which reject (w.h.p) received words far from every codeword (like the current "standard" testers) and accept (w.h.p) close-by received words (unlike the current ones which only need to accept codewords). We will refer to codes that admit a tolerant tester as a tolerant LTCs. In the general context of property testing, the notion of tolerant testing was introduced by Parnas *et al* [13] along with the related notion of distance approximation. Parnas *et al* also give tolerant testers for clustering. We feel that codeword-testing is a particularly natural instance to study tolerant testing. (In fact, if LTCs were defined solely from a coding-theoretic viewpoint, without their relevance and applications to PCPs in mind, we feel that it is likely that the original definition itself would have required tolerant testers.)

For any vectors $u, v \in \mathbb{F}_q^n$, the relative distance between u and v, denoted $\text{dist}(u, v)$, equals the fraction of positions where u and v differ. For any subset $A \subset \mathbb{F}_q^n$, $\text{dist}(v, A) = \min_{u \in A} \text{dist}(u, v)$. An $[n, k, d]_q$ linear code $\mathcal{C}$ is a k-dimensional subspace of $\mathbb{F}_q^n$ such that every pair of distinct elements $x, y \in \mathcal{C}$ differ in at least d locations, i.e., $\text{dist}(x, y) \geq d/n$. The ratio $\frac{k}{n}$ is called the rate of the code and d is the (minimum) distance of the code. We now formally define a *tolerant* tester.

Definition 1. *For any linear code $\mathcal{C}$ over $\mathbb{F}_q$ of block length n and distance d, and $0 \leq c_1 \leq c_2 \leq 1$, a (c_1, c_2)-tolerant tester T for $\mathcal{C}$ with query complexity $p(n)$ (or simply p when the argument is clear from the context) is a probabilistic polynomial time oracle Turing machine such that for every vector $v \in \mathbb{F}_q^n$:*

1. *If $\text{dist}(v, \mathcal{C}) \leq \frac{c_1 d}{n}$, T upon oracle access to v accepts with probability at least $\frac{2}{3}$ (tolerance),*
2. *If $\text{dist}(v, \mathcal{C}) > \frac{c_2 d}{n}$, T rejects with probability at least $\frac{2}{3}$ (soundness),*
3. *T makes $p(n)$ probes into the string (oracle) v.*

A code is said to be (c_1, c_2, p)-testable if it admits a (c_1, c_2)-tolerant tester of query complexity $p(\cdot)$.

We will be interested in asymptotics and thus we implicitly are interested in a family of codes with the stated properties in the above definition (and so, the notion of the tester being a polynomial time machine, in particular, makes sense). We usually hide this for notational simplicity.

A tester has *perfect completeness* if it accepts any codeword with probability 1. As pointed out earlier, the existing literature just consider $(0, c_2)$-tolerant testers with perfect completeness. We will refer to these as *standard* testers henceforth. Note that our definition of tolerant testers is per se not a generalization of standard testers since we do not require perfect completeness for the case when the input v is a codeword. However, all our constructions will inherit this property from the standard testers we obtain them from.

Recall one of the applications of tolerant testers mentioned earlier– a tolerant tester is used to decide if the expensive decoding algorithm should be used. In this scenario, one would like to set the parameters c_1 and c_2 such that the tester is tolerant up to the decoding radius. For example, if we have an unique decoding algorithm which can correct up to $\frac{d}{2}$ errors, a particularly appealing setting of parameters would be $c_1 = \frac{1}{2}$ and c_2 as close to $\frac{1}{2}$ as possible. However, we would not be able to achieve such large c_1. In general we will aim for positive constants c_1 and c_2 with $\frac{c_2}{c_1}$ being as small as possible while minimizing $p(n)$.

One might hope that the existing standard testers could also be tolerant testers. We give a simple example to illustrate the fact that this is not the case in general. Consider the tester for the Reed-Solomon (RS) codes of dimension $k+1$– pick $k+2$ points uniformly at random and check if the degree k univariate polynomial obtained by interpolating on the first $k + 1$ points agrees with the input on the last point. It is well known that this is a standard tester [16]. However, this is not a tolerant tester. Assume we have an input which differs from a degree k polynomial in only one point. Thus, for $\binom{n-1}{k+1}$ choices of $k + 2$ points, the tester would reject, that is, the rejection probability is $\frac{\binom{n-1}{k+1}}{\binom{n}{k+2}} = \frac{k+2}{n}$ which is greater than $\frac{1}{3}$ for high rate RS codes.

Another pointer towards the inherent difficulty in coming up with a tolerant tester is the recent work of Fischer and Fortnow [8] which shows that there are certain boolean properties which have a standard tester with constant number of queries but for which every tolerant tester requires at least $n^{\Omega(1)}$ queries.

In this work, we examine existing standard testers and convert some standard testers into tolerant ones. In Section 2 we record a few general facts which will be useful in performing this conversion. The ultimate goal, if this can be realized at all, would be to construct tolerant LTCs of constant rate which can be tested using $O(1)$ queries (we remark that such a construction has not been obtained even without the requirement of tolerance). In this work, we show that we can achieve either constant number of queries with slightly sub-constant rate (Section 3) as well as constant rate with sub-linear number of queries (Section 4). That is, something non-trivial is possible in both the domains: (a) constant rate, and (b) constant number of queries. Specifically, in Section 3 we discuss binary codes which encode k bits into codewords of length $n = k \cdot \exp(\log^{\varepsilon} k)$ for any $\varepsilon > 0$, and can be tolerant tested using $O(1/\varepsilon)$ queries. In Section 4, following [5], we study the simple construction of LTCs using products of codes – this yields asymptotically good codes which are tolerant testable using a sub-linear number n^γ of queries for any desired $\gamma > 0$. An interesting common feature of the codes in Section 3 and 4 is that they can be constructed from any code that has good distance properties and which in particular need not admit a local tester with sub-linear query complexity. The overall message from our work is that a lot of the work on locally testable code constructions extends fairly easily to also yield tolerant locally testable codes. However, there does not seem to be a generic way to "compile" a standard tester to a tolerant tester for an arbitrary code.

Due to the lack of space most of the proofs are omitted and will appear in the full version of the paper.

2   General Observations

In this section we will fix some notations and spell out some general properties of tolerant testers and subsequently use them to design tolerant testers for some existing codes. We will denote the set $\{1, \cdots, n\}$ by $[n]$. All the testers we refer to are non-adaptive testers which decide on the locations to query all at once based only on the random choices. In the sequel, we use n to denote the block length and d the distance of the code under consideration. The motivation for the definition below will be clear in Section 3.

Definition 2. *Let $0 < \alpha \leq 1$. A tester T is $(\langle s_1, q_1 \rangle, \langle s_2, q_2 \rangle, \alpha)$-smooth if there exists a set $A \subseteq [n]$ where $|A| = \alpha n$ with the following properties:*

- *T queries at most q_1 points in A, and for every $x \in A$, the probability that each of these queries equals location x is at most $\frac{s_1}{|A|}$, and*
- *T queries at most q_2 points in $[n] - A$, and for every $x \in [n] - A$, the probability that each of these queries equals location x is at most $\frac{s_2}{n-|A|}$.*

As a special case a $(\langle 1, q \rangle, \langle 0, 0 \rangle, 1)$-smooth tester makes a total of q queries each of them distributed uniformly among the n possible probe points. The following lemma follows easily by an application of the union bound.

Lemma 1. *For any $0 < \alpha < 1$, a $(\langle s_1, q_1 \rangle, \langle s_2, q_2 \rangle, \alpha)$-smooth $(0, c_2)$-tolerant tester T with perfect completeness is a (c_1, c_2)-tolerant tester T', where*
$$c_1 = \frac{n\alpha(1-\alpha)}{3d \max\{q_1 s_1(1-\alpha),\ q_2 s_2 \alpha\}}.$$

The above lemma is not useful for us unless the relative distance and the number of queries are constants. Next we sketch how to design tolerant testers from existing *robust* testers with certain properties. We first recall the definition of robust testers from [5].

A standard tester T has two inputs– an oracle for the received word v and a random string s. Depending on s, T generates q query positions $i_1, \cdots, i_q$, fixes a circuit C_s and then accepts if $C_s(v_f(s)) = 1$ where $v_f(s) = \langle v_{i_1}, \cdots, v_{i_q} \rangle$. The robustness of T on inputs v and s, denoted by $\rho^T(v, s)$, is defined to be the minimum, over all strings y such that $C_s(y) = 1$, of $\text{dist}(v_f(s), y)$. The expected robustness of T on v is the expected value of $\rho^T(v, s)$ over the random choices of s and would be denoted by $\rho^T(v)$.

A standard tester T is said to be c-robust for $\mathcal{C}$ if for every $v \in \mathcal{C}$, the tester accepts with probability 1, and for every $v \in \mathbb{F}_q^n$, $\text{dist}(v, \mathcal{C}) \leq c \cdot \rho^T(v)$.

The tolerant version T' of the standard c-robust tester T is obtained by accepting an oracle v on random input s, if $\rho^T(v, s) \leq \tau$ for some threshold τ. (Throughout the paper τ will denote the threshold.) We will sometimes refer to such a tester as one with threshold τ. Recall that a standard tester T accepts if $\rho^T(v, s) = 0$. We next show that T' is sound.

The following lemma follows from a simple averaging argument and the fact that T is c-robust:

Lemma 2. *Let $0 \leq \tau \leq 1$, and let $c_2 = \frac{(\tau+2)cn}{3d}$. For any $v \in \mathbb{F}_q^n$, if $\text{dist}(v, \mathcal{C}) > \frac{c_2 d}{2^n}$, then the tolerant tester T' with threshold τ rejects v with probability at least $\frac{2}{3}$.*

We next mention a property of the query pattern of T which would make T' tolerant. Let S be the set of all possible choices for the random string s. Further for each s, let $p^T(s)$ be the set of positions queried by T.

Definition 3. *A tester T has a partitioned query pattern if there exists a partition $S_1 \cup \cdots \cup S_m$ of the random choices of T for some m, such that for every i,*

- $\cup_{s \in S_i} p^T(s) = \{1, 2, \cdots, n\}$, *and*
- *For all $s, s' \in S_i$, $p^T(s) \cap p^T(s') = \emptyset$ if $s \neq s'$.*

Lemma 3. *Let T have a partitioned query pattern. For any $v \in \mathbb{F}_q^n$, if $\text{dist}(v, \mathcal{C}) \leq \frac{c_1 d}{n}$, where $c_1 = \frac{n\tau}{3d}$, then the tolerant test T' with threshold τ rejects with probability at most $\frac{1}{3}$.*

Proof. Let $S_1, \cdots, S_m$ be the partition of S, the set of all random choices of the tester T. For each j, by the properties of S_j, $\sum_{s \in S_j} \rho^T(v, s) \leq \text{dist}(v, \mathcal{C})$. By an averaging argument and by the assumption on $\text{dist}(v, \mathcal{C})$ and the value of c_1, at least $\frac{2}{3}$ fraction of the choices of s in S_j have $\rho^T(v, s) \leq \tau$ and thus, T' accepts. Recalling that $S_1, \cdots, S_m$ was a partition of S, for at least $\frac{2}{3}$ of the choices of s in S, T' accepts. This completes the proof.

3   Tolerant Testers for Binary Codes

One of the natural goals in the study of tolerant codes is to design explicit tolerant binary codes with constant relative distance and as large a rate as possible. In the case of standard testers, Ben-Sasson et al [4] give binary locally testable codes which map k bits to $k \cdot \exp(\log^\varepsilon k)$ bits for any $\varepsilon > 0$ and which are testable with $O(1/\varepsilon)$ queries. Their construction uses objects called PCPs of Proximity (PCPP) which they also introduce in [4]. In this section, we show that a simple modification to their construction yields tolerant testable binary codes which map k bits to $k \cdot \exp(\log^\varepsilon k)$ bits for any $\varepsilon > 0$. We note that a similar modification is used by Ben-Sasson et al to give a relaxed locally decodable codes [4] but with worse parameters (specifically they gives codes with block length $k^{1+\varepsilon}$).

3.1   PCP of Proximity

We start with the definition[1] of of a Probabilistic Checkable proof of Proximity (PCPP). A pair language is simply a language whose elements are naturally a pair of strings, i.e., it is some collection of strings (x, y). A notable example is CIRCUITVAL $= \{\langle C, a \rangle \mid$ Boolean circuit C evaluates to 1 on assignment $a\}$.

Definition 4. *Fix $0 \le \delta \le 1$. A probabilistic verifier V is a PCPP for a pair language L with proximity parameter δ and query complexity $q(\cdot)$ if the following conditions hold:*

 - *(Completeness) If $(x, y) \in L$ then there exists a proof π such that V accepts by accessing the oracle $y \circ \pi$ with probability 1.*
 - *(Soundness) If y is δ-far from $L(x) = \{y | (x, y) \in L\}$, then for all proofs π, V accepts by accessing the oracle $y \circ \pi$ with probability strictly less than $\frac{1}{4}$.*
 - *(Query complexity) For any input x and proof π, V makes at most $q(|x|)$ queries in $y \circ \pi$.*

Note that a PCPP differs from a standard PCP in that it has a more relaxed soundness condition but its queries into part of the input y are also counted in its query complexity.

Ben-Sasson et. al. give constructions of PCPPs with the following guarantees:

Lemma 4. *([4]) Let $\varepsilon > 0$ be arbitrary. There exists a PCP of proximity for the pair language CIRCUITVAL $= \{(C, x) | C$ is a boolean circuit and $C(x) = 1\}$ whose proof length, for input circuits of size s, is at most $s \cdot \exp(\log^{\varepsilon/2} s)$ and for $t = \frac{2 \log \log s}{\log \log \log s}$ the verifier of proximity has query complexity $O(\max\{\frac{1}{\delta}, \frac{1}{\varepsilon}\})$ for any proximity parameter δ that satisfies $\delta \ge \frac{1}{t}$. Furthermore, the queries of the verifier are non-adaptive and each of the queries which lie in the input part x are uniformly distributed among the locations of x.*

The fact that the queries to the input part are uniformly distributed follows by an examination of the verifier construction in [4]. In fact, in the extended version of that paper, the authors make this fact explicit and use it in their construction of relaxed locally decodable codes (LDCs). To achieve a tolerant LTC using the PCPP, we will need all queries of the verifier to be somewhat uniformly or smoothly distributed. We will now proceed to make the queries of the PCPP verifier that fall into the "proof part" π near-uniform. This will follow

---

[1] The definition here is a special case of the general PCPP defined in [4] which would be sufficient for our purposes.

a fairly general method suggested in [4] to smoothen out the query distribution, which the authors used to obtain relaxed locally decodable codes from the PCPP. We will obtain tolerant LTCs instead, and in fact will manage to do so without a substantial increase in the encoding length (i.e., the encoding length will remain $k \cdot 2^{\log^\varepsilon k}$). On the other hand, the best encoding length achieved for relaxed LDCs in [4] is $k^{1+\varepsilon}$ for constant $\varepsilon > 0$. We begin with the definition of a mapping that helps smoothen out the query distribution.

Definition 5. *Given any $v \in \mathbb{F}_q^n$ and $\boldsymbol{p} = \langle p_i \rangle_{i=1}^n$ with $p_i \geq 0$ for all $i \in [n]$ and $\sum_{i=1}^n p_i = 1$, we define the mapping $\mathsf{Repeat}(\cdot, \cdot)$ as follows: $\mathsf{Repeat}(v, \boldsymbol{p}) \in \mathbb{F}_q^{n'}$ such that v_i is repeated $\lfloor 4np_i \rfloor$ times in $\mathsf{Repeat}(v, \boldsymbol{p})$ and $n' = \sum_{i=1}^n \lfloor 4np_i \rfloor$.*

The foloowing lemma shows why the mapping is useful. A similar fact appears in [4].

Lemma 5. *For any $v \in \mathbb{F}_q^n$ let a non-adaptive verifier T (with oracle access to v) make $q(n)$ queries and let p_i be the probability that each of these queries probes location $i \in [n]$. Let $c_i = \frac{1}{2n} + \frac{p_i}{2}$ and $\boldsymbol{c} = \langle c_i \rangle_{i=1}^n$. Consider the map $\mathsf{Repeat}(v, \boldsymbol{c}) : \mathbb{F}_q^n \to \mathbb{F}_q^{n'}$. Then there exists another tester T' for strings of length n' with the following properties:*

1. *T' makes $2q(n)$ queries on $v' = \mathsf{Repeat}(v, \boldsymbol{c})$ each of which probes location j, for any $j \in [n']$, with probability at most $\frac{2}{n'}$, and*
2. *for every $v \in \mathbb{F}_q^n$, the decision of T' on v' is identical to that of T on v. Further, $3n < n' \leq 4n$.*

Applying the transformation of Lemma 5 to the proximity verifier and proof of proximity of Lemma 4, we conclude the following.

Proposition 1. *Let $\varepsilon > 0$ be arbitrary. There exists a PCP of proximity for the pair language CIRCUITVAL $= \{(C, x) | C \text{ is a boolean circuit and } C(x) = 1\}$ with the following properties:*

1. *The proof length, for input circuits of size s, is at most $s \cdot \exp(\log^{\varepsilon/2} s)$, and*
2. *for $t = \frac{2 \log \log s}{\log \log \log s}$ the verifier of proximity has query complexity $O(\max\{\frac{1}{\delta}, \frac{1}{\varepsilon}\})$ for any proximity parameter δ that satisfies $\delta \geq \frac{1}{t}$.*

Furthermore, the queries of the verifier are non-adaptive with the following properties:

1. *Each query made to one of the locations of the input x is uniformly distributed among the locations of x, and*
2. *each query to one of the locations in the proof of proximity π probes each location with probability at most $2/|\pi|$ (and thus is distributed nearly uniformly among the locations of π).*

3.2 The Code

We now outline the construction of the locally testable code from [4]. The idea behind the construction is to make use of a PCPP to aid in checking if the received word is a codeword is far away from being one. Details follow.

Suppose we have a binary code $C_0 : \{0,1\}^k \rightarrow \{0,1\}^m$ of distance d defined by a parity check matrix $H \in \{0,1\}^{(m-k)\times m}$ that is sparse, i.e., each of whose rows has only an absolute constant number of 1's. Such a code is referred to as a low-density parity check code (LDPC). For the construction below, we will use any such code which is asymptotically good (i.e., has rate k/m and relative distance d/m both positive as $m \rightarrow \infty$). Explicit constructions of such codes are known using expander graphs [15]. Let V be a verifier of a PCP of proximity for membership in C_0; more precisely, the proof of proximity of an input string $w \in \{0,1\}^m$ will be a proof that $\tilde{C}_0(w) = 1$ where $\tilde{C}_0$ is a linear-sized circuit which performs the parity checks required by H on w (the circuit will have size $O(m) = O(k)$ since H is sparse and C_0 has positive rate). Denote by $\pi(x)$ be the proof of proximity guaranteed by Proposition 1 for the claim that the input $C_0(x)$ is a member of C_0 (i.e., satisfies the circuit $\tilde{C}_0$). By Proposition 1 and fact that the size of $\tilde{C}_0$ is $O(k)$, the length of $\pi(x)$ can be made at most $k \exp(\log^{\varepsilon/2} k)$.

The final code is defined as $C_1(x) = (C_0(x)^t, \pi(x))$ where $t = \frac{(\log k - 1)|\pi(x)|}{|C_0(x)|}$. The repetition of the code part $C_0(x)$ is required in order to ensure good distance, since the length of the proof part $\pi(x)$ typically dominates and we have no guarantee on how far apart $\pi(x_1)$ and $\pi(x_2)$ for $x_1 \neq x_2$ are.

For the rest of this section let ℓ denote the proof length. The tester T_1 for C_1 on an input $w = (w_1, \cdots, w_t, \pi) \in \{0,1\}^{tm+\ell}$ picks $i \in [t]$ at random and runs the PCPP verifier V on $w_i \circ \pi$. It also performs a few rounds of the following consistency checks: pick $i_1, i_2 \in [t]$ and $j_1, j_2 \in [m]$ at random and check if $w_{i_1}(j_1) = w_{i_2}(j_2)$. Ben-Sasson et al in [4] show that T_1 is a standard tester. However, T_1 need not be a tolerant tester. To see this, note that the proof part of C_1 forms a $\frac{1}{\log k}$ fraction of the total length. Now consider a received word $w_{rec} = (w_0, \cdots, w_0, \pi')$ where $w_0 \in C_0$ but π' is not a correct proof for w_0 being a valid codeword in c_0. Note that w_{rec} is close to C_1. However, T_1 is not guaranteed to accept w_{rec} with high probability.

The problem with the construction above was that the proof part was too small: a natural fix is to make the proof part a constant fraction of the codeword. We will show that this is sufficient to make the code tolerant testable. We also remark that a similar idea was used by Ben-Sasson et. al. to give efficient constructions for relaxed locally decodable codes [4].

Construction 1 *Let $0 < \beta < 1$ be a parameter, $C_0 : \{0,1\}^k \rightarrow \{0,1\}^m$ be a good [2] binary code and V be a PCP of proximity verifier for membership in C_0. Finally let $\pi(x)$ be the proof corresponding to the claim that $C_0(x)$ is a codeword in C_0. The final code is defined as $C_2(x) = (C_0(x)^{r_1}, \pi(x)^{r_2})$ with $r_1 = \frac{(1-\beta)\log k |\pi(x)|}{m}$ and $r_2 = \beta \log k$.*

[2] This means that $m = O(k)$ and the encoding can be done by circuits of nearly linear size $s_0 = \tilde{O}(k)$.

For the rest of the section the proof length $|\pi(x)|$ will be denoted by ℓ. Further the proximity parameter and the number of queries made by the PCPP verifier V would be denoted by δ_p and q_p respectively. Finally let ρ_0 denote the relative distance of the code $\mathcal{C}_0$.

The tester T_2 for $\mathcal{C}_2$ is also the natural generalization of T_1. For a parameter q_r (to be instantiated later) and input $w = (w_1, \cdots, w_{r_1}, \pi_1, \cdots, \pi_{r_2}) \in \{0,1\}^{r_1 m + r_2 \ell}$, T_2 does the following:

1. Repeat the next two steps twice.
2. Pick $i \in [r_1]$ and $j \in [r_2]$ randomly and run V on $w_i \circ \pi_j$.
3. Do q_r repetitions of the following: pick $i_1, i_2 \in [r_1]$ and $j_1, j_2 \in [m]$ randomly and check if $w_{i_1}(j_1) = w_{i_2}(j_2)$.

The following lemma captures the properties of the code $\mathcal{C}_2$ and its tester T_2.

Lemma 6. *The code $\mathcal{C}_2$ in Construction 1 and the tester T_2 (with parameters β and q_r respectively) above have the following properties:*

1. *The code $\mathcal{C}_2$ has block length $n = \log k \cdot \ell$ with minimum distance d lower bounded by $(1 - \beta)\rho_0 n$.*
2. *T_2 makes a total of $q = 2q_p + 4q_r$ queries.*
3. *T_2 is $(\langle 1, q \rangle, \langle 2, 2q_p \rangle, 1 - \beta)$-smooth.*
4. *T_2 is a (c_1, c_2)-tolerant tester with $c_1 = \frac{n\beta(1-\beta)}{6d \max\{(2q_r + q_p)\beta,\, 2(1-\beta)q_p\}}$ and $c_2 = \frac{n}{d}(\delta_p + \frac{4}{q_r} + \beta)$.*

Fix any $0 < \delta < 1$ and let $\beta = \frac{\delta}{2}$, $\delta_p = \frac{\delta}{6}$, $q_r = \frac{12}{\delta}$. With these settings we get $\delta_p + \frac{4}{q_r} + \beta = \delta$ and $q_p = O(\frac{1}{\delta})$ from Proposition 1 with the choice $\varepsilon = 2\delta$. Finally, $q = 2q_p + 4q_r = O(\frac{1}{\delta})$. Substituting the parameters in c_2 and c_1, we get $c_2 = \frac{\delta n}{d}$ and

$$\frac{c_1 d}{n} = \frac{\delta}{24 \max\{\delta(q_r + q_p/2),\, (2 - \delta)q_p\}} = \Omega(\delta^2).$$

Also note that the minimum distance $d \geq (1 - \beta)\rho_0 n = (1 - \frac{\delta}{2})\rho_0 n \geq \frac{\rho_0}{2} n$. Thus, we have the following result for tolerant testable binary codes.

Theorem 1. *There exists an absolute constant $\alpha_0 > 0$ such that for every δ, $0 < \delta < 1$, there exists an explicit binary linear code $\mathcal{C} : \{0,1\}^k \to \{0,1\}^n$ where $n = k \cdot \exp(\log^\delta k)$ with minimum distance $d \geq \alpha_0 n$ which admits a (c_1, c_2)-tolerant tester with $c_2 = O(\delta)$, $c_1 = \Omega(\delta^2)$ and query complexity $O(\frac{1}{\delta})$.*

The claim about explicitness follows from the fact that the PCPP of Lemma 4 and hence Proposition 1 has an explicit construction. The claim about linearity follows from the fact that the PCPP for CIRCUITVAL is a linear function of the input when the circuit computes linear functions – this aspect of the construction is discussed in detail in Section 8.4 of the extended version of [4].

4   Tolerant Testers for Tensor Products of Codes

Tensor product of codes is simple way to construct new codes from any existing codes such that the constructed codes have testers with sub-linear query complexity even though the original code need not admit a sub-linear complexity tester [5]. We first briefly define product of codes and then outline the tester of product of codes from [5].

Given an $[n, k, d]_q$ code $\mathcal{C}$, the product of $\mathcal{C}$ with itself, denoted by $\mathcal{C}^2$, is a $[n^2, k^2, d^2]_q$ code such that a codeword (viewed as a $n \times n$ matrix) restricted to any row or column is a codeword in $\mathcal{C}$. More formally, given the $n \times k$ generator matrix M of $\mathcal{C}$, $\mathcal{C}^2$ is precisely the set of matrices in the set $\{M \cdot X \cdot M^T \mid X \in F_q^{k \times k}\}$. A very natural test for $\mathcal{C}^2$ is to randomly choose a row or a column and then check if the restriction of the received word on that row or column is a codeword in $\mathcal{C}$ (which can be done for example by querying all the n points in the row or column). Unfortunately, it is not known if this test is robust in general (see the discussion in [5]).

Ben-Sasson and Sudan in [5] considered the more general product of codes $\mathcal{C}^t$ for $t \geq 3$ along with the following general tester– Choose at random $b \in \{1, \cdots, t\}$ and $i \in \{1, \cdots, n\}$ and check if b^{th} coordinate of the received word (which is an element of $\mathbb{F}_q^{n^t}$) when restricted[3] to i is a codeword in $\mathcal{C}^{t-1}$. It is shown in [5] that this test is robust, in that if a received word is far from $\mathcal{C}^t$, then many of the tested substrings will be far from $\mathcal{C}^{t-1}$. This tester lends itself to recursion: the test for $\mathcal{C}^{t-1}$ can be reduced to a test for $\mathcal{C}^{t-2}$ and so on till we need to check whether a word in $\mathbb{F}_q^{n^2}$ is a codeword of $\mathcal{C}^2$. This last check can done by querying all the n^2 points, out of the n^t points in the original received word, thus leading to a sub-linear query complexity. As shown in [5], the reduction can be done in $\log t$ stages by the standard halving technique.

We now give a tolerant version of the test for product of codes given by Ben-Sasson and Sudan [5]. In what follows $t \geq 4$ will be a power of two. As mentioned above the tester T for the tensor product $\mathcal{C}^t$ reduces the test to checking if some restriction of the given string belong to $\mathcal{C}^2$. For the rest of this section, with a slight abuse of notation let $v_f \in \mathbb{F}_q^{n^2}$ denote the final restriction being tested. In what follows we assume that by looking at all points in any $v \in \mathbb{F}_q^{n^2}$ one can determine if $\text{dist}(v, \mathcal{C}^2) \leq \tau$ in time polynomial in n^2.

The tolerant version of the test of [5] is a simple modification as mentioned in Section 2: reduce the test on $\mathcal{C}^t$ to $\mathcal{C}^2$ as in [5] and then accept if v_f is τ-close to $\mathcal{C}^2$.

First we make the following observation about the test in [5]. The test recurses $\log t$ times to reduce the test to $\mathcal{C}^2$. At step l , the test chooses an random coordinate b_l (this will just be a random bit) and fixes the value of the b_l^{th} coordinate of the current $\mathcal{C}^{\frac{t}{2^l}}$ to an index i_l (where i_l takes values in the range $1 \leq i_l \leq n^{t/2^l}$). The key observation here is that for each fixed choice of $b_1, \cdots, b_{\log t}$, distinct choices of $i_1, \cdots, i_{\log t}$ correspond to querying disjoint sets n^2 points in

---

[3] For the $t = 2$ case b signifies either row or column and i denotes the row/column index.

the original $v \in \mathbb{F}_q^{n^t}$ string, which together form a partition of all coordinates of v. In other words, T has a *partitioned* query pattern, which will be useful to argue tolerance. For soundness, we use the results in [5], which show that their tester is $C^{\log t}$-robust for $C = 2^{32}$.

Applying Lemmas 2 and 3, therefore, we have the following result:

Theorem 2. *Let $t \geq 4$ be a power of two and $0 < \tau \leq 1$. There exist $0 < c_1 < c_2 \leq 1$ with $\frac{c_2}{c_1} = C^{\log t}(1 + 2/\tau)$ such that the proposed tolerant tester for $\mathcal{C}^t$ is a (c_1, c_2)-tolerant tester with query complexity $N^{2/t}$ where N is the block length of $\mathcal{C}^t$. Further, c_1 and c_2 are constants (independent of N) if t is a constant and $\mathcal{C}$ has constant relative distance.*

Thus, Theorem 2 achieves the goal of a simple construction of tolerant testable codes with sub-linear query complexity, as the following corollary records:

Corollary 1 *For every $\gamma > 0$, there is an explicit family of asymptotically good binary linear codes which are tolerant testable using n^γ queries, where n is the block length of the concerned code. (The rate, relative distance and thresholds c_1, c_2 for the tolerant testing depend on γ.)*

5   Concluding Remarks

Obtaining non-trivial lower bounds on the the block length of codes that are locally testable with very few (even 3) queries is an extremely interesting question. This problem has remained open and resisted even moderate progress despite all the advancements in constructions of LTCs. The requirement of having a tolerant local tester is a stronger requirement. While we have seen that we can get tolerance with similar parameters to the best known LTCs, it remains an interesting question whether the added requirement of tolerance makes the task of proving lower bounds more tractable. This seems like a good first step in making progress towards understanding whether asymptotically good locally testable codes exist, a question which is arguably one of the grand challenges in this area. For interesting work in this direction which proves that such codes, if they exist, cannot also be *cyclic*, see [3].

We also investigated whether Reed-Muller codes admit tolerant testers. For large fields, the existing results on low-degree testing of multivariate polynomials [9, 14] immediately imply results on tolerant testing for these codes. The details were omitted due to the lack of space and would appear in the full version of the paper.

Very recently, (standard) locally testable codes with blocklength $n = k \cdot$ polylog$k$ have been constructed [6, 7]. In the full version of the paper we hope to investigate whether this new code admits a tolerant tester as well.

References

1. S. Arora, C. Lund, R. Motwani, M. Sudan, and M. Szegedy. Proof verification and the intractibility of approximation problems. *Journal of the ACM*, 45(3):501–555, 1998.

2. S. Arora and S. Safra. Probabilistic checking of proofs: A new characterization of NP. *Journal of the ACM*, 45(1):70–122, 1998.
3. L. Babai, A. Shpilka, and D. Stefankovic. Locally testable cyclic codes. In *Proceedings of 44th Annual Symposium on Foundations of Computer Science (FOCS)*, pages 116–125, 2003.
4. E. Ben-Sasson, O. Goldreich, P. Harsha, M. Sudan, and S. Vadhan. Robust PCPs of proximity, shorter PCPs and application to coding. In *Proceedings of the 36th Annual ACM Symposium on Theory of Computing (STOC)*, pages 1–10, 2004.
5. E. Ben-Sasson and M. Sudan. Robust locally testable codes and products of codes. In *Proceedings of the 8th International Workshop on Randomization and Computation (RANDOM)*, pages 286–297, 2004.
6. E. Ben-Sasson and M. Sudan. Simple PCPs with poly-log rate and query complexity. In *Proceedings of 37th ACM Symposium on Theory of Computing (STOC)*, pages 266–275, 2005.
7. I. Dinur. The PCP theorem by gap amplification. In *ECCC Technical Report TR05-046*, 2005.
8. E. Fischer and L. Fortnow. Tolerant versus intolerant testing for boolean properties. In *Proceedings of the 20th IEEE Conference on Computational Complexity*, 2005. To appear.
9. K. Friedl and M. Sudan. Some improvements to total degree tests. In *Proceedings of the 3rd Israel Symp. on Theory and Computing Systems (ISTCS)*, pages 190–198, 1995.
10. O. Goldreich. Short locally testable codes and proofs (Survey). *ECCC Technical Report TR05-014*, 2005.
11. O. Goldreich and M. Sudan. Locally testable codes and PCPs of almost linear length. In *Proceedings of 43rd Symposium on Foundations of Computer Science (FOCS)*, pages 13–22, 2002.
12. T. Kaufman and D. Ron. Testing polynomials over general fields. In *Proceedings of the 45th Annual IEEE Symposium on Foundations of Computer Science (FOCS)*, pages 413–422, 2004.
13. M. Parnas, D. Ron, and R. Rubinfeld. Tolerant property testing and distance approximation. In *ECCC Technical Report TR04-010*, 2004.
14. A. Polishchuk and D. A. Spielman. Nearly-linear size holographic proofs. In *Proceedings of the 26th Annual ACM Symposium on Theory of Computing (STOC)*, pages 194–203, 1994.
15. M. Sipser and D. Spielman. Expander codes. *IEEE Transactions on Information Theory*, 42(6):1710–1722, 1996.
16. M. Sudan. *Efficient Checking of Polynomials and Proofs and the Hardness of Approximation Problems*. ACM Distinguished Theses Series. Lecture Notes in Computer Science, no. 1001, Springer, 1996.

A Lower Bound on List Size for List Decoding

Venkatesan Guruswami[1,*] and Salil Vadhan[2,**]

[1] Department of Computer Science & Engineering
University of Washington, Seattle, WA
venkat@cs.washington.edu
[2] Division of Engineering & Applied Sciences
Harvard University, Cambridge, MA
salil@eecs.harvard.edu

Abstract. A q-ary error-correcting code $C \subseteq \{1, 2, \ldots, q\}^n$ is said to be *list decodable* to radius ρ with list size L if every Hamming ball of radius ρ contains at most L codewords of C. We prove that in order for a q-ary code to be list-decodable up to radius $(1 - 1/q)(1 - \varepsilon)n$, we must have $L = \Omega(1/\varepsilon^2)$. Specifically, we prove that there exists a constant $c_q > 0$ and a function f_q such that for small enough $\varepsilon > 0$, if C is list-decodable to radius $(1 - 1/q)(1 - \varepsilon)n$ with list size c_q/ε^2, then C has at most $f_q(\varepsilon)$ codewords, independent of n. This result is asymptotically tight (treating q as a constant), since such codes with an exponential (in n) number of codewords are known for list size $L = O(1/\varepsilon^2)$.
A result similar to ours is implicit in Blinovsky [Bli] for the binary ($q = 2$) case. Our proof works for all alphabet sizes, and is technically and conceptually simpler.

1 Introduction

List decoding was introduced independently by Elias [Eli1] and Wozencraft [Woz] as a relaxation of the classical notion of error-correction by allowing the decoder to output a *list* of possible answers. The decoding is considered successful as long as the correct message is included in the list. We point the reader to the paper by Elias [Eli2] for a good summary of the history and context of list decoding.

The basic question raised by list decoding is the following: How many errors can one recover from, when constrained to output a list of small size? The study of list decoding strives to (1) understand the combinatorics underlying this question, (2) realize the bounds with explicit constructions of codes, and (3) list decode those codes with efficient algorithms. This work falls in the combinatorial facet of list decoding. Combinatorially, an error-correcting code has "nice" list-decodability properties if every Hamming ball of "large" radius has a "small" number of codewords in it. In this work, we are interested in exposing some combinatorial *limitations* on the performance of list-decodable codes.

* Supported in part by NSF Career Award CCF-0343672.
** Supported by NSF grant CCF-0133096, ONR grant N00014-04-1-0478, and a Sloan Research Fellowship. Work done in part while a Fellow at the Radcliffe Institute for Advanced Study.

C. Chekuri et al. (Eds.): APPROX and RANDOM 2005, LNCS 3624, pp. 318–329, 2005.
© Springer-Verlag Berlin Heidelberg 2005

Specifically, we seek lower bounds on the list size needed to perform decoding up to a certain number of errors, or in other words, lower bounds on the number of codewords that must fall inside *some* ball of specified radius centered at some point. We show such a result by picking the center in a certain probabilistic way. We now give some background definitions and terminology, followed by a description of our main result.

1.1   Preliminaries

We denote the set $\{1, 2, \ldots, m\}$ by the shorthand $[m]$. For $q \geq 2$, a q-ary *code* of *block length* n is simply a subset of $[q]^n$. The elements of the code are referred to as *codewords*. The high-level property of a code that makes it useful for error-correction is its sparsity — the codewords must be well spread-out, so they are unlikely to distort into one another. One way to insist on sparsity is that the Hamming distance between every pair of distinct codewords is at least d. Note that this is equivalent to requiring that every Hamming ball of radius $\lfloor (d-1)/2 \rfloor$ has at most one codeword. Generalizing this, one can allow up to a small number, say L, of codewords in Hamming balls of certain radius. This leads to the notion of list decoding and a good list-decodable code. Since the expected Hamming distance of a random string of length n from any codeword is $(1 - 1/q) \cdot n$ for a q-ary code, the largest fraction of errors one can sensibly hope to correct is $(1 - 1/q)$. This motivates the following definition of a list-decodable code.

Definition 1. *Let $q \geq 2$, $0 < \rho < 1$, and L be a positive integer. A q-ary code C of block length n is said to be (ρ, L)-list-decodable if for every $y \in [q]^n$, the Hamming ball of radius $\rho \cdot (1 - 1/q) \cdot n$ centered at y contains at most L codewords of C.*

We will study (ρ, L)-list-decodable codes for $\rho = 1 - \varepsilon$ in the limit of $\varepsilon \to 0$. This setting is the one where list decoding is most beneficial, and is a clean setting to initially study the asymptotics. In particular, we will prove that, except for trivial codes whose size does not grow with n, $(1 - \varepsilon, L)$-list-decodable codes require list size $L = \Omega(1/\varepsilon^2)$ (hiding dependence on q).

1.2   Context and Related Results

Before stating our result, we describe some of the previously known results to elucidate the broader context where our work fits. The *rate* of a q-ary code of block length n is defined to be $\frac{\log_q |C|}{n}$. For $0 \leq x \leq 1$, we denote by $H_q(x)$ the q-ary entropy function, $H_q(x) = x \log_q(q - 1) - x \log_q x - (1 - x) \log_q(1 - x)$.

   Using the probabilistic method, it can be shown that (ρ, L)-list-decodable q-ary codes of rate $1 - H_q((1 - 1/q)\rho) - 1/L$ exist [Eli2, GHSZ]. In particular, in the limit of large L, we can achieve a rate of $1 - H_q((1 - 1/q)\rho)$, which equals both the Hamming bound and the Shannon capacity of the q-ary channel that changes a symbol $\alpha \in [q]$ to a uniformly random element of $[q]$ with probability ρ and leaves α unchanged with probability $1 - (1 - 1/q)\rho$. When $\rho = 1 - \varepsilon$ for small ε, we have $H_q((1 - 1/q)\rho) = 1 - \Omega(q\varepsilon^2/\log q)$. Therefore, there exist $(1 - \varepsilon, L(q, \varepsilon))$-list-decodable q-ary codes with $2^{\Omega(q\varepsilon^2 n)}$ codewords and $L(q, \varepsilon) = O(\frac{\log q}{q\varepsilon^2})$. In particular, for constant q, list size of $O(1/\varepsilon^2)$ suffices for non-trivial list decoding up to radius $(1 - 1/q) \cdot (1 - \varepsilon)$.

We are interested in whether this quadratic dependence on $1/\varepsilon$ in the list size is inherent. The quadratic bound is related to the $2 \log(1/\varepsilon) - O(1)$ lower bound due to [RT] for the amount of "entropy loss" in *randomness extractors*, which are well-studied objects in the subject of pseudorandomness. In fact a lower bound of $\Omega(1/\varepsilon^2)$ on list size will implies such an entropy loss bound for ("strong") randomness extractors. However, in the other direction, the argument loses a factor of ε in the lower bound, yielding only a lower of $\Omega(1/\varepsilon)$ for list size (cf. [Vad]).

For the model of erasures, where up to a fraction $(1 - \varepsilon)$ of symbols are erased by the channel, optimal bounds of $\Theta(\log(1/\varepsilon))$ are known for the list size required for binary codes [Gur]. This can be compared with the $\log \log(1/\varepsilon) - O(1)$ lower bound on entropy loss for dispersers [RT].

A lower bound of $\Omega(1/\varepsilon^2)$ for list size L for $(1 - \varepsilon, L)$-list-decodable *binary* codes follows from the work of Blinovsky [Bli]. We discuss more about his work and how it compares to our results in Section 1.5.

1.3   Our Result

Our main result is a proof of the following fact: the smallest list size that permits list decoding up to radius $(1 - 1/q)(1 - \varepsilon)$ is $\Theta(\varepsilon^{-2})$ (hiding constants depending on q in the Θ-notation). The formal statement of our main result is below.

Theorem 2 (Main). *For every integer $q \geq 2$ there exists $c_q > 0$ and $d_q < \infty$ such that for all small enough $\varepsilon > 0$, the following holds. If C is a q-ary $(1 - \varepsilon, c_q/\varepsilon^2)$-list-decodable code, then $|C| \leq 2^{d_q \cdot \varepsilon^{-2} \log(1/\varepsilon)}$.*

1.4   Overview of Proof

We now describe the high-level structure of our proof. Recall that our goal is to exhibit a center z that has several (specifically $\Omega(1/\varepsilon^2)$) codewords of C with large correlation, where we say two codewords have correlation ε if they agree in $(1/q + \varepsilon) \cdot n$ locations. (The actual definition we use, given in Definition 3, is slightly different, but this version suffices for the present discussion.) Using the probabilistic method, it is not very difficult to prove the existence of such a center z and $\Omega(1/\varepsilon^2)$ codewords whose *average* correlation with z is at least $\Omega(\varepsilon)$. (This is the content of our Lemma 6.) This step is closely related to (and actually follows from) the known lower bound of Radhakrishnan and Ta-Shma [RT] on the "entropy loss" of "randomness extractors," by applying the known connection between randomness extractors and list-decodable error-correcting codes (see [Tre, TZ, Vad]).

However, this large average could occur due to about $1/\varepsilon$ codewords having a $\Omega(1)$ correlation with z, whereas we would like to find many more (i.e., $\Omega(1/\varepsilon^2)$)

codewords with smaller (i.e., $\Omega(\varepsilon)$) correlation. We get around this difficulty by working with a large subcode C' of C where such a phenomenon cannot occur. Roughly speaking, we will use the probabilistic method to prove the existence of a large "L-pseudorandom" subcode C', for which looking at any set of L codewords of C *never* reveals any significant overall bias in terms of the most popular symbol (out of $[q]$). More formally, all ℓ-tuples, $\ell \leq L$, the average "plurality" (i.e., frequency of most frequent symbol) over all the coordinates isn't much higher than ℓ/q. (This is the content of our Lemma 7.) This in turn implies that for every center z, the sum of the correlations of z with all codewords that have "large" correlation (say at least $D\varepsilon$, for a sufficiently large constant D) is small. Together with the high average correlation bound, this means several codewords must have "intermediate" correlation with z (between ε and $D\varepsilon$). The number of such codewords is our lower bound on list size.

1.5 Comparison with Blinovsky [Bli]

As remarked earlier, a lower bound of $L = \Omega(1/\varepsilon^2)$ for *binary* $(1 - \varepsilon, L)$-list-decodable codes follows from the work of Blinovsky [Bli]. He explores the tradeoff between ρ, L, and the relative rate γ of a (ρ, L)-list-decodable code, when all three of these parameters are constants and the block length n tends to infinity. A special case of his main theorem shows that if $\rho = 1 - \varepsilon$ and $L \leq c/\varepsilon^2$ for a certain constant $c > 0$, then the rate γ must be zero asymptotically, which means that the code can have at most $2^{o(n)}$ codewords for block length n. A careful inspection of his proof, however, reveals an $f(\varepsilon)$ bound (independent of n) on the number of codewords in any such code. This is similar in spirit to our Theorem 2. However, our work compares favorably with [Bli] in the following respects.

1. Our result also holds for q-ary codes for $q > 2$. The result in [Bli] applies only to binary codes, and it is unclear whether his analysis can be generalized to q-ary codes.
2. Our result is quantitatively stronger. The dependence $f(\varepsilon)$ of the bound on the size of the code in [Bli] is much worse than the $(1/\varepsilon)^{O(\varepsilon^{-2})}$ that we obtain. In particular, $f(\varepsilon)$ is at least an exponential tower of height $\Theta(1/\varepsilon^2)$ (and is in fact bigger than the Ackermann function of $1/\varepsilon$).
3. Our proof seems significantly simpler and provides more intuition about why and how the lower bound arises.

We now comment on the proof method in [Bli]. As with our proof, the first step in the proof is a bound for the case when the average correlation (w.r.t every center) for every set of $L+1$ codewords is small (this is Theorem 2 in [Bli]). Note that this is a more stringent condition than requiring no set of $L + 1$ codewords lie within a small ball. Our proof uses the probabilistic method to show the existence of codewords with large average correlation in any reasonable sized code. The proof in [Bli] is more combinatorial, and uses a counting argument to bound the size of the code when all subsets of $L + 1$ codewords have low average correlation (with every center). But the underlying technical goal of the first step in both the approaches is the same.

The second step in Blinovsky's proof is to use this bound to obtain a bound for list-decodable codes. The high-level idea is to pick a subcode of the list-decodable code with certain nice properties so that the bound for average correlation can be turned into one for list decoding. This is also similar in spirit to our approach (Lemma 7). The specifics of how this is done are, however, quite different. The approach in [Bli] is to find a large subcode which is $(L+1)$-*equidistant*, i.e., for every $k \leq L+1$, all subsets of k codewords have the same value for their k'th order scalar product, which is defined as the sum over all coordinates of the product of the k symbols (from $\{0,1\}$) in that coordinate.[1] Such a subcode has the following useful property: in each subset of $L+1$ codewords, all codewords in the subset have the same agreement with the best center, i.e., the center obtained by taking their coordinate-wise majority, and moreover this value is independent of the choice of the subset of $L+1$ codewords. This in turn enables one to get a bound for list decoding from one for average correlation. The requirement of being $(L+1)$-equidistant is a rather stringent one, and is achieved iteratively by ensuring k-equidistance for $k = 1, 2, \ldots, L+1$ successively. Each stage incurs a rather huge loss in the size of the code, and thus the bound obtained on the size of the original code is an enormously large function of $1/\varepsilon$. We make do with a much weaker property than $(L+1)$-equidistance, letting us pick a much larger subcode with the property we need. This translates into a good upper bound on the size of the original list-decodable code.

2   Proof of Main Result

We first begin with convenient measures of closeness between strings, the agreement and the correlation.

Definition 3 (Agreement and Correlation). *For strings* $x, y \in [q]^n$, *define their agreement, denoted* $\mathsf{agr}(x,y) = \frac{1}{n} \cdot \#\{i : x_i = y_i\}$. *Their correlation is the value* $\mathsf{corr}(x,y) \in [-1/(q-1), 1]$ *such that* $\mathsf{agr}(x,y) = \frac{1}{q} + \left(1 - \frac{1}{q}\right) \cdot \mathsf{corr}(x,y)$.[2]

The standard notion of correlation between two strings in $\{1, -1\}^n$ is simply their dot product divided by n; the definition above is a natural generalization to larger alphabets.

A very useful notion for us will be the plurality of a set of codewords.

Definition 4 (Plurality). *For symbols* $a_1, \ldots, a_k \in [q]$, *we define their plurality* $\mathsf{plur}(a_1, \ldots, a_k) \in [q]$ *to be the most frequent symbol among* $a_1, \ldots, a_k$, *breaking ties arbitrarily. We define the plurality count* $\#\mathsf{plur}(a_1, \ldots, a_k) \in \mathbb{N}$ *to be the number of times that* $\mathsf{plur}(a_1, \ldots, a_k)$ *occurs among* $a_1, \ldots, a_k$.

---

[1] A slight relaxation of the $(L+1)$-equidistance property is actually what is used in [Bli], but this description should suffice for the discussion here.

[2] Note that we find it convenient to work with agreement and correlation that are normalized by dividing by the length n.

For vectors $c_1, \ldots, c_k \in [q]^n$, we define $\mathsf{plur}(c_1, \ldots, c_k) \in [q]^n$ to be the componentwise plurality, i.e. $\mathsf{plur}(c_1, \ldots, c_k)_i = \mathsf{plur}(c_{1i}, \ldots, c_{ki})$.

We define $\#\mathsf{plur}(c_1, \ldots, c_k)$ to be the average plurality count over all coordinates; that is, $\#\mathsf{plur}(c_1, \ldots, c_k) = (1/n) \cdot \sum_{i=1}^{n} \#\mathsf{plur}(c_{1i}, \ldots, c_{ki})$.

The reason pluralities will be useful to us is that they capture the maximum average correlation any vector has with a set of codewords:

Lemma 5. *For all $c_1, \ldots, c_k \in [q]^n$,*

$$\max_{z \in [q]^n} \sum_{i=1}^{k} \mathsf{agr}(z, c_i) = \sum_{i=1}^{k} \mathsf{agr}(\mathsf{plur}(c_1, \ldots, c_k), c_i) = \#\mathsf{plur}(c_1, \ldots, c_k)$$

Note that our goal of proving lower bound on list size is the same as proving that in every not too small code, there must be some center z that has several (i.e. $\Omega(1/\varepsilon^2)$) close-by codewords, or in other words several codewords with large (i.e., at least ε) correlation. We begin by showing the existence of a center which has a large *average* correlation with a collection of several codewords. By Lemma 5, this is equivalent to finding a collection of several codewords whose total plurality count is large.

Lemma 6. *For all integers $q \geq 2$ and $t \geq 37q$, there exists a constant $b_q > 0$ such that for every positive integer t and every code $C \subseteq [q]^n$ with $|C| \geq 2t$, there exist t distinct codewords $c_1, c_2, \ldots, c_t \in C$ such that*

$$\#\mathsf{plur}(c_1, \ldots, c_t) \geq \frac{t}{q} + \Omega\left(\sqrt{\frac{t}{q}}\right).$$

Equivalently, there exists a $z \in [q]^n$ such that

$$\sum_{i=1}^{t} \mathsf{corr}(z, c_i) \geq \Omega\left(\sqrt{\frac{t}{q}}\right). \tag{1}$$

Proof. Without loss of generality, assume $|C| = 2t$. Pick a subset $\{c_1, c_2, \ldots, c_t\}$ from C, chosen uniformly at random among all t-element subsets of C. For $j = 1, \ldots, n$, define the random variable $P_j = \#\mathsf{plur}(c_{1j}, \ldots, c_{tj})$ to be the plurality of the j'th coordinates. By definition, $\#\mathsf{plur}(c_1, \ldots, c_t) = (1/n) \cdot \sum_{j=1}^{n} P_j$. Notice that P_j is always at least t/q, and we would expect the plurality to occasionally deviate from the lower bound. Indeed, it can be shown that for any sequence of $2t$ elements of $[q]$, if we choose a random subset of half of them, the expected plurality count is $t/q + \Omega(\sqrt{t/q})$. Thus, $\mathbf{E}[P_j] = t/q + \Omega(\sqrt{t/q})$. So $\mathbf{E}[(1/n) \cdot \sum_j P_j] = t/q + \Omega(\sqrt{t/q})$, and thus the lemma follows by taking any $c_1, \ldots, c_t$ that achieves the expectation. The equivalent reformulation in terms of correlation follows from Lemma 5 and the definition of correlation in terms of agreement (Definition 3). ∎

For any $\varepsilon > 0$, the above lemma gives a center z and $t = \Omega(1/q\varepsilon^2)$ codewords $c_1, \ldots, c_t$ such that the average correlation between z and $c_1, \ldots, c_t$ is at least ε. This implies that at least $(\varepsilon/2) \cdot t$ of the c_i's have correlation at least $\varepsilon/2$ with z. Thus we get a list-size lower bound of $(\varepsilon/2) \cdot t = \Omega(1/q\varepsilon)$ for decoding from correlation $\varepsilon/2$. (This argument has also appeared in [LTW].)

Now we would like to avoid the ε factor loss in list size in the above argument. The reason it occurs is that the average correlation can be ε due to the presence of $\approx \varepsilon t$ of the c_i's having extremely high correlation with z. This is consistent with the code being list-decodable with list size $o(1/(q\varepsilon^2))$ for correlation ε, but it means that it code has very poor list-decoding properties at some higher correlations — e.g., having $\varepsilon t = \Omega(1/(q\varepsilon))$ codewords at correlation $\Omega(1)$, whereas we'd expect a "good" code to have only $O(1)$ such codewords. In our next (and main) lemma, we show that we can pick a subcode of the code where this difficulty does not occur. Specifically, if C has good list-decoding properties at correlation ε, we get a subcode that has good list-decoding properties at every correlation larger than ε.

Lemma 7 (Main technical lemma). *For all positive integers L, t, $m \geq 2t$ and $q \geq 2$, and all small enough $\varepsilon > 0$, the following holds. Let C be a $(1-\varepsilon, L)$-list-decodable q-ary code of block length n with $|C| > 2L \cdot t \cdot m!/(m-t)!$. Then there exists a subcode $C' \subseteq C$, $|C'| \geq m$, such that for all positive integers $\ell \leq t$ and every $c_1, c_2, \ldots, c_\ell \in C'$,*

$$\#\mathsf{plur}(c_1, \ldots, c_\ell) \leq \left(\frac{1}{q} + \left(1 - \frac{1}{q}\right) \cdot \varepsilon + O\left(\frac{q^{3/2}}{\sqrt{\ell}}\right) \right) \cdot \ell.$$

Equivalently, for every $z \in [q]^n$ and every $c_1, \ldots, c_\ell \in C'$, we have

$$\sum_{i=1}^{\ell} \mathsf{corr}(z, c_i) \leq \left(\varepsilon + O\left(\frac{q^{3/2}}{\sqrt{\ell}}\right) \right) \cdot \ell. \tag{2}$$

Notice that the lemma implies a better upper bound on list size for correlations much larger than ε. More precisely, for every $\delta > 0$, it implies that the number of codewords having correlation at least $\varepsilon + \delta$ with a center z is at most $\ell = O(q^3/\delta^2)$. In fact, any ℓ codewords must even have *average* correlation at most $\varepsilon + \delta$.

Proof. We will pick a subcode $C' \subseteq C$ of size m at random from all m-element subsets of C, and prove that C' will fail to have the claimed property with probability less than 1.

For now, however, think of the code C' as being fixed, and we will reduce proving the desired properties above to bounding some simpler quantities. Let $(c_1, c_2, \ldots, c_\ell)$ be an arbitrary ℓ-tuple of codewords in C'. We will keep track of the average plurality count $\#\mathsf{plur}(c_1, \ldots, c_i)$ as we add each codeword to this sequence. To describe how this quantity can change at each step, we need a couple of additional definitions. We say a sequence $(a_1, \ldots, a_i) \in [q]^i$ has a *plurality tie* if at least two symbols occur $\#\mathsf{plur}(a_1, \ldots, a_i)$ times among $a_1, \ldots, a_i$. For vectors

$c_1, \ldots, c_i \in [q]^n$, we define $\#\mathsf{ties}(c_1, \ldots, c_i)$ to be the fraction of coordinates $j \in [n]$ such that $(c_{1j}, \ldots, c_{ij})$ has a plurality tie. Then:

Claim 8. *For every* $c_1, \ldots, c_i \in [q]^n$, $\#\mathsf{plur}(c_1, \ldots, c_i) \leq \#\mathsf{plur}(c_1, \ldots, c_{i-1}) + \mathsf{agr}(c_i, \mathsf{plur}(c_1, \ldots, c_{i-1})) + \#\mathsf{ties}(c_1, \ldots, c_{i-1})$.

Proof of Claim. Consider each coordinate $j \in [n]$ separately. Clearly,

$$\#\mathsf{plur}(c_{1j}, \ldots, c_{ij}) \leq \#\mathsf{plur}(c_{1j}, \ldots, c_{(i-1)j}) + 1 \ .$$

Moreover, if $(c_{1j}, \ldots, c_{(i-1)j})$ does not have a plurality tie, then the plurality increases iff c_{ij} equals the unique symbol $\mathsf{plur}(c_{1j}, \ldots, c_{(i-1)j})$ achieving the plurality. Thus,

$$\#\mathsf{plur}(c_{1j}, \ldots, c_{ij}) \leq \#\mathsf{plur}(c_{1j}, \ldots, c_{(i-1)j}) + A_j + T_j,$$

where T_j is the indicator variable for $(c_{1j}, \ldots, c_{(i-1)j})$ having a plurality tie, and A_j for c_{ij} agreeing with $\mathsf{plur}(c_{1j}, \ldots, c_{(i-1)j})$. The claim follows by averaging over $j = 1, \ldots, n$. ∎

Therefore, our task of bounding $\#\mathsf{plur}(c_1, \ldots, c_\ell)$ reduces to bounding $\mathsf{agr}(c_i, \mathsf{plur}(c_1, \ldots, c_{i-1}))$ and $\#\mathsf{ties}(c_1, \ldots, c_{i-1})$ for each $i = 1, \ldots, \ell$. The first term we bound using the list-decodability of C and the random choice of the subcode C'.

Claim 9. *There exists a choice of the subcode C' such that $|C'| = m$ and for every $i \leq t$ and every (ordered) sequence $c_1, \ldots, c_i \in C'$, we have*

$$\mathsf{agr}(c_i, \mathsf{plur}(c_1, \ldots, c_{i-1})) \leq 1/q + (1 - 1/q) \cdot \varepsilon \ .$$

Proof of Claim. We choose the subcode C' uniformly at random from all m-subsets of C. We view C' as a sequence of m codewords selected randomly from C without replacement. Consider any i of the codewords $c_1, \ldots, c_i$ in this sequence. By the list-decodability of the C, for any $c_1, \ldots, c_{i-1}$, there are at most L choices for c_i having agreement larger than $(1/q + (1 - 1/q) \cdot \varepsilon)$ with $\mathsf{plur}(c_1, \ldots, c_{i-1})$. Conditioned on $c_1, \ldots, c_{i-1}$, c_i is distributed uniformly on the remaining $|C| - i + 1$ elements of C, so the probability of c_i being one of the $\leq L$ bad codewords is at most $L/(|C| - i + 1)$.

By a union bound, the probability that the claim fails for at least one subsequence $c_1, \ldots, c_i$ of at most t codewords in C' is at most

$$\sum_{i=1}^{t} \frac{m!}{(m-i)!} \cdot \frac{L}{|C| - i + 1} \leq t \cdot \frac{m!}{(m-t)!} \cdot \frac{L}{|C| - t + 1} < 1.$$

Thus, there exists a choice of subcode C' satisfying the claim. ∎

For the $\#\mathsf{ties}(c_1, \ldots, c_{i-1})$ terms, we consider the codewords $c_1, \ldots, c_\ell$ in a random order.

Claim 10. *For every sequence of $c_1, \ldots, c_\ell \in [q]^n$, there exists a permutation $\sigma : [\ell] \to [\ell]$ such that*

$$\sum_{i=1}^{\ell} \#\mathsf{ties}(c_{\sigma(1)}, \ldots, c_{\sigma(i)}) = O(q^{3/2} \cdot \sqrt{\ell}).$$

Proof of Claim. We choose σ uniformly at random from all permutations $\sigma :$ $[\ell] \to [\ell]$ and show that the expectation of the left side is at most $O(q^{3/2} \cdot \sqrt{\ell})$. By linearity of expectations, it suffices to consider the expected number of plurality ties occurring in each coordinate $j \in [n]$. That is, we read the symbols $c_{1j}, \ldots, c_{\ell j} \in [q]$ in a random order σ and count the number of prefixes $c_{\sigma(1)j}, \ldots, c_{\sigma(i)j}$ having a plurality tie. If this prefix were i symbols chosen independently according to some (arbitrary) distribution, then it is fairly easy to show that the probability of a tie is $O(1/\sqrt{i})$ (ignoring the dependence on q), and summing this from $i = 1, \ldots, \ell$ gives $O(\sqrt{\ell})$ expected ties in each coordinate. Since they are not independently chosen, but rather i distinct symbols from a fixed sequence of ℓ symbols, the analysis becomes a bit more involved, but nevertheless the bound remains essentially the same. Specifically, by Lemma 11 stated below, the expected number of ties is at most $O(q^{3/2} \cdot \sqrt{\ell})$, yielding the claim. ∎

Lemma 11. *Let $b_1, b_2, \ldots, b_k$ be a sequence of elements from the universe $[q]$. Recall that a prefix of such a sequence has a plurality tie if there are at least two elements of $[q]$ that occur the same number of times in the prefix, and no other element occurs a strictly greater number of times in the prefix. Let Y be the random variable counting the number of prefixes with a plurality tie in a random permutation of the b_i's. Then $\mathbf{E}[Y] = O(q^{3/2}\sqrt{k})$.*

A full proof of the above lemma is omitted for reasons of space and can be found in the full version of the paper. Our approach is as follows. For $\alpha \in [q]$ and $i \in [k]$, let $Y_{\alpha,i}$ be the indicator random variable (over the choice of the permutation π of the sequence) for the event that the prefix $(b_{\pi(1)}, \ldots, b_{\pi(i)})$ has a plurality tie, α achieves the plurality, and $b_{\pi(i)} \neq \alpha$. Then $Y \leq \sum_{\alpha,i} Y_{\alpha,i}$. (For every prefix with a plurality tie, at least one of the two symbols achieving the plurality must be different from the last symbol in the prefix.) Essentially what we prove is that $\mathbf{E}[Y_{\alpha,i}] = O(\sqrt{q/\min\{i, k-i\}})$. Then summing over all $i \in [k]$ and $\alpha \in [q]$ gives the desired bound of $O(\sqrt{qk})$.

Now to complete the proof of Lemma 7, let C' be as in Claim 9, and let $c_1, \ldots, c_\ell$ be an arbitrary sequence of distinct codewords in C'. Let σ be permutation guaranteed by Claim 10. Then, by Claim 8, we have

$$\#\mathsf{plur}(c_1, \ldots, c_\ell) = \#\mathsf{plur}(c_{\sigma(1)}, \ldots, c_{\sigma(\ell)})$$

$$\leq \sum_{i=1}^{\ell} \left[\mathsf{agr}(c_{\sigma(i)}, \mathsf{plur}(c_{\sigma(1)}, \ldots, c_{\sigma(i-1)})) + \#\mathsf{ties}(c_{\sigma(1)}, \ldots, c_{\sigma(i-1)}) \right]$$

$$\leq \ell \cdot (1/q + (1 - 1/q) \cdot \varepsilon) + O(q^{3/2} \cdot \sqrt{\ell}),$$

as desired. The equivalent reformulation in terms of correlation again follows from Lemma 5 and the definition of correlation in terms of agreement (Definition 3). ∎

The following corollary of Lemma 7 will be useful in proving our main result.

Corollary 12. *Let L, t, m, q, ε, C, and C' be as in Lemma 7 for a choice of parameters satisfying $t \geq L$. Then for all $z \in [q]^n$ and all $D \geq 2$,*

$$\sum_{\substack{c \in C' \\ \mathrm{corr}(z,c) \geq D\varepsilon}} \mathrm{corr}(z,c) \leq O\left(\frac{q^3}{D\varepsilon}\right). \tag{3}$$

Proof. Let $c_1, c_2, \ldots, c_r$ be all the codewords of C' that satisfy $\mathrm{corr}(z,c) \geq D\varepsilon$. Since C, and hence C', is $(1-\varepsilon, L)$-list-decodable, we have $r \leq L \leq t$. Using (2) for the codewords $c_1, c_2, \ldots, c_r$, we have

$$D\varepsilon \leq \frac{1}{r}\sum_{i=1}^{r} \mathrm{corr}(z,c_i) \leq \varepsilon + O\left(\frac{q^{3/2}}{\sqrt{r}}\right)$$

which gives $r = O(q^3/((D-1)^2\varepsilon^2)) = O(q^3/(D^2\varepsilon^2))$, since $D \geq 2$. Applying (2) again,

$$\sum_{\substack{c \in C' \\ \mathrm{corr}(z,c) \geq D\varepsilon}} \mathrm{corr}(z,c) = \sum_{i=1}^{r} \mathrm{corr}(z,c_i)$$

$$\leq \varepsilon r + O(q^{3/2}\sqrt{r})$$

$$\leq O\left(\frac{q^3}{D^2\varepsilon}\right) + O\left(\frac{q^3}{D\varepsilon}\right)$$

$$\leq O\left(\frac{q^3}{D\varepsilon}\right).$$

∎

We are now ready to prove our main result, Theorem 2, which restate (in slightly different form) below.

Theorem 13 (Main). *There exist constants $c > 0$, $d < \infty$, such that for all small enough $\varepsilon > 0$, the following holds. Suppose C is a q-ary $(1-\varepsilon, L)$-list-decodable code with $|C| > 1/(q\varepsilon^2)^{d/(q\varepsilon^2)}$. Then $L \geq c/(q^5\varepsilon^2)$.*

Proof. Let T be a large enough constant to be specified later. Let $t = \lfloor \frac{1}{Tq\varepsilon^2}\rfloor$. If $L > t$, then there is nothing to prove. So assume that $t \geq L \geq 1$ and set $m = 2t$. Then

$$2L \cdot t \cdot \frac{m!}{(m-t)!} \leq 2t^2 \cdot (2t)^t = \left(\frac{1}{q\varepsilon^2}\right)^{O(1/(q\varepsilon^2))} < |C|,$$

for a sufficiently large choice of the constant d. Let C' be a subcode of C of size $m = 2t$ guaranteed by Lemma 7.

By Lemma 6, there exist t codewords c_i, $1 \le i \le t$, in C', and a center $z \in [q]^n$ such that

$$\sum_{i=1}^{t} \mathrm{corr}(z, c_i) = \Omega\left(\sqrt{\frac{t}{q}}\right). \tag{4}$$

Also, for any $D \ge 2$, we have $\sum_{i=1}^{t} \mathrm{corr}(z, c_i)$ equals

$$\sum_{i:\mathrm{corr}(z,c_i)<\varepsilon} \mathrm{corr}(z, c_i) + \sum_{i:\varepsilon \le \mathrm{corr}(z,c_i)<D\varepsilon} \mathrm{corr}(z, c_i) + \sum_{i:\mathrm{corr}(z,c_i)\ge D\varepsilon} \mathrm{corr}(z, c_i)$$

$$\le \varepsilon t + D\varepsilon L + O\left(\frac{q^3}{D\varepsilon}\right) \tag{5}$$

where to bound the second part we used that C' is $(1 - \varepsilon, L)$-list-decodable, and to bound the third part we used the fact that C' satisfies (3).

Putting these together, and setting $D = q^4 \cdot T$, we have

$$\Omega\left(\sqrt{\frac{t}{q}}\right) \le \varepsilon t + D\varepsilon L + O\left(\frac{q^3}{D\varepsilon}\right)$$

$$\le \sqrt{\frac{t}{Tq}} + D\varepsilon L + O\left(\sqrt{\frac{t}{Tq}}\right).$$

For a sufficiently large choice of the constant T, this gives

$$L \ge \frac{1}{D\varepsilon} \cdot \Omega\left(\sqrt{\frac{t}{q}}\right) = \Omega\left(\frac{1}{\sqrt{T} \cdot q^5 \cdot \varepsilon^2}\right).$$

as desired.   ∎

3   Open Questions

Several questions are open in the general direction of exhibiting limitations on the performance of list-decodable codes. We mention some of them below.

- We have not attempted to optimize the dependence on the alphabet size q in our bound on list size (i.e. the constant c_q in Theorem 2), and this leaves a gap between the upper and lower bounds. The probabilistic code construction of [Eli2, GHSZ] achieves a nearly linear dependence on q (specifically, list size $L = O(\log q/(q\varepsilon^2))$), whereas our lower bound (Theorem 13) has a polynomial dependence on q (namely, it shows $L = \Omega(1/(q^5\varepsilon^2))$).
- It should be possible to use our main result, together with an appropriate "filtering" argument (that focuses, for example, on a subcode consisting of all codewords of a particular Hamming weight) to obtain upper bounds on rate of list-decodable q-ary codes. In particular, can one confirm that for

each fixed L, the maximum rate achievable for list decoding up to radius p with list size L is strictly less than the capacity $1 - H_q(p)$? Such a result is known for binary codes [Bli]. Also, it is an interesting question whether some of the ideas in this paper can be used to improve the rate upper bounds of Blinovsky [Bli] for the binary case.

- Can one prove a lower bound on list size as a function of distance from "capacity"? In particular, does one need list size $\Omega(1/\gamma)$ to achieve a rate that is within γ of capacity?
- Can one prove stronger results for linear codes?

Acknowledgments

We thank Michael Mitzenmacher, Madhu Sudan, and Amnon Ta-Shma for helpful conversations.

References

[Bli]   V. M. Blinovsky. Bounds for codes in the case of list decoding of finite volume. *Problems of Information Transmission*, 22(1):7–19, 1986.

[Eli1]   P. Elias. List decoding for noisy channels. *Technical Report 335, Research Laboratory of Electronics, MIT*, 1957.

[Eli2]   P. Elias. Error-correcting codes for list decoding. *IEEE Transactions on Information Theory*, 37:5–12, 1991.

[Gur]   V. Guruswami. List decoding from erasures: Bounds and code constructions. *IEEE Transactions on Information Theory*, 49(11):2826–2833, 2003.

[GHSZ]   V. Guruswami, J. Hastad, M. Sudan, and D. Zuckerman. Combinatorial bounds for list decoding. *IEEE Transactions on Information Theory*, 48(5):1021–1035, 2002.

[LTW]   C.-J. Lu, S.-C. Tsai, and H.-L. Wu. On the complexity of hardness amplification. In *Proceedings of the 20th Annual IEEE Conference on Computational Complexity*, San Jose, CA, June 2005. To appear.

[RT]   J. Radhakrishnan and A. Ta-Shma. Bounds for dispersers, extractors, and depth-two superconcentrators. *SIAM Journal on Discrete Mathematics*, 13(1):2–24 (electronic), 2000.

[TZ]   A. Ta-Shma and D. Zuckerman. Extractor codes. *IEEE Transactions on Information Theory*, 50(12):3015–3025, 2004.

[Tre]   L. Trevisan. Extractors and Pseudorandom Generators. *Journal of the ACM*, 48(4):860–879, July 2001.

[Vad]   S. P. Vadhan. Randomness Extractors and their Many Guises. Tutorial at IEEE Symposium on Foundations of Computer Science, November 2002. Slides available at http://eecs.harvard.edu/~salil.

[Woz]   J. M. Wozencraft. List Decoding. *Quarterly Progress Report, Research Laboratory of Electronics, MIT*, 48:90–95, 1958.

A Lower Bound
for Distribution-Free Monotonicity Testing

Shirley Halevy and Eyal Kushilevitz

Department of Computer Science, Technion, Haifa 3200, Israel
{shirleyh,eyalk}@cs.technion.ac.il

Abstract. We consider monotonicity testing of functions $f : [n]^d \to \{0,1\}$, in the property testing framework of Rubinfeld and Sudan [23] and Goldreich, Goldwasser and Ron [14]. Specifically, we consider the framework of *distribution-free* property testing, where the distance between functions is measured with respect to a *fixed but unknown* distribution D on the domain and the testing algorithms have an oracle access to random sampling from the domain according to this distribution D. We show that, though in the uniform distribution case, testing of boolean functions defined over the boolean hypercube can be done using query complexity that is polynomial in $\frac{1}{\epsilon}$ and in the dimension d, in the distribution-free setting such testing requires a number of queries that is exponential in d. Therefore, in the high-dimensional case (in oppose to the low-dimensional case), the gap between the query complexity for the uniform and the distribution-free settings is exponential.

1 Introduction

Property testing is a relaxation of decision problems, where algorithms are required to distinguish objects having some property $\mathcal{P}$ from those objects which are at least "ϵ-far" from every such object. The notion of property testing was introduced by Rubinfeld and Sudan [23] and since then attracted a considerable amount of attention. Property testing algorithms (or *property testers*) were introduced for monotonicity testing (e.g. [2, 6–8, 10, 11, 13, 18]), problems in graph theory (e.g. [1, 4, 14–16, 20]) and other properties (e.g. [3, 5]; the reader is referred to excellent surveys by Ron [22], Goldreich [12], and Fischer [9] for a presentation of some of this work, including some connections between property testing and other topics). The main goal of property testing is to avoid "reading" the whole object (which requires complexity at least linear in the size of its representation); i.e., to make the decision by reading a small (possibly, selected at random) fraction of the input (e.g., a fraction of size polynomial in $1/\epsilon$ and poly-logarithmic in the size of the representation) and still having a good (say, at least $2/3$) probability of success.

A crucial component in the definition of property testing is that of the *distance* between two objects. For the purpose of this definition, it is common to think of objects as being *functions* over some domain $\mathcal{X}$. The distance between functions f and g is then measured by considering the set $\mathcal{X}_{f \neq g}$ of all points

C. Chekuri et al. (Eds.): APPROX and RANDOM 2005, LNCS 3624, pp. 330–341, 2005.

x where $f(x) \neq g(x)$ and comparing the size of this set $\mathcal{X}_{f \neq g}$ to that of $\mathcal{X}$; equivalently, one may introduce a uniform distribution over $\mathcal{X}$ and measure the probability of picking $x \in \mathcal{X}_{f \neq g}$. Note that property testers access the input function (object) via *membership queries* (i.e., the algorithm gives a value x and gets $f(x)$).

It is natural to generalize the above definition of distance between two functions, to deal with arbitrary probability distributions D over $\mathcal{X}$, by measuring the probability of $\mathcal{X}_{f \neq g}$ according to D. Ideally, one would hope to get *distribution-free* property testers. Such a tester, for a given property $\mathcal{P}$, accesses the function using membership queries, as above, and by randomly sampling the **fixed but unknown** distribution D (this mimics similar definitions from *learning theory* and is implemented via an oracle access to D; see, e.g., [21] [1]). As before, the tester is required to accept the given function f, with high probability, if f satisfies the property $\mathcal{P}$, and to reject it, with high probability, if f is at least ϵ-far from $\mathcal{P}$ with respect to the distribution D.

This notion of testing, termed *distribution-free testing*, has been introduced by Goldreich, Goldwasser and Ron in [14], who also showed that it is impossible to test a variety of partition problems (for which they showed testers with respect to the uniform distribution) in a distribution-free manner. Halevy and Kushilevitz [17] presented the first non-trivial distribution-free testers for two properties: monotonicity and low-degree polynomials. In [19], distribution-free testers were presented for testing connectivity of sparse graphs.

A natural question that arises when dealing with distribution-free testing is whether it can be done using the same query complexity as in the uniform setting. We answer this question with respect to monotonicity testing, by showing an exponential gap between the query complexity required for the uniform and the distribution-free settings.

Monotonicity is one of the central problems studied in the property testing literature (e.g. [2, 6–8, 10, 11, 13, 18]). In monotonicity testing, the domain $\mathcal{X}$ is usually the d-dimensional cube $[n]^d$. A partial order is defined on this domain in the natural way (for $\boldsymbol{y}, \boldsymbol{z} \in [n]^d$, we say that $\boldsymbol{y} \leq \boldsymbol{z}$ if each coordinate of $\boldsymbol{y}$ is bounded by the corresponding coordinate of $\boldsymbol{z}$). A function f over the domain $[n]^d$ is *monotone* if whenever $\boldsymbol{z} \geq \boldsymbol{y}$ then $f(\boldsymbol{z}) \geq f(\boldsymbol{y})$. Testers were developed to deal with both the low-dimensional and the high-dimensional cases.

The low-dimensional case: In this case, d is considered to be small compared to n (e.g., d may be a constant); a successful algorithm for this case is typically one that is polynomial in $1/\epsilon$ and in $\log n$. The first paper to deal with this case in the uniform setting is by Ergün et al. [7] which presented an $O(\frac{\log n}{\epsilon})$ algorithm for the line (i.e., the case $d = 1$), and showed that this query complexity cannot be achieved without using membership queries. This algorithm was generalized for any fixed d in [2]. For the case $d = 1$, there is a lower bound showing that

---

[1] More precisely, distribution-free property testing is the analogue of the PAC+MQ model of learning (that was studied by the learning-theory community mainly via the EQ+MQ model); standard property testing is the analogue of the uniform+MQ model.

testing monotonicity (for some constant ϵ) indeed requires $\Omega(\log n)$ queries [8]. In [18], the authors further analyzed a tester presented in [6], and showed an upper bound of $O(\frac{d \cdot 4^d \cdot \log n}{\epsilon})$ on the query complexity required for the low-dimensional case. In the distribution-free setting, [17] presented a distribution-free tester with query complexity $O(\frac{\log^d n \cdot 2^d}{\epsilon})$; by this, showing that in the low-dimensional case the overhead caused by the distribution-free requirement is relatively small (both tests have polylogarithmic query complexity and, in fact, for $d = 1$ they have the same query complexity).

The high-dimensional case: In this case, d is considered as the main parameter (and n might be as low as 2); a successful algorithm is typically one that is polynomial in $1/\epsilon$ and d. This case was first considered by Goldreich et al. [13] that showed an algorithm for testing monotonicity of functions over the boolean ($n = 2$) d-dimensional hyper-cube to a boolean range using $O(\frac{d}{\epsilon})$ queries. This result was generalized in [6] to arbitrary values of n, showing that $O(\frac{d \cdot \log^2 n}{\epsilon})$ queries suffice for testing monotonicity of general functions over $[n]^d$. In the distribution-free setting, the only known tester is the one presented by [17] using $O(\frac{\log^d n \cdot 2^d}{\epsilon})$ queries (this query complexity is non-trivial whenever $n \geq 8$, but is exponential in d). However, the question whether distribution-free testing can be done using similar query complexity to the one used in the uniform setting remained open.

Our Contribution: We show that, while $O(\frac{d}{\epsilon})$ queries suffice for testing monotonicity of boolean functions over the boolean hypercube in the uniform case [13], in the distribution-free case it requires a number of queries that is exponential in the dimension d (that is, there exists a constant c such that $\Omega(2^{cd})$ queries are required for the testing). Hence, such testing is infeasible in the high-dimensional case. Our bound can be trivially extended to monotonicity testing of boolean functions defined over the domain $[n]^d$ for general values of n. By this, showing a gap between the query complexity of uniform and distribution-free testers for a natural testing problem.

Other related work: Lower bounds for monotonicity testing of functions $f : \{0,1\}^d \to \{0,1\}$ were shown in [10]: $\Omega(\sqrt{d})$ lower bound for non-adaptive, one-sided error algorithms, and an $\Omega(\log \log d)$ lower bound for two-sided error algorithms. The authors also considered graphs other than the hyper-cube, proving, for example, that testing monotonicity with a constant (depending on $\frac{1}{\epsilon}$ only) number of queries is possible for certain classes of graphs.

Organization: In Section 2, we formally define the notions we are about to use in this work. In Section 3, we give an overview of the lower-bound proof and present some families of functions used for this proof. Then, in Section 4, we prove the lower bound for one-sided error testing and, in Section 5, we extend our lower bound to the two-sided error case.

2   Preliminaries

In this section, we formally define the notion of a function $f : \{0,1\}^d \rightarrow \{0,1\}$ being ϵ-far from monotone with respect to a given distribution D defined over $\{0,1\}^d$, and of monotonicity distribution-free testing. Then, we define some notation that will be used throughout this work. Denote by $\mathcal{P}_{mon}$ the set of all monotone functions $f : \{0,1\}^d \rightarrow \{0,1\}$.

Definition 1. *Let D be a distribution defined over $\{0,1\}^d$. The* distance *between functions $f, g : \{0,1\}^d \rightarrow \{0,1\}$, measured with respect to D, is defined by $dist_D(f,g) \overset{def}{=} \Pr_{x \sim D}\{f(x) \neq g(x)\}$.*
The distance *of a function f from being monotone, measured with respect to D, is $dist_D(f, \mathcal{P}_{mon}) \overset{def}{=} \min_{g \in \mathcal{P}_{mon}} dist_D(f,g)$.*
We say that f is ϵ-far from monotone with respect to D if $dist_D(f, \mathcal{P}_{mon}) \geq \epsilon$.

Definition 2. *A* distribution-free monotonicity tester *is a probabilistic oracle machine M, which is given a distance parameter, $\epsilon > 0$, and an oracle access to an arbitrary function $f : \{0,1\}^d \rightarrow \{0,1\}$ and to sampling of a fixed but unknown distribution D over $\{0,1\}^d$, and satisfies the following two conditions:*
1. If f is monotone, then $\Pr\{M^{f,D} = Accept\} \geq \frac{2}{3}$.
2. If f is ϵ-far from monotone with respect to D, then $\Pr\{M^{f,D} = Accept\} \leq \frac{1}{3}$.

We will also be interested in testers that allow one-sided error; such testers always accept any monotone function.

Hereafter, we identify any point $x \in \{0,1\}^d$ with the corresponding set $x \subseteq [d]$. This allows us to apply set theory operations, such as union and intersection, to points. In addition, for any two points $p, p' \in \{0,1\}^{\frac{d}{2}}$, denote by $p||p'$ the point $x \in \{0,1\}^d$ that is the concatenation of p and p' (i.e., it is identical to p in its first $\frac{d}{2}$ coordinates and to p' in its last $\frac{d}{2}$ coordinates). Given a point $x \in \{0,1\}^d$, we say that $p \in \{0,1\}^{\frac{d}{2}}$ is the prefix of x if $x = p||p'$ for some p'. Given two points $x, y \in \{0,1\}^d$, denote by $H(x,y)$ the hamming distance between x and y.

Denote by B_λ^d the set of all the λd weight points in $\{0,1\}^d$, that is $B_\lambda^d = \{x \in \{0,1\}^d : |x| = \lambda d\}$.

3   The Two Families

To prove the lower bound on the query complexity, we show that any monotonicity distribution-free tester with sub-exponential query complexity is unable to distinguish between monotone functions to functions that are $\frac{1}{2}$-far from monotone. To this aim, we consider two families $\mathcal{F}_1$ and $\mathcal{F}_2$ of pairs (f, D_f), where f is a boolean function and D_f is a probability distribution corresponding to f, both defined over $\{0,1\}^d$, such that the following holds:

1. Every function in $\mathcal{F}_1$ is monotone, hence every tester has to accept every pair $(f, D_f) \in \mathcal{F}_1$ with probability of at least $\frac{2}{3}$.

2. Every function f in $\mathcal{F}_2$ is $\frac{1}{2}$-far from monotone with respect to D_f. Therefore, every tester has to reject every pair $(f, D_f) \in \mathcal{F}_2$ with probability of at least $\frac{2}{3}$ (regardless of the choice of the distance parameter ϵ).
3 There exists a constant c such that no algorithm A, that asks less than 2^{cd} membership queries and samples the distribution less than 2^{cd} times, accepts every pair (f, D_f) in $\mathcal{F}_1$ with probability of at least $\frac{2}{3}$, and rejects every pair (f, D_f) in $\mathcal{F}_2$ with probability of at least $\frac{2}{3}$.

Hence, there exists no two-sided error distribution-free tester for monotonicity of boolean functions defined over $\{0, 1\}^d$ with sub-exponential query complexity. (A similar approach was previously used for lower bound proofs in property testing, e.g., [15].)

In the rest of this section, we describe the two families of functions (and distributions) $\mathcal{F}_1$ and $\mathcal{F}_2$ defined over $\{0, 1\}^d$. Let $\alpha < \frac{1}{16}$ be a parameter used for the construction of both families and define $m \stackrel{\text{def}}{=} 2^{d\alpha}$. In the construction, we use a set $\mathcal{M} \subset B^{d/2}_{1/2 - 2\alpha}$ of size $2m$ such that every two points in $\mathcal{M}$ are "far apart". This property of $\mathcal{M}$ will be used in the proof of the third requirement. We first define the exact requirements from such a set $\mathcal{M}$ and claim its existence. The proof of the following lemma will appear in the full version of the paper.

Lemma 1. *There exists a set $\mathcal{M} \subset B^{d/2}_{1/2 - 2\alpha}$ such that: (a) $|\mathcal{M}| = 2m$, and (b) $H(p, p') > 2\alpha d$ for every two points $p, p' \in \mathcal{M}$ such that $p \neq p'$.*

Using such $\mathcal{M}$, we define the two families. For each pair (f, D_f), we first define the function $f : \{0, 1\}^d \to \{0, 1\}$ and then, based on the definition of f, define the corresponding distribution D_f over $\{0, 1\}^d$.

3.1   The Family $\mathcal{F}_1$

Each function f is defined by first choosing (in all possible ways) two sets X_1 and X_2, each of size m such that $X_1 \subset B^d_{1/2 - \alpha}$ and $X_2 \subset B^d_{1/2 + \alpha}$. The choice of the two sets is done as follows:

- Choose m points from $\mathcal{M}$, denote the set of m selected points by $\mathcal{M}'$.
- For each point p in $\mathcal{M}'$, choose a point $y \in B^{d/2}_{1/2}$, and add the point $x = p||y$ to X_1. ($|x|$ is exactly $(1/2 - \alpha)d$.)
- For each point p in $\mathcal{M} \setminus \mathcal{M}'$, choose a point $y \in B^{d/2}_{1/2 + 4\alpha}$ and add the point $x = p||y$ to the set X_2. ($|x|$ is exactly $(1/2 + \alpha)d$.)

The function f is now defined in the following manner:

- For every point $x_1 \in X_1$ and every point y such that $y \geq x_1$, set $f(y) = 1$.
- For every point $x_2 \in X_2$ and every point y such that $y \leq x_2$, set $f(y) = 0$.
- For every point y, such that $f(y)$ was not defined above, if $|y| \leq \frac{d}{2}$ then $f(y) = 0$, otherwise $f(y) = 1$.

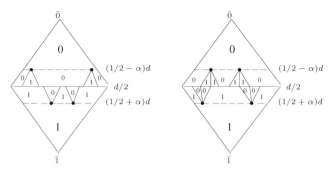

Fig. 1. An example for functions in $\mathcal{F}_1$ (left) and in $\mathcal{F}_2$ (right) for $m = 2$. (The bullets represent points in X_1 and X_2)

See left side of Figure 1 for an example of such a function for $m = 2$.

The distribution D_f corresponding to the function f is the uniform distribution over the $2m$ points in $X_1 \cup X_2$. That is, $D_f(x) = \frac{1}{2m}$ for every $x \in X_1 \cup X_2$ and $D_f(x) = 0$ for any other point in $\{0,1\}^d$.

The fact that every function in $\mathcal{F}_1$ is monotone follows the definition. The fact that $\mathcal{F}_1$ is not empty follows the existence of such a set $\mathcal{M}$. To see that the function f is well defined, and hence the family $\mathcal{F}_1$ satisfies the first requirement of the construction, we observe the following simple lemma.

Lemma 2. $\{y : y \geq x_1\} \cap \{y : y \leq x_2\} = \phi$, for every $x_1 \in X_1$ and $x_2 \in X_2$.

Proof. If there exists a point $z \in \{y : y \geq x_1\} \cap \{y : y \leq x_2\}$, then $x_1 < x_2$. By the construction of f, there exists two different points $p_1, p_2 \in \mathcal{M}$ such that $x_1 = p_1 || y_1$ and $x_2 = p_2 || y_2$ for some y_1 and y_2. Thus, $p_1 \leq p_2$, contradicting the fact that $p_1 \neq p_2$ and $|p_1| = |p_2|$. $\square$

3.2   The Family $\mathcal{F}_2$

Each function f in $\mathcal{F}_2$ is also defined by first choosing two sets X_1 and X_2, each of size m such that $X_1 \subset B^d_{1/2-\alpha}$ and $X_2 \subset B^d_{1/2+\alpha}$. The choice of the two sets is done as follows:

- Choose m points from $\mathcal{M}$, denote the set of m selected points by $\mathcal{M}'$.
- For each point p in $\mathcal{M}'$, choose a pair of points (y_1, y_2) such that: (a) $y_1 \in B^{d/2}_{1/2}$, (b) $y_2 \in B^{d/2}_{1/2+4\alpha}$, and (c) $y_1 < y_2$.
- Add the point $x_1 = p || y_1$ to X_1, and the point $x_2 = p || y_2$ to X_2.
- The two points x_1 and x_2 will be referred to as a *couple* in f.

The function f is now defined as follows:

- For every point $x_1 \in X_1$ and every point y such that $y \in U(x_1)$, set $f(y) = 1$, where $U(x) \stackrel{\text{def}}{=} \{y : y \geq x, |y| \leq \frac{d}{2}\}$.
- For every point $x_2 \in X_2$ and every point y such that $y \in L(x_2)$, set $f(y) = 0$, where $L(x) \stackrel{\text{def}}{=} \{y : y \leq x, |y| > \frac{d}{2}\}$.

- For every point y such that $f(y)$ was not defined above, if $|y| \leq \frac{d}{2}$ then $f(y) = 0$, otherwise, $f(y) = 1$.

See right side of Figure 1 for an example of such a function for $m = 2$.

The distribution D_f is again the uniform distribution over $X_1 \cup X_2$, as in the definition of $\mathcal{F}_1$.

The fact that $\mathcal{F}_2$ is not empty follows immediately the existence of the set $\mathcal{M}$. However, we have to argue that every function f in $\mathcal{F}_2$ is $\frac{1}{2}$-far from monotone with respect to D_f.

Lemma 3. f is $\frac{1}{2}$-far from monotone with respect to D_f, for every $(f, D_f) \in \mathcal{F}_2$.

Proof. To turn f into a monotone function, we have to alter the value of f either in x_1 or in x_2, for every couple (x_1, x_2) in f. By this, increasing the distance by $\frac{1}{2m}$. Since there are m different couples, the total distance is at least $\frac{1}{2}$. $\quad\square$

4   Lower Bound for One-Sided Error Testing

In this section, we prove that there exists no one-sided error distribution-free monotonicity tester with sub-exponential query complexity that accepts every pair (f, D_f) in $\mathcal{F}_1$, with probability 1, and rejects every pair (f, D_f) in $\mathcal{F}_2$, with high probability. To this aim, we will show that for every tester A, there exists a pair $(f, D_f) \in \mathcal{F}_2$, such that with high probability the run of A on (f, D_f) is also consistent with a monotone function from $\mathcal{F}_1$. Since A has to accept every monotone function, then A has to accept f with high probability.

The above claim is simple if the tester is not allowed to use membership queries, but only to sample the distribution D_f. In this case, it is clear that distinguishing between the two families requires an exponential number of queries. However, the difference between one tester to the other is in the tester's choice of membership queries, and dealing with the tester's membership queries is where the difficulty of this proof lays.

To state our claim formally, we introduce some notation. Let A be an algorithm with query and sample complexity $n = n(d, \frac{1}{\epsilon})$. Such an algorithm can also be viewed as a mapping from a sequence $\{(p_i, v_i)\}_{i=1}^{j}$ of labelled points to either "sample the distribution" or "query p_{j+1}" if $j < n$, and to "accept" or "reject" if $j = n$. We refer to such a sequence of labelled points obtained by A as a *knowledge sequence*. Given a pair (f, D_f), denote by S_f the knowledge sequence learnt by A during its run on (f, D_f) (for some possible run of A on (f, D_f)). We say that a function f is consistent with a knowledge sequence S if $f(p_i) = v_i$, for every $1 \leq i \leq j$.

As stated above, we show that for some constant c, for every algorithm A with query complexity $n < 2^{cd}$, there exists a pair $(f, D_f) \in \mathcal{F}_2$ such that with high probability, over the possible runs of A and the sampling of the domain according to D_f, the sequence S_f is consistent with some monotone function $f' \in \mathcal{F}_1$. Hence, with high probability S_f causes A to accept, contradicting the requirement that A rejects (f, D_f) with probability of at least $\frac{2}{3}$.

For this, we show that for every algorithm A and for every possible choice of random coins for A, the probability that, for a randomly drawn pair (f, D_f) from $\mathcal{F}_2$, the sequence S_f is consistent with some monotone function from $\mathcal{F}_1$ is very high (where the probability is taken over the choice of (f, D_f) and the sampling of D_f). In other words, for every algorithm A and for every choice of random coins for A, most pairs in $\mathcal{F}_2$ are such that, with high probability (over the sampling of D_f), S_f is consistent with a monotone function. Hence, for every algorithm A, there exists a pair $(f, D_f) \in \mathcal{F}_2$ such that for most choices of random coins for A, the probability that S_f is consistent with a monotone function is very high.

Lemma 4. *For every algorithm A, there exits a pair $(f, D_f) \in \mathcal{F}_2$ such that $Pr\{S_f$ is consistent with a monotone function$\} \geq 1 - \frac{2n^2}{m}$, where the probability is over the choice of random coins for A and the samplings of D_f.*

Based on the above lemma, we can state our lower bound.

Theorem 1. *Let $c = \alpha/3$. For every one-sided tester A that asks less than 2^{cd} membership queries and samples the distribution less than 2^{cd} times, there exists a pair $(f, D_f) \in \mathcal{F}_2$ such that $Pr\{A$ accepts $(f, D_f)\} > \frac{1}{3}$.*

It remains to prove Lemma 4. We first state the condition that a knowledge sequence has to satisfy in order to be consistent with some function in $\mathcal{F}_1$. We observe that in order for a knowledge sequence to be extendable to a monotone function in $\mathcal{F}_1$, it is not enough that S_f is consistent with some monotone function. If for some couple (x_1, x_2) with prefix p, the sequence S_f contains points $z \in U(x_1)$ and $z' \in L(x_2)$ both with the same prefix p, then we can deduce that $f \in \mathcal{F}_2$ regardless of whether z and z' are comparable or not. Therefore, we have to define a relaxed notion of a witness for non-monotone functions for our case, such that every sequence that does not contain a witness is indeed extendable to a function in $\mathcal{F}_1$. (In general, it is enough to show that there exists an *arbitrary* monotone function that is consistent with the knowledge sequence, and not necessarily a function in $\mathcal{F}_1$. However, we will use this fact later on when extending the proof of the lower bound to the two-sided error case.)

Definition 3. *Let S_f be a knowledge sequence learnt by A while running on the pair $(f, D_f) \in \mathcal{F}_2$. We say that S_f contains a witness if there exist points y_1, y_2 in S_f such that for some couple (x_1, x_2) in f, $y_1 \in U(x_1)$ and $y_2 \in L(x_2)$.*

Lemma 5. *If the knowledge sequence S_f, learnt by A while running on a pair $(f, D_f) \in \mathcal{F}_2$, does not contain a witness, then there exists a function $f' \in \mathcal{F}_1$ that is consistent with S_f.*

To prove the existence of such a function f', we show that it is possible construct the sets X'_1 and X'_2 such that f', the function induced by X'_1 and X'_2, is in $\mathcal{F}_1$ and is consistent with S_f. The proof of the lemma is omitted.

Next, we prove that the probability that the knowledge sequence S_f, learnt by A during its run on a randomly chosen pair $(f, D_f) \in \mathcal{F}_2$, contains a witness, is very small. To do so, we prove that the probability for any step of A, that the knowledge sequence will contain a witness after that step is very small.

We start with the following technical lemma. It shows that, for every query that A may ask and sample it may receive, there can be only one prefix in $\mathcal{M}$ such that this query or sample belongs to $U(x_1)$ or $L(x_2)$ for some couple (x_1, x_2) with that prefix. The proof of the lemma relies on the fact that the points in $\mathcal{M}$ are far apart and is omitted for lack of space.

Lemma 6. *Let f be a function in $\mathcal{F}_2$. Then, for every point $z \in \{y \in \{0,1\}^d :$ $(\frac{1}{2} - \alpha)d \leq |y| \leq \frac{d}{2}\}$, there exists at most one prefix $p \in \mathcal{M}$ such that $z \in U(x)$ for some point $x \in B_{1/2-\alpha}^d$ with prefix p. Similarly, for every point $z \in \{y \in \{0,1\}^d : \frac{d}{2} < |y| \leq (\frac{1}{2} + \alpha)d\}$, there exists at most one prefix $p \in \mathcal{M}$ such that $z \in L(x)$ for some point $x \in B_{1/2+\alpha}^d$ with prefix p.*

Given $(f, D_f) \in \mathcal{F}_2$ and a knowledge sequence S_f, denote by $S_f(i)$ the length i prefix of S_f.

Lemma 7. *For every algorithm A and every possible sequence of random coins for A, the following holds:*
$Pr\{S_f(i+1)$ *contains a witness* $\mid S_f(i)$ *does not contain a witness*$\} \leq \frac{i+2}{m}$
(where the probability is over the random choice of (f, D_f) from $\mathcal{F}_2$ that is consistent with $S_f(i)$, and over the possible samplings of D_f for the chosen pair (f, D_f)).

Proof. Distinguish between the two possible types of steps for A. The first possibility is that in step $i + 1$ A asks for a sample of the distribution D_f, and the second possibility is that A chooses to query the function. In the first case, for every possible choice of (f, D_f) from $\mathcal{F}_2$, we show that the probability that $S_f(i+1)$ contains a witness is at most $\frac{i}{2m}$. Let (f, D_f) be a pair in $\mathcal{F}_2$ that is consistent with $S_f(i)$ and let X_1 and X_2 be the sets used for its construction. The sampled point can cause $S_f(i)$ to contain a witness only if $S_f(i)$ already contains a point that is related to its couple. That is, if for some couple (x_1, x_2) in f, the sampled point is either x_1 and there are points in $S_f(i)$ from $L(x_2)$, or that the sampled point is x_2 and $S_f(i)$ contains points from $U(x_1)$. By Lemma 6 and the fact that every two different points in X_1 have different prefixes, every point in $S_f(i)$ can be in $U(x_1)$ for at most one point $x_1 \in X_1$ and similarly, every point can be in $L(x_2)$ for at most one point $x_2 \in X_2$. Hence, for every point in $S_f(i)$, only sampling of the relevant point x_1 or x_2 can cause the knowledge sequence to contain a witness. (In general, if the points in $\mathcal{M}$ were not "far apart", it is possible that the couples are very close to one another", and hence almost every sampled point causes the knowledge sequence to contain a witness). The probability for this event is $\frac{1}{2m}$. Therefore, the probability that the sampled point, that was returned by the oracle, causes $S_f(i)$ to contain a witness can be bounded by $\frac{i}{2m}$.

In the second case, let q_{i+1} be the query asked by A in step $i+1$. If $|q_{i+1}| < (\frac{1}{2} - \alpha)d$ or $|q_{i+1}| > (\frac{1}{2} + \alpha)d$, then $S_f(i+1)$ cannot contain a witness. Assume w.l.o.g. that $|q_{i+1}| \leq \frac{d}{2}$ (the proof for the case that $|q_{i+1}| > \frac{d}{2}$ is similar). Based on Lemma 6, there can be at most one relevant prefix $p \in \mathcal{M}$ such that a couple (x_1, x_2) with prefix p can include q_{i+1} in $U(x_1)$. If there is no 0-labelled point z in $S_f(i)$ that can possibly be in $L(x_2)$, for a point x_2 with prefix p, then q_{i+1} cannot cause $S_f(i)$ to contain a witness. Hence, assume there exists a 0-labelled point z in $S_f(i)$ that can possibly be in $L(x_2)$ for a point x_2 with prefix p. The sequence $S_f(i+1)$ will contain a witness only if $q_{i+1} \in U(x_1)$ for the point x_1 corresponding to x_2. Note that S_f does not necessarily include x_2 itself, but only points in $L(x_2)$. We will bound the probability, over the possible choices of a pair (f, D_f) such that f is consistent with $S_f(i)$, that indeed $q_{i+1} \in U(x_1)$, and show that this probability is at most $\frac{2}{m}$.

The proof is by showing an upper bound on the probability, for a randomly drawn point x_1 that is consistent with S_f, that indeed $q_{i+1} \in U(x_1)$. Before proving this bound, we explain why such a bound implies a bound on the probability that a randomly drawn pair in $\mathcal{F}_2$ that is consistent with S_f satisfies $q_{i+1} \in U(x_1)$.

We first observe that, based on Lemma 6, there exists a unique prefix $p \in \mathcal{M}$ such that all the functions in $\mathcal{F}_2$ that are consistent with S_f contain a couple with the prefix p. Clearly, once the prefix of a couple is determined and we consider only functions in $\mathcal{F}_2$ that include a couple with that prefix, the distribution induced over the possible choices of this couple in f is uniform over all prefix p couples (x_1, x_2) such that $x_1 < x_2$. In addition, for every such function f, the question whether indeed $q_{i+1} \in U(x_1)$ is fully determined by the choice of the prefix p couple (x_1, x_2). Hence, it is enough to bound the probability over the possible choices of the couple (x_1, x_2) in f with the prefix p such that the couple is consistent with $S_f(i)$, that $q_{i+1} \in U(x_1)$. Following the fact that every $(\frac{1}{2} - \alpha)d$-weight point x plays the role of x_1 in the same number of couples (note that the weight of x_1 and x_2 is identical in all the couples), we conclude that the distribution induced by a random choice of a couple with prefix p over the choice of x_1 (or x_2) is uniform. Thus, bounding the probability that a random couple that is consistent with S_f satisfies $q_{i+1} \in U(x_1)$ is done by considering the probability that a random choice of x_1, that is consistent with $S_f(i)$, satisfies $q_{i+1} \in U(x_1)$, and showing that this probability is at most $\frac{2}{m}$. The technical details are omitted and will appear in the full version of the paper. □

Based on the above lemma, it is possible to prove the following proposition. Lemma 4 follows.

Proposition 1. *For every algorithm A and every possible choice of random coins for A, the following holds: $Pr\{S_f$ contains a witness$\} \leq \frac{2n^2}{m}$ (where the probability is taken over the random choice of the pair (f, D_f) from $\mathcal{F}_2$ and over the possible sampling of D_f for the chosen pair (f, D_f)).*

Proof. Given an algorithm A and a choice of random coins for A, the probability that S_f contains a witness for a randomly chosen pair $(f, D_f) \in \mathcal{F}_2$ can be bounded by $\sum_{i=1}^{n} Pr\{S_f(i+1)$ contains a witness $| S_f(i)$ does not contain a witness$\}$. By Lemma 7, this probability can be bounded by $\sum_{i=1}^{n} \frac{i+2}{m} = \frac{n(n+1)+2n}{m} < \frac{2n^2}{m}$. □

5   Lower Bound for Two-Sided Error Testing

In this section, we extend Theorem 1 to the two-sided error case, and prove that there exists no two-sided error distribution-free monotonicity tester with subexponential query complexity that accepts every pair $(f, D_f) \in \mathcal{F}_1$, with high probability, and rejects every pair $(f, D_f) \in \mathcal{F}_2$, with high probability. To do so, we show that there exists a constant c, such that for every algorithm A that asks less than 2^{cd} membership queries and samples D_f less than 2^{cd} times, the distributions induced on the set of possible knowledge sequences by running A on a randomly chosen pair from $\mathcal{F}_1$ and from $\mathcal{F}_2$ are indistinguishable.

In the previous section, we saw that for every algorithm A, the probability that a knowledge sequence obtained by A, while running on a randomly chosen pair from $\mathcal{F}_2$, contains a witness, is very small. At first glance, it may seem that this is enough to show that the two distributions are indistinguishable. However, this is not true. In the case of functions from $\mathcal{F}_1$, for every prefix $p \in \mathcal{M}$ there exists a point either in X_1 or in X_2 with the prefix p (m of the prefixes in $\mathcal{M}$ are used as prefixes for points in X_1, while the other m are used as prefixes for points in X_2). This does not hold in the case of functions from $\mathcal{F}_2$, where we choose only m out of the $2m$ prefixes, and select a couple (x_1, x_2) with each prefix. Hence, one can suggest a testing approach that looks for unused prefixed from $\mathcal{M}$. (Such an approach is not relevant in the case of one-sided error tester, since the tester has to accept every function in $\mathcal{F}_1$ with probability 1.) Clearly, if the tester is only allowed to sample the distribution D_f, then this testing approach requires an exponential number of queries. However, it is possible that we can use membership queries to significantly reduce the required query complexity. Hence, showing that with high probability the knowledge sequence S_f does not contain a witness is not enough, and there are other undesired events we have to eliminate. The full details of the two-sided testing lower bound proof will appear in the full version of this paper.

Acknowledgment

This work was initiated by a discussion with Nader Bshouty. We wish to thank Nader for sharing his ideas with us; they played a significant role in the development of the results presented in this paper.

References

1. N. Alon, E. Fischer, M. Krivelevich, and M. szegedy, *Efficient testing of large graphs*, FOCS 1999, pages 656–666.
2. T. Batu, R. Rubinfeld, and P. White, *Fast approximation PCPs for multidimensional bin-packing problems*, RANDOM-APPROX 1999, pages 246–256,.
3. M. Blum, M. Luby, and R. Rubinfeld, *Self testing/correcting with applications to numerical problems*, Journal of Computer and System Science 47:549–595, 1993.
4. A. Bogdanov, K. Obata, and L. Trevisan, *A lower bound for testing 3-colorability in bounded-degree graphs*, FOCS, 2002, pages 93-102.

5. A. Czumaj and C. Sohler, *Testing hypergraph coloring, ICALP* 2001, pp. 493–505.
6. Y. Dodis, O. Goldreich, E. Lehman, S. Raskhodnikova, D. Ron, and A. Samorodnitsky, *Improved testing algorithms for monotonicity, RANDOM-APPROX* 1999, pages 97–108.
7. E. Ergün, S. Kannan, R. Kumar, R. Rubinfeld, and M. Viswanathan, *Spotcheckers, Journal of Computing and System Science*, 60:717–751, 2000 (a preliminary version appeared in STOC 1998).
8. E. Fischer, *On the strength of comparisons in property testing*, manuscript (available at ECCC 8(8): (2001)).
9. E. Fischer, *The art of uninformed decisions: A primer to property testing, The Computational Complexity Column of The bulletin of the European Association for Theoretical Computer Science*, 75:97–126, 2001.
10. E. Fischer, E. Lehman, I. Newman, S. Raskhodnikova, R. Rubinfeld and, A. Samorodnitsky, *Monotonicity testing over general poset domains, STOC* 2002, pages 474–483.
11. E. Fischer and I. Newman, *Testing of matrix properties, STOC* 2001, pages 286–295.
12. O. Goldreich, *Combinatorical property testing – a survey, In: Randomized Methods in Algorithms Design*, AMS-DIMACS pages 45–61, 1998 .
13. O. Goldreich, S. Goldwasser, E. Lehman, D. Ron, and A. Samorodnitsky, *Testing Monotonicity, Combinatorica*, 20(3):301–337, 2000 (a preliminary version appeared in FOCS 1998).
14. O. Goldreich, S. Goldwasser, and D. Ron, *Property testing and its connection to learning and approximation, Journal of the ACM*, 45(4):653–750, 1998 (a preliminary version appeared in FOCS 1996).
15. O. Goldreich and D. Ron, *Property testing in bounded degree graphs, STOC* 1997, pages 406–415.
16. O. Goldreich and L. Trevisan, *Three theorems regarding testing graph properties, FOCS* 2001, pages 302–317.
17. S. Halevy and E. Kushilevitz, *Distribution-free property testing. In RANDOM-APPROX* 2003, pages 341–353.
18. S. Halevy and E. Kushilevitz, *Testing monotonicity over graph products, ICALP* 2004, pages 721–732.
19. S. Halevy and E. Kushilevitz, *Distribution-free connectivity testing. In RANDOM-APPROX* 2004, pages 393–404.
20. T. Kaufman, M. Krivelevich, and D. Ron, *Tight bounds for testing bipartiteness in general graphs, RANDOM-APPROX* 2003, pages 341–353.
21. M. J. Kearns and U. V. Vzirani, *An introduction to Computational Learning Theory*, MIT Press, 1994.
22. D. Ron, *Property testing (a tutorial), In: Handbook of Randomized Computing* (S.Rajasekaran, P. M. Pardalos, J. H. Reif and J. D. P. Rolin eds), Kluwer Press (2001).
23. R. Rubinfeld and M. Sudan, *Robust characterization of polynomials with applications to program testing, SIAM Journal of Computing*, 25(2):252–271, 1996. (first appeared as a technical report, Cornell University, 1993).

On Learning Random DNF Formulas
Under the Uniform Distribution

Jeffrey C. Jackson[1] and Rocco A. Servedio[2]

[1] Dept. of Math. and Computer Science
Duquesne University
Pittsburgh, PA 15282
jackson@mathcs.duq.edu
[2] Dept. of Computer Science
Columbia University
New York, NY 10027, USA
rocco@cs.columbia.edu

Abstract. We study the average-case learnability of DNF formulas in the model of learning from uniformly distributed random examples. We define a natural model of random monotone DNF formulas and give an efficient algorithm which with high probability can learn, for any fixed constant $\gamma > 0$, a random t-term monotone DNF for any $t = O(n^{2-\gamma})$. We also define a model of random nonmonotone DNF and give an efficient algorithm which with high probability can learn a random t-term DNF for any $t = O(n^{3/2-\gamma})$. These are the first known algorithms that can successfully learn a broad class of polynomial-size DNF in a reasonable average-case model of learning from random examples.

1   Introduction

1.1   Motivation and Background

A *disjunctive normal form* formula, or DNF, is an AND of ORs of Boolean literals. A question that has been open since Valiant's initial paper on computational learning theory [22] is whether or not efficient algorithms exist for learning polynomial size DNF formulas in various learning models. The only positive result to date is the Harmonic Sieve [11], which is a membership-query algorithm that efficiently learns DNF with respect to the uniform distribution (and certain related distributions). The approximating function produced by the Sieve is not itself a DNF; thus, the Sieve is an *improper* learning algorithm.

There has been little progress on polynomial-time algorithms for learning arbitrary DNF since the discovery of the Sieve. There are two obvious relaxations of the uniform-plus-membership model that can be pursued: learn with respect to arbitrary distributions using membership queries, and learn with respect to uniform without membership queries. Given standard cryptographic assumptions, the first direction is essentially as difficult as showing that DNF is learnable with respect to arbitrary distributions without membership queries

C. Chekuri et al. (Eds.): APPROX and RANDOM 2005, LNCS 3624, pp. 342–353, 2005.

[3]. However, there are also substantial known obstacles to learning DNF in the second model of uniform distribution without membership queries. In particular, no algorithm which can be recast as a Statistical Query algorithm can learn arbitrary polynomial-size DNF under the uniform distribution in $n^{o(\log n)}$ time [7]. Since nearly all non-membership learning algorithms can be recast as SQ algorithms [15], a major conceptual shift seems necessary to obtain an algorithm for efficiently learning arbitrary DNF formulas from uniform examples alone.

An apparently simpler question is whether *monotone* DNF formulas can be learned efficiently. Angluin showed that monotone DNF can be properly learned with respect to arbitrary distributions using membership queries [2]. It has also long been known that with respect to arbitrary distributions without membership queries, monotone DNF are no easier to learn than arbitrary DNF. [16]. This leaves the following enticing question (posed in [5, 6, 14]): are monotone DNF efficiently learnable from uniform examples alone?

In 1990, Verbeurgt [23] gave an algorithm that can properly learn any poly(n)-size (arbitrary) DNF from uniform examples in time $n^{O(\log n)}$. More recently, the algorithm of [21] learns any $2^{\sqrt{\log n}}$-term monotone DNF in poly(n) time. However, despite significant interest in the problem, no algorithm faster than that of [23] is known for learning arbitrary poly(n)-size monotone DNF from uniform examples, and no known hardness result precludes such an algorithm (the SQ result of [7] is at its heart a hardness result for low-degree parity functions).

Since worst-case versions of several DNF learning problems have remained stubbornly open for a decade or more, it is natural to ask about DNF learning from an average-case perspective, i.e., about learning *random* DNF formulas. In fact, this question has been considered before: Aizenstein & Pitt [1] were the first to ask whether random DNF formulas are efficiently learnable. They proposed a model of random DNF in which each of the t terms is selected independently at random from all possible terms, and gave a membership and equivalence query algorithm which with high probability learns a random DNF generated in this way. However, as noted in [1], a limitation of this model is that with very high probability all terms will have length $\Omega(n)$. The learning algorithm itself becomes quite simple in this situation. Thus, while this is a "natural" average-case DNF model, from a learning perspective it is not a particularly interesting model. To address this deficiency, they also proposed another natural average-case model which is parameterized by the expected length k of each term as well as the number of independent terms t, but left open the question of whether or not random DNF can be efficiently learned in such a model.

1.2   Our Results

We consider an average-case DNF model very similar to the latter Aizenstein and Pitt model, although we simplify slightly by assuming that k represents a fixed term length rather than an expected length. We show that, in the model of learning from uniform random examples only, random monotone DNF are properly and efficiently learnable for many interesting values of k and t. In particular, for $t = O(n^{2-\gamma})$ where $\gamma > 0$, and for $k = \log t$, our algorithm can achieve any error rate $\epsilon > 0$ in poly$(n, 1/\epsilon)$ time with high probability (over both the selection of the target DNF and the selection of examples). In addition, we obtain slightly weaker results for arbitrary DNF: our algorithm can properly and efficiently learn random t-term DNF for t such that $t = O(n^{\frac{3}{2}-\gamma})$. This algorithm cannot achieve arbitrarily small error but can achieve error $\epsilon = o(1)$ for any $t = \omega(1)$. For detailed result statements see Theorems 3 and 5.

While our results would clearly be stronger if they held for any $t = $ poly(n) rather than the specific polynomials given, they are a marked advance over the previous state of affairs for DNF learning. (Recall that in the standard worst-case model, poly(n)-time uniform-distribution learning of $t(n)$-term DNF for any $t(n) = \omega(1)$ is an open problem with an associated cash prize [4].)

We note that taking $k = \log t$ is a natural choice when learning with respect to the uniform distribution. (We actually allow a somewhat more general choice of k.) This choice leads to target DNFs that, with respect to uniform, are roughly *balanced* (0 and 1 values are equally likely). From a learning perspective balanced functions are generally more interesting than unbalanced functions, since a constant function is trivially a good approximator to a highly unbalanced function.

Our results shed some light on which cases are *not* hard in the worst-case model. While "hard" cases were previously known for arbitrary DNF [4], our findings may be particularly helpful in guiding future research on monotone DNF. In particular, our algorithm learns any monotone DNF which (i) is near-balanced, (ii) has every term uniquely satisfied with reasonably high probability, (iii) has every pair of terms jointly satisfied with much smaller probability, and (iv) has no variable appearing in significantly more than a $1/\sqrt{t}$ fraction of the t terms (this is made precise in Lemma 6). So in order to be "hard," a monotone DNF must violate one or more of these criteria.

Our algorithms work in two stages: they first identify pairs of variables which cooccur in some term of the target DNF, and then use these pairs to reconstruct terms via a specialized clique-finding algorithm. For monotone DNF we can with high probability determine for every pair of variables whether or not the pair cooccurs in some term. For nonmonotone DNF, with high probability we can identify most pairs of variables which cooccur in some term; as we show, this enables us to learn to fairly (but not arbitrarily) high accuracy.

We give preliminaries in Section 2. Sections 3 and 4 contain our results for monotone and nonmonotone DNF respectively. Section 5 concludes. Because of space constraints, many proofs are omitted and will appear in the full version.

2   Preliminaries

We first describe our models of random monotone and nonmonotone DNF. Let $\mathcal{M}_n^{t,k}$ be the probability distribution over monotone t-term DNF induced by the following random process: each term is independently and uniformly chosen at random from all $\binom{n}{k}$ monotone ANDs of size exactly k over variables $v_1, \ldots, v_n$. For nonmonotone DNF, we write $\mathcal{D}_n^{t,k}$ to denote the following distribution over t-term DNF: first a monotone DNF is selected from $\mathcal{M}_n^{t,k}$, and then each occurrence of each variable in each term is independently negated with probability $1/2$. (Equivalently, a draw from $\mathcal{D}_n^{t,k}$ is done by independently selecting t terms from the set of all terms of length exactly k).

Given a Boolean function $\phi : \{0,1\}^n \to \{0,1\}$, we write $\Pr[\phi]$ to denote $\Pr_{x \sim U_n}[\phi(x) = 1]$, where U_n denotes the uniform distribution over $\{0,1\}^n$. We write log to denote $\log_2$ and ln to denote natural log.

In the uniform distribution learning model which we consider, the learner is given a source of labeled examples $(x, f(x))$ where each x is uniformly drawn from $\{0,1\}^n$ and f is the unknown function to be learned. The goal of the learner is to efficiently construct a hypothesis h which with high probability (over the choice of labeled examples used for learning) has low error relative to f under the uniform distribution, i.e. $\Pr_{x \sim U_n}[h(x) \neq f(x)] \leq \epsilon$ with probability $1 - \delta$. This model has been intensively studied in learning theory, see e.g. [9, 10, 12, 18, 19, 21, 23]. In our average case framework, the target function f will be drawn randomly from either $\mathcal{M}_n^{t,k}$ or $\mathcal{D}_n^{t,k}$, and (as in [13]) our goal is to construct a low-error hypothesis h for f with high probability over both the random examples used for learning and the random draw of f.

3 Learning Random Monotone DNF

3.1 Interesting Parameter Settings

Consider a random draw of f from $\mathcal{M}_n^{t,k}$. It is intuitively clear that if t is too large relative to k then a random $f \in \mathcal{M}_n^{t,k}$ will likely have $\Pr[f] \approx 1$; similarly if t is too small relative to k then a random $f \in \mathcal{M}_n^{t,k}$ will likely have $\Pr[f] \approx 0$. Such cases are not very interesting from a learning perspective since a trivial algorithm can learn to high accuracy. We are thus led to the following definition:

Definition 1. *For $0 < \alpha \leq 1/2$, a pair of values (k,t) is said to be* monotone α-interesting *if $\alpha \leq \mathbf{E}_{f \in \mathcal{M}_n^{t,k}}[\Pr[f]] \leq 1 - \alpha$.*

Throughout the paper we will assume that $0 < \alpha \leq 1/2$ is a fixed constant independent of n and that $t \leq p(n)$, where $p(\cdot)$ is a fixed polynomial (and we will also make assumptions about the degree of p). The following lemma gives necessary conditions for (k,t) to be monotone α-interesting. (As Lemma 1 indicates, we may always think of k as being roughly $\log t$.)

Lemma 1. *If (k,t) is monotone α-interesting then $\alpha 2^k \leq t \leq 2^{k+1} \ln \frac{2}{\alpha}$.*

3.2 Properties of Random Monotone DNF

Throughout the rest of Section 3 we assume that $\alpha > 0$ is fixed and (k,t) is a monotone α-interesting pair where $t = O(n^{2-\gamma})$ for some $\gamma > 0$. In this section we state some useful lemmas regarding $\mathcal{M}_n^{t,k}$.

Let f^i denote the projected function obtained from f by first removing term T_i from the monotone DNF for f and then restricting all of the variables which were present in term T_i to 1. For $\ell \neq i$ we write T_ℓ^i to denote the term obtained by setting all variables in T_i to 1 in T_ℓ, i.e. T_ℓ^i is the term in f^i corresponding to T_ℓ. Note that if $T_\ell^i \not\equiv T_\ell$ then T_ℓ^i is smaller than T_ℓ.

The following lemma shows that each variable appears in a limited number of terms; intuitively this means that not too many terms T_ℓ^i in f^i are smaller than their corresponding terms T_ℓ in f. In this and later lemmas, "n sufficiently large" means n is larger than a constant which depends on α but not on k or t.

Lemma 2. Let $\delta_{\mathrm{many}} := n(\frac{ekt^{3/2}\log t}{n2^{k-1}\alpha^2})^{2^{k-1}\alpha^2/\sqrt{t}\log t}$. For n sufficiently large, with probability at least $1 - \delta_{\mathrm{many}}$ over the draw of f from $\mathcal{M}_n^{t,k}$ we have that every variable v_j, $1 \leq j \leq n$, appears in at most $2^{k-1}\alpha^2/\sqrt{t}\log t$ terms of f.

Note that since (k,t) is a monotone α-interesting pair and $t = O(n^{2-\gamma})$ for some fixed $\gamma > 0$, for sufficiently large n this probability bound is non-trivial.

Using straightforward probabilistic arguments, we can show that for f drawn from $\mathcal{M}_n^{t,k}$, with high probability each term is "uniquely satisfied" by a noticeable fraction of assignments. More precisely, we have:

Lemma 3. Let $\delta_{\mathrm{usat}} := tk(\frac{ek(t-1)\log t}{n2^k})^{2^k/(\log t)} + t^2(\frac{k^2}{n})^{\log\log t}$. For n sufficiently large, with probability at least $1 - \delta_{\mathrm{usat}}$ over the random draw of f from $\mathcal{M}_n^{t,k}$, f is such that for all $i = 1,\ldots,t$ we have $\mathrm{Pr}_x[T_i$ is satisfied by x but no other T_j is satisfied by $x] \geq \frac{\alpha^3}{2^{k+2}}$.

On the other hand, we can upper bound the probability that two terms of a random DNF f will be satisfied simultaneously:

Lemma 4. Let $\delta_{\mathrm{shared}} := t^2(\frac{k^2}{n})^{\log\log t}$. With probability at least $1 - \delta_{\mathrm{shared}}$ over the random draw of f from $\mathcal{M}_n^{t,k}$, for all $1 \leq i < j \leq t$ we have $\mathrm{Pr}[T_i \wedge T_j] \leq \frac{\log t}{2^{2k}}$.

3.3   Identifying Cooccurring Variables

We now show how to identify pairs of variables that cooccur in some term of f.

First, some notation. Given a monotone DNF f over variables $v_1,\ldots,v_n$, define DNF formulas g_{**}, g_{1*}, g_{*1} and g_{11} over variables $v_3,\ldots,v_n$ as follows:

- g_{**} is the disjunction of the terms in f that contain neither v_1 nor v_2;
- g_{1*} is the disjunction of the terms in f that contain v_1 but not v_2 (but with v_1 removed from each of these terms);
- g_{*1} is defined similarly as the disjunction of the terms in f that contain v_2 but not v_1 (but with v_2 removed from each of these terms);
- g_{11} is the disjunction of the terms in f that contain both v_1 and v_2 (with both variables removed from each term).

We thus have $f = g_{**} \vee (v_1 g_{1*}) \vee (v_2 g_{*1}) \vee (v_1 v_2 g_{11})$. Note that any of g_{**}, g_{1*}, g_{*1}, g_{11} may be an empty disjunction which is identically false.

We can empirically estimate each of the following using uniform random examples $(x, f(x))$:

$$p_{00} := \Pr_x[g_{**}] = \Pr_{x \in U_n}[f(x) = 1 \mid x_1 = x_2 = 0]$$

$$p_{01} := \Pr_x[g_{**} \vee g_{*1}] = \Pr_{x \in U_n}[f(x) = 1 \mid x_1 = 0, x_2 = 1]$$

$$p_{10} := \Pr_x[g_{**} \vee g_{1*}] = \Pr_{x \in U_n}[f(x) = 1 \mid x_1 = 1, x_2 = 0]$$

$$p_{11} := \Pr_x[g_{**} \vee g_{*1} \vee g_{1*} \vee g_{11}] = \Pr_{x \in U_n}[f(x) = 1 \mid x_1 = 1, x_2 = 1].$$

It is clear that g_{11} is nonempty if and only if v_1 and v_2 cooccur in some term of f; thus we would ideally like to obtain $\Pr_{x \in U_n}[g_{11}]$. While we cannot obtain this probability from p_{00}, p_{01}, p_{10} and p_{11}, the following lemma shows that we can estimate a related quantity:

Lemma 5. *Let* P *denote* $p_{11} - p_{10} - p_{01} + p_{00}$. *Then* $P = \Pr[g_{11} \wedge \overline{g}_{1*} \wedge \overline{g}_{*1} \wedge \overline{g}_{**}] - \Pr[g_{1*} \wedge g_{*1} \wedge \overline{g}_{**}]$.

More generally, let P_{ij} be defined as P but with v_i, x_i, v_j, and x_j substituted for v_1, x_1, v_2, and x_2, respectively, throughout the definitions of the g's and p's above. The following lemma shows that, for most random choices of f, for all $1 \leq i, j \leq n$, the value of P_{ij} is a good indicator of whether or not v_i and v_j cooccur in some term of f:

Lemma 6. *For* n *sufficiently large and* $t \geq 4$, *with probability at least* $1 - \delta_{\text{usat}} - \delta_{\text{shared}} - \delta_{\text{many}}$ *over the random draw of* f *from* $\mathcal{M}_n^{t,k}$, *we have that for all* $1 \leq i, j \leq n$ *(i) if* v_i *and* v_j *do not cooccur in some term of* f *then* $P_{ij} \leq 0$; *(ii) if* v_i *and* v_j *do cooccur in some term of* f *then* $P_{ij} \geq \frac{\alpha^4}{8t}$.

Proof. Part (i) holds for any monotone DNF by Lemma 5. For (ii), we first note that with probability at least $1 - \delta_{\text{usat}} - \delta_{\text{shared}} - \delta_{\text{many}}$, a randomly chosen f has all the following properties:

1. Each term in f is uniquely satisfied with probability at least $\alpha^3/2^{k+2}$ (by Lemma 3);
2. Each pair of terms T_i and T_j in f are both satisfied with probability at most $\log t / 2^{2k}$ (by Lemma 4); and
3. Each variable in f appears in at most $2^{k-1}\alpha^2/\sqrt{t} \log t$ terms (by Lemma 2).

We call such an f *well-behaved*. For the sequel, assume that f is well-behaved and also assume without loss of generality that $i = 1$ and $j = 2$. We consider separately the two probabilities $\rho_1 = \Pr[g_{11} \wedge \overline{g}_{1*} \wedge \overline{g}_{*1} \wedge \overline{g}_{**}]$ and $\rho_2 = \Pr[g_{1*} \wedge g_{*1} \wedge \overline{g}_{**}]$ whose difference defines $P_{12} = P$. By property (1) above, $\rho_1 \geq \alpha^3/2^{k+2}$, since each instance x that uniquely satisfies a term T_j in f containing both v_1 and v_2 also satisfies g_{11} while falsifying all of g_{1*}, g_{*1}, and g_{**}. Since (k, t) is monotone α-interesting, this implies that $\rho_1 \geq \alpha^4/4t$. On the other hand, clearly $\rho_2 \leq \Pr[g_{1*} \wedge g_{*1}]$. By property (2) above, for any pair of terms consisting of one term from g_{1*} and the other from g_{*1}, the probability that both terms are satisfied is at most $\log t/2^{2k}$. Since each of g_{1*} and g_{*1} contains at most $2^{k-1}\alpha^2/\sqrt{t}\log t$ terms by property (3), by a union bound we have $\rho_2 \leq \alpha^4/(4t\log t)$, and the lemma follows given the assumption that $t \geq 4$.

Thus, our algorithm for finding all of the cooccurring pairs of a randomly chosen monotone DNF consists of estimating P_{ij} for each of the $n(n-1)/2$ pairs (i, j) so that all of our estimates are—with probability at least $1 - \delta$—within an additive factor of $\alpha^4/16t$ of their true values. The reader familiar with Boolean Fourier analysis will readily recognize that P_{12} is just $\hat{f}(110_{n-2})$ and that in general all of the P_{ij} are simply second-order Fourier coefficients. Therefore, by the standard Hoeffding bound, a uniform random sample of size $O(t^2 \ln(n^2/\delta)/\alpha^8)$ is sufficient to estimate all of the P_{ij}'s to the specified tolerance with overall probability at least $1 - \delta$. We thus have:

Theorem 1. *For n sufficiently large and any $\delta > 0$, with probability at least $1 - \delta_{\mathrm{usat}} - \delta_{\mathrm{shared}} - \delta_{\mathrm{many}} - \delta$ over the choice of f from $\mathcal{M}_n^{t,k}$ and the choice of random examples, our algorithm runs in $O(n^2 t^2 \log(n/\delta))$ time and identifies exactly those pairs (v_i, v_j) which cooccur in some term of f.*

3.4   Forming a Hypothesis from Pairs of Cooccurring Variables

Here we show how to construct an accurate DNF hypothesis for a random f drawn from $\mathcal{M}_n^{t,k}$.

Identifying all k-cliques. By Theorem 1, with high probability we have complete information about which pairs of variables (v_i, v_j) cooccur in some term of f. We thus may consider the graph G with vertices $v_1, \ldots, v_n$ and edges for precisely those pairs of variables (v_i, v_j) which cooccur in some term of f. This graph is a union of t randomly chosen k-cliques from $\{v_1, \ldots, v_n\}$ which correspond to the t terms in f. We will show how to efficiently identify (with high probability over the choice of f and random examples of f) all of the k-cliques in G. Once these k-cliques have been identified, as we show later it is easy to construct an accurate DNF hypothesis for f.

The following lemma (whose proof is omitted) shows that with high probability over the choice of f, each pair (v_i, v_j) cooccurs in at most a constant number of terms:

Lemma 7. *Fix $1 \leq i < j \leq n$. For any $C \geq 0$ and all sufficiently large n, we have $\Pr_{f \in \mathcal{M}_n^{t,k}}[$some pair of variables (v_i, v_j) cooccur in more than C terms of $f] \leq \delta_C := (\frac{tk^2}{n^2})^C$.*

By Lemma 7 we know that, for any given pair (v_i, v_j) of variables, with probability at least $1 - \delta_C$ there are at most Ck other variables v_ℓ such that (v_i, v_j, v_ℓ) all cooccur in some term of f. Suppose that we can efficiently (with high probability) identify the set S_{ij} of all such variables v_ℓ. Then we can perform an exhaustive search over all $(k-2)$-element subsets S' of S_{ij} in at most $\binom{Ck}{k} \leq (eC)^k = n^{O(\log C)}$ time, and can identify exactly those sets S' such that $S' \cup$

$\{v_i, v_j\}$ is a clique of size k in G. Repeating this over all pairs of variables (v_i, v_j), we can with high probability identify all k-cliques in G.

Thus, to identify all k-cliques in G it remains only to show that for every pair of variables v_i and v_j, we can determine the set S_{ij} of those variables v_ℓ that cooccur in at least one term with both v_i and v_j. Assume that f is such that all pairs of variables cooccur in at most C terms, and let T be a set of variables of cardinality at most C having the following properties:

- In the projection $f_{T \leftarrow 0}$ of f in which all of the variables of T are fixed to 0, v_i and v_j do not cooccur in any term; and
- For every set $T' \subset T$ such that $|T'| = |T| - 1$, v_i and v_j do cooccur in $f_{T' \leftarrow 0}$.

Then T is clearly a subset of S_{ij}. Furthermore, if we can identify all such sets T, then their union will be S_{ij}. There are only $O(n^C)$ possible sets to consider, so our problem now reduces to the following: given a set T of at most C variables, determine whether v_i and v_j cooccur in $f_{T \leftarrow 0}$.

The proof of Lemma 6 shows that f is well-behaved with probability at least $1 - \delta_{\text{usat}} - \delta_{\text{shared}} - \delta_{\text{many}}$ over the choice of f. Furthermore, if f is well-behaved then it is easy to see that for every $|T| \leq C$, $f_{T \leftarrow 0}$ is also well-behaved, since $f_{T \leftarrow 0}$ is just f with $O(\sqrt{t})$ terms removed (by Lemma 2). That is, removing terms from f can only make it more likely that the remaining terms are uniquely satisfied, does not change the bound on the probability of a pair of remaining terms being satisfied, and can only decrease the bound on the number of remaining terms in which a remaining variable can appear. Furthermore, Lemma 5 holds for any monotone DNF f. Therefore, if f is well-behaved then the proof of Lemma 6 also shows that for every $|T| \leq C$, the P_{ij}'s of $f_{T \leftarrow 0}$ can be used to identify the cooccurring pairs of variables within $f_{T \leftarrow 0}$. It remains to show that we can efficiently simulate a uniform example oracle for $f_{T \leftarrow 0}$ so that these P_{ij}'s can be accurately estimated.

In fact, for a given set T, we can simulate a uniform example oracle for $f_{T \leftarrow 0}$ by filtering the examples from the uniform oracle for f so that only examples setting the variables in T to 0 are accepted. Since $|T| \leq C$, the filter accepts with constant probability at least $1/2^C$. A Chernoff argument shows that if all P_{ij}'s are estimated using a single sample of size $2^{C+10} t^2 \ln(2(C + 2)n^C/\delta)/\alpha^8$ (filtered appropriately when needed) then all of the estimates will have the desired accuracy with probability at least $1 - \delta$. This gives us the following:

Theorem 2. *For n sufficiently large, any $\delta > 0$, and any fixed $C \geq 2$, with probability at least $1 - \delta_{\text{usat}} - \delta_{\text{shared}} - \delta_{\text{many}} - \delta_C - \delta$ over the random draw of f from $\mathcal{M}_n^{t,k}$ and the choice of random examples, all of the k-cliques of the graph G can be identified in time $n^{O(C)} t^3 k^2 \log(n/\delta)$.*

The main learning result for monotone DNF. We now have a list $T'_1, \ldots, T'_N$ (with $N = O(n^C)$) of length-k monotone terms which contains all t true terms $T_1, \ldots, T_t$ of f. Now note that the target function f is simply an OR of some subset of these N "variables" $T_1, \ldots, T_N$, so the standard elimination algorithm for learning disjunctions (see e.g. Chapter 1 of [17]) can be used to PAC learn the target function.

Call the above described entire learning algorithm A. In summary, we have proved the following:

Theorem 3. *Fix γ, $\alpha > 0$ and $C \geq 2$. Let (k,t) be a monotone α-interesting pair. For any $\epsilon > 0$, $\delta > 0$, and $t = O(n^{2-\gamma})$, algorithm A will with probability at least $1 - \delta_{\mathrm{usat}} - \delta_{\mathrm{shared}} - \delta_{\mathrm{many}} - \delta_C - \delta$ (over the random choice of DNF from $\mathcal{M}_n^{t,k}$ and the randomness of the example oracle) produce a hypothesis h that ϵ-approximates the target with respect to the uniform distribution. Algorithm A runs in time polynomial in n, $\log(1/\delta)$, and $1/\epsilon$.*

4   Nonmonotone DNF

Because of space constraints here we only sketch our results and techniques for nonmonotone DNF.

As with $\mathcal{M}_n^{t,k}$, we are interested in pairs (k,t) for which $\mathbf{E}_{f \in \mathcal{D}_n^{t,k}}[\Pr[f]]$ is between α and $1 - \alpha$. For $\alpha > 0$, we now say that the pair (k,t) is α-*interesting* if $\alpha \leq \mathbf{E}_{f \in \mathcal{D}_n^{t,k}}[\Pr[f]] \leq 1 - \alpha$. For any fixed $x \in \{0,1\}^n$ we have $\Pr_{f \in \mathcal{D}_n^{t,k}}[f(x) = 0] = (1 - \frac{1}{2^k})^t$, and thus by linearity of expectation we have $\mathbf{E}_{f \in \mathcal{D}_n^{t,k}}[\Pr[f]] = 1 - (1 - \frac{1}{2^k})^t$; this formula will be useful later.

Throughout the rest of Section 4 we assume that $\alpha > 0$ is fixed and (k,t) is an α-interesting pair where $t = O(n^{3/2-\gamma})$ for some $\gamma > 0$.

4.1   Identifying (Most Pairs of) Cooccurring Variables

Recall that in Section 3.3 we partitioned the terms of our monotone DNF into four disjoint groups depending on what subset of $\{v_1, v_2\}$ was present in each term. In the nonmonotone case, we will partition the terms of f into nine disjoint groups depending on whether each of v_1, v_2 is unnegated, negated, or absent:

$$f = g_{**} \vee (v_1 g_{1*}) \vee (\overline{v_1} g_{0*}) \vee (v_2 g_{*1}) \vee (v_1 v_2 g_{11}) \vee (\overline{v_1} v_2 g_{01}) \vee (\overline{v_2} g_{*0}) \vee (v_1 \overline{v_2} g_{10}) \vee (\overline{v_1} \overline{v_2} g_{00})$$

Thus g_{**} contains those terms of f which contain neither v_1 nor v_2 in any form; g_{0*} contains the terms of f which contain $\overline{v_1}$ but not v_2 in any form (with $\overline{v_1}$ removed from each term); g_{*1} contains the terms of f which contain v_2 but not v_1 in any form (with v_2 removed from each term); and so on. Each $g_{\cdot,\cdot}$ is thus a DNF (possibly empty) over literals formed from $v_3, \ldots, v_n$.

For all four possible values of $(a,b) \in (0,1)^2$, we can empirically estimate

$$p_{ab} := \Pr_x[g_{**} \vee g_{a*} \vee g_{*b} \vee g_{ab}] = \Pr_x[f(x) = 1 \mid x_1 = a, x_2 = b].$$

It is easy to see that $\Pr[g_{11}]$ is either 0 or else at least $\frac{4}{2^k}$ depending on whether g_{11} is empty or not. Ideally we would like to be able to accurately estimate each of $\Pr[g_{00}], \Pr[g_{01}], \Pr[g_{10}]$ and $\Pr[g_{11}]$; if we could do this then we would have complete information about which pairs of literals involving variables v_1 and v_2 cooccur in terms of f. Unfortunately, the probabilities $\Pr[g_{00}], \Pr[g_{01}], \Pr[g_{10}]$ and $\Pr[g_{11}]$ cannot in general be obtained from p_{00}, p_{01}, p_{10} and p_{11}. However, we will show that we can efficiently obtain some partial information which enables us to learn to fairly high accuracy.

As before, our approach is to accurately estimate the quantity $P = p_{11} - p_{10} - p_{01} + p_{00}$. We have the following lemmas:

Lemma 8. *If all four of g_{00}, g_{01}, g_{10} and g_{11} are empty, then P equals*

$$\Pr[g_{1*} \wedge g_{*0} \wedge (\text{ no other } g_{.,.})] + \Pr[g_{0*} \wedge g_{*1} \wedge (\text{ no other } g_{.,.})]$$
$$- \Pr[g_{1*} \wedge g_{*1} \wedge (\text{ no other } g_{.,.})] - \Pr[g_{0*} \wedge g_{*0} \wedge (\text{ no other } g_{.,.})]. \quad (1)$$

Lemma 9. *If exactly one of g_{00}, g_{01}, g_{10} and g_{11} is nonempty (say g_{11}), then P equals (1) plus $\Pr[g_{11} \wedge g_{1*} \wedge g_{*0} \wedge (\text{ no other } g_{.,.})] + \Pr[g_{11} \wedge g_{0*} \wedge g_{*1} \wedge (\text{ no other } g_{.,.})] - \Pr[g_{11} \wedge g_{1*} \wedge g_{*1} \wedge (\text{ no other } g_{.,.})] - \Pr[g_{11} \wedge g_{0*} \wedge g_{*0} \wedge (\text{ no other } g_{.,.})] + \Pr[g_{11} \wedge g_{0*} \wedge (\text{ no other } g_{.,.})] + \Pr[g_{11} \wedge g_{*0} \wedge (\text{ no other } g_{.,.})] + \Pr[g_{11} \wedge (\text{ no other } g_{.,.})].$*

Using the above two lemmas we can show that the value of P is a good indicator for distinguishing between all four of $g_{00}, g_{01}, g_{10}, g_{11}$ being empty versus exactly one of them being nonempty:

Lemma 10. *For n sufficiently large and $t \geq 4$, with probability at least $1 - \delta'_{usat} - \delta_{shared} - \delta_{many}$ over a random draw of f from $\mathcal{D}_n^{t,k}$, we have that: (i) if v_1 and v_2 do not cooccur in any term of f then $P \leq \frac{\alpha^2}{8t}$.; (ii) if v_1 and v_2 do cooccur in some term of f and exactly one of g_{00}, g_{01}, g_{10} and g_{11} is nonempty, then $P \geq \frac{3\alpha^2}{16t}$.*

The proof (omitted, as are the exact definitions of the various δ's) uses analogues of the properties established in Section 3.2 for nonmonotone DNF.

It is clear that an analogue of Lemma 10 holds for any pair of variables v_i, v_j in place of v_1, v_2. Thus, for each pair of variables v_i, v_j, if we decide whether v_i and v_j cooccur (negated or otherwise) in any term on the basis of whether P_{ij} is large or small, we will err only if two or more of $g_{00}, g_{01}, g_{10}, g_{11}$ are nonempty.

We now show that for $f \in \mathcal{D}_n^{t,k}$, with very high probability there are not too many pairs of variables (v_i, v_j) which cooccur (with any sign pattern) in at least two terms of f. Note that this immediately bounds the number of pairs (v_i, v_j) which have two or more of the corresponding $g_{00}, g_{01}, g_{10}, g_{11}$ nonempty.

Lemma 11. *Let $d > 0$ and $f \in \mathcal{D}_n^{t,k}$. The probability that more than $(d + 1)t^2k^4/n^2$ pairs of variables (v_i, v_j) each cooccur in two or more terms of f is at most $\exp(-d^2t^3k^4/n^4)$.*

Taking $d = n^2/(t^{5/4}k^4)$ in the above lemma (note that $d > 1$ for n sufficiently large since $t^{5/4} = O(n^{15/8})$), we have $(d + 1)t^2k^4/n^2 \leq 2t^{3/4}$ and the failure probability is at most $\delta_{cooccur} := \exp(-\sqrt{t}/k^4)$. The results of this section (together with a standard analysis of error in estimating each P_{ij}) thus yield:

Theorem 4. *For n sufficiently large and for any $\delta > 0$, with probability at least $1 - \delta_{cooccur} - \delta'_{usat} - \delta_{shared} - \delta_{many} - \delta$ over the random draw of f from $\mathcal{D}_n^{t,k}$*

and the choice of random examples, the above algorithm runs in $O(n^2 t^2 \log(n/\delta))$ time and outputs a list of pairs of variables (v_i, v_j) such that: (i) if (v_i, v_j) is in the list then v_i and v_j cooccur in some term of f; and (ii) at most $N_0 = 2t^{3/4}$ pairs of variables (v_i, v_j) which do cooccur in f are not on the list.

4.2   Reconstructing an Accurate DNF Hypothesis

It remains to construct a good hypothesis for the target DNF from a list of pairwise cooccurrence relationships as provided by Theorem 4. As in the monotone case, we consider the graph G with vertices $v_1, \ldots, v_n$ and edges for precisely those pairs of variables (v_i, v_j) which cooccur (with any sign pattern) in some term of f. As before this graph is a union of t randomly chosen k-cliques $S_1, \ldots, S_t$ which correspond to the t terms in f, and as before we would like to find all k-cliques in G. However, there are two differences now: the first is that instead of having the true graph G, we instead have access only to a graph G' which is formed from G by deleting some set of at most $N_0 = 2t^{3/4}$ edges. The second difference is that the final hypothesis must take the signs of literals in each term into account. To handle these two differences, we use a different reconstruction procedure than we used for monotone DNF in Section 3.4; this reconstruction procedure only works for $t = O(n^{3/2 - \gamma})$ where $\gamma > 0$.

Because of space constraints we do not describe the reconstruction procedure here. All in all the following is our main learning result for nonmonotone DNF:

Theorem 5. *Fix γ, $\alpha > 0$ and $C \geq 2$. Let (k, t) be a monotone α-interesting pair. For f randomly chosen from $\mathcal{D}_n^{t,k}$, with probability at least $1 - \delta_{\text{cooccur}} - \delta'_{\text{usat}} - \delta_{\text{shared}} - \delta_{\text{many}} - \delta_{\text{clique}} - 1/n^{\Omega(C)}$ the above algorithm runs in $\tilde{O}(n^2 t^2 + n^{O(\log C)})$ time and outputs a hypothesis h whose error rate relative to f under the uniform distribution is at most $1/\Omega(t^{1/4})$.*

It can be verified from the definitions of the various δ's that for any $t = \omega(1)$ as a function of n, the failure probability is $o(1)$ and the accuracy is $1 - o(1)$.

5   Future Work

We can currently only learn random DNFs with $o(n^{3/2})$ terms ($o(n^2)$ terms for monotone DNF); can stronger results be obtained which hold for all polynomial-size DNF? A natural approach here for learning n^c-term DNF might be to first try to identify all c'-tuples of variables which cooccur in a term, where c' is some constant larger than c.

Acknowledgement

Avrim Blum suggested to one of us (JCJ) the basic strategy that learning monotone DNF with respect to uniform might be reducible to finding the cooccurring pairs of variables in the target function. This material is based upon work supported by the National Science Foundation under Grant No. CCR-0209064 (JCJ) and CCF-0347282 (RAS).

References

1. H. Aizenstein and L. Pitt. On the learnability of disjunctive normal form formulas. *Machine Learning*, 19:183–208, 1995.
2. D. Angluin. Queries and concept learning. *Machine Learning*, 2:319–342, 1988.
3. D. Angluin and M. Kharitonov. When won't membership queries help? *J. Comput. & Syst. Sci.*, 50:336–355, 1995.
4. A. Blum. Learning a function of r relevant variables (open problem). In *Proc. 16th COLT*, pages 731–733, 2003.
5. A. Blum. Machine learning: a tour through some favorite results, directions, and open problems. FOCS 2003 tutorial slides, available at http://www-2.cs.cmu.edu/~avrim/Talks/FOCS03/tutorial.ppt, 2003.
6. A. Blum, C. Burch, and J. Langford. On learning monotone boolean functions. In *Proc. 39th FOCS*, pages 408–415, 1998.
7. A. Blum, M. Furst, J. Jackson, M. Kearns, Y. Mansour, and S. Rudich. Weakly learning DNF and characterizing statistical query learning using Fourier analysis. In *Proc. 26th STOC*, pages 253–262, 1994.
8. B. Bollobas. *Combinatorics: Set Systems, Hypergraphs, Families of Vectors and Combinatorial Probability.* Cambridge University Press, 1986.
9. M. Golea, M. Marchand, and T. Hancock. On learning μ-perceptron networks on the uniform distribution. *Neural Networks*, 9:67–82, 1994.
10. T. Hancock. Learning $k\mu$ decision trees on the uniform distribution. In *Proc. Sixth COLT*, pages 352–360, 1993.
11. J. Jackson. An efficient membership-query algorithm for learning DNF with respect to the uniform distribution. *J. Comput. & Syst. Sci.*, 55:414–440, 1997.
12. J. Jackson, A. Klivans, and R. Servedio. Learnability beyond AC^0. In *Proc. 34th STOC*, 2002.
13. J. Jackson and R. Servedio. Learning random log-depth decision trees under the uniform distribution. In *Proc. 16th COLT*, pages 610–624, 2003.
14. J. Jackson and C. Tamon. Fourier analysis in machine learning. ICML/COLT 1997 tutorial slides, available at http://learningtheory.org/resources.html, 1997.
15. M. Kearns. Efficient noise-tolerant learning from statistical queries. *Journal of the ACM*, 45(6):983–1006, 1998.
16. M. Kearns, M. Li, L. Pitt, and L. Valiant. Recent results on Boolean concept learning. In *Proc. Fourth Int. Workshop on Mach. Learning*, pages 337–352, 1987.
17. M. Kearns and U. Vazirani. *An Introduction to Computational Learning Theory.* MIT Press, Cambridge, MA, 1994.
18. A. Klivans, R. O'Donnell, and R. Servedio. Learning intersections and thresholds of halfspaces. In *Proc. 43rd FOCS*, pages 177–186, 2002.
19. L. Kucera, A. Marchetti-Spaccamela, and M. Protassi. On learning monotone DNF formulae under uniform distributions. *Inform. and Comput.*, 110:84–95, 1994.
20. C. McDiarmid. On the method of bounded differences. In *Surveys in Combinatoric 1989*, pages 148–188. London Mathematical Society Lecture Notes, 1989.
21. R. Servedio. On learning monotone DNF under product distributions. In *Proc. 14th COLT*, pages 473–489, 2001.
22. L. Valiant. A theory of the learnable. *CACM*, 27(11):1134–1142, 1984.
23. K. Verbeurgt. Learning DNF under the uniform distribution in quasi-polynomial time. In *Proc. Third COLT*, pages 314–326, 1990.

Derandomized Constructions
of k-Wise (Almost) Independent Permutations

Eyal Kaplan[1], Moni Naor[2,*], and Omer Reingold[2,**]

[1] Tel-Aviv University
kaplaney@post.tau.ac.il
[2] Department of Computer Science and Applied Math, Weizmann Institute of Science
{moni.naor,omer.reingold}@weizmann.ac.il

Abstract. Constructions of k-wise almost independent permutations have been receiving a growing amount of attention in recent years. However, unlike the case of k-wise independent functions, the size of previously constructed families of such permutations is far from optimal. This paper gives a new method for reducing the size of families given by previous constructions. Our method relies on pseudorandom generators for space-bounded computations. In fact, all we need is a generator, that produces "pseudorandom walks" on undirected graphs with a consistent labelling. One such generator is implied by Reingold's log-space algorithm for undirected connectivity [21, 22]. We obtain families of k-wise almost independent permutations, with an optimal description length, up to a constant factor. More precisely, if the distance from uniform for any k tuple should be at most δ, then the size of the description of a permutation in the family is $O(kn + \log \frac{1}{\delta})$.

1 Introduction

In explicit constructions of pseudorandom objects, we are interested in simulating a large random object using a succinct one and would like to capture some essential properties of the former. A natural way to phrase such a requirement is via limited access. Suppose the object that we are interested in simulating is a random function $f : \{0,1\}^n \mapsto \{0,1\}^n$ and we want to come up with a small family of functions G that simulates it. The k-wise independence requirement in this case is that a function g chosen at random from G be completely *indistinguishable* from a function f chosen at random from the set of all functions, for any process that receives the value of either f or g at any k points. We can also relax the requirement and talk about *almost k-wise independence* by requiring that the advantage of a distinguisher be limited by some δ.

Families of functions that are k-wise independent (or almost independent) were constructed and applied extensively in the computer science literature (see [1, 14]).

Suppose now that the object we are interested in constructing is a *permutation*, i.e. a 1-1 function $g : \{0,1\}^n \mapsto \{0,1\}^n$, which is indistinguishable from a random permutation for a process that examines at most k points (a variant also allows examining

* Incumbent of the Judith Kleeman Professorial Chair. Research supported in part by a grant from the Israel Science Foundation.
** Incumbent of the Walter and Elise Haas Career Development Chair. Research supported by US-Israel Binational Science Foundation Grant 2002246.

C. Chekuri et al. (Eds.): APPROX and RANDOM 2005, LNCS 3624, pp. 354–365, 2005.

the inverse). In other words, we are interested in families of permutations such that restricted to k inputs their output is identical (or statistically close, up to distance δ), to that of a random permutation. For $k = 2$ the set of linear permutations ($ax + b$ where $a \neq 0$) over $GF[2^n]$ constitutes such a family. Similarly, there is an algebraic trick when $k = 3$ (Schulman, private communication). For $k > 3$ no explicit (non-trivial) construction is known for k-wise *exactly* independent permutations.

Once we settle on k-wise *almost* dependent permutations, with error parameter δ, then we can hope for permutations with description length $O(kn + \log(\frac{1}{\delta}))^1$; this is what a random (non-explicit) construction gives (see Section 3.2) . There are a number of proposals in the literature of constructing k-wise almost independent permutations (see Section 4), but the description length they obtain is in general significantly higher than this asymptotically optimal value. This paper obtains the first construction of k-wise almost independent permutations, with description length $O(kn + \log(\frac{1}{\delta}))$, for *every* value of k.

Motivation: Given the simplicity of the question, and given how fundamental k-wise independent functions are, we feel that it is well motivated in its own right. Indeed, k-wise independent permutations have been receiving a growing amount of attention with various motivations and applications in mind (e.g. [5]). One motivation for this study, is the relation between k-wise independent permutations and block ciphers [7, 15].

In block-ciphers, modelled by pseudorandom permutations, the distinguisher is not limited by the *number of calls* to the permutations but rather by its computational power. Still, the two notions are related (see [7, 10, 15]).

Our Technique and Main Results: We give a method for "derandomizing" essentially all previous constructions of k-wise almost independent permutations. It is most effective, and easiest to describe for permutation families obtained by composition of simpler permutations. As most previous constructions fall into this category, this is a rather general method. In particular, based on any one of a few previous constructions, we obtain k-wise almost independent permutations with optimal description length, up to a constant factor.

Consider a family of permutations $\mathcal{F}$, with rather small description length s. We denote by $\mathcal{F}^t$ the family of permutations obtained by composing any t permutations $f_1, f_2, \ldots, f_t$ in $\mathcal{F}$. Now assume that $\mathcal{F}^t$ is a family of k-wise almost independent permutations. The description length of $\mathcal{F}^t$ is $t \cdot s$, as we need to describe t independent permutations from $\mathcal{F}$. We will argue that such constructions can be derandomized in the sense that *it is sufficient to consider a subset of the t-tuples of $\mathcal{F}$ functions*. This will naturally reduce the overall description length.

Our first idea uses generators that fool bounded space computations for the task of choosing the subset of $\mathcal{F}^t$. Pseudorandomness for space-bounded computation has been a very productive area, see [16, 17].

Such pseudorandomness has been used before in the context of combinatorial constructions where space is not an *explicit* issue by Indyk [8] and by Sivakumar [26].

---

[1] The lower bound of kn trivially follows as in the case of functions (simply since the output of a random permutation on k fixed inputs has entropy close to kn). If for no other reason, $\log(\frac{1}{\delta})$ bits are needed to reduce roundoff errors. This lower bound also follows for more significant reasons, unless k-wise *exactly* independent permutations can be constructed.

The key observation is, that if the composition of permutations, chosen according to the generator, is not k-wise almost independent, then there is a distinguisher, using only space kn, which contradicts the security of the generator.

Given an "ideal" generator for space bounded computations with optimal parameters we could expect the method above to give k-wise almost independent permutations with description length $O(nk + \log(\frac{1}{\delta}) + s + \log t)$. Based on previous constructions this implies description length $O(nk + \log(\frac{1}{\delta}))$ as desired. However, applying this derandomization method with currently known generators (which are not optimal) implies description length $(nk + \log(\frac{1}{\delta}))$ *times poly-logarithmic factors*.

This leads us to our second idea: to obtain families with description length $O(nk + \log(\frac{1}{\delta}))$ we revise the above method to use a more restricted derandomization tool: we use *pseudorandom generators for walks on undirected labelled graphs*. That is walks which are indistinguishable from a random walk for any 'consistently labelled graph' and sufficient length. Such generators with sufficiently good parameters are implied by the proof that undirected connectivity is in logspace of Reingold [21], and made explicit by Reingold, Trevisan and Vadhan [22].

Adaptive vs. Static Distinguishers: A natural issue is whether the distinguisher, trying to guess whether the permutation it has is random or from the family G has to decide on its k queries ahead of time (statically) or adaptively, as a function of the responses the process receives. Here we shall consider the static case for at least two reasons: (i) static indistinguishability up to distance $\delta 2^{-nk}$ implies adaptive indistinguishability up to distance δ. (ii) A result of Maurer and Pietrzak [12] shows that composing two independently chosen k-wise almost independent permutations in the static case yields k-wise almost independent permutations with adaptive queries of similar parameters.

Related Work: There are several lines of constructions that are of particular relevance to our work. We describe them in more detail in Section 4. The information is summarized in Table 1.

Table 1. Summary of Results and Previous Work on k-wise δ-dependent Permutations

Family	Description Length	Range of Queries
Feistel[2] (Luby Rackoff)	$nk + O(n)$	$k < 2^{\frac{n}{4}-O(1)}, \delta = \frac{k^2}{2^{n/2}}$
	$O(nk \cdot \log \frac{\delta}{\delta_0})$	$k < 2^{\frac{n}{4}-O(1)}$, any δ, $\delta_0 = \frac{k^2}{2^{n/2}}$
Simple 3-Bit Permutations [4, 6, 7]	$O(n^2 k(nk + lg(\frac{1}{\delta})) \lg(n))$	$k \leq 2^n - 2$
Thorp Shuffle [13, 15, 23]	$O(n^{45} k \log(\frac{1}{\delta}))$	$k \leq 2^n$
Non-Explicit Constructions:		
Probabilistic (Thm. 3)	$O(nk + \log(\frac{1}{\delta}))$	$k \leq 2^n$
Sample space existence (Thm. 4)	$O(nk)$	$k \leq 2^n$
This Work (Theorem 10)	$O(nk + \log(\frac{1}{\delta}))$	$k \leq 2^n$

---

[2] The first row is based on 4 rounds with the first and last being pair-wise independent [10, 15]. The second row is obtained by the composition theorem (Theorem 5). Analysis of related constructions [12, 15, 19, 20] allows k that approaches $k = 2^{n/2}$, but does not go beyond.

Organization

In Section 2 we provide notation and some basic information regarding random walks and the spectral gap of graphs. In Section 3 we define k-wise δ-dependent permutations, argue the (non-constructive) existence of small families of such permutations and study the composition of such permutations. In Section 4 we discuss some known families of permutations. Section 5 describes our general construction of a permutation family, and proves our main result. In Section 6 we describe possible extensions for future research.

2 Preliminaries and Notation

- Let P_n be the set of all permutations over $\{0,1\}^n$. We will use $N = 2^n$.
- Let x and y be two bit strings of equal length, then $x \oplus y$ denotes their bit-by-bit exclusive-or.
- For any $f, g \in P_n$ denote by $f \circ g$ their composition (i.e., $f \circ g(x) = f(g(x))$).
- For a set Ω, Denote by $\mathcal{D}(\Omega)$ the set of distributions over Ω. Denote by U_Ω the uniform distribution on the elements of Ω.
- Denote by $[N_k]$ the set of all k-tuples of distinct n-bit strings.

2.1 Random Walks

A random walk on a graph starting at a vertex v is a sequence of vertices, $u_0, u_1, \ldots$ where $u_0 = v$ and for $i > 0$ the vertex u_i is obtained by selecting an edge (u_{i-1}, u_i), uniformly from the edges leaving u_{i-1}. Regular, connected, undirected graphs, with self-loops, have the property that a random walk on the graph (starting at an arbitrary vertex) converges to the uniform distribution on the vertices. The rate of convergence is governed by the second largest (in absolute value) eigenvalue of the graph. Below we formalize these notions.

Definition 1 (Spectral Gap). *Let $G = (V, E)$ be a connected, d-regular undirected graph on n vertices. The normalized adjacency matrix of G is its adjacency matrix divided by d. Denote this matrix by $M \in M_n(\mathbb{R})$. Denote by $1 = \lambda_1 \geq \lambda_2 \geq \ldots \geq \lambda_n$ its eigenvalues. We denote by $\lambda(G)$ the second eigenvalue in absolute value. Namely, $\lambda(G) \doteq \max\{|\lambda_2|, |\lambda_n|\}$. The spectral gap of G, is defined by $gap(G) \doteq 1 - \lambda(G)$.*

Definition 2 (Mixing Time). *Let $G = (V, E)$ be a connected, regular, undirected graph with self-loops, on n vertices. Let $M \in M_n(\mathbb{R})$ be the normalized adjacency matrix of G. A random walk on this graph is an ergodic Markov chain, whose transition matrix is M. Its stationary distribution π is the uniform distribution on the vertices. For $x \in V$, define the mixing time of the walk starting from x, by $\tau_x(\epsilon) = \min\{n \mid \|M^n 1_x - \pi\| \leq \epsilon\}$, where 1_x is the distribution concentrated on x. The mixing time of the walk is defined by $\tau(\epsilon) = \max_{x \in V} \tau_x(\epsilon)$.*

We have the following claims, relating the mixing time of a walk with the spectral gap of the graph.

Claim 1 *[24] Let $G = (V, E)$, M, π be as in Definition 2. Let $\epsilon > 0$. Let λ be the second largest eigenvalue of G. Then*

$$\frac{1}{2}\frac{\lambda}{1-\lambda}\ln(\frac{1}{2\epsilon}) \le \tau(\epsilon) \le \frac{1}{1-\lambda}\ln(\frac{|V|}{\epsilon}).$$

Usually, such a claim is used to bound the mixing time. However, we will be using constructions with a proven mixing time. The construction itself may also provide a bound on the spectral gap. In case it does not, we will be able to use Claim 1, to bound the gap of the graph from below. A simple calculation using Claim 1, shows that

$$gap(G) = \Omega(\frac{\ln(\frac{1}{2\epsilon})}{\tau(\epsilon)}).$$

Another useful Claim is the following.

Claim 2 *[7] For $\ell \ge 1$, $\tau(2^{-\ell-1}) \le \ell \cdot \tau(\frac{1}{4})$.*

3   The Existence of k-Wise δ-Dependent Permutations

In this section we define k-wise δ-dependent permutations, discuss their existence, and show that the distance parameter δ is reduced by the composition of such permutations. For simplicity of presentation this paper concentrates on permutations over bit strings (rather than considering more general domains).

3.1   Definitions

The output of a k-wise almost independent permutation on any k inputs is δ-close to random, where "closeness" is measured by statistical variation distance between distributions.

Definition 3. *Let $n, k \in \mathbb{N}$, and let $\mathcal{F} \subseteq P_n$ be a family of permutations. Let $\delta \ge 0$. The family $\mathcal{F}$ is k-wise δ-dependent if for every k-tuple of distinct elements $(x_1, \ldots, x_k) \in [N_k]$, the distribution $(f(x_1), f(x_2), \ldots, f(x_k))$, for $f \in \mathcal{F}$ chosen uniformly at random is δ-close to $U_{[N_k]}$. We refer to a k-wise 0-dependent family of permutations as k-wise independent.*

We are mostly interested in *explicit* families of permutations, meaning that both sampling uniformly at random from $\mathcal{F}$ and evaluating permutations from $\mathcal{F}$ can be done in polynomial time. The parameters we will be interested in analyzing are the following:

Description Length. The description length of a family $\mathcal{F}$ is the number of random bits, used by the algorithm for sampling permutations uniformly at random from $\mathcal{F}$. Alternatively, we may consider the **size** of $\mathcal{F}$, which is the number of permutations in $\mathcal{F}$, denoted $|\mathcal{F}|$. In all of our applications, the description length of a family $\mathcal{F}$ equals $O(\log(|\mathcal{F}|))$. By allowing $\mathcal{F}$ to be a multi-set we can assume without loss of generality that the description length is exactly $\log(|\mathcal{F}|)$.

Time Complexity. The time complexity of a family $\mathcal{F}$ is the running time of the algorithm for evaluating permutations from $\mathcal{F}$.

Our main goal would be to reduce the *description length* of constructions of k-wise δ-dependent permutations. Still, we would take care to keep time complexity as efficient as possible. See additional discussion in Section 6.

3.2 Non-explicit Constructions

We note the following *non-explicit* families of permutations. Our goal would be to obtain families of size which is as close as possible to that obtained by the non-explicit arguments below. The following theorem follows by a standard application of the probabilistic method.

Theorem 3 (Non-explicit Construction). *Let $n \in \mathbb{N}$. For all $1 \leq k \leq 2^n$ and $\delta > 0$, there exists a k-wise δ-dependent family $\mathcal{F} \subseteq P_n$, of size $|\mathcal{F}| = \frac{2^{(2+\epsilon)nk}}{\delta^2}$, for some constant $0 < \epsilon < 1$.*

The existence (even with a non-explicit construction) of exact k-wise family of permutations is unknown. Nonetheless, using an approach due to Koller and Megiddo [9], we can show that there exists a distribution on permutations, which is k-wise independent and has a small support. This follows, by observing that such a requirement on the permutations defines a set of $\binom{N}{k}^2$ constraints, in the terminology of Koller and Megiddo [9].

Theorem 4 (Existence of k-Wise Independent Distribution). *There exists a distribution on permutations which is k-wise independent (i.e. for any k points the value of the chosen permutation is uniform in $[N_k]$) and the size of the support of the distribution is at most 2^{2nk}.*

3.3 Composition of Permutations

Some of the permutations families we will inspect, require several compositions, to get a distribution close to uniform. In fact, as we argue below, composing permutations is an effective method for reducing the distance parameter δ. This motivates the following definition.

Definition 4. *Let $\mathcal{F} \subseteq P_n$. The tth power of $\mathcal{F}$, denoted by $\mathcal{F}^t \subseteq P_n$, is $\{ f_1 \circ \ldots \circ f_t \mid f_1, \ldots, f_t \in \mathcal{F} \}$.*

Remark 1. Let $\mathcal{F} \subseteq P_n$. Observe that $|\mathcal{F}^t| = |\mathcal{F}|^t$, and that the time complexity of $\mathcal{F}^t$ is essentially t times the time complexity of $\mathcal{F}$.

We now state a composition theorem, that is proven in the full version.

Theorem 5. *1. Let $\mathcal{F}$ be a k-wise δ-dependent family. Then, $\mathcal{F}^2$ is a k-wise $2\delta^2$-dependent family.*

2. Let $\mathcal{F}_1$ and $\mathcal{F}_2$ be k-wise δ_1-dependent and δ_2-dependent families respectively. Then, $\mathcal{F}_1 \circ \mathcal{F}_2$ is a k-wise $2\delta_1\delta_2$-dependent family.

Theorem 5 has the following corollary.

Corollary 1. *Let $\mathcal{F}$ be a k-wise δ-dependent family. Then, for any $\ell \in \mathbb{N}$, $\mathcal{F}^\ell$ is a k-wise $(\frac{1}{2}(2\delta)^\ell)$-dependent family.*

4 Short Survey of Explicit Constructions

We now survey some known constructions yielding k-wise almost independent permutations with reasonable parameters.

4.1 Feistel Based Constructions

In their famed work Luby and Rackoff [10] showed how to construct pseudorandom permutations from pseudorandom functions. The construction is based on the *Feistel Permutation*: For any function $f \in \{0,1\}^{n/2} \mapsto \{0,1\}^{n/2}$ the Feistel Permutation is defined by $(L, R) \mapsto (R, L \oplus f(R))$, where $|L| = |R| = n/2$. The construction uses a composition of several such permutations.

There are Feistel constructions of k-wise δ-dependent permutations, for k up to $2^{n/2}$ (see Naor and Reingold [15], Patarin [18–20], and Maurer and Pietrzak [11]).

Feistel permutations approach yields succinct k-wise δ-dependent permutation as long as k is not too large and δ is not too small, and is probably the method of choice for this range. To reduce the dependency δ one can use Theorem 5 and obtain a permutation with description size $O(kn \log(1/\delta))$ (or even $O(k \log(1/\delta))$ for certain ranges of k and δ). The Feistel method is not known to be useful for k larger than $2^{n/2}$.

4.2 Card Shuffling

Consider a process for shuffling cards. Each round (shuffle) in such a procedure selects a permutation on the locations of the N cards of a deck (selected from some collection of basic permutations). Starting at an arbitrary ordering of the cards, we are interested at how long does it take to get the deck into a (close to) random position. In other words, a card shuffling defines a Markov chain on the state of the deck, and the goal is to bound its mixing time.

An "old" proposal by the second author [23, page 17], [15] for the construction of k-wise almost independent permutations was to utilize "oblivious" card shuffling procedure. Briefly, a shuffle is oblivious if the location of a card, after each round, is easy to trace and is determined by only a few random bits, say $O(1)$. An excellent example is the Thorp Shuffle [27], defined below.

Definition 5 (Thorp Shuffle). *Let $n \in \mathbb{N}$. Given a deck of 2^n cards, one stage of the shuffle is determined by 2^{n-1} bits that we will view as a random function $g : \{0,1\}^{n-1} \mapsto \{0,1\}$. View the location of each card as an n-bit string according to the lexical order. Card at location (σ, x) where $\sigma \in \{0,1\}$ and $x \in \{0,1\}^{n-1}$ moves to location $(x, \sigma \oplus g(x))$.*

Theorem 6. *[13] The mixing time for the Thorp shuffle is $O(n^{44})$.*

It can be seen, that the Thorp Shuffle is oblivious. When using such a card shuffle to construct an k-wise almost independent permutation, all we care for is the final locations of k cards. If we replace the random function g by a k-wise independent function, then this will not change the distribution on the k final locations.

It is also possible to utilize *non*-oblivious shuffles such as riffle shuffle using *range-summable* k-wise independent shuffles; details will be given in the full version.

4.3 Simple 3-Bit Permutations

A very intriguing method for generating k-wise δ-dependent permutation was explored first by Gowers [6] and then (with some variation) by Hoory et al. [7] and Brodsky and Hoory [4]. The idea is to pick a few bit positions (actually 3) and chose a permutation on the resulting small cube. In the Hoory et al. variation only a single bit is changed as a function of the other bits. This is reminiscent of a shuffle, but there is no chance that the shuffle will converge in reasonable time (as we invest too few bits in each shuffle). This approach is treated more formally in the Section 5.4 and it works very well with the derandomized walk approach, since the underlying set of permutations considered is the simplest and hence the description length of simple permutations is quite short. What this line of research shows is that a composition of not too many simple permutations yields a k-wise almost independent permutation.

5 Main Results

In this section we give a method for reducing the description length of previous constructions of k-wise δ-dependent permutations.

5.1 Permutation Families and Random Walks on Graphs

We associate with a family $\mathcal{F}$ of permutations a graph as follows:

Definition 6 (Companion Graph). *Let $\mathcal{F} \subseteq P_n$ be a family of permutations. For $k \in \mathbb{N}$, define the companion (multi-)graph of $\mathcal{F}$, $G_{\mathcal{F},k} = (V, E)$ by:*

- *$V = [N_k]$.*
- *$E = \{ (i, \sigma(i)) \mid i \in [N_k], \sigma \in \mathcal{F} \}$.*
- *Each edge $(i, \sigma(i)) \in E$ is labelled by σ.*

All of our families of permutations of Section 4 closed under taking an inverse of a permutation and always include the identity permutation. We summarize the properties of the companion graph in the following proposition:

Proposition 1. *Let $\mathcal{F} \subseteq P_n$ be a family of permutations, which is closed under taking an inverse and contains the identity permutation. Let $k \in \mathbb{N}$. Then, the companion graph $G_{\mathcal{F},k}$, is an undirected, $|\mathcal{F}|$-regular, consistently labelled graph, with self-loops.*

Remark 2. A consistently labelled graph has the property that for any vertex w, any two incoming edges to w are labelled with different labels.

Assume that $\mathcal{F}$ is such that $\mathcal{F}^t$ is a family of k-wise δ-dependent permutations. This means that the distribution over the vertices we reach, by taking a walk of length t, starting at any vertex of $G_{\mathcal{F},k}$, is δ-close to uniform. Simply, traversing an edge is the same as applying the permutation that is the label of this edge. Taking t random edges is the same as applying the composition of t randomly chosen permutations.

Derandomizing the family $\mathcal{F}^t$ will mean that instead of composing independently chosen permutations from $\mathcal{F}$, we will select the permutations with some dependencies. Equivalently, we will take a pseudorandom walk instead of a random one. We will use a pseudorandom generator, to generate this walk. Such a generator was given by Reingold, Trevisan and Vadhan [21, 22].

5.2 Pseudorandom Walk Generators

We now discuss generators for pseudorandom walks on graphs. We will refer to graphs with the following parameters:

Definition 7 (Parameters for a Graph). *Let $G = (V, E)$ be a connected, undirected d-regular graph, on m vertices. Then G is an (m, d, λ)-graph if $\lambda(G) \leq \lambda$.*

Definition 8 (Pseudorandom Walk). *Let $G = (V, E)$ be a d-regular graph where each node labels its adjacent edges in $[d]$. Let $\mathcal{A}$ be a distribution over*

$$a = a_1, a_2, \ldots a_\ell \in [d]^\ell.$$

We say that $\mathcal{A}$ is δ-pseudorandom for G, if for every $u \in V$, the distribution on the possible end vertices of a walk in G, which starts from u, and follows the edge labels in a is δ-close to uniform when a is distributed according to $\mathcal{A}$.

Note that if G is an (m, d, λ) graph, λ is sufficiently smaller than 1 and the walk is sufficiently long, then we expect a (truly) random walk to end in vertex that is close to being uniformly distributed no matter where the walk started. We are now ready to state the parameters of the best known construction of pseudorandom walk generators.

Theorem 7. *[21, 22][Pseudorandom Walk Generator] For every $m, d \in \mathbb{N}$, $\delta, \epsilon > 0$, there is a pseudorandom walk generator PRG where $PRG_{m,d,\delta,\epsilon} : \{0,1\}^r \to [d]^\ell$, with the following parameters:*

- *Seed length $r = O(\log(md/\epsilon\delta))$.*
- *Walk length $\ell = poly(1/\epsilon) \cdot \log(md/\delta)$.*
- *Computable in space $O(\log(md/\epsilon\delta))$ and time $poly(1/\epsilon, \log(md/\delta))$.*

such that for every consistently labelled $(m, d, 1 - \epsilon)$-graph G, the output of $PRG(U_r)$ is δ-pseudorandom for G, where U_r is the uniform distribution on $\{0,1\}^r$.

5.3 Derandomizing Compositions of Permutation Families

By Proposition 1, the companion graph $G_{\mathcal{F},k}$, is regular and consistently labelled. As argued above, if $\mathcal{F}^t$ (for t not too large) is k-wise almost independent then the random walk on $G_{\mathcal{F},k}$ has small mixing time. By Claim 1, this implies a bound on the eigenvalue gap ε. Therefore, Theorem 7 gives a way to generate a pseudorandom walk for $G_{\mathcal{F},k}$ with $PRG_{m,d,\delta,\epsilon}$ with $m = |[N]_k|$ and $d = |\mathcal{F}|$. The idea is to use each seed $s \in \{0,1\}^r$ of the pseudorandom generator PRG, to define a new permutation σ_s, which is the composition of permutations from $\mathcal{F}$. Theorem 8 formalizes this approach (though we assume for simplicity here the bound on the eigenvalue gap, rather than deducing it by Claim 1 as in the discussion above).

An advantage we have, which affects the parameters of our results (especially the time complexity), is that the efficiency of the generator of [22] depends on the spectral gap of the *initial graph*. Since we are using families of permutations for which the companion graph is known to be of good expansion, we manage to achieve non-trivial parameters in the families we construct.

The following theorem describes the family of permutations we achieve.

Theorem 8. *Let* $\mathcal{F} \subseteq P_n$ *be a family of size* $d = |\mathcal{F}|$, *and* $G_{\mathcal{F},k}$ *be its companion graph. Suppose that* $gap(G_{\mathcal{F},k}) = \epsilon$, *where* ϵ *may be a function of* n *and* k. *Then, there exists* $\mathcal{F}' \subseteq P_n$, *such that* $\mathcal{F}'$ *is a* k-*wise* δ-*dependent family, with following properties.*

- *The description length of* $\mathcal{F}'$ *is* $O(nk + \log(\frac{d}{\epsilon\delta}))$.
- *If the time complexity of any permutation in* $\mathcal{F}$ *is bounded by* $\xi(n,k)$, *then the time complexity of* $\mathcal{F}'$ *is* $poly(1/\epsilon, n, k, \log(\frac{d}{\delta})) \cdot \xi(n,k)$.

Proof. We apply Theorem 7 on the companion graph of $\mathcal{F}$. Following Proposition 1 we know that $G_{\mathcal{F},k}$ fits the requirements there. Let $r = O(\log(\frac{2^{nk} \cdot d}{\epsilon\delta}))$ and $\ell = poly(1/\epsilon) \cdot \log(\frac{2^{nk} \cdot d}{\delta})$ be as in Theorem 7. For a string $s \in \{0,1\}^r$, we define $\sigma_s \in P_n$ as follows. Let $\boldsymbol{w} = PRG_{2^{nk},d,\delta,\epsilon}(s) \in [d]^\ell$. Then $\boldsymbol{w} = \tau_1, \tau_2, \ldots, \tau_\ell$, where for all $1 \le i \le \ell$, $\tau_i \in \mathcal{F}$. We let $\sigma_s = \tau_\ell \circ \ldots \circ \tau_1$.

Next define a permutations family $\mathcal{F}' \subseteq P_n$ by

$$\mathcal{F}' = \{ \sigma_s \mid s \in \{0,1\}^r \}.$$

We now show that $\mathcal{F}'$ is a k-wise δ-dependent family. By Theorem 7, for any starting vertex $u \in V(G_{\mathcal{F},k})$, the pseudorandom walk starting at u and following the labels of $PRG_{2^{nk},d,\delta,\epsilon}(U_r)$ reaches a vertex that is δ-close to uniform. Observe that picking a random $\sigma_s \in \mathcal{F}'$ and applying it to any value $A \in V(G_{\mathcal{F},k}) = [N_k]$ is exactly as taking a random walk on $G_{\mathcal{F},k}$ according to the output of $PRG_{2^{nk},d,\delta,\epsilon}$ with a random seed s. Therefore, the output of a uniform σ_s on any such $A \in [N_k]$, is δ-close to uniform. We can conclude that $\mathcal{F}'$ is k-wise δ-dependent.

The description length of $\mathcal{F}'$ is $|r| = O(\log(\frac{2^{nk}d}{\epsilon\delta})) = O(nk + \log(\frac{d}{\epsilon\delta}))$. The time complexity of $\mathcal{F}'$ depends on the time complexity of running the generator, and of running permutations from $\mathcal{F}$. This can be bounded by $poly(1/\epsilon, n, k, \log(\frac{d}{\delta})) \cdot \xi(n,k)$.

5.4 Particular Derandomization – 3-Bit Permutations

We now provide a formal definition and analysis of simple 3-bit permutations, mentioned in Section 4.3.

Definition 9 (Simple Permutations). *[7] Let* $w \le n$. *For* $i \in [n]$, $J = \{j_1, \ldots, j_w\} \subseteq [n] \smallsetminus \{i\}$, *and a function* $f \in \{0,1\}^w \to \{0,1\}$, *denote by* $\sigma_{i,J,f}$ *the permutation*

$$\sigma_{i,J,f}(x_1, \ldots, x_n) \doteq (x_1, \ldots, x_{i-1}, x_i \oplus f(x_{j_1}, \ldots, x_{j_w}), x_{i+1} \ldots, x_n)$$

The following simple permutations family $\mathcal{F}_w$ *is defined by*

$$\mathcal{F}_w = \{\sigma_{i,J,f} | i \in [n], J \subseteq [n] \smallsetminus \{i\}, |J| = w, f \in \{0,1\}^w \to \{0,1\}\}$$

Theorem 9. *[4] For all* $2 \le k \le 2^n - 2$, $\mathcal{F}_2^t$ *is* k-*wise* δ-*dependent , for* $t = O(n^2k(nk + \log(\frac{1}{\delta})))$. *Furthermore,* $gap(G_{\mathcal{F},k}) = \Omega(\frac{1}{n^2k})$.

Evaluating $\sigma_{i,J,f} \in \mathcal{F}_2$ takes $O(n)$ time. The size of $\mathcal{F}_2$ is $O(n^3)$, and the size of $\mathcal{F}_2^t$ is $O(n^3)^t = n^{O(n^2k(nk+\log(\frac{1}{\delta})))}$. It follows that $\mathcal{F}_2^t$ has description length $O(n^2k(nk + \log(\frac{1}{\delta})) \log(n))$, and time complexity $O(n^3k(nk + \log(\frac{1}{\delta})))$.

Combining Theorems 9 and 8 we obtain the main result of this paper:

Theorem 10. *There exists* $\mathcal{F} \subseteq P_n$, *such that* $\mathcal{F}$ *is* k-*wise* δ-*dependent.* $\mathcal{F}$ *has description length* $O(nk + \log(\frac{1}{\delta}))$, *and time complexity* $poly(n,k)$.

Proof. Apply Theorem 8, with $d = |\mathcal{F}_2| = O(n^3)$, $\epsilon = \Omega(\frac{1}{n^2 k})$ and $\xi(n, k) = O(n)$.

6   Discussion and Further Work

One issue that we have not resolved is coming up with k-wise permutations where the time complexity of evaluating at a given point is small. Note that even for k-wise independent functions this issue is not completely resolved; the basic construction based on polynomials is expensive and some lower and upper bounds are given by Siegel [25]. In general the transformation we propose via the random walks does not preserve the time complexity of evaluating permutations in $\mathcal{F}$: when the composed permutation is stored in its succinct form we do not know how to evaluate it at a given point without first 'decompressing' and representing explicitly as a composition of ℓ permutations in $\mathcal{F}$.

In order to maintain the complexity of evaluation, we need a generator with 'random access' properties. In such a generator, evaluating the ith bit of its output, does not entail computing all bits up to i. The Nisan generator [16] has some aspects of this nature, but is sub-optimal. Also note an advantage of the general space bounded pseudorandomness over the random walk pseudorandomness: the former preserves the number of rounds whereas the latter may increase them.

One interesting question is whether it is possible to 'scale down' a construction for k-wise dependent permutations on n bits to one on $n' \leq n$ bits. This is most relevant in the computational pseudorandomness setting: is it possible to obtain from a block-cipher on large blocks (e.g. 128 bits) a block-cipher on small blocks (e.g. 40 bits), while maintaining the security of the former.

An issue that we did not explore so far is constructing k-wise independent permutations over domains that are not powers of 2. This problem was raised by Bar-Noy and S. Naor inspired by the needs of [2]. Black and Rogaway [3] suggested several methods for obtaining a pseudo-random permutation on domain size M that is not a power of 2 from a pseudo-random permutation on domain size N that is a power of 2 (say $N = 2^{\lceil \log M \rceil}$). The most relevant method for our purposes is the 'cycle walking' one, where the idea is to construct a permutation on $[M]$ elements by iterating a permutation on $[N]$ until it lands in the first M values of $[N]$. It is possible to use this method in the k-wise independent case, but it introduces an additional error. The derandomized walk method is applicable for reducing the error with no significant penalty, since Theorem 8 does not require the domain size to be a power of 2. For details, see the full version.

Finally, there is no strong reason to suppose that explicit small families (or distributions) of *exact* k-wise independent permutation do not exist and Theorem 4 hints to their existence. So how about finding them?

Acknowledgments

The authors are grateful to Ronen Shaltiel for his invaluable collaboration during the early stages of this work and thank Danny Harnik and Adam Smith for useful comments.

References

1. N. Alon and J. Spencer, **The Probabilistic Method**, Wiley, 1992.

2. A. Bar-Noy, J. Naor and B. Schieber, *Pushing Dependent Data in Clients-Providers-Servers Systems*, Wireless Networks 9(5), 2003, pp. 421-430.
3. J. Black and P. Rogaway, *Ciphers with Arbitrary Finite Domains*. Topics in Cryptology - CT-RSA 2002, Lecture Notes in Computer Science, vol. 2271, Springer, 2002, 114–130.
4. A. Brodsky and S. Hoory, *Simple Permutations Mix Even Better*, Arxiv math.CO/0411098.
5. Y. Z. Ding, D. Harnik, A. Rosen and R. Shaltiel, *Constant-Round Oblivious Transfer in the Bounded Storage Model*, First Theory of Cryptography Conference, TCC 2004, LNCS vol. 2951, Springer, pp. 446-472
6. W. T. Gowers, *An almost m-wise independent random permutation of the cube*, Combinatorics, Probability and Computing, vol. 5(2), 1996, pp. 119-130.
7. S. Hoory, A. Magen, S. Myers, and C. Rackoff, *Simple permutations mix well*, The 31st International Colloquium on Automata, Languages and Programming (ICALP), 2004.
8. P. Indyk, *Stable Distributions, Pseudorandom Generators, Embeddings and Data Stream Computation*, FOCS 2000, pp. 189–197.
9. D. Koller and N. Megiddo, Constructing small sample spaces satisfying given constraints, *SIAM J. Discrete Math.* , vol. 7(2), 1994, pp. 260-274.
10. M. Luby and C. Rackoff, *How to construct pseudorandom permutations and pseudorandom functions*, SIAM J. Comput., vol. 17, 1988, pp. 373-386.
11. U. M. Maurer and K. Pietrzak, *The Security of Many-Round Luby-Rackoff Pseudo-Random Permutations*, EUROCRPYT 2003, LNCS vol. 2656, Springer, pp. 544–561.
12. U. M. Maurer and K. Pietrzak, *Composition of Random Systems: When Two Weak Make One Strong*, First Theory of Cryptography Conference, TCC 2004, LNCS vol. 2951, Springer, pp. 410–427.
13. B. Morris, *On the mixing time for the Thorp shuffle*, STOC 2005, pp. 403–412.
14. R. Motwani and P. Raghavan, **Randomized Algorithms**, Cambridge University Press, New York (NY), 1995.
15. M. Naor and O. Reingold, *On the Construction of Pseudorandom Permutations: Luby-Rackoff Revisited*, J. of Cryptology, vol. 12(1), Springer-Verlag, 1999, pp. 29-66.
16. N. Nisan, *Pseudorandom generators for space-bounded computation*, Combinatorica 12(4), 1992, 449–461.
17. N. Nisan and D. Zuckerman, *Randomness is Linear in Space*, J. Comput. Syst. Sci. vol. 52(1), 1996, pp. 43–52.
18. J. Patarin, *Improved security bounds for pseudorandom permutations*, 4th ACM Conference on Computer and Communications Security, 1997, pp. 142–150.
19. J. Patarin, *Luby-Rackoff: 7 Rounds Are Enough for $2^{n(1-epsilon)}$) Security*, CRYPTO 2003: pp. 513-529.
20. J. Patarin *Security of Random Feistel Schemes with 5 or More Rounds*, CRYPTO 2004, pp. 106–122.
21. O. Reingold, *Undirected ST-Connectibvity in Log-Space*, STOC 2005, pp. 376-385.
22. O. Reingold, L. Trevisan, S. Vadhan, *Pseudorandom Walks in Biregular Graphs and the RL vs. L Problem*, ECCC, TR05-22, February 2005.
23. S. Rudich, *Limits on the provable consequences of one-way functions*, PhD Thesis, U. C. Berkeley.
24. A. Sinclair, *Improved bounds for mixing rates of Markov chains and multicommodity flow*, Combinatorics, Probability and Computing, vol. 1(4), 1992, pp. 351-370.
25. A. Siegel, *On Universal Classes of Extremely Random Constant-Time Hash Functions*, SIAM Journal on Computing 33(3), 2004, pp. 505–543.
26. D. Sivakumar, *Algorithmic derandomization via complexity theory*, STOC 2002, pp. 619-626
27. E. Thorp, *Nonrandom shuffling with applications to the game of Faro*, Journal of the American Statistical Association, vol. 68, 1973, pp. 842-847.

Testing Periodicity*

Oded Lachish and Ilan Newman

Haifa University, Haifa, 31905 Israel
{loded,ilan}@cs.haifa.ac.il

Abstract. A string $\alpha \in \Sigma^n$ is called *p-periodic*, if for every $i, j \in \{1, \ldots, n\}$, such that $i \equiv j \bmod p$, $\alpha_i = \alpha_j$, where α_i is the i-th place of α. A string $\alpha \in \Sigma^n$ is said to be *period*($\leq g$), if there exists $p \in \{1, \ldots, g\}$ such that α is *p-periodic*.

An ϵ-property tester for *period*($\leq g$) is a randomized algorithm, that for an input α distinguishes between the case that α is in *period*($\leq g$) and the case that one needs to change at least ϵ-fraction of the letters of α, so that it will become *period*($\leq g$). The complexity of the tester is the number of letter-queries it makes to the input. We study here the complexity of ϵ-testers for *period*($\leq g$) when g varies in the range $1, \ldots, \frac{n}{2}$. We show that there exists a surprising exponential phase transition in the query complexity around $g = \log n$. That is, for every $\delta > 0$ and for each g, such that $g \geq (\log n)^{1+\delta}$, the number of queries required and sufficient for testing *period*($\leq g$) is polynomial in g. On the other hand, for each $g \leq \frac{\log n}{4}$, the number of queries required and sufficient for testing *period*($\leq g$) is only poly-logarithmic in g.

We also prove an exact asymptotic bound for testing general periodicity. Namely, that 1-sided error, non adaptive ϵ-testing of periodicity (*period*($\leq \frac{n}{2}$)) is $\Theta(\sqrt{n \log n})$ queries.

1 Introduction

Periodicity in strings plays an important role in several branches of CS and engineering applications. It is being used as a measure of 'self similarity' in many application regarding string algorithms (e.g. pattern matching), computational biology, data analysis and planning (e.g. analysis of stock prices, communication patterns etc.), signal and image processing and others. On the other hand, sources of very large streams of data are now common inputs for strategy-planning or trend detection algorithms. Typically, such streams of data are either too large to store entirely in the computer memory, or so large that even linear processing time is not feasible. Thus it would be of interest to develop very fast (sub-linear time) algorithms that test whether a long sequence is periodic or approximately periodic, and in particular, that test if it has a very short period. This calls for algorithms in the framework of Combinatorial Property Testing [1]. In this framework, introduced initially by Rubinfeld and Sudan [2]

* This research was supported by THE ISRAEL SCIENCE FOUNDATION (grant number 55/03).

C. Chekuri et al. (Eds.): APPROX and RANDOM 2005, LNCS 3624, pp. 366–377, 2005.

and formalized by Goldreich et al. [1], one uses a randomized algorithm that queries the input at very few locations and based on this, decides whether it has a given property or it is 'far' from having the property. Indeed related questions to periodicity have already been investigated [3–5], although the focus here is somewhat different.

In [6], the authors constructs an algorithm that approximates in a certain sense the DFT (Discrete Fourier Transform) of a finite sequence in sub linear time. This is quite related but not equivalent to testing how close is a sequence to being periodic. In [3], the authors study some alternative parametric definitions of periodicity that intend to 'capture the distance' of a sequence to being periodic. They mainly relate the different definitions of periodicity. They also show that there is a tolerant tester for periodicity. That is, they show a simple algorithm, that given $0 \leq \epsilon_1 < \epsilon_2 \leq 1$, decides whether a sequence is ϵ_1-close to periodic or ϵ_2-far from being periodic, using $O(\sqrt{n} \cdot poly(\log n))$ queries.

There are other works on sequences sketching [4] etc. but none of those seems to address directly periodicity testing.

The property of being periodic is formalized here in a very general form: Let Σ be a finite alphabet. A string $\alpha \in \Sigma^n$ is said to be p-periodic for an integer $p \in [n]$, if for every $i, j \in [n]$, $i \equiv j (\mathrm{mod} p)$, $\alpha_i = \alpha_j$, where α_i is the i-th character of α. We say that α is in $period(\leq g)$, where $g \in [\frac{n}{2}]$, if it is p-periodic for some $p \leq g$. We say that a string is periodic if it is in $period(\leq \frac{n}{2})$.

We study here the complexity of ϵ-testing the property $period(\leq g)$ when g varies in the range $1, \ldots, \frac{n}{2}$. An ϵ-tester for $period(\leq g)$ is a randomized algorithm, that for an input $\alpha \in \Sigma^n$ distinguishes between the case that α is in $period(\leq g)$ and the case that one needs to change at least an ϵ-fraction of the letters of α, so that it will be in $period(\leq g)$. The complexity of the tester is the number of letter-queries it makes to the input.

We show that there exists a surprising exponential phase transition in the complexity of testing $period(\leq g)$ around $g = \log n$. That is, for every $\delta > 0$ and for each g, such that $g \geq (\log n)^{1+\delta}$, the number of queries required and sufficient for testing $period(\leq g)$ is polynomial in g. On the other hand, for each $g \leq \frac{\log n}{4}$, the number of queries required and sufficient for testing $period(\leq g)$ is only poly-logarithmic in g. We also settle the exact complexity of non-adaptive 1-sided error test for general periodicity (that is, $period(\leq n/2)$). We show that the exact complexity in this case is $\theta(\sqrt{n \log n})$. The upper bound that we prove is an improvement over the result of [3] and uses a construction of a small random set $A \subseteq [n]$ of size $\sqrt{n \log n}$ for which the multi-set $A - A = \{a - b | a, b \in A\}$ contains at least $\log n$ copies of each member of $[n/2]$. This can be trivially done with a set A of size $\sqrt{n} \log n$. We improve on the previous bound by giving up the full independence of the samples, but still retain the property that the probability of the $\log n$ copies of each number in $A - A$ behave as if being not too far from independent. A similar problem has occurred in various other situation, (e.g. [7, 8]) and thus could be interesting in its own.

The rest of the paper is organized as follows. In section 2 we introduce the necessary notations and some very basic observations. Section 3 contains an ϵ-

368   Oded Lachish and Ilan Newman

test for $period(\leq g)$ that uses $\theta(\sqrt{g \log g})$ queries. In section 4 we construct an ϵ-test for $period(\leq g)$, $g \leq \frac{\log n}{4}$ that uses only $\tilde{O}((\log g)^6)$ queries. Sections 5 and 6 contain the corresponding lower bounds, thus showing the claimed phase transition. Finally, the special case of $g = \theta(n)$ is treated in Section 7 which contains a proof that any 1-sided error non adaptive tester for periodicity ($period(\leq \frac{n}{2})$) requires $\Omega(\sqrt{n \log n})$ queries.

2   Preliminaries

In the following Σ is a fixed size alphabet that contains $0, 1$. For a string $\alpha \in \Sigma^n$ and an integer $i \in [n]$, ($[n] = \{1, \ldots, n\}$) we denote by α_i the i-th symbol of α, that is $\alpha = \alpha_1 \ldots \alpha_n$. Given a set $S \subseteq [n]$ such that $S = \{i_1, i_2, \ldots, i_m\}$ and $i_1 < i_2 < \ldots < i_m$ we define $\alpha_S = \alpha_{i_1} \alpha_{i_2} \ldots \alpha_{i_m}$. In the following Σ will be fixed. For short unless otherwise stated, all strings for which the length is not specified are in Σ^n.

For two strings α, β we denote by $dist(\alpha, \beta)$ the Hamming distance between α and β. Namely, $dist(\alpha, \beta) = |\{i| \ \alpha_i \neq \beta_i\}|$. For a property $\mathcal{P} \subseteq \Sigma^n$ and a string $\alpha \in \Sigma^n$, $dist(\alpha, \mathcal{P}) = min\{dist(\alpha, \beta)| \ \beta \in \mathcal{P}\}$ denotes the distance from α to $\mathcal{P}$. We say that α is ϵ-far from $\mathcal{P}$ if $dist(\alpha, \mathcal{P}) \geq \epsilon n$, otherwise we say that α is ϵ-close to $\mathcal{P}$.

Definition 1. *For a string α and a subset $S \subseteq [n]$ we say that α_S is homogeneous if for all $i, j \in S$, $\alpha_i = \alpha_j$.*

Property Testing

The type of algorithms that we consider here are 'property-testers' [1, 9, 10]. An ϵ-test is a randomized algorithms that accesses the input string via a 'location-oracle' which it can query: A query is done by specifying one place in the string to which the answer is the value of the string in the queried location. The complexity of the algorithm is the amount of queries it makes in the worst case. Such an algorithm is said to be an ϵ-test for a property $\mathcal{P} \subseteq \Sigma^n$ if it distinguishes with success probability at least $2/3$ between the case that the input string belongs to $\mathcal{P}$ and the case that it is ϵ-far from $\mathcal{P}$.

Periodicity

Definition 2. *A string $\alpha \in \Sigma^n$ has period p (denoted p-periodic), if $\alpha_i = \alpha_j$ for every $i, j \in [n]$ such that $i \equiv j \bmod p$.*

Note that a string is homogeneous if and only if it is *1-periodic*.

Definition 3. *A string has the property $period(\leq g)$ if it is p-periodic for some $p \leq g$. We say that it is periodic if it has the property $period(\leq \frac{n}{2})$.*

Definition 4. *A witness that a string α is not p-periodic, denoted as p-witness, is an unordered pair $\{i, j\} \subseteq [n]$ such that $i \equiv j (\bmod p)$ and $\alpha_i \neq \alpha_j$.*

According to the definition of periodicity a string has a period $p \leq \frac{n}{2}$ if and only if there does not exist a *p-witness*.

In the same manner a witness for not having $period(\leq g)$ is defined as follows:

Definition 5. *A witness that a string is not in period($\leq g$) is a set of integers $Q \subseteq [n]$ such that for every $p \leq g$ there are two integers $i, j \in Q$ that form a p-witness.*

Fact 1. *A string has period($\leq g$) if and only if there does not exist a witness that the string is not in period($\leq g$).*

Fact 2. *If a string $\alpha \in \Sigma^n$ does not have a period in $(\frac{g}{2}, g]$, where $g \leq \frac{n}{2}$, then it has no period of at most $\frac{g}{2}$. If a string $\alpha \in \Sigma^n$ is ϵ-far from having a period in $(\frac{g}{2}, g]$, where $g \leq \frac{n}{2}$, then it is ϵ-far from having the property period($\leq g$).*

Proof. Observe that if a string $\alpha \in \Sigma^n$ has period $p \leq \frac{g}{2}$ then it also has period q for every q that is a multiple of p. Note also that there must exist such $q \in (\frac{g}{2}, g]$. □

Definition 6. *For $\alpha \in \Sigma^n$, $p \in [n]$ and $0 \leq i \leq p-1$, let $Z(p, i) = \{j|\ j \equiv i(mod\ p)\}$. We call $\alpha_{Z(p,i)}$ the i-th p-section of α.*

The following obvious fact relates the distance of a string to *p-periodic* and the homogeneity of its *p-sections*.

Fact 3. *For each α, $dist(\alpha, p - periodic) = \Sigma_{i=0}^{p-1} dist(\alpha_{Z(p,i)}, homogeneous)$.*

In further sections we use the following basic ϵ-test for *p-period*.

Algorithm p-Test

Input: a string $\alpha \in \Sigma^n$, the string length n, a period $p \in [\frac{n}{2}]$ and a distance parameter $0 < \epsilon < 1$;

1. Select $\frac{1}{\epsilon}$ random unordered pairs (with repetitions) $\{i, j\} \subset [n]$ such that $i \equiv j(mod p)$.
2. Reject if one of the selected pairs is a *p-witness* (namely a witness for being *non-p-periodic*). Otherwise accept.

Proposition 1. *Algorithm p-test is a 1-sided error, non-adaptive ϵ-test for p-periodic. Its query complexity is $\frac{2}{\epsilon}$.*

Proof. The query complexity is obvious. The test is 1-sided error since it rejects only if it finds a *p-witness*. To estimate its error probability let α be such that $dist(\alpha, p - periodic) \geq \epsilon n$.

Assume that p divides n (the general case is essentially the same). With this assumption $|Z(p, i)| = n/p = m$. For every $0 \leq i \leq p-1$ set $dist(\alpha_{Z(p,i)}, homogeneous) = d_i$. For fixed i and every $\sigma \in \Sigma$ let n_σ be the number of occurrences of σ in $Z(p, i)$. Assume also that we have renumbered the letters in Σ so that $n_1 \geq n_2, \ldots \geq n_k$. Then $d_i = n - n_1$ as it is easy to see that the closest homogeneous string is obtained by changing all letters different from σ_1 to σ_1 (exactly $m - n_1$ letters). Hence the number of *p-witnesses* in $Z(p, i)$,

W_i, is $W_i = \frac{1}{2}(\Sigma n_j)^2 - \Sigma n_j^2$. It is easy to see that $W_i \geq m \cdot d_i/2$. Now, according to Fact 3 we get that,

$$dist(\alpha, p - periodic) = \Sigma_{i=0}^{p-1} dist(\alpha_{Z(p,i)}, homogeneous) =$$

$$= \Sigma_{i=0}^{p-1} d_i \leq \frac{2}{m}\Sigma W_i \leq \frac{2}{m}W$$

where W is the total number of p-witnesses.

It follows that $W \geq \frac{m}{2}dist(\alpha, p - periodic) \geq \frac{\epsilon n^2}{2p}$. Note however that the total number of unordered pairs i, j such that $j \equiv i(mod\ p)$ is $\frac{n^2}{2p}$. Thus we conclude that the probability that a random such pair is a p-witness is at least ϵ and the result follows. □

We will need the following proposition in a number of lower bound proofs. Let $m < n/8$, $S \subset [n]$, $|S| = m$ and a be an assignment $a : S \longrightarrow \{0,1\}$. We define the following distribution U/S on strings of length n. We choose every letter $\alpha_i = a(i)$ if $i \in S$. For all places $i \notin S$ we choose α_i to be '1' with probability $1/2$ and '0' with probability $1/2$, independently between different i's. Note that U/S is just the uniform distribution on all binary strings conditioned on the event that the projection on S is a. We have,

Proposition 2. *Let m, S, a be as above and let $\mathcal{G}(g)$ be the event that a string selected according to U/S is $\frac{1}{16}$-far from having $period(\leq g)$. Then $Prob_{U/S}(\mathcal{G}(g)) \geq 1 - \frac{1}{n}$.*

Proof: We present the proof for $g = n/2$ this immedialty implies the result for every $g \leq n/2$ by Fact 2. Let U be the uniform distribution over Σ^n. Again according to Fact 2, it is enough to prove that the probability that a string selected according to U/S is $\frac{1}{16}$-far from having a period in $\left(\frac{n}{4}, \frac{n}{2}\right]$, is at least $1 - \frac{1}{n}$. We prove that for each $p \in \left(\frac{n}{4}, \frac{n}{2}\right]$ the probability that a string selected according to U, is $\frac{1}{16}$-close to being p-periodic, is at most $\frac{4}{n^2}$, and therefore by the union bound we are done.

Indeed let $p \in \left(\frac{n}{4}, \frac{n}{2}\right]$, and let α be a string selected according to U/S. Since $p \in \left(\frac{n}{4}, \frac{n}{2}\right]$ there are at least $\frac{n}{4}$ p-sections. Because $m \leq \frac{n}{8}$ and the size of each p-section is at least 2, at least $\frac{3n}{16}$ of these p-sections contain at least 1 location not in S. We call such a location a *free location*, and define B to be the set of all integers x such that the x p-section contains a free location.

Then for every $i \in B$, $Prob_{U/S}(dist(\alpha_{Z(p,i)}, homogeneous) \geq 1) \geq \frac{1}{2}$ (the *free location* is different than some other location in the p-sections). Thus, since the p-sections are mutually disjoint and $|B| \geq \frac{3n}{16}$, Chernoff bound implies that, $Prob_{U/S}\left(\Sigma_{i \in B} dist(\alpha_{Z(p,i)}, homogeneous) \leq \frac{n}{16}\right) \leq e^{-\frac{n}{24}} \leq \frac{4}{n^2}$. □

3   Upper Bound for Testing $period(\leq g)$

In this section we construct a 1-sided error, non adaptive ϵ-test for $period(\leq g)$ that uses $O(\sqrt{g \log g}/\epsilon^2)$ queries. The construction consists of two stages. We first show a 1-sided error, non adaptive algorithm, PT, for testing periodicity ($period(\leq \frac{n}{2})$), that uses $O(\frac{\sqrt{n \log n}}{\epsilon^2})$ queries. Then we show how to use algorithm PT in order to construct the claimed algorithm for testing $period(\leq g)$.

The intuition behind the algorithm PT is simple. We first explain it by hinting to a test for $period(\leq \frac{n}{2})$ that uses $O(\frac{\sqrt{n \log n}}{\epsilon^2})$ queries. By Fact 2 α is ϵ-far from $period(\leq n)$ if and only if α is ϵ-far from being p-periodic for every $p \in (\frac{n}{4}, \frac{n}{2}]$. We show that if α is ϵ-far from $period(\leq \frac{n}{2})$, we can find a p-witness for each $p \in (\frac{n}{4}, \frac{n}{2}]$ with probability at least $\frac{1}{n}$. This is sufficient since by the union bound we will be done.

Suppose that we can construct a random set $Q \subseteq [n]$ of size $O(\sqrt{n} \log n)$, such that Q contains at least $\log n$ independent potential p-witnesses for each $p \in (\frac{n}{4}, \frac{n}{2}]$. Then obviously we achieve the above goal. Indeed such a set Q can be constructed e.g. by choosing Q uniformly between all sets of the required cardinality.

To reduce the size of Q, we construct such a random set Q of size $\sqrt{n} \log n$ with the above properties while giving up the independence between the $\log n$ pairs for each p. In turn we will have to show that the probability that none will be a p-witness is still low enough.

We use the following definition.

Definition 7. *Let $\ell = \sqrt{n \log n}$ and let $J = \{I_i\}_{i=1}^{\ell}$ be the set of pairwise disjoint intervals $I_i = \left((i-1)\sqrt{\frac{n}{\log n}}, i\sqrt{\frac{n}{\log n}}\right]$.*

We now present the ϵ-test for $period(\leq \frac{n}{2})$.

Algorithm PT

Input: a string $\alpha \in \Sigma^n$, the string length n and a distance parameter $0 < \epsilon < 1$;
Let ℓ, J be as in Definition 7.

1. Select a set of integers $T = \{t_1, \ldots, t_\ell\}$ by choosing one random point from each $I \in J$.
2. Repeat the following independently for $m = \frac{2^{10} \cdot \log n}{\epsilon^2}$ times: uniformly select an interval $I \in J$. Let J^* be the set of intervals that where selected and let $H = \cup_{I \in J^*} I$.
3. Reject if for every $p \in (\frac{n}{4}, \frac{n}{2}]$ the set $H \cup T$ contains a p-witness. Otherwise accept.

Theorem 1. *Algorithm PT is a 1-sided error, non-adaptive ϵ-test for periodicity. Its query complexity is $O(\sqrt{n \log n}/\epsilon^2)$.*

The proof is in omitted.

Theorem 2. *For any $g \leq \frac{n}{2}$ there is a 1-sided error, non-adaptive ϵ-test for $period(\leq g)$. Its query complexity is $O(\sqrt{g \log g}/\epsilon^2)$.*

Proof Idea: Let $\alpha \in \Sigma^n$. We think of α as being composed of $\frac{n}{2g}$ pieces of length $n' = 2g$ each. We now run the ϵ-test for periodicity for strings of length n' and for each query $q \in [n']$ we query q in one the pieces that is chosen randomly and independently for each query. We avoid further details here.   □

4   Upper Bound for Testing $period(\leq g)$, Where $g \leq \frac{\log n}{4}$

Let $g \leq \frac{\log n}{4}$, we describe an algorithm for testing whether a string has $period(\leq g)$ that has query complexity $poly(\log g)$. The technicalities are somewhat involved, however, the intuition is simple and motivated by the following reasoning: Our goal is to find a witness for not being r-*periodic* for every $r \in [g]$. One thing that we could do easily is to check whether α is q-*periodic* where q is the product of all numbers in $[g]$ (this number would be a actually too big, but for understanding the following intuition this should suffice). Now if α is not q-*periodic* then it is certainly not in $period(\leq g)$ and we are done. Yet as far as we know α may be far from having $period(\leq g)$ but is q-*periodic*. Our goal is to show that if α is far from p-*periodic*, where $p \leq g$, then there exists $q' >> g$, such that, α is far from q'-*periodic* and p divides q'. Indeed it follows form Lemma 3 that if α is far form p then there exists such a q' as above. This together with the following concept enables us to construct an ϵ-*test*.

Definition 8.   *For integers $n, g \in [n]$, a gcd-cover of $[g]$ is a set $E \subseteq [n]$ such that for every $\ell \leq g$, there exists a subset $I \subseteq E$, that satisfies $\ell = \gcd(I)$.*

The importance of a *gcd-cover of $[g]$* is the following: We prove that if $\ell = gcd(I)$ for an integer ℓ and a subset I, then, the assumption that α is far from being ℓ-*periodic* implies that it is also sufficiently far from being t-*periodic* for some $t \in I$. Using this, let E be a *gcd-cover of $[g]$*. Then we are going to query for each $t \in E$ enough pairs $i, j \in [n]$ such that $i \equiv j \pmod{t}$. We say that such pairs i, j cover t. Now for each $\ell \leq g$ there is a set $I_\ell \subseteq E$ such that $\ell = gcd(I_\ell)$. The pairs of queries that cover I_ℓ cover ℓ. Thus if α is 'far' from being ℓ-*periodic* then, as $gcd(I_\ell) = \ell$, there exists a period t, $t \in I_\ell$, such that the string is far from t-*periodic*. We thus expect that this t will distinguish between strings that are ℓ-*periodic* and those that are $k\ell$-*periodic* but far from being ℓ-*periodic*.

As the complexity of the test will depend crucially on the size of the *gcd-cover*, we need to guarantee the existence of a small one. This is done in the following Lemma, which also brings in an additional technical requirement.

Lemma 1.   *For every integers n and $g \in \left[\frac{\log n}{4}\right]$, there exists a gcd-cover of $[g]$, $E \subseteq [\sqrt{n}]$, of size $O((\log g)^3)$.*

The proof of Lemma 1 is omitted.

A *gcd-cover of $[g]$* as in the Lemma is called an *efficient gcd-cover of $[g]$*.
We next describe our algorithm.

Algorithm SPT

Input: a string $\alpha \in \Sigma^n$, a distance parameter $0 < \epsilon < 1$ and a threshold period $g \le \frac{\log n}{4}$;

1. Set E to be an *efficient gcd-cover of $[g]$*.
2. Let $M = \frac{|E| \cdot \log(8|E|)}{2\epsilon}$. For each $\ell \in E$ select M pairs of integers uniformly and with repetitions from the set $\{\{x, y\} \mid x \equiv y \bmod \ell \text{ and } x, y \in [n]\}$. Let Q be the resulting (multi)set.
3. Reject if for each $\ell \in [g]$ the set Q contains a witness that α is not ℓ-*periodic*. Otherwise accept.

Theorem 3. *Algorithm SPT is a 1-sided error, non-adaptive ϵ-test for period($\le g$), where $g \le \frac{\log n}{4}$. Its query complexity is $\tilde{O}\left(\frac{(\log g)^6}{\epsilon}\right)$.*

Proof: Since for every member of E we used M queries and $|E| = O((\log g)^3)$ the estimate on the query complexity follows. The fact that the algorithm never rejects a ℓ-*periodic* string, for $\ell \le g$ is immediate, as the algorithm rejects only when it finds a witness that the input string is not in *period($\le g$)*.

In order to compute the success probability of algorithm *SPT* we first need the following definition.

Definition 9. *A string $\alpha \in \Sigma^n$ is called ϵ-bad if for every $\ell \le g$ and every $S \subseteq E$, where $\ell = gcd(S)$, there exists $s \in S$ for which α is $\frac{\epsilon}{2|E|}$-far from being s-periodic.*

Note: if α is ϵ-*bad* and s, ℓ are as in the definition, then a witness for it showing that it is not s-*periodic* is also a witness that it is not ℓ-*periodic*.

The proof of the Theorem now follows from Lemma 2 and Lemma 3.   □

Lemma 2. *Algorithm SPT rejects every α that is ϵ-bad with probability at least $\frac{3}{4}$.*

Proof. Let $\alpha \in \Sigma^n$ be ϵ-*bad*. Let $B \subseteq E$ be the set that contains every $b \in E$ such that α is $\frac{\epsilon}{2|E|}$-*far* from being b-*periodic*. By Definition 9, for every $\ell \le g$ there exists a $b \in B$ such that $\ell \mid b$, and hence if we find a b-*witness* for each $b \in B$ - this is also a witness that α is not in *period($\le g$)*. Thus, it is enough to prove that for each $b \in B$, the probability that there is no b-*witness* in Q is at most $\frac{1}{8|E|}$, as then by the union bound we are done.

Let $\delta = \frac{\epsilon}{2|E|}$ and let b be a member of B, namely α is δ-*far* from being b-*periodic*. By Corollary 1, for a random $x, y \in [n]$, such that $x \equiv y(\bmod b)$ we have that $Prob(\alpha_x \ne \alpha_y) \ge \delta$. Since M random pairs are being queried for each $\ell \in E$ and in particular for b, the probability that a b-*witness* is not found is at most $(1 - \delta)^M \le \frac{1}{(8|E|)}$.   □

Lemma 3. *If a string $\alpha \in \Sigma^n$ is ϵ-far from having short period($\le g$), where $g \le (\log n)/4$ then it is a ϵ-bad string.*

The proof of Lemma 3 is omitted.

5 Lower Bound for Testing *period*(≤ *g*)

In this section we prove that every $g \leq n/2$ every adaptive, 2-sided error, $\frac{1}{32}$-test for $period(\leq g)$ uses $\Omega(\sqrt{g/(\log g \cdot \log n)})$ queries. We prove this for $\Sigma = \{0,1\}$. This implies the same bound for every alphabet that contains at least two symbols.

Theorem 4. *Any adaptive 2-sided, error $\frac{1}{32}$-test for $period(\leq g)$, uses $\Omega(\sqrt{g/(\log g \cdot \log n)})$ queries.*

Proof. Fix $g \leq n/2$. We prove the theorem by using Yao's principle. That is, we construct a distribution $\mathcal{D}$ over legitimate instances (strings that are either in $period(\leq g)$, or strings that are $\frac{1}{32}$-far from $period(\leq g)$) and prove that any adaptive deterministic tester, that uses $m = o(\sqrt{g/(\log g \cdot \log n)})$ queries, gives an incorrect answer with probability greater than $\frac{1}{3}$.

In order to define $\mathcal{D}$ we use auxiliary distributions $\mathcal{D}_P, \mathcal{D}_N$ and the following notation. Let $Primes = \{p \mid p \text{ is a prime such that } p \leq g\}$. According to the *Prime Number Theorem* [11–13], $|Primes| = \theta(\frac{g}{\log g})$.

We now define distributions $\mathcal{D}_P, \mathcal{D}_N$.

- $\mathcal{D}_N$ is simply the uniform distribution over Σ^n.
- An instance α of length n is selected according to distribution $\mathcal{D}_P$ as follows. Uniformly select $p \in Primes$, then uniformly select $\omega \in \Sigma^p$ and finally set α to be the concatenation of ω to itself enough times until a total length of n (possibly concatinating a prefix of ω at the end if p does not divide n).

We next define distribution $\mathcal{D}$. Let $\mathcal{G}$ be the event that α is $\frac{1}{16}$-far from having $period(\leq g)$. Let $\mathcal{D}_{N/\mathcal{G}}$ be the distribution D_N given that event $\mathcal{G}$ is true. A string α is selected according to distribution $\mathcal{D}$ by choosing one of the distributions $\mathcal{D}_P, \mathcal{D}_{N/\mathcal{G}}$ with equal probability and then selecting α according to the distribution chosen. Namely $\mathcal{D} = \frac{1}{2}\mathcal{D}_P + \frac{1}{2}\mathcal{D}_{N/\mathcal{G}}$.

We don't work directly with $\mathcal{D}$ but rather with a simpler distribution $\mathcal{D}'$ which approximates $\mathcal{D}$ well enough and is defined as follows $\mathcal{D}' = \frac{1}{2}\mathcal{D}_P + \frac{1}{2}\mathcal{D}_N$. Let $\mathcal{B}$ be the event that the tester gives an incorrect answer. We prove that $Prob_{\mathcal{D}'}(\mathcal{B}) \geq \frac{2}{5}$. This is indeed sufficient as $Prob_{\mathcal{D}}(\mathcal{B}) \geq Prob_{\mathcal{D}'}(\mathcal{B}) - Prob_{\mathcal{D}_N}(\overline{\mathcal{G}})$. Using Proposition 2, $Prob_{\mathcal{D}}(\mathcal{B}) \geq \frac{1-o(1)}{2} - \frac{1}{g} > \frac{1}{3}$.

We assume with out loss of generality that for any string of length n the tester uses the same number of queries. Hence we can view the tester as a full binary decision tree of depth m which is labeled as follows. Each node of the tree represents a query location, for each internal node one of the outgoing edges is labeled by 1 and the other by 0, where $0, 1$ represent the answers to the query, and each leaf is labeled either by "accept" or by "reject" according to the decision of the algorithm.

For each leaf l in the tree we associate a pair Q_l, f_l, where Q_l is the set of queries on the path from the root to the leaf l and $f_l : Q_l \longrightarrow \{0,1\}$ is a mapping between each query and its answer, that is the labellings on the edges of the path from the root to the leaf l. Let L be the set of all leaves, L_0 be the set of all leaves that are labeled by "reject" and let L_1 be the set of all leaves that are labeled by "accept".

Let $h_\ell : \Sigma^n \longrightarrow \{0,1\}$ be a function that is 1 only on strings $\alpha \in \Sigma^n$ such that for every $q \in Q_\ell$ we have $\alpha_q = f_\ell(q)$. That is $h_\ell(\alpha) = 1$ if and only if α is consistent with Q_ℓ, f_ℓ. Let $far : \Sigma^n \longrightarrow \{0,1\}$ be a function that is 1 only on strings $\alpha \in \Sigma^n$ such that are $\frac{1}{16}$-far from $period(\leq g)$). Thus the $\text{Prob}_{\mathcal{D}'}(\mathcal{B})$ is at least

$$\frac{1}{2}\Sigma_{\ell \in L_0} prob_{\mathcal{D}_P}[h_\ell(\alpha) = 1] + \frac{1}{2}\Sigma_{\ell \in L_1} prob_{\mathcal{D}_N}[far(\alpha) = 1 \bigwedge h_\ell(\alpha) = 1] \quad (1)$$

We will prove that for each $\ell \in L_0$, $prob_{\mathcal{D}_P}[h_\ell(\alpha) = 1] \geq \frac{1-o(1)}{2^m}$, and for each $\ell \in L_1$, $prob_{\mathcal{D}_N}[far(\alpha) = 1 \bigwedge h_\ell(\alpha) = 1] \geq \frac{1-o(1)}{\cdot 2^m}$. This implies that $\text{Prob}_{\mathcal{D}'}(\mathcal{B}) \geq \frac{1}{2}\Sigma_{\ell \in L}\frac{1-o(1)}{\cdot 2^m} \geq \frac{1-o(1)}{2}$.

Indeed, recall that a string is selected according to D_P by first selecting $z \in Primes$ and then selecting a z-periodic string. For $Q \subset [n]$, $|Q| = m$ let $\mathcal{A}(Q)$ be the event that for α selected according to $\mathcal{D}_P$, there exists no $j, k \in Q$ such that $j \equiv k \mod z$. For any fixed $i < j \in Q$ there are at most $\log n$ prime divisors of $j - i$. Hence for each $\ell \in L_0$, $prob_{\mathcal{D}_P}[\mathcal{A}(Q_\ell)] \geq 1 - \frac{m^2 \cdot \log n}{|Primes|} \geq 1 - o(1)$. Observe that for any fixed ℓ if $\mathcal{A}(Q_\ell)$ occurs then according to the definition of $\mathcal{D}_P$, for each $q \in Q_l$, α_q is selected uniformly and independently of any other q'. Hence, $prob_{\mathcal{D}_P}[h_\ell(\alpha) = 1] \geq prob_{\mathcal{D}_P}[h_\ell(\alpha) = 1 \wedge \mathcal{A}(Q_\ell)] \geq prob_{\mathcal{D}_P}[h_\ell(\alpha) = 1 \mid \mathcal{A}(Q_\ell)] \cdot Prob_{\mathcal{D}_P}[\mathcal{A}(Q_\ell)] \geq \frac{1}{2^m}(1 - o(1))$.

Observe that for each $\ell \in L_1$,

$$\begin{aligned}\text{Prob}_{\mathcal{D}_N}[far(\alpha) = 1 \bigwedge h_\ell(\alpha) = 1] \geq \\ \text{Prob}_{\mathcal{D}_N}[far(\alpha) = 1 \mid h_\ell(\alpha) = 1] \cdot \text{Prob}_{\mathcal{D}_N}[h_\ell(\alpha) = 1]\end{aligned} \quad (2)$$

By the definition of $\mathcal{D}_N$, $\text{Prob}_{\mathcal{D}_N}[h_\ell(\alpha) = 1] = \frac{1}{2^m}$. Also, using the definition just before Proposition 2 we see that $\text{Prob}_{\mathcal{D}_N}[far(\alpha) = 1 \mid h_\ell(\alpha) = 1] = prob_{U(Q_\ell)}[far(\alpha) = 1] \geq 1 - \frac{1}{n}$ (where the last inequality is by Proposition 2). $\square$

6   Lower Bound for Testing $period(\leq g)$, Where $g \leq \frac{\log n}{4}$

In this section we prove that every 2-sided error, $\frac{1}{16}$-test for $period(\leq g)$, where $g \leq \frac{\log n}{4}$, uses $\Omega((\log g)^{\frac{1}{4}})$ queries. We prove this for $\Sigma = \{0,1\}$. This implies the same bound for every alphabet that contains at least two symbols. This also shows that the test presented in Section 4 cannot be dramatically improved.

Theorem 5. *Any 2-sided error $\frac{1}{16}$-test for $period(\leq g)$, where $g \leq \frac{\log n}{4}$, requires $\Omega((\log g)^{\frac{1}{4}})$ queries.*

Proof. The proof is very similar to the proof of Theorem 4. We just describe here the two probabilities $\mathcal{D}_P$ and $\mathcal{D}_N$ that are concentrated on positive inputs (those that have $period(\leq g)$) and negative inputs (those that are $\frac{1}{16}$-far from being $period(\leq g)$) respectively.

Let $S = \{p_1, \ldots, p_k\}$ where $k = 1 + \sqrt{\log g}$ and each p_i is a prime such that $2^{\sqrt{\log g} - 0.9} \leq p_i \leq 2^{\sqrt{\log g}}$. According to the *Prime Number Theorem* [11–13] there exists at least $\dfrac{1}{8\sqrt{\log g}} \cdot 2^{\sqrt{\log g}} > 2 + \sqrt{\log g}$ such primes, hence the set S is well defined. Let $t = \prod_{p \in S} p$ and let $Z = \{\frac{t}{p} \mid p \in S\}$. Note that $t > g$ while $z \leq g$ for every $z \in Z$.

We now define distributions $\mathcal{D}_\mathcal{P}$, $\mathcal{D}_\mathcal{N}$ (we assume w.l.o.g that t divides n).

- We use an auxiliary distribution $\mathcal{U}$ in order to define $\mathcal{D}_\mathcal{N}$. To select an instance according to $\mathcal{U}$, select a random string $\omega \in \Sigma^t$ and then set $\alpha = (\omega)^{\frac{n}{t}}$. Let G be the event that α is $\frac{1}{16}$-far from being $period(\leq g)$. Then $\mathcal{D}_\mathcal{N}$ is defined as the distribution $\mathcal{U}|\mathcal{G}$, namely $\mathcal{U}$ given $\mathcal{G}$.
- An instance α of length n is selected according to distribution $\mathcal{D}_\mathcal{P}$ as follows. Uniformly a select $z \in Z$, select uniformly a string $\omega \in \Sigma^z$ and then set α to be the concatination of ω to itself enough times until a total length of n.

We omit further details.   □

7   Lower Bound for Testing Periodicity

Let $\Sigma = \{0, 1\}$, we prove the following lower bound on the number of queries that is needed for testing $period(\leq \frac{n}{2})$ over Σ. This clearly implies the same lower bound for any alphabet that contains at least two letter. It also shows that the test presented in Section 3 is asymptotically optimal.

Theorem 6.   *Any non-adaptive 1-sided error $\frac{1}{16}$-test for periodicity requires $\Omega(\sqrt{n \log n})$ queries.*

Proof: A 1-sided error test rejects only when the input string is not periodic. Therefore according to corollary 1 such a test rejects only if the set of queries it uses contains a witness that the string in not periodic. To prove the Theorem we use Yao's principle (the easy direction). Namely, we construct a distribution on $1/16$-*far* inputs and show that any *deterministic* non-adaptive test that queries at most $\dfrac{\sqrt{n \log n}}{100}$ queries finds a witness for non-periodicity with probability at most $1/3$.

Let U be the uniform distribution over Σ^n, and let $\mathcal{G}$ be the event that a string selected according to U is $\frac{1}{16}$-*far* from being periodic. Let D be the distribution $U/\mathcal{G}$ (U given $\mathcal{G}$). Namely, D is uniform over strings in Σ^n that are $1/16$-*far* from being periodic. Let $\mathcal{B}$ be the event that the set of queries Q contains a witness. Proposition 2 $Prob_U(\mathcal{G}) \geq 1 - \frac{1}{n}$. Thus, it is enough to prove that $prob_U(\mathcal{B}) < \frac{1}{4}$, since $prob_D(\mathcal{B}) \leq \mathrm{Prob}_U(\mathcal{B}) + \mathrm{Prob}_U(\overline{\mathcal{G}}) \leq \frac{1}{4} + \frac{1}{n}$. The proof follows from Lemma 4 below.   □

Lemma 4.   *Let $Q \subseteq [n]$ be a set of $\dfrac{\sqrt{n \log n}}{100}$ queries. Then for α chosen according to U, the probability that Q contains a witness that α is not periodic is at most $\frac{1}{4}$.*

The proof is omitted.

References

1. S. Goldwasser O. Goldreich and D. Ron. Property testing and its connection to learning and approximation. *Journal of the ACM*, 45:653–750, 1998.
2. R. Rubinfeld and M. Sudan. Robust characterization of polynomials with applications to program testing. *SIAM Journal of Computing*, 25:252–271, 1996.
3. F. Ergun, S. Muthukrishnan, and C. Sahinalp. Sub-linear methods for detecting periodic trends in data streams. In *LATIN 2004, Proc. of the 6th Latin American Symposium on Theoretical Informatics*, pages 16–28, 2004.
4. P. Indyk, N. Koudas, and S. Muthukrishnan. Identifying representative trends in massive time series data sets using sketches. In *VLDB 2000, Proceedings of 26th International Conference on Very Large Data Bases, September 10-14, 2000, Cairo, Egypt*, pages 363–372. Morgan Kaufmann, 2000.
5. R. Krauthgamer O. Sasson. Property testing of data dimensionality. In *Proceedings of the fourteenth annual ACM-SIAM symposium on Discrete algorithms*, pages 18–27. Society for Industrial and Applied Mathematics, 2003.
6. A. C. Gilbert, S. Guha, P. Indyk, S. Muthukrishnan, and M. Strauss. Near-optimal sparse fourier representations via sampling. In *STOC 2002, Proceedings of the thirty-fourth annual ACM symposium on Theory of computing*, pages 152–161, 2002.
7. Alex Samorodnitsky and Luca Trevisan. A PCP characterization of NP with optimal amortized query complexity. In *Proc. of the 32 ACM STOC*, pages 191–199, 2000.
8. Johan Håstad and Avi Wigderson. Simple analysis of graph tests for linearity and pcp. *Random Struct. Algorithms*, 22(2):139–160, 2003.
9. E. Fischer. The art of uninformed decisions: A primer to property testing. *The computational complexity column of The Bulletin of the European Association for Theoretical Computer Science*, 75:97–126, 2001.
10. D. Ron. Property testing (a tutorial). In *Handbook of Randomized computing*, pages 597–649. Kluwer Press, 2001.
11. J. Hadamard. Sur la distribution des zéros de la fonction $\zeta(s)$ et ses conséquences arithmétiques. *Bull. Soc. Math. France*, 24:199–220, 1896.
12. V. Poussin. Recherces analytiques sur la théorie des nombres premiers. *Ann. Soc. Sci. Bruxelles*, 1897.
13. D.J. Newman. Simple analytic proof of the prime number theorem. *Amer. Math. Monthly*, 87:693–696, 1980.

The Parity Problem in the Presence of Noise, Decoding Random Linear Codes, and the Subset Sum Problem*

(Extended Abstract)

Vadim Lyubashevsky

University of California at San Diego, La Jolla CA 92093, USA
vlyubash@cs.ucsd.edu

Abstract. In [2], Blum *et al.* demonstrated the first sub-exponential algorithm for learning the parity function in the presence of noise. They solved the length-n parity problem in time $2^{O(n/\log n)}$ but it required the availability of $2^{O(n/\log n)}$ labeled examples. As an open problem, they asked whether there exists a $2^{o(n)}$ algorithm for the length-n parity problem that uses only $poly(n)$ labeled examples. In this work, we provide a positive answer to this question. We show that there is an algorithm that solves the length-n parity problem in time $2^{O(n/\log\log n)}$ using $n^{1+\epsilon}$ labeled examples. This result immediately gives us a sub-exponential algorithm for decoding $n \times n^{1+\epsilon}$ random binary linear codes (i.e. codes where the messages are n bits and the codewords are $n^{1+\epsilon}$ bits) in the presence of random noise. We are also able to extend the same techniques to provide a sub-exponential algorithm for dense instances of the random subset sum problem.

1 Introduction

In the length-n parity problem with noise, there is an unknown to us vector $c \in \{0,1\}^n$ that we are trying to learn. We are also given access to an oracle that generates examples a_i and labels l_i where a_i is uniformly distributed in $\{0,1\}^n$ and l_i equals $c \cdot a_i \pmod 2$ with probability $\frac{1}{2} + \eta$ and $1 - c \cdot a_i \pmod 2$ with probability $\frac{1}{2} - \eta$. The problem is to recover c. In [2], Blum, Kalai, and Wasserman demonstrated the first sub-exponential algorithm for solving this problem. They gave an algorithm that recovers c in time $2^{O(n/\log n)}$ using $2^{O(n/\log n)}$ labeled examples for values of η is greater than $2^{-n^{\delta}}$ for any constant $\delta < 1$. An open problem was whether it was possible to have an algorithm with a sub-exponential running time when only given access to a polynomial number of labeled examples. In this work, we show that by having access to only $n^{1+\epsilon}$ labeled examples, we can recover c in time $2^{O(n/\log\log n)}$ for values of η greater than $2^{-(\log n)^{\delta}}$. So the penalty we pay for using fewer examples is both in the time and the error tolerance.

* Research supported in part by NSF grant CCR-0093029.

C. Chekuri et al. (Eds.): APPROX and RANDOM 2005, LNCS 3624, pp. 378–389, 2005.
© Springer-Verlag Berlin Heidelberg 2005

The parity problem in the presence of noise is equivalent to the problem of decoding random binary linear codes in the presence of random noise. Suppose that A is a random $n \times m$ boolean matrix, and let $l = cA$ for some binary string c of length n. Now flip each bit of l with probability $\frac{1}{2} - \eta$, and call the resulting bit string l'. The goal is to recover c given the matrix A and the string l'. Notice that we can just view every column of A as an example a_i and view the ith bit of l' as the value of $c \cdot a_i \pmod 2$ which is correct with probability $\frac{1}{2} + \eta$. So this is exactly the length n parity problem in the presence of noise where we are given m labeled examples. In this application, being able to solve the length n parity problem with fewer examples is crucial because we don't really have an oracle that will provide us with as many labeled examples as we want. The number of labeled examples we get is exactly m, the number of columns of A. Our result for learning the parity function provides the first algorithm which can decode $n \times n^{1+\epsilon}$ random binary linear codes in sub-exponential time ($2^{O(n/\log\log n)}$) in the presence of random noise. Unfortunately, the techniques do not extend to providing a sub-exponential algorithm for decoding $n \times \alpha n$ random binary linear codes for constant $\alpha > 1$, and we leave that as an open problem.

We then show how to apply essentially the same techniques to obtain a sub-exponential time algorithm for high density random instances of the subset sum problem. In the random subset sum problem, we are given n random integers $(a_1, \ldots, a_n)$ in the range $[0, M)$, and a target t. We are asked to produce a subset of the a_i's whose sum is t modulo M. We say that the instance of the problem is high density if $M < 2^n$. For such values of M, a solution to the problem exists with extremely high probability. It has been shown that solving worst case instances of some lattice problems reduces to solving random instances of high density subset sum [1],[10]. Thus it is widely conjectured that solving random high density instances of subset sum is indeed a hard problem.

Random subset sum instances where $M = 2^{\Omega(n^2)}$ have been shown to be solvable in polynomial time [8],[4], and the same techniques extend to provide better algorithms for values of $M > 2^{n^{1+\epsilon}}$. But the techniques do not carry over to high density instances where $M < 2^n$. A recent algorithm by Flaxman and Przydatek [3] gives a polynomial time solution for instances of random subset sum where $M = 2^{O(\log^2 n)}$. But for all other values of M between $2^{\log^2 n}$ and 2^n, the best algorithms work in time $O(\sqrt{M})$[11]. In our work, we provide an algorithm that works in time $M^{O(\frac{1}{\log n})}$ for all values of $M < 2^{n^\epsilon}$ for any $\epsilon < 1$. So our algorithm is a generalization of [3]. Technical difficulties do not allow us to extend the techniques to provide an algorithm that takes time $M^{O(\frac{1}{\log n})}$ when $M = 2^{\epsilon n}$ for $\epsilon < 1$, and we leave this as an open problem.

1.1   Conventions and Terminology

We will use the phrase "bit b gets corrupted by noise of rate $\frac{1}{2} - \eta$" to mean that the bit b retains its value with probability $\frac{1}{2} + \eta$ and gets flipped with probability $\frac{1}{2} - \eta$. And this probability is completely independent of anything else. Also, all logarithms are base 2.

1.2   Main Ideas and Main Results

In order for the Blum, Kalai, Wasserman algorithm to work, it must have access to a very large (slightly sub-exponential) number of labeled examples. What we will do is build a function to provide this very large number of labeled examples to their algorithm, and use their algorithm as a black box. We are given $n^{1+\epsilon}$ pairs (a_i, l_i) where a_i is uniformly distributed on $\{0,1\}^n$ and l_i is $c \cdot a_i (\mathrm{mod}\ 2)$ corrupted by noise of rate $\frac{1}{2} - \eta$. Instead of treating this input as labeled examples, we will treat it as a random element from a certain family of functions that maps elements from some domain X to the domain of labeled examples. To be a little more precise, say we are given ordered pairs $(a_1, l_1), \ldots, (a_{n^{1+\epsilon}}, l_{n^{1+\epsilon}})$. Now consider the set $X = \{x \in \{0,1\}^{n^{1+\epsilon}} | \sum_i x_i = \lceil \frac{2n}{\epsilon \log n} \rceil \}$ (i.e all bit strings of length $n^{1+\epsilon}$ that have exactly $\lceil \frac{2n}{\epsilon \log n} \rceil$ ones). Our function works by selecting a random element $x \in X$ and outputting $a' = \bigoplus x_i a_i$ and $l' = \bigoplus x_i l_i$ (where $\bigoplus$ is the xor operation, and x_i is the ith bit of x).

What we will need to show is that the distribution of the a' produced by our function is statistically close to uniform on $\{0,1\}^n$, that l' is the correct value of $c \cdot a' (\mathrm{mod}\ 2)$ with probability approximately $\frac{1}{2} + \eta^{\frac{2n}{\epsilon \log n}}$, and that the correctness of the label is almost independent of a'. This will allow us to use the function to produce a sub-exponential number of random looking labeled examples that the Blum, Kalai, Wasserman algorithm needs.

The idea of using a small number of examples to create a large number of examples is somewhat reminiscent of the idea used by Goldreich and Levin in [5]. In that work, they showed how to learn the parity function in polynomial time when given access to a "membership oracle" that gives the correct label for a $\frac{1}{2} + \frac{1}{poly(n)}$ fraction of the 2^n possible examples. A key step in their algorithm was guessing the correct labels for a logarithmic number of examples and then taking the xor of every possible combination of them to create a polynomial number of correctly labeled examples. The main difference between their problem and ours is that they were allowed to pick the examples with which to query the oracle, while in our formulation of the problem, the examples are provided to us randomly.

We prove the following theorem in Section 3:

Theorem 1. *We are given $n^{1+\epsilon}$ ordered pairs (a_i, l_i) where a_i are chosen uniformly and independently at random from the set $\{0,1\}^n$ and for some $c \in \{0,1\}^n$, $l_i = c \cdot a_i (\mathrm{mod}\ 2)$ corrupted by noise of rate $\frac{1}{2} - \eta$. If $\eta > 2^{-(\log n)^\delta}$ for any constant $\delta < 1$, then there is an algorithm that can recover c in time $2^{O(n/\log \log n)}$. The algorithm succeeds with error exponentially small in n.*

In Section 4, we move on to the random subset sum problem. We can no longer use the Blum, Kalai, Wasserman algorithm as a black box, but we do use an algorithmic idea that was developed by Wagner [13], and is in some sense a more general version of an algorithm in [2]. Wagner considers the following problem: given n lists each containing $M^{\frac{1}{\log n}}$ independent integers uniformly distributed in the interval $[0, M)$ and a target t, find one element from each list such that the sum of the elements is $t(mod\,M)$. Wagner shows that there exists an algorithm that in time $nM^{O(\frac{1}{\log n})}$, returns a list of solutions. The number of solutions that the algorithm returns is a random variable with expected value 1. That is, we expect to find one solution. Notice that this does not imply that the algorithm finds a solution with high probability because, for example, it can find 2^n solutions with probability 2^{-n} and find 0 solutions every other time. By inspecting Wagner's algorithm, there is no reason to assume such pathological behavior, and thus the algorithm works well as a cryptanalytic tool (as was intended by the author). We make some modifications to the parameters of the algorithm and are able to obtain a proof that the algorithm indeed succeeds with high probability in finding a solution. So if $M = 2^{n^\epsilon}$ and we are given n lists each containing $2^{O(n^\epsilon/\log n)}$ numbers in the range $[0, M)$, then we can find one number from each list such that their sum is the target t in time $2^{O(n^\epsilon/\log n)}$. We will be using this algorithm as a black box to solve the random subset sum problem.

We will give a sketch of the proof for the below theorem in Section 4. For complete details, the reader is referred to [9] for a full, self-contained version of the paper describing the algorithm.

Theorem 2. *Given n numbers, $a_1, \ldots, a_n$, that are independent and uniformly distributed in the range $[0, M)$ where $M = 2^{n^\epsilon}$ for some $\epsilon < 1$, and a target t, there exists an algorithm that in time $2^{O(n^\epsilon/\log n)}$ and with probability at least $1 - 2^{-\Omega(n^\epsilon)}$ will find $x_1, \ldots, x_n \in \{0,1\}$ such that $\sum_{i=1}^{n} a_i x_i = t(mod\,M)$.*

2   Preliminaries

2.1   Statistical Distance

Statistical distance is a measure of how far apart two probability distributions are. In this subsection, we review the definition and some of the basic properties of statistical distance. The reader may refer to [12] for reference.

Definition 1. *Let X and Y be random variables over a countable set A. The statistical distance between X and Y, denoted $\Delta(X, Y)$, is*

$$\Delta(X, Y) = \frac{1}{2} \sum_{a \in A} |Pr[X = a] - Pr[Y = a]|$$

Proposition 1. *Let $X_1, \ldots, X_k$ and $Y_1, \ldots, Y_k$ be two lists of independent random variables. Then*

$$\Delta((X_1, \ldots, X_k), (Y_1, \ldots, Y_k)) \leq \sum_{i=1}^{k} \Delta(X_i, Y_i)$$

Proposition 2. *Let X, Y be two random variables over a set A. For any predicate $f : A \rightarrow \{0,1\}$,*

$$|Pr[f(X) = 1] - Pr[f(Y) = 1]| \leq \Delta(X, Y)$$

2.2   Leftover Hash Lemma

In this sub-section, we state the Leftover Hash Lemma [7]. We refer the reader to [12] for more information on the material in this sub-section, and to [7] for the proof of the Leftover Hash Lemma.

Definition 2. *Let $\mathcal{H}$ be a family of hash functions from X to Y. Let H be a random variable whose distribution is uniform on $\mathcal{H}$. We say that $\mathcal{H}$ is a* **universal** *family of hash functions if for all $x, x' \in X$ with $x \neq x'$,*

$$Pr_{H \in \mathcal{H}}[H(x) = H(x')] \leq \frac{1}{|Y|}$$

Proposition 3. *(Leftover Hash Lemma) Let $X \subset \{0,1\}^n, |X| \geq 2^l$ and $|Y| = \{0,1\}^{l-2e}$ for some $e > 0$. Let U be the uniform distribution over Y. If $\mathcal{H}$ is a universal family of hash functions from X to Y, then for all but a $2^{-\frac{e}{2}}$ fraction of the possible $h_i \in \mathcal{H}$, $\Delta(h_i(x), U) \leq 2^{-\frac{e}{2}}$ where x is chosen uniformly at random from X.*

2.3   Learning Parity

The below result was proved in [2] and we will be using it as a black box.

Proposition 4. *(Blum, Kalai, Wasserman) For any integers a and b such that $ab \geq n$, if we are given $poly\left(\left(\frac{1}{2\gamma}\right)^{2^a}, 2^b\right)$ labeled examples uniformly and independently distributed in $\{0,1\}^n$, each of whose labels is corrupted by noise of rate $\frac{1}{2} - \gamma$, then there exists an algorithm that in time $poly\left(\left(\frac{1}{2\gamma}\right)^{2^a}, 2^b\right)$ solves the length-n parity problem. The algorithm succeeds with error exponentially small in n.* □

For $\gamma = 2^{-n^{\delta}}$ where δ is any positive constant less than 1, we can set a to be any value less than $(1 - \delta) \log n$ and set b to any value grater than $\frac{n}{(1-\delta)\log n}$ in order to get an algorithm with running time $2^{O(n/\log n)}$. The examples that we will be constructing in our algorithm will be corrupted by noise of rate more than $\frac{1}{2} - 2^{-n^{\delta}}$, that is why we will not be able to get the same running time.

3   Solving the Parity Problem with Fewer Examples

The main goal of this section will be to prove Theorem 1. But first, we will prove a lemma which will be used several times in the proof of the theorem.

Lemma 1. *Let $X \subseteq \{0,1\}^{n^{1+\epsilon}}$ such that $|X| > 2^{2n}$ and let $Y = \{0,1\}^n$. Let $\mathcal{H}$ be the family of hash functions from X to Y, where $\mathcal{H} = \{h_a | a = (a_1, \ldots, a_{n^{1+\epsilon}}), a_i \in \{0,1\}^n\}$, and where $h_a(x) = x_1 a_1 \oplus \ldots \oplus x_{n^{1+\epsilon}} a_{n^{1+\epsilon}}$. If h_a is chosen uniformly at random from $\mathcal{H}$, then with probability at least $1 - 2^{-\frac{n}{4}}$, $\Delta(h_a(x), U) \leq 2^{-\frac{n}{4}}$ where x is chosen uniformly at random from X and U is the uniform distribution over Y.*

Proof. First we will show that $\mathcal{H}$ is a universal family of hash functions from X to Y. Let $x, x' \in X$ such that $x \neq x'$ and H be a random variable whose distribution is uniform on $\mathcal{H}$. Since x has to differ from x' in some coordinate, without loss of generality, assume that $x_j = 1$ and $x'_j = 0$ for some j. Then,

$$
Pr_{H \in \mathcal{H}}[H(x) = H(x')] = Pr\left[\bigoplus_i x_i a_i = \bigoplus_i x'_i a_i\right]
$$

$$
= Pr\left[a_j \oplus \bigoplus_{i \neq j} x_i a_i = \bigoplus_{i \neq j} x'_i a_i\right]
$$

$$
= Pr\left[a_j = \bigoplus_{i \neq j}(x_i a_i) \oplus \bigoplus_{i \neq j}(x'_i a_i)\right] = \frac{1}{2^n}
$$

with the last equality being true because a_j is chosen uniformly at random and is independent of all the other a_i's.

Now we will apply the leftover hash lemma. Recall that $|X| > 2^{2n}$ and $|Y| = 2^n$. So in the application of Proposition 3, if we set $l = 2n$, and $e = \frac{n}{2}$, then we obtain the claim in the lemma. ☐

Now we are ready to prove the theorem.

Proof of Theorem 1

Proof. As we alluded to earlier, the way our algorithm works is by generating random-looking labeled examples and using them in the Blum, Kalai, Wasserman algorithm. We now describe the function $h(x)$ for outputting the labeled examples. We are given $n^{1+\epsilon}$ labeled examples (a_i, l_i). Define the set $X \subset \{0, 1\}^{n^{1+\epsilon}}$ as $X = \{x \in \{0, 1\}^{n^{1+\epsilon}} | \sum_i x_i = \lceil \frac{2n}{\epsilon \log n} \rceil\}$. In other words, the set X consists of all the bit strings of length $n^{1+\epsilon}$ that have exactly $\lceil \frac{2n}{\epsilon \log n} \rceil$ ones. The input to our function will be a random element from X.

$h(x)\{$
$\quad a' = \bigoplus_i x_i a_i$
$\quad l' = \bigoplus_i x_i l_i$
$\quad$ Output (a', l')
$\}$

So an element $x \in X$ essentially chooses which examples and labels to xor together.

Now we will show that with high probability, the function h produces labeled examples such that the a' are distributed statistically close to uniform, and that the l' are the correct values of $a' \cdot c \pmod 2$ corrupted by noise of rate at most $\frac{1}{2} - \frac{1}{2}\left(\frac{\eta}{2}\right)^{\frac{2n}{\epsilon \log n}}$.

To prove that the distribution of a' is statistically close to uniform on the set $\{0, 1\}^n$, we need to think of the $n^{1+\epsilon}$ a_i given to us as a random element of a family of hash functions $\mathcal{H}$ defined in Lemma 1. And this hash function is mapping elements from X to $\{0,1\}^n$. Notice that,

$$|X| = \binom{n^{1+\epsilon}}{\lceil \frac{2n}{\epsilon \log n} \rceil} \geq \left(\frac{n^{1+\epsilon}}{\lceil \frac{2n}{\epsilon \log n} \rceil} \right)^{\lceil \frac{2n}{\epsilon \log n} \rceil} > 2^{2n}$$

so now we can apply Lemma 1 and conclude that with high probability the distribution of a' generated by h is statistically close to uniform.

Now we need to find the probability that l' is the correct value of $c \cdot a' (\mathrm{mod}\ 2)$. For that, we first need to show that with high probability enough of the $n^{1+\epsilon}$ examples we are given are labeled correctly.

Lemma 2. *If we are given $n^{1+\epsilon}$ pairs (a_i, l_i) where l_i is the correct label of a_i with probability $\frac{1}{2} + \eta$, then with probability at least $1 - e^{-(n^{1+\epsilon})\frac{\eta^2}{2}}$, there will be more than $\frac{n^{1+\epsilon}}{2}(1 + \eta)$ pairs (a_i, l_i) where l_i is the correct label of a_i.* □

Notice that l' will equal $c \cdot a' (\mathrm{mod}\ 2)$ if $l' = \bigoplus x_i l_i$ is the xor of an even number of incorrectly labeled examples. Using the next lemma, we will be able to obtain a lower bound for the probability that our random x chooses an even number of mislabeled examples.

Lemma 3. *If a bucket contains m balls, $(\frac{1}{2} + p)m$ of which are colored white, and the rest colored black, and we select k balls at random without replacement, then the probability that we selected an even number of black balls is at least*

$$\frac{1}{2} + \frac{1}{2} \left(\frac{2mp - k + 1}{m - k + 1} \right)^k.$$ □

Lemma 2 tells us that out of $n^{1+\epsilon}$ examples, at least $\frac{n^{1+\epsilon}}{2}(1 + \eta)$ are correct. Combining this with the statement of Lemma 3, the probability when we pick $\lceil \frac{2n}{\epsilon \log n} \rceil$ random examples, an even number of them will be mislabeled is at least

$$\frac{1}{2} + \frac{1}{2} \left(\frac{n^{1+\epsilon}\eta - \lceil \frac{2n}{\epsilon \log n} \rceil + 1}{n^{1+\epsilon} - \lceil \frac{2n}{\epsilon \log n} \rceil + 1} \right)^{\lceil \frac{2n}{\epsilon \log n} \rceil} > \frac{1}{2} + \frac{1}{2} \left(\eta - \frac{\lceil \frac{2n}{\epsilon \log n} \rceil}{n^{1+\epsilon}} \right)^{\lceil \frac{2n}{\epsilon \log n} \rceil}$$

$$> \frac{1}{2} + \frac{1}{2} \left(\eta - \frac{2}{\epsilon n^\epsilon} \right)^{\frac{2n}{\epsilon \log n}}$$

$$> \frac{1}{2} + \frac{1}{2} \left(\frac{\eta}{2} \right)^{\frac{2n}{\epsilon \log n}}$$

with the last inequality being true since $\eta > \frac{4}{\epsilon n^\epsilon}$.

So far we have shown that our function produces examples a' which are statistically close to uniform and labels l' which are correct with probability at least $\frac{1}{2} + \frac{1}{2} \left(\frac{\eta}{2} \right)^{\frac{2n}{\epsilon \log n}}$. All that is left to show is that the correctness of l' is almost independent of a'. We do this by showing that the distribution of the a' that are labeled correctly and the distribution of the a' that are labeled incorrectly are both statistically close to the uniform distribution over the set $\{0,1\}^n$.

Consider the sets X_{even} and X_{odd} where X_{even} consists of all the $x \in X$ that choose an even number of mislabeled examples, and X_{odd} consists of all the $x \in X$ that choose an odd number of mislabeled examples. While we of course do not know the contents of X_{even} or X_{odd}, we can lower bound their sizes.

X_{even} contains all the x that choose exactly zero mislabeled examples. Since Lemma 2 says that there are at least $\frac{n^{1+\epsilon}}{2}(1+\eta)$ correctly labeled examples,

$$|X_{even}| > \left(\genfrac{}{}{0pt}{}{\frac{n^{1+\epsilon}}{2}}{\lceil\frac{2n}{\epsilon\log n}\rceil}\right) > 2^{2n}$$

and similarly, X_{odd} contains all the x that choose exactly one mislabeled example. So,

$$|X_{odd}| > \left(\genfrac{}{}{0pt}{}{\frac{n^{1+\epsilon}}{2}}{\lceil\frac{2n}{\epsilon\log n}\rceil - 1}\right) > 2^{2n}$$

Now we can use Lemma 1 to conclude that the distribution of the a' generated by function $h(x)$ is statistically close to the uniform distribution over the set $\{0,1\}^n$ when x is chosen uniformly from the set X_{even} (and similarly from X_{odd}).

So we have shown that with high probability, our function h outputs labeled examples with distribution statistically close to the distribution of labeled examples chosen uniformly from $\{0,1\}^n$ whose labels are corrupted by noise of rate at most $\frac{1}{2} - \frac{1}{2}\left(\frac{\eta}{2}\right)^{\frac{2n}{\epsilon\log n}}$. Now we can apply the parity function learning algorithm from Proposition 4. If our noise rate is $\eta = 2^{-(\log n)^\delta}$ for some constant $\delta < 1$, then we use the algorithm with the following parameters: $a = \kappa(\log\log n)$, $b = \frac{n}{\kappa\log\log n}$, $\gamma = \frac{1}{2}\left(\frac{\eta}{2}\right)^{\frac{2n}{\epsilon\log n}}$, where $\delta + \kappa < 1$. The algorithm will run in time,

$$poly\left(\left(\frac{1}{2\gamma}\right)^{2^a}, 2^b\right) = poly\left(\left(\frac{1}{\left(\frac{\eta}{2}\right)^{\frac{2n}{\epsilon\log n}}}\right)^{(\log n)^\kappa}, 2^{\frac{n}{\kappa\log\log n}}\right)$$

$$= poly\left(2^{O\left((\log n)^\delta\left(\frac{2n}{\epsilon\log n}\right)(\log n)^\kappa\right)}, 2^{\frac{n}{\kappa\log\log n}}\right)$$

$$= poly\left(2^{O\left(n(\log n)^{\delta+\kappa-1}\right)}, 2^{\frac{n}{\kappa\log\log n}}\right)$$

$$= 2^{O\left(\frac{n}{\log\log n}\right)}$$

if given access to $2^{O(n/\log\log n)}$ labeled examples. Since our function h generates labeled examples whose distribution is statistically close $(2^{-\Omega(n/4)})$ to the correct distribution, if we generate a sub-exponential number of such pairs, Proposition 1 tells us that the statistical distance is still exponentially small. And Proposition 2 says that if we run the Blum, Kalai, Wasserman algorithm with the labeled examples produced by our function, the success probability of the algorithm only goes down by at most an exponentially small amount. And this concludes the proof of Theorem 1.   □

4   The Subset Sum Problem

In this section, we will sketch the proof for Theorem 2. All the details of the proof may be found in [9].

First, we state a theorem about an algorithm that will be used as a black box analogous to the way the Blum, Kalai, Wasserman algorithm was used as a black box in the previous section.

Theorem 3. *Given b independent lists each consisting of $kM^{\frac{2}{\log b}}$ independent uniformly distributed integers in the range $[-\frac{M}{2}, \frac{M}{2})$, there exists an algorithm that with probability at least $1 - be^{-\frac{k}{4b^2}}$ returns one element from each of the b lists such that the sum of the b elements is 0. The running time of this algorithm is $O(b \cdot kM^{\frac{2}{\log b}} \cdot \log(kM^{\frac{2}{\log b}}))$ arithmetic operations.*

Proof. (sketch) The algorithm will be building a tree whose nodes are lists of numbers. The initial b lists will make up level 0 of the tree. Define intervals

$$
I_i = \left[-\frac{M^{1-\frac{i}{\log b}}}{2}, \frac{M^{1-\frac{i}{\log b}}}{2} \right)
$$

for integers $0 \le i \le \log b$. Notice that all the numbers in the lists at level 0 are in the range I_0. Level 1 of the tree will be formed by pairing up the lists on level 0 in any arbitrary way, and for each pair of lists, L_1 and L_2 (there are $\frac{b}{2}$ such pairs), create a new list L_3 whose elements are in the range I_1 and of the form $a_1 + a_2$ where $a_1 \in L_1$ and $a_2 \in L_2$. So level 1 will have half the number of lists as level 0, but the numbers in the lists will be in a smaller range. We construct level 2 in the same way; that is, we pair up the lists in level 1 and for each pair of lists L_1 and L_2, create a new list L_3 whose elements are in the range I_2 and of the form $a_1 + a_2$ where $a_1 \in L_1$ and $a_2 \in L_2$. Notice that if we continue in this way, then level $\log b$ will have one list of numbers in the interval $I_{\log b}$ where each number is the sum of b numbers, one from each of the original b lists at level 0. And since the only possible integer in $I_{\log b}$ is 0, if the list at level $\log b$ is not empty, then we are done. So what we need to prove is that the list at level $\log b$ is not empty with high probability.

Claim: *With probability at least $1 - be^{-\frac{k}{4b^2}}$, for $0 \le i \le \log b$, level i consists of $\frac{b}{2^i}$ lists each containing $\frac{k}{4^i} \cdot M^{\frac{2}{\log b}}$ independent, uniformly distributed numbers in the range I_i.*

Notice that proving this claim immediately proves the theorem. To prove the claim, we essentially have to show that when we combine two lists in order to obtain a list containing numbers in a smaller range, the new list still consists of many random, independently distributed numbers. The key to the proof is the following technical lemma:

Lemma 4. *Let L_1 and L_2 be lists of numbers in the range $\left[-\frac{R}{2}, \frac{R}{2} \right)$ and let p and c be positive reals such that $e^{-\frac{c}{12}} < p < \frac{1}{8}$. Let L_3 be a list of numbers $a_1 + a_2$ such that $a_1 \in L_1, a_2 \in L_2$, and $a_1 + a_2 \in \left[-\frac{Rp}{2}, \frac{Rp}{2} \right)$. If L_1 and L_2 each contain* at least $\frac{c}{p^2}$ independent, uniformly distributed numbers in $\left[-\frac{R}{2}, \frac{R}{2}\right)$, then with probability greater than $1 - e^{-\frac{c}{4}}$, L_3 contains at least $\frac{c}{4p^2}$ independent, uniformly distributed numbers in the range $\left[-\frac{Rp}{2}, \frac{Rp}{2}\right)$.   □

Using this lemma, it's not too hard to finish the proof of the theorem by induction.   □

Corollary 1. *Given b independent lists each consisting of $kM^{\frac{2}{\log b}}$ independent uniformly distributed integers in the range $[0, M)$, and a target t, there exists an algorithm that with probability at least $1 - be^{-\frac{k}{4b^2}}$ returns one element from each of the b lists such that the sum of the b elements is $t \pmod M$. The running time of this algorithm is $O(b \cdot kM^{\frac{2}{\log b}} \cdot \log(kM^{\frac{2}{\log b}}))$.*

Proof. First, we subtract the target t from every element in the first list. Then we transform every number (in every list) in the interval $[0, M)$ into an equivalent number modulo M in the interval $[-\frac{M}{2}, \frac{M}{2})$. We do this by simply subtracting M from every number greater or equal to $\frac{M}{2}$. Now we apply Theorem 3. Since exactly one number from every list got used, it means that $-t$ got used once. And since the sum is 0, we found an element from each list such that their sum is $t \pmod M$.   □

Trying to solve the random subset sum problem by using the algorithm implicit in the above corollary is very similar to the position we were in trying to decode random linear codes by using the Blum, Kalai, Wasserman algorithm. In both cases, we had algorithms that solved almost the same problem but required a lot more samples than we had. So just as in the case of random linear codes, the idea for the algorithm for random subset sum will be to combine the small number of elements that we are given in order to create a slightly subexponential number of elements that are almost indistinguishable from uniform, independently distributed ones.

Now we state a lemma which is analogous and will be used analogously to Lemma 1. The proof is very similar to the one for Lemma 1, and something very similar was actually previously shown by Impagliazzo and Naor in [6].

Lemma 5. *Let $X = \{0, 1\}^n$, and $Y = \{0, 1, \ldots, M\}$ for some $M \leq 2^{n/2}$. Let $\mathcal{H}$ be a family of hash functions from X to Y, where $\mathcal{H} = \{h_a | a = (a_1, \ldots, a_n), a_i \in Y\}$, and where $h_a(x) = \sum x_i a_i \pmod M$. If h_a is chosen uniformly at random from $\mathcal{H}$, then with probability at least $1 - 2^{-\frac{n}{4}}$, $\Delta(h_a(x), U) \leq 2^{-\frac{n}{4}}$ where x is chosen uniformly at random from X and U is the uniform distribution over Y.*   □

Now we can finally sketch the idea to prove Theorem 2. The idea for our algorithm will be to first take the n given numbers modulo $M = 2^{n^\epsilon}$ and break them up into $\frac{1}{2}n^{1-\epsilon}$ groups each containing $2n^\epsilon$ numbers. Then for each group, we generate a list of $n^2 M^{\overline{\log \frac{1}{2} n^{1-\epsilon}}}$ numbers by taking sums of random subsets of the elements in the group. By Lemma 5 and Proposition 1, we can say that with high probability the lists are not too far from being uniformly distributed. Now that we have lists that are big enough, we can apply Corollary 1 which states that we can find one number from each list such that the sum of the numbers is the target. And since every number in the list is some subset of the numbers in the group from which the list was generated, we have found a subset of the original n numbers which sums to the target.

Proof of Theorem 2

Proof. We will show that the below *SubsetSum* algorithm satisfies the claim in the theorem.

$SubsetSum(a_1, \ldots, a_n, t, M)$ /** Here $M = 2^{n^\epsilon}$
(1) Break up the n numbers into $\frac{1}{2}n^{1-\epsilon}$ groups each containing $2n^\epsilon$ numbers.
(2) For group $i = 1$ to $\frac{1}{2}n^{1-\epsilon}$ do
(3)   List $L_i = GenerateListFromGroup(\{a_j | a_j \in group\ i\}, M)$
(4) Apply the algorithm from Corollary 1 to $L_1, \ldots, L_{.5n^{1-\epsilon}}, t, M$.

$GenerateListFromGroup(\{a_1, \ldots, a_m\}, M)$
(5) Initialize list L to be an empty list.
(6) For $i = 1$ to $n^2 2^{\frac{2n^\epsilon}{\log .5n^{1-\epsilon}}}$ do /** Notice that $n^2 2^{\frac{2n^\epsilon}{\log .5n^{1-\epsilon}}} = n^2 M^{\frac{2}{\log \frac{1}{2}n^{1-\epsilon}}}$
(7)   Generate m random $x_j \in \{0, 1\}$
(8)   Add the number $\sum_{j=1}^{m} a_j x_j (mod\ M)$ to list L.
(9) Return L.

If we assume for a second that the lists $L_1, \ldots L_{.5n^{1-\epsilon}}$ in line (4) consist of independent, uniformly distributed numbers, then we are applying the algorithm from Corollary 1 with parameters $b = \frac{1}{2}n^{1-\epsilon}$ and $k = n^2$. So with probability

$$1 - be^{-\frac{k}{4b^2}} = 1 - \frac{1}{2}n^{1-\epsilon}e^{-\frac{n^2}{4(.5n^{1-\epsilon})^2}} = 1 - e^{-\Omega(n^{2\epsilon})}$$

the algorithm will return one element from each list such that the sum of the elements is $t(mod\ M)$. And this gives us the solution to the subset sum instance. The running time of the algorithm is

$$O(b \cdot kM^{\frac{2}{\log b}} \cdot \log(kM^{\frac{2}{\log b}})) = O((b \cdot kM^{\frac{2}{\log b}})^2)$$
$$= O\left(\left(\frac{1}{2}n^{1-\epsilon}n^2 M^{\frac{2}{\log .5n^{1-\epsilon}}}\right)^2\right)$$
$$= O\left(\left(\frac{1}{2}n^{1-\epsilon}n^2 2^{\frac{2n^\epsilon}{\log .5n^{1-\epsilon}}}\right)^2\right)$$
$$= 2^{O\left(\frac{n^\epsilon}{(1-\epsilon)\log n}\right)} = 2^{O\left(\frac{n^\epsilon}{\log n}\right)}$$

The problem is that each list in line (4) is not generated uniformly and independently from the interval $[0, M)$. But it is not too hard to finish off the proof by showing that the distribution of each list is statistically close to the uniform distribution, and thus our algorithm still has only a negligible probability of failure.   □

Acknowledgements

I am very grateful to Daniele Micciancio for his many comments and corrections of an earlier version of this work. I am also very grateful to Russell Impagliazzo for some extremely useful conversations about some issues related to the technical parts of the paper. I would also like to thank the anonymous members of the program committee for many useful suggestions.

References

1. Miklos Ajtai. "Generating hard instances of lattice problems" Proc. 28th Annual Symposium on Theory of Computing, pp. 99-108, 1996
2. Avrim Blum, Adam Kalai, and Hal Wasserman. "Noise-tolerant learning, the parity problem, and the statistical query model" JACM., vol. 50, iss. 4 2003 pp. 506-519
3. Abraham Flaxman and Bartosz Przydatek. "Solving medium-density subset sum problems in expected polynomial time" Proc. of the 22nd Symposium on Theoretical Aspects of Computer Science (STACS) (2005) pp. 305-314
4. Alan Frieze. "On the Lagarias-Odlyzko algorithm for the subset sum problem" SIAM J. Comput., vol. 15 1986 pp. 536-539
5. Oded Goldreich and Leonid A. Levin. "Hard-core predicates for any one-way function." Proc. 21st Annual Symposium on Theory of Computing, pp. 25-32, 1989
6. Russell Impagliazzo and Moni Naor. "Efficient cryptographic schemes provably as secure as subset sum" Journal of Cryptology Volume 9, Number 4, 1996, pp. 199-216
7. Russell Impagliazzo and David Zuckerman. "How to recycle random bits" Proc. 30th Annual Symposium on Foundations of Computer Science, 1989, pp. 248-253
8. Jeffrey C. Lagarias and Andrew M. Odlyzko. "Solving low density subset sum problems", J. of the ACM, Volume 32, 1985 pp. 229-246
9. Vadim Lyubashevsky "On random high density subset sums", Technical Report TR05-007, Electronic Coloquium on Computational Complexity (ECCC), 2005
10. Daniele Micciancio and Oded Regev. "Worst-case to average-case reductions based on gaussian measure" Proc. 45th Annual Symposium on Foundations of Computer Science, pp. 372-381, 2004.
11. Richard Schroeppel and Adi Shamir. "A $T = O(2^{(n/2)})$, $S = O(2^{(n/4)})$ algorithm for certain NP-complete problems." SIAM J. Comput. vol. 10 1981 pp. 456-464
12. Victor Shoup. *A Computational Introduction to Number Theory and Algebra.* Cambridge University Press 2005
13. David Wagner. "A generalized birthday problem" CRYPTO 2002, LNCS, Springer-Verlag, pp. 288-303

The Online Clique Avoidance Game on Random Graphs

Martin Marciniszyn, Reto Spöhel, and Angelika Steger

Institute of Theoretical Computer Science, ETH Zürich
8092 Zürich, Switzerland
{mmarcini,rspoehel,steger}@inf.ethz.ch

Abstract. Consider the following one player game on an empty graph with n vertices. The edges are presented one by one to the player in a random order. One of two colors, red or blue, has to be assigned to each edge immediately. The player's object is to color as many edges as possible without creating a monochromatic clique K_ℓ of some fixed size ℓ. We prove a threshold phenomenon for the expected duration of this game. We show that there is a function $N_0 = N_0(\ell, n)$ such that the player can asymptotically almost surely survive up to $N(n) \ll N_0$ edges by playing greedily and that this is best possible, i.e., there is no strategy such that the game would last for $N(n) \gg N_0$ edges.

1 Introduction

In the classical edge coloring problem, we are given a graph G, and our goal is to color the edges of G with as few colors as possible such that no two incident edges get the same color, that is avoiding a monochromatic path of length two. It is well known that the minimum number of colors is either $\Delta(G)$ or $\Delta(G)+1$, where $\Delta(G)$ denotes the maximum degree in G. Deciding which one of these two bounds is tight is an NP-complete problem. A well-studied generalization of this is the problem of finding edge colorings that avoid instead of a monochromatic path of length two a monochromatic copy of a given fixed graph F. It follows from Ramsey's celebrated result [1] that *every* two-coloring of the edges of a complete graph on n vertices contains a monochromatic clique on $\Theta(\log n)$ vertices and thus, for n sufficiently large, in particular a monochromatic copy of F. While this seems to rely on the fact that the K_n is a very dense graph, Folkman [2] and, in a more general setting, Nešetřil and Rödl [3] showed that there also exist locally sparse graphs $G = G(F)$ with the property that every two-coloring of the edges of G contains a monochromatic copy of F. By transferring the problem into a random setting, Rödl and Ruciński [4] showed that in fact such graphs G are quite frequent. More precisely, Rödl and Ruciński showed that the existence of a two-coloring without a monochromatic copy of F a.a.s. depends only on the density p of the underlying random graph $G_{n,p}$.

While the above results concern only the *existence* of an appropriate two-coloring, Friedgut, Kohayakawa, Rödl, Ruciński, and Tetali transfered these problems into an *algorithmic* setting. They studied the following one player game.

C. Chekuri et al. (Eds.): APPROX and RANDOM 2005, LNCS 3624, pp. 390–401, 2005.

The board is a graph with n vertices, which initially contains no edges. These are presented to the player one by one in an order chosen uniformly at random among all permutations of the underlying complete graph. One of two colors, red or blue, has to be assigned to each edge immediately. The player's object is to color as many edges as possible without creating a monochromatic copy of some fixed graph F. As soon as the first monochromatic copy of F is closed, the game ends. We call this the online F-avoidance game and refer to the number of properly colored edges as its *duration*. Friedgut et al. [5] showed that for triangles there is a threshold that differs dramatically from the one in the offline case. While in the offline setting a.a.s. every random graph with $cn^{3/2}$ edges can be properly colored, the online case will a.a.s. stop whenever the painter has seen substantially more than $n^{4/3}$ edges.

In this paper, we shall prove that if F is a clique of size ℓ, there exists a function $N_0 = N_0(\ell, n)$ and a strategy such that the player a.a.s. 'survives' with this strategy as long as the number of edges is substantially below N_0. We analyze this strategy and show that it is best possible, that is, there is no strategy such that a.a.s. the player survives substantially more than N_0 edges.

Theorem 1 (Clique-avoidance games). *For $\ell \geq 2$, the threshold for the online K_ℓ-avoidance game with two colors is*

$$N_0(\ell, n) = n^{\left(2 - \frac{2}{\ell-1}\right)\left(1 + \frac{1}{\binom{\ell}{2}}\right)} = n^{\left(2 - \frac{2}{\ell+1}\right)\left(1 - \frac{1}{\binom{\ell}{2}^2}\right)} .$$

The player will a.a.s. succeed with the greedy strategy if the number of edges $N \ll N_0$, and she will fail with any strategy if $N \gg N_0$.

The greedy strategy is as simple as it could be: always use one color, say red, if this does not close a monochromatic copy of K_ℓ, and use blue otherwise. We give the formulas in two different forms since there are two interpretations of our result. Clearly, the game cannot end before the first copy of F has appeared. The well-known theorem of Bollobás [6], which is a generalization of a theorem found by Erdős and Rényi in [7] to arbitrary graphs F, states a threshold for this event.

Theorem 2 (Bollobás, 1981). *Let F be a non-empty graph. Then the threshold for the property $\mathcal{P} = `G_{n,m}$ contains a copy of F' is $m_0(n) = n^{2-1/m(F)}$, where*

$$m(F) := \max \left\{ \frac{e_H}{v_H} : H \subseteq F \right\} .$$

Thus, the first clique of size ℓ will not appear before $n^{2-2/(\ell-1)}$ edges have been inserted into the graph. The first formula of Theorem 1 relates the threshold of the clique avoidance game to this obvious lower bound.

The second formula emphasizes the connection between the threshold and a trivial upper bound for the duration of the game. Obviously, knowing all edges in advance would ease the player's situation. Using a result of Rödl and Ruciński [4], we can derive an upper bound for the online game from a Ramsey-type property of random graphs.

Theorem 3 (Rödl, Ruciński, 1995). *Let $r \geq 2$ and F be a non-empty graph with $v_F \geq 3$ which is not a star forest and in the case $r = 2$ not a forest of stars and P_3's. Moreover, let $\mathcal{P} = $ 'every r-edge-coloring of G contains a monochromatic copy of F'. Then there exist constants $c = c(F, r)$ and $C = C(F, r)$ such that*

$$\lim_{n \to \infty} \mathbb{P}[G_{n,m} \in \mathcal{P}] = \begin{cases} 1 & \text{if } m > Cn^{2 - \frac{1}{m_2(F)}} \\ 0 & \text{if } m < cn^{2 - \frac{1}{m_2(F)}}, \end{cases}$$

where

$$m_2(F) := \max\left\{ \frac{e_H - 1}{v_H - 2} : H \subset F \text{ and } v_H \geq 3 \right\}.$$

Thus, there is no hope of finding a 2-coloring avoiding a monochromatic K_ℓ, whenever $N \gg n^{2-2/(l+1)}$. In fact, it was the search for a sharp threshold for the Ramsey property studied in Theorem 3 which led to the notion of the online graph avoidance game. It is a major open problem in random graph theory and motivated many research activities (see [8], [9], [10], [11]). Recently these efforts culminated in a sharp threshold for 2-edge colorings avoiding monochromatic triangles, see [12].

In the remainder of this paper, we shall prove Theorem 1.

Organization of this Paper We conclude this introduction with some notation and a summary of known results. In Sect. 2 we present a general approach to the analysis of the online F-avoidance game for arbitrary graphs F. We apply these results in the proof Theorem 1 in Sect. 3. In Sect. 4 we state a generalization of Theorem 1, the proof of which is more technical and beyond the scope of this paper.

Preliminaries and Notation Our notation is fairly standard and mostly adopted from [13]. The expression $f \asymp g$ means that the functions f and g differ at most by a multiplicative constant. We denote a clique on ℓ vertices by K_ℓ and a path with $\ell + 1$ vertices by P_ℓ. By v_G or $v(G)$ we denote the number of vertices of a graph G and similarly the number of edges by e_G or $e(G)$. We define abbreviations for the following ratios:

$$\varepsilon(G) := \frac{e_G}{v_G} \quad \text{and} \quad \varepsilon_2(G) := \frac{e_G - 1}{v_G - 2}$$

We say that a graph G is *balanced* if $m(G) = \varepsilon(G)$, where $m(G)$ is defined as in Theorem 2. Analogously we call G *2-balanced* if $m_2(G) = \varepsilon_2(G)$ for $m_2(G)$ as defined in Theorem 3. By a *proper coloring* we mean an edge coloring without a monochromatic copy of a fixed graph F. We will use the two standard models for random graphs. $\mathcal{G}_{n,p}$ denotes the probability space of all graphs $G_{n,p}$ on n labeled vertices, where each edge is present with probability p independently of all other edges. The probability p usually depends on the number of vertices n. In the model $\mathcal{G}_{n,m}$, the graph $G_{n,m}$ is chosen uniformly at random among all graphs on n vertices and exactly m edges. Since both models are equivalent in terms of asymptotic properties if $m \asymp pn^2$ and p sufficiently large, we may switch from one to the other to simplify the presentation. We assume that the reader is familiar with the notion of thresholds for (monotone) graph properties. A property $\mathcal{P}$ holds asymptotically almost surely (a.a.s.) if $\lim_{n\to\infty} \mathbb{P}[\mathcal{P}] = 1$.

We will consider the model of the random graph process $(G(n, N))_{0 \le N \le \binom{n}{2}}$, where the edges appear uniformly at random one after the other, i.e., in one of $\binom{n}{2}!$ possible permutations. In this setting we consider the random variable $\hat{N}$ that measures the number of inserted edges until $G(n, N)$ satisfies some increasing property $\mathcal{P}$. This is called the *hitting time* of $\mathcal{P}$. There is a straightforward relation between the hitting time and a threshold function m_0 in the uniform model $\mathcal{G}_{n,m}$. It follows from the observation that $G(n, N)$ and $\mathcal{G}_{n,N}$ have the same distribution that m_0 is also a threshold for the random process $(G(n, N))_{0 \le N \le \binom{n}{2}}$.

Lemma 4. *Let $\mathcal{P}$ be an increasing graph property with threshold m_0. Then*

$$\lim_{n\to\infty} \mathbb{P}[\hat{N} \le m] = \begin{cases} 0 & \text{if } m \ll m_0 \\ 1 & \text{if } m \gg m_0 . \end{cases}$$

The following theorem is a counting version of Theorem 3, which also appeared in [4].

Theorem 5 (Rödl, Ruciński, 1995). *Let $r \ge 1$ and F be a non-empty graph with $v_F \ge 3$. Then there exist constants $C = C(F, r)$ and $a = a(F, r)$ such that for*

$$p(n) \ge Cn^{-1/m_2(F)}$$

a.a.s. in every r-edge-coloring of a random graph $G_{n,p}$, one color contains at least $an^{v_F}p^{e_F}$ copies of F.

2   Bounds for Online Graph Avoidance Games

This section is divided into three subsections: In Sect. 2.1 we establish a lower bound $\underline{N}(n)$ on the duration of the game by analyzing the greedy strategy for arbitrary graphs F. By identifying a family of dangerous graphs, we reduce this problem to the appearance of small subgraphs in a random graph. Sect. 2.2 is devoted to the proof of an upper bound $\overline{N}(n)$ for graphs F with certain properties. We will subdivide the game into two rounds and argue as follows: for a particular family of graphs, the player will a.a.s. create a lot of monochromatic copies of some member of this family regardless of her strategy. We shall prove by means of the second moment method that then these substructures will force the player to create a monochromatic copy of F in the second round of the game. In Sect. 2.3 we combine both bounds to a criterion which asserts a threshold for the F-avoidance game.

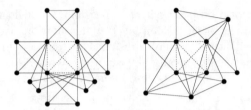

Fig. 1. Two representatives of the family $\mathcal{F}'$ in the K_4-avoidance game

2.1   Lower Bound

The greedy player uses one color, say red, in every move if this does not create a monochromatic red copy of F. Hence, the process will stop with a blue copy of F which was forced by a surrounding red structure. More precisely, when the game is over, $G(n, N)$ contains a blue copy of F, each edge of which is the missing link of a red subgraph isomorphic to F.

In [5] the case $F = K_3$ is analyzed. If this game is played greedily, eventually the graph will contain a blue triangle whose edges cannot be recolored due to the surrounding red edges. This entire structure will form either a K_4 or a pyramid graph, which consists of an inner triangle each edge of which is embedded into a disjoint outer triangle. Since both graphs do a.a.s. not appear in $G_{n,m}$ for $m \ll m_0 = n^{4/3}$ by Theorem 2, it follows from Lemma 4 that the greedy strategy a.a.s. ensures survival up to any $N \ll n^{4/3}$ edges.

In our general setting, we let $\mathcal{F}'$ denote the class of all graphs which have an 'inner' (blue) copy of F, every edge of which would close an 'outer' (red) copy of F. Note that here the colors should only provide the intuitive connection to the greedy strategy, but the members of the family $\mathcal{F}'$ are actually uncolored. We say that the inner copy of F is formed by inner edges and vertices, and refer to the surrounding elements as outer edges and vertices respectively. Observe that in this terminology, every outer copy contains exactly one inner edge and two or more inner vertices. This is illustrated in Fig. 1 for the case $F = K_4$. The inner edges are dashed. Formally we define this family of graphs $\mathcal{F}'$ as follows:

Definition 6. *For every graph $F = (V, E)$, let*

$$\mathcal{F}' := \Big\{ F' = (V \,\dot\cup\, U, E \,\dot\cup\, D) \big| F' \text{ is a minimal } graph \text{ such that}$$
$$\text{for all } e = \{u, v\} \in E \text{ there are sets } U_e \subseteq V \,\dot\cup\, U \text{ and } D_e \subseteq D \text{ with}$$
$$(\{u, v\} \cup U_e, e \cup D_e) \cong F \Big\} .$$

The inner vertices and edges V and E form the inner copy of F. Every edge $e \in E(F)$ is completed to a graph isomorphic to F by U_e and D_e. Hence, $|U_e| = v_F - 2$ and $|D_e| = e_F - 1$. We take F' as a minimal element with respect to subgraph inclusion, i.e., F' is not a proper subgraph of any other graph in $\mathcal{F}'$ that satisfies the same properties. In particular, this ensures that $\mathcal{F}'$ is finite.

With these definitions at hand, we can state our main lemma for the lower bound.

Lemma 7. *Consider the online F-avoidance game with two colors for a non-empty graph F. Playing greedily one can a.a.s. color any $N(n) \ll n^{2-1/m^*(F)}$ edges, where*

$$m^*(F) := \min\{m(F') : F' \in \mathcal{F}'\} \ .$$

Proof. The player is safe as long as no member of $\mathcal{F}'$ appears in $G(n, N)$, which by Theorem 2 and Lemma 4 does a.a.s. not happen for any $N \ll n^{2-1/m^*(F)}$. ∎

2.2   Upper Bound

For each color $\mathcal{C} \in \{R, B\}$, let $\mathcal{C}$ denote the subgraph of $G = G(n, N)$ spanned by all $\mathcal{C}$-colored edges, and let $\text{Base}(\mathcal{C})$ be the set of all vertex-pairs that joined by an edge would complete a subgraph isomorphic to F in $\mathcal{C}$. Thus, if an edge from the set $\text{Base}(\mathcal{C})$ is added to the graph, there is at most one proper coloring of it. Observe that edges in $\text{Base}(\mathcal{C})$ are induced by $\mathcal{C}$-colored copies of graphs $F_- \subseteq F$ with exactly v_F vertices and $e_F - 1$ edges. We refer to $\text{Base}(\mathcal{C})$ as the *base graph* of $\mathcal{C}$.

We relax the game and grant a mercy period of $N_1(n)$ to the player. She may wait until the end of this phase to decide on the edge coloring. Then the game proceeds normally, and another $N_2(n) \gg N_1(n)$ edges are added to the graph with online coloring. We argue that regardless of her strategy, the player will a.a.s. create many 'forbidden' copies of F either in $\text{Base}(R)$ or in $\text{Base}(B)$ after the first period. We will call these forbidden copies *threats*. Then for any number $N_2(n) \gg N_1(n)$ of additional edges in the second round, one of these threats will be added to G completely, which finishes the game.

The key to the upper bound for the F-avoidance game is to observe that the threats in the base graphs are induced by copies of graphs $F' \in \mathcal{F}'$, in which the inner edges are missing. We define a family of graphs $\mathcal{F}'_-$ similarly to the family $\mathcal{F}'$ reflecting this fact.

Definition 8. *For every graph $F = (V, E)$, let*

$$\mathcal{F}'_- := \Big\{ F'_- = (V \,\dot\cup\, U, D) \big| F'_- \text{ is a minimal graph such that}$$

$$\text{for all } e = \{u, v\} \in E \text{ there are sets } U_e \subseteq V \,\dot\cup\, U \text{ and } D_e \subseteq D \text{ with}$$

$$(\{u, v\} \cup U_e, e \cup D_e) \cong F \Big\} \ .$$

We say that $F'_- \in \mathcal{F}'_-$ and $F' \in \mathcal{F}'$ are *associated* if one can obtain F'_- by removing the inner copy of F from F'. Note that only edges are removed, the vertex set of F' remains unchanged.

Our proof strategy for the upper bound on the duration of the game is as follows. In the first round we generate a random graph G with $N_1(n)$ edges. Subsequently, the player has to provide an edge coloring of G. If $N_1(n) \geq$

$Cn^{2-1/m_2(F'_-)}$ for some graph $F'_- \in \mathcal{F}'_-$ and some large constant C, then Theorem 5 asserts that a positive fraction of all subgraphs in G isomorphic to F'_- is monochromatically colored. We shall show that almost every such subgraph induces a different copy of F in the base graph. Hence, there will be approximately $n^{v(F'_-)}(N_1 n^2)^{e(F'_-)}$ threats in G after the first round.

In the second round another $N_2 \gg N_1$ edges are presented to the player one by one. We count the number of encountered threats Z. In order to apply the second moment method, the expectation of Z has to be unbounded, i.e.,

$$\mathbb{E}[Z] \asymp n^{v(F'_-)} \left(\frac{N_1}{n^2} \right)^{e(F'_-)} \left(\frac{N_2}{n^2} \right)^{e_F} = n^{v(F')} \left(\frac{N_1}{n^2} \right)^{e(F')} \left(\frac{N_2}{N_1} \right)^{e_F} \to \infty \ . \quad (1)$$

Note that $v(F') = v(F'_-)$. Clearly, (1) holds provided

$$n^{v(F')} \left(\frac{N_1}{n^2} \right)^{e(F')} = \Omega(1) \Leftrightarrow N_1 = \Omega \left(n^{2-1/\varepsilon(F')} \right) \ .$$

These ideas motivate the following lemma.

Lemma 9. *Let F be a non-empty graph such that there exists $F' \in \mathcal{F}'$ and an associated $F'_- \in \mathcal{F}'_-$ satisfying $m_2(F'_-) \leq \varepsilon(F')$. Moreover, suppose that every proper subgraph $H \subset F$ with $v_H \geq 2$ satisfies*

$$v_F - \frac{e_F}{\varepsilon(F')} < v_H - \frac{e_H}{\varepsilon(F')} \ . \quad (2)$$

Then in the online F-avoidance game with two colors the player cannot color any $N(n) \gg n^{2-1/\varepsilon(F')}$ edges a.a.s. regardless of her strategy.

Proof. Let $F' \in \mathcal{F}'$ and $F'_- \in \mathcal{F}'_-$ be fixed. We tacitly switch between the models $G_{n,p}$ and $G(n, N)$, exploiting their asymptotic equivalence via $p \asymp N/n^2$.

Claim 10. *There exists $C > 0$ such that a.a.s. in every 2-edge-coloring of a random graph $G(n, N_1)$ with $N_1 = Cn^{2-1/\varepsilon(F')}$, the base graph of one color contains $\Omega((N_1/n^2)^{-e_F})$ copies of F.*

Proof. Set $C = C(F'_-, 2)$ according to Theorem 5, and let Y denote the number of copies of F'_- in $G(n, N_1)$. The expected number of such copies is

$$\mathbb{E}[Y] \asymp n^{v(F'_-)} \left(\frac{N_1}{n^2} \right)^{e(F'_-)} \asymp n^{v(F'_-) - \frac{e(F'_-)v(F')}{e(F')}} = n^{\frac{e_F}{\varepsilon(F')}} \asymp \left(\frac{N_1}{n^2} \right)^{-e_F} \ .$$

Here we used that $v(F'_-) = v(F')$ and $e(F') - e(F'_-) = e_F$. Theorem 5 yields asymptotically the same number of copies of monochromatic F'_- in (w.l.o.g.) R since $N_1 = Cn^{2-1/\varepsilon(F')} \geq Cn^{2-1/m_2(F'_-)}$ due to the assumption of Lemma 9.

Every such copy induces one copy of F in $\mathrm{Base}(R)$. To conclude the proof of Claim 10, we need to show that only few of these copies of F are counted multiple times, i.e., are induced by more than one copy of F'_-.

Observe that any multiply counted copy of F has at least one of its edges e induced by two or more copies of graphs isomorphic to F_-. Hence such a copy of F is induced by a subgraph $(F'_-)_H$ of the following structure: $(F'_-)_H$ looks like F'_- with one inner edge covered by an additional F_-, which overlaps with the original F'_- such that their intersection complemented with e is a copy of a proper subgraph $H \subset F$ (see Fig. 2 for two examples). Recall that F_- denotes a graph isomorphic to F with one missing edge. For any graph $(F'_-)_H$, we have $e\left((F'_-)_H\right) = e(F'_-) + e(F_-) - (e_H - 1) = e(F'_-) + e_F - e_H$ and $v\left((F'_-)_H\right) = v(F'_-) + v_F - v_H$.

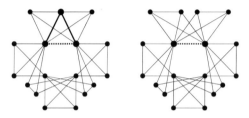

Fig. 2. Two graphs $(K'_{4-})_H$ that induce the same K_4 in the base graph twice. The bold edges and large dots form $H = K_3$ and $H = K_2$ respectively

We denote the number of subgraphs isomorphic to $(F'_-)_H$ in $G(n, N_1)$ by Y_H. It follows that

$$\mathbb{E}[Y_H] \asymp n^{v(F'_-)+v_F-v_H} \left(\frac{N_1}{n^2}\right)^{e(F'_-)+e_F-e_H} \asymp \mathbb{E}[Y] \frac{n^{v_F-e_F/\varepsilon(F')}}{n^{v_H-e_H/\varepsilon(F')}} \overset{(2)}{=} o(\mathbb{E}[Y]) \ .$$

The Markov inequality now yields $\mathbb{P}[Y_H \geq c\mathbb{E}[Y]] = o(1)$ for any $c > 0$. Since the number of copies of F induced by a fixed occurrence of a graph $(F'_-)_H$ is bounded by a constant only depending on F, the multiply counted copies of F are a.a.s. of lower order of magnitude than $\mathbb{E}[Y]$ and thus negligible. This concludes the proof of Claim 10.   □

We fix N_1 as in Claim 10. Let X denote the duration of the game. Continuing the proof of Lemma 9, we apply the claim to show that $\mathbb{P}[X \geq N] = o(1)$ for $N \gg N_1$. Let $R \dot{\cup} B$ be any coloring assigned by the player to the first N_1 edges. For $N \gg N_1$, we have $N_2 := N - N_1 \gg N_1$ edges in the second round. By Claim 10, there are $M = \Omega\left((N_1/n^2)^{-e_F}\right)$ copies of F in (w.l.o.g) $\text{Base}(R)$. All the edges in $\text{Base}(R)$ have to be colored blue if presented to the player, and hence she loses if the N_2 remaining edges contain one of those copies. Since there are still $\binom{n}{2} - N_1 = (1 + o(1))\binom{n}{2}$ edges missing, the probability that this happens is asymptotically the same as the probability that $G(n, N_2)$ contains one of the M forbidden copies of F.

For a fixed copy $F_i \subseteq \text{Base}(R)$, let the random variable Z_i denote the event that F_i is in $G(n, N_2)$, and let $Z := \sum_{F_i \subseteq \text{Base}(R)} Z_i$. We obtain as in (1)

$$\mathbb{E}[Z] \asymp M \left(\frac{N_2}{n^2}\right)^{e_F} = \Omega\left(\left(\frac{N_2}{N_1}\right)^{e_F}\right) \to \infty \ .$$

Standard variance calculations, which are omitted here due to space restrictions, yield $\mathrm{Var}[Z] = \mathbb{E}[Z^2] - \mathbb{E}[Z]^2 = o\left(\mathbb{E}[Z]^2\right)$. The second moment method now yields $\mathbb{P}[X \geq N] \leq \mathbb{P}[Z = 0] = o(1)$, and Lemma 9 is proved.   □

2.3   Threshold

Concluding this section, we restate our findings for graphs for which the existence of a threshold follows. Combining Lemmas 7 and 9, we obtain the following corollary.

Corollary 11. *Let F be a non-empty graph such that there exists $F' \in \mathcal{F}'$ and an associated $F'_- \in \mathcal{F}'_-$ satisfying $m_2(F'_-) \leq \varepsilon(F') \leq m^*(F)$. Moreover, suppose that every proper subgraph $H \subset F$ with $v_H \geq 2$ satisfies*

$$v_F - \frac{e_F}{\varepsilon(F')} < v_H - \frac{e_H}{\varepsilon(F')} \ .$$

Then $m^*(F) = \varepsilon(F')$, and the threshold for the online F-avoidance game with two colors is $n^{2-1/\varepsilon(F')}$.

As an example, we give the threshold for the online P_3-avoidance game. Simple but somewhat tedious calculations show that $m^*(P_3) = 7/8$. A graph F' that attains $\varepsilon(F') = m^*(P_3)$ is a path of length three with one additional edge emitting from each vertex. Hence, the perfect matching with exactly four edges is associated with this graph. Due to $m_2(4K_2) = 1/2 \leq m^*(P_3) = 7/8$, Corollary 11 is applicable, and the threshold is $n^{2-8/7} = n^{6/7}$.

3   Proof of Theorem 1

As a concrete application of Corollary 11, we calculate the threshold for clique avoidance games. Let $\mathcal{K}'_\ell$ denote the class of potentially dangerous graphs for the greedy player of the K_ℓ-avoidance game as in Definition 6. $\mathcal{K}'_\ell$ contains a unique member denoted by K^*_ℓ in which the outer copies do not overlap as in the left drawing of Fig. 1. K^*_ℓ has the edge density

$$\varepsilon(K^*_\ell) = \frac{\binom{l}{2}^2}{1 + \binom{l}{2}(l-2)} \ . \tag{3}$$

The formulas in Theorem 1 are equivalent to the claim that the threshold for the online K_ℓ-avoidance game is given by $N_0(n) = n^{2-1/\varepsilon(K^*_\ell)}$. To conclude this from Corollary 11, one needs to show the following statements.

Claim 12. *For all $\ell \geq 3$, every proper subgraph $H \subset K_\ell$ with $v_H \geq 2$ satisfies*

$$v_F - \frac{e_F}{\varepsilon(K^*_\ell)} < v_H - \frac{e_H}{\varepsilon(K^*_\ell)} \ .$$

The proof of this claim involves rather simple calculations, which we omit in this extended abstract. Let $K_{\ell-}^*$ denote the graph associated with K_ℓ^* in the class $\mathcal{K}_{\ell-}'$. It is obtained by removing all inner edges from K_ℓ^*.

Claim 13. *For all $\ell \geq 3$, $m_2(K_{\ell-}^*) \leq \varepsilon(K_\ell^*)$.*

Proof (Sketch). The proof idea is to show first that a subgraph $H \subseteq K_{\ell-}^*$ with maximal 2-density must be a union of complete outer copies of K_ℓ with one missing edge. Once this is established, the remaining candidates for H bijectively correspond to subgraphs H_0 of the inner K_ℓ. Similarly to (3), the 2-density of such a graph H can be expressed as a function of H_0 and l, and the claim follows with elementary calculations. □

The next theorem yields the lower bound on the expected duration of the clique avoidance game if the player plays greedily. In order to conclude that K_ℓ^* is the *most dangerous* graph for the player, it suffices to prove that any other graph $K_\ell' \in \mathcal{K}_\ell'$ is at least as dense as K_ℓ^*. This claim seems to be obvious: we expect that the density of K_ℓ^* increases if several outer copies share vertices. A formal proof, however, seems not to be straightforward. In particular, there are graphs in $\mathcal{K}_\ell'$ such that their density decreases when outer vertices are merged. Our proof therefore relies on an amortized analysis of a process transforming K_ℓ^* into K_ℓ'. Essentially we show that within this process there may be steps that decrease the density, but that these are compensated by earlier increasing steps.

Theorem 14. *For all $\ell \geq 3$, $m^*(K_\ell) \geq \varepsilon(K_\ell^*)$.*

Proof (Sketch). We refer to a graph $K_\ell' \in \mathcal{K}_\ell'$ not isomorphic to K_ℓ^* as a *contraction* of K_ℓ^* since it can be obtained by merging outer vertices and edges of K_ℓ^*. In order to prove that $m(K_\ell') \geq \varepsilon(K_\ell^*)$, it suffices to show that

$$\varepsilon(K_\ell') \geq \varepsilon(K_\ell^*) = \frac{\binom{l}{2}^2}{l + \binom{l}{2}(l-2)} = \frac{l+1}{2} - \frac{2}{l^2 - 3l + 4}$$

for any contraction K_ℓ' of K_ℓ^*. Clearly, this holds if

$$\frac{\Delta E}{\Delta V} \leq \varepsilon(K_\ell^*) \ , \tag{4}$$

where $\Delta E := e(K_\ell^*) - e(K_\ell')$ and $\Delta V := v(K_\ell^*) - v(K_\ell')$.

Number the $\binom{l}{2}$ outer copies of K_ℓ such that the inner edges of $K_\ell^1, \ldots, K_\ell^{\binom{k}{2}}$ form a k-clique for all $1 \leq k \leq \ell$. K_ℓ' can be constructed from K_ℓ^* in $\binom{l}{2} - 1$ steps: For $i = 2, \ldots, \binom{l}{2}$, merge the outer edges and vertices of K_ℓ^i with those of $K_\ell^1, \ldots, K_\ell^{i-1}$ with which they coincide. At the same time, keep track of all (outer) edges and vertices in K_ℓ^i that vanish in the process – i.e., are merged with edges or vertices of $K_\ell^1, \ldots, K_\ell^{i-1}$ – by marking them in a separate (non-contracted) copy of K_ℓ^*.

In general the marked edges and vertices do not form a well-defined graph unless we add some or all of the (unmarked) inner vertices. For an outer copy

K_ℓ^i, let $H^i \subset K_\ell$ denote the graph that is formed by the marked edges and vertices in K_ℓ^i and the two inner vertices of K_ℓ^i. Furthermore, let $e_i = e(H^i)$ and $v_i = v(H^i) - 2$ denote the numbers of marked edges and vertices in K_ℓ^i and call K_ℓ^i *empty* if $v_i = 0$; note that this implies $e_i = 0$. Clearly, K_ℓ^1 is always empty. For every non-empty copy K_ℓ^i, set $\Delta_i := e_i/v_i$. The main idea of the proof is to view

$$\frac{\Delta E}{\Delta V} = \frac{1}{\Delta V} \sum_i e_i = \frac{1}{\Delta V} \sum_{v_i \neq 0} v_i \Delta_i \tag{5}$$

as a weighted average of the Δ_i of all non-empty copies K_ℓ^i. Let us thus call a non-empty outer copy K_ℓ^i *good* if $\Delta_i \leq \varepsilon(K_\ell^*)$, *bad* otherwise. A simple calculation shows that K_ℓ^i is bad only if all its outer $\binom{l}{2} - 1$ edges and $l - 2$ vertices are marked, i.e., if

$$\Delta_i = \frac{\binom{l}{2} - 1}{l - 2} = \frac{l + 1}{2} .$$

Moreover, one can prove that for $k = 0, \ldots, l-1$, the first non-empty H^i among $H^{\binom{k}{2}+1}, \ldots, H^{\binom{k+1}{2}}$ is disconnected. For the disconnected copy one obtains the strong bound

$$\Delta_i \leq \frac{\binom{v_i+1}{2}}{v_i} = \frac{v_i + 1}{2} .$$

We call it *very good*. Now, (4) follows from the fact that this single very good copy outweighs all of the at most $\ell - 2$ bad copies among $K_\ell^{\binom{k}{2}+1}, \ldots, K_\ell^{\binom{k+1}{2}}$ in (5). Set

$$E_k := \sum_{i=\binom{k}{2}+1}^{i=\binom{k+1}{2}} e_i \quad \text{and} \quad V_k := \sum_{i=\binom{k}{2}+1}^{i=\binom{k+1}{2}} v_i .$$

In the worst case we have

$$\frac{E_k}{V_k} \leq \frac{(l-2)(\binom{l}{2} - 1) + \binom{v+1}{2}}{(l-2)^2 + v}$$

for some $v \in \{1, \ldots, l-2\}$. By elementary calculations this is at most $\varepsilon(K_\ell^*)$, and (4) follows with

$$\frac{\Delta E}{\Delta V} = \frac{1}{\Delta V} \sum_{k=0}^{l-1} E_k = \frac{1}{\Delta V} \sum_{V_k \neq 0} V_k \cdot \frac{E_k}{V_k} \leq \varepsilon^*(K_\ell) .$$

$\square$

4   Generalization

With a similar approach one can prove the threshold of the online F-avoidance with two colors for a larger family of graphs. This includes for example all cycles of fixed length. A graph F is 2-balanced if $m_2(F) = \varepsilon_2(F)$.

Theorem 15. *Let F be a 2-balanced graph which is not a tree, and from which one edge can be removed to obtain a graph F_- which satisfies $m_2(F_-) \leq \varepsilon(F^*)$. Then the threshold for the online F-avoidance game with 2 colors is*

$$N_0(n) = n^{\left(2 - \frac{1}{m(F)}\right)\left(1 + \frac{1}{e_F}\right)} = n^{\left(2 - \frac{1}{m_2(F)}\right)\left(1 - \frac{1}{e_F^2}\right)}.$$

The proof for this general family is more involved than for cliques. In particular, generalizing Claim 13 and Theorem 14 requires technical details beyond the scope of this extended abstract.

References

1. Ramsey, F.P.: On a problem of formal logic. Proceedings of the London Mathematical Society **30** (1930) 264–286
2. Folkman, J.: Graphs with monochromatic complete subgraphs in every edge coloring. SIAM J. Appl. Math. **18** (1970) 19–24
3. Nešetřil, J., Rödl, V.: The Ramsey property for graphs with forbidden complete subgraphs. J. Combinatorial Theory Ser. B **20** (1976) 243–249
4. Rödl, V., Ruciński, A.: Threshold functions for Ramsey properties. J. Amer. Math. Soc. **8** (1995) 917–942
5. Friedgut, E., Kohayakawa, Y., Rödl, V., Ruciński, A., Tetali, P.: Ramsey games against a one-armed bandit. Combinatorics, Probability and Computing **12** (2003) 515–545
6. Bollobás, B.: Threshold functions for small subgraphs. Math. Proc. Cambridge Philos. Soc. **90** (1981) 197–206
7. Erdős, P., Rényi, A.: On the evolution of random graphs. Publ. Math. Inst. Hung. Acad. **5** (1960) 17–61
8. Friedgut, E., Kalai, G.: Every monotone graph property has a sharp threshold. Proc. Amer. Math. Soc. **124** (1996) 2993–3002
9. Friedgut, E.: Sharp thresholds of graph properties, and the k-sat problem. J. Amer. Math. Soc. **12** (1999) 1017–1054 With an appendix by Jean Bourgain.
10. Friedgut, E., Krivelevich, M.: Sharp thresholds for certain Ramsey properties of random graphs. Random Structures & Algorithms **17** (2000) 1–19
11. Friedgut, E.: Hunting for sharp thresholds. Random Structures & Algorithms **26** (2005) 37–51
12. Friedgut, E., Rödl, V., Ruciński, A., Tetali, P.: A sharp threshold for random graphs with monochromatic triangle in every edge coloring. Memoirs of the AMS (to appear)
13. Janson, S., Łuczak, T., Ruciński, A.: Random graphs. Wiley-Interscience, New York (2000)

A Generating Function Method
for the Average-Case Analysis of DPLL

Rémi Monasson

CNRS-Laboratoire de Physique Théorique, ENS, Paris, France
monasson@lpt.ens.fr

Abstract. A method to calculate the average size of Davis-Putnam-Loveland-Logemann (DPLL) search trees for random computational problems is introduced, and applied to the satisfiability of random CNF formulas (SAT) and the coloring of random graph (COL) problems. We establish recursion relations for the generating functions of the average numbers of (variable or color) assignments at a given height in the search tree, which allow us to derive the asymptotics of the expected DPLL tree size, $2^{N\omega + o(N)}$, where N is the instance size. ω is calculated as a function of the input distribution parameters (ratio of clauses per variable for SAT, average vertex degree for COL), and the branching heuristics.

1 Introduction and Main Results

Many efforts have been devoted to the study of the performances of the Davis-Putnam-Loveland-Logemann (DPLL) procedure [19], and more generally, resolution proof complexity for combinatorial problems with randomly generated instances. Two examples are random k-Satisfiability (k-SAT), where an instance $\mathcal{F}$ is a uniformly and randomly chosen set of $M = \alpha N$ disjunctions of k literals built from N Boolean variables and their negations (with no repetition and no complementary literals), and random graph k-Coloring (k-COL), where an instance $\mathcal{F}$ is an Erdős-Rényi random graph from $G(N, p = c/N)$ *i.e.* with average vertex degree c.

Originally, efforts were concentrated on the random width distribution for k-SAT, where each literal appear with a fixed probability. Franco, Purdom and collaborators showed that simplified versions of DPLL had polynomial average-case complexity in this case, see [14, 24] for reviews. It was then recognized that the fixed clause length ensemble might provide harder instances for DPLL [12]. Chvátal and Szemerédi indeed showed that DPLL proof size is w.h.p. exponentially large (in N at fixed ratio α) for an unsatisfiable instance [9]. Later on, Beame *et al.* [4] showed that the proof size was w.h.p. bounded from above by $2^{cN/\alpha}$ (for some constant c), a decreasing function of α. As for the satisfiable case, Frieze and Suen showed that backtracking is irrelevant at small enough ratios α (≤ 3.003 with the Generalized Unit Clause heuristic, to be defined below) [15], allowing DPLL to find satisfying assignment in polynomial (linear) time. Achlioptas, Beame and Molloy proved that, conversely, at ratios smaller

C. Chekuri et al. (Eds.): APPROX and RANDOM 2005, LNCS 3624, pp. 402–413, 2005.
© Springer-Verlag Berlin Heidelberg 2005

than the generally accepted satisfiability threshold, DPLL takes w.h.p. exponential time to find a satisfying assignment [3]. Altogether these results provide explanations for the 'easy-hard-easy' (or, more precisely, 'easy-hard-less hard') pattern of complexity experimentally observed when running DPLL on random 3-SAT instances [23].

A precise calculation of the average size of the search space explored by DPLL (and #DPLL, a version of the procedure solving the enumeration problems #SAT and #COL) as a function of the parameters N and α or c is difficult due to the statistical correlations between branches in the search tree resulting from backtracking. Heuristic derivations were nevertheless proposed by Cocco and Monasson based on a 'dynamic annealing' assumption [10, 11, 13]. Hereafter, using the linearity of expectation, we show that 'dynamic annealing' turns not to be an assumption at all when the expected tree size is concerned.

We first illustrate the approach, based on the use of recurrence relations for the generating functions of the number of nodes at a given height in the tree, on the random k-SAT problem and the simple Unit Clause (UC) branching heuristic where unset variables are chosen uniformly at random and assigned to True or False uniformly at random [7, 8]. Consider the following counting algorithm

Procedure #DPLL-UC[$\mathcal{F}$,A,S]
Call $\mathcal{F}_A$ what is left from instance $\mathcal{F}$ given partial variable assignment A;
  1. If $\mathcal{F}_A$ is empty, $S \to S + 2^{N-|A|}$, Return; *(Solution Leaf)*
  2. If there is an empty clause in $\mathcal{F}_A$, Return; *(Contradiction Leaf)*
  3. If there is no empty clause in $\mathcal{F}_A$, let $\Gamma_1 = \{$1-clauses $\in \mathcal{F}_A\}$,
      if $\Gamma_1 \neq \emptyset$, pick any 1-clause, say, ℓ, and call DPLL[$\mathcal{F}$,A∪ℓ]; *(unit-propagation)*
      if $\Gamma_1 = \emptyset$, pick up an unset literal uniformly at random, say, ℓ, and call
            DPLL[$\mathcal{F}$,A∪ℓ], then DPLL[$\mathcal{F}$,A∪$\bar{\ell}$] ; *(variable splitting)*
End;

#DPLL-UC, called with $A = \emptyset$ and $S = 0$, returns the number S of solutions of the instance $\mathcal{F}$; the history of the search can be summarized as a search tree with leaves marked with solution or contradiction labels. As the instance to be treated and the sequence of operations done by #DPLL-UC are stochastic, so are the numbers L_S and L_C of solution and contradiction leaves respectively.

Theorem 1. *Let $k \geq 3$ and $\Omega(t, \alpha, k) = t + \alpha \ \log_2 \left(1 - \frac{k}{2^k} t^{k-1} + \frac{k-1}{2^k} t^k\right)$. The expectations of the numbers of solution and contradiction leaves in the #DPLL-UC search tree of random k-SAT instances with N variables and αN clauses are, respectively, $L_S(N, \alpha, k) = 2^{N\omega_S(\alpha,k)+o(N)}$ with $\omega_S(\alpha, k) = \Omega(1, \alpha, k)$ and $L_C(N, \alpha, k) = 2^{N\omega_C(\alpha,k)+o(N)}$ with $\omega_C(\alpha, k) = \max_{t \in [0;1]} \Omega(t, \alpha, k)$.*

An immediate consequence of Theorem 1 is that the expectation value of the total number of leaves, $L_S + L_C$, is $2^{N\omega_C(\alpha,k)+o(N)}$. This result was first found by Méjean, Morel and Reynaud in the particular case $k = 3$ and for ratios $\alpha > 1$ [22]. Our approach not only provides a much shorter proof, but can also be easily extended to other problems and more sophisticated heuristics, see Theorems 2 and 3 below. In addition, Theorem 1 provides us with some information about the expected search tree size of the decision procedure DPLL-UC, corresponding to #DPLL-UC with Line 1 replaced with: If $\mathcal{F}_A$ is empty, output Satisfiable; Halt.

Corollary 1. *Let* $\alpha > \alpha_u(k)$, *the root of* $\omega_C(\alpha, k) = 2 + \alpha \log_2(1 - 2^{-k})$ *e.g.* $\alpha_u(3) = 10.1286....$ *The average size of DPLL-UC search trees for random k-SAT instances with N variables and* αN *clauses equals* $2^{N\omega_C(\alpha,k)+o(N)}$.

Functions ω_S, ω_C are shown in Figure 1 in the $k = 3$ case. They coincide and are equal to $1 - \alpha \log_2(8/7)$ for $\alpha < \alpha^* = 4.56429...$, while $\omega_C > \omega_S$ for $\alpha > \alpha^*$. In other words, for $\alpha > \alpha^*$, most leaves in #DPLL-UC trees are contradiction leaves, while for $\alpha < \alpha^*$, both contradiction and solution leaf numbers are (to exponential order in N) of the same order. As for DPLL-UC trees, notice that $\omega_C(\alpha, k) \asymp \dfrac{2 \ln 2}{3 \alpha} = \dfrac{0.46209...}{\alpha}$. This behaviour agrees with Beame et al.'s result $(\Theta(1/\alpha))$ for the average resolution proof complexity of unsatisfiable instances [4]. Corollary 1 shows that the expected DPLL tree size can be estimated for a whole range of α; we conjecture that the above expression holds for ratios smaller than α_u *i.e.* down to α^* roughly. For generic $k \geq 3$, we have $\omega_C(\alpha, k) \asymp \dfrac{k-2}{k-1} \left(\dfrac{2^k \ln 2}{k(k-1)\alpha} \right)^{1/(k-2)}$; the decrease of ω_C with α is therefore slower and slower as k increases.

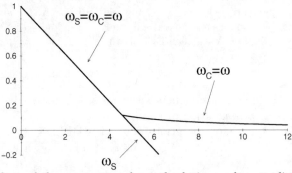

Fig. 1. Logarithms of the average numbers of solution and contradiction leaves, respectively ω_S and ω_C, in #DPLL-UC search trees versus ratio α of clauses per variable for random 3-SAT. Notice that ω_C coincides with the logarithm of the expected size of #DPLL-UC at all ratios α, and with the one of DPLL-UC search trees for $\alpha \geq 10.1286...$

So far, no expression for ω has been obtained for more sophisticated heuristics than UC. We consider the Generalized Unit Clause (GUC) heuristic [2, 8] where the shortest clauses are preferentially satisfied. The associated decision procedure, DPLL-GUC, corresponds to DPLL-UC with Line 3 replaced with: Pick a clause uniformly at random among the shortest clauses, and a literal, say, ℓ, in the clause; call DPLL[$\mathcal{F}$,A∪ℓ], then DPLL[$\mathcal{F}$,A∪$\bar{\ell}$].

Theorem 2. *Define* $m(x_2) = \frac{1}{2}\left(1 + \sqrt{1 + 4x_2}\right) - 2x_2$, $y_3(y_2)$ *the solution of the ordinary differential equation* $dy_3/dy_2 = 3(1 + y_2 - 2\ y_3)/(2m(y_2))$ *such that* $y_3(1) = 1$, *and*

$$\omega^g(\alpha) = \max_{\frac{3}{4} < y_2 \leq 1} \left[\int_{y_2}^1 \frac{dz}{m(z)} \log_2\left(2z + m(z)\right) \exp\left(-\int_z^1 \frac{dw}{m(w)}\right) + \alpha \log_2 y_3(y_2) \right]. \tag{1}$$

Let $\alpha > \alpha_u^g = 10.2183...$, *the root of* $\omega^g(\alpha) + \alpha \log_2(8/7) = 2$. *The expected size of DPLL-GUC search tree for random 3-SAT instances with N variables and αN clauses is* $2^{N\,\omega^g(\alpha) + o(N)}$.

Notice that, at large α, $\omega^g(\alpha) \asymp \frac{3 + \sqrt{5}}{6 \ln 2}\left[\ln\left(\frac{1 + \sqrt{5}}{2}\right)\right]^2 \frac{1}{\alpha} = \frac{0.29154...}{\alpha}$ in agreement with the $1/\alpha$ scaling established in [4]. Furthermore, the multiplicative factor is smaller than the one for UC, showing that DPLL-GUC is more efficient than DPLL-UC in proving unsatisfiability.

A third application is the analysis of the counterpart of GUC for the random 3-COL problem. The version of DPLL we have analyzed operates as follows [1]. Initially, each vertex is assigned a list of 3 available colors. In the course of the procedure, a vertex, say, v, with the smallest number of available colors, say, j, is chosen at random and uniformly. DPLL-GUC then removes v, and successively branches to the j color assignments corresponding to removal of one of the j colors of v from the lists of the neighbors of v. The procedure backtracks when a vertex with no color left is created (contradiction), or no vertex is left (a proper coloring is found).

Theorem 3. *Define* $\omega^h(c) = \max_{0 < t < 1} \left[\frac{c}{6}t^2 - \frac{c}{3}t - (1 - t)\ln 2 + \ln\left(3 - e^{-2ct/3}\right)\right]$.
Let $c > c_u^h = 13.1538...$, *the root of* $\omega^h(c) + \frac{c}{6} = 2\ln 3$. *The expected size of DPLL-GUC search tree for deciding 3-COL on random graphs from $G(N, c/N)$ with N vertices is* $e^{N\,\omega^h(c) + o(N)}$.

Asymptotically, $\omega^h(c) \asymp \frac{3\ln 2}{2\,c^2} = \frac{1.0397...}{c^2}$ in agreement with Beame et al.'s scaling $(\Theta(1/c^2))$ [5]. An extension of Theorem 3 to higher values of the number k of colors gives $\omega^h(c, k) \asymp \frac{k(k - 2)}{k - 1}\left[\frac{2\ln 2}{k - 1}\right]^{1/(k-2)} c^{-(k-1)/(k-2)}$. This result is compatible with the bounds derived in [5], and suggests that the $\Theta(c^{-(k-1)/(k-2)})$ dependence could hold w.h.p. (and not only in expectation).

2   Recurrence Equation for #DPLL-UC Search Tree

Let $\mathcal{F}$ be an instance of the 3-SAT problem defined over a set of N Boolean variables X. A partial assignment A of length $T(\leq N)$ is the specification of the truth values of T variables in X. We denote by $\mathcal{F}_A$ the residual instance given A. A clause $c \in \mathcal{F}_A$ is said to be a ℓ-clause with $\ell \in \{0, 1, 2, 3\}$ if the number of false literals in c is equal to $3 - \ell$. We denote by $C_\ell(\mathcal{F}_A)$ the number of ℓ-clauses in $\mathcal{F}_A$. The instance $\mathcal{F}$ is said to be satisfied under A if $C_\ell(\mathcal{F}_A) = 0$ for $\ell = 0, 1, 2, 3$, unsatisfied (or violated) under A if $C_0(\mathcal{F}_A) \geq 1$, undetermined under A otherwise. The clause vector of an undetermined or satisfied residual instance $\mathcal{F}_A$ is the three-dimensional vector $\boldsymbol{C}$ with components $C_1(\mathcal{F}_A), C_2(\mathcal{F}_A), C_3(\mathcal{F}_A)$. The search tree associated to an instance $\mathcal{F}$ and a run of #DPLL is the tree whose nodes carry the residual assignments A considered in the course of the search. The height T of a node is the length of the attached assignment.

It was shown by Chao and Franco [7, 8] that, during the first descent in the search tree *i.e.* prior to any backtracking, the distribution of residual instances remains uniformly random conditioned on the numbers of ℓ-clauses. This statement remains correct for heuristics more sophisticated than UC *e.g.* GUC, SC_1 [2, 8], and was recently extended to splitting heuristics based on variable occurrences by Kaporis, Kirousis and Lalas [18]. Clearly, in this context, uniformity is lost after backtracking enters into play (with the exception of Suen and Frieze's analysis of a limited version of backtracking [15]). Though this limitation appears to forbid (and has forbidden so far) the extension of average-case studies of backtrack-free DPLL to full DPLL with backtracking, we point out here that it is not as severe as it looks. Indeed, let us forget about how #DPLL or DPLL search tree is built and consider its final state. We refer to a branch (of the search tree) as the shortest path from the root node (empty assignment) to a leaf. The two key remarks underlying the present work can be informally stated as follows. First, the expected size of a #DPLL search tree can be calculated from the knolwedge of the statistical distribution of (residual instances on) a single branch; no characterization of the correlations between distinct branches in the tree is necessary. Secondly, the statistical distribution of (residual instances on) a single branch is simple since, along a branch, uniformity is preserved (as in the absence of backtracking). More precisely,

Lemma 1 (from Chao & Franco [7]). *Let $\mathcal{F}_A$ be a residual instance attached to a node A at height T in a #DPLL-UC search tree produced from an instance $\mathcal{F}$ drawn from the random 3-SAT distribution. Then the set of ℓ-clauses in $\mathcal{F}_A$ is uniformly random conditioned on its size $C_\ell(\mathcal{F}_A)$ and the number $N - T$ of unassigned variables for each $\ell \in \{0, 1, 2, 3\}$.*

Proof. the above Lemma is an immediate application of Lemma 3 in Achlioptas' Card Game framework which establishes uniformity for algorithms (*a*) 'pointing to a particular card (clause)', or (*b*) 'naming a variable that has not yet been assigned a value' (Section 2.1 in Ref. [2]). The operation of #DPLL-UC along a branch precisely amounts to these two operations: unit-propagation relies on action (*a*), and variable splitting on (*b*).   □

Lemma 1 does not address the question of uniformity among different branches. Residual instances attached to two (or more) nodes on distinct branches in the search tree are correlated. However, these correlations can be safely ignored in calculating the average number of residual instances, in much the same way as the average value of the sum of correlated random variables is simply the sum of their average values.

Proposition 1. *Let $L(\boldsymbol{C}, T)$ be the expectation of the number of undetermined residual instances with clause vector $\boldsymbol{C}$ at height T in #DPLL-UC search tree, and $G(x_1, x_2, x_3; T) = \sum_{\boldsymbol{C}} x_1^{C_1} x_2^{C_2} x_3^{C_3} L(\boldsymbol{C}, T)$ its generating function. Then, for $0 \leq T < N$,*

$$G(x_1, x_2, x_3; T+1) = \frac{1}{f_1} G(f_1, f_2, f_3; T) + \left(2 - \frac{1}{f_1}\right) G(0, f_2, f_3; T)$$
$$- 2 G(0, 0, 0; T) \tag{2}$$

where f_1, f_2, f_3 stand for the functions $f_1^{(T)}(x_1) = x_1 + \frac{1}{2}\mu(1-2x_1)$, $f_2^{(T)}(x_1, x_2) = x_2 + \mu(x_1 + 1 - 2x_2)$, $f_3^{(T)}(x_2, x_3) = x_3 + \frac{3}{2}\mu(x_2 + 1 - 2x_3)$, and $\mu = 1/(N-T)$. The generating function G is entirely defined from recurrence relation (2) and the initial condition $G(x_1, x_2, x_3; 0) = (x_3)^{\alpha N}$.

Proof. Let δ_n denote the Kronecker function ($\delta_n = 1$ if $n = 0$, $\delta_n = 0$ otherwise), $B_n^{m,q} = \binom{m}{n} q^n (1-q)^{m-n}$ the binomial distribution. Let A be a node at height T, and $\mathcal{F}_A$ the attached residual instance. Call $\boldsymbol{C}$ the clause vector of $\mathcal{F}_A$. Assume first that $C_1 \geq 1$. Pick up one 1-clause, say, ℓ. Call z_j the number of j-clauses that contain $\bar{\ell}$ or ℓ (for $j = 1, 2, 3$). From Lemma 1, the z_j's are binomial variables with parameter $j/(N-T)$ among $C_j - \delta_{j-1}$ (the 1-clause that is satisfied through unit-propagation is removed). Among the z_j clauses, w_{j-1} contained $\bar{\ell}$ and are reduced to $(j-1)$-clauses, while the remaining $z_j - w_{j-1}$ contained ℓ and are satisfied and removed. From Lemma 1 again, w_{j-1} is a binomial variable with parameter $1/2$ among z_j. The probability that the instance produced has no empty clause ($w_0 = 0$) is $B_0^{z_1, \frac{1}{2}} = 2^{-z_1}$. Thus, setting $\mu = \frac{1}{N-T}$,

$$M_P\ [\boldsymbol{C}', \boldsymbol{C}; T] = \sum_{z_3=0}^{C_3} B_{z_3}^{C_3, 3\mu} \sum_{w_2=0}^{z_3} B_{w_2}^{z_3, \frac{1}{2}} \sum_{z_2=0}^{C_2} B_{z_2}^{C_2, 2\mu} \sum_{w_1=0}^{z_2} B_{w_1}^{z_2, \frac{1}{2}}$$
$$\times \sum_{z_1=0}^{C_1-1} B_{z_1}^{C_1-1, \mu} \frac{1}{2^{z_1}} \delta_{C_3' - (C_3 - z_3)} \delta_{C_2' - (C_2 - z_2 + w_2)} \delta_{C_1' - (C_1 - 1 - z_1 + w_1)}$$

expresses the probability that a residual instance at height T with clause vector $\boldsymbol{C}$ gives rise to a (non-violated) residual instance with clause vector $\boldsymbol{C}'$ at height $T+1$ through unit-propagation. Assume now $C_1 = 0$. Then, a yet unset variable is chosen and set to True or False uniformly at random. The calculation of the new vector $\boldsymbol{C}'$ is identical to the unit-propagation case above, except that: $z_1 = w_0 = 0$ (absence of 1-clauses), and two nodes are produced (instead of one). Hence,

$$M_{UC}[\boldsymbol{C}', \boldsymbol{C}; T] = 2 \sum_{z_3=0}^{C_3} B_{z_3}^{C_3, 3\mu} \sum_{w_2=0}^{z_3} B_{w_2}^{z_3, \frac{1}{2}} \sum_{z_2=0}^{C_2} B_{z_2}^{C_2, 2\mu} \sum_{w_1=0}^{z_2} B_{w_1}^{z_2, \frac{1}{2}}$$
$$\times \delta_{C_3' - (C_3 - z_3)} \delta_{C_2' - (C_2 - z_2 + w_2)} \delta_{C_1' - w_1}$$

expresses the expected number of residual instances at height $T + 1$ and with clause vector C' produced from a residual instance at height T and with clause vector C through UC branching.

Now, consider all the nodes A_i at height T, with $i = 1, \ldots, \mathcal{L}$. Let o_i be the operation done by #DPLL-UC on A_i. o_i represents either unit-propagation (literal ℓ_i set to True) or variable splitting (literals ℓ_i set to T and F on the descendent nodes respectively). Denoting by $\mathbf{E}_Y(X)$ the expectation value of a quantity X over variable Y, $L(C'; T + 1) = \mathbf{E}_{\mathcal{L}, \{A_i, o_i\}} \left(\sum_{i=1}^{\mathcal{L}} \mathcal{M}[C'; A_i, o_i] \right)$ where $\mathcal{M}$ is the number (0, 1 or 2) of residual instances with clause vector C' produced from A_i after #DPLL-UC has carried out operation o_i. Using the linearity of expectation,

$$L(C'; T + 1) = \mathbf{E}_{\mathcal{L}} \left(\sum_{i=1}^{\mathcal{L}} \mathbf{E}_{\{A_i, o_i\}} \big(\mathcal{M}[C'; A_i, o_i] \big) \right) = \mathbf{E}_{\mathcal{L}} \left(\sum_{i=1}^{\mathcal{L}} M[C', C_i; T] \right)$$

where C_i is the clause vector of the residual instance attached to A_i, and $M[C', C; T] = \left(1 - \delta_{C_1} \right) M_P[C', C; T] + \delta_{C_1} M_{UC}[C', C; T]$. Gathering assignments with identical clause vectors gives the reccurence relation $L(C', T + 1) = \sum_C M[C', C; T] \, L(C, T)$. Recurrence relation (2) for the generating function is an immediate consequence. The initial condition over G stems from the fact that the instance is originally drawn from the random 3-SAT distribution, $L(C; 0) = \delta_{C_1} \delta_{C_2} \delta_{C_3 - \alpha N}$. □

3   Asymptotic Analysis and Application to DPLL-UC

The asymptotic analysis of G relies on the following technical lemma:

Lemma 2. *Let* $\gamma(x_2, x_3, t) = (1 - t)^3 x_3 + \frac{3t}{2}(1 - t)^2 x_2 + \frac{t}{8}(12 - 3t - 2t^2)$, *with* $t \in]0; 1[$ *and* $x_2, x_3 > 0$. *Define* $S_0(T) \equiv \sum_{H=0}^{T} 2^{T-H} G(0, 0, 0; H)$. *Then, in the large N limit,* $S_0([tN]) \leq 2^{N(t + \alpha \log_2 \gamma(0,0,t)) + o(N)}$ *and* $G\big(\frac{1}{2}, x_2, x_3; [tN]\big) = 2^{N(t + \alpha \log_2 \gamma(x_2, x_3, t)) + o(N)}$.

Due to space limitations, we give here only some elements of the proof. The first step in the proof is inspired by Knuth's kernel method [20]: when $x_1 = \frac{1}{2}$, $f_1 = \frac{1}{2}$ and recurrence relation (2) simplifies and is easier to handle. Iterating this equation then allows us to relate the value of G at height T and coordinates $(\frac{1}{2}, x_2, x_3)$ to the (known) value of G at height 0 and coordinates $(\frac{1}{2}, y_2, y_3)$ which are functions of x_2, x_3, T, N, and α. The function γ is the value of y_3 when T, N are sent to infinity at fixed ratio t. The asymptotic statement about $S_0(T)$ comes from the previous result and the fact that the dominant terms in the sum defining S_0 are the ones with H close to T.

Proposition 2. *Let* $L_C(N, T, \alpha)$ *be the expected number of contradiction leaves of height T in the #DPLL-UC resolution tree of random 3-SAT instances with*

N variables and αN clauses, and $\epsilon > 0$. Then, for $t \in [\epsilon; 1 - \epsilon]$ and $\alpha > 0$,
$\Omega(t, \alpha, 3) \leq \frac{1}{N} \log_2 L_C(N, [tN], \alpha) + o(1) \leq \max_{h \in [\epsilon,;t]} \Omega(h, \alpha, 3)$ where Ω is defined
in Theorem 1.

Observe that a contradiction may appear with a positive (and non–exponentially small in N) probability as soon as two 1-clauses are present. These 1-clauses will be present as a result of 2-clause reduction when the residual instances include a large number ($\Theta(N)$) of 2-clauses. As this is the case for a finite fraction of residual instances, $G(1, 1, 1; T)$ is not exponentially larger than $L_C(T)$. Use of the monotonicity of G with respect to x_1 and Lemma 2 gives the announced lower bound (recognize that $\Omega(t, \alpha, 3) = t + \alpha \log_2 \gamma(1, 1; t)$). To derive the upper bound, remark that contradictions leaves cannot be more numerous than the number of branches created through splittings; hence $L_C(T)$ is bounded from above by the number of splittings at smaller heights H, that is, $\sum_{H<T} G(0, 1, 1; H)$.
Once more, we use the monotonicity of G with respect to x_1 and Lemma 2 to obtain the upper bound. The complete proof will be given in the full version.

Proof. (Theorem 1) By definition, a solution leaf is a node in the search tree where no clauses are left; the average number L_S of solution leaves is thus given by $L_S = \sum_{H=0}^{N} L(0, 0, 0; H) = \sum_{H=0}^{N} G(\mathbf{0}; H)$. A straightforward albeit useful upper bound on L_S is obtained from $L_S \leq S_0(N)$. By definition of the algorithm #DPLL, $S_0(N)$ is the average number of solutions of an instance with αN clauses over N variables drawn from the random 3-SAT distribution, $S_0(N) = 2^N (7/8)^{\alpha N}$ [12]. This upper bound is indeed tight (to within terms that are subexponential in N), as most solution leaves have heights equal, or close to N. To show this, consider $\epsilon > 0$, and write

$$L_S \geq \sum_{H=N(1-\epsilon)}^{N} G(\mathbf{0}; H) \geq 2^{-N\epsilon} \sum_{H=N(1-\epsilon)}^{N} 2^{N-H} G(\mathbf{0}; H) = 2^{-N\epsilon} S_0(N) \left[1 - A\right]$$

with $A = 2^{N\epsilon} S_0(N(1 - \epsilon))/S_0(N)$. From Lemma 2, $A \leq (\kappa + o(1))^{\alpha N}$ with $\kappa = \frac{\gamma(0, 0, 1 - \epsilon)}{7/8} = 1 - \frac{9}{7} \epsilon^2 + \frac{2}{7} \epsilon^3 < 1$ for small enough ϵ (but $\Theta(1)$ with respect to N). We conclude that A is exponential small in N, and $-\epsilon + 1 - \alpha \log_2 \frac{8}{7} + o(1) \leq \frac{1}{N} \log_2 L_S \leq 1 - \alpha \log_2 \frac{8}{7}$. Choosing arbitrarily small ϵ allows us to establish the statement about the asymptotic behaviour of L_S in Theorem 1.

Proposition 2, with arbitrarily small ϵ, immediately leads to Theorem 1 for $k = 3$, for the average number of contradiction leaves, L_C, equals the sum over all heights $T = tN$ (with $0 \leq t \leq 1$) of $L_C(N, T, \alpha)$, and the sum is bounded from below by its largest term and, from above, by N times this largest term. The statement on the number of leaves following Theorem 1 comes from the observation that the expected total number of leaves is $L_S + L_C$, and $\omega_S(\alpha, 3) = \Omega(1, \alpha, 3) \leq \max_{t \in [0;1]} \Omega(t, \alpha, 3) = \omega_C(\alpha, 3)$. □

Proof. (*Corollary 1*) Let P_{sat} be the probability that a random 3-SAT instance with N variables and αN clauses is satisfiable. Define $\#L_{sat}$ and $\#L_{unsat}$ (respectively, L_{sat} and L_{unsat}) the expected numbers of leaves in #DPLL-UC (resp. DPLL-UC) search trees for satisfiable and unsatisfiable instances respectively. All these quantities depend on α and N. As the operations of #DPLL and DPLL coincide for unsatifiable instances, we have $\#L_{unsat} = L_{unsat}$. Conversely, $\#L_{sat} \geq L_{sat}$ since DPLL halts after having encountered the first solution leaf. Therefore, the difference between the average sizes #L and L of #DPLL-UC and DPLL-UC search trees satisfies $0 \leq \#L - L = P_{sat} (\#L_{sat} - L_{sat}) \leq P_{sat} \#L_{sat}$. Hence, $1 - P_{sat} \#L_{sat}/\#L \leq L/\#L \leq 1$. Using $\#L_{sat} \leq 2^N$, $P_{sat} \leq 2^N (7/8)^{\alpha N}$ from the first moment theorem and the asymptotic scaling for $\#L$ given in Theorem 1, we see that the left hand side of the previous inequality tends to 1 when $N \to \infty$ and $\alpha > \alpha_u$. □

Proofs for higher values of k are identical, and will be given in the full version.

4   The GUC Heuristic for Random SAT and COL

The above analysis of the DPLL-UC search tree can be extended to the GUC heuristic [8], where literals are preferentially chosen to satisfy 2-clauses (if any). The outlines of the proofs of Theorems 2 and 3 are given below; details will be found in the full version.

3-SAT. The main difference with respect to the UC case is that the two branches issued from the split are not statistically identical. In fact, the literal ℓ chosen by GUC satisfies at least one clause, while this clause is reduced to a shorter clause when ℓ is set to False. The cases $C_2 \geq 1$ and $C_2 = 0$ have also to be considered separately. With f_1, f_2, f_3 defined in the same way as in the UC case, we obtain

$$G(x_1, x_2, x_3 \; ; \; T+1) = \frac{1}{f_1} G(f_1, f_2, f_3; T) + \left(\frac{1+f_1}{f_2} - \frac{1}{f_1}\right) G(0, f_2, f_3; T)$$
$$+ \left(\frac{1+f_2}{f_3} - \frac{1+f_1}{f_2}\right) G(0, 0, f_3; T) - \frac{1+f_2}{f_3} G(0, 0, 0; T) . \quad (3)$$

The asymptotic analysis of G follows the lines of Section 3. Choosing $f_2 = f_1 + f_1^2$ i.e. $x_1 = (-1 + \sqrt{1 + 4x_2})/2 + O(1/N)$ allows us to cancel the second term on the r.h.s. of (3). Iterating relation (3), we establish the counterpart of Lemma 2 for GUC: the value of G at height $[tN]$ and argument x_2, x_3 is equal to its (known) value at height 0 and argument y_2, y_3 times the product of factors $\frac{1}{f_1}$, up to an additive term, A, including iterates of the third and fourth terms on the right hand side of (3). y_2, y_3 are the values at 'time' $\tau = 0$ of the solutions of the ordinary differential equations (ODE) $dY_2/d\tau = -2m(Y_2)/(1-\tau)$, $dY_3/d\tau = -3((1+Y_2)/2 - Y_3)/(1-\tau)$ with 'initial' condition $Y_2(t) = x_2$, $Y_3(t) = x_3$ (recall that function m is defined in Theorem 2). Eliminating 'time' between Y_2, Y_3 leads to the ODE in Theorem 2. The first term on the r.h.s. in the expression of ω^g (1) corresponds to the logarithm of the product of factors $\frac{1}{f_1}$ between heights 0 and

T. The maximum over y_2 in expression (1) for ω^g is equivalent to the maximum over the reduced height t appearing in ω_C in Theorem 1 (see also Proposition 2). Finally, choosing $\alpha > \alpha_u^g$ ensures that, from the one hand, the additive term A mentioned above is asymptotically negligible and, from the other hand, the ratio of the expected sizes of #DPLL-GUC and DPLL-GUC is asymptotically equal to unity (see proof of Corollary 1).

3-COL. The uniformity expressed by Lemma 1 holds: the subgraph resulting from the coloring of T vertices is still Erdős-Rényi-like with edge probability $\frac{c}{N}$, conditioned to the numbers C_j of vertices with j available colors [1]. The generating function G of the average number of residual asignments equals $(x_3)^N$ at height $T = 0$ and obeys the reccurence relation, for $T < N$,

$$G(x_1, x_2, x_3; T+1) = \frac{1}{f_1} \, G\big(f_1, f_2, f_3; T\big) + \left(\frac{2}{f_2} - \frac{1}{f_1}\right) G\big(0, f_2, f_3; T\big)$$

$$+ \left(\frac{3}{f_3} - \frac{2}{f_2}\right) G\big(0, 0, f_3; T\big) \tag{4}$$

with $f_1 = (1-\mu)x_1$, $f_2 = (1-2\mu)x_2 + 2\mu x_1$, $f_3 = (1-3\mu)x_3 + 3\mu x_2$, and $\mu = c/(3N)$. Choosing $f_1 = \frac{1}{2}f_2$ i.e. $x_1 = \frac{1}{2}x_2 + O(1/N)$ allows us to cancel the second term on the r.h.s. of (4). Iterating relation (3), we establish the counterpart of Lemma 2 for GUC: the value of G at height $[tN]$ and argument x_2, x_3 is equal to its (known) value at height 0 and argument y_2, y_3 respectively, times the product of factors $\frac{1}{f_1}$, up to an additive term, A, including iterates of the last term in (4). An explicit calculation leads to $G(\frac{1}{2}x_2, x_2, x_3; [tN]) = e^{N\gamma^h(x_2, x_3, t) + o(N)} + A$ for $x_2, x_3 > 0$, where $\gamma^h(x_2, x_3, t) = \frac{c}{6}t^2 - \frac{c}{3}t + (1-t)\ln(x_2/2) + \ln[3 + e^{-2ct/3}(2x_2/x_3 - 3)]$. As in Proposition 2, we bound from below (respectively, above) the number of contradiction leaves in #DPLL-GUC tree by the exponential of (N times) the value of function γ^h in $x_2 = x_3 = 1$ at reduced height t (respectively, lower than t). The maximum over t in Theorem 3 is equivalent to the maximum over the reduced height t appearing in ω_C in Theorem 1 (see also Proposition 2). Finally, we choose c_u^h to make the additive term A negligible. Following the notations of Corollary 1, we use $L_{sat} \le 3^N$, and $P_{sat} \le 3^N e^{-Nc/6 + o(N)}$, the expected number of 3-colorings for random graphs from $G(N, c/N)$.

5   Conclusion and Perspectives

We emphasize that the average #DPLL tree size can be calculated for even more complex heuristics e.g. making decisions based on literal degrees [18]. This task requires, in practice, that one is able: first, to find the correct conditioning ensuring uniformity along a branch (as in the study of DPLL in the absence of backtracking); secondly, to determine the asymptotic behaviour of the associated generating function G from the recurrence relation for G.

To some extent, the present work is an analytical implementation of an idea put forward by Knuth thirty years ago [11, 21]. Knuth indeed proposed to estimate the average computational effort required by a backtracking procedure through successive runs of the non–backtracking counterpart, each weighted in an appropriate way [21]. This weight is, in the language of Section II.B, simply the probability of a branch (given the heuristic under consideration) in #DPLL search tree times 2^S where S is the number of splits [11].

Since the amount of backtracking seems to have a heavy tail [16, 17], the expectation is often not a good predictor in practice. Knowledge of the second moment of the search tree size would be very precious; its calculation, currently under way, requires us to treat the correlations between nodes attached to distinct branches. Calculating the second moment is a step towards the distant goal of finding the expectation of the logarithm, which probably requires a deep understanding of correlations as in the replica theory of statistical mechanics.

Last of all, #DPLL is a complete procedure for enumeration. Understanding its average-case operation will, hopefully, provide us with valuable information not only on the algorithm itself but also on random decision problems e.g. new bounds on the sat/unsat or col/uncol thresholds, or insights on the statistical properties of solutions.

Acknowledgments

The present analysis is the outcome of a work started four years ago with S. Cocco to which I am deeply indebted [10, 11]. I am grateful to C. Moore for numerous and illuminating discussions, as well as for a critical reading of the manuscript. I thank J. Franco for his interest and support, and the referee for pointing out Ref. [6], the results of which agree with the $\alpha^{-1/(k-2)}$ asymptotic scaling of ω found here.

References

1. Achlioptas, D. and Molloy, M. Analysis of a List-Coloring Algorithm on a Random Graph, in *Proc. Foundations of Computer Science (FOCS)*, vol 97, 204 (1997).
2. Achlioptas, D. Lower bounds for random 3-SAT via differential equations, *Theor. Comp. Sci.* **265**, 159–185 (2001).
3. Achlioptas, D., Beame, P. and Molloy, M. A Sharp Threshold in Proof Complexity. in *Proceedings of STOC 01*, p.337-346 (2001).
4. Beame, P., Karp, R., Pitassi, T. and Saks, M. On the complexity of unsatisfiability of random k-CNF formulas. In *Proceedings of the Thirtieth Annual ACM Symposium on Theory of Computing (STOC98)*, p. 561–571, Dallas, TX (1998).
5. Beame, P., Culberson, J., Mitchell, D. and Moore, C. The resolution complexity of random graph k-colorability. To appear in *Discrete Applied Mathematics* (2003).
6. Ben-Sasson, E. and Galesi, N. Space Complexity of Random Formulae in Resolution, *Random Struct. Algorithms* **23**, 92 (2003).
7. Chao, M.T. and Franco, J. Probabilistic analysis of two heuristics for the 3-satisfiability problem, *SIAM Journal on Computing* **15**, 1106-1118 (1986).
8. Chao, M.T. and Franco, J. Probabilistic analysis of a generalization of the unit clause literal selection heuristics for the k-satisfiability problem, *Information Science* **51**, 289–314 (1990).

9. Chvátal, V. and Szemerédi, E. Many hard examples for resolution, *J. Assoc. Comput. Mach.* **35**, 759–768 (1988).
10. Cocco, S. and Monasson, R. Analysis of the computational complexity of solving random satisfiability problems using branch and bound search algorithms, *Eur. Phys. J. B* **22**, 505 (2001).
11. Cocco, S. and Monasson, R. Heuristic average-case analysis of the backtrack resolution of random 3-Satisfiability instances. *Theor. Comp. Sci.* **320**, 345 (2004).
12. Franco, J. and Paull, M. Probabilistic analysis of the Davis-Putnam procedure for solving the satisfiability problem. *Discrete Appl. Math.* **5**, 77–87 (1983).
13. Ein-Dor, L. and Monasson, R. The dynamics of proving uncolourability of large random graphs: I. Symmetric colouring heuristic. *J. Phys. A* **36**, 11055-11067 (2003).
14. Franco, J. Results related to thresholds phenomena research in satisfiability: lower bounds. *Theor. Comp. Sci.* **265**, 147–157 (2001).
15. Frieze, A. and Suen, S. Analysis of two simple heuristics on a random instance of k-SAT. *J. Algorithms* **20**, 312–335 (1996).
16. Gent, I. and Walsh, T. Easy problems are sometimes hard. *Artificial Intelligence* **70**, 335 (1993).
17. Jia, H. and Moore, C. How much backtracking does it take to color random graphs? Rigorous results on heavy tails. In *Proc. 10th International Conference on Principles and Practice of Constraint Programming* (CP '04) (2004).
18. Kaporis, A.C., Kirousis, L.M. and Lalas, E.G. The Probabilistic Analysis of a Greedy Satisfiability Algorithm. *ESA*, p. 574-585 (2002).
19. Karp, R.M. The probabilistic analysis of some combinatorial search algorithms, in J.F. Traub, ed. Algorithms and Complexity, Academic Press, New York (1976).
20. Knuth, D.E. The art of computer programming, vol. 1: Fundamental algorithms, Section 2.2.1, Addison-Wesley, New York (1968).
21. Knuth, D.E. Estimating the efficiency of backtrack programs. *Math. Comp.* **29**, 136 (1975).
22. Méjean, H-M., Morel, H. and Reynaud, G. A variational method for analysing unit clause search. *SIAM J. Comput.* **24**, 621 (1995).
23. Mitchell, D., Selman, B. and Levesque, H. Hard and Easy Distributions of SAT Problems, *Proc. of the Tenth Natl. Conf. on Artificial Intelligence (AAAI-92)*, 440-446, The AAAI Press / MIT Press, Cambridge, MA (1992).
24. Purdom, P.W. A survey of average time analyses of satisfiability algorithms. *J. Inform. Process.* **13**, 449 (1990).

A Continuous-Discontinuous Second-Order Transition in the Satisfiability of Random Horn-SAT Formulas

Cristopher Moore[1], Gabriel Istrate[2], Demetrios Demopoulos[3], and Moshe Y. Vardi[4]

[1] Department of Computer Science, University of New Mexico, Albuquerque NM, USA
moore@cs.unm.edu
[2] CCS-5, Los Alamos National Laboratory, Los Alamos NM, USA
istrate@lanl.gov
[3] Archimedean Academy, 12425 SW Sunset Drive, Miami FL, USA
demetrios_demopoulos@archimedean.org
[4] Department of Computer Science, Rice University, Houston TX, USA
vardi@rice.edu

Abstract. We compute the probability of satisfiability of a class of random Horn-SAT formulae, motivated by a connection with the nonemptiness problem of finite tree automata. In particular, when the maximum clause length is three, this model displays a curve in its parameter space along which the probability of satisfiability is discontinuous, ending in a second-order phase transition where it becomes continuous. This is the first case in which a phase transition of this type has been rigorously established for a random constraint satisfaction problem.

1 Introduction

In the past decade, phase transitions, or *sharp thresholds*, have been studied intensively in combinatorial problems. Although the idea of thresholds in a combinatorial context was introduced as early as 1960 [15], in recent years it has been a major subject of research in the communities of theoretical computer science, artificial intelligence and statistical physics. Phase transitions have been observed in numerous combinatorial problems in which they the probability of satisfiability jumps from 1 to 0 when the density of constraints exceeds a critical threshold.

The problem at the center of this research is, of course, 3-SAT. An instance of 3-SAT is a Boolean formula, consisting of a conjunction of clauses, where each clause is a disjunction of three literals. The goal is to find a truth assignment that satisfies all the clauses and thus the entire formula. The *density* of a 3-SAT instance is the ratio of the number of clauses to the number of variables. We call the number of variables the *size* of the instance. Experimental studies [9, 28, 29] show a dramatic shift in the probability of satisfiability of random 3-SAT instances, from 1 to 0, located at a critical density $r_c \approx 4.26$. In spite of compelling arguments, however, from statistical physics [25, 26], and rigorous upper and lower bounds on the threshold if it exists [11, 18, 23], there is still no mathematical proof that a phase transition takes place at that density. For a view variants of SAT the existence and location of phase transitions have been established rigorously, in particular for 2-SAT, 3-XORSAT, and 1-in-k SAT [2, 7, 8, 12, 17].

C. Chekuri et al. (Eds.): APPROX and RANDOM 2005, LNCS 3624, pp. 414–425, 2005.

In this paper we prove the existence of a more elaborate type of phase transition, where a curve of discontinuities in a two-dimensional parameter space ends in a *second-order* transition, where the probability of satisfiability becomes continuous. We focus on a particular variant of 3-SAT, namely Horn-SAT. A Horn clause is a disjunction of literals of which *at most one* is positive, and a Horn-SAT formula is a conjunction of Horn clauses. Unlike 3-SAT, Horn-SAT is a tractable problem; the complexity of Horn-SAT is linear in the size of the formula [10]. This tractability allows one to study random Horn-SAT formulae for much larger input sizes that we can achieve using complete algorithms for 3-SAT.

An additional motivation for studying random Horn-SAT comes from the fact that Horn formulae are connected to several other areas of computer science and mathematics [24]. In particular, Horn formulae are connected to automata theory, as the transition relation, the starting state, and the set of final states of an automaton can be described using Horn clauses. For example, if we consider automata on binary trees, then Horn clauses of length three can be used to describe its transition relation, while Horn clauses of length one can describe the starting state and the set of the final states of the automaton. (we elaborate on this below). Then the question of the emptiness of the language of the automaton can be translated to a question about the satisfiability of the formula. Since automata-theoretic techniques have recently been applied in automated reasoning [30, 31], the behavior of random Horn formulae might shed light on these applications.

Threshold properties of random Horn-SAT problems have recently been actively studied. The probability of satisfiability of random Horn formulae under two related *variable*-clause-length models was fully characterized in [20, 21]; in those models random Horn formulae have a *coarse* threshold, meaning that the probability of satisfiability is a continuous function of the parameters of the model. The variable-clause-length model used there is ideally suited to studying Horn formulae in connection with knowledge-based systems [24]. Bench-Capon and Dunne [4] studied a *fixed*-clause-length model, in which each Horn clause has precisely k literals, and proved a sharp threshold with respect to assignments that have at least $k - 1$ variables assigned to true.

Motivated by the connection between the automata emptiness problem and Horn satisfiability, Demopoulos and Vardi [14] studied the satisfiability of two types of fixed-clause-length random Horn-SAT formulae. They considered 1-2-Horn-SAT, where formulae consist of clauses of length one and two only, and 1-3-Horn-SAT, where formulae consist of clauses of length one and three only. These two classes can be viewed as the Horn analogue of 2-SAT and 3-SAT. For 1-2-Horn-SAT, they showed experimentally that there is a coarse transition (see Figure 4), which can be explained and analyzed in terms of random digraph connectivity [22]. The situation for 1-3-Horn-SAT is less clear cut. On one hand, recent results on random undirected hypergraphs [13] fit the experimental data of [14] quite well. On the other, a scaling analysis of the data suggested that transition between the mostly-satisfiable and mostly-unsatisfiable regions (the "waterfall" in Figure 1) is steep but continuous, rather than a step function. It was therefore not clear if the model exhibits a phase transition, in spite of experimental data for instances with tens of thousands of variables.

In this paper we generalize the fixed-clause-length model of [14] and offer a complete analysis of the probability of satisfiability in this model. For a finite $k > 0$ and a vector $\mathbf{d}$ of k nonnegative real numbers $d_1, d_2, \ldots, d_k$, $d_1 < 1$, let the random Horn-SAT formula $H_{n,\mathbf{d}}^k$ be the conjunction of

- a single negative literal $\overline{x}_1$,
- $d_1 n$ positive literals chosen uniformly without replacement from $x_2, \ldots, x_n$, and
- for each $2 \leq j \leq k$, $d_j n$ clauses chosen uniformly from the $j\binom{n}{j}$ possible Horn clauses with j variables where one literal is positive.

Thus, the classes studied in [14] are H_{n,d_1,d_2}^2 and $H_{n,d_1,0,d_3}^3$ respectively.

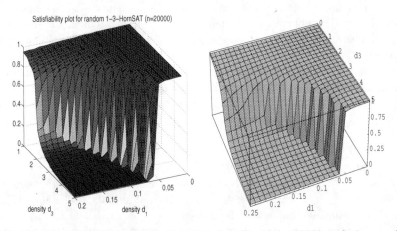

Fig. 1. Satisfiability probability of random 1-3-Horn formulae of size 20000. Left, the experimental results of [14]; right, our analytic results

With this model in hand, we settle the question of sharp thresholds for 1-3-Horn-SAT. In particular, we show that there are sharp thresholds in some regions of the (d_1, d_3) plane in the probability of satisfiability, although not from 1 to 0. We start with the following general result for the $H_{n,\mathbf{d}}^k$ model.

Theorem 1. *Let t_0 be the smallest positive root of the equation*

$$\ln \frac{1-t}{1-d_1} + \sum_{j=2}^{k} d_j t^{j-1} = 0 \; . \tag{1}$$

Call t_0 simple if it is not a double root of (1), i.e., if the derivative of the left-hand-side of (1) with respect to t is nonzero at t_0. If t_0 is simple, the probability that a random formula from $H_{n,\mathbf{d}}^k$ is satisfiable in the limit $n \to \infty$ is

$$\Phi(\mathbf{d}) := \lim_{n \to \infty} \Pr[H_{n,\mathbf{d}}^k \text{ is satisfiable}] = \frac{1 - t_0}{1 - d_1} \; . \tag{2}$$

Specializing this result to the case $k = 2$ yields an exact formula that matches the experimental results in [14]:

Proposition 1. *The probability that a random formula from H^2_{n,d_1,d_2} is satisfiable in the limit $n \to \infty$ is*

$$\Phi(d_1, d_2) := \lim_{n \to \infty} \Pr[H^2_{n,d_1,d_2} \text{ is satisfiable}] = -\frac{W\left(-(1 - d_1)d_2 e^{-d_2}\right)}{(1 - d_1)d_2} . \tag{3}$$

Here $W(\cdot)$ is Lambert's function, defined as the principal root y of $ye^y = x$.

For the case $k = 3$ and $d_2 = 0$, we do not have a closed-form expression for the probability of satisfiability, though numerically Figure 1 shows a very good fit to the experimental results of [14]. In addition, we find an interesting phase transition behavior in the (d_1, d_3) plane, described by the following proposition.

Proposition 2. *The probability of satisfiability $\Phi(d_1, d_3)$ that a random formula from $H^3_{n,d_1,0,d_3}$ is satisfiable is continuous for $d_3 < 2$ and discontinuous for $d_3 > 2$. Its discontinuities are given by a curve Γ in the (d_1, d_3) plane described by the equation*

$$d_1 = 1 - \frac{\exp\left(\frac{1}{4}\left(\sqrt{d_3} - \sqrt{d_3 - 2}\right)^2\right)}{d_3 - \sqrt{d_3(d_3 - 2)}} . \tag{4}$$

This curve consists of the points (d_1, d_3) at which t_0 is a double root of (1), and ends at the critical point

$$(1 - \sqrt{e}/2, 2) = (0.1756..., 2) .$$

The curve Γ of discontinuities described in Proposition 2 can be seen in the right part of Figure 1. The drop at the "waterfall" decreases as we approach the critical point $(0.1756..., 2)$, where the probability of satisfiability becomes continuous (although its derivatives at that point are infinite). We can also see this in Figure 2, which shows a contour plot; the contours are bunched at the discontinuity, and "fan out" at the critical point. In both cases our calculates closely match the experimental results of [14].

In statistical physics, we would say that Γ is a curve of *first-order* transitions, in which the order parameter is discontinuous, ending in a *second-order* transition at the tip of the curve, at which the order parameter is continuous, but has a discontinuous derivative (see e.g. [5]). A similar transition takes place in the Ising model, where the two parameters are the temperature T and the external field H; the magnetization is discontinuous at the line $H = 0$ for $T < T_c$ where T_c is the transition temperature, but there is a second-order transition at $(T_c, 0)$ and the magnetization is continuous for $T \geq T_c$.

To our knowledge, this is the first time that a continuous-discontinuous transition of this type has been established rigorously in a model of random constraint satisfaction problems. We note that a similar phenomenon is believed to take place for $(2 + p)$-SAT model at the triple point $p = 2/5$; here the order parameter is the size of the backbone, i.e., the number of variables that take fixed values in every truth assignment [3, 27]. Indeed, in our model the probability of satisfiability is closely related to the size of the backbone, as we see below.

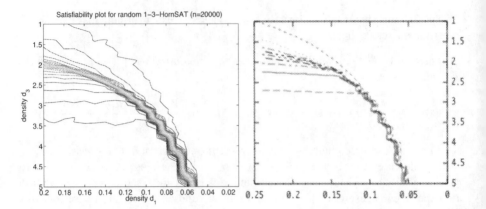

Fig. 2. Contour plots of the probability of satisfiability of random 1-3-Horn formulae. Left, the experimental results of [14]. Right, our analytical results

2   Horn-SAT and Automata

Our main motivation for studying the satisfiability of Horn formulae is the unusually rich type of phase transition described above, and the fact that its tractability allows us to perform experiments on formulae of very large size. The original motivation of [14] that led to the present work is the fact that Horn formulae can be used to describe finite automata on words and trees.

A finite automaton A is a 5-tuple $A = (S, \Sigma, \delta, s, F)$, where S is a finite set of states, Σ is an alphabet, s is a starting state, $F \subseteq S$ is the set of final (accepting) states and δ is a transition relation. In a word automaton, δ is a function from $S \times \Sigma$ to 2^S, while in a binary-tree automaton δ is a function from $S \times \Sigma$ to $2^{S \times S}$. Intuitively, for word automata δ provides a set of successor states, while for binary-tree automata δ provides a set of successor state pairs. A run of an automaton on a word $a = a_1 a_2 \cdots a_n$ is a sequence of states $s_0 s_1 \cdots s_n$ such that $s_0 = s$ and $(s_{i-1}, a_i, s_i) \in \delta$. A run is succesful if $s_n \in F$: in this case we say that A accepts the word a. A run of an automaton on a binary tree t labeled with letters from Σ is a binary tree r labeled with states from S such that $\text{root}(r) = s$ and for a node i of t, $(r(i), t(i), r(\text{left-child-of-}i), r(\text{right-child-of-}i)) \in \delta$. Thus, each pair in $\delta(r(i), t(i))$ is a possible labeling of the children of i. A run is succesful if for all leaves l of r, $r(l) \in F$: in this case we say that A accepts the tree t. The language $L(A)$ of a word automaton A is the set of all words a for which there is a successful run of A on a. Likewise, the language $L(A)$ of a tree automaton A is the set of all trees t for which there is a successful run of A on t. An important question in automata theory, which is also of great practical importance in the field of formal verification [30, 31], is: given an automaton A, is $L(A)$ non-empty? We now show how the problem of non-emptiness of automata languages translates to Horn satisfiability. Thus, getting a better understanding of the satisfiability of Horn formulae would tell us about the expected answer to automata nonemptiness problems.

Consider first a word automaton $A = (S, \Sigma, \delta, s_0, F)$. Construct a Horn formula ϕ_A over the set S of variables as follows: create a clause $(\bar{s}_0)$, for each $s_i \in F$ create a

---

Algorithm PUR:
1. while (ϕ contains positive unit clauses)
2.   choose a random positive unit clause x
3.   remove all other clauses in which x occurs positively in ϕ
4.   shorten all clauses in which x appears negatively
5.   label x as "implied" and call the algorithm recursively.
6. if no contradictory clause was created
7.   accept ϕ
8. else
9.   reject ϕ.

Fig. 3. Positive Unit Resolution

clause (s_i), for each element (s_i, a, s_j) of δ create a clause $(\bar{s}_j, s_i)$, where $(s_i, \cdots, s_k)$ represents the clause $s_i \vee \cdots \vee s_k$ and $\bar{s}_j$ is the negation of s_j. Note that the formula ϕ_A consists of unary and binary clauses. Similarly to the word automata case, we can show how to construct a Horn formula from a binary-tree automaton. Let $A = (S, \Sigma, \delta, s_0, F)$ be a binary-tree automaton. Then we can construct a Horn formula ϕ_A using the construction above with the only difference that since δ in this case is a function from $S \times \{a\}$ to $S \times S$, for each element (s_i, α, s_j, s_k) of δ we create a clause $(\bar{s}_j, \vec{s}_k, s_i)$. Note that the formula ϕ_A now consists of unary and ternary clauses. This explains why the formula classes studied in [14] are H^2_{n,d_1,d_2} and $H^3_{n,d_1,0,d_3}$.

Proposition 3 ([14]). *Let A be a word or binary tree automaton and ϕ_A the Horn formula constructed as described above. Then $L(A)$ is non-empty if and only if ϕ_A is unsatisfiable.*

3 Main Result

Consider the positive unit resolution algorithm PUR applied to a random formula ϕ (Figure 3). The proof of Theorem 1 follows immediately from the following theorem, which establishes the size of the backbone of the formula with the single negative literal $\bar{x}_1$ removed: that is, the set of positive literals implied by the positive unit clauses and the clauses of length greater than 1. Then ϕ is satisfiable as long as x_1 is not in this backbone.

Lemma 1. *Let ϕ be a random Horn-SAT formula $H^k_{n,\mathbf{d}}$ with $d_1 > 0$. Denote by t_0 the smallest positive root of Equation (1), and suppose that t_0 is simple. Then, for any $\epsilon > 0$, the number $N_{\mathbf{d},n}$ of implied positive literals, including the $d_1 n$ initially positive literals, satisfies w.h.p. the inequality*

$$(t_0 - \epsilon) \cdot n < N_{\mathbf{d},n} < (t_0 + \epsilon) \cdot n, \tag{5}$$

Proof. First, we give a heuristic argument, analogous to the branching process argument for the size of the giant component in a random graph. The number m of clauses of length j with a given literal x as their implicate (i.e., in which x appears positively) is Poisson-distributed with mean d_j. If any of these clauses have the property that all their literals whose negations appear are implied, then x is implied as well. In addition, x is implied if it is one of the $d_1 n$ initially positive literals. Therefore, the probability that x is *not* implied is the probability that it is not one of the initially positive literals, and that, for all j, for all m clauses c of length j with x as their implicate, at least one of the $j - 1$ literals whose negations appear in c is not implied. Assuming all these events are independent, if t is the fraction of literals that are implied, we have

$$1 - t = (1 - d_1) \prod_{j=2}^{k} \left(\sum_{m=0}^{\infty} \frac{e^{-d_j} d_j^m}{m!} (1 - t^{j-1})^m \right)$$

$$= (1 - d_1) \prod_{j=2}^{k} \exp(-d_j t^{j-1})) = (1 - d_1) \exp \left(-\sum_{j=2}^{k} d_j t^{j-1} \right)$$

yielding Equation (1).

To make this rigorous, we use a standard technique for proving results about threshold properties: analysis of algorithms via differential equations [32] (see [1] for a review). We analyze the while loop of PUR shown in Figure 3; specifically, we view PUR as working in *stages*, indexed by the number of literals that are labeled "implied." After T stages of this process, T variables are labeled as implied. At each stage the resulting formula consists of a set of Horn clauses of length j for $1 \leq j \leq k$ on the $n - T$ unlabeled variables. Let the number of distinct clauses of length j in this formula be $S_j(T)$; we rely on the fact that, at each stage T, conditioned on the values of $S_j(T)$ the formula is uniformly random. This follows from an easy induction argument which is standard for problems of this type (see e.g. [21]).

Now, the variables appearing in the clauses present at stage T are chosen uniformly from the $n - T$ remaining variables, so the probability that the chosen variable x appears in a given clause of length j is $j/(n - T)$, and the probability that a given clause of length $j + 1$ is shortened to one of length j (as opposed to removed) is $j/(n - T)$. A newly shortened clause is distinct from all the others with probability $1 - o(1)$ unless $j = 1$, in which case it is distinct with probability $(n - T - S_1)/(n - T)$. Finally, each stage labels the variable in one of the $S_1(T)$ unit clauses as implied. Thus the expected effect of each step is

$$E[S_j(T + 1)] = S_j(T) + j \frac{S_{j+1}(T) - S_j(T)}{n - T} + o(1) \quad \text{for all } 2 \leq j \leq k$$

$$E[S_1(T + 1)] = S_1(T) + \left(\frac{n - T - S_1}{n - T} \right) \left(\frac{S_2(T)}{n - T} \right) - 1 + o(1)$$

Wormald's theorem [32] allows us to rescale these difference equations by setting $T = t \cdot n$ and $S_j(T) = s_j(t) \cdot n$, obtaining the following system of differential equations:

$$\frac{ds_j}{dt} = j \frac{s_{j+1}(t) - s_j(t)}{1 - t} \quad \text{for all } 2 \leq j \leq k$$

$$\frac{ds_1}{dt} = \left(\frac{1 - t - s_1(t)}{1 - t} \right) \left(\frac{s_2(t)}{1 - t} \right) - 1 . \tag{6}$$

Then for any constant $\delta > 0$, for all t such that $s_1 > \delta$, w.h.p. we have $S_j(t \cdot n) = s_j(t) \cdot n + o(n)$ where $s_j(t)$ is the solution to the system (6). With the initial conditions

$s_j(0) = d_j$ for $1 \leq j \leq k$, a little work shows that this solution is

$$s_j(t) = (1-t)^j \sum_{\ell=j}^{k} \binom{\ell-1}{j-1} d_\ell t^{\ell-j} \quad \text{for all } 2 \leq j \leq k$$

$$s_1(t) = 1 - t - (1-d_1)\exp\left(-\sum_{j=2}^{k} d_j t^{j-1}\right) . \tag{7}$$

Note that $s_1(t)$ is continuous, $s_1(0) = d_1 > 0$, and $s_1(1) < 0$ since $d_1 < 1$. Thus $s_1(t)$ has at least one root in the interval $(0,1)$. Since PUR halts when there are no unit clauses, i.e., when $S_1(T) = 0$, we expect the stage at which it halts. Thus the number of implied positive literals, to be $T = t_0 n + o(n)$ where t_0 is the smallest positive root of $s_1(t) = 0$, or equivalently, dividing by $1-d_1$ and taking the logarithm, of Equation (1).

However, the conditions of Wormald's theorem do not allow us to run the differential equations all the way up to stage $t_0 n$. We therefore choose small constants $\epsilon, \delta > 0$ such that $s_1(t_0 - \epsilon) = \delta$ and run the algorithm until stage $(t_0 - \epsilon)n$. At this point $(t_0 - \epsilon)n$ literals are already implied, proving the lower bound of (5).

To prove the upper bound of (5), recall that by assumption t_0 is a simple root of (1), i.e., the second derivative of the left-hand size of (1) with respect to t is nonzero at t_0. It is easy to see that this is equivalent to $ds_1/dt < 0$ at t_0. Since ds_1/dt is analytic, there is a constant $c > 0$ such that $ds_1/dt < 0$ for all $t_0 - c \leq t \leq t_0 + c$. Set $\epsilon < c$; the number of literals implied during these stages is bounded above by a subcritical branching process whose initial population is w.h.p. $\delta n + o(n)$, and by standard arguments we can bound its total progeny to be $\epsilon' n$ for any $\epsilon' > 0$ by taking δ small enough.

It is easy to see that the backbone of implied positive literals is a uniformly random subset of $\{x_1, \ldots, x_n\}$ of size $N_{\mathbf{d},n}$. Since x_1 is guaranteed to not be among the $d_1 n$ initially positive literals, the probability that x_1 is not in this backbone is

$$\frac{n - N_{\mathbf{d},n}}{(1-d_1) \cdot n}.$$

By completeness of positive unit resolution for Horn satisfiability, this is precisely the probability that the ϕ is satisfiable. Applying Lemma 1 and taking $\epsilon \to 0$ proves equation (2) and completes the proof of Theorem 1.

We make several observations. First, if we set $k = 2$ and take the limit $d_1 \to 0$, Theorem 1 recovers Karp's result [22] on the size of the reachable component of a random directed graph with mean out-degree $d = d_2$, the root of $\ln(1-t) + dt = 0$.

Secondly, as we will see below, the condition that t_0 is simple is essential. Indeed, for the 1-3-Horn-SAT model studied in [14], the curve Γ of discontinuities, where the probability of satisfiability drops in the "waterfall" of Figure 1, consists exactly of those (d_1, d_3) where t_0 is a double root, which implies $ds_1/dt = 0$ at t_0.

Finally, we note that Theorem 1 is very similar to Darling and Norris's work on identifiability in random undirected hypergraphs [13], where the number of hyperedges of length j is Poisson-distributed with mean β_j. Their result reads

$$\ln(1-t) + \sum_{j=1}^{k} j\beta_j t^{j-1} = 0 .$$

We can make this match (1) as follows. First, since each hyperedge of length j corresponds to j Horn clauses, we set $d_j = j\beta_j$ for all $j \geq 2$. Then, since edges are chosen with replacement in their model, the expected number of distinct clauses of length 1 (i.e., positive literals) is $d_1 n$ where $d_1 = 1 - e^{-\beta_1}$.

4   Application to H^2_{n,d_1,d_2}

For H^2_{n,d_1,d_2}, Equation (1) can rewritten as $1 - t = (1 - d_1) \cdot e^{-d_2 \cdot t}$. Denoting $y = d_2(t - 1)$, this implies

$$y \cdot e^y = d_2(t - 1) \cdot e^{d_2 \cdot (t-1)} = -d_2(1 - d_1) \cdot e^{-d_2 \cdot t} \cdot e^{d_2 \cdot (t-1)} = -(1 - d_1)d_2 e^{-d_2}.$$

Solving this yields

$$t_0 = 1 + \frac{1}{d_2} W\left(-(1 - d_1)d_2 e^{-d_2}\right)$$

and substituting this into (2) proves Equation (3) and Proposition 1. In Figure 4 we plot the probability of satisfiability $\Phi(d_1, d_2)$ as a function of d_2 for $d_1 = 0.1$ and compare it with the experimental results of [14]; the agreement is excellent.

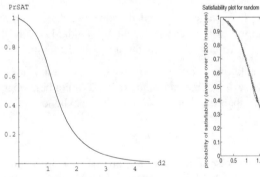
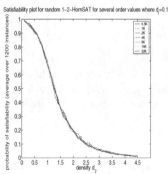

Fig. 4. The probability of satisfiability for 1-2-Horn formulae as a function of d_2, where $d_1 = 0.1$. Left, our analytic results; right, the experimental data of [14]

5   A Continuous-Discontinuous Phase Transition for $H^3_{n,d_1,0,d_3}$

For the random model $H^3_{n,d_1,0,d_3}$ studied in [14], an analytic solution analogous to (3) does not seem to exist. Let us, however, rewrite (1) as

$$1 - t = f(t) := (1 - d_1)e^{-d_3 t^2} . \tag{8}$$

We claim that for some values of d_1 and d_3 there is a phase transition in the roots of (8). For instance, consider the plot of $f(t)$ shown in Figure 5 for $d_1 = 0.1$ and $d_3 = 3$. Here $f(t)$ is tangent to $1 - t$, so there is a bifurcation as we vary either parameter; for $d_3 = 2.9$, for instance, $f(t)$ crosses $1 - t$ three times and there is a root of (8) at $t =$

0.185, but for $d_3 = 3.1$ the unique root is at $t = 0.943$. This causes the probability of satisfiability to jump discontinuously but from 0.905 to 0.064. The set of pairs (d_1, d_3) for which $f(t)$ is tangent to $1 - t$ is exactly the curve Γ on which the smallest positive root t_0 of (1) or (8) is a double root, giving the "waterfall" of Figure 1.

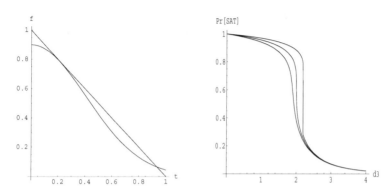

Fig. 5. Left, the function $f(t)$ of (8) when $d_1 = 0.1$ and $d_3 = 3$. Right, the probability of satisfiability $\Phi(d_1, d_3)$ with d_1 equal to 0.15 (continuous), 0.1756 (critical), and 0.2 (discontinuous)

To find where this transition takes place, we set $f' = -1$, yielding $f(t) = 1/(2d_3 t)$. Setting this to $1 - t$ and solving for t gives

$$d_1 = 1 - \frac{e^{d_3 t^2}}{2d_3 t} \tag{9}$$

where

$$t = \frac{1}{2}\left(1 - \sqrt{1 - \frac{2}{d_3}}\right) \tag{10}$$

Substituting (10) in (9) and simplifying gives (4), proving Proposition 2.

The fact that d_1 is only real for $d_3 \geq 2$ explains why Γ ends at $d_3 = 2$. At this extreme case we have

$$d_1 = 1 - \frac{\sqrt{e}}{2} \approx 0.1756 \text{ and } \frac{\partial d_1}{\partial d_3} = -\frac{\sqrt{e}}{8} \ .$$

What happens for $d_3 < 2$? In this case, there are no real t for which $f'(t) = -1$, so the kind of tangency displayed in Figure 5 cannot happen. In that case, (8) (and equivalently (1)) has a unique solution t, which varies continuously with d_1 and d_3, and therefore the probability of satisfiability $\Phi(d_1, d_3)$ is continuous as well. To illustrate this, in the right part of Figure 5 we plot $\Phi(d_1, d_3)$ as a function of d_3 with three values of d_1. For $d_1 = 0.15$, Φ is continuous; for $d_1 = 0.2$, it is discontinuous; and for $d_1 = 0.1756...$, the critical value at the second-order transition, it is continuous but has an infinite derivative at $d_3 = 2$.

Acknowledgments

We thank Mark Newman for helpful conversations. CM would also like to acknowledge the Big Island of Hawai'i where some of this work was done. This work has been supported by the U.S. Department of Energy under contract W-705-ENG-36, by NSF grants PHY-0200909, EIA-0218563, CCR-0220070, CCR-0124077, CCR-0311326, IIS-9908435, IIS-9978135, eIA-0086264, and ANI-0216467, by BSF grant 9800096, and by Texas ATP grant 003604-0058-2003.

References

1. D. Achlioptas. Lower Bounds for Random 3-SAT via Differential Equations. *Theoretical Computer Science* 265 (1-2), 159–185, 2001.
2. D. Achlioptas, A. Chtcherba, G. Istrate, and C. Moore. The phase transition in 1-in-k SAT and NAE 3-SAT. In *Proc. 12th ACM-SIAM Symp. on Discrete Algorithms*, 721–722, 2001.
3. D. Achlioptas, L.M. Kirousis, E. Kranakis, and D. Krizanc. Rigorous results for random $(2 + p)$-SAT. *Theor. Comput. Sci.* 265(1-2): 109-129 (2001).
4. T. Bench-Capon and P. Dunne. A sharp threshold for the phase transition of a restricted satisfiability problem for Horn clauses. *Journal of Logic and Algebraic Programming*, 47(1):1–14, 2001.
5. J. J. Binney, N. J. Dowrick, A. J. Fisher and M. E. J. Newman. *The Theory of Critical Phenomena.* Oxford University Press (1992).
6. B. Bollobás. *Random Graphs.* Academic Press, 1985.
7. S. Cocco, O. Dubois, J. Mandler, and R. Monasson. Rigorous decimation-based construction of ground pure states for spin glass models on random lattices. *Phys. Rev. Lett.*, 90, 2003.
8. V. Chvátal and B. Reed. Mick gets some (the odds are on his side). In *Proc. 33rd IEEE Symp. on Foundations of Computer Science*, IEEE Comput. Soc. Press, 620–627, 1992.
9. J. M. Crawford and L. D. Auton. Experimental results on the crossover point in random 3-SAT. *Artificial Intelligence*, 81(1-2):31–57, 1996.
10. W. F. Dowling and J. H. Gallier. Linear-time algorithms for testing the satisfiability of propositional Horn formulae. *Logic Programming. (USA) ISSN: 0743-1066*, 1(3):267–284, 1984.
11. O. Dubois, Y. Boufkhad, and J. Mandler. Typical random 3-SAT formulae and the satisfiability theshold. In *Proc. 11th ACM-SIAM Symp. on Discrete Algorithms*, 126–127, 2000.
12. O. Dubois and J. Mandler. The 3-XORSAT threshold. In *Proc. 43rd IEEE Symp. on Foundations of Computer Science*, 769–778, 2002.
13. R. Darling and J.R. Norris. Structure of large random hypergraphs. *Annals of Applied Probability*, 15(1A), 2005.
14. D. Demopoulos and M. Vardi. The phase transition in random 1-3 Hornsat problems. In A. Percus, G. Istrate, and C. Moore, editors, *Computational Complexity and Statistical Physics*, Santa Fe Institute Lectures in the Sciences of Complexity. Oxford University Press, 2005. Available at http://www.cs.rice.edu/~vardi/papers/.
15. P. Erdös and A. Rényi. On the evolution of random graphs. *Publications of the Mathematical Institute of the Hungarian Academy of Science*, 5:17–61, 1960.
16. E. Friedgut. Necessary and sufficient conditions for sharp threshold of graph properties and the k-SAT problem. *J. Amer. Math. Soc.*, 12:1917–1054, 1999.
17. A. Goerdt. A threshold for unsatisfiability. *J. Comput. System Sci.*, 53(3):469–486, 1996.
18. M. Hajiaghayi and G.B. Sorkin. The satisfiability threshold for random 3-SAT is at least 3.52. IBM Technical Report , 2003.

19. T. Hogg and C. P. Williams. The hardest constraint problems: A double phase transition. *Artificial Intelligence*, 69(1-2):359–377, 1994.
20. G. Istrate. The phase transition in random Horn satisfiability and its algorithmic implications. *Random Structures and Algorithms*, 4:483–506, 2002.
21. G. Istrate. On the satisfiability of random k-Horn formulae. In P. Winkler and J. Nesetril, editors, *Graphs, Morphisms and Statistical Physics*, volume 64 of *AMS-DIMACS Series in Discrete Mathematics and Theoretical Computer Science*, 113–136, 2004.
22. R. Karp. The transitive closure of a random digraph. *Random Structures and Algorithms*, 1:73–93, 1990.
23. A. Kaporis, L. M. Kirousis, and E. Lalas. Selecting complementary pairs of literals. In *Proceedings of LICS'03 Workshop on Typical Case Complexity and Phase Transitions*, June 2003.
24. J. A. Makowsky. Why Horn formulae matter in Computer Science: Initial structures and generic examples. *JCSS*, 34(2-3):266–292, 1987.
25. M. Mézard and R. Zecchina. Random k-satisfiability problem: from an analytic solution to an efficient algorithm. *Phys. Rev. E*, 66:056126, 2002.
26. M. Mézard, G. Parisi, and R. Zecchina. Analytic and algorithmic solution of random satisfiability problems. *Science*, 297:812–815, 2002.
27. R. Monasson, R. Zecchina, S. Kirkpatrick, B. Selman, and L. Troyansky. $2+p$-SAT: Relation of typical-case complexity to the nature of the phase transition. *Random Structures and Algorithms*, 15(3–4):414–435, 1999.
28. B. Selman, D. G. Mitchell, and H. J. Levesque. Generating hard satisfiability problems. *Artificial Intelligence*, 81(1-2):17–29, 1996.
29. B. Selman and S. Kirkpatrick. Critical behavior in the computational cost of satisfiability testing. *Artificial Intelligence*, 81(1-2):273–295, 1996.
30. M.Y. Vardi and P. Wolper. Automata-Theoretic Techniques for Modal Logics of Programs *J. Computer and System Science* 32:2(1986), 181–221, 1986.
31. M.Y. Vardi and P. Wolper. An automata-theoretic approach to automatic program verification (preliminary report). In *Proc. 1st IEEE Symp. on Logic in Computer Science*, 332–344, 1986.
32. N. Wormald. Differential equations for random processes and random graphs. *Annals of Applied Probability*, 5(4):1217–1235, 1995.

Mixing Points on a Circle

Dana Randall[1,*] and Peter Winkler[2,**]

[1] College of Computing, Georgia Institute of Technology, Atlanta GA 30332-0280
[2] Dept. of Mathematics, Dartmouth College, Hanover NH 03755-3551

Abstract. We determine, up to a log factor, the mixing time of a Markov chain whose state space consists of the successive distances between n labeled "dots" on a circle, in which one dot is selected uniformly at random and moved to a uniformly random point between its two neighbors. The method involves novel use of auxiliary discrete Markov chains to keep track of a vector of quadratic parameters.

1 Introduction

Randomized approximation algorithms using Markov chains have proven to be powerful tools with an astounding variety of applications. Most notably, when the state space consists of configurations of a physical system, random sampling provides a method for estimating the thermodynamic properties of the model. Showing that these chains converge quickly to their stationary distributions is typically the vital step in establishing the efficiency of these algorithms.

Over the last 15 years there has been substantial progress in developing methods for bounding the mixing rates of finite Markov chains, including coupling, canonical paths, and various isoperimetric inequalities (see, e.g., [4], [6], and [11] for surveys). As a consequence, there are now provably efficient algorithms for many discrete sampling problems, such as matchings, independents sets and colorings. However, there has been a scarcity of results for problems defined on continuous spaces, mostly because methods used to analyze discrete chains seem to break down in the continuous setting. This is likely an artifact of the appearance of the "minimum stationary probability" parameter which often arises in the discrete cases, and not anything inherent in continuous chains, which often seem to be efficient. Thus, an obvious challenge to the mixing community is to develop new tools amenable to Markov chains on continuous spaces.

For example, the hard-core lattice gas model from statistical physics examines the possible configurations of n balls of radius r in a finite region R of $\mathbf{R}^n$, where balls are required to lie in non-overlapping positions. Although the centers of the balls can occupy any real point in $\mathbf{R}^n$, the model is typically discretized so that they lie on lattice points in $\Lambda_n \subseteq \mathbf{Z}^n$. In this setting, if r is chosen to be slightly more than $1/2$, then configurations correspond to independent sets on the lattice

* Supported in part by NSF grant CCR-0105639.
** Completed during a term as Clay Senior Fellow at the Mathematical Sciences Research Institute.

C. Chekuri et al. (Eds.): APPROX and RANDOM 2005, LNCS 3624, pp. 426–435, 2005.

graph. While much analysis has focused on rigorous means of efficiently sampling discrete models such as independent sets in Λ_n (e.g., [3], [8]) very little is known about sampling in the continuous setting.

One powerful tool for sampling points in a continuous state space is a walk used to estimate the volume of a convex body in $\mathbf{R}^n$ [2]. Notice that this chain provides a way to sample sets consisting of n real points on the unit interval because these configurations can be mapped to points in the unit simplex. However, the moves arising from walks in the simplex are less natural in the context of independent sets since they move all the points in the independent set in a single move.

Alternative chains for sampling lattice gas configurations were considered in [5]. The first variant is a single particle, global chain that connects all pairs of states that differ by the position of one point. It is shown that the chain is rapidly mixing in $\mathbf{R}^n$ if the density of points is sufficiently low. The second variant examined is a single particle, local chain where steps involve moving one point to a new position in its current neighborhood. Although it is believed that this chain mixes rapidly in many settings, it was only shown that a discretized version of the chain mixes rapidly in one dimension. In this chain, the unit interval is subdivided into m discrete points and at each step one of the n particles is chosen and moved to an adjacent point if that point is currently unoccupied. The mixing time for this chain was shown to be $O(n^3 m^2)$.

In [12], the authors of this paper previously considered a natural chain whose stationary distribution is the set of configurations of n dots on the real interval $[0, 1]$ and showed that the mixing time is $\Theta(n^3 \log n)$. The analysis was based on a "two phase coupling": the first phase brings the dots of one chain close to the corresponding dots of the other; the second aligns them exactly.

Here, we tackle another simple and natural chain whose stationary state is uniform on the unit n-simplex. A state of the chain consists of the successive distances (along the circle) between n sequentially-numbered points on a unit-circumference circle. Moves consist of moving a random such point to a spot on the circle chosen uniformly at random between its two neighbors. We formalize the moves in the next section. The mixing time τ of the chain is then shown to satisfy

$$cn^3 \le \tau \le c'n^3 \log n,$$

for constants c, c'.

This chain seems superficially similar to the chain "n dots on an interval." However, the methods used there required a lattice structure for the state space and an eigenfunction for the chain, neither of which is available here. Instead we use an entirely different and apparently novel approach involving auxiliary discrete Markov chains, borrowing only the two phase coupling from [12].

2 n Dots on a Circle

To avoid confusion with other points on the circle, the n designated points defining the chain will be called *dots* and the chain itself will be called "n dots on a circle." The circle will be assumed to have circumference 1 and its points are permanently identified with points in the half-open interval $[0, 1)$, modulo 1.

The *configuration* $\bar{x} = (x_1, \ldots, x_n)$ of chain $\mathbf{X}$ at any particular time is the vector of locations of dots 1 through n. The dots are assumed to be labeled clockwise around the circle.

A step of the chain is determined by a uniformly random integer k between 1 and n, and an independent real λ chosen from the Lebesgue distribution on $[0, 1]$. If the previous configuration of the chain was $\bar{x}$, the new configuration will be

$$\bar{x}' = (x_1, \ldots, x_{k-1}, (1-\lambda)x_{k-1} + \lambda x_{k+1}, x_{k+1}, \ldots, x_n)$$

where the subscripts are taken modulo n.

For the *state* of $\mathbf{X}$ we take not $\bar{x}$, but the difference vector $\bar{u} = (u_1, \ldots, u_n)$ with $u_i := x_{i+1} - x_i \in [0, 1)$. One reason for this reduction is that otherwise, the mixing time of the chain is dominated by the less interesting issue of the location of the dots' "center of gravity" along the circumference. It is not difficult to show that this point executes a driftless random walk around the cycle which takes time $\Theta(n^4)$ to mix.

However, it is convenient for us, at times, to consider the unreduced chain $\mathbf{X}^+$ whose state is the configuration $\bar{x}$.

Theorem 1. *The stationary state of the unreduced "n dots on a circle" chain $\mathbf{X}^+$ is uniform on $[0, 1)^n$ subject to clockwise labeling modulo 1.*

Proof. This follows from detail balance (or simply noticing that the position of x_k in a uniform point is indeed uniformly random between x_{k-1} and x_{k+1}). $\quad\blacksquare$

It is convenient here to define the *mixing time* of a Markov chain $\mathbf{X}$ to be the least number t of steps such that, beginning at any state, the total variation distance $\frac{1}{2}\|\mathbf{X}(t) - \sigma\|_1$ between the state of $\mathbf{X}$ at time t and the stationary distribution σ of $\mathbf{X}$ is less than $\frac{1}{4}$.

Theorem 2. *The mixing time of the reduced "n dots on an circle" chain $\mathbf{X}$ is $O(n^3 \log n)$.*

Proof. For the upper bound, it suffices to exhibit a coupling which will achieve coalescence of any chain $\mathbf{X}$ with a stationary chain $\mathbf{Y}$ in time $O(n^3 \log n)$, with probability at least $\frac{3}{4}$. Our coupling proceeds in two phases, first bringing the dots of one chain close to the dots of the other, then aligning them exactly.

Our second ("exact alignment") phase is much like the corresponding phase in "n dots on a line." The first phase employs the familiar linear coupling, but our argument will rest on quite a different eigenfunction. Let $\bar{z} := \bar{x} - \bar{y}$, where $\bar{y}$ is the (current) configuration of the $\mathbf{Y}$ chain. Coupling will be achieved exactly when $\bar{z} = c\mathbf{1}$ for some $c \in [0, 1)$, i.e. when $z_1 = z_2 = \cdots = z_n$.

The first-stage "linear" coupling of $\mathbf{X}$ and $\mathbf{Y}$ is achieved simply by using the same random k and λ at each step.

Let $s_0 := 3\sum_{i=1}^{n} z_i^2$, $s_1 = \sum_{i=1}^{n} z_i z_{i-1}$, $s_m := 2\sum_{i=1}^{n} z_i z_{i-m}$ for $m = 2, \ldots, q-1$ where $q = \lfloor n/2 \rfloor$, and $s_q := \beta \sum_{i=1}^{n} z_i z_{i-q}$, where $\beta = 2$ when n is odd but only 1 when n is even. Let us observe what becomes of $\bar{s} := (s_0, \ldots, s_q)$, in expectation, as a step of the chain is taken.

In case dot k is chosen, z_k will fall uniformly between z_{k-1} and z_{k+1}, hence

$$
Ez_k^2 = \frac{1}{z_{k+1} - z_{k-1}} \int_{z_{k-1}}^{z_{k+1}} z^2 dz = \frac{1}{3}\left(z_{k-1}^2 + z_{k-1}z_{k+1} + z_{k+1}^2\right).
$$

It follows that over all choices of k (using primes to denote future values),

$$
Es_0' = \frac{n-1}{n}s_0 + 3\frac{1}{3n}\left(\frac{s_0}{3} + \frac{s_2}{2} + \frac{s_0}{3}\right) = \left(1 - \frac{1}{3n}\right)s_0 + \frac{1}{2n}s_2.
$$

Similarly,

$$
Es_1' = \left(1 - \frac{2}{n}\right)s_1 + \frac{1}{n}\frac{s_0}{3} + \frac{1}{n}\frac{s_2}{2} = \frac{1}{3n}s_0 + \left(1 - \frac{2}{n}\right)s_1 + \frac{1}{2n}s_2,
$$

$$
Es_2' = \left(1 - \frac{2}{n}\right)s_2 + 2\frac{1}{n}s_1 + 2\frac{1}{n}\frac{s_3}{2} = \frac{2}{n}s_1 + \left(1 - \frac{2}{n}\right)s_2 + \frac{1}{n}s_3,
$$

and for $2 < m < q-1$,

$$
Es_m' = \frac{1}{n}s_{m-1} + \left(1 - \frac{2}{n}\right)s_m + \frac{1}{n}s_{m+1}.
$$

Finally,

$$
Es_{q-1}' = \frac{1}{n}s_{q-2} + \left(1 - \frac{2}{n}\right)s_{q-1} + \frac{2}{\beta n}s_q
$$

and

$$
Es_q' = \frac{1}{n}s_{q-1} + \left(1 - \frac{2}{\beta n}\right)s_q.
$$

Let us now define a completely separate, discrete Markov chain on the state space $\{0, 1, \ldots, q\}$, which behaves according to the following transition matrix **D**:

$$
\begin{pmatrix}
1 - \frac{1}{3n} & \frac{1}{3n} & 0 & 0 & 0 & \cdot & \cdot & \cdot & 0 \\
0 & 1 - \frac{2}{n} & \frac{2}{n} & 0 & 0 & \cdot & \cdot & \cdot & 0 \\
\frac{1}{2n} & \frac{1}{2n} & 1 - \frac{2}{n} & \frac{1}{n} & 0 & \cdot & \cdot & \cdot & 0 \\
0 & 0 & \frac{1}{n} & 1 - \frac{2}{n} & \frac{1}{n} & 0 & \cdot & \cdot & 0 \\
0 & 0 & 0 & \frac{1}{n} & 1 - \frac{2}{n} & \frac{1}{n} & 0 & \cdot & 0 \\
\cdot & \cdot & \cdot & \cdot & \cdot & \cdot & \cdot & \cdot & \cdot \\
\cdot & \cdot & \cdot & \cdot & \cdot & \cdot & \cdot & \cdot & \cdot \\
0 & 0 & 0 & 0 & 0 & 0 & \frac{1}{n} & 1 - \frac{2}{n} & \frac{1}{n} \\
0 & 0 & 0 & 0 & 0 & 0 & 0 & \frac{2}{\beta n} & 1 - \frac{2}{\beta n}
\end{pmatrix}
$$

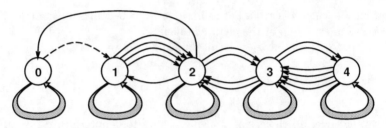

Fig. 1. The chain $\mathbf{D}$ for $n = 8$. The dotted arrow represents a transition of probability $1/3n$, other thin arrows $1/2n$

See Fig. 1 for a schematic of the state space of $\mathbf{D}$.

We see that $\mathbf{D}$ is designed so that $\bar{s} \cdot \mathbf{D} = E\bar{s}'$. The stationary distribution of $\mathbf{D}$ is

$$\bar{s}(\infty) := \frac{1}{n+1}(3, 1, 2, 2, 2, \ldots, 2, 2, \beta) \ .$$

If at time 0 the chains $\mathbf{X}$ and $\mathbf{Y}$ yield a vector $\bar{s}(0)$ whose coordinates sum to ρ, then after t steps we have $Es = \bar{s}(0) \cdot \mathbf{D}^t \to \frac{\rho}{n+1}(3, 1, 2, 2, 2, \ldots, 2, 2, \beta)$ and in particular $f(t) := s_0(t)/3 - s_1(t) \to 0$. But $f(t) = \frac{1}{2}\sum_{i=1}^{n}(z_{i+1}(t) - z_i(t))^2$ so the z_i's are becoming equal; thus, we see that the mixing time of $\mathbf{X}$ is intimately tied to the mixing time of $\mathbf{D}$.

Lemma 1. *The discrete chain $\mathbf{D}$ mixes in time $O(n^3)$.*

Proof. We employ a simple coupling on the state space. The coupling is defined so that pairs of points $(i, j) \in \{0, \ldots, q\}^2$ are updated simultaneously, ensuring that the two points never move in opposite directions. We now define a distance function on pairs of points by letting $d(i, j) = |i - j|$ if $\min(i, j) \geq 2$ or if $\max(i, j) \leq 1$ and letting $d(i, j) = |i - j| - 1/2$ otherwise. An upper bound on this distance function is $B = q$. For this distance function and the above coupling, it is easy to see that for any $(i, j) \in \{0, \ldots, q\}^2$, the distance between i and j will be non-increasing in expectation under moves of the coupled chain. Moreover, $E(\Delta d(i, j))^2) \geq V = 1/n$, whenever $i \neq j$. Applying the coupling theorem 4.1 from [7], we find that the coupling time $T \leq d(i_0, j_0)(2B - d(i_0, j_0))/V = O(n^3)$, and consequently the mixing time is $O(n^3)$.

Observing that $\rho < n^2$, we deduce from Lemma 1 that

$$\|\bar{s}(0) \cdot \mathbf{D}^t, \bar{s}(\infty)\|_{\mathrm{TV}} < an^3 \left(1 - \frac{1}{bn^3}\right)^t$$

for appropriate constants a and b. Thus, after $10bn^3 \log n$ steps,

$$Ef(t) < \|E\bar{s}(t), \bar{s}(\infty)\|_{\mathrm{TV}} < an^3 \left(1 - \frac{1}{(bn^3)}\right)^{10bn^3 \log n} = O(n^{-8}) \ .$$

Since $f(t)$ is non-negative, we deduce from Markov's inequality that for some constant c, there is a z so that with probability at least .9, $|z_i - z| < cn^{-4}$ for every i.

We may now assume that $\mathbf{Y}$ is shifted so that $z = 0$; the rest of the argument proceeds just as in the "n dots on a line" chain, using the second coupling to get $x_k = y_k$ with high probability when dot k is moved, and running $n(\log n)^2$ steps to ensure with high probability that every k has been called.

To lower bound the mixing time of $\mathbf{X}$, we will use a somewhat different auxiliary chain.

Theorem 3. *The mixing time of the reduced "n dots on an circle" chain $\mathbf{X}$ is $\Omega(n^3)$.*

Proof. We now have only one chain, $\mathbf{X}$, and are directly concerned with its state $\bar{u}$. Recall that $u_i := x_{i+1} - x_i$ is the distance between dots i and $i+1$.

Put $q = \lfloor n/2 \rfloor$ as before, again with $\beta = 2$ for n odd and $\beta = 1$ for n even. Define $r_0 := \sum_{i=1}^{n} u_i^2$, $r_q = \beta \sum_{i=1}^{n} u_i u_{i-q}$, and $r_m := 2 \sum_{i=1}^{n} u_i u_{i-m}$ for $0 < m < q$. Note that $\sum_{m=0}^{q} r_m = (\sum u_i)^2 = 1$.

When dot k is moved, u_{k-1} and u_k are both affected. Each lands uniformly between 0 and $u_{k-1} + u_k$, thus $E(u'_{k-1})^2 = E(u'_k)^2 = \frac{1}{3}(u_{k-1} + u_k)^2$ and consequently, over all choices of k,

$$\mathrm{E}r'_0 = (1 - \frac{2}{n})r_0 + 2\frac{1}{3n}(r_0 + 2(r_1/2) + r_0) = (1 - \frac{2}{3n})r_0 + \frac{2}{3n}r_1 .$$

Similarly,

$$\mathrm{E}r'_1 = \frac{2}{3n}r_0 + \left(1 - \frac{5}{3n}\right)r_1 + \frac{1}{n}r_2 ,$$

for $2 < m < q-1$

$$\mathrm{E}r'_m = \frac{1}{n}r_{m-1} + \left(1 - \frac{2}{n}\right)r_m + \frac{1}{n}r_{m+1} .$$

for $1 < m < q-1$,

$$\mathrm{E}r'_{q-1} = \frac{1}{n}r_{q-2} + \left(1 - \frac{2}{n}\right)r_{q-1} + \frac{2}{\beta n}r_q ,$$

and

$$\mathrm{E}r'_q = \frac{1}{n}r_{q-1} + \left(1 - \frac{2}{\beta n}\right)r_q .$$

We now define another discrete, auxiliary chain on $\{0, 1, \ldots, q\}$ by the following transition matrix $\widehat{\mathbf{D}}$:

$$
\begin{pmatrix}
1 - \frac{2}{3n} & \frac{2}{3n} & 0 & 0 & 0 & \cdot & \cdot & \cdot & 0 \\
\frac{2}{3n} & 1 - \frac{5}{3n} & \frac{1}{n} & 0 & 0 & \cdot & \cdot & \cdot & 0 \\
0 & \frac{1}{n} & 1 - \frac{2}{n} & \frac{1}{n} & 0 & \cdot & \cdot & \cdot & 0 \\
0 & 0 & \frac{1}{n} & 1 - \frac{2}{n} & \frac{1}{n} & 0 & \cdot & \cdot & 0 \\
0 & 0 & 0 & \frac{1}{n} & 1 - \frac{2}{n} & \frac{1}{n} & 0 & \cdot & 0 \\
\cdot & \cdot & \cdot & \cdot & \cdot & \cdot & \cdot & \cdot & \cdot \\
\cdot & \cdot & \cdot & \cdot & \cdot & \cdot & \cdot & \cdot & \cdot \\
0 & 0 & 0 & 0 & 0 & 0 & \frac{1}{n} & 1 - \frac{2}{n} & \frac{1}{n} \\
0 & 0 & 0 & 0 & 0 & 0 & 0 & \frac{2}{\beta n} & 1 - \frac{2}{\beta n}
\end{pmatrix}
$$

See Fig. 2 below, for a schematic of the state space of $\widehat{\mathbf{D}}$.

Fig. 2. The chain $\widehat{\mathbf{D}}$ for $n = 8$. Each dotted arrow represents a transition of probability $1/3n$, other thin arrows $1/2n$

As before, $\widehat{\mathbf{D}}$ has been designed so that $\widehat{\mathbf{D}} \cdot \bar{r} = \mathrm{E}\bar{r}'$. Note that $\widehat{\mathbf{D}}$ is reversible and very close to being a simple random walk on a path. Its stationary distribution is

$$
\bar{r}(\infty) := \frac{1}{n+1}(2, 2, 2, 2, 2, \ldots, 2, 2, \beta) \ .
$$

and, as in the case of $\mathbf{D}$:

Lemma 2. *The discrete chain* $\widehat{\mathbf{D}}$ *mixes in time* $\Theta(n^3)$.

Proof. For the upper bound, we follow the approach in Lemma 1 and define the same coupling on the state space. This time we use the standard distance function $d(i, j) = |i - j|$ and again find that $E(\Delta d(i, j)) \geq 0$ and $E(\Delta d(i, j))^2 \geq 1/n$. This shows that the mixing time is $O(n^3)$.

To establish the matching lower bound, we turn to the spectral gap of the chain. Lemma 3.1 in [10] states that if $\widehat{\mathbf{D}}$ and $\widetilde{\mathbf{D}}$ are Markov chains on the same state space with the same stationary distribution, and if there are constants c_1 and c_2 such that

$$
c_1 \widehat{\mathbf{D}}(i, j) \leq \widetilde{\mathbf{D}}(i, j) \leq c_2 \widehat{\mathbf{D}}(i, j) \tag{1}
$$

for all $i \neq j$, then

$$c_1 \operatorname{Gap}(\widehat{\mathbf{D}}) \leq \operatorname{Gap}(\widetilde{\mathbf{D}}) \leq c_2 \operatorname{Gap}(\widehat{\mathbf{D}}).$$

Let $\widetilde{D}$ be the standard symmetric random walk with transitions $\widetilde{D}(i, i \pm 1) = 1/n$, and self-loops elsewhere. This well-studied chain has spectral gap $\operatorname{Gap}(\widetilde{\mathbf{D}}) = O(\frac{1}{n^3})$. Comparing the two chains and noting that the condition (1) is satisfied with constants c_1 and c_2, we find that $\operatorname{Gap}(\widehat{\mathbf{D}}) = O(\frac{1}{n^3})$ as well. Using upper bounds on the spectral gap to lower bound the mixing rate (see, e.g., [13]), we can conclude that the chain mixes in time $\Omega(n^3)$.

Now we start the chain $\mathbf{X}$ with all dots coincident, aiming to show that after $t = cn^3$ steps (for some constant c) the total variation distance of its state distribution from the stationary distribution will still exceed $1/4$. To do this we exhibit a state-dependent event whose probability differs from a stationary chain's by more than $1/4$. This event will itself be based on a total variation distance, namely the value $d_t := \|\bar{r}(t), \bar{r}(\infty)\|_{\mathrm{TV}}$.

Let us consider the behavior of the auxiliary chain $\widehat{\mathbf{D}}$ while the chain $\mathbf{X}$ is running. When $\mathbf{X}$ has all dots coincident, the chain $\widehat{\mathbf{D}}$ starts at state $r_0 = (1, 0, 0, \ldots, 0)$ and has total variation distance from $\bar{r}(\infty)$ close to 1. We know from Lemma 2 that for some constant c, $\|\bar{r}(0)\widehat{\mathbf{D}}^{cn^3}, \bar{r}(\infty)\|_{\mathrm{TV}}$ still exceeds, say, 0.6.

Then

$$\mathrm{E}d_t = \mathrm{E}\|\bar{r}(t), \bar{r}(\infty)\|_{\mathrm{TV}} \geq \|\mathrm{E}\bar{r}(t), \bar{r}(\infty)\|_{\mathrm{TV}} = \|\bar{r}(0)\widehat{\mathbf{D}}^t, \bar{r}(\infty)\|_{\mathrm{TV}} > .6$$

for $t \leq cn^3$. Since d_t can never exceed 1, it follows that $\Pr(d_t > .2) > .5$. It therefore suffices to show that for the stationary chain $\mathbf{Y}$, $\Pr(d_t > .2) < .25$.

In $\mathbf{Y}$, however, the distribution of $\bar{r} := \bar{r}(t)$ does not depend on t, and is instead determined by a uniformly random vector $\bar{u} = (u_1, \ldots, u_n)$ in the simplex

$$S := \{\bar{u} : \bar{u}_1, \ldots, u_n \geq 0, \ \sum_{i=1}^n u_i = 1\}.$$

We know that $\mathrm{E}\bar{r} = \bar{r}(\infty)$, so we need only that $\bar{r}$ becomes concentrated about its mean as $n \to \infty$. To establish this, we represent $\bar{u}$ by a sequence of independent exponentially distributed random variables $y_1, \ldots, y_n$ with $u_i = y_i/S$ where $S = \sum_{j=1}^n y_j$. We now consider the random variables $w_{i,m} := y_i y_{i-m}, \ 0 \leq m \leq q = \lfloor n/2 \rfloor$.

For each fixed $m > 0$, we define a "dependence graph" G_m on $V = \{1, \ldots, n\}$ by $i \sim j$ iff $|i - j| = m$; since G_m is just a union of cycles, it has an equitable 3-colorable, i.e., there is a partition $V = A_m \cup B_m \cup C_m$ of the vertices into three sets of the same size (give or take one vertex) none of which contains an edge. But then no variable y_i appears in more than one term of the sum $a_m = \sum_{i \in A_m} w_{i,m}$, so the terms in a_m are i.i.d.; thus from the Central Limit Theorem, we have that $a_m/\sqrt{n/3}$ is normally distributed in the limit. The same applies of course to b_m and c_m, and it follows that for n larger than some n_0,

$a_m + b_m + c_m = \sum_{i=1}^{n} w_{i,m}$ lies within, say, $n^{2/3}$ of its mean (n) with probability greater than $1 - 1/10n$.

For $m = 0$, the terms $w_{i,0} = y_i^2$ are already i.i.d. so again by CLT, there is some $n_1 > n_0$ such that for $n > n_1$, $\sum_{i=1}^{n} w_{i,0}$ lies within $n^{2/3}$ of its mean $(2n)$ with probability greater than $1 - 1/10n$.

We have

$$\bar{r} = \left(\sum_{i=1}^{n} w_{i,0} + 2 \sum_{i=1}^{n} w_{i,1} + \cdots + 2 \sum_{i=1}^{n} w_{i,q-1} + \beta \sum_{i=1}^{n} w_{i,q} \right) / S^2$$

with $E(S^2) = n(n+1)$ and $\sigma(S^2) = \sqrt{2n(n+1)(2n+3)}$. For $n > n_1$, the numerator lies within $n \times n^{2/3} = n^{5/3}$ of its mean with probability at least $9/10$, and by Chebyshev's inequality the denominator lies within distance $3n^{3/2}$ of its mean with probability at least $8/9$.

Putting this all together, we have that $\Pr\left(\bar{r}, \bar{r}(\infty) \| _{\mathrm{TV}} > .2\right) < .25$ with plenty to spare, and the proof is (finally) complete.

We remark that it is equally easy to use just $|r_0(t) - r_0(\infty)|$ instead of the total variation distance to prove the lower bound, using special properties of the Markov chain $\widehat{\mathbf{D}}$. The above method was chosen since it appears to be more generally applicable.

Acknowledgments

This work began at the Banff International Research Station and has benefited from a conversation with Graham Brightwell and several with David Aldous.

References

1. P. Diaconis and M. Shashahani, Time to reach stationarity in the Bernoulli-Laplace diffusion model, *SIAM J. Math. Anal.* **81** #1 (1987), 208–218.
2. M. Dyer, A. Frieze, and R. Kannan, A random polynomial time algorithm for approximating the volume of convex sets, *J. for the Association for Computing Machinery* **38** (1991), 1–17.
3. M. Dyer and C. Greenhill, On Markov Chains for Independent Sets, *J. Algorithms*, **35** #1 (2000), 17–49.
4. R. Kannan, Markov chains and polynomial time algorithms. *Proc. 35th FOCS*, (1994), 656–671.
5. R. Kannan, M.W. Mahoney, and R. Montenegro, Rapid mixing of several Markov chains for a hard-core model, *Proc. 14th Annual ISAAC* (2003), 663–675.
6. L. Lovász and P. Winkler, Mixing times, *Microsurveys in Discrete Probability*, D. Aldous and J. Propp, eds., *DIMACS Series in Discrete Math. and Theoretical Computer Science*, (1998) **41**: 85-134.
7. M. Luby, D. Randall and A. Sinclair, Markov chain algorithms for planar lattice structures, *SIAM Journal on Computing.* **31** (2001), 167–192.
8. M. Luby and E. Vigoda, Fast convergence of the Glauber dynamics for sampling independent sets, *Random Structures and Algorithms*, **15** #3-4 (1999), 229–241.

9. T.-Y. Lee and H.-T. Yau, Logarithmic Sobolev inequalities for some models of random walks, *Ann. Prob.* **26** #4 (1998), 1855–1873.
10. N. Madras and D. Randall, Markov chain decomposition for convergence rate analysis, *Ann. Appl. Prob* **12** (2002), 581–606.
11. D. Randall, Mixing, *Proc. 44th Annual FOCS* (2004), 4–15.
12. D. Randall and P. Winkler, Mixing points in an interval, *Proc. ANALCO 2005*, Vancouver BC.
13. A. Sinclair, *Algorithms for Random Generation & Counting: A Markov Chain Approach.* Birkhäuser, Boston, 1993.
14. D.B. Wilson, Mixing times of lozenge tiling and card shuffling Markov chains, *Annals of Appl. Probab.* **14** (2004), 274–325.
15. D.B. Wilson, Mixing time of the Rudvalis shuffle, *Elect. Comm. in Probab.* **8** (2003), 77–85.

Derandomized Squaring of Graphs

Eyal Rozenman[1] and Salil Vadhan[2,*]

[1] Division of Engineering & Applied Sciences
Harvard University, Cambridge, Massachusetts
[2] Department of Computer Science & Applied Mathematics
Weizmann Institute, Rehovot, Israel

Abstract. We introduce a "derandomized" analogue of graph squaring. This operation increases the connectivity of the graph (as measured by the second eigenvalue) almost as well as squaring the graph does, yet only increases the degree of the graph by a constant factor, instead of squaring the degree.

One application of this product is an alternative proof of Reingold's recent breakthrough result that S-T Connectivity in Undirected Graphs can be solved in deterministic logspace.

1   Introduction

"Pseudorandom" variants of graph operations have proved to be useful in a variety of settings. Alon, Feige, Wigderson, and Zuckerman [AFWZ] introduced "derandomized graph products" to give a more illuminating deterministic reduction from approximating clique to within relatively small (eg constant) factors to approximating clique to within relatively large (eg n^ϵ) factors. Reingold, Vadhan, and Wigderson [RVW] introduced the "zig-zag graph product" to give a new construction of constant-degree expander graphs. The zig-zag product found many applications, the most recent and most dramatic of which is Reingold's deterministic logspace algorithm [Rei] for connectivity in undirected graphs.

In this paper, we present a pseudorandom analogue of graph squaring. The *square* X^2 of a graph X is the graph on the same vertex set whose edges are paths of length 2 in the original graph. This operation improves many connectivity properties of the graph, such as the diameter and mixing time of random walks of the graph (both of which roughly halve). However, the degree of the graph squares. In terms of random walks on the graph, this means that although half as many steps are needed to mix, each step costs twice as many random bits. Thus, there is no savings in the amount of randomness needed for mixing.

Our derandomized graph squaring only increases the degree by a constant factor rather than squaring it. Nevertheless, it improves the connectivity almost as much as the standard squaring operation. The measure of connectivity for which we prove this is the second eigenvalue of the graph, which is well-known to be a good measure of the mixing time of random walks, as well as of graph

* Supported by NSF grant CCF-0133096 and ONR grant N00014-04-1-0478.

C. Chekuri et al. (Eds.): APPROX and RANDOM 2005, LNCS 3624, pp. 436–447, 2005.

expansion. The standard squaring operation squares the second eigenvalue; we prove that the derandomized squaring does nearly as well.

The main new application of our derandomized squaring is a new logspace algorithm for connectivity in undirected graphs, thereby giving a new proof of Reingold's theorem [Rei]. Our algorithm, while closely related to Reingold's algorithm, is arguably more natural. In particular, it can be viewed as applying a natural pseudorandom generator, namely that of Impagliazzo, Nisan, and Wigderson [INW], to random walks on the input graph. This makes the analysis of the space requirements of the algorithm simpler. Reingold's algorithm is based on the *zig-zag product*, and constructs a sequence of graphs with an increasing number of vertices. Our analysis, based on derandomized squaring, only works on the vertex set of the original input graph.

Below we describe the derandomized squaring and its application to undirected s-t connectivity in more detail.

1.1   Derandomized Graph Squaring

Let X be an undirected regular graph of degree K.[1] The square X^2 of X has an edge for every path of length 2 in X. One way to visualize it is that for every vertex v in X, we place a clique on its K neighbours (this connects every pair of vertices that has a length 2 path through v). The degree thus becomes K^2. (Throughout the paper, we allow multiple edges and self-loops.)

In derandomized squaring, we use an auxiliary graph G on K vertices and place it instead of a clique on the K neighbours of every vertex v (thus connecting only *some* of the pairs of vertices which have a length 2 path through v). We denote the resulting graph by $X \circledS G$.

If the degree of G is D, the derandomized square will have degree KD, which will be smaller than K^2. We will see, however, that if G is an expander, then even if D is much smaller than K, the derandomized square of X with respect to G improves connectivity similarly to standard squaring.

Our measure of connectivity is the second eigenvalue $\lambda \in [0,1]$ of (the random walk on) the graph; small λ indicates that the random walk mixes rapidly and that the graph has good expansion (i.e. is highly connected). If the second eigenvalue of X is λ then the second eigenvalue of X^2 is λ^2. The second eigenvalue of the derandomized product is not very far. For example, we prove that it is at most $\lambda^2 + 2\mu$ where μ is the second eigenvalue of G.

1.2   A New Logspace Algorithm for Undirected Connectivity

Recall that the problem of undirected st-connectivity is: given an undirected graph G and two vertices s, t, decide whether there is a path from s to t in G. While the time complexity of this problem is well-understood, the space

---

[1] Actually, following [RTV], we actually work with regular *directed* graphs in the technical sections of the paper, but thinking of undirected graphs suffices for the informal discussion here.

complexity was harder to tackle. A long line of research, starting with Savitch's deterministic $\log^2$-space algorithm [Sav], and the log-space randomized algorithm [AKL+], culminated in Reingold's optimal deterministic log-space algorithm [Rei] (see his paper and its references for more on the history and applications of this problem). We now shortly describe Reingold's algorithm, then present our algorithm and compare the two.

Reingold's Algorithm. Notice that undirected connectivity is solvable in log-space on bounded-degree graphs with *logarithmic diameter* (simply enumerate all paths of logarithmic length in the graph out of the origin vertex). Examples of graphs with logarithmic diameter are expander graphs, i.e. graphs whose second eigenvalue is bounded away from 1. Reingold's idea is to transform the input graph into a bounded-degree expander by gradually decreasing its second eigenvalue.

A natural attempt would be to square the graph. This indeed decreases the second eigenvalue, but increases the degree. To decrease the degree, Reingold uses the *zig-zag graph product* of [RVW], or the related *replacement product*. We describe his algorithm in terms of the latter product.

Given a K-regular graph X on N vertices, and an auxiliary D-regular graph G on K vertices, the replacement product $X \circledr G$ is a $D + 1$-regular graph on NK vertices. On each edge (v, w) in X put two vertices, one called e_v "near" v and another called e_w "near" w, for a total of NK vertices. This can be thought of as splitting each vertex v into K vertices forming a "cloud" near v. Place the graph G on each cloud. Now for each edge $e = (v, w)$ of X, put an edge between e_v and e_w. The result is a $(D+1)$-regular graph. Notice that $X \circledr G$ is connected if and only if both X and G are.

The replacement product reduces the degree from K to $D + 1$. It is proven in [RVW] (and also follows from [MR]) that when G is a good enough expander, replacement product roughly preserves the second eigenvalue of X. Suppose that X is $(D + 1)$ regular and G has $(D + 1)^2$ vertices and degree D. Then $X^2 \circledr G$ is again a $(D + 1)$-regular graph, whose second eigenvalue is roughly the square of the second eigenvalue of X. Iterating this procedure $\log N$ times leads to a constant-degree expander on *polynomially many vertices*, since at each iteration the number of vertices grows by a factor of about D^2. On the resulting expander we can therefore solve connectivity in logarithmic space. (One also must confirm that the iterations can be computed in logarithmic space as well).

Our Algorithm. Our approach also follows from this idea of increasing connectivity by squaring the graph. However, instead of squaring, and then reducing the degree by a zigzag product (and thus increasing the number of vertices) we will replace the squaring by derandomized squaring, which maintains the vertex set (but increases the degree). Iterating the derandomized squaring operation yields highly connected graphs with relatively small degree compared to doing the same iterations with standard squaring. In the next two paragraphs we compare the resulting graphs in each case.

Let X be a regular graph on N vertices. Squaring the graph $\log N$ times, results in the graph $X^{2^{\log N}} = X^N$ (whose edges are all paths of length N in X). This graph is extremely well connected; it contains an edge between every two vertices which are connected by a path in X. The degree however, is huge — exponential in N. We want to simulate the behavior of X^N with a graph that has much smaller degree.

Suppose that instead of standard squaring at each step we apply derandomized squaring to obtain a sequence of graphs $X_1, X_2, \ldots$. At each step the degree increases by a constant factor (instead of the degree squaring at each step). For $m = O(\log N)$ the degree of X_m is only *polynomial* in N. But we will show that is as well-connected as X^N (as measured by the second eigenvalue). In particular, X_m will contain an edge between every pair of vertices s, t that are connected by a path in X. Deciding whether s, t are connected therefore reduces to enumerating all neighbors of s in X_m and looking for t. There are only polynomially many neighbors, so the search can be done in logarithmic space. We will show that computing neighbors in X_m can also be done in logarithmic space. These two facts yield a logarithmic space algorithm for undirected connectivity.

Comparing our approach to Reingold's original solution, the main way in which our algorithm differs from (and is arguably more natural than) Reingold's algorithm is that all the graphs we construct are on the *same* vertex set. Edges in the graph X_m correspond to paths of length 2^m in X. The price we pay is that the degree increases, but, thanks to the use of *derandomized* squaring, only by a constant factor (which we can afford). In contrast, each step of Reingold's algorithm creates a graph that is larger than the original graph (but maintains constant degree throughout).

1.3   Derandomized Squaring as a Pseudo-random Generator

Impagliazzo, Nisan, and Wigderson [INW] proposed the following pseudorandom generator. Let G be an expander graph with K vertices and degree D. Choose a random vertex $x \leftarrow [K]$, a random edge label $a \leftarrow [D]$, and output $(x, x[a]) \in [K] \times [K]$. This pseudorandom generator is at the heart of derandomized squaring. Notice that using this pseudorandom generator to generate a pseudorandom walk of length 2 in a graph X of degree K is *equivalent* to taking a random step in the derandomized square of X using auxiliary graph G.

Impagliazzo, Nisan, and Wigderson [INW] suggested to increase the stretch of the above generator by recursion. They proved that when the graphs G used in the construction are sufficiently good expanders of relatively *large degree*, this construction fools various models of computation (including randomized logspace algorithms)[2]. However, the resulting generator has seed length $O(\log^2 n)$, and hence does not prove that RL=L.

Our construction of the graph X_m in our st-connectivity algorithm is precisely the graph corresponding to using the INW generator to derandomize random

---

[2] Specifically, to fool an algorithm running in space $\log n$, they use expanders of degree poly(n).

walks of length 2^m in X. However, we are able to use *constant-degree* expanders for G (for most levels of recursion), thereby obtaining seed length $O(\log n)$ and hence a logspace algorithm (albeit for undirected st-connectivity rather than all of RL).

Moreover, it follows from our analysis that taking the pseudorandom walk in X corresponding to a random step in X_m (equivalently, according to the INW generator with appropriate parameters) will end at an almost-uniformly distributed vertex. A pseudorandom generator with such a property was already given in [RTV] based on Reingold's algorithm and the zig-zag product, but again it is more natural in terms of derandomized squaring.

1.4   Relation to the Zig-Zag Product

The reader may have noticed a similarity between the derandomized squaring and the zig-zag product of [RVW] (which we define precisely later in the paper). Indeed, they are very closely related. When we use a square graph G^2 as auxiliary graph, the derandomized square $X \text{Ⓢ} G^2$ turns out to be a "projection" of the square of the zigzag product $(X \text{Ⓩ} G)^2$. This observation allows us to prove the expansion properties of the derandomized squaring by reducing to the zig-zag product case.

We note that the derandomized squaring has complementary properties to the zigzag product. In the zigzag product we are given a graph X and can decrease its degree while (nearly) maintaining its second eigenvalue. We must pay by slightly increasing the number of vertices. In the derandomized squaring we manage to decrease the second eigenvalue while maintaining the number of vertices, and we pay by slightly increasing the degree.

2   Preliminaries

Reingold, Trevisan, and Vadhan [RTV] generalized Reingold's algorithm and the zig-zag product to (regular) *directed* graphs, and working in this more general setting turns out to be useful for us, too (even if we are only interested in solving st-connectivity for undirected graphs). We present the necessary background on such graphs in this section.

Let X be a directed graph (*digraph* for short) on N vertices. We say that X is K-*outregular* if every node has outdegree K, K-*inregular* if every node has indegree K, and K-*regular* if both conditions hold. Graphs may have self-loops and multiple edges, where a self-loop is counted as both an outgoing and incoming edge. All graphs in this paper are outregular directed graphs (and most are regular).

For a K-regular graph X on N vertices, we denote by M_X the transition matrix of the random walk on X, i.e. the adjacency matrix divided by K. Let $u = (1/N, \ldots, 1/N) \in \mathbb{R}^N$ be the uniform distribution on the vertices of X. Then, by regularity, $M_X u = u$ (so u is an eigenvector of eigenvalue 1).

Following [Mih], we consider the following measure of the rate at which the random walk on X converges to the stationary distribution u:

$$\lambda(X) = \max_{v \perp u} \frac{\|M_X(v)\|}{\|v\|} \in [0,1]$$

where $v \perp u$ refers to orthogonality with respect to the standard dot product $\langle x, y \rangle = \sum_i x_i y_i$ on $\mathbb{R}^N$ and $\|x\| = \sqrt{\langle x, x \rangle}$ is the L_2 norm. The smaller $\lambda(X)$, the faster the random walk converges to the stationary distribution and the better "expansion" properties X has. Hence, families of graphs with $\lambda(X) \le 1 - \Omega(1)$ are referred to as *expanders*.

In case X is undirected, $\lambda(X)$ equals the second largest eigenvalue of the symmetric matrix M_X in absolute value. In directed graphs, it equals the square root of the second largest eigenvalue of $M_X^T M_X$.

A K-regular directed graph X on N vertices with $\lambda(X) \le \lambda$ will be called an (N, K, λ)-graph. We define $g(X) = 1 - \lambda(X)$ to be the *spectral gap* of X.

A *labelling* of a K-outregular graph X is an assignment of a number in $[K]$ to every edge of X, such that the edges exiting every vertex have K distinct labels. For a vertex v of X and an edge label $x \in [K]$ we denote by $v[x]$ the neighbor of v via the outgoing edge labelled x. We say that a labelling is *consistent* if for every vertex all incoming edges have distinct labels. Notice that if a graph has a consistent labeling, then it is K-inregular (and hence K-regular). We will work with consistently labelled graphs in this extended abstract for simplicity and to make the connection between derandomized squaring and the INW pseudorandom generator [INW] more apparent. But this condition condition can be relaxed (eg in the definition and analysis of the derandomized square) by allowing each edge (u, v) to have two labels, one as an outgoing edge from u and one as an incoming edge to v, as formalized using the "rotation maps" of [RVW, RTV].

The notion of consistent labelling we use is the same as in [HW] and [RTV]. For undirected graphs, one can consider a stronger notion of consistent labelling, where an edge (u, v) is required to have the same label leaving from u as it does when leaving v, but this has the disadvantage that it is not preserved under the operations we perform (such as the squaring operation below).

The square X^2 of a graph X is the graph whose edges are paths of length 2 in X. The square of a K-regular graph is K^2-regular, and a consistent labelling of X induces a consistent labelling of X^2 in a natural way. Specifically, for a label $(x, y) \in [K]^2$, we define $v[x, y] = v[x][y]$. Notice that $\lambda(X^2) \le \lambda(X)^2$. (This is always an equality for undirected graphs, but not necessarily so for directed graphs). We similarly define the n-th power X^n using paths of length n in X.

Like undirected graphs, the spectral gap of *regular* connected directed graphs is always at least an inverse polynomial. This holds provided the graph is connected and aperiodic (i.e. the gcd of all cycle-lengths is 1). (If either of these conditions doesn't hold than $\lambda(X) = 1$.) The following can be proven by reduction to the corresponding lemma in the undirected case [AS].

Lemma 2.1. *For every K-regular, connected, aperiodic graph. Then $\lambda(X) \le 1 - 1/(2D^2N^2)$.*

The next proposition shows that when the second eigenvalue is very small, the graph is very well connected - it contains a clique (we omit the proof).

Proposition 2.2. *An $(N, D, 1/2N^{1.5})$-graph X contains an edge between any pair of vertices. Indeed, for a pair of vertices v, w the probability that a random neighbor of v is equal w is at least $1/N - 1/N^2$.*

3   Derandomized Squaring

After giving a formal definition of derandomized squaring, we will show that it decreases the second eigenvalue of a graph in a way comparable to squaring it. The proof is by reduction to the zigzag product $X \, \textcircled{z} \, G$ (also defined below), of X with an auxiliary graph G. We shall see that the derandomized square of X with the auxiliary squared graph G^2 is a "projection" (defined precisely later) of the squared graph $(X \, \textcircled{z} \, G)^2$. We then show that projection does not increase the second eigenvalue. We can therefore use the known bounds on the second eigenvalue from [RTV].

Definition 3.1. *Let X be a labelled K-regular graph on vertex set $[N]$, let G be a labelled D-regular graph on vertex set $[K]$. The* derandomized square graph *$X \, \textcircled{s} \, G$ has vertex set $[N]$ and is KD-outregular. The edges exiting a vertex v are paths $v[x][y]$ of length two in X such that y is a neighbor of x in G. Equivalently, when $x \in [K]$ is an edge label in X and $a \in [D]$ is an edge label in G, the neighbor of $v \in [N]$ via the edge labelled (x, a) is $v[x][x[a]]$.*

The derandomized square may, in general, not produce an in-regular graph. However, it will do so provided that X is consistently labelled.

Proposition 3.2. *If X is consistently labelled, then $X \, \textcircled{s} \, G$ is KD-regular. If, in addition, G is consistently labelled, then $X \, \textcircled{s} \, G$ is consistently labelled.* ∎

Notice that even if X and G are consistently labelled and undirected, i.e. for every edge (u, v) there is a corresponding reverse edge (v, u), then the derandomized square $X \, \textcircled{s} \, G$ need not be undirected[3].

Our main result is that when G is a good expander, then the expansion of $X \, \textcircled{s} \, G$ is close to that of X^2.

Theorem 3.3. *If X is an (N, K, λ)-graph and G is a (K, D, μ)-graph, then $X \, \textcircled{s} \, G^2$ is an $(N, KD^2, f(\lambda, \mu))$-graph, for a function f, monotone increasing in λ and μ, and satisfying*

- $f(\lambda, \mu) \leq \lambda^2 + 2\mu^2$,
- $f(1 - \gamma, 1/100) \leq 1 - (8/7) \cdot \gamma$, *when* $\gamma < 1/4$.

We will prove the theorem via reduction to the zigzag product $X \, \textcircled{z} \, G$ which we now turn to define[4].

---

[3] If X satisfied the stronger notion of consistent labeling where (u, v) and (v, u) are required to have the same label, then $X \, \textcircled{s} \, G$ would be undirected. Alas, this stronger notion is not preserved under the derandomized square (or even the standard squaring).

[4] In the full version, we will provide a direct proof, which proceeds along similar lines to the zig-zag analysis of [RVW, RTV], but is simpler and provides a better bound.

Definition 3.4 ([RVW]). *Let X be an labelled K-regular graph on vertex set $[N]$. Let G be a labelled D-regular graph on vertex set $[K]$. The* zig-zag product *$X \textcircled{z} G$ is a graph on vertex set $[N] \times [K]$ of outdegree D^2, where the edge label $(a,b) \in [D] \times [D]$ connects vertex $(v,x) \in [N] \times [K]$ to $(v[x[a]], x[a][b])$.*

Again, consistent labelling of X ensures that $X \textcircled{z} G$ is a regular graph. We will prove the following lemma.

Lemma 3.5. *Let X be consistently labelled. Then $\lambda(X \textcircled{s} G^2) \leq \lambda(X \textcircled{z} G)^2$.*

Theorem 3.3 then follows by plugging in the bounds on $\lambda(X \textcircled{z} G)$ given by [RVW, RTV].

Proof of Lemma 3.5: We will show that $X \textcircled{s} G^2$ is a projection (defined next) of $(X \textcircled{z} G)^2$, and prove that projection cannot increase the second eigenvalue.

Definition 3.6. *Let X be a D-regular graph on vertex set $[N] \times [K]$. The* **projection graph** *$\mathrm{P}X$ has vertex set $[N]$ and edge labels $[K] \times [D]$. The neighbor labelled (x,a) of a vertex v is defined by $v[x,a] = (v,x)[a]_1$, where the right-hand side refers to the first component of the a'th neighbor of (v,x) in X. The projection $\mathrm{P}X$ is a KD-regular graph, but the labeling above is not necessarily consistent.*

Proposition 3.7. *The projection graph $\mathrm{P}(X \textcircled{z} G)^2$ is equal to the graph obtained from $X \textcircled{s} G^2$ by duplicating each edge K^2 times.*

We omit the proof, which is a straightforward verification from the definitions.

Proposition 3.8. $\lambda(\mathrm{P}X) \leq \lambda(X)$.

Proof. Let $f : [N] \to \mathbb{R}$ be a vector of $\mathrm{P}X$ that is orthogonal to the uniform distribution (i.e. $\sum_i f(i) = 0$) and $\|A_{\mathrm{P}X}f\| = \lambda \cdot \|f\|$, where $\lambda = \lambda(\mathrm{P}X)$.

Define $\hat{f} : [N] \times [K] \to \mathbb{R}$ by $\hat{f}(v,x) = f(v)$. Observe that $f(v[x,a]) = \hat{f}(v,x)[a]$. So

$$\frac{1}{K} \sum_{x \in [K]} M_X \hat{f}(v,x) = \frac{1}{KD} \sum_{\substack{a \in [D] \\ x \in [K]}} \hat{f}((v,x)[a]) = \frac{1}{KD} \sum_{\substack{a \in [D] \\ x \in [K]}} f(v[x,a]) = M_{\mathrm{P}X} f(v).$$

The leftmost expression above is a vector on $[N] \times [K]$. It is obtained by applying an average operator to the vector $M_X \hat{f}$, which cannot increase the L_2 norm. Therefore $\|M_X \hat{f}\|$ is at least $\lambda \|\hat{f}\|$. The sum of coordinates of $\hat{f}$ is zero, so $\lambda(X) \geq \lambda$. ∎

4   A Log-Space Algorithm for Undirected Connectivity

We describe how to solve undirected st-connectivity on an undirected graph X with N vertices in logarithmic space.

Overview. We will assume that the input graph X is 3-regular and consistently labelled. (in the full version we will show that this assumption does not lose generality). By Lemma 2.1, every 3-regular connected graph has second eigenvalue $1 - \Omega(1/N)$. Our goal is to use derandomized squaring to decrease the second eigenvalue (of each connected component) to less than $1/2N^3$ (we will need to square $O(\log N)$ times). By Prop. 2.2, the resulting graph must contain a clique on every connected component of X. We can therefore go over all the neighbors of s in the resulting graph and look for t.

Starting with (some power of) X, we define a sequence of graphs X_m, each of which is a derandomized square of its predecessor using a suitable auxiliary graph. The algorithm works in two phases. Phase one works for $m \le 100 \log N$, and reduces the second eigenvalue to a constant $(3/4)$, by using as auxiliary graphs a sequence G_m of *fixed-degree* expanders. We will see that the spectral gap $g(X_m)$ grows by at least a factor of $8/7$ at each step. Therefore, after $m_0 = O(\log N)$ steps, we obtain an expander X_{m_0} with second eigenvalue at most $3/4$ and degree polynomial in N.

At this point we cannot use fixed-degree expanders as auxiliary graphs any more. If we did, the second eigenvalue of the derandomized square would be dominated by the second eigenvalue of the auxiliary graph, which is constant. Thus we would not be able to decrease the eigenvalue to $1/2N^3$. In phase two, we therefore use auxiliary graphs G_m with non-constant degrees. Specifically, for $m > m_0$, the auxiliary graph G_m will have degree doubly-exponential in $m - m_0$. The fast growth of the degree allows the eigenvalue of the auxiliary graph to remain small enough to imply that $\lambda(X_{m+1}) \le c \cdot \lambda(X_m)^2$ for some $c > 1$ quite close to 1. Therefore, after an additional $\log \log N + O(1)$ steps we obtain a graph X_{m_1} with second eigenvalue at most $1/2N^3$.

Since the graph X_{m_1} has degree polynomial in N, we can enumerate all the neighbors of s in logarithmic space. We will show (in Prop. 4.4) that neighbors in X_{m_1} are log-space computable, making the whole algorithm work in logarithmic space.

The Auxiliary Expanders. We will need a family of logspace-constructible constant-degree expanders with the following parameters, (which can be obtained from e.g. [GG] or [RVW].

Lemma 4.1. *For some constant $Q = 3^q$, there exists a sequence H_m of consistently labeled $(Q^{2m}, Q, 1/100)$-graphs. Neighbors in H_m are computable in space $O(m)$ (i.e. given a vertex name $v \in [Q^{2m}]$ and an edge label $x \in [Q]$, we can compute $v[x]$ in space $O(m)$).*

Definition 4.2. *Let H_m be the graph sequence of lemma 4.1. For a positive integer N, we set $m_0 = \lceil 100 \log N \rceil$, we define a graph sequence G_m by $G_m = (H_m)^2$ for $m \le m_0$, and $G_m = (H_{m_0 + 2^{m-m_0-1}})^{2^{m-m_0}}$ for $m > m_0$. Neighbors in G_m are computable in space $O(m + 2^{m-m_0})$.*

The Algorithm. Let X be a 3-regular consistently labelled graph. Given two vertices s, t connected in X, we describe a log-space algorithm that outputs a path between s and t. For simplicity, assume that X is connected (else carry out the analysis below on each connected component of X).

Define $X_1 = X^{2q}$, where $Q = 3^q$ is from lemma 4.1. Define inductively $X_{m+1} = X_m \circledS G_m$. It can be verified by induction that the degree D_m of X_m is equal to the number of vertices of G_m, so the operation $X_m \circledS G_m$ is indeed well-defined. Specifically, we have $D_m = Q^{2m}$ for $m \leq m_0$, and $D_m = Q^{2 \cdot (m_0 + 2^{m-m_0})}$ for $m > m_0$.

Phase One. By lemma 2.1 we have $g(X_1) \geq 1/3N$. From the second inequality in theorem 3.3 $g(X_m) \geq g(X_{m-1}) \cdot (8/7)$ as long as $g(X_{m-1}) \leq 1/4$. Therefore for some $m < 100 \log N$ we will get $g(X_m) \leq 1/4$. Because of the monotonicity mentioned in theorem 3.3 the gap does not decrease in the proceeding iterations. Therefore, for $m_0 = \lceil 100 \log N \rceil$ we have $\lambda(X_{m_0}) < 3/4$.

Phase Two. We now decrease the second eigenvalue from $3/4$ to $1/2N^3$. For $m \geq m_0$ define $\lambda_m = (64/65) \cdot (7/8)^{2^{(m-m_0)}}$, $\mu_m = (1/100)^{2^{m-m_0}}$. Suppose that $\lambda(X_m) \leq \lambda_m$. Since $2\mu_m^2 < \lambda_m^2/64$ we can use the first inequality in theorem 3.3 to deduce that $\lambda(X_{m+1}) \leq \lambda_m^2(1 + 1/64)$. Since for $m = m_0$ indeed $\lambda(X_m) \leq \lambda_m$ we deduce this holds for all $m \geq m_0$. The next proposition is a direct consequence.

Proposition 4.3. Let $m_1 = m_0 + \log \log N + 10$. Then $\lambda(X_{m_1}) \leq 1/2N^3$. ∎

By Prop. 2.2 the graph X_{m_1} contains a clique on the N vertices. Moreover, it has degree $D_{m_1} = Q^{2 \cdot (100 \log N + 2^{\log \log N + 10})} = \text{poly}(N)$. If we could compute neighbors in X_{m_1} in space $O(\log N)$ we could find a path from s to t in logarithmic space.

Proposition 4.4. *Neighborhoods in X_{m_1} are computable in space $O(\log N)$.*

Proof. Edge labels in X_m are vectors $y_m = (y_1, a_1, \ldots, a_{m-1})$ where y_1 is an edge label in X_1 and a_i is an edge label on G_i. Given a vertex v and an edge label y_m in X_m we wish to compute the neighbor $v[y_m]$ in X_m.

Every edge in X_m corresponds to a path of length 2^m in X. It suffices to give a (log-space) algorithm that, given v, y and an integer b in the range $[1, 2^m]$, returns the edge label in X of the b-th edge in this path of length 2^m. As we will see below, this edge label is actually independent of the vertex v (and thus can be computed given only y and b).

The path of length 2^m originating from v corresponding to the edge label y_m consists of two paths of length 2^{m-1} corresponding to two edges in X_{m-1}. These two edges in X_{m-1} have labels $y_{m-1} = (y_1, a_1, \ldots, a_{m-2})$ and $y_{m-1}[a_{m-1}]$, where the latter is a neighbor computation in G_{m-1}.

From these observations the algorithm is simple. If $b \leq 2^{m-1}$ then solve the problem encoded by y_{m-1}, b in X_{m-1}. If $b > 2^{m-1}$ then instead set $y_{m-1} \leftarrow y_{m-1}[a_{m-1}]$, $b \leftarrow b - 2^{m-1}$, and now solve the problem encoded by y_{m-1}, b on X_{m-1}.

Here is a pseudo code for the algorithm. Write $b - 1$ as a binary string $(b_{m-1}, \ldots, b_0)$, and let y_i be the string $y_1, a_1, \ldots, a_{i-1}$.

for $i = m - 1$ to 0 **do**
   if $b_i = 1$ **then**
      set $y_i = y_i[a_i]$ (this is a computation in G_i).
   end if
end for
output y_0

Now we argue that this can be computed in space $O(\log N)$ when $m = m_1$. Notice that the input length to the algorithm is $m + \log D_{m_1} = O(\log N)$. By lemma 4.1, the computation in G_i steps in the loop described in the code can be performed in space $O(m + 2^{m-m_0}) = O(\log N)$, and we are done. $\blacksquare$

A Pseudo-random Generator for Walks on X. Picking a random edge in X_{m_1} yields a walk of length $2^{m_1} = \text{poly}(N)$ in X. This walk is "pseudo-random" - it ends at an almost-uniformly distributed vertex (by prop. 2.2 the probability to reach any vertex of X is at least $1/N - 1/N^2$), but is generated by only $\log D_{m_1} = O(\log N)$ random bits (compare with $\text{poly}(N)$ bits required to generate a standard random walk of this length). Moreover, the edge labels in this walk do not depend on the initial vertex v, but only on the edge label chosen in X_{m_1}. Indeed, the algorithm given above describes how to compute the labels in the walk given the edge label y_{m_1} in X_{m_1}. In fact, the map from y_{m_1} to the sequence of edge labels in the walk is precisely the pseudorandom generator constructed in [INW] from the expanders $G_1, \ldots, G_{m-1}$. This pseudo-random walk generator will have the above property (producing walks that end at almost-uniformly distributed vertices) in any consistently labeled graph (of specified parameters). A pseudorandom walk generator with similar properties was given in [RTV] based on Reingold's algorithm (which uses the zig-zag product). However, the generator does not have as simple a description as above. In particular, computing the b'th label produced by the walk seems to require computing all the previous $b - 1$ labels of the walk (taking time up to $2^{m_1} = \text{poly} N$, rather than being computable directly as above (in time $\text{poly}(\log N)$).

Acknowledgments

This work emerged from of our collaborations with Omer Reingold, Luca Trevisan, and Avi Wigderson. We are deeply grateful to them for their insights on this topic and their encouragement in writing this paper.

References

[AKL+] R. Aleliunas, R. M. Karp, R. J. Lipton, L. Lovász, and C. Rackoff. Random walks, universal traversal sequences, and the complexity of maze problems. In *20th Annual Symposium on Foundations of Computer Science (San Juan, Puerto Rico, 1979)*, pages 218–223. IEEE, New York, 1979.

[AFWZ] N. Alon, U. Feige, A. Wigderson, and D. Zuckerman. Derandomized graph products. *Comput. Complexity*, 5(1):60–75, 1995.

[AS]   N. Alon and B. Sudakov. Bipartite subgraphs and the smallest eigenvalue. *Combin. Probab. Comput.*, 9(1):1–12, 2000.

[GG]   O. Gabber and Z. Galil. Explicit Constructions of Linear-Sized Superconcentrators. *J. Comput. Syst. Sci.*, 22(3):407–420, June 1981.

[HW]   S. Hoory and A. Wigderson. Universal Traversal Sequences for Expander Graphs. *Inf. Process. Lett.*, 46(2):67–69, 1993.

[INW]   R. Impagliazzo, N. Nisan, and A. Wigderson. Pseudorandomness for Network Algorithms. In *Proceedings of the Twenty-Sixth Annual ACM Symposium on the Theory of Computing*, pages 356–364, Montréal, Québec, Canada, 23–25 May 1994.

[MR]   R. A. Martin and D. Randall. Sampling Adsorbing Staircase Walks Using a New Markov Chain Decomposition Method. In *Proceedings of the 41st Annual Symposium on Foundations of Computer Science*, pages 492–502, Redondo Beach, CA, 17–19 Oct. 2000. IEEE.

[Mih]   M. Mihail. Conductance and convergence of markov chains: a combinatorial treatment of expanders. In *In Proc. of the 37th Conf. on Foundations of Computer Science*, pages 526–531, 1989.

[RTV]   Reingold, Trevisan, and Vadhan. Pseudorandom Walks in Biregular Graphs and the RL vs. L Problem. In *ECCCTR: Electronic Colloquium on Computational Complexity, technical reports*, 2005.

[Rei]   O. Reingold. Undirected ST-connectivity in log-space. In *STOC '05: Proceedings of the thirty-seventh annual ACM symposium on Theory of computing*, pages 376–385, New York, NY, USA, 2005. ACM Press.

[RTV]   O. Reingold, L. Trevisan, and S. Vadhan. Pseudorandom Walks in Biregular Graphs and the RL vs. L Problem. *Electronic Colloquium on Computational Complexity* Technical Report TR05-022, February 2005. http://www.eccc.uni-trier.de/eccc.

[RVW]   O. Reingold, S. Vadhan, and A. Wigderson. Entropy waves, the zig-zag graph product, and new constant-degree expanders. *Ann. of Math. (2)*, 155(1):157–187, 2002.

[Sav]   W. J. Savitch. Relationships between nondeterministic and deterministic tape complexities. *J. Comput. System. Sci.*, 4:177–192, 1970.

Tight Bounds for String Reconstruction Using Substring Queries

Dekel Tsur

University of California, San Diego
dtsur@cs.ucsd.edu

Abstract. We resolve two open problems presented in [8]. First, we consider the problem of reconstructing an unknown string T over a fixed alphabet using queries of the form "does the string S appear in T?" for some query string S. We show that $\Omega(\epsilon^{-1/2}n^2)$ queries are needed in order to reconstruct a $1 - \epsilon$ fraction of the strings of length n. This lower bound is asymptotically optimal since it is known that $O(\epsilon^{-1/2}n^2)$ queries are sufficient. The second problem is reconstructing a string using queries of the form "does a string from $\mathcal{S}$ appear in T?", where $\mathcal{S}$ is a set of strings. We show that a $1 - \epsilon$ fraction of the strings of length n can be reconstructed using $O(n)$ queries, where the maximum length of a string in the queries is $2\log_\sigma n + \log_\sigma \frac{1}{\epsilon} + O(1)$. This construction is optimal up to constants.

1 Introduction

Consider the following problem: There is some unknown string T of length n over a fixed alphabet. The goal is to reconstruct T by making queries of the form "does the string S appear in T?" for a *query string* S, where all the queries are asked simultaneously. Besides the theoretical interest, this problem is motivated by *Sequencing by Hybridization (SBH)* which is a method for sequencing unknown DNA molecules [2]. In this method, the target string is hybridized to a chip containing known strings. For each string in the chip, if its reverse complement appears in the target, then the two strings will bind (or hybridize), and this hybridization can be detected. Thus, SBH can be modeled by the string reconstruction problem described above.

An algorithm for the string reconstruction problem is composed of two parts: The first part is designing a set of queries Q depending on the value of n, and the second part is reconstructing the unknown string given the answers to the queries in Q. In this work, we study the design of the query set. In other words, we are interested whether a given query set provides enough information to solve the reconstruction problem.

Denote by Σ the alphabet of the string T, and let $\sigma = |\Sigma|$. Margaritis and Skiena [6] showed that at least $\sigma^{n/2}/n$ queries are needed in order to reconstruct all strings of length n, and $O(\sigma^{n/2})$ queries are sufficient.

The string reconstruction problem becomes easier when relaxing the requirement to reconstruct all strings of length n. For a given query set Q, the *resolution power* $p(Q, n)$ of Q is the fraction of the strings of length n that are

C. Chekuri et al. (Eds.): APPROX and RANDOM 2005, LNCS 3624, pp. 448–459, 2005.

unambiguously reconstructed from the answers to the queries in Q. The query set containing all the strings of length $2 \log_\sigma n + \frac{1}{2} \log_\sigma \epsilon^{-1} + O(1)$, called the *uniform query set*, has a resolution power $1 - \epsilon$ [1, 3, 7, 10]. The number of queries in this set is $O(\epsilon^{-1/2} n^2)$ For simplicity, we assume throughout the paper that σ is constant and $\epsilon \geq \frac{1}{n}$.

The question whether the uniform query set is optimal remained open. This problem was posed explicitly in [8, problem 36]. In this paper we settle this open problem by showing that the uniform query set is asymptotically optimal. More precisely, we show that every query set with resolution power $1 - \epsilon$ contains $\Omega(\epsilon^{-1/2} n^2)$ queries.

Our lower bound also applies to a variant of the problem in which the queries are quantitative, namely the queries are of the form "how many times does the string S appear in T?". Clearly, the $O(\epsilon^{-1/2} n^2)$ upper bound for the uniform query set also applies to this problem. Our $\Omega(\epsilon^{-1/2} n^2)$ lower bound which was mentioned above also applies to this model.

Set Queries. Consider the reconstruction problem in which every query is a set of strings, and the answer to a query is whether at least one of the strings in the set appears in T. The *length* of a query set Q is the maximum length of a string that appears in the sets of Q. Frieze et al. [4] showed that every query set Q with resolution power at least $\frac{1}{2}$ contains $\Omega(n)$ queries. Moreover, the results of [1, 3, 7] imply that the length of Q must be at least $2 \log_\sigma n + \frac{1}{2} \log_\sigma \epsilon^{-1}$.

Pevzner and Waterman [8, problem 37] presented the problem of designing a query set which is optimal in its length and in the number of queries. Frieze et al. [4] gave a construction of a query set with $O(n)$ queries, but the length of the set is $\Theta(\log_\sigma^2 n)$. Preparata and Upfal [9] gave an improved analysis for the construction of [4]. From this analysis, it follows that there is a construction of a query set with $O(n)$ queries whose length is $3 \log_\sigma n + O(1)$ (we note that this result is not explicitly mentioned in [9]). The second result of this paper is a construction of a query set that is asymptotically optimal in the length and in the number of queries, thus settling the open question of Pevzner and Waterman. To prove this result we use a probabilistic construction. A similar approach was used by Halperin et al. [5], but their construction and analysis are quite different from ours.

Due to lack of space, some proofs are omitted from this abstract. We note that this work focuses on the asymptotic behavior, so the constants in the theorems below were not optimized. We finish this section with some definitions. For a string A, we denote its letters by $a_1, a_2, \ldots$, and we denote by $A[i: j]$ the substring $a_i a_{i+1} \cdots a_j$. For two strings A and B, AB is the concatenation of A and B.

2   The Lower Bound

In this section, we show the $\Omega(\epsilon^{-1/2} n^2)$ lower bound for the quantitative queries This lower bound implies the other lower bound mentioned in the introduction.

Theorem 1. *For every $\epsilon \leq 1/(500\sigma^2)$, every query set Q with $p(Q,n) \geq 1 - \epsilon$ satisfies $|Q| = \Omega(\epsilon^{-1/2}n^2)$.*

Proof. Consider some $\epsilon \leq 1/(500\sigma^2)$, and let Q be some query set with $p(Q,n) \geq 1 - \epsilon$. The main idea of the proof is to partition a subset of the strings of length n into pairs. The paired strings will constitute at least 2ϵ fraction of the strings of length n, and thus Q reconstructs at least half of these strings. Every paired string that is reconstructed by Q needs to be distinguished from the string it was paired to by some query. By showing that a single query can distinguish between only a small part of the pairs, we will obtain a lower bound on the size of Q. The reason we pair only part of the strings of length n is that some strings have a "complex" structure, and thus handling them is difficult. Pairing only "simple" strings simplifies the analysis.

Let $k = \lfloor 2\log_\sigma n + \log_\sigma(1/20\sqrt{\epsilon}) \rfloor$. We say that two indices i and j are *far* if $|i - j| \geq 3k + 1$, and they are *close* if $|i - j| \leq k$. For a string A of length n, a pair of indices (i,j) with $i < j$ is called a *repeat* if $A[i: i + k - 1] = A[j: j + k - 1]$. A repeat (i,j) is called *rightmost* if $j \neq n - k + 1$ and $(i + 1, j + 1)$ is not a repeat (i.e., $a_{i+k} \neq a_{j+k}$).

A string u of length k will be called *repetitive* if it has a substring of length $\lceil k/2 \rceil$ that appears at least twice in u. For example, for $k = 6$, the string 'ababac' is repetitive as the string 'aba' appears twice in it. We say that two strings u and v of length k are *similar* if they have a common substring of length $\lceil k/2 \rceil$. An ordered pair (u,v) of dissimilar non-repetitive strings of length k will be called a *simple pair*.

For every simple pair (u,v), let $B_{u,v}$ be the set of all strings A of length n for which there are indices i, i', j, and j' such that

1. $i < i' < j < j'$.
2. Every two indices from $\{i, i', j, j'\}$ are far.
3. (i,j) and (i',j') are rightmost repeats, and there are no other rightmost repeats in A.
4. $A[i: i + k - 1] = u$ and $A[i': i' + k - 1] = v$.

From conditions 3 and 4, we have that the sets $B_{u,v}$ are disjoint. We denote by B the union of the sets $B_{u,v}$ for all simple pairs (u,v).

For a string $A \in B_{u,v}$ whose rightmost repeats are (i,j) and (i',j'), let $\hat{A}$ be the string obtained from A by exchanging the substrings $A[i + k: i' - 1]$ and $A[j + k: j' - 1]$, namely,

$$\hat{A} = a_1 \cdots a_{i+k-1}a_{j+k} \cdots a_{j'-1}a_{i'} \cdots a_{j+k-1}a_{i+k} \cdots a_{i'-1}a_{j'} \cdots a_n.$$

It is easy to verify that $\hat{A} \in B_{u,v}$. Furthermore, $\hat{\hat{A}} = A$.

Lemma 1. *The number of simple pairs is $(1 - o(1))\sigma^{2k}$.*

Proof. Let u be a random string of length k. For fixed indices $i < j$, the probability that $u[i: i + \lceil k/2 \rceil - 1] = u[j: j + \lceil k/2 \rceil - 1]$ is $1/\sigma^{\lceil k/2 \rceil}$. There are $\binom{\lfloor k/2 \rfloor + 1}{2}$

ways to choose the indices i and j. Therefore, the probability that u is repetitive is at most $\binom{\lfloor k/2 \rfloor + 1}{2}/\sigma^{\lceil k/2 \rceil} = O(\log_\sigma^2 n/n) = o(1)$. Similarly, for two random strings u and v of length k, the probability that u and v are similar is at most $(\lfloor k/2 \rfloor + 1)^2/\sigma^{\lceil k/2 \rceil} = o(1)$. □

Lemma 2. *For every simple pair* (u, v), $|B_{u,v}| \geq \frac{1-o(1)}{48}\left(\frac{\sigma-1}{\sigma}\right)^2 n^4 \sigma^{n-4k}$.

Proof. Fix a simple pair (u, v). Let A be a random string of length n, and let Y be the event that $A \in B_{u,v}$. Our goal is to show that $\mathrm{P}[Y] \geq \frac{1-o(1)}{48}\left(\frac{\sigma-1}{\sigma}\right)^2 n^4/\sigma^{4k}$.

Let I be the set of all pairs $(\alpha = (i,j), \beta = (i',j'))$ such that i, i', j, j' satisfy conditions 1 and 2 in the definition of $B_{u,v}$. For a pair of indices $\alpha = (i,j)$, let Z_α be the event that (i,j) is a rightmost repeat in A, and let Z_α^w be the event that (i,j) is a rightmost repeat and $A[i:i+k-1] = w$. For $(\alpha, \beta) \in I$, let $Y_{\alpha,\beta} = Z_\alpha^u \wedge Z_\beta^v \wedge \bigwedge_{\gamma \neq \alpha, \beta} \overline{Z_\gamma}$ (note that the indices of γ can be close). The events $\{Y_{\alpha,\beta}\}_{\alpha,\beta \in I}$ are disjoint, so $\mathrm{P}[Y] = \mathrm{P}\left[\bigvee_{(\alpha,\beta)\in I} Y_{\alpha,\beta}\right] = \sum_{(\alpha,\beta)\in I} \mathrm{P}[Y_{\alpha,\beta}]$. Clearly,

$$\mathrm{P}[Y_{\alpha,\beta}] = \mathrm{P}\left[Z_\alpha^u \wedge Z_\beta^v\right] \mathrm{P}\left[\left.\bigwedge_\gamma \overline{Z_\gamma}\right| Z_\alpha^u \wedge Z_\beta^v\right]$$

$$= \mathrm{P}\left[Z_\alpha^u \wedge Z_\beta^v\right]\left(1 - \mathrm{P}\left[\left.\bigvee_\gamma Z_\gamma\right| Z_\alpha^u \wedge Z_\beta^v\right]\right)$$

$$\geq \mathrm{P}\left[Z_\alpha^u \wedge Z_\beta^v\right]\left(1 - \sum_\gamma \mathrm{P}\left[Z_\gamma | Z_\alpha^u \wedge Z_\beta^v\right]\right).$$

We now estimate the probabilities in the last expression. As every two indices from α and β are far, we have that the events Z_α^u and Z_β^v are independent, so $\mathrm{P}\left[Z_\alpha^u \wedge Z_\beta^v\right] = ((\sigma - 1)/\sigma^{2k+1})^2$.

To estimate $\mathrm{P}\left[Z_\gamma | Z_\alpha^u \wedge Z_\beta^v\right]$, consider some pair of indices $\gamma = (\gamma_1, \gamma_2)$ with $\gamma_1 < \gamma_2$. An index in γ can be close to at most one index in α or β. We consider four cases:

Case 1. If at most one index from γ is close to an index in α or β, then the events Z_γ and $Z_\alpha^u \wedge Z_\beta^v$ are independent, and $\mathrm{P}\left[Z_\gamma | Z_\alpha^u \wedge Z_\beta^v\right] = \mathrm{P}[Z_\gamma] = (\sigma - 1)/\sigma^{k+1}$ (note that $\mathrm{P}[Z_\gamma] = (\sigma - 1)/\sigma^{k+1}$ even if the indices of γ are close).

For the rest of the cases, we assume that γ_1 is close to an index $j_1 \in \alpha \cup \beta$, and γ_2 is close to an index $j_2 \in \alpha \cup \beta$.

Case 2. Suppose that $j_1 \neq j_2$ and j_1, j_2 are both from α or both from β. If $\gamma_2 - \gamma_1 = j_2 - j_1$ then let $l = \min\{j_1, \gamma_1\}$. The letters $A[l+k]$ and $A[l+j_2-j_1]$ are required to be equal by one of the events Z_γ and $Z_\alpha^u \wedge Z_\beta^v$, and are required to be unequal by the other event. Therefore, $\mathrm{P}\left[Z_\gamma | Z_\alpha^u \wedge Z_\beta^v\right] = 0$. If $\gamma_2 - \gamma_1 \neq j_2 - j_1$ then we have from [1, p. 437] that the events Z_γ and $Z_\alpha^u \wedge Z_\beta^v$ are independent, so $\mathrm{P}\left[Z_\gamma | Z_\alpha^u \wedge Z_\beta^v\right] = (\sigma - 1)/\sigma^{k+1}$.

Case 3. Suppose that $j_1 = j_2$. W.l.o.g. assume that j_1 is from α. The event Z_γ consists of k letter equality events $A[\gamma_1 + l] = A[\gamma_2 + l]$ for $l = 0, \dots, k - 1$. Let S be the set of all indices l such that the letters $A[\gamma_1 + l]$ and $A[\gamma_2 + l]$ are inside the substring $A[j_1 : j_1 + k - 1]$. For every $l \leq k - 1$, if $l \in S$ then the event $A[\gamma_1 + l] = A[\gamma_2 + l]$ depends on the event $Z_\alpha^u \wedge Z_\beta^u$, and otherwise these events are independent. More precisely, if event Z_α^u happens, then for every $l \in S$ we have that $A[\gamma_1 + l] = A[\gamma_2 + l]$ if and only if $u[\gamma_1 + l - j_1 + 1] = u[\gamma_2 + l - j_1 + 1]$. Thus, $\mathrm{P}\left[A[\gamma_1 + l] = A[\gamma_2 + l]|Z_\alpha^u\right]$ is either 0 or 1. If the latter probability is 0 for some l, then $\mathrm{P}\left[Z_\gamma|Z_\alpha^u \wedge Z_\beta^v\right] = 0$. Otherwise, we have that there is a substring of u of length $|S|$ that appears twice in u. Since u is non-repetitive, we have that $|S| \leq \lceil k/2 \rceil - 1$. The events $A[\gamma_1 + l] = A[\gamma_2 + l]$ for $l \notin S$ are independent, and they are independent of the event $Z_\alpha^u \wedge Z_\beta^u$. Therefore, $\mathrm{P}\left[Z_\gamma|Z_\alpha^u \wedge Z_\beta^v\right] = 1/\sigma^{k - |S|} \leq 1/\sigma^{\lfloor k/2 \rfloor + 1}$.

Case 4. If one of the indices j_1 and j_2 is from α and the other is from β, we can use similar argument to the one used in case 3, using the fact that u and v are dissimilar. In this case, $\mathrm{P}\left[Z_\gamma|Z_\alpha^u \wedge Z_\beta^v\right] \leq 1/\sigma^{\lfloor k/2 \rfloor + 1}$.

The number of pairs γ for which cases 1 or 2 occur is at most $\binom{n}{2}$. The number of pairs γ for which cases 3 or 4 occur is at most $8(2k + 1)^2$: There are 4 ways to choose the indices j_1 and j_2 (from $\alpha \cup \beta$) for case 3, and 4 ways to choose these indices in case 4. After choosing j_1 and j_2, there are at most $2k + 1$ ways to choose each of the two indices of γ. We conclude that

$$\sum_\gamma \mathrm{P}\left[Z_\gamma|Z_\alpha^u \wedge Z_\beta^v\right] \leq \binom{n}{2}\frac{\sigma - 1}{\sigma^{k+1}} + 8(2k + 1)^2\frac{1}{\sigma^{\lfloor k/2 \rfloor + 1}} \leq \frac{1}{2} \cdot \frac{n^2}{\sigma^k} + o(1) \leq \frac{1}{2}.$$

Since $|I| = \binom{n - 3 \cdot 3k}{4} = (1 - o(1))n^4/24$, it follows that $\mathrm{P}[Y] \geq \frac{1 - o(1)}{48}\left(\frac{\sigma - 1}{\sigma}\right)^2 n^4/\sigma^{4k}$. $\square$

From Lemmas 1 and 2 we have that

$$|B| \geq \frac{1 - o(1)}{48}\left(\frac{\sigma - 1}{\sigma}\right)^2 \frac{n^4}{\sigma^{2k}} \cdot \sigma^n \geq \frac{1 - o(1)}{48} \cdot \frac{1}{4} \cdot \frac{n^4}{\sigma^{2k}} \cdot \sigma^n \geq 2\epsilon \cdot \sigma^n.$$

Since $p(Q, n) \geq 1 - \epsilon$, it follows that Q reconstructs at least half of the strings in B.

If a string $A \in B$ is uniquely reconstructible by Q, then there must be a query $q \in Q$ that distinguishes between A and $\hat{A}$, that is, q appears a different number of times in A and $\hat{A}$. If q appears different number of times in A and $\hat{A}$ we say that q *separates* A and $\hat{A}$. Let M denote the maximum over all $q \in Q$, of the number of pairs of strings that q separates. From the definition of M we have that the number of strings in Q is at least $\frac{1}{2}|B|/M$. To complete the proof of the theorem, we will give an upper bound on M.

Lemma 3. $M \leq \frac{1}{6}n^4 \cdot \frac{(\sigma - 1)^2}{\sigma^{3k + 4}} \cdot \sigma^n$.

Proof. If q is a string of length at most $k+1$, then for every string A, q appears the same number of times in A and $\hat{A}$. Thus, q does not separate any pair of strings in B. Consider some string $q \in Q$ of length $k+l$ where $l \geq 2$.

In order to bound the number of pairs that q separates, consider some random string A. If $A \in B$ and the rightmost repeats of A are (i,j) and (i',j'), then q appear more time in A than in $\hat{A}$ if and only if one appearance of q in A is a superstring of one of the following substrings of A: $A[i-1:i+k]$, $A[i'-1:i'+k]$, $A[j-1:j+k]$, or $A[j'-1:j'+k]$. In other words, q separates A if and only if $q = A[h:h+k+l-1]$ for $h \in \{i-l+1,\ldots,i-1\} \cup \{i'-l+1,\ldots,i'-1\} \cup \{j-l+1,\ldots,j-1\} \cup \{j'-l+1,\ldots,j'-1\}$, and denote the latter event by $X_{(i,j),(i',j')}$.

Consider some fixed h. Recall that Z_α denotes the event that α is a rightmost repeat in A. If $l \leq 2k$, then $\mathrm{P}\left[q = A[h:h+k+l-1]\big|Z_{(i,j)} \wedge Z_{(i',j')}\right] = 1/\sigma^{k+l}$ since the substring $A[h:h+k+l-1]$ intersects only one of the substrings $A[i:i+k-1]$, $A[i':i'+k-1]$, $A[j:j+k-1]$, and $A[j':j'+k-1]$. If $l > 2k$, then consider the event that the prefix of q of length $3k$ is equal to the prefix of $A[h:h+k+l-1]$ of length $3k$. The probability of this event, conditioned on the event $Z_{(i,j)} \wedge Z_{(i',j')}$ is $1/\sigma^{3k}$, and therefore, $\mathrm{P}\left[q = A[h:h+k+l-1]\big|Z_{(i,j)} \wedge Z_{(i',j')}\right] \leq 1/\sigma^{3k}$. Thus, for every l,

$$\mathrm{P}\left[q = A[h:h+k+l-1]\big|Z_{(i,j)} \wedge Z_{(i',j')}\right] \leq 1/\sigma^{k+\min(l,2k)}.$$

Since there are $4(l-1)$ ways to choose h for fixed i, i', j, and j', it follows that

$$\mathrm{P}\left[X_{(i,j),(i',j')}\big|Z_{(i,j)} \wedge Z_{(i',j')}\right] \leq \frac{4(l-1)}{\sigma^{k+\min(l,2k)}} \leq \frac{4}{\sigma^{k+2}}.$$

Since the inequality above is true for every i, i', j, and j', we conclude that the number of strings that q separates is at most

$$\frac{4}{\sigma^{k+2}} \cdot \sum_{(\alpha,\beta)\in I} \mathrm{P}\left[Z_\alpha \wedge Z_\beta\right] \cdot \sigma^n \leq \frac{4}{\sigma^{k+2}} \cdot \frac{n^4}{24} \cdot \frac{(\sigma-1)^2}{\sigma^{2k+2}} \cdot \sigma^n. \quad \square$$

From Lemma 3, the number of strings in Q is at least

$$\frac{|B|}{2M} \geq \frac{\frac{1-o(1)}{48}\left(\frac{\sigma-1}{\sigma}\right)^2 \frac{n^4}{\sigma^{2k}} \cdot \sigma^n}{2 \cdot \frac{1}{6}n^4 \cdot \frac{(\sigma-1)^2}{\sigma^{3k+4}} \cdot \sigma^n} = \Omega(\sigma^k) = \Omega(\epsilon^{-1/2}n^2). \quad \square$$

3   Set Queries Model

We now show a construction of optimal query set in the set queries model. We will give our construction in a slightly different model, called the *gapped queries model*, in which each query is a single word q that contains *gaps*, namely don't care symbols, which are denoted by ϕ. The answer to a query q for a string A is "yes" if and only if q matches to a substring of A, where a don't care symbol matches to every symbol in Σ. Such a query q can be translated to the set queries model by creating a set q' containing all the words of length $|q|$ that matches to q. For example, if $q = a\phi\phi b$ and $\Sigma = \{a, b\}$, then $q' = \{aaab, aabb, abab, abbb\}$. Due to lack of space, we will consider a simplified version of the reconstruction problem in which a short prefix and suffix of the string A are known. The full version of paper will show how to solve the full reconstruction problem.

We will construct the query set in the following way: We will choose k and k', and build a set $I \subseteq \{1, \ldots, k'\}$ of size k. Then, Q_I is the set of size σ^k that contains every string q of length k' such that the i-th letter of q is a regular character if $i \in I$ and ϕ if $i \notin I$. Our goal is to choose a set I that maximizes the resolution power of Q_I.

In order to understand the intuition for building I, consider the following generic reconstruction algorithm that receives as input the answers to the queries of a set Q on the string A. We assume that for every $q \in Q$, all the strings in q have length l_Q, and that the first and last $l_Q - 1$ letters of A are known. Let l be some positive integer. The algorithm for reconstructing the first $\lceil n/2 \rceil$ letters of A is as follows (reconstructing the last $\lfloor n/2 \rfloor$ letters is performed in a similar manner):

1. Let $s_1, s_2, \ldots, s_{l_Q-1}$ be the first $l_Q - 1$ letters of A.
2. For $t = l_Q, l_Q + 1, \ldots, \lceil n/2 \rceil$ do:
   (a) Let $\mathcal{B}_t$ be the set of all strings B of length l, such that the string $s_1 \cdots s_{t-1}B$ is consistent with the answers for Q on A (i.e., for every $q \in Q$, if the answer to q on $s_1 \cdots s_{t-1}B$ is "yes", then the answer to q on A is "yes").
   (b) If all the strings in $\mathcal{B}_t$ have a common first letter a, then set $s_t \leftarrow a$. Otherwise, stop.
3. Return $s_1 \cdots s_{\lceil n/2 \rceil}$.

Clearly, for every t, the set $\mathcal{B}_t$ contains the string $a_t \cdots a_{t+l-1}$. Thus, after iteration t, $s_1 \cdots s_t = a_1 \cdots a_t$. Moreover, the algorithm stops at iteration t if and only if there is a string $B \in \mathcal{B}_t$ whose first letter is not equal to a_t. Such a string B will be called a *bad string* (w.r.t. t).

Now, suppose that we run the algorithm above with Q being the uniform query set of size k (that is, all the strings of length k without don't care symbols). Then, the algorithm stops at iteration t if and only if there is a bad string $B' \in \mathcal{B}_t$, that is, every substring of $s_1 \cdots s_{t-1}B'$ of length k is also a substring A. The latter events happen if and only if every substring of $B = s_{t-k+1} \cdots s_{t-1}B'$ of length k is a substring A. Now, the event that the first substring of B of length k is a substring of A has small probability. However, if this event happens, then probability that second substring of B of length k is a substring of A is at least $1/\sigma$: If the first event happens then $B[1:k] = A[i: i + k - 1]$ for some i. Therefore, $A[i + k] = B[k + 1]$ is a sufficient condition for the second event, and the probability that $A[i + k] = B[k + 1]$ is $1/\sigma$. This "clumping" phenomenon is also true for the other substrings of B, and therefore, the algorithm fails with relatively large probability.

In order to reduce the failure probability, we will use gapped queries as described above, and our goal is to build a set I which will reduce the "clumping" phenomenon. Here, the algorithm fails at iteration t if and only if there is bad a string $B' \in \mathcal{B}_t$ such that every substring of $B = s_{t-k'+1} \cdots s_{t-1} B'$ of length k' is equal to some substring of A on the letters of I. The event that the first substring of B is equal to some substring of A on the letters of I has small probability (note that we need to assume that $k' \in I$ otherwise this event will always occur). The event that the second substring of B is equal to a substring of A on the letters of I still depends on the first event, but now the conditional probability is small: To see this, assume that $B[1:k']$ is equal to $A[i:i+k'-1]$ on the letters of I. Then, the event that $B[2:k'+1]$ is equal to $A[i+1:i+k']$ on the letters of I consists of k letters equalities. Some of this equalities are satisfied due to the fact that $B[1:k']$ is equal to $A[i:i+k'-1]$ on the letters of I, while the other equalities are satisfied with probability $1/\sigma$ each. To have a small probability for the event that $B[2:k'+1]$ is equal to a substring of A on the letters of I, we need the number of letter equalities of the first kind to be small. This implies the following requirement on I: The intersection of I and $\{x+1 : x \in I\}$ should be small.

We now formalize the idea above. We will define what is a good set I, show that such set exists using the probabilistic method, and then show that if I is a good set, then the set Q_I has sufficiently large resolution power.

Define $\alpha = \frac{1}{50}$, $\beta = \frac{1}{20}$, $b = 500$, $b' = 1002 + \log_\sigma \frac{1}{\epsilon}$, $k = \lceil \log_\sigma n + b \rceil$ and $k' = \lceil 2\log_\sigma n + b' \rceil$. Note that $\frac{1}{3} \le \frac{k}{k'} \le \frac{1}{2}$ as we assumed that $\epsilon \ge \frac{1}{n}$. For an integer x, we denote $I + x = \{y + x : y \in I\}$. We say that a set $I \subseteq \{1, \ldots, k'\}$ is *good* if it satisfies the following requirements:

1. $\{k' - \alpha k + 1, \ldots, k'\} \subseteq I$.
2. $|I| = k$.
3. For every $i \in \{\alpha k + 1, \ldots, k'\}$, $|\{x \in \{0, \ldots, \alpha k - 1\} : i \in I + x\}| > (\frac{1}{3} - \beta)\alpha k$.
4. If $X \subseteq \{0, \ldots, \alpha k - 1\}$ and $|X| \ge (\frac{2}{3} + \beta)\alpha k$ then $\bigcup_{x \in X}(I + x) = \{1 + \min(X), \ldots, k' + \max(X)\}$.
5. For every integer $x \ne 0$, $|I \setminus (I + x)| \ge (\frac{1}{2} - \beta)k$.
6. For every integers $x \ne 0$ and $x' \ne 0$, $|I \setminus ((I + x) \cup (I + x'))| \ge (\frac{1}{4} - \beta)k$.

Lemma 4. *There exists a good set I.*

Proof. Let I be a subset of $\{1, \ldots, k'\}$ which is built by taking the set $\hat{I} = \{1, \ldots, \alpha k\} \cup \{k' - \alpha k + 1, \ldots, k'\}$ and then, for each j in $\{\alpha k + 1, \ldots, k' - \alpha k\}$, j is added to I with probability $p = (k - 2\alpha k)/(k' - \alpha k)$. We will show that I is good with positive probability. Requirement 1 is clearly satisfied.

Let $m = k' - 2\alpha k$ and $l = k - 2\alpha k$. Using Stirling's formula, we obtain that

$$P[|I| = k] = \binom{m}{l}\left(\frac{l}{m}\right)^l\left(1 - \frac{l}{m}\right)^{m-l} \ge \frac{1}{3\sqrt{l}} \ge \frac{1}{3\sqrt{k}},$$

namely, the set I satisfies requirement 2 with probability at least $1/(3\sqrt{k})$. We will show that for each other requirement, the probability that it is not satisfied is $2^{-\Omega(k)}$. Therefore, there is a positive probability that all the requirements are satisfied.

Requirement 3. Consider some fixed $i \in \{\alpha k + 1, \ldots, k'\}$ and let $Y = |\{x \in \{0, \ldots, \alpha k - 1\} : i \in I + x\}|$. For every $x \in \{0, \ldots, \alpha k - 1\}$, the probability that $i \in I + x$ is either 1 (if $i \in \hat{I} + x$) or p. Thus, $\mathrm{E}\,[Y] \geq p\alpha k \geq (\frac{k}{k'} - \frac{1}{2}\beta)\alpha k \geq (\frac{1}{3} - \frac{1}{2}\beta)\alpha k$. By Chernoff bounds, the probability that $Y \leq (\frac{1}{3} - \beta)\alpha k$ is $2^{-\Omega(k)}$. Therefore, the probability that requirement 3 is not satisfied is $(k' - \alpha k) \cdot 2^{-\Omega(k)} = 2^{-\Omega(k)}$.

Requirement 4. For every integers m and M with $0 \leq m < M \leq \alpha k - 1$, and for every set $X \subseteq \{0, \ldots, \alpha k - 1\}$ such that $\min(X) = m$ and $\max(X) = M$, we have that

$$\bigcup_{x \in X} (I + x) \supseteq \bigcup_{x \in X} (\hat{I} + x) = \{1 + m, \ldots, \alpha k + M\} \cup \{k' - \alpha k + 1 + m, \ldots, k' + M\}.$$

Moreover, if requirement 3 is satisfied then $\{\alpha k + 1, \ldots, k'\} \subseteq \bigcup_{x \in X}(I + x)$. Therefore, $\bigcup_{x \in X}(I + x) = \{1 + m, \ldots, k' + M\}$.

Requirement 5. Clearly, if $|x| \geq k'$ then $|I \setminus (I + x)| = |I| = k$. Consider some fixed integer x with $|x| < k'$. Let Y be the number of elements of $I \cap \{\alpha k + 1, \ldots, k' - \alpha k\}$ which are not in $I + x$. At least $k' - 3\alpha k$ elements of $\{\alpha k + 1, \ldots, k' - \alpha k\}$ are not contained in $\hat{I} + x$. For each such element, the probability that it appears in I and not in $I + x$ is $p(1 - p)$. Therefore, $\mathrm{E}\,[Y] \geq p(1-p)(k'-3\alpha k) \geq (1 - \frac{k}{k'} - \frac{1}{2}\beta)k \geq (\frac{1}{2} - \frac{1}{2}\beta)k$. Using Chernoff bounds, we obtain that the probability that $Y < (\frac{1}{2} - \beta)k$ is $2^{-\Omega(k)}$. Therefore, the probability that requirement 5 is not satisfied is $2k' \cdot 2^{-\Omega(k)} = 2^{-\Omega(k)}$.

Requirement 6. Using the same arguments as above, we obtain that the probability that $|I \setminus ((I + x) \cup (I + x'))| < (\frac{1}{4} - \beta)k$ is $2^{-\Omega(k)}$.   □

For the rest of the section, we assume that I is good. We use the reconstruction algorithm that was given in the beginning of this section with $Q = Q_I$, $l_Q = k'$, and $l = \alpha k$.

Lemma 5. *The probability that the algorithm fails is at most $\epsilon/2$.*

Proof. Fix an iteration t. Recall that a bad string (w.r.t. t) is a string in $\mathcal{B}_t$ whose first letter is not equal to a_t. We will bound the probability that there is a bad string w.r.t. t. We generate a random string $B' = b'_1 \cdots b'_{\alpha k}$ as follows: b'_1 is selected uniformly at random from $\Sigma - \{a_t\}$, and for $i > 1$, b'_i is selected uniformly from Σ.

Let $B = s_{t-k'+1} \cdots s_{t-1} B' = a_{t-k'+1} \cdots a_{t-1} B'$, and denote $B = b_1 \cdots b_{k'+\alpha k-1}$. By definition, B' is a bad string if and only if there are indices $r_1, \ldots, r_{\alpha k}$, called *probes*, such that $B[i: i + k' - 1]$ is equal to $A[r_i: r_i + k' - 1]$ on the letters of I for all i. We denote by $E(r_i)$ the event corresponding to the probe r_i (i.e. the event $B[i: i + k' - 1]$ is equal to $A[r_i: r_i + k' - 1]$ on the letters of I). Since $b'_1 \neq a_t$ and I satisfies requirement 1, we have that $r_i \neq t - 1 + i$ for all i.

For a probe r_i, the index i is called the *offset* of the probe, and the value r_i is called the *position* of the probe. A probe r_i is called *close* if $r_i \in \{t - k' + 2, \ldots, t\}$.

Two probes r_i and $r_{i'}$ will be called *adjacent* if $r_{i'} - r_i = i' - i$, and they will be called *overlapping* if $|r_{i'} - r_i| < k'$ and they are not adjacent. The adjacency relation is an equivalence relation. We say that an equivalence class of this relation is *close* if it contains a close probe, and we say that two equivalence classes are overlapping if they contain two probes that are overlapping.

Our goal is to estimate the probability that all the events $E(r_i)$ happen. This is easy if these events are independent, but this is not necessarily true. The cause of dependency between the events are close, adjacent, and overlapping probes. We consider several cases:

Case 1. There is an equivalence class which overlaps with at least two different equivalence classes, and all these classes are not close. Let r_{i_1}, r_{i_2} and r_{i_3} be probes from these classes, where r_{i_1} is from the first class, and r_{i_2}, r_{i_3} are from the other two classes. From [1], the events $E(r_{i_1})$ and $E(r_{i_2})$ are independent, so the probability that both events happen is σ^{-2k}. Event $E(r_{i_3})$ consists of k equalities between a letter of B and a letter of A. Since I satisfies requirement 6, at least $(\frac{1}{4} - \beta)k$ of the letter equalities involve a letter of A that does not participate in the equalities of the events $E(r_{i_1})$ and $E(r_{i_2})$. Therefore, the probability that the events $E(r_{i_1})$, $E(r_{i_2})$, and $E(r_{i_3})$ happen is at least $\sigma^{-(2+\frac{1}{4}-\beta)k}$. The number of ways to choose the positions of the probes r_{i_1}, r_{i_2}, and r_{i_3} is at most $n \cdot (2 \cdot (k' + \alpha k) - 1)^2$, and the number of ways to choose the offsets of these probes is $\binom{\alpha k}{3} \le (\alpha k)^3$. Furthermore, there are at most $n/2$ ways to choose t, and $(\sigma - 1) \cdot \sigma^{\alpha k - 1}$ ways to choose the string B'. Therefore, the overall probability that case 1 happens during the run of the algorithm is at most

$$\frac{n}{2}\sigma^{\alpha k} \cdot \frac{n \cdot (2k' + 2\alpha k)^2 (\alpha k)^3}{\sigma^{(2+\frac{1}{4}-\beta)k}} = O\left(\frac{k^5}{\sigma^{(\frac{1}{4}-\alpha-\beta)k}}\right) = o(1).$$

Case 2. There are two pairs of overlapping equivalence classes, and all these classes are not close. Let $r_{i_1}, r_{i_2}, r_{i_3}$, and r_{i_4} be probes from these classes, where r_{i_1} and r_{i_2} are from one pair of overlapping classes, and r_{i_3} and r_{i_4} are from the other pair. The events $E(r_{i_1})$ and $E(r_{i_2})$ are independent. Moreover, since r_{i_3} does not overlap with r_{i_1} or r_{i_2} (otherwise, case 1 occurs), the event $E(r_{i_3})$ is independent of the events $E(r_{i_1})$ and $E(r_{i_2})$. Therefore, $\mathrm{P}\left[E(r_{i_1}) \wedge E(r_{i_2}) \wedge E(r_{i_3})\right] = \sigma^{-3k}$. From the fact that I satisfies requirement 5, it follows that

$$\mathrm{P}\left[E(r_{i_4})|E(r_{i_1}) \wedge E(r_{i_2}) \wedge E(r_{i_3})\right] \ge \sigma^{-(\frac{1}{2}-\beta)k}.$$

Therefore, the probability that case 2 happens is at most

$$\frac{n}{2}\sigma^{\alpha k} \cdot \frac{n^2 \cdot (2k' + 2\alpha k)^2 (\alpha k)^4}{\sigma^{(3+\frac{1}{2}-\beta)k}} = O\left(\frac{k^6}{\sigma^{(\frac{1}{2}-\alpha-\beta)k}}\right) = o(1).$$

Case 3. There are two close equivalence classes. Let r_{i_1} and r_{i_2} be probes from these classes. Then, $\mathrm{P}\left[E(r_{i_1})\right] = \sigma^{-k}$, and $\mathrm{P}\left[E(r_{i_2})|E(r_{i_1})\right] \ge \sigma^{-(\frac{1}{2}-\beta)k}$. Thus, the probability of this case is bounded by

$$\frac{n}{2}\sigma^{\alpha k} \cdot \frac{(k' + \alpha k)^2 (\alpha k)^2}{\sigma^{(\frac{3}{2}-\beta)k}} = O\left(\frac{k^4}{\sigma^{(\frac{1}{2}-\alpha-\beta)k}}\right) = o(1).$$

Case 4. There is a pair of overlapping classes, and a close equivalence class. The close class can be either one of the two classes of the overlapping pair, or a third class. In both cases, using the same arguments as above, the probability of such an event is $o(1)$.

If cases 1–4 do not happen, then there is at most one pair of overlapping equivalence classes or one close equivalence class. However, these cases do not affect the analysis of the next cases, so we assume in the sequel that there are no overlapping or close probes.

Case 5. The adjacency relation contains 3 equivalence classes each of size at least 2. Let $r_{i_1}, \ldots, r_{i_6}$ be probes from these classes, where $r_{i_{2j-1}}$ and $r_{i_{2j}}$ are adjacent for $j = 1, 2, 3$. Since these probe do not overlap, we have that the events $E(r_{i_1}) \wedge E(r_{i_2})$, $E(r_{i_3}) \wedge E(r_{i_4})$, and $E(r_{i_5}) \wedge E(r_{i_6})$ are independent. From the fact that I satisfies requirement 5, it follows that an event $E(r_{i_{2j-1}}) \wedge E(r_{i_{2j}})$ contains at least $(\frac{3}{2} - \beta)k$ distinct letter equalities, so the probability that all the events happen is at least $\left(\sigma^{-(\frac{3}{2}-\beta)k}\right)^3$. Hence, the probability that case 5 happens is

$$\frac{n}{2}\sigma^{\alpha k} \cdot \frac{n^3(\alpha k)^3}{\sigma^{3(\frac{3}{2}-\beta)k}} = O\left(\frac{k^3}{\sigma^{(\frac{1}{2}-\alpha-3\beta)k}}\right) = o(1).$$

Case 6. The adjacency relation contains 2 equivalence classes each of size at least 3. Using similar arguments to the ones in case 5, the probability of this case is at most

$$n\sigma^{\alpha k} \cdot \frac{n^2(\alpha k)^2}{\sigma^{2(\frac{7}{4}-2\beta)k}} = O\left(\frac{k^2}{\sigma^{(\frac{1}{2}-\alpha-2\beta)k}}\right) = o(1).$$

In the following, we assume that cases 5 and 6 do not happen, so there is at most one equivalence class of size at least 3. All the over equivalence classes, except perhaps one class, have size 1.

Case 7. The adjacency equivalence relation contains at least $\frac{1}{4}\alpha k$ equivalence classes of size 1. Let $r_{i_1}, \ldots, r_{i_{\alpha k/4}}$ be the corresponding probes. We have that the events $E(r_{i_j})$ for $j = 1, \ldots, \frac{1}{4}\alpha k$ are independent, so for fixed probes, the probability of the event $\bigwedge_{j=1}^{\alpha k/4} E(r_{i_j})$ is $\sigma^{-k \cdot \alpha k/4}$. For each probe r_{i_j}, there are at most n ways to choose its position. The number of ways to choose the offsets of the probes is $\binom{\alpha k}{\alpha k/4} \leq 2^{\alpha k}$. Therefore, the probability of this case is

$$\frac{n}{2}\sigma^{\alpha k} \cdot 2^{\alpha k} \cdot \left(\frac{n}{\sigma^k}\right)^{\frac{\alpha k}{4}} \leq n\left(\frac{16\sigma^4}{\sigma^b}\right)^{\frac{\alpha k}{4}} \leq \frac{n}{\sigma^{(b-8)\alpha k/4}} \leq \frac{1}{4n} \leq \frac{\epsilon}{4}.$$

Case 8. If case 7 does not happen, then there is an equivalence class of size at least $\frac{3}{4}\alpha k - 1 \geq (\frac{2}{3} + \beta)\alpha k$. Let $r_{i_1}, \ldots, r_{i_s}$ be the probes of this class, and let $r_{j_1}, \ldots, r_{j_{s'}}$ be the rest of the probes. Denote by m and M the minimal and maximal offsets in $r_{i_1}, \ldots, r_{i_s}$, respectively, and let $l = \alpha k + m - M$. By the fact that I is satisfies requirement 4, we have that the letter equality event of $E(r_{i_1}) \wedge \cdots \wedge E(r_{i_s})$ involves continuous segments of letters in B and A. In other words, $E(r_{i_1}) \wedge \cdots \wedge E(r_{i_s})$ happens if and only if $B[m \colon M + k' - 1] = A[p \colon p + k' + M - m - 1]$, where p is the minimal position in $r_{i_1}, \ldots, r_{i_s}$. Therefore, $\mathrm{P}\left[E(r_{i_1}) \wedge \cdots \wedge E(r_{i_s})\right] = \sigma^{-(k'+M-m)} = \sigma^{-(k'+\alpha k - l)}$. Moreover, the event $E(r_{i_1}) \wedge \cdots \wedge E(r_{i_s})$ depends only on the values of m and M and not on the choice of the probes $r_{i_1}, \ldots, r_{i_s}$.

The events $E(r_{j_1}), \ldots, E(r_{j_{s'}})$, perhaps except one, are independent, and each event has probability σ^{-k}. In order to avoid multiplying the probability of these events by the number of ways to choose the offsets of the probes $r_{j_1}, \ldots, r_{j_{s'}}$, we will consider only the events of probes with offset either less than m or greater than M. The number of such probes is l. For a fixed value of l, there are $l + 1$ ways to choose m and M. Therefore, the probability of this case is

$$\sum_{l=1}^{\frac{1}{4}\alpha k + 1} \frac{n}{2} \sigma^{\alpha k} \frac{n}{\sigma^{k'+\alpha k - l}} \cdot (l+1) \cdot \left(\frac{n}{\sigma^k}\right)^{l-1} \le \frac{1}{\sigma^{b'}} \cdot \sum_{l=1}^{\infty} \frac{l+1}{\sigma^{(b-1)l}} \le \frac{1}{\sigma^{b'}} \cdot \sum_{l=1}^{\infty} \frac{l+1}{2^l} \le \frac{\epsilon}{4}. \quad \square$$

We obtain the following theorem:

Theorem 2. *For every $\epsilon > 0$ and n, there is a query set Q with $O(n)$ queries such that $p(Q, n) \ge 1 - \epsilon$, and the length of Q is $2\log_\sigma n + \log_\sigma \frac{1}{\epsilon} + O(1)$.*

References

1. R. Arratia, D. Martin, G. Reinert, and M. S. Waterman. Poisson process approximation for sequence repeats, and sequencing by hybridization. *J. of Computational Biology*, 3(3):425–463, 1996.
2. W. Bains and G. C. Smith. A novel method for nucleic acid sequence determination. *J. Theor. Biology*, 135:303–307, 1988.
3. M. E. Dyer, A. M. Frieze, and S. Suen. The probability of unique solutions of sequencing by hybridization. *J. of Computational Biology*, 1:105–110, 1994.
4. A. Frieze, F. Preparata, and E. Upfal. Optimal reconstruction of a sequence from its probes. *J. of Computational Biology*, 6:361–368, 1999.
5. E. Halperin, S. Halperin, T. Hartman, and R. Shamir. Handling long targets and errors in sequencing by hybridization. In *Proc. 6th Annual International Conference on Computational Molecular Biology (RECOMB '02)*, pages 176–185, 2002.
6. D. Margaritis and S. Skiena. Reconstructing strings from substrings in rounds. In *Proc. 36th Symposium on Foundation of Computer Science (FOCS 95)*, pages 613–620, 1995.
7. P. A. Pevzner, Y. P. Lysov, K. R. Khrapko, A. V. Belyavsky, V. L. Florentiev, and A. D. Mirzabekov. Improved chips for sequencing by hybridization. *J. Biomolecular Structure and Dynamics*, 9:399–410, 1991.
8. P. A. Pevzner and M. S. Waterman. Open combinatorial problems in computational molecular biology. In *Proc. 3rd Israel Symposium on Theory of Computing and Systems, (ISTCS 95)*, pages 158–173, 1995.
9. F. Preparata and E. Upfal. Sequencing by hybridization at the information theory bound: an optimal algorithm. *J. of Computational Biology*, 7:621–630, 2000.
10. R. Shamir and D. Tsur. Large scale sequencing by hybridization. *J. of Computational Biology*, 9(2):413–428, 2002.

Reconstructive Dispersers
and Hitting Set Generators

Christopher Umans*

Computer Science Department
California Institute of Technology
Pasadena CA 91125
umans@cs.caltech.edu

Abstract. We give a generic construction of an optimal hitting set generator (HSG) from any good "reconstructive" disperser. Past constructions of optimal HSGs have been based on such disperser constructions, but have had to modify the construction in a complicated way to meet the stringent efficiency requirements of HSGs. The construction in this paper uses existing disperser constructions with the "easiest" parameter setting in a black-box fashion to give new constructions of optimal HSGs without any additional complications.

Our results show that a straightforward composition of the Nisan-Wigderson pseudorandom generator that is similar to the composition in works by Impagliazzo, Shaltiel and Wigderson in fact yields optimal HSGs (in contrast to the "near-optimal" HSGs constructed in those works). Our results also give optimal HSGs that do not use any form of hardness amplification or implicit list-decoding – like Trevisan's extractor, the only ingredients are combinatorial designs and any good list-decodable error-correcting code.

1   Introduction

Ever since Trevisan [1] showed that certain pseudorandom generator (PRG) constructions yield extractors, many of the results in these two areas have been intertwined. In this paper we are concerned with the one-sided variant of a PRG, called a *hitting set generator* (HSG), and the one-sided variant of an extractor, called a *disperser*.

Informally, a HSG construction takes an n-bit truth table of a hard function f and converts it into a collection of $\text{poly}(n)$ shorter m-bit strings, with the property that every small circuit D that accepts at least $1/2$ of its inputs, also accepts one of these m-bit strings. The proof that a construction is indeed a HSG typically gives an efficient way to convert a small circuit D on which the construction fails to meet the definition into a small circuit computing f, thus contradicting the hardness of f.

Informally, a disperser takes an n-bit string x sampled from a weak random source with sufficient min-entropy and converts it into a collection of $\text{poly}(n)$ shorter m-bit strings, with the property that every circuit D that accepts at least

* Supported by NSF grant CCF-0346991 and an Alfred P. Sloan Research Fellowship.

C. Chekuri et al. (Eds.): APPROX and RANDOM 2005, LNCS 3624, pp. 460–471, 2005.
© Springer-Verlag Berlin Heidelberg 2005

$1/2$ of its inputs, also accepts one of these m-bit strings. Trevisan's insight is that a HSG construction whose proof uses D in a black-box fashion *is* a disperser, for the following reason: if there is a circuit D on which the construction fails to meet the disperser definition, then we have a small circuit *relative to D* that describes input x, and it cannot be that every string in a source with sufficiently high min-entropy has such a short description.

Thus we can produce a formal statement to the effect that "every black-box HSG construction yields a disperser with similar parameters." In this paper we consider the reverse question, namely: "under what conditions does a disperser construction yield a HSG construction?"

We will limit ourselves to so-called "reconstructive" dispersers which means, roughly, that the associated proof has the same outline as the one sketched above, and that the conversion in the proof is efficient. At first glance this may seem to be such a strong constraint that the question becomes uninteresting. However, there is an important issue related to the precise meaning of "efficient." It turns out that there are (at least) two possible notions of "efficient;" one is satisfied naturally in several disperser constructions, and the other – which is the one that is actually required for the construction to be a HSG – is far more stringent. The distinction between the two is analogous to the distinction between an error-correcting code being efficiently decodable in the usual sense, and being efficiently *implicitly* decodable in the sense of [2].

To be precise, consider a function $E : \{0,1\}^n \times \{0,1\}^t \to \{0,1\}^m$ (where $m < n$ may be as small as poly $\log n$) and a circuit $D : \{0,1\}^m \to \{0,1\}$ relative to which E fails to be a disperser; i.e. D accepts at least half of its inputs, but for every x in the weak random source, $E(x,\cdot)$ fails to hit an accepting input of D. If E is equipped with a "reconstructive" proof then the proof should give an efficient way to reconstruct x from D and short advice (the advice may depend on x and auxiliary randomness w used in the proof). The two notions of reconstructivity for E that we discuss are:

Explicit Reconstructivity. Given oracle access to circuit D, and advice $A(x,w)$, compute in time poly(n) the string x with non-negligable probability over the choice of w.

Implicit Reconstructivity. Given oracle access to circuit D, and advice $A(x,w)$, and an index i, compute in time poly(m) the i-th bit of x with non-negligable probability over the choice of w.

Explicit reconstructivity is naturally satisfied by relatively simple disperser constructions[1] [1, 3, 4], and these construction possess two additional useful features as well (as observed in [5–7]): (1) w has length $O(\log n)$ and (2) A is computable in time poly(n).

In contrast, obtaining *implicit reconstructivity* has always required significant extra effort: in [4], one needs to use multiple copies of the original disperser with multiple "strides" and to piece them together in a complex manner; in [8], similar

---

[1] Of course these are all actually extractor constructions, but an extractor is a disperser.

ideas are used, together with an "augmented" low-degree extension; and in [5, 6], the construction of [1] is modified by repeated composition, but at the price of a super-polynomial degradation in the output length.

In this paper, we show that the weaker notion of explicit reconstructivity (together with the two additional properties satisfied by known constructions) is in fact sufficient to construct *optimal* HSGs, thus avoiding the complications of past work.

Our result is shown by analyzing a composition of reconstructive dispersers, similar to the one used in [5, 6]. Our composition has the advantage of being simpler (we believe) and generic (it works for any reconstructive disperser). Most importantly, it produces optimal HSGs for all hardnesses, starting only with reconstructive dispersers for a particular "easy" setting of parameters.

Our results also shed light on two issues: the first concerns the Nisan-Wigderson pseudorandom generator (PRG) [2, 9, 10], which has a non-optimal seed length for sub-exponential hardness assumptions. Two works [5, 6] addressed this deficiency by composing the PRG with itself in a clever way. The result came close to an optimal construction, but fell short. One might have guessed that there was some inherent loss associated with the composition-based approach, or that the combinatorial-design-based constructions were too weak to obtain the optimal result (as subsequent solutions to this problem employed different, algebraic constructions [4, 8]). Our results show that in the end, composing the Nisan-Wigderson PRG with itself *can* be made to work to obtain HSGs, by using a somewhat different composition than those used previously.

Second, until this paper, known constructions of optimal HSGs [4, 8] have made crucial use of implicit list-decoding (cf. [2]) of Reed-Muller codes. Implicit list-decoding of Reed-Muller codes also underlies the hardness amplification results that were a component of earlier (non-optimal) constructions [2, 5, 6]. One may wonder whether some form of implicit list-decoding is in fact *necessary* to construct HSGs. Our results show that it is not, and indeed we can construct optimal HSGs using only combinatorial designs and any good list-decodable code (which, not accidentally, are also the two ingredients needed for Trevisan's extractors).

We remark that "reconstructivity" of various disperser and extractor constructions has emerged as a crucial property of these constructions in a number of applications: error-correcting codes [11, 12], data structures [13], and complexity theory [14]. This suggests that it is worthwhile to formalize and study notions of reconstructivity as we do in this paper.

2   Preliminaries

We use $[n]$ as shorthand for the set $\{1, 2, 3 \ldots n\}$. We use U_m for the random variable uniformly distributed on $\{0, 1\}^m$.

Definition 1. *Let Z be a random variable distributed on $\{0, 1\}^m$. We say that a function $D : \{0, 1\}^m \to \{0, 1\}$ "ϵ-catches" Z if $|\Pr[D(U_m) = 1] - \Pr[D(Z) = 1]| > \epsilon$. In the special case that $\Pr[D(Z) = 1] = 0$, we say that D "ϵ-avoids Z".*

In this paper we will always be in the aforementioned special case, since we are discussing one-sided objects (HSGs and dispersers). Replacing "ϵ-avoids" with "ϵ-catches" everywhere in the paper yields the two-sided version of all of the definitions, theorems and proofs, with the exception of one place in the proof of Theorem 4 where we use the one-sidedness critically.

Definition 2. *Let x be a n-bit truth table of a function that requires circuits of size k. An ϵ-HSG is a function $H_x : \{0,1\}^t \to \{0,1\}^m$ such that no size m circuit $D : \{0,1\}^m \to \{0,1\}$ ϵ-avoids $H_x(U_t)$.*

This means that every size m circuit D that accepts more than an ϵ fraction of its inputs, also accepts $H_x(y)$ for some y. It has become customary to refer to families of ϵ-HSGs that for every $k = k(n)$ have parameters $t \leq O(\log n)$ and $m \geq k^\delta$ for some constant $\delta > 0$ as *optimal* (because they give rise to hardness vs. randomness tradeoffs that are optimal up to a polynomial).

For reference, we give the standard definition of a (k, ϵ)-disperser before stating the more complicated definition of a reconstructive disperser that we will need for this paper.

Definition 3. *A (k, ϵ)-disperser is a function $E : \{0,1\}^n \times \{0,1\}^t \to \{0,1\}^m$ such that for all subsets $X \subseteq \{0,1\}^n$ with $|X| \geq 2^k$, no circuit $D : \{0,1\}^m \to \{0,1\}$ ϵ-avoids $E(X, U_t)$.*

2.1   Reconstructive Dispersers

We now define the central object, which we call a reconstructive disperser. It has more parameters than one would prefer, but keeping track of all of these parameters will make the composition much easier to state.

Definition 4. *A $(n, t, m, d, a, b, \epsilon, \delta)$-reconstructive disperser is a triple of functions:*

- *the "disperser" function $E : \{0,1\}^n \times \{0,1\}^t \to \{0,1\}^m$*
- *the "advice" function $A : \{0,1\}^n \times \{0,1\}^d \to \{0,1\}^a$*
- *the randomized[2] oracle "reconstruction" procedure $R : \{0,1\}^a \times [n/b] \to \{0,1\}^b$*

that satisfy the following property: for every $D : \{0,1\}^m \to \{0,1\}$ and $x \in \{0,1\}^n$ for which D ϵ-avoids $E(x, U_t)$, we have

$$\forall i \in [n/b] \quad \Pr_w[R^D(A(x,w), i) = x_i] \geq \delta. \tag{1}$$

Here x_i refers to the i-th b-bit block in x. When $b = n$ we drop the second argument to R.

---

[2] When we refer to a "randomized" function, we mean a function f that takes an extra argument which is thought of as random bits. We refrain from explicitly writing the second argument; however whenever f occurs within a probability, we understand that the probability space includes the randomness of f.

Note that in this definition there is no reference to the parameter "k" that occurs in Definitions 2 and 3. The idea is that if (E, A, R) is a reconstructive disperser, then E must be a (k, ϵ)-disperser for k slightly larger than a, because relative to D, many strings x in the source X have descriptions of size approximately a (via R). Similarly, $E(x, \cdot)$ is an ϵ-HSG when x is the truth table of a function that does not have size k circuits, for k slightly larger than a plus the running time of R. This is because relative to D, and given a bits of advice, R can be used to compute any specified bit of x. See Theorems 2 and 3 for formal statements of these assertions.

The parameter b interpolates between the two types of reconstructivity. When $b = n$, the reconstruction procedure outputs all of x. A reconstruction procedure of this type running in time $\text{poly}(n)$ is *explicit* and is often implied by disperser constructions whose proofs use the so-called "reconstruction proof paradigm."

To obtain an optimal HSG, we need b to be small and the running time of the reconstruction procedure to be $\text{poly}(m)$ (where $m < n$ may be as small as $\text{poly} \log n$). Such a reconstruction procedure is necessarily *implicit*, and it satisfies a much more stringent efficiency requirement that typically does *not* follow from disperser constructions without modification. Note however that it is trivial to decrease b by a multiplicative factor without changing the running time of R (and we will use this fact). The challenge is to decrease b while decreasing the running time of R simultaneously.

There are three quite clean constructions of reconstructive dispersers known for a certain "easy" setting of the parameters.

Theorem 1 ([1, 3, 4]). *For every n and $\epsilon \geq n^{-1}$, and pair of constants $\gamma > \beta > 0$ (and $\gamma > \beta + 1/2$ in the case of [3]), there is a*

$$(n, t = O(\log n), m = n^\beta, d = O(\log n), a = n^\gamma, b = n, \epsilon, \delta = 1/4)$$

reconstructive disperser (E, A, R) with the running time of E, A, R at most n^c for a universal constant c.

Proof. (sketch) The proofs associated with these constructions all conform to the following outline: (1) given D that ϵ-avoids $E(x, U_t)$, convert it into a randomized *next-bit predictor* with success rate $1/2 + \epsilon/m$; (2) use the next-bit predictor together with the advice $A(x, w)$ to recover x with probability $1/4$ over the choice of w. In some cases the original proof concludes by obtaining a short *list* containing x; in this case we add a random hash of x to the advice string to allow us to correctly select x from the list with probability $1/4$.   □

Note that these constructions are not sufficient to produce HSGs directly, because the running time of the reconstruction procedure R is by itself far greater than the trivial upper bound on the circuit complexity of a function whose truth table has size n, so we cannot get the required contradiction.

However, if the running time of R is much smaller than the input length n, reconstructive dispersers *are* hitting set generators when their input is fixed to be a hard function.

Theorem 2. *Let (E, A, R) be a $(n, t, m, d, a, b, \epsilon, \delta = 2/3)$ reconstructive dispersers, for which R runs in time T. Let $x \in \{0,1\}^n$ be the truth table of a function that cannot be computed by circuits of size k. There is a universal constant c for which: if $k > c(\log n)(Tm + a)$, then $H_x(\cdot) = E(x, \cdot)$ is an ϵ-HSG.*

Proof. If $E(x, \cdot)$ is not the claimed HSG, then there is a size m circuit $D :$ $\{0,1\}^m \to \{0,1\}$ that ϵ-avoids $E(x, U_t)$. By the definition of reconstructive dispersers we have: $\forall i \in [n/b]$ $\mathrm{Pr}_w[R^D(A(x, w), i) = x_i] \geq 2/3$. We repeat the reconstruction $\Theta(\log n)$ times with independent random w's and take the majority outcome. For a given i, the probability that this fails to produce x_i is less than $1/n$ by Chernoff bounds. By a union bound, the probability that we fail on any i is strictly less than one. Thus we can fix the random bits used in this procedure so that we correctly produce x_i for all i. We hardwire $A(x, w)$ for the chosen w's. The resulting circuit has size $c(\log n)(Tm + a)$ which contradicts the hardness of x.   □

We remark that the same argument shows that reconstructive dispersers are indeed a special case of ordinary dispersers (a more efficient conversion is possible but that is not important for this paper):

Theorem 3. *Let (E, A, R) be a $(n, t, m, d, a, b, \epsilon, \delta = 2/3)$ reconstructive disperser. There is a universal constant c for which E is a $(ca(\log n), \epsilon)$-disperser.*

3   Intuition for the Composition

In this discussion we focus almost entirely on the advice and reconstruction functions. Getting the parameters of these "right" in the composition gives a reconstructive disperser that is an optimal HSG. From Theorem 2 we can see that the key is for the advice length a to be as short as possible, and for the running time of the reconstruction procedure R to be as small as possible. Specifically, our goal will be to obtain (after several compositions) a reconstructive disperser with input length $N \gg n$, advice length poly(n), and R's running time poly(n), while maintaining an output length of at least $n^{\Omega(1)}$.

Our starting point is the "simple" constructions of Theorem 1. Note that $b = n$ in those constructions so we drop the second argument of the reconstruction procedure R, and then it simply maps $a = n^\gamma$ bits to n bits.

Let as also assume for the purpose of this simplified exposition that $d = 0$, i.e., the advice function A just maps n bits to $a = n^\gamma$ bits. It is useful to think of the advice function A as a procedure that "compresses" an arbitrary n-bit string x into n^γ bits and the reconstruction procedure R as a procedure that "decompresses" the n^γ bit string back to the original n-bit string x. Of course this is information-theoretically impossible, but a slightly relaxed version of this (in which $d = O(\log n)$ rather than 0, and R is given oracle access to "distinguishing" function D) is exactly the definition of A and R in Definition 4.

So we have the ability to compress from n bits down to n^γ bits. Suppose we want to be able to compress from $N \gg n$ down to n^γ bits. A natural thing to do is to divide the N-bit string into N/n substrings of size n, and use A to compress each of the substrings. The resulting string has length $(N/n)n^\gamma$; we can repeat the process $O(\log N/\log n)$ times until we get down to length n^γ. To decompress, we use R to reverse the process. A nice side-effect of this scheme is that when we know *which* n-bit substring of the original N bit string we wish to recover, we only need to invoke R *once* at each level (as opposed to, e.g., N/n times at the bottom level, if we insist on recovering the entire original N bit string). In the formalism of Definition 4, this means that we can take $b = n$, and have R only recover the specified b-bit block of the input string.

The above description specifies an advice function that maps an N-bit string x down to n^γ bits, and a reconstruction procedure that recovers any specified n-bit substring of x from those n^γ bits. A crucial observation is that the new reconstruction procedure has $b = n \ll N$, and it runs in time $\text{poly}(n, \log N/\log n) \ll N$, since it invokes R once for each of the $O(\log N/\log n)$ levels.

We have made only two simplifying assumptions. First, we have assumed that $d = 0$, when in fact it must be $O(\log n)$. When $d = O(\log n)$ the "compression" function A produces $2^d = \text{poly}(n)$ *candidate* compressed versions of x, with the property that with high probability over a choice of candidates, the "decompression" procedure R succeeds. We will carry out the above scheme for all candidates at each level, and ensure that the decompression at each level works with sufficiently high probability so that the overall decompression works with constant probability.

Second, we have ignored the fact that R needs oracle access to a function D with certain properties to succeed. To deal with this, we simply run the disperser E on every n-bit string x that we "compress" in the entire process, and define the disperser E' of the composed object to be the union of these. The reconstruction procedure R' for the composed object is only required to work relative to D that ϵ-avoids the output of E'. Such a D also ϵ-avoids the output of each invocation of E we have performed[3], and so D is exactly what is required to actually allow the "decompressions" to work at all levels.

Both the disperser function E' and the advice function A' of the composed object have asymptotically optimal seed lengths of $O(\log N)$. This is because at each level we need a fresh $O(\log n)$ bit seed for E and A coming from Theorem 1, and there are $O(\log N/\log n)$ levels.

The formal analysis of a single level of this composition is the content of Theorem 4. Corollary 1 and Theorem 5 apply this composition $O(\log N/\log n)$ times to the reconstructive dispersers of Theorem 1 to obtain the final result.

Comparison with [5, 6]. As noted, our composition is similar to the one used in [5, 6], which can also be understood in the language of reconstructive dispersers. If one interprets their composition this way, the crucial difference is that our

---

[3] This is the single place where we rely critically on the fact that we are dealing with "ϵ-avoids," rather than "ϵ-catches."

reconstruction procedure runs on strings of length n at all levels, while in their work, the reconstruction procedure runs on inputs whose lengths increase from one level to the next. In addition, [5, 6] have a layer of hardness amplification at each level, which our composition avoids. The result is that they incur a loss that is superpolynomial if the number of levels is super-constant (and to minimize this loss, they vary parameters from level to level in a sophisticated way). We incur only a loss that is *multiplicative* in the number of levels, which for us is logarithmic in the input length, so we can ignore it altogether.

4   Analysis of the Composition

We begin by showing that a simple pairwise independent repetition can increase the success probability of the reconstruction procedure:

Lemma 1 (increasing δ). *Suppose (E, A, R) is a $(n, t, m, d, a, b = n, \epsilon, \delta)$ reconstructive extractor. For every $\alpha > 0$, we can convert (E, A, R) into a*

$$(n, t, m, d' = O(d + \log(pn)), a' = O(pa), b = n, \epsilon, \delta' = 1 - \alpha)$$

reconstructive extractor (E, A', R'), where $p = \delta^{-1}\alpha^{-1}$. The running time of A' is at most $poly(p, n)$ times the running time of A, and the running time of R' is at most $poly(p, n)$ times the running time of R.

Proof. (sketch) A' uses its random bits to output $2p$ invocations of A using pairwise independent seeds, and a random hash of the input x; R' runs R on each of the advice strings, using the hash to select which reconstruction to output.   □

Our main composition operation for reconstructive dispersers increases n. More specifically, it roughly squares n, while multiplying each of the two seed lengths (t and d) by a constant. The final object inherits the advice length a from the first reconstructive disperser, and the block length b from the second.

Theorem 4 (composition of reconstructive dispersers). *Suppose (E_1, A_1, R_1) is a $(n_1, t_1, m, d_1, a_1, b_1, \epsilon, \delta_1)$ reconstructive disperser and that (E_2, A_2, R_2) is a $(n_2, t_2, m, d_2, a_2, b_2, \epsilon, \delta_2)$ reconstructive disperser, with $a_2 = b_1$. Set $r = n_1/b_1$. Then (E, A, R) defined as:*

$$E(x_1, \ldots, x_r; w_2, y, p \in [r+1]) = \begin{cases} E_1(A_2(x_1, w_2) \circ \cdots \circ A_2(x_r, w_2), y) & \text{if } p = r+1 \\ E_2(x_p, y) & \text{otherwise} \end{cases}$$

$$A(x_1, \ldots, x_r; w_1, w_2) = A_1(A_2(x_1, w_2) \circ \cdots \circ A_2(x_r, w_2), w_1)$$

$$R^D(z; i_1 \in [r], i_2 \in [n_2/b_2]) = R_2^D(R_1^D(z, i_1), i_2)$$

is a

$$(n = (n_1 n_2)/a_2, t = \max(t_1, t_2) + d_2 + \log(r+1), m,$$
$$d = d_1 + d_2, a = a_1, b = b_2, \epsilon, \delta = \delta_1 + \delta_2 - 1)$$

reconstructive disperser.

A few words of explanation are in order. The input for the composed object is $x = x_1, x_2, \ldots, x_r$, where the x_i coincide with the blocks R will need to output. The function A is very simple: it just concatenates the output of A_2 run on each of $x_1, x_2, \ldots, x_r$, and runs A_1 on the concatenated string. The function R is similarly simple; it reverses the process: it takes the output of A and first uses R_1 to extract the advice associated with x_{i_1}, and then uses R_2 to actually recover (the i_2-th block of) x_{i_1}. Finally E uses part of its seed, p, to decide either to run E_2 on input x_i for some i, or to run E_1 on the concatenated advice string on which A_1 is run.

Proof. (of Theorem 4) Fix D and $x = x_1, x_2, \ldots, x_r$ for which D ϵ-avoids $E(x_1, \ldots, x_r; U_t)$. Also fix $i = i_1 i_2$, where $i_1 \in [r]$ and $i_2 \in [n_2/b_2]$.

From the fact that D ϵ-avoids $E(x_1, \ldots, x_r; U_t)$, we know that

$$\forall w_2 \quad D \ \epsilon\text{-avoids } E_1(A_2(x_1, w_2) \circ \cdots \circ A_2(x_r, w_2), U_{t_1}) \tag{2}$$

$$\forall p \in [r] \quad D \ \epsilon\text{-avoids } E_2(x_p, U_{t_2}) \tag{3}$$

From (2) and the definition of reconstructive dispersers, we get that for all w_2:

$$\Pr_{w_1}[R_1^D(A_1(A_2(x_1, w_2) \circ \cdots \circ A_2(x_r, w_2), w_1), i_1) = A_2(x_{i_1}, w_2)] \geq \delta_1. \tag{4}$$

From (3) and the definition of reconstructive dispersers, we get that:

$$\Pr_{w_2}[R_2^D(A_2(x_{i_1}, w_2), i_2) = (x_{i_1})_{i_2}] \geq \delta_2. \tag{5}$$

The probability over a random choice of w_1 and w_2 that both events occur is at least $\delta_1 + \delta_2 - 1$. If both events occur, then:

$$R^D(A(x_1, \ldots, x_r; w_1, w_2); i_1, i_2) = R_2^D(R_1^D(A(x_1, \ldots, x_r; w_1, w_2), i_1), i_2) \tag{6}$$

$$= R_2^D(R_1^D(A_1(A_2(x_1, w_2) \circ \cdots \circ A_2(x_r, w_2), w_1), i_1), i_2) \tag{7}$$

$$= R_2^D(A_2(x_{i_1}, w_2), i_2) = (x_{i_1})_{i_2}. \tag{8}$$

where (6) just applies the definition of R, (7) applies the definition of A, and (8) follows from our assumption that the events in (4) and (5) both occur.

We conclude that $\Pr_{w_1, w_2}[R^D(A(x; w_1, w_2), i) = x_i] \geq \delta_1 + \delta_2 - 1$, which is what was to be shown. $\qquad\square$

We now apply Theorem 4 repeatedly:

Corollary 1. *Fix N, n, ϵ, and constant $\gamma < 1$. Let (E, A, R) be a*

$$(n, t_1 = O(\log n), m, d_1 = O(\log n), a = n^\gamma, b = a, \epsilon, \delta_1 = 1 - (1/\log N))$$

reconstructive disperser with the running time of E, A, R at most n^{c_1} for a universal constant c_1. Then (E', A', R') obtained by $\left(\frac{\log N}{(1-\gamma)\log n}\right)$ applications of Theorem 4 is a $(N, t = O(\log N), m, d = O(\log N), a = n^\gamma, b = a, \epsilon, \delta = 2/3)$ reconstructive disperser with the running time of E' and A' at most N^c, and the running time of R' at most n^c for a universal constant c.

Proof. We claim that after i compositions of (E, A, R) with itself, we obtain a

$$\left(n_i = n^{i-(i-1)\gamma}, t_i = t_1 + (i-1)(d_1 + \log(n^{1-\gamma} + 1)), m,\right.$$
$$d_i = id_1, a, b, \epsilon, \delta_i = i(\delta_1 - 1) + 1)$$

reconstructive disperser (E_i, A_i, R_i). This clearly holds for $i = 1$.

To see that it holds for arbitrary i, consider composing $(E_{i-1}, A_{i-1}, R_{i-1})$ with (E, A, R). By Theorem 4 and our inductive assumption, we get (E_i, A_i, R_i) with parameters:

$$n_i = n_{i-1}n/a = n^{i-1-(i-2)\gamma}n/n^\gamma = n^{i-(i-1)\gamma}$$
$$t_i = \max(t_{i-1}, t_1) + d_1 + \log(n^{1-\gamma} + 1)$$
$$= [t_1 + (i-1)(d_1 + \log(n^{1-\gamma} + 1))] + d_1 + \log(n^{1-\gamma} + 1)$$
$$= t_1 + i(d_1 + \log(n^{1-\gamma} + 1))$$
$$d_i = d_{i-1} + d_1 = id_1$$
$$\delta_i = \delta_{i-1} + \delta_1 - 1 = (i-1)(\delta_1 - 1) + \delta_1 = i(\delta_1 - 1) + 1$$

as claimed. Note that $n_i > n^{i(1-\gamma)}$. Thus when $i = \frac{\log N}{(1-\gamma)\log n}$, we have $n_i \geq N$, and also $d_i = O(\log N)$ and $t_i = O(\log N)$. Also note that for sufficiently large n, $\delta_i = 1 - i(1/\log N) = 1 - \frac{1}{(1-\gamma)\log n} > 2/3$.

Finally, let $T(E_i), T(A_i), T(R_i)$ denote the running times of the functions E_i, A_i, R_i respectively. From the specification of the composition in Theorem 4, we see that

$$T(E_i) \leq \max(T(E_{i-1}) + n_{i-1}T(A), T(E))$$
$$T(A_i) \leq T(A_{i-1}) + n_{i-1}T(A)$$
$$T(R_i) \leq T(R_{i-1}) + T(R),$$

and then it is easy to verify by induction that $T(E_i), T(A_i) \leq in^{i+c_1}$ and $T(R_i) \leq in^{c_1}$. Plugging in $i = O(\log N/\log n)$ gives the claimed running times.   □

5   Optimal Hitting Set Generators

Now, we can use any one of the constructions of Theorem 1 in Corollary 1 to obtain an optimal HSG for arbitrary hardness. Specifically, our HSG is built from the N-bit truth table of a function that is hard for circuits of size k, has a seed length of $O(\log N)$, and outputs $k^{\Omega(1)}$ bits, while running in time poly(N). As usual, we assume $N < 2^{k^\eta}$ for any constant η; as otherwise we could just output the seed.

Theorem 5. *Let x be the N-bit truth table of a function that cannot be computed by circuits of size k, and set $n = k^{1/c}$ for a sufficiently large constant c. Let (E, A, R) be a*

$$(n, t = O(\log n), m = n^\beta, d = O(\log n), a = n^\gamma, b = n, \epsilon, \delta = 1/4)$$

reconstructive disperser from Theorem 1. If (E', A', R') is the reconstructive disperser obtained by applying Lemma 1 with $\alpha = 1/\log N$ and then applying Corollary 1, then $H_x(\cdot) = E'(x, \cdot)$ is an ϵ-HSG against circuits of size $m = k^{\beta/c}$.

Proof. After applying Lemma 1 with $\alpha = 1/\log N$, we have a

$$\left(n, t = O(\log n), m = n^\beta, d' = O(\log n), a' = (\log N)n^\gamma, b = n, \epsilon, \delta' = 1 - \frac{1}{\log N} \right)$$

reconstructive disperser. By the assumption on N stated before the theorem, we have that $a' \leq n^{\gamma'}$ for some constant $\gamma' < 1$. As noted following Definition 4, we can decrease b from n to a' trivially. Then, after applying Corollary 1 we obtain a $(N, O(\log N), m, O(\log N), n^{\gamma'}, a', \epsilon, 2/3)$ reconstructive disperser (E', A', R'), with the running time of E' and A' at most $\text{poly}(N)$ and the running time of R' at most $\text{poly}(n)$.

To satisfy Theorem 2 we need $c_1 \log N(n^{c_2} n^\beta + n^{\gamma'}) < k$, where c_1 and c_2 are universal constants; we can set c to ensure that this holds. Theorem 2 then states that $E'(x, \cdot)$ is an ϵ-HSG against circuits of size $m = n^\beta = k^{\beta/c}$. $\qquad\square$

As noted in the introduction, if we use Trevisan's extractor [1] as our starting object, then the entire construction requires only two ingredients: (1) any good list-decodable error-correcting code and (2) combinatorial designs (to obtain the original Trevisan extractor). In particular there is *no* hardness amplification or implicit list-decoding hidden in the construction, precisely because we are able to work with a starting object that only has *explicit reconstructivity* rather than implicit reconstructivity (the latter type of reconstructivity typically has required implicit list-decoding in some form or another).

6   Open Problems

We mention briefly two interesting open problems related to this work.

First, is it possible to extend these results to two-sided objects, by giving a similar composition for reconstructive *extractors*? Because implicit reconstructivity of extractors is closely related to efficient *implicit* list-decodability, it is possible that such a result would give a new generic construction of implicitly list-decodable codes (in the sense of [2]) from *any* good list-decodable codes.

Second, is it possible to extend our result to the non-deterministic setting? Here there is an important technical issue of *resiliency* of HSGs discussed in [15]; obtaining a resilient HSG construction would lead to so-called "low-end" uniform hardness vs. randomness tradeoffs for the class AM. One possible route to constructing low-end resilient HSGs against nondeterministic circuits is to construct high-end resilient HSGs (typically an easier task) that possess the features needed to apply the composition in this paper – namely an associated advice function $A(x, w)$ computable in polynomial time, with w having length $O(\log n)$. In the currently known resilient construction [16], w has length much larger than $O(\log n)$.

Acknowledgements

We thank Ronen Shaltiel for his comments on an early draft of this paper.

References

1. Trevisan, L.: Extractors and pseudorandom generators. Journal of the ACM **48** (2002) 860–879
2. Sudan, M., Trevisan, L., Vadhan, S.: Pseudorandom generators without the XOR lemma. JCSS: Journal of Computer and System Sciences **62** (2001)
3. Ta-Shma, A., Zuckerman, D., Safra, S.: Extractors from Reed-Muller codes. In: Proceedings of the 42nd Annual IEEE Symposium on Foundations of Computer Science. (2001)
4. Shaltiel, R., Umans, C.: Simple extractors for all min-entropies and a new pseudo-random generator. In: Proceedings of the 42nd Annual IEEE Symposium on Foundations of Computer Science. (2001)
5. Impagliazzo, R., Shaltiel, R., Wigderson, A.: Near-optimal conversion of hardness into pseudo-randomness. In: Proceedings of the 40th Annual IEEE Symposium on Foundations of Computer Science. (1999) 181–190
6. Impagliazzo, R., Shaltiel, R., Wigderson, A.: Extractors and pseudo-randomn generators with optimal seed-length. In: Proceedings of the Thirty-second Annual ACM Symposium on the Theory of Computing. (2000)
7. Ta-Shma, A., Umans, C., Zuckerman, D.: Loss-less condensers, unbalanced expanders, and extractors. In: Proceedings of the 33rd Annual ACM Symposium on Theory of Computing. (2001) 143–152
8. Umans, C.: Pseudo-random generators for all hardnesses. In: Proceedings of the 34th Annual ACM Symposium on Theory of Computing. (2002) 627–634
9. Nisan, N., Wigderson, A.: Hardness vs randomness. Journal of Computer and System Sciences **49** (1994) 149–167
10. Impagliazzo, R., Wigderson, A.: P = BPP if E requires exponential circuits: Derandomizing the XOR lemma. In: Proceedings of the 29th Annual ACM Symposium on Theory of Computing. (1997) 220–229
11. Ta-Shma, A., Zuckerman, D.: Extractor codes. IEEE Transactions on Information Theory **50** (2004) 3015–3025
12. Guruswami, V.: Better extractors for better codes? In: STOC. (2004) 436–444
13. Ta-Shma, A.: Storing information with extractors. Inf. Process. Lett. **83** (2002) 267–274
14. Buhrman, H., Lee, T., van Melkebeek, D.: Language compression and pseudo-random generators. In: IEEE Conference on Computational Complexity. (2004) 15–28
15. Gutfreund, D., Shaltiel, R., Ta-Shma, A.: Uniform hardness vs. randomness trade-offs for Arthur-Merlin games. In: 18th Annual IEEE Conference on Computational Complexity. (2003)
16. Miltersen, P.B., Vinodchandran, N.V.: Derandomizing Arthur-Merlin games using hitting sets. In: Proceedings of the 40th Annual IEEE Symposium on Foundations of Computer Science. (1999) 71–80

The Tensor Product of Two Codes
Is Not Necessarily Robustly Testable

Paul Valiant

Massachusetts Institute of Technology
pvaliant@mit.edu

Abstract. There has been significant interest lately in the task of constructing codes that are testable with a small number of random probes. Ben-Sasson and Sudan show that the repeated tensor product of codes leads to a general class of locally testable codes. One question that is not settled by their work is the local testability of a code generated by a single application of the tensor product. Special cases of this question have been studied in the literature in the form of "tests for bivariate polynomials", where the tensor product has been shown to be locally testable for certain families of codes. However the question remained open for the tensor product of generic families of codes. Here we resolve the question negatively, giving families of codes whose tensor product does not have good local testability properties.

1 Introduction

Sub-linear algorithms in coding theory have become an increasingly important area of study, both for their own sake, and because of their strong connection with the techniques of the PCP program. A key problem is that of constructing and understanding Locally Testable Codes (LTCs), namely codes for which membership can be estimated probabilistically using a sub-linear number of queries (cf. [1, 7, 8, 11]).

The tensor product code of two linear codes C_1, C_2 of length n, denoted $C_1 \otimes C_2$, is the code consisting of the set of all $n \times n$ matrices in which each row belongs to C_1 and each column to C_2. The prime example of a tensor code is the set of bivariate polynomials of individual degree k over a field of size n, which is the tensor product of a pair of (univariate) Reed-Solomon codes of degree k.

A basic result of Arora and Safra says that if an $n \times n$ matrix is δ-far from the above-mentioned bivariate code then the expected distance of a random row/column of the matrix from a univariate degree k polynomial for $k = O(n^{\frac{1}{3}})$ is $\Omega(\delta)$ [3]. (A similar result with linear relationship between k and n was shown by Polishchuk and Spielman [10].) Formally, this means that the transformation from the univariate codes to the bivariate product code is *robust*, as distance from the product code implies similar expected distance from the component Reed-Solomon codes.

Recently, Ben-Sasson and Sudan, using ideas from Raz and Safra [12], showed that the slightly more complicated three-wise tensor product code is robustly locally testable under much more general conditions than the above polynomial restriction [5].

Robust locally testable codes are crucial in efficient constructions of LTCs and in fact many PCPs (see [5, 6]). In light of the above two results, it is natural to ask whether the tensor product of pairs of general codes is robustly locally testable. The current paper gives a negative answer to this basic question.

2 Definitions

We begin with some basic notions from coding theory. The first is the *Hamming metric* on strings.

Definition 1 (Hamming Distance). *Given a set Σ called the* alphabet, *and two strings $x, y \in \Sigma^n$ for some n, denote the ith entry of each string by x_i, y_i respectively. The Hamming distance between x and y, denoted $\Delta(x, y)$, is the number of indices i for which $x_i \neq y_i$.*

A *code* is a means of transforming strings into (typically) longer strings, such that the original message is recoverable from the *encoded* message even when the encoded message has been tampered with by some bounded adversary. This notion is made precise with the following definition.

Definition 2 (Code). *Given an alphabet Σ, and integers (n, k, d), referred to respectively as the* block length, *the* message length, *and the* distance, *an $[n, k, d]_\Sigma$ code is defined by an encoding function*

$$f : \Sigma^k \to \Sigma^n$$

such that Hamming distance between any two elements of the image of f is at least d.

A code C is often identified with the image of its encoding function

$$C \stackrel{\text{def}}{=} \{f(x) | x \in \Sigma^k\}.$$

The elements of C are referred to as codewords.

We define a few more notions related to the Hamming distance that are frequently used in coding theory.

Definition 3 (Distance Notions). *The ratio of the Hamming distance to the length of strings n is called the* relative distance, *and is denoted by $\delta(x, y) \stackrel{\text{def}}{=} \Delta(x, y)/n$. Given a code—or alternatively any set of strings—C, we define $\Delta_C(x), \delta_C(x)$ to be respectively, the minimum Hamming distance and relative distance between x and any codeword of C. Similarly, let $\Delta(C)$ be the minimum Hamming distance between any two distinct codewords of C, and let $\delta(C) \stackrel{\text{def}}{=} \Delta(C)/n$.*

In this paper we consider *linear codes*, defined as follows:

Definition 4 (Linear Code). *An $[n, k, d]_\Sigma$ linear code is an $[n, k, d]_\Sigma$ code where Σ is a field, and the encoding function f is linear. We represent f by a $k \times n$ matrix M, known as the* generator *matrix, with the property that $f(x) = x^T M$.*

Alternatively, an $[n, k, d]_\Sigma$ linear code C is a k-dimensional linear subspace of Σ^n such that $\Delta(C) \geq d$.

Given a code C, and a string $x \in \Sigma^n$, we often wish to determine if x is a codeword of C and, if not, what the closest codeword of C to x is. Recently, the focus has changed from deterministic tests that look at x in its entirety, to probabilistic tests that sample x at only a few locations. Such tests are called *local tests* (cf. [5]).

Definition 5 (Local Tester). *A tester T with query complexity q is a probabilistic oracle machine that when given oracle access to a string $r \in \Sigma^n$, makes q queries to the oracle for r and returns an accept/reject verdict. We say that T tests a code C of minimum distance d if whenever $r \in C$, T accepts with probability one; and when $r \notin C$, the tester rejects with probability at least $\delta_C(r)/2$. A code C is said to be locally testable with q queries if C has minimum distance $d > 0$ and there is a tester for C with query complexity q.*

A generic way to create codes that have non-trivial local tests is via the tensor product. The next two definitions describe the tensor product and a natural local test for codes constructed by this product (cf. [9], [13, Lecture 6, Sect. 2.4]).

Definition 6 (Tensor Product Code). *Given an $[n_1, k_1, d_1]_\Sigma$ code C_1, and an $[n_2, k_2, d_2]_\Sigma$ code C_2, define their* tensor product, *denoted $C_1 \otimes C_2 \subseteq \Sigma^{n_2 \times n_1}$, to be the set of $n_2 \times n_1$ matrices each of whose rows is an element of C_1, and each of whose columns is an element of C_2. It is well known that $C_1 \otimes C_2$ is an $[n_1 n_2, k_1 k_2, d_1 d_2]_\Sigma$ code.*

We note that if C_1, C_2 are linear codes, with corresponding generator matrices G_1, G_2, then the set of matrices with rows in the row-space of G_1 and columns in the row-space of G_2 is just the set

$$C_1 \otimes C_2 \stackrel{\text{def}}{=} \{M_2^T X M_1 | X \in \Sigma^{k_2 \times k_1}\}.$$

The natural probabilistic test for membership in a product code C is the following.

Definition 7 (Product Tester). *Given a product code $C = C_1 \otimes C_2$, test a matrix r for membership in C as follows: flip a coin; if it is heads, test whether a random row of r is a codeword of C_1; if it is tails, test whether a random column of r is a codeword of C_2.*

It is straightforward to show that in fact this product tester meets the criteria of Definition 5 for a local tester (see the appendix).

Thus we see that the tensor product of two codes of length n has a local test with query complexity $\sqrt{N}$, where $N = n^2$ is the length of the tensored code.

We now ask whether the local tests for the tensor product may be recursively applied. Specifically, if we take the tensor product of two tensor product codes $C^4 = C^2 \otimes C^2 = C \otimes C \otimes C \otimes C$, we know we can locally test for membership in C^4 by randomly testing for membership in C^2; could we now save time on this test by applying a local test to C^2 instead of a more expensive deterministic test?

It turns out that locally testable codes do not necessarily compose in this way without an additional *robustness* property. Motivated by the notion of Robust PCP verifiers introduced in [4], Ben-Sasson and Sudan introduce the notion of a *robust* local tester (cf. [5]). Informally a test is said to be robust if words that are far from codewords are not only rejected by the tester with high probability, but in fact, with high probability, the *view* of the tester is actually far from any accepting *view*. This is defined formally as follows.

Definition 8 (Robust Local Tester). *A tester T has two inputs: an oracle for a received vector r, and a random string s. On input the string s the tester generates queries $i_1, \ldots, i_q \in [n]$ and fixes circuit $C = C_s$ and accepts if $C(r[i_1], \ldots, r[i_q]) = 1$. For oracle r and random string s, define the robustness of the tester T on r, s, denoted $\rho^T(r, s)$, to be the minimum, over strings x satisfying $C(x) = 1$, of relative distance of $\langle r[i_1], \ldots, r[i_q] \rangle$ from x. We refer to the quantity $\rho^T(r) \stackrel{\text{def}}{=} \mathbf{E}_s[\rho^T(r, s)]$ as the expected robustness of T on r.*

A tester T is said to be α-robust for a code C if for every $r \in C$, the tester accepts w.p. one, and for every $r \in \Sigma^n$, $\rho^T(r) \geq \alpha \cdot \delta_C(r)$.

This definition is especially meaningful for the product tester of Definition 7. In this case, we may interpret the above definition of robustness as follows. Given a tensor product code $C = C_1 \otimes C_2$, the queries $i_1, \ldots, i_q$ comprise either a row or a column of the matrix r. For random seed s, denote this row or column by r^s. We note that if r^s is a row, then the robustness $\rho^T(r, s)$ is just the relative distance of r^s from the nearest codeword of C_1, namely $\delta_{C_1}(r^s)$, while if r^s is a column, then $\rho^T(r, s) = \delta_{C_2}(r^s)$. Thus the expected robustness of T on r is just the average of the following two quantities: the average relative distance of rows of r from C_1; and the average relative distance of columns of r from C_2. A robust tester signifies that any time r is close to C_1 row-wise, and close to C_2 column-wise, then it is also close to a single element of C with rows in C_1 and columns in C_2.

One of the few known examples of robust locally testable codes are codes based on multivariate polynomials. Codes based on multivariate polynomials may in turn be viewed as tensor products of (univariate) Reed-Solomon codes, which are defined as follows. Given an alphabet Σ that is a finite field $\mathbb{F}_q$, we encode degree k polynomials over this field by their values on $n > k$ fixed elements $\alpha_1, \ldots, \alpha_n$ of $\mathbb{F}_q$.

The tensor product of two Reed-Solomon codes is a *bivariate* code, which represents a bivariate polynomial by its values on a "rectangle" in $\mathbb{F}_q \times \mathbb{F}_q$.

Local testability—and in fact robustness—of the product tester for the bivariate polynomial codes is well-studied in the literature on PCPs. Results of

Polishchuk and Spielman (cf.[10, Theorem 9]) imply that for bivariate codes $C = C_1 \otimes C_1$, there exist $\epsilon > 0, c < \infty$ such that the product tester of above is in fact an ϵ-robust local tester provided the parameters n, k of C_1 satisfy $n \geq ck$. Similar results are implied by Lemma 5.2.1 of [3], specialized to the case $m = 2$. The resulting robustness analysis of the bivariate test is critical to many PCPs in the literature including those of [2, 3, 6, 10].

The robustness of the product tester for the product of Reed-Solomon codes raises the natural question of whether the product tester may be robust for the product of any pair of codes. We note that for this question to be meaningful, it must be asked of families of codes with asymptotically good distance properties. Formally, this question takes the following form.

> **Question:** Is the product tester always robust for general families of tensor product codes? Specifically, is it the case that there exists a function $\alpha(\delta) > 0$ such that for every infinite family of pairs of codes C_1, C_2 of relative distance δ, the tensor product $C_1 \otimes C_2$ is $\alpha(\delta)$ robust?

A somewhat more complicated tester for the tensor product of more than two codes is analyzed and shown to be robust in [5]. A positive resolution of the above question would have been sufficient for their purposes. At the same time, it would generalize the analyses of Arora and Safra [3] and Polishchuk and Spielman [10]. To date, however the status of the above question was open. Here, we provide a negative answer to this question in the form of the following theorem.

Theorem 1. *There exists an infinite family of pairs of codes $C_1 = \{C_1^{(i)}\}$ and $C_2 = \{C_2^{(i)}\}$ of relative distances δ_1, δ_2 at least $\frac{1}{10}$ such that the robustness of $C_1^{(i)} \otimes C_2^{(i)}$ converges to 0 as i increases.*

We note that it remains an interesting open problem whether the tensor product of a code with itself is always robust. Our counterexample involves the tensor product of *different* codes.

3   Proofs

As noted in the previous section, when we apply the definition of a robust local tester to the product tester of code $C = C_1 \otimes C_2$, the robustness $\rho^T(r)$ equals the average of the relative distance of r from the closest matrix with each row in C_1 and the relative distance of r from the closest matrix with each column in C_2. Defining $\delta_{C_1}(r)$ and $\delta_{C_2}(r)$ respectively to be these distances, we find the robustness of the product tester on $C = C_1 \otimes C_2$ is

$$\alpha_{C_1,C_2} = \min_r \frac{\delta_{C_1}(r) + \delta_{C_2}(r)}{2\delta_C(r)}. \tag{1}$$

Our goal then, is to find a matrix r whose rows and columns lie close to codewords of C_1 and C_2 respectively, but which lies far from any codeword of

$C_1 \otimes C_2$. Specifically, we show a general construction for generating pairs of codes C_1, C_2 whose tensor product codes have arbitrarily bad robustness. This construction implies the theorem.

To motivate the following construction we note that we seek a matrix r whose rows and columns, when taken individually, are very simple, but when taken together are complicated (the rows and columns are very close to codewords of C_1 and C_2 respectively, but the whole matrix is far from any code of the tensor product). A natural matrix that fits this general rubric is the *identity* matrix. Specifically, each row or column has exactly one nonzero entry, and is thus "simple", however, when considered as a whole, the identity matrix has *full rank* which, in linear algebra terms, is as complicated as a matrix can get. We use these properties of the identity matrix in a fundamental way in the following construction.

Construction. Given an $[n, k, d]_2$ code C_1 such that the dual space $C_1^{\perp}$ is an $[n, n - k, d']_2$ code for some d' and an $[m, k, D]$ code C_g with the additional property that the distance of 1^m from any codeword of C_g, namely $\Delta_{C_g}(1^m)$ is also at least D, we construct a tensor product code as follows.

Let G_1 be a generator matrix for code C_1, i.e., let its rows be some basis for the subspace C_1. Similarly, let G_g be a generator for C_g. We note that G_1 will be $k \times n$ and G_g will be $k \times m$. Define the $n \times m$ matrix H to be their product $H = G_1^T G_g$.

Next, consider the $n \times mn$ matrix obtained by horizontally concatenating n copies of H, which we denote by H^n. Let I_n^m be the $n \times mn$ matrix consisting of the $n \times n$ identity matrix with each column duplicated to appear m times consecutively.

Let

$$G_2 = H^n + I_n^m$$

be the $n \times mn$ matrix that is the sum of these two matrices, and let C_2 be the code generated by the rows of G_2. Define the tensor product code $C = C_1 \otimes C_2$.
□

We claim the following lemma.

Lemma 1. *The distance of the code C_2 constructed above is at least $\min\{md', nD\}$, and the robustness of the code $C = C_1 \otimes C_2$, which we denote α_{C_1, C_2}, satisfies*

$$\alpha_{C_1, C_2} \le \frac{nm}{2(n - k) \min\{md', nD\}}. \qquad (2)$$

We first show how the lemma implies our main result, and then prove the lemma.

Proof (Theorem 1). We produce a family of codes $C_1^{(i)}, C_g^{(i)}$ and show how, by Lemma 1, our construction transforms these codes into a family that satisfies the conditions of this theorem.

Let the parameters of $C_1^{(i)}$ be defined as

$$[n^{(i)}, k^{(i)}, d^{(i)}]_2 \overset{\text{def}}{=} [i, \frac{i}{2}, \frac{i}{10}]_2.$$

We note that from standard coding theory arguments if the $i/2 \times i$ generator matrix $G_1^{(i)}$ is filled in *at random*, then for large enough i the ensuing code will have distance $d^{(i)} \geq i/10$ with overwhelming probability. Take $C_1^{(i)}$ to be such a code. From standard linear algebra, the dual space $C_1^{\perp(i)}$ will also be a randomly generated subspace of $\mathbb{R}^i$, and thus the dual code will also have distance at least $i/10$ with overwhelming probability.

Similarly, let the $[m^{(i)}, k^{(i)}, D^{(i)}]_2$ code $C_g^{(i)}$ have parameters $[i, \frac{i}{2}, \frac{i}{10}]_2$ and be constructed from a random $i/2 \times i$ generator. Similar arguments show that for large enough i, the code C_g will also satisfy the additional property of codewords being far from 1^i with overwhelming probability.

We note that from Lemma 1, the distance of code C_2 is at least $\min\{md', nD\}$ $= \min\{i^2/10, i^2/10\} = i^2/10$. Thus C_2 is an $[i^2, i, i^2/10]_2$ code, and the *relative* distance of C_2 is $\frac{1}{10}$, as desired.

Explicitly evaluating (2), we have that

$$\alpha_{C_1, C_2} \leq \frac{nm}{2(n-k)\min\{md', nD\}} = \frac{i^2}{2(i - \frac{i}{2})\frac{i^2}{10}} = \frac{10}{i},$$

which converges to 0 as i increases. Thus the families $C_1^{(i)}, C_2^{(i)}$ satisfy the desired properties of the theorem. □

We now prove the lemma.

Proof (Lemma 1). Consider the above construction. We first note that since both G_1 and G_g have rank k, both G_1 and G_g must have full-rank $k \times k$ submatrices. Further, the product of these submatrices will be a submatrix of their product $H = G_1^T G_g$. Thus H has rank k.

We prove now that

$$\Delta(C_2) \geq \min\{md', nD\}.$$

Consider a nonzero codeword $c \in C_2$. Since G_2 generates C_2, we have by definition that for some nonzero vector $v \in \mathbb{R}^n$, $c = v^T G_2$. We consider two cases.

In the first case, suppose $v^T H = 0$. Since H has rank k and its columns are linear combinations of rows of G_1, its column space must equal the row space of G_1, which consists of the elements of C_1. Thus $G_1 v = 0$, and we conclude that $v \in C_1^\perp$, the dual code of C_1.

Since the dual code $C_1^\perp$ has minimum distance d' by assumption, v must differ from the codeword 0^n in at least d' places. Recall that $G_2 \overset{\text{def}}{=} H^n + I_n^m$. Thus we have $c \overset{\text{def}}{=} v^T G_2 = v^T I_n^m$, from which we conclude that c differs from 0^{nm} in at least md' places.

In the second case, suppose $v^T H \neq 0$. Then since the rows of H are linear combinations of the rows of G_g, $v^T H$ is a nonzero codeword of C_g, which we denote by $x \overset{\text{def}}{=} v^T H$.

We note that $v^T I_n^m$ consists of n consecutive $1 \times m$ vectors that are either uniformly zero or uniformly one. Thus our codeword $c = v^T G_2 = v^T(H^n + I_n^m)$ consists of n consecutive chunks containing either x or $1 - x$. From our assumptions on C_g, both x and $1 - x$ differ from 0^m in at least D places. Thus c has at least nD nonzero entries.

Thus, in either case, a nonzero codeword of C_2 differs from 0^{mn} in least $\min\{md', nD\}$ places. Since codes are linear, the minimum distance of 0^{nm} to another codeword equals the minimum distance of the code, and thus $\Delta(C_2) \geq \min\{md', nD\}$, as desired.

Recall from (1) that to demonstrate the low robustness of the product tester on $C_1 \otimes C_2$ we are trying to find a matrix whose rows are very close to codes in C_1 and whose columns are very close to codes in C_2, yet which lies far from any code in the tensor product $C = C_1 \otimes C_2$. The matrix we consider here is in fact G_2^T, the transpose of the generator matrix of C_2. Note that by definition,

$$\delta_{C_2}(G_2^T) = 0,$$

since G_2 generates C_2, and thus all the columns of G_2^T are in C_2. Also, we have that

$$\delta_{C_1}(G_2^T) = 1/n,$$

since all the columns of H^n are in C_1 and each column of I_n^m has relative weight $1/n$. Thus rows and columns of G_2^T lie "close" to codewords of C_1 and C_2 respectively.

We now show that G_2^T is far from any codeword of the tensor product $C_1 \otimes C_2$, specifically, that

$$\delta_{C_1 \otimes C_2}(G_2^T) \geq \frac{(n-k)\Delta(C_2)}{n^2 m}.$$

Consider the difference $R - G_2^T$ for any codeword $R \in C_1 \otimes C_2$. We note that each column of $R - G_2^T$ is a codeword of C_2. Also, we note that each row of $R - (G_2^T - I_n^{mT})$ is a codeword of C_1, implying that $R - (G_2^T - I_n^{mT})$ has rank at most k. We now invoke the fact that the identity matrix has full rank, and conclude that the difference $[R - (G_2^T - I_n^{mT})] - I_n^{mT} = R - G_2^T$ must therefore have has rank at least $n - k$. Thus there are at least $n - k$ nonzero columns in $R - G_2^T$, each of which is a codeword of C_2. Since each nonzero codeword of C_2 has weight at least $\Delta(C_2)$, we conclude that

$$\delta_{C_1 \otimes C_2}(G_2^T) \geq \frac{(n-k)\Delta(C_2)}{n^2 m}.$$

Thus we have constructed a matrix G_2^T for which

$$\alpha_{C_1,C_2} \leq \frac{\delta_{C_1}(G_2^T) + \delta_{C_2}(G_2^T)}{2\delta_C(G_2^T)} \leq \frac{nm}{2(n-k)\Delta(C_2)},$$

as desired, and the lemma is proven.   □

Acknowledgement

I am grateful to Madhu Sudan for introducing this problem to me, and for many valuable discussions along the way to solving it and writing it up. I would also like to acknowledge the suggestions of an anonymous referee with respect to my introduction.

References

1. Sanjeev Arora. *Probabilistic checking of proofs and the hardness of approximation problems.* PhD thesis, University of California at Berkeley, 1994.
2. Sanjeev Arora, Carsten Lund, Rajeev Motwani, Madhu Sudan, and Mario Szegedy. Proof verification and the hardness of approximation problems. *Journal of the ACM*, 45(3):501–555, May 1998.
3. Sanjeev Arora and Shmuel Safra. Probabilistic checking of proofs: A new characterization of NP. *Journal of the ACM*, 45(1):70–122, January 1998.
4. Eli Ben-Sasson, Oded Goldreich, Prahladh Harsha, Madhu Sudan, and Salil Vadhan. Robust PCPs of proximity, shorter PCPs and applications to coding. In *Proceedings of the 36th Annual ACM Symposium on Theory of Computing*, pages 1-10, 2004.
5. E. Ben-Sasson and M. Sudan. Robust Locally Testable Codes and Products of Codes. RANDOM 2004, pages 286–297, 2004.
6. E. Ben-Sasson and M. Sudan. Simple PCPs with Poly-log Rate and Query Complexity. 37th STOC (to appear), 2005.
7. Katalin Friedl and Madhu Sudan. Some improvements to total degree tests. In *Proceedings of the 3rd Annual Israel Symposium on Theory of Computing and Systems*, pages 190–198, Tel Aviv, Israel, 4-6 January 1995. Corrected version available online at http://theory.csail.mit.edu/~madhu/papers/friedl.ps.
8. Oded Goldreich and Madhu Sudan. Locally testable codes and PCPs of almost-linear length. In *Proceedings of the 43rd Annual IEEE Symposium on Foundations of Computer Science*, Vancouver, Canada, 16-19 November 2002.
9. F. J. MacWilliams and Neil J. A. Sloane. *The Theory of Error-Correcting Codes.* Elsevier/North-Holland, Amsterdam, 1981.
10. Alexander Polishchuk and Daniel A. Spielman. Nearly linear-size holographic proofs. In *Proceedings of the Twenty-Sixth Annual ACM Symposium on the Theory of Computing*, pages 194–203, Montreal, Quebec, Canada, 23-25 May 1994.
11. Ronitt Rubinfeld and Madhu Sudan. Robust characterizations of polynomials with applications to program testing. *SIAM Journal on Computing*, 25(2):252–271, April 1996.
12. Ran Raz and Shmuel Safra. A Sub-Constant Error-Probability Low-Degree Test, and a Sub-Constant Error-Probability PCP Characterization of NP. In *Proceedings of the 29th STOC*, pages 475–484, 1997.
13. Madhu Sudan. Algorithmic introduction to coding theory. Lecture notes, Available from http://theory.csail.mit.edu/~madhu/FT01/, 2001.

Appendix

We prove here the fact mentioned in the body of the article that the product tester meets the criteria for a (non-robust) local tester.

Proof. Suppose we have a tensor product code $C = C_1 \otimes C_2$ and a matrix r for which the product tester accepts with probability α. This implies that for some $\beta + \gamma = 2\alpha$, β-fraction of the rows of r are in C_1 and γ-fraction of the columns of r are in C_2.

Consider the submatrix $r_{\beta,\gamma}$ at the intersection of these β rows and γ columns. Let G_1, G_2 be the generators of C_1, C_2 respectively. Let G_1^β and G_2^γ be the restrictions of G_1, G_2 respectively to those columns that correspond to the β rows of r and γ columns of r respectively. Since $r_{\beta,\gamma}$ is an element of the tensor product of the codes generated by G_1^β and G_2^γ, we may express $r_{\beta,\gamma}$ as

$$r_{\beta,\gamma} = G_2^{\gamma T} X G_1^\beta,$$

for some matrix X. We now extend $r_{\beta,\gamma}$ to a full matrix r', with the property that *every* row of r' is in C_1 and every column is in C_2. Specifically, let

$$r' \stackrel{\text{def}}{=} G_2^T X G_1,$$

for the full matrices G_1, G_2. Clearly r' is a codeword of C.

We note that by definition, r' agrees with r on the submatrix $r_{\beta,\gamma}$. Thus, $\delta_C(r)$, the distance of r from the nearest codeword of C is at most the distance from r to r', which is at most $1 - \beta\gamma$.

Recall that we are trying to prove that the probability of r failing the test, namely $1 - \alpha$, is at least half this distance $\delta_C(r)$. The algebra from here is straightforward. Recall that $1 - \alpha = 1 - \frac{\beta+\gamma}{2}$, and further that

$$\delta_C(r)/2 \leq \frac{1}{2} - \frac{\beta\gamma}{2} \leq \frac{1}{2} - \frac{\beta\gamma}{2} + \frac{(1-\beta)(1-\gamma)}{2} = 1 - \frac{\beta+\gamma}{2}.$$

Comparing these two equations yields the desired result.

Fractional Decompositions
of Dense Hypergraphs

Raphael Yuster

Department of Mathematics
University of Haifa
Haifa 31905, Israel
raphy@research.haifa.ac.il

Abstract. A seminal result of Rödl (the *Rödl nibble*) asserts that the edges of the complete r-uniform hypergraph K_n^r can be packed, almost completely, with copies of K_k^r, where k is fixed. This result is considered one of the most fruitful applications of the probabilistic method. It was not known whether the same result also holds in a dense hypergraph setting. In this paper we prove that it does. We prove that for every r-uniform hypergraph H_0, there exists a constant $\alpha = \alpha(H_0) < 1$ such that every r-uniform hypergraph H in which every $(r-1)$-set is contained in at least αn edges has an H_0-packing that covers $|E(H)|(1 - o_n(1))$ edges. Our method of proof is via fractional decompositions. Let H_0 be a fixed hypergraph. A *fractional H_0-decomposition* of a hypergraph H is an assignment of nonnegative real weights to the copies of H_0 in H such that for each edge $e \in E(H)$, the sum of the weights of copies of H_0 containing e is precisely one. Let k and r be positive integers with $k > r > 2$. We prove that there exists a constant $\alpha = \alpha(k, r) < 1$ such that every r-uniform hypergraph with n (sufficiently large) vertices in which every $(r-1)$-set is contained in at least αn edges has a fractional K_k^r-decomposition. We then apply a recent result of Rödl, Schacht, Siggers and Tokushige to obtain our integral packing result. The proof makes extensive use of probabilistic arguments and additional combinatorial ideas.

1 Introduction

A *hypergraph* H is an ordered pair $H = (V, E)$ where V is a finite set (the *vertex set*) and E is a family of distinct subsets of V (the *edge set*). A hypergraph is *r-uniform* if all edges have size r. In this paper we only consider r-uniform hypergraphs where $r \geq 2$ is fixed. Let H_0 be a fixed hypergraph. For a hypergraph H, the *H_0-packing number*, denoted $\nu_{H_0}(H)$, is the maximum number of pairwise edge-disjoint copies of H_0 in H. A function ψ from the set of copies of H_0 in H to $[0, 1]$ is a *fractional H_0-packing* of H if $\sum_{e \in H_0} \psi(H_0) \leq 1$ for each $e \in E(H)$. For a fractional H_0-packing ψ, let $|\psi| = \sum_{H_0 \in \binom{H}{H_0}} \psi(H_0)$. The *fractional H_0-packing number*, denoted $\nu_{H_0}^*(H)$, is defined to be the maximum value of $|\psi|$ over all fractional H_0-packings ψ. Notice that, trivially, $e(H)/e(H_0) \geq \nu_{H_0}^*(H) \geq$

C. Chekuri et al. (Eds.): APPROX and RANDOM 2005, LNCS 3624, pp. 482–493, 2005.

$\nu_{H_0}(H)$. In case $\nu_{H_0}(H) = e(H)/e(H_0)$ we say that H has an H_0-decomposition.
In case $\nu^*_{H_0}(H) = e(H)/e(H_0)$ we say that H has a *fractional H_0-decomposition*.
It is well known that computing $\nu_{H_0}(H)$ is NP-Hard already when H_0 is a
2-uniform hypergraph (namely, a graph) with more than two edges in some
connected component [4]. It is well known that computing $\nu^*_{H_0}(H)$ is solvable
in polynomial time for every fixed hypergraph H_0 as this amounts to solving a
(polynomial size) linear program.

For fixed integers k and r with $k > r \geq 2$, let K^r_k denote the complete
r-uniform hypergraph with k vertices. For $n > k$ it is trivial that K^r_n has a
fractional K^r_k-decomposition. However, it is far from trivial (and unknown for
$r > 2$) whether this fractional decomposition can be replaced with an integral
one, even when necessary divisibility conditions hold. In the graph-theoretic case
this is known to be true (for n sufficiently large), following the seminal result
of Wilson [10]. Solving an old conjecture of Erdős and Hanani, Rödl proved
in [8] that K^r_n has a packing with $(1 - o_n(1))\binom{n}{r}/\binom{k}{r}$ copies of K^r_k (namely,
an asymptotically optimal K^r_k-packing). In case we replace K^r_n with a dense
and large n-vertex r-uniform hypergraph H, it was not even known whether a
fractional K^r_k-decomposition of H exists, or whether an asymptotically optimal
K^r_k-packing exists. In this paper we answer both questions affirmatively. We note
that the easier graph theoretic case has been considered by the author in [12].

In order to state our density requirements we need a few definitions. Let
$H = (V, E)$ be an r-uniform hypergraph. For $S \subset V$ with $1 \leq |S| \leq r - 1$,
let $deg(S)$ be the number of edges of H that contain S. For $1 \leq d \leq r - 1$ let
$\delta_d(H) = \min_{S \subset V, |S|=d} deg(S)$ be the *minimum d-degree* of H. Usually, $\delta_1(H)$ is
also called the *minimum degree* and $\delta_2(H)$ is also called the *minimum co-degree*.
The analogous maximum d-degree is denoted by $\Delta_d(H)$. For $0 \leq \alpha \leq 1$ we say
that H is α-dense if $\delta_d(H) \geq \alpha\binom{n-d}{r-d}$ for all $1 \leq d \leq r - 1$. Notice that K^r_n is
1-dense and that H is α-dense if and only if $\delta_{r-1}(H) \geq \alpha(n - r + 1)$.

Our first main result is given in the following theorem.

Theorem 1. *Let k and r be integers with $k > r \geq 3$. There exists a positive
$\alpha = \alpha(k,r)$ and an integer $N = N(k,r)$ such that if H is a $(1 - \alpha)$-dense
r-uniform hypergraph with more than N vertices then H has a fractional K^r_k-
decomposition.*

We note that the constant $\alpha = \alpha(k,r)$ that we obtain is *only* exponential in k
and r. It is not difficult to show that our proof already holds for $\alpha(k,r) = 6^{-kr}$
although we make no effort to optimize the constant. We note the the proof
in the graph-theoretic case given in [12] yields $\alpha(k,2) \leq 1/9k^{10}$. However, the
proof in the graph-theoretic case is quite different for the most part and cannot
be easily generalized to the hypergraph setting.

Although Theorem 1 is stated only for K^r_k, it is easy to see that a similar
theorem also holds for any k-vertex r-uniform hypergraph H_0. Indeed, if H_0
has k vertices then, trivially, K^r_k has a fractional H_0-decomposition. Thus, any
hypergraph which has a fractional K^r_k-decomposition also has a fractional H_0-
decomposition. We note that in the very special case where H_0 is an r-uniform
simple hypertree then exact decomposition results are known [11].

Our second result is, in fact, a corollary obtained from Theorem 1 and a theorem of Rödl, Schacht, Siggers and Tokushige [9] who proved that the H_0-packing number and the fractional H_0-packing number are very close for dense r-uniform hypergraphs (an earlier result of Haxell, Nagle and Rödl [6] asserted this for the case $r = 3$). The exact statement of their result is the following.

Theorem 2. *[Rödl, Schacht, Siggers and Tokushige [9].] For a fixed r-uniform hypergraph H_0, if H is an n-vertex r-uniform hypergraph then $\nu^*_{H_0}(H) - \nu_{H_0}(H) = o(n^r)$.* ∎

From Theorem 1 and the comments after it, and from Theorem 2, we immediately obtain the following.

Theorem 3. *Let H_0 be a fixed r-uniform hypergraph. There exists a positive constant $\alpha = \alpha(H_0)$ such that if $H = (V, E)$ is a $(1 - \alpha)$-dense r-uniform hypergraph with n vertices then H has an H_0-packing that covers $|E|(1 - o_n(1))$ edges.* ∎

In the next section we prove Theorem 1. The final section contains some concluding remarks and open problems.

2   Proof of Theorem 1

Let $\mathcal{F}$ be a fixed family of r-uniform hypergraphs. An $\mathcal{F}$-*decomposition* of an r-uniform hypergraph H is a set L of subhypergraphs of H, each isomorphic to an element of $\mathcal{F}$, and such that each edge of H appears in precisely one element of L. Let $H(t, r)$ denote the complete r-uniform hypergraph with t vertices and with one missing edge. For the remainder of this section we shall use $t = k(r+1)$. Let $\mathcal{F}(k, r) = \{K_k^r, K_t^r, H(t, r)\}$. The proof of Theorem 1 is a corollary of the following stronger theorem.

Theorem 4. *For all $k > r \geq 3$ there exists a positive $\alpha = \alpha(k, r)$ and an integer $N = N(k, r)$ such that every r-uniform hypergraph with $n > N$ vertices which is $(1 - \alpha)$-dense has an $\mathcal{F}(k, r)$-decomposition.*

Clearly K_t^r has a fractional K_k^r-decomposition, since $t > k$. Thus, in order to prove that Theorem 1 is a corollary of Theorem 4 is suffices to prove that $H(t, r)$ has a fractional K_k^r-decomposition. This is done in the following two lemmas.

Lemma 1. *Let A be an upper triangular matrix of order r satisfying $A_{j,j} > 0$ for $j = 1, \ldots, r$, $A_{i,j} \geq 0$ for all $1 \leq i < j \leq r$, and $A_{i,j} \geq A_{i-1,j}$ for all $2 \leq i \leq j \leq r$. Let J be the all-one column vector of length r. Then, in the unique solution of $Ax = J$ all coordinates of x are nonnegative.*

Proof. Clearly $Ax = J$ has a unique solution since A is upper triangular and the diagonal consists of nonzero entries. Let $x^t = (x_1, \ldots, x_r)$ be the unique solution. Clearly, $x_r = 1/A_{r,r} > 0$. Assuming $x_{i+1} \geq 0$ we prove $x_i \geq 0$. Indeed,

$$x_i = \frac{1}{A_{i,i}}\left(1 - \sum_{j=i+1}^r A_{i,j}x_j\right) \geq \frac{1}{A_{i,i}}\left(1 - \sum_{j=i+1}^r A_{i+1,j}x_j\right) = 0. \qquad ∎$$

Lemma 2. *For all $k \geq r \geq 2$, $H(t,r)$ has a fractional K_k^r-decomposition.*

Proof. Let $A = \{u_1, \ldots, u_r\}$ be the unique set of vertices of $H(t,r)$ for which A is not an edge, and let B denote the set of the remaining $t - r$ vertices. For $i = 0, \ldots, r - 1$, we say that an edge of $H(t,r)$ is of *type* i if it intersects i elements of A. For $j = 0, \ldots, r - 1$ we say that a copy of K_k^r in $H(t,r)$ is of *type* j if it intersects j elements of A. For $j \geq i$, each edge of type i lies on precisely

$$f(i,j) = \binom{r-i}{j-i}\binom{t-2r+i}{k-r-j+i}$$

copies of K_k^r of type j. We now prove that there are nonnegative real numbers $x_0, \ldots, x_{r-1}$ such that by assigning the value x_j to each copy of K_k^r of type j, we obtain a fractional K_k^r decomposition, namely we must show that for each $i = 0, \ldots, r - 1$,

$$\sum_{j=i}^{r-1} x_j f(i,j) = 1.$$

Indeed, consider the upper triangular matrix A of order r with $A_{i,j} = f(i - 1, j - 1)$. By Lemma 1 it suffices to show that $f(j,j) > 0$ and $f(i,j) \geq 0$ for all $0 \leq i \leq j \leq r - 1$ and $f(i,j) \geq f(i - 1, j)$ for all $1 \leq i \leq j \leq r - 1$. Indeed, by definition $f(i,j) \geq 0$. Furthermore,

$$f(j,j) = \binom{t - 2r + j}{k - r} > 0$$

and

$$\frac{f(i,j)}{f(i-1,j)} = \frac{(t - 2r + i)(j - i + 1)}{(r - i + 1)(k - r - j + i)} \geq \frac{t - 2r}{(r + 1)(k - r)} = \frac{kr + k - 2r}{kr + k - r - r^2} \geq 1.$$

∎

Our goal in the remainder of this section is to prove Theorem 4. Our first tool is the following powerful result of Kahn [7] giving an upper bound for the minimum number of colors in a proper edge-coloring of a uniform hypergraph (his result is, in fact, more general than the one stated here).

Lemma 3 (Kahn [7]). *For every $r^* \geq 2$ and every $\gamma > 0$ there exists a positive constant $\rho = \rho(r^*, \gamma)$ such that the following statement is true:*
If U is an r^*-uniform hypergraph with $\Delta_1(U) \leq D$ and $\Delta_2(U) \leq \rho D$ then there is a proper coloring of the edges of U with at most $(1 + \gamma)D$ colors. ∎

Our second Lemma quantifies the fact that in a dense r-uniform hypergraph every edge appears on many copies of K_t^r.

Lemma 4. *Let $t \geq r \geq 3$ and let $\zeta > 0$. Then, for all sufficiently large n, if H is a $(1 - \zeta)$-dense r-uniform hypergraph with n vertices then every edge of H appears on at least $\frac{1}{(t-r)!} n^{t-r}(1 - \zeta t 2^t)$ copies of K_t^r.*

Proof. Fix an edge $e = \{u_1, \ldots, u_r\}$. We prove the lemma by induction on t. Our base cases are $t = r, \ldots, 2r - 1$ for which we prove the lemma directly. The case $t = r$ is trivial. If $r + 1 \leq t \leq 2r - 1$, then for any $(t - r)$-subset S of $V(H) - e$, the set of t-vertices $S \cup e$ is *not* a K_t^r if and only if there exists some $f \subset e$ with $2r - t \leq |f| \leq r - 1$ and some $g \subset S$ with $|g| = r - |f|$ such that $f \cup g$ is not an edge. For any $f \subset e$ with $2r - t \leq |f| \leq r - 1$, the number of non-edges containing f is at most $\zeta \binom{n - |f|}{r - |f|}$. For each such non-edge e', if $g = e' - f$ then g appears in at most $\binom{n}{t - r - |g|} = \binom{n}{t - 2r + |f|}$ possible $(t - r)$-subsets S of $V(H) - e$. It follows that e appears on at least

$$\binom{n - r}{t - r} - \sum_{d = 2r - t}^{r - 1} \binom{r}{d} \zeta \binom{n - d}{r - d} \binom{n}{t - 2r + d} > \frac{n^{t - r}}{(t - r)!} (1 - \zeta 2^t)$$

copies of K_t^r.

Assume the lemma holds for all $t' < t$ and that $t \geq 2r$. Let H^* be the subhypergraph of H induced on $V(H) - e$. H^* has $n - r$ vertices. Since n is chosen large enough, the deletion of a constant (namely r) vertices from a $(1 - \zeta)$-dense n-vertex hypergraph has a negligible affect on the density. In particular, the density of H^* is larger than $(1 - 2\zeta)$. By the induction hypothesis, each edge of H^* appears in at least

$$\frac{(n - r)^{t - 2r}}{(t - 2r)!} (1 - 2\zeta(t - r)2^{t - r})$$

copies of $K_{t - r}^r$ in H^*. Since H^* is $(1 - 2\zeta)$-dense it has at least $\binom{n - r}{r}(1 - 2\zeta)$ edges. As each copy of $K_{t - r}^r$ has $\binom{t - r}{r}$ edges, we have that H^* contains at least

$$\frac{(n - r)^{t - 2r}}{(t - 2r)!} (1 - 2\zeta(t - r)2^{t - r}) \binom{n - r}{r} (1 - 2\zeta) \frac{1}{\binom{t - r}{r}}$$

$$> \frac{n^{t - r}}{(t - r)!} (1 - 2\zeta(t - r)2^{t - r})(1 - 3\zeta)$$

copies of $K_{t - r}^r$. If S is the set of vertices of some $K_{t - r}^r$ in H^* we say that S is *good* if $S \cup e$ is the set of vertices of a K_t^r, otherwise S is *bad*. We can estimate the number of bad S in a similar fashion to the estimation in the base cases of the induction. Indeed, S is bad if and only if there exists some $f \subset e$ with $1 \leq |f| \leq r - 1$ and some $g \subset S$ with $|g| = r - |f|$ such that $f \cup g$ is not an edge. It follows that the number of bad S is at most

$$\sum_{d = 1}^{r - 1} \binom{r}{d} \zeta \binom{n - d}{r - d} \binom{n}{t - 2r + d} < \frac{n^{t - r}}{(t - r)!} \zeta 2^t.$$

It follows that the number of good S, and hence the number of K_t^r of H containing e, is at least

$$\frac{n^{t - r}}{(t - r)!} ((1 - 2\zeta(t - r)2^{t - r})(1 - 3\zeta) - \zeta 2^t) > \frac{n^{t - r}}{(t - r)!} (1 - \zeta t 2^t)$$

as required.   ∎

Proof of Theorem 4. Let $k > r \geq 3$ be fixed integers. We must prove that there exists $\alpha = \alpha(k, r)$ and $N = N(k, r)$ such that if H is an r-uniform hypergraph with $n > N$ vertices and $\delta_d(H) \geq \binom{n-d}{r-d}(1 - \alpha)$ for all $1 \leq d \leq r - 1$ then H has an $\mathcal{F}(k, r)$-decomposition.

Let $\epsilon = \epsilon(k, r)$ be a constant to be chosen later (in fact, it suffices to take $\epsilon = (2kr)^{-2r}$ but we make no attempt to optimize ϵ). Let $\eta = (2^{-H(\epsilon)}0.9)^{1/\epsilon}$ where $H(x) = -x \log x - (1 - x) \log(1 - x)$ is the entropy function. Let $\alpha = \min\{(\eta/2)^2, \ \epsilon^2/(t^2 4^{t+1})\}$. Let $\gamma > 0$ be chosen such that $(1 - \alpha t 2^t)(1 - \gamma)/(1 + \gamma)^2 > 1 - 2\alpha t 2^t$. Let $r^* = \binom{t}{r}$. Let $\rho = \rho(r^*, \gamma)$ be the constant from Lemma 3. In the proof we shall assume, whenever necessary, that N is sufficiently large as a function of these constants.

Let $H = (V, E)$ be an r-uniform hypergraph with $n > N$ vertices and $\delta_d(H) \geq \binom{n-d}{r-d}(1 - \alpha)$ for all $1 \leq d \leq r - 1$.

Our first step is to color the edges of H such that the spanning subhypergraph on each color class has some "nice" properties. We shall use q colors where $q = n^{1/(4\binom{t}{r}-4)}$ (for convenience we ignore floors and ceilings as they do not affect the asymptotic nature of our result). Each $e \in E$ selects a color from $[q]$ uniformly at random. The choices are independent. Let $H_i = (V, E_i)$ denote the subhypergraph whose edges received the color i. Let $S \subset V$ with $1 \leq |S| \leq r - 1$. Clearly, the degree of S in H_i, denoted $deg_i(S)$, has binomial distribution $B(deg(S), 1/q)$. Thus, $E[deg_i(S)] = deg(S)/q$. By a large deviation inequality of Chernoff (cf. [2], Appendix A) it follows that the probability that $deg_i(S)$ deviates from its mean by more than a constant fraction of the mean is exponentially small in n. In particular, for n sufficiently large,

$$\Pr\left[\left|deg_i(S) - \frac{deg(S)}{q}\right| > \gamma \frac{deg(S)}{q}\right] < \frac{1}{4qrn^r}. \tag{1}$$

Let $e \in E_i$. Let $C(e)$ denote the set of K_t^r copies of H that contain e and let $c(e) = |C(e)|$. Trivially, $c(e) \leq \binom{n-r}{t-r}$. Thus, by Lemma 4 with $\zeta = \alpha$ we have

$$\frac{1}{(t-r)!} n^{t-r} \geq c(e) \geq \frac{1}{(t-r)!} n^{t-r}(1 - \alpha t 2^t).$$

Let $C_i(e)$ denote the set of K_t^r copies of H_i containing e, and put $c_i(e) = |C_i(e)|$. Clearly, $E[c_i(e)] = c(e)q^{-\binom{t}{r}+1} = c(e)n^{-1/4}$. Therefore,

$$\frac{1}{(t-r)!} n^{t-r-1/4} \geq E[c_i(e)] \geq \frac{1}{(t-r)!} n^{t-r-1/4}(1 - \alpha t 2^t).$$

However, this time we cannot simply use Chernoff's inequality to show that $E[c_i(e)]$ is concentrated around its mean, since, given that $e \in E_i$, two elements of $C(e)$ are *dependent* if they contain another common edge in addition to e. However, we can overcome this obstacle using the fact that the dependence is *limited*. This is done as follows. Consider a graph G whose vertex set is $C(c)$ and whose edges connect two elements of $C(e)$ that share at least one edge (in addition to e). For $X \in C(e)$ the degree of X in G is clearly at most $\left(\binom{t}{r} - \right.$

1)$\binom{n-r-1}{t-r-1}$ since given $f \in E(X)$ with $f \neq e$ we have $|f \cup e| \geq r+1$ and thus there are at most $\binom{n-|f\cup e|}{t-|f\cup e|}$ copies of K_t^r containing both f and e. In particular, $\Delta(G) = O(n^{t-r-1})$. On the other hand $|V(G)| = c(e) = \Theta(n^{t-r})$. Notice also that the chromatic number of G is $\chi = \chi(G) = O(n^{t-r-1})$. Consider a coloring of G with $\chi(G)$ colors. If X and X' are in the same color class then, given that $e \in E_i$, the event that $X \in C_i(e)$ is *independent* of the event that $X' \in C_i(e)$. For $z = 1, \ldots, \chi(G)$, let $C^z(e)$ denote the elements of $C(e)$ colored with z and put $c^z(e) = |C^z(e)|$. Put $C_i^z(e) = C^z(e) \cap C_i(e)$ and let $c_i^z(e) = |C_i^z(e)|$. Clearly, $c_i(e) = \sum_{z=1}^{\chi} c_i^z(e)$ and $E[c_i^z(e)] = c^z(e)n^{-1/4}$. Whenever $|c^z(e)| > n^{1/2}$ we can use Chernoff's inequality to show that $c_i^z(e)$ is highly concentrated around its mean (that, is, the probability that it deviates from its mean by any given constant fraction of the mean is exponentially small in n). Whenever $|c^z(e)| \leq n^{1/2}$ we simply notice that the overall number of elements of $C(e)$ belonging to these small color classes is at most $\chi n^{1/2} = O(n^{t-r-1/2}) << n^{t-r-1/4}$. We therefore have that for n sufficiently large,

$$\Pr[c_i(e) < (1-\gamma)\frac{1}{(t-r)!}n^{t-r-1/4}(1-\alpha t2^t)] < \frac{1}{4\binom{n}{r}}, \qquad (2)$$

$$\Pr[c_i(e) > (1+\gamma)\frac{1}{(t-r)!}n^{t-r-1/4}] < \frac{1}{4\binom{n}{r}}. \qquad (3)$$

Since the overall number of subsets S with $1 \leq |S| \leq r-1$ is less than rn^r, and since $|E| \leq \binom{n}{r}$ we have, by (1), (2) and (3) that with probability at least $1 - qrn^r/(4qrn^r) - 2\binom{n}{r}/(4\binom{n}{r}) \geq 1/4$, a random q-coloring of the edges of H satisfies the following:

A. For all $S \subset V$ with $1 \leq |S| \leq r-1$, and for all $i = 1, \ldots, q$, $|deg_i(S) - \frac{deg(S)}{q}| \leq \gamma\frac{deg(S)}{q}$.

B. For each $e \in E$, if $e \in E_i$ then $c_i(e) \geq (1-\gamma)\frac{1}{(t-r)!}n^{t-r-1/4}(1-\alpha t2^t)$ and $c_i(e) \leq (1+\gamma)\frac{1}{(t-r)!}n^{t-r-1/4}$.

We therefore fix an edge coloring and the resulting spanning subhypergraphs $H_1, \ldots, H_q$ satisfying properties A and B.

For each $H_i = (V, E_i)$ we create another hypergraph, denoted U_i, as follows. The vertex set of U_i is E_i. The edges of U_i are the sets of edges of copies of K_t^r in H_i. Notice that U_i is a $\binom{t}{r}$-uniform hypergraph. Let $D = (1+\gamma)((t-r)!)^{-1}n^{t-r-1/4}$. By Property B, $\Delta_1(U_i) \leq D$. Also, we trivially have that for all n sufficiently large, $\Delta_2(U_i) \leq n^{t-r-1} < \rho D$. It follows from Lemma 3 that the set of K_t^r copies of H_i can be partitioned into at most $(1+\gamma)D$ packings. Denote these packings by $L_i^1, \ldots, L_i^{z_i}$ where $z_i \leq (1+\gamma)D$.

We now choose a K_t^r-packing of H as follows. For each $i = 1, \ldots, q$ we select, uniformly at random, one of the packings $\{L_i^1, \ldots, L_i^{z_i}\}$. Denote by L_i the randomly selected packing. All q selections are performed independently. Notice that $L = L_1 \cup \cdots \cup L_q$ is a K_t^r-packing of H. Let M denote the set of edges of H that do not belong to any element of L, and let $H[M]$ be the spanning subhypergraph of H consisting of the edges of M. Let $p = \binom{k}{r} - 1$. We say that a p-subset $S = \{S_1, \ldots, S_p\}$ of L is *good for* $e \in M$ if we can select edges $f_i \in E(S_i)$ such that $\{f_1, \ldots, f_p, e\}$ is the set of edges of a K_k^r in H. We say that L is *good* if for each $e \in M$ there exists a p-subset $S(e)$ of L such that $S(e)$ is good for e and such that if $e \neq e'$ then $S(e) \cap S(e') = \emptyset$.

Lemma 5. *If L is good then H has an $\mathcal{F}(k,r)$-decomposition.*

Proof. For each $e \in M$, pick a copy of K_k^r in H containing e and precisely one edge from each element of $S(e)$. As each element of $S(e)$ is a K_t^r, deleting one edge from such an element results in an $H(t,r)$. We therefore have $|M|$ copies of K_k^r and $|M|(\binom{k}{r} - 1)$ copies of $H(t,r)$, all being edge disjoint. The remaining element of L not belonging to any of the $S(e)$ are each a K_t^r, and they are edge-disjoint from each other and from the previously selected K_k^r and $H(t,r)$. ∎

Our goal in the remainder of this section is to show that there exists a good L. We will show that with positive probability, the random selection of the q packings $L_1, \ldots, L_q$ yields a good L. We begin by showing that with high probability, $H[M]$ has a relatively small maximum d-degree, for all $1 \leq d \leq r-1$.

Lemma 6. *With positive probability, for all $d = 1, \ldots, r - 1$, $\Delta_d(H[M]) \leq 2\epsilon\binom{n-d}{r-d}$.*

Proof. Let $S \subset V$ with $1 \leq |S| \leq r - 1$. Let $F_i(S) \subset E_i$ denote the edges of H_i containing S and let $L_i(S) \subset F_i(S)$ denote those edges of $F_i(S)$ that are covered by L_i. For $e \in F_i(S)$, the probability that e is covered by L_i is $c_i(e)/z_i$. By Property B and since $z_i \leq (1 + \gamma)D$ we have

$$\frac{c_i(e)}{z_i} \geq \frac{(1 - \gamma)(1 - \alpha t 2^t)}{(1 + \gamma)^2} \geq 1 - 2\alpha t 2^t.$$

It follows that $E[|L_i(S)|] \geq (1 - 2\alpha t 2^t)|F_i(S)| = (1 - 2\alpha t 2^t)deg_i(S)$ and that

$$\Pr[|L_i(S)| \leq (1 - \alpha^{1/2} t 2^t)deg_i(S)] \leq 2\alpha^{1/2} \leq \eta.$$

Since $|L_1(S)|, \ldots, |L_q(S)|$ are independent random variables it follows that the probability that at least ϵq of them have cardinality at most $(1 - \alpha^{1/2} t 2^t)deg_i(S)$ is at most

$$\binom{q}{\epsilon q}\eta^{\epsilon q} < 0.9^q << \frac{1}{qrn^r}$$

where in the last inequality we used the fact that $\eta = (2^{-H(\epsilon)}0.9)^{1/\epsilon}$. It follows that there exists a choice of $L_1, \ldots, L_q$ such that for all S, at most ϵq of the packings have $|L_i(S)| \leq (1 - \alpha^{1/2} t 2^t)deg_i(S)$. Let $deg^M(S)$ denote the degree of S in $H[M]$. By Property A, $deg_i(S) \leq (1+\gamma)deg(S)/q$. Thus, since $\sum_{i=1}^{q} deg_i(S) = deg(S)$ we have

$$deg^M(S) = deg(S) - \sum_{i=1}^{q} |L_i(S)| \leq deg(S) - (1 - \alpha^{1/2} t 2^t)deg(S) + \epsilon q(1+\gamma)\frac{deg(S)}{q}$$

$$\leq deg(S)(\alpha^{1/2}t2^t + \epsilon(1+\gamma)) \leq 2\epsilon deg(S) \leq 2\epsilon\binom{n-|S|}{r-|S|}$$

where in the last inequality we used the fact that $\alpha \leq \epsilon^2/(t^2 4^{t+1})$. It follows that there is a choice of $L_1, \ldots, L_q$ such that for all $d = 1, \ldots, r-1$, $\Delta_d(H[M]) \leq 2\epsilon\binom{n-d}{r-d}$. ∎

By Lemma 6, we may *fix* L and M such that $\Delta_d(H[M]) \leq 2\epsilon\binom{n-d}{r-d}$ for $d = 1, \ldots, r-1$. Let $M = \{e_1, \ldots, e_m\}$. Notice that, in particular, $m \leq 2\epsilon\binom{n}{r}$. Let $\mathcal{U} = \{U_1, \ldots, U_m\}$ be the family of p-uniform hypergraphs defined as follows. The vertex set of each U_i is L. The edges of U_i are the p-subsets of L that are good for e_i. A *system of disjoint representatives* (SDR) for $\mathcal{U}$ is a set of m edges $S(e_i) \in E(U_i)$ for $i = 1, \ldots, m$ such that $S(e_i) \cap S(e_j) = \emptyset$ whenever $i \neq j$. Thus, L is good if and only if $\mathcal{U}$ has an SDR. Generalizing Hall's Theorem, Aharoni and Haxell [1] gave a sufficient condition for the existence of an SDR.

Lemma 7. *[Aharoni and Haxell [1]] Let $\mathcal{U} = \{U_1, \ldots, U_m\}$ be a family of p-uniform hypergraphs. If for every $\mathcal{W} \subset \mathcal{U}$ there is a matching in $\cup_{U \in \mathcal{W}} U$ of size greater than $p(|\mathcal{W}| - 1)$ then $\mathcal{U}$ has an SDR.* ∎

We use Lemma 7 to prove:

Lemma 8. *If $\Delta_d(H[M]) \leq 2\epsilon\binom{n-d}{r-d}$ for $d = 1, \ldots, r-1$ then $\mathcal{U}$ has an SDR.*

Proof. Let R_i denote the set of K_k^r copies of H that contain e_i and whose remaining p edges are each from a distinct element of L. We establishing a lower bound for $|R_i|$. Let a_i denote the number of copies of K_k^r containing e_i, let b_i denote the number of copies of K_k^r containing e_i and at least two edges from the same element of L. Let c_i denote the number of copies of K_k^r containing e_i and at least another edge of M. Clearly, $|R_i| = a_i - b_i - c_i$.

A similar proof to that of Lemma 4 where we use k instead of t and $\zeta = \alpha$ immediately gives

$$a_i \geq \frac{1}{(k-r)!}n^{k-r}(1 - \alpha k 2^k). \tag{4}$$

Consider a pair of edges f_1, f_2 that belong to the same element of L. Suppose $|(f_1 \cup f_2) \cap e_i| = d$ then we must have $0 \leq d \leq r-1$. The overall number of choices for f_1, f_2 for which $|(f_1 \cup f_2) \cap e_m| = d$ is $O(n^{r-d})$ (there are $O(n^{r-d})$ choices for f_1, and given f_1 there are only $\binom{t}{r} - 1$ choices for f_2 in the same element of L). Given f_1, f_2, the number of K_k^r containing f_1, f_2, e_i is at most $O(n^{k-(2r+1-d)})$, since $|f_1 \cup f_2 \cup e_m| \geq 2r + 1 - d$. Thus, in total, we get,

$$b_i = \sum_{d=0}^{r-1} O(n^{r-d}n^{k-(2r+1-d)}) = O(n^{k-r-1}). \tag{5}$$

Consider an edge $f \in M$ with $f \neq e_i$. If f and e_i are independent then there are at most $\binom{n-2r}{k-2r}$ copies of K_k^r containing both of them. Overall, there are less than $m\binom{n-2r}{k-2r}$ such copies. If f and e_i intersect in d vertices then there are at most $\binom{n-2r+d}{k-2r+d}$ copies of K_k^r containing both of them. However, the maximum d-degree of M is at most $2\epsilon\binom{n-d}{r-d}$ and hence there are at most $\binom{r}{d}2\epsilon\binom{n-d}{r-d}$ choices for f. We therefore have that

$$c_i \le m\binom{n-2r}{k-2r} + \sum_{d=1}^{r-1}\binom{r}{d}2\epsilon\binom{n-d}{r-d}\binom{n-2r+d}{k-2r+d} \tag{6}$$

$$\le \sum_{d=0}^{r-1}2\epsilon\binom{r}{d}\binom{n-d}{r-d}\binom{n-2r+d}{k-2r+d}.$$

We now get, using (4), (5) and (6), that for $\epsilon = \epsilon(k,r)$ sufficiently small and for n sufficiently large,

$$|R_i| \ge \frac{1}{(k-r)!}n^{k-r}(1-\alpha k2^k) - O(n^{k-r-1}) - \sum_{d=0}^{r-1}2\epsilon\binom{r}{d}\binom{n-d}{r-d}\binom{n-2r+d}{k-2r+d}$$

$$\ge \frac{1}{2(k-r)!}n^{k-r}.$$

Let $W \subset \mathcal{U}$ with $w = |W|$. Without loss of generality assume $W = \{U_1,\ldots,U_w\}$. Put $M(W) = \{e_1,\ldots,e_w\}$. We must show that the condition in Lemma 7 holds. Assume that this is not the case. Consider a maximum matching T in $U_1 \cup \cdots \cup U_w$. Thus, $|T| \le p(w-1)$. In particular, $|T|$ contains at most $p^2(w-1)$ vertices (recall that the vertices are element of L). Let $L' \subset L$ denote the vertices contained in T. Thus, $|L'| \le p^2(w-1)$. The overall number of copies of K_k^r that contain precisely one edge from $M(W)$ and whose other edges are in p distinct elements of L is

$$|R_1| + \cdots + |R_w| \ge w\frac{1}{2(k-r)!}n^{k-r}.$$

Let F be the set of edges in the elements of L'. Hence, $|F| = |L'|\binom{t}{r}$. Let $f \in F$. Let $c(f)$ denote the number of copies of K_k^r containing f and precisely one edge from $M(W)$. For $Y \subset f$, let $M_f(Y) = \{e_i \mid e_i \cap f = Y, \ i = 1,\ldots,w\}$. This partitions $M(W)$ into $2^r - 1$ classes according to the choice of Y. Let $c(f, Y)$ denote the number of copies of K_k^r containing f and precisely one edge from $M_f(Y)$. Given $e \in M_f(Y)$, the number of K_k^r containing both f and e is at most $\binom{n-2r+|Y|}{k-2r+|Y|}$. On the other hand, since $deg^M(Y) \le 2\epsilon\binom{n-|Y|}{r-|Y|}$ we have $|M_f(Y)| \le 2\epsilon\binom{n-|Y|}{r-|Y|}$. Thus,

$$c(f, Y) \le 2\epsilon\binom{n-|Y|}{r-|Y|}\binom{n-2r+|Y|}{k-2r+|Y|} < 2\epsilon n^{k-r}.$$

It follows that

$$c(f) < 2^{r+1}\epsilon n^{k-r}.$$

Now, for $\epsilon = \epsilon(k,r)$ sufficiently small

$$\sum_{f \in F} c(f) < |L'| \binom{t}{r} 2^{r+1} \epsilon n^{k-r} \leq p^2(w-1)\binom{t}{r} 2^{r+1}\epsilon n^{k-r}$$

$$\leq w\frac{1}{2(k-r)!} n^{k-r} \leq |R_1| + \cdots + |R_w|.$$

It follows that there exists a K_k^r containing precisely one edge from $M(\mathcal{W})$, say, e_i, and whose other edges are in p distinct elements of $L - L'$. The p distinct elements form an edge $\{S_1, \ldots, S_p\}$ of U_i and hence $\{S_1, \ldots, S_p\}$ is an edge of $\cup_{U \in \mathcal{W}} U$. Since $\{S_1, \ldots, S_p\}$ is independent of all the edges of T we have that T is *not* a maximal matching of $\cup_{U \in \mathcal{W}} U$, a contradiction.   ∎

We have now completed the proof of Theorem 4.   ∎

3   Concluding Remarks and Open Problems

- A simpler version of Theorem 1 holds in case we assume that every edge of K_n^r lies on approximately the same number of copies of K_k^r (such is the case in, say, the random r-uniform hypergraph). In this case the statement of Theorem 1 follows quite easily from the result given in [3] and the result of [5]. However, our Theorem 1 does *not* assume these regularity conditions. It only assumes a minimum density threshold.
- Theorem 4 gives a nontrivial minimum density requirement which guarantees the existence of an $\mathcal{F}$-decomposition for the family $\mathcal{F} = \{K_k^r, K_t^r, H(t,r)\}$. It is interesting to find other more general families $\mathcal{F}$ for which nontrivial density conditions guarantee an $\mathcal{F}$-decomposition.

References

1. R. Aharoni and P. Haxell, *Hall's theorem for hypergraphs*, Journal of Graph Theory 35 (2000), 83–88.
2. N. Alon and J. H. Spencer, The Probabilistic Method, *Second Edition*, Wiley, New York, 2000.
3. N. Alon and R. Yuster, *On a hypergraph matching problem*, submitted.
4. D. Dor and M. Tarsi, *Graph decomposition is NPC - A complete proof of Holyer's conjecture*, Proc. 20th ACM STOC, ACM Press (1992), 252–263.
5. P. Frankl and V. Rödl, *Near perfect coverings in graphs and hypergraphs*, European J. Combinatorics 6 (1985), 317–326.
6. P. E. Haxell, B. Nagle and V. Rödl, *Integer and fractional packings in dense 3-uniform hypergraphs*, Random Structures and Algorithms 22 (2003), 248–310.
7. J. Kahn, *Asymptotically good list colorings*, J. Combin. Theory, Ser. A 73 (1996), 1–59.
8. V. Rödl, *On a packing and covering problem*, Europ. J. of Combin. 6 (1985), 69–78.

9. V. Rödl, M. Schacht, M. H. Siggers and N. Tokushige, *Integer and fractional packings of hypergraphs*, submitted.

10. R. M. Wilson, *Decomposition of complete graphs into subgraphs isomorphic to a given graph*, Congressus Numerantium XV (1975), 647–659.

11. R. Yuster, *Decomposing hypergraphs with simple hypertrees*, Combinatorica 20 (2000), 119–140.

12. R. Yuster, *Asymptotically optimal K_k-packings of dense graphs via fractional K_k-decompositions*, J. Combin. Theory, Ser. B, to appear.

Author Index

Lecture Notes in Computer Science

For information about Vols. 1–3523

please contact your bookseller or Springer